W9-AXT-937

ADFRIIA 1
University
Bookstore

4th Edition

CONTEMPORARY MATHEMATICS

Bruce E. Meserve

Professor of Mathematics, Emeritus, University of Vermont

Max A. Sobel

Professor of Mathematics, Montclair State College

John A. Dossey

Professor of Mathematics, Illinois State University

Prentice-Hall, Inc., Englewood Cliffs, New Jersey 07632

Library of Congress Cataloging-in-Publication Data

MESERVE, BRUCE ELWYN, 1917-
Contemporary mathematics.

Includes index.
1. Mathematics—1961- . I. Sobel, Max A.
II. Dossey, John A. III. Title.
QA39.2.M47 1987 510 86-25358
ISBN 0-13-170127-4

Editorial/production supervision:
Zita de Schauensee
Interior and cover design: Jayne Conte
Photo research: Kay Dellosa
Cover photo by Michel Tchervkoff
Manufacturing buyer: John B. Hall

Chapter-opening illustration credits. Chapter 1: Courtesy of Stanford University, CA. Chapters 2, 3, 6: The Granger Collection. Chapters 4, 7, 12: New York Public Library-Picture Collection. Chapter 5: Courtesy of American Mathematical Society, R.I. Chapters 8, 10: The Bettmann Archive. Chapter 9: Courtesy of The New York Historical Society. Chapter 11: Courtesy of Harvard University, Dept. of Health. Chapter 13: Courtesy of Dr. Sylvia Wiegand, University of Wisconsin. Chapter 14: The Bettmann Archive/BBC Hulton.

Credits for other illustrations. P. 67: From left to right, U.S. Fish and Wildlife Service, photo by David Klinger; Dr. Malcolm T. Sanford; U.S. Air Force. Pp. 68, 77: New York Public Library. Pp. 253, 425, 450; Teri Leigh Stratford. P. 309: NASA. Pp. 367, 519, 527: Scott Publishing Co. Pp. 456, 591: IBM.

A Division of Simon & Schuster
Englewood Cliffs, New Jersey 07632

Printed in the United States of America

10 9 8 7 6 5 4 3 2

ISBN 0-13-170127-4 01

Prentice-Hall International (UK) Limited, *London*
Prentice-Hall of Australia Pty. Limited, *Sydney*
Prentice-Hall Canada Inc., *Toronto*
Prentice-Hall Hispanoamericana, S.A., *Mexico*
Prentice-Hall of India Private Limited, *New Delhi*
Prentice-Hall of Japan, Inc., *Tokyo*
Prentice-Hall of Southeast Asia Pte. Ltd., *Singapore*
Editora Prentice-Hall do Brasil, Ltda., *Rio de Janeiro*

Contents

3 Numeration and Whole Numbers 93

4 Integers 145

5 Elements of Number Theory 172

6 Rational Numbers 203

7 Real Numbers 261

8 An Introduction to Geometry 307

9 Measurement and Geometry 361

10 An Introduction to Probability 418

Computer Applications Sections

Selected Explorations

Preface

Contemporary Mathematics, fourth edition, has been written to serve the needs of both liberal arts students and prospective teachers of elementary mathematics in grades K through 8. It is designed primarily as a one-year text at the undergraduate level, although it may be used effectively for in-service courses as well.

Users of the three previous editions of this text, teachers and students alike, have made valuable suggestions that have been incorporated into this fourth edition. An emphasis on skills and techniques of problem solving provides the basis for a new first chapter and is continued throughout the book. The rapidly increasing role of computers is recognized in new sections on computer applications following the chapter review at the end of each of the first thirteen chapters and in the new final chapter, Chapter 14: Computers in Elementary Mathematics.

Throughout this text there is an emphasis on the basic mathematical concepts and skills that a prospective teacher needs to know in order to teach effectively using contemporary elementary school textbooks. Such a background enables the teacher to enjoy the pleasures that come from teaching with confidence and competence. However, this text does not neglect those aspects of mathematics that remain important regardless of the changing curricula. Careful attention is given to the development of the number system and to the various algorithms that represent the fundamental operations of arithmetic.

A one-semester course with an emphasis on problem solving and number systems can be based effectively on Chapters 1 through 7 or selections from their 41 sections and nearly 1700 exercises. The extensive sets of exercises provide ample practice for students who need to spend extra time on these important topics. There are also challenging exercises and about 250 explorations for motivating the very capable students. Similarly, the 40 sections of Chapters 8 through 14 provide an effective basis for a one-semester course with an emphasis on geometry, probability, statistics, algebra, and the use of computers. As for the first half of this text there are ample numbers and levels of exercises and explorations to serve the needs of, and challenge, all students and to provide a basis for courses based upon selections from the topics available.

Prior editions of *Contemporary Mathematics* have been cited for their special attention to features that assist student learning. These include the following distinctive features that have been further expanded in this edition.

Illustrative Examples Each section of the text contains illustrative examples with detailed solutions that provide the student with a model for the correct solutions of problems.

Exercises Each section of the text is followed by a comprehensive set of exercises, with answers to the odd-numbered problems given at the back of the book. In all, there are more than 3100 exercises in the text so that choices may be readily made for classes of students of varying abilities.

Explorations Explorations are included at the end of each section of the text. These are often discovery exercises that allow the reader to pursue independently open-ended explorations that supplement knowledge of the material under discussion. They also serve to provide vistas of possible future extensions of such material.

Chapter Review A Chapter Review appears at the end of each chapter of the text. Answers to all the review exercises are given at the back of the book. The Chapter Review enables the student to demonstrate a comprehensive mastery of the basic concepts and skills of the chapter.

Each of the authors has participated for many years in the activities of national committees as guidelines and recommendations for the mathematical preparation of teachers have been formulated. Accordingly, as should be expected, the present text contains the mathematical background for courses designed for prospective elementary school teachers as suggested in the latest reports. These include the *Guidelines for the Preparation of Teachers of Mathematics* developed by the National Council of Teachers of Mathematics and the *Recommendations on the Mathematical Preparation of Teachers* of the Committee on the Undergraduate Program in Mathematics (CUPM) of the Mathematical Association of America.

The authors wish to express their appreciation to all those who have submitted suggestions for the revision of this text, as well as to the many students throughout the country who have both directly and indirectly contributed to the conception and formulation of this material. We are particularly grateful to the reviewers for their detailed comments and suggestions:

James R. Boone, Texas A & M University
Martha C. Jordan, Oskaloosa-Walton Junior College
Judy Kasabian, El Camino College
Patricia S. Kusimo, West Virginia State College
J. G. McCann, Surry Community College

Special recognition is given to Dorothy Meserve for her careful analysis of the manuscript, her detailed work in checking answers, and her preparation of the Instructor's Manual that accompanies this text.

The famous French mathematician René Descartes concludes his famous *La Géométrie* with the statement: "I hope that posterity will judge me kindly, not only as to the things which I have explained, but also as to those which I have intentionally omitted so as to leave to others the pleasure of discovery." The authors have attempted to provide a great deal of exposition in this text. They have, however, left many opportunities for the reader to experience the true beauty and excitement of mathematics through discovery.

Bruce E. Meserve

Max A. Sobel

John A. Dossey

1

An Introduction to Problem Solving

George Polya
(1887–1985)

George Polya was an outstanding mathematician who came to the United States from Hungary and served for many years on the faculty of Stanford University. He made contributions to both mathematics and the teaching of mathematics through his insightful works on problem solving. His 1945 text How to Solve It *is a classic in this field. In addition to writing, Dr. Polya spent a great deal of his career in working with teachers of mathematics encouraging them to teach in a problem-solving style.*

1-1 What Is Mathematics?

Mathematics has been described as a set of games played according to specified rules. Mathematics has also been described as a set of tools for solving specified problems. In this sense, arithmetic, geometry, and algebra are specific tools. These are the basic tools from which other tools have evolved. Any activity in which arithmetic, geometry, or algebra is used is either a *form of mathematics* or an *application of mathematics.* In **applied mathematics** the problems to be solved arise from other areas, including real-life situations, physics, astronomy, and practically all areas of systematic study. In **pure mathematics** the problems arise from the study of the abstract (mathematical) system under consideration.

The essence of mathematics is solving problems. The problems may be of any imaginable type—easy or very difficult, realistic or assumed, practical or theoretical. There are so many types of problems that we look for general strategies rather than rules for solving specific types of problems. Often several strategies are needed for a single problem. Several useful strategies are indicated in the margin as guides to the processes used to solve the problems in the following examples.

Example 1

Read the problem carefully.

(a) How many telephone poles are needed to reach the moon?
(b) Assume that one rabbit eats 2 pounds of food each week. There are 52 weeks in a year. How much food do 5 rabbits eat in one week?

Solution

(a) One, if it is long enough. (Avoid making unstated assumptions.)
(b) 10 pounds. (Use only the needed information. Do not be distracted by superfluous data.) ■

Example 2

Find the next three terms in each sequence.

(a) 1, 3, 5, 7, __, __, __, ...
(b) 1, 1, 2, 4, 7, __, __, __, ...

Solution A **sequence** is a list of elements called **terms** given in a particular order so that one can tell which term is first, second, and so forth.

Look for a pattern.

(**a**) The terms of this sequence are all odd numbers. Each term is larger than its predecessor; indeed, each can be obtained from its predecessor by adding 2. The next three terms are 9, 11, and 13. The corresponding points on a number line provide a picture, or model, of this pattern.

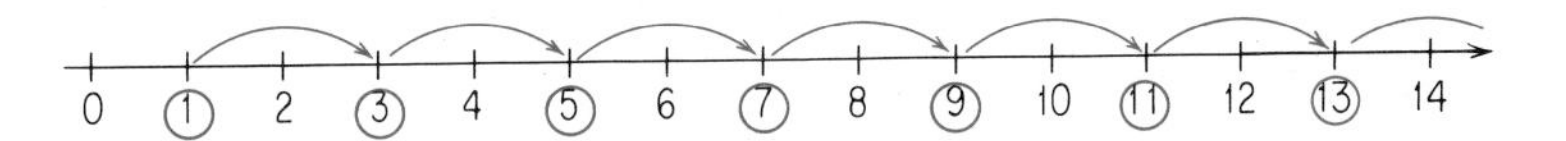

(**b**) After the second term each term is larger than its predecessor. Thus the differences may be helpful. We write the differences below the spaces between the given terms.

Terms 1, 1, 2, 4, 7, _, _, _, ...

Differences 0, 1, 2, 3, ...

For the differences the pattern of increasing by one at each step is easily observed. We use this pattern to extend the sequence of differences and add each difference to the last known term of the given sequence to obtain the next term of that sequence.

$$7 + 4 = 11 \qquad 11 + 5 = 16 \qquad 16 + 6 = 22$$

The next three terms of the given sequence are 11, 16, and 22.

Terms 1, 1, 2, 4, 7, 11, 16, 22, ...

Differences 0, 1, 2, 3, 4, 5, 6, ... ■

Example 3 Suppose that a down coat and a ski cap cost \$110 and the coat costs \$100 more than the cap. What did they each cost?

Solution

Guess and test.

We are tempted to guess that the price for the coat is \$100 and the price for the cap is \$10. But the coat would then cost only \$90 more than the cap. The difference is not large enough by \$10. If we change each price by \$5 to make them further apart, we have \$105 and \$5. These prices check in the given statements.

$$105 + 5 = 110 \qquad 105 - 5 = 100$$

The coat costs \$105 and the cap costs \$5.

Alternative Solution

State an equation.

Let the variable S stand for the cost of the ski cap. Then the cost of the coat is $S + 100$ and

$$S + (S + 100) = 110$$

Twice a number S increased by 100 is 110. Twice the number S is then 10. Thus S must equal 5 and the cost of the ski cap is \$5. Since $S + 100 = 105$, the cost of the coat is \$105. ■

An approach in which equations and variables are used to represent unknown quantities is an **algebraic approach** to solving the problem.

Example 4 How many games are needed to determine a winner in a basketball tournament having ten teams if a team is eliminated upon losing a game?

Solution All possible cases may be listed in a tournament pairing table.

List all possibilities.

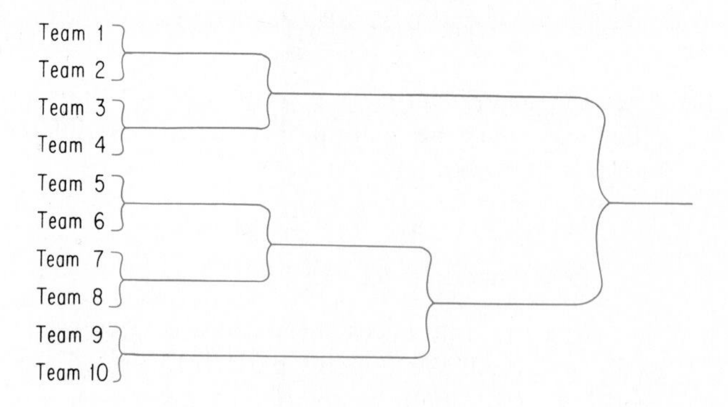

Each brace represents a game played; each horizontal line segment, the winner of a game. The tournament pairing table provides a model (representation) for the problem. There are nine braces and nine games are needed.

Alternative Solution Example 4 may be solved also by reasoning that if there is to be one winner in the tournament, then there must be nine losers. The only way one can be a loser is to play a game and lose it. Hence there must be nine games to determine nine losers and one winner. ■

Use logical reasoning.

There are many useful problem-solving strategies. For the previous examples we have noted:

Read the problem carefully.
Look for a pattern.
Guess and test.
State an equation.
List all possibilities.
Use logical reasoning.

Among other useful strategies are:

Make a model.
Draw a picture or diagram.
Attempt to solve a simpler but related problem.
Work backward from the answer.
Try to look at the problem from a different point of view.

There are four stages in solving a given problem. Each problem may involve one or more of the suggested strategies.

Understanding the Problem

Read the problem carefully.
List the details you know.
Look for the key phrases.
Pick out the important information.
State the problem in your own words.

Analyzing the Problem

Look for a pattern.
Draw a picture.
Use models or act out the problem.
Make an organized list.
Guess and test.
State and solve an equation.
Use logical reasoning.
Consider a simpler form of the problem.
Work the problem backward.

Doing the Necessary Computations

Simplify any arithmetic expressions.
Solve any algebraic equations.
Use a calculator or computer as appropriate.

Providing the Desired Results

Answer all questions.
Provide all requested information.
Be sure all results (solutions) are reasonable.
Check all solutions.
Look for extensions or generalizations of the solutions.
State the problem and its solution in sentence form.

Mathematics, both pure and applied, involves not only the solution of problems but also the identification, formulation, communication, and generalization of the problems. **Recognition** refers to the discovery that there is a problem, that something is not quite correct or some information is needed to understand a given situation. **Formulation** refers to the structuring of the problem, the collecting of information, the identification of relationships and desired unknown information, and—ultimately—a careful statement of the problem. **Analysis** refers to the process of understanding and studying the problem, performing the necessary computations, and providing the desired

Recognition.
Formulation.
Analysis.

Communication. results. **Communication** refers to the frequent need to explain in words or in writing the results of the previous stages so that others can understand the problem and its solution. This aspect is particularly important for prospective teachers. Generalization. **Generalization** refers to the extension of the problem to provide insights into the solutions of other, usually more general, problems. This aspect provides a basis for the development of new mathematical concepts.

Mathematics is not a spectator sport even though it may be considered as a set of games. The following exercises provide you with opportunities for participation. Read each problem carefully, select one or more strategies, and become involved in this sport.

EXERCISES

Note that many of the problems considered here are included "just for fun." Do not feel discouraged if you are unable to solve some at first glance. Also, you should not feel disturbed if you discover that some are trick questions!

In Exercises 1 *and* 2 *find the next three terms in each sequence.*

1. **(a)** 1, 4, 9, 16, ...
   (b) 2, 5, 8, 11, ...
2. **(a)** 101, 98, 95, 92, ...
   (b) 2, 5, 10, 17, ...
3. All the following puzzles have logical answers, but they are not strictly mathematical. See how many you can answer.
   (a) How many ten-cent stamps are there in a dozen?
   (b) How far can you walk into a forest?
   (c) Two American coins total 55¢ in value, yet one of them is not a nickel. Can you explain this?
   (d) How much dirt is there in a hole which is 3 feet wide, 4 feet long, and 2 feet deep?
   (e) There was a blind beggar who had a brother, but this brother had no brother. What was the relationship between the two?
4. A bottle and cork cost $2.00 together. The bottle costs $1.00 more than the cork. How much does each cost?
5. Three cannibals and three missionaries need to cross the river in a boat big enough only for two. The cannibals are fine if they are left alone or if they are with the same number or with a larger number of missionaries. They are dangerous if they are left alone in a situation where they outnumber the missionaries. How do they all get across the river without harm?
6. A farmer has to get a fox, a goose, and a bag of corn across a river in a boat which is only large enough for him and one of these three items. Now if he leaves the fox alone with the goose, the fox will eat the goose. If he leaves the goose alone with the corn, the goose will eat the corn. How does he get all items across the river?
7. A cat is at the bottom of an 18-foot well. Each day it climbs up 3 feet; each night it slides back 2 feet. How long will it take for the cat to get out of the well?
8. If a cat and a half eats a rat and a half in a day and a half, how many days will it take for 50 cats to eat 50 rats?
9. Ten coins are arranged to form a triangle, as shown in the accompanying figure. By rearranging only three of the coins, form a new triangle showing

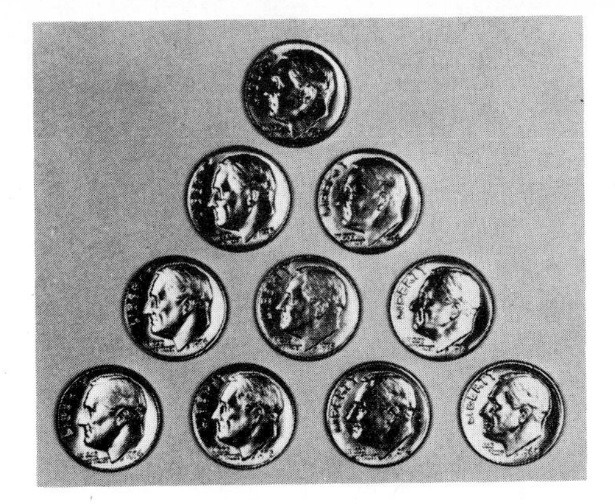

how the given triangle would appear with a vertex at the bottom instead of the top.

10. A woman goes to a well with three cans whose capacities are 3 liters, 5 liters, and 8 liters. Explain how she can obtain exactly 4 liters of water from the well
(**a**) using all the cans
(**b**) using only two of the cans.

11. Here is a mathematical trick you can try on a friend. Ask someone to place a penny in one hand and a dime in the other. Then tell your friend to multiply the value of the coin in the right hand by 6, multiply the value of the coin held in the left hand by 3, and add. Ask for the result. If the number given is an even number, you then announce that the penny is in the right hand; if the result is an odd number, then the penny is in the left hand and the dime is in the right. Explain why this trick works.

12. Four sticks are arranged to form a "cup" with a coin contained within three of the segments. Move only two sticks so that the figure formed by the sticks has the same shape and size as the given figure but the coin is no longer in the cup.

13. Three men enter a hotel and rent a room for \$60. After they are taken to their room the manager discovers he overcharged them; the room rents for only \$50. He thereupon sends a bellhop upstairs with the \$10 change. The dishonest bellhop decides to keep \$4 and returns only \$6 to the men. Now the room originally cost \$60, but the men had \$6 returned to them. This means that they paid only \$54 for the room. The bellhop kept \$4. \$54 + \$4 = \$58. What happened to the extra two dollars?

14. Find at least one way of using four 4's and ordinary arithmetic operations to write each of the numbers 1 through 10. Here are possible solutions for 1, 2, and 3.

$$\frac{44}{44} = 1 \qquad \frac{4}{4} + \frac{4}{4} = 2 \qquad \frac{4 + 4 + 4}{4} = 3$$

15. Use six matchsticks, all the same size, to form four equilateral triangles.

16. Rearrange the eight segments shown in the following figure to form three congruent squares. Each of the four smaller segments is one-half of the length of one of the larger segments.

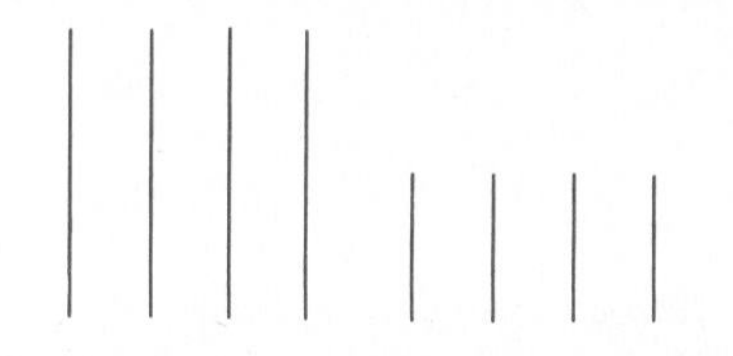

*17. A sailor lands on an island inhabited by two types of people. The *A*'s always lie, and the *B*'s always tell the truth. The sailor meets three inhabitants on the beach and asks the first of these: "Are you an *A* or a *B*?" The man answers, but the sailor doesn't understand him and asks the second person what the first man said. The man replies: "He said that he was a *B*. He is, and so am I." The third inhabitant then says: "That's not true. The first man is an *A* and I'm a *B*." Can you tell who was lying and who was telling the truth?

Cabinet		Desk
Television set	Sofa	Bookcase

18. Consider a house with six rooms and furniture arranged as in the accompanying figure. We wish to interchange the desk and the bookcase, but in such a way that there is never more than one piece of furniture in a room at a time. The other three pieces of furniture do not need to return to their original places. Can you do this? Try it using coins or other objects to represent the furniture.

19. Arrange two pennies *P* and two dimes *D* as in the accompanying figure. Try to interchange the coins so that the pennies are at the right and the dimes at the left. A coin may be moved to an adjacent empty square. A coin may also be jumped over a single occupied square to an empty square. You may move only one coin at a time, and no two coins may occupy the same space at the same time. Pennies may be moved only to the right, whereas dimes may be moved only to the left. What is the minimum number of moves required to complete the game?

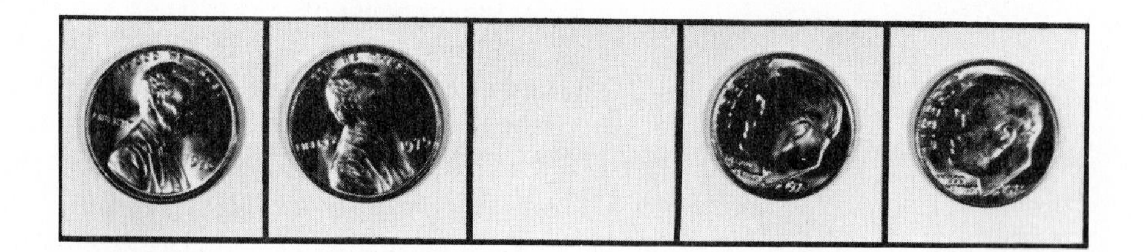

20. Repeat Exercise 19 for three pennies and three dimes, using seven squares. What is the minimum number of moves required to complete the game?

21. Think of the decimal digits and identify the next two letters in the sequence: *O*, *T*, *T*, *F*, *F*, *S*, *S*, ...

* An asterisk preceding an exercise indicates that the exercise is more challenging than the others.

22. As in Exercise 21 identify the next two letters in the sequence: $E, F, F, N, O, S, S, \ldots$.

23. Place a half-dollar, a quarter, and a nickel in one position, A, as in the figure. Then try to move these coins, one at a time, to position C. Coins may be placed in any available position. At no time may a larger coin be placed on a smaller coin, This can be accomplished in $2^3 - 1$, that is, 7, moves.

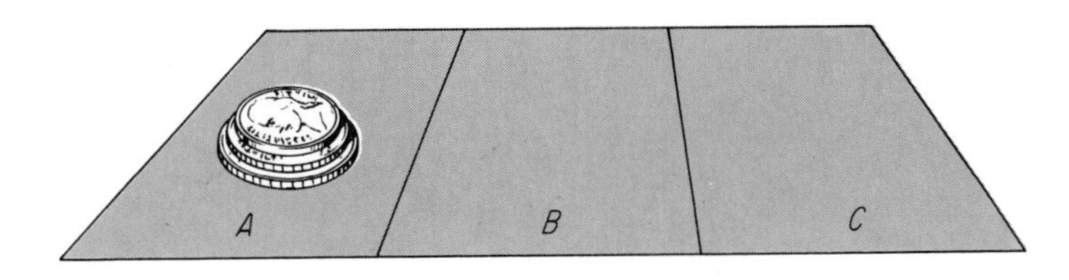

Next add a penny to the pile and try to make the change in $2^4 - 1$, that is, 15, moves.

This is an example of a famous problem called the **Tower of Hanoi**. The ancient Brahman priests were to move a pile of 64 such disks of decreasing size, after which the world would end. This would require $2^{64} - 1$ moves. Estimate how long this would take at the rate of one move per second.

24. Here is a game that must be played by two persons. Two players alternate in selecting one of the numbers 1, 2, 3, 4, 5, or 6. After each number is selected, it is added to the sum of those previously selected. For example, if player A selects 3 and player B selects 5, then the total is 8. If A selects 3 again, then the total is 11 and player B takes his turn. The object of the game is to be the first one to reach 50. There is a way to win at all times if you are permitted to go first. See if you can discover this method for winning, and then try to play the game with a classmate.

25. Many tricks of magic have their basis in elementary mathematics and may be found in books on mathematical recreations. Here is one example of such a trick.

Have someone place three dice on top of one another while you turn your back. Then instruct that person to look at and find the sum of the values shown on the two faces that touch each other for the top and middle dice, the two faces that touch each other for the middle and bottom dice, and the value of the bottom face of the bottom die. You then turn around and at a glance tell the sum. The trick is this: You merely subtract the value showing on the top face of the top die from 21. Stack a set of three dice in the manner described and try to figure out why the trick works as it does.

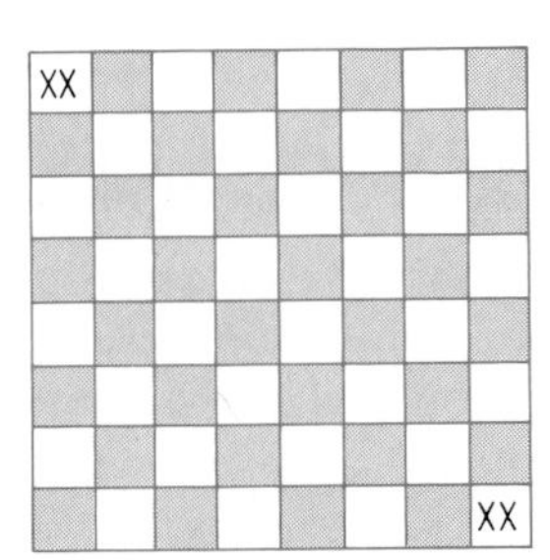

26. You are given a checkerboard and a set of dominoes. The size of each domino is such that it is able to cover two squares on the board. Can you arrange the dominoes in such a way that all of the board is covered with the exception of two squares in opposite corners? (That is, you are to leave uncovered the two squares marked XX in the adjacent figure.) Try to

explain why you should or should not be able to arrange the dominoes in this way.

27. Find the sum.

$$\frac{1}{2}+\frac{1}{2^2}+\frac{1}{2^3}+\frac{1}{2^4}+\cdots+\frac{1}{2^n}$$

To help you discover this sum, complete these partial sums and search for a pattern.

(a) $\frac{1}{2}+\frac{1}{2^2}=\frac{1}{2}+\frac{1}{4}=?$

(b) $\frac{1}{2}+\frac{1}{2^2}+\frac{1}{2^3}=\frac{1}{2}+\frac{1}{4}+\frac{1}{8}=?$

(c) $\frac{1}{2}+\frac{1}{2^2}+\frac{1}{2^3}+\frac{1}{2^4}=\frac{1}{2}+\frac{1}{4}+\frac{1}{8}+\frac{1}{16}=?$

28. Consider the case of the carpenter making a walnut jewelry box for a customer. The customer asked for a box whose interior dimensions are such that the length is twice the width and the height is half the width. If the length is 16 centimeters, what is the volume of the interior of the box?

29. Suppose that someone gave you a puzzle with the six clues A, B, C, D, E and F.

A. There are four students in a room. Their names are Fred, Joe, Kathy, and Linda.
B. Each student has one and only one favorite color among the colors red, pink, green, and yellow.
C. Kathy does not like pink.
D. The name of each student has a different number of letters than the name of the student's favorite color.
E. One of the boys, Joe and Fred, likes red and the other likes green.
F. No two students have the same favorite colors.

Which colors are which students' favorites?

***30.** If each different letter in the adjacent problem stands for a different digit, what replacements for the letters by digits will result in a correct statement?

```
   S E N D
+  M O R E
 M O N E Y
```

EXPLORATIONS Consider a triangular array of circles as shown in the figure. We are asked to place the numbers 1 through 9 in the circles so that the same sum is obtained for the four numbers along each side of the figure. In addition, we would like to obtain the smallest possible sum in figure (a) and the largest possible sum in figure (b).

(*a*) The sum is 17 along each side. (*b*) The sum is 23 along each side.

The explorations found at the end of every section of this book are designed to provide the basis for further study and understanding of the topic under discussion.

To obtain the smallest sum it seems logical to place the smallest numbers (1, 2, 3) in the corner circles since they will each be added along two different directions. Similar reasoning indicates that we should place the largest numbers (7, 8, 9) in the corner circles to obtain the largest sum. The remaining numbers are placed in the circles by a trial-and-error procedure to obtain the results.

The following explorations provide guidelines for trying to obtain triangular arrays with sums 17, 18, 19, 20, 21, 22, and 23, that is, for all numbers from the smallest possible to the largest. As an aid to finding patterns that might be useful, copy and extend this table as you complete the explorations.

Sum along each side	17	18	19	20	21	22	23
Sum of numbers in corners	6						24

1. Construct a triangle with the sum of 19 along each side. (*Hint*: Try the numbers 2, 3, 7 or the numbers 1, 4, 7 in the corners.)
2. Repeat Exploration 1 for the sum of 21. (*Hint*: The sum of the numbers in the corners will be 18.)
3. Study the entries in the table for the triangles completed thus far. What appears to be the pattern for the numbers that represent the sums of the corner numbers?
4. Look at the apparent pattern in the table. What do you conjecture for the sum of the corner numbers for a triangle whose sum will be 20 along each side? Construct such a triangle. Try to find more than one solution. Very often there may be several different solutions possible for a problem.
5. Unfortunately, it is impossible to construct triangles with the sums of 18 or 22 along each side. Experiment with a few cases to convince yourself that such is the case. Although this may shock some readers, it is well to note that the solutions of some problems in mathematics are proofs that the problems are impossible, that is, the problems do not have numerical or other ordinary answers as solutions.
6. Read *Problem Solving in School Mathematics*, the 1980 Yearbook of the National Council of Teachers of Mathematics. Prepare a report on three of the chapters of the book.

1-2 Problem Solving with Arithmetic Patterns

A search for patterns is often a significant way to solve problems or to establish conjectures that can lead to the solution of problems. Furthermore, many individuals find the exploration of patterns to be one of the most fascinating aspects of the study of mathematics. Patterns appear in many branches of mathematics, especially in the study of arithmetic. We shall begin by exploring one of the first tables that a student encounters in elementary arithmetic, a table of the **natural numbers** through 100.

Natural numbers are often called **counting numbers**.

Before reading on see how many different patterns in this number table you can discover by yourself.

1	2	3	4	5	6	7	8	9	10
11	12	13	14	15	16	17	18	19	20
21	22	23	24	25	26	27	28	29	30
31	32	33	34	35	36	37	38	39	40
41	42	43	44	45	46	47	48	49	50
51	52	53	54	55	56	57	58	59	60
61	62	63	64	65	66	67	68	69	70
71	72	73	74	75	76	77	78	79	80
81	82	83	84	85	86	87	88	89	90
91	92	93	94	95	96	97	98	99	100

There are many different patterns that can be found in this table; we shall explore a few of these and suggest others in the exercises at the end of this section. For example, the first ten natural numbers are on the horizontal line at the top (the first row). The *multiples* of 10 are on the vertical line at the right (the last column). Can you locate the multiples of 11? These begin with the first entry in the second row and continue down to the right:

11, 22, 33, 44, 55, 66, 77, 88, 99

Strategy: Search for a pattern.

Do you see why the multiples of 11 are on a line through the first entry in the second row and extending down diagonally to the right? As you move to a position one place to the right of a number in the table you are effectively adding one to the number; as you move one space down, you are adding 10.

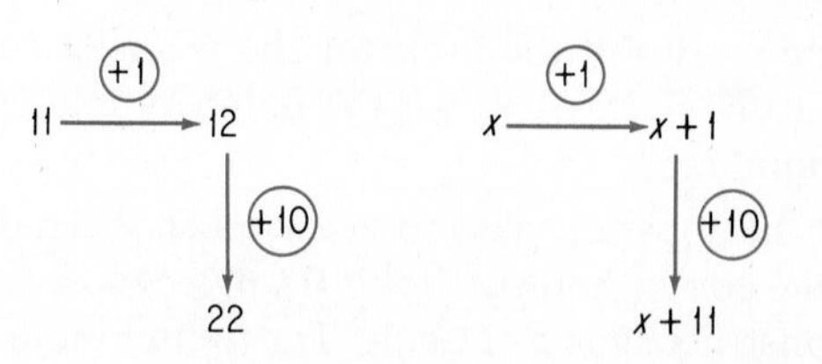

Next consider the pattern of multiples of 9. The multiples of 9 are on a line through 9 in the first row and extending down diagonally to the left:

9, 18, 27, 36, 45, 54, 63, 72, 81

Each movement one space to the left on the table has the effect of subtracting 1 from a number; each movement down adds 10 to a number. The adjacent diagrams show why the multiples of 9 are on the designated line.

Before leaving the number table, let us explore several more patterns. For example, consider any rectangular array of numbers and find the sums of the numbers in opposite corners. Here are two examples.

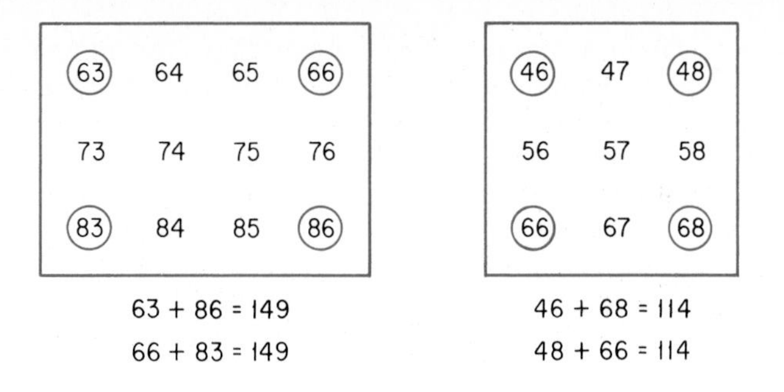

Strategy: Make an educated guess or conjecture. To establish a proof, generalize by using x, $x + 1$, $x + 2$, and so on.

Do you think that the sums of the numbers in the opposite corners of such rectangular arrays will *always* be equal? Try a few more of your own and then conjecture your answer. Finally, see if you can offer a reasonable *proof* for a particular array that you select.

Earlier in this section we discovered a diagonal line in the number table that contained the multiples of 9. We now consider these multiples again in a vertical array.

Most people speak of "adding digits," "subtracting from the units digit." and so on, as we have done. Many teachers are more precise in their terminology and recognize that digits are *numerals*, that is, symbols for numbers, rather than numbers. Numerals can be written. Only numbers can be added or subtracted. However, unless the more precise terminology is needed to avoid major confusion, we use the commonly accepted phraseology.

$$\begin{aligned} 1 \times 9 &= 9 \\ 2 \times 9 &= 18 \\ 3 \times 9 &= 27 \\ 4 \times 9 &= 36 \\ 5 \times 9 &= 45 \\ 6 \times 9 &= 54 \\ 7 \times 9 &= 63 \\ 8 \times 9 &= 72 \\ 9 \times 9 &= 81 \end{aligned}$$

Before reading on, see what patterns you can find in this column of multiples.

You may observe that the sum of the digits in each case is always 9. You should observe that the units digit decreases (9, 8, 7, ...), whereas the tens digit increases (1, 2, 3, ...). What lies behind this pattern?

Consider the product

$$2 \times 9 = 18$$

This is another explanation for the appearance of the multiples of 9 along a diagonal line in the number table shown at the start of this section.

To find 3×9 we need to add 9 to 18. Instead of adding 9, we may add 10 and subtract 1.

$$\begin{array}{r} 18 \\ +\ 10 \\ \hline 28 \end{array} \qquad \begin{array}{r} 28 \\ -\ 1 \\ \hline 27 \end{array} = 3 \times 9$$

That is, by adding 1 to the tens digit, 1, of 18, we are really adding 10 to 18. We then subtract 1 from the units digit, 8, of 18 to obtain 27 as our product.

The number 9, incidentally, has other convenient properties such as procedures for multiplying by 9 on one's fingers. For example, to multiply 9 by 3, place both hands together as in the adjacent figure, and bend the third finger from the left. The result is read as 27.

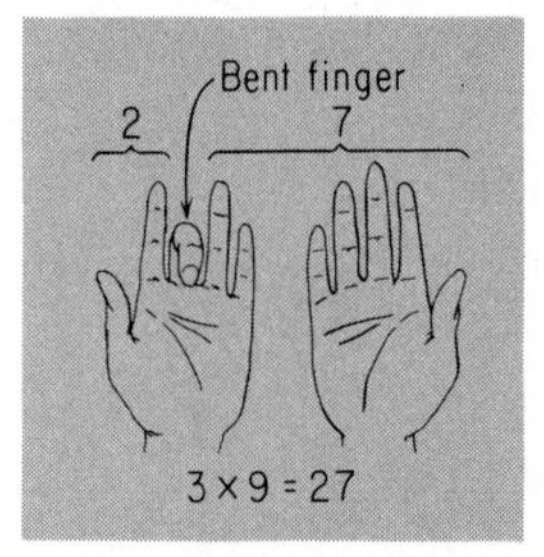

The next figure shows the procedure for finding the product 7×9. Note that the seventh finger from the left is bent, and the result is read in terms of the tens digit on the left of the bent finger and the units digit to the right. (Note that a thumb is considered to be a finger.) What number fact is shown in the figure on the right?

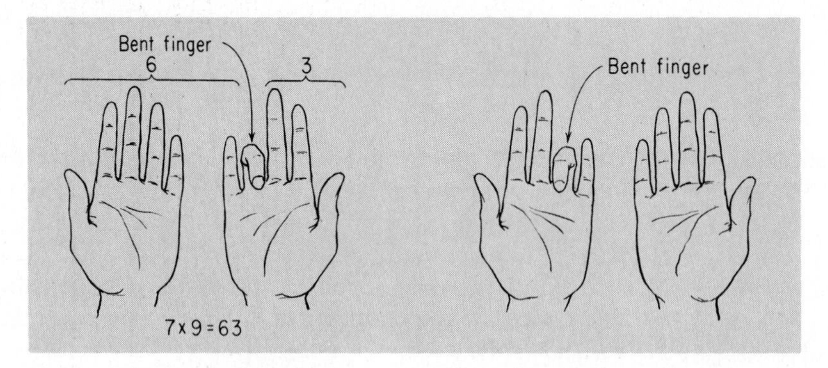

Here is one more pattern related to the number 9.

Use a calculator to verify that each of these statements is correct.

$$1 \times 9 + 2 = 11$$
$$12 \times 9 + 3 = 111$$
$$123 \times 9 + 4 = 1{,}111$$
$$1{,}234 \times 9 + 5 = 11{,}111$$
$$12{,}345 \times 9 + 6 = 111{,}111$$

Try to find a correspondence of the number of 1's in the number symbol on the right with one of the numbers used on the left. Now see if, without computation, you can supply the answers to the following questions.

$$123{,}456 \times 9 + 7 = ?$$
$$1{,}234{,}567 \times 9 + 8 = ?$$

Let us see *why* this pattern works. To do so we shall examine just one of the statements. A similar explanation can be offered for each of the other statements. Consider the statement

$$12{,}345 \times 9 + 6 = 111{,}111$$

Note here that we discover a pattern, extend it to other cases, and then attempt to verify the pattern through use of a specific case.

We express 12,345 as a sum of five numbers.

$$\begin{array}{r} 11{,}111 \\ 1{,}111 \\ 111 \\ 11 \\ 1 \\ \hline 12{,}345 \end{array}$$

We multiply each of these five numbers by 9.

$$\begin{aligned} 11{,}111 \times 9 &= 99{,}999 \\ 1{,}111 \times 9 &= 9{,}999 \\ 111 \times 9 &= 999 \\ 11 \times 9 &= 99 \\ 1 \times 9 &= 9 \end{aligned}$$

Finally, we add 6 by adding six 1's as in the following array, and find the total sum.

$$
\begin{aligned}
99{,}999 + 1 &= 100{,}000 \\
9{,}999 + 1 &= 10{,}000 \\
999 + 1 &= 1{,}000 \\
99 + 1 &= 100 \\
9 + 1 &= 10 \\
1 &= \underline{1} \\
 & \ 111{,}111
\end{aligned}
$$

Study the pattern in the following table. Then try to write the next four lines without doing any computation. Finally, verify your entries by completing the necessary multiplications; use a calculator if one is available.

$$
\begin{aligned}
1 \times 1 &= 1 \\
11 \times 11 &= 121 \\
111 \times 111 &= 12{,}321 \\
1{,}111 \times 1{,}111 &= 1{,}234{,}321 \\
11{,}111 \times 11{,}111 &= 123{,}454{,}321
\end{aligned}
$$

Do you think that the pattern displayed will continue indefinitely? Compute the product $1{,}111{,}111{,}111 \times 1{,}111{,}111{,}111$ to help you answer this question.

Interesting discoveries can often be made by studying arithmetic patterns. A famous German mathematician by the name of Carl Gauss (1777–1855) is said to have been a precocious child who would often drive his teachers to despair. The story is told that on one occasion his teacher asked him to add the first 100 natural numbers, hoping to keep him suitably occupied for some time. Instead, young Gauss recognized a pattern, and gave the answer immediately. He is said to have found the desired sum as indicated in the following array.

Germany honored this famous mathematician by producing a stamp in his honor. Find out which mathematician has been pictured on a U.S. postage stamp.

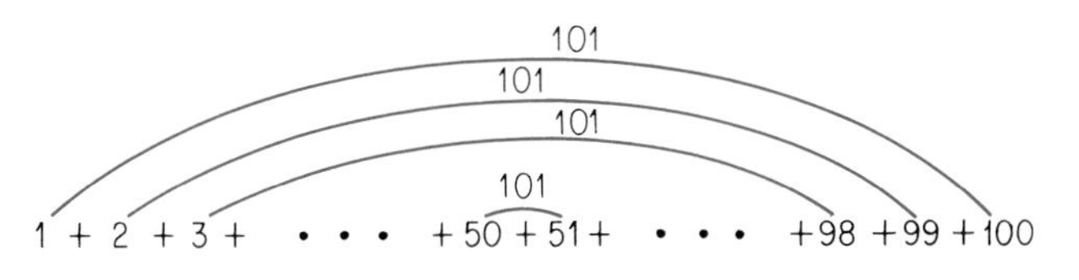

Gauss reasoned that there would be 50 pairs of numbers, each with a sum of 101 (consider $100 + 1, 99 + 2, 98 + 3, \ldots, 50 + 51$). Thus the sum is 50×101, that is, 5050.

Example Use a method similar to that of Gauss to find the sum of the first 200 natural numbers.

Solution Consider the sum $1 + 2 + 3 + \cdots + 198 + 199 + 200$. The sum of the first and last numbers is 201, and there will be 100 pairs of numbers, each with this sum. Thus the total sum is 100×201, that is, 20,100. ■

We conclude this section with the following problem. Suppose that you are offered a job for a month of 30 days with the option of being paid \$1000 per day or at the rate of 1¢ for the first day, 2¢ for the second day, then 4¢, 8¢, 16¢, and so forth. That is, your salary would double each day. Which option would you choose?

Strategy: Solve a simpler but related problem. This problem-solving strategy will often provide you with the clues to solve the more difficult problem.

Of course, one could list the 30 daily salaries and add, but this would be a tedious job indeed. Let us, rather, rely on a very helpful tool used in problem solving: when possible, consider a similar but smaller task first. Thus we will consider total salaries in cents for 5, for 6, for 7, and for 8 days first.

For 5 days	*For 6 days*	*For 7 days*	*For 8 days*
1	1	1	1
2	2	2	2
4	4	4	4
8	8	8	8
16	16	16	16
31	32	32	32
	63	64	64
		127	128
			255

Do you see a pattern emerging? Compare the total for the first 5 days with the salary for the sixth day; compare the total for 6 days with the salary for the seventh day. Notice that the total for 5 days is 1¢ less than the salary for the sixth day; the total for 6 days is 1¢ less than the salary for the seventh day; and so forth. Thus your total salary for 10 days will be 1¢ less than your salary for the eleventh day. That is,

$$1 + 2 + 4 + 8 + 16 + 32 + 64 + 128 + 256 + 512 = 2(512) - 1 = 1023$$

Your salary for the eleventh day would be 1024¢; in 10 days you will earn a *total* of 1023¢, or \$10.23.

To use this approach to answer the original question will require a good deal more work but will still be easier than the addition of 30 amounts. By a doubling process we first need to find your salary for the thirtieth day. Your total salary for all 30 days can then be found by doubling this amount (to find your salary for the thirty-first day), and subtracting 1.

To generalize, we note that the doubling process starting with 1 gives **powers of 2**.

$$2^0 = 1 \qquad 2^1 = 2 \qquad 2^2 = 4 \qquad 2^3 = 8 \qquad 2^4 = 16 \qquad 2^5 = 32 \qquad \ldots$$

Then, by our discovery of a pattern, we may say that

$$2^0 + 2^1 + 2^2 + 2^3 + \cdots + 2^n = 2^{n+1} - 1$$

When $n = 5$, we have

$$2^0 + 2^1 + 2^2 + 2^3 + 2^4 + 2^5 = 1 + 2 + 4 + 8 + 16 + 32 = 2^6 - 1 = 63$$

To find your total salary for 30 working days, we need to find the sum

$$\underset{\text{1st day}}{2^0} + \underset{\text{2nd day}}{2^1} + \underset{\text{3rd day}}{2^2} + \cdots + \underset{\text{30th day}}{2^{29}}$$

From the preceding discussion we know that this sum is equal to $2^{30} - 1$. We can use a calculator, can compute 2^{30} by a doubling process, or can use a shortcut such as the following.

$$2^{10} = 2 \times 2^9 = 2 \times 512 = 1024$$

Alternatively, we may say that

$$2^{10} = 2^5 \times 2^5 = 32 \times 32 = 1024$$

In a similar manner we compute

$$2^{30} = 2^{10} \times 2^{10} \times 2^{10} = 1024 \times 1024 \times 1024 = 1{,}073{,}741{,}824$$

and thus $2^{30} - 1 = 1{,}073{,}741{,}823$. In 30 days you would earn a total of 1,073,741,823¢, that is, $10,737,418.23, which is substantially more than the $30,000 that you would earn at $1000 per day.

EXERCISES

In the exercises the reader will have opportunities to solve problems and make discoveries through careful explorations of patterns.

1. Verify that the process for finger multiplication shown in this section will work for each of the multiples of 9 from 1×9 through 9×9. For each product, count from the left to the bent finger and specify its position as first, second, third, and so on.
2. Follow the procedure outlined in this section and show that

   $$1234 \times 9 + 5 = 11{,}111$$

3. Study the following pattern and use it to express the squares of 6, 7, 8, and 9 in the same manner.

   $$1^2 = 1$$
   $$2^2 = 1 + 2 + 1$$
   $$3^2 = 1 + 2 + 3 + 2 + 1$$
   $$4^2 = 1 + 2 + 3 + 4 + 3 + 2 + 1$$
   $$5^2 = 1 + 2 + 3 + 4 + 5 + 4 + 3 + 2 + 1$$

4. Study the entries that follow and use the pattern that is exhibited to complete the last four rows.

   $$1 + 3 = 4, \quad \text{that is,} \quad 2^2$$
   $$1 + 3 + 5 = 9, \quad \text{that is,} \quad 3^2$$
   $$1 + 3 + 5 + 7 = 16, \quad \text{that is,} \quad 4^2$$
   $$1 + 3 + 5 + 7 + 9 = ?$$
   $$1 + 3 + 5 + 7 + 9 + 11 = ?$$
   $$1 + 3 + 5 + 7 + 9 + 11 + 13 = ?$$
   $$1 + 3 + 5 + \cdots + (2n - 1) = ?$$

5. An addition problem can be checked by a process called **casting out nines**. To do this, you first find the sum of the digits of each of the **addends** (numbers that are added), divide by 9, and record the remainder. Digits may be added again and again until a one-digit remainder is obtained. The sum of these remainders is then divided by 9 to find a final remainder. This should be equal to the remainder found by considering the sum of the addends (the answer), adding its digits, dividing the sum of these digits by 9, and finding the remainder. Here is an example.

Addends	*Sum of digits*	*Remainders*
4,378	22	4
2,160	9	0
3,872	20	2
1,085	14	5
11,495		11

When the sum of the remainders is divided by 9, the final remainder is 2. This corresponds to the remainder obtained by dividing the sum of the digits in the answer ($1 + 1 + 4 + 9 + 5 = 20$) by 9.

Try this procedure for several other examples and verify that it works in each case.

6. Try to discover a procedure for checking multiplication by casting out nines. Verify for several cases that this procedure works.

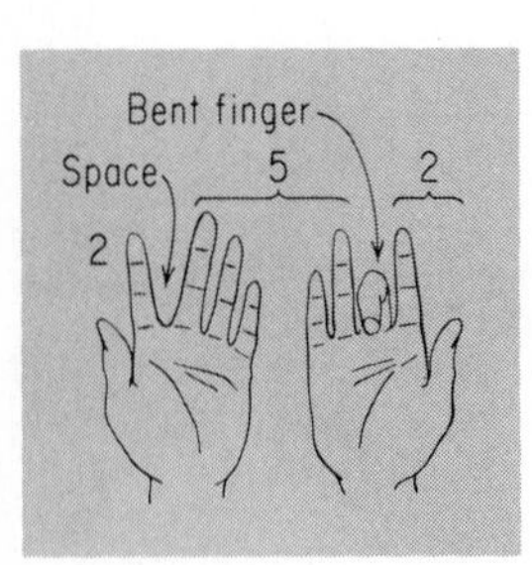

7. There is a procedure for multiplying a two-digit number by 9 on one's fingers provided that the tens digit is smaller than the units digit. The accompanying diagram shows how to multiply 28 by 9. Reading from the left, put a space after the second finger and bend the eighth finger. Read the product in groups of fingers as 252. Use this procedure to find
(a) 9×47 **(b)** 9×39 **(c)** 9×18 **(d)** 9×27.
Check each of the numerical answers that you have obtained.

8. John offered to work for 1¢ the first day, 2¢ the second day, 4¢ the third day, and so on, doubling the amount each day. Bill offered to work for $30 a day. Which boy would receive more money for a job that lasted
(a) 10 days **(b)** 15 days **(c)** 16 days?

9. Using a method similar to that of Gauss, find
(a) The sum of the first 90 natural numbers.
(b) The sum of the first 300 natural numbers.
(c) The sum of all the odd numbers from 1 through 39.
(d) The sum of all the odd numbers from 1 through 299.
(e) The sum of all the even numbers from 2 through 500.

10. Use the results obtained in Exercise 9 and try to find a formula for the sum of
(a) The first n natural numbers, that is,

$$1 + 2 + 3 + \cdots + (n - 1) + n$$

(**b**) The first n odd numbers, that is,

$$1 + 3 + 5 + \cdots + (2n - 3) + (2n - 1)$$

For Exercises 11 *through* 15 *use the number table given at the beginning of this section.*

11. Consider any square array of nine numbers. Find the sum of the numbers and compare this with the number in the center of the array. Form a conjecture and test this by using several other such square arrays.

12. Draw a rectangle around any four numbers in a row. Compare the sum of the two outer numbers with the sum of the two inner numbers. State what you notice about these sums and check your conjecture with several other sets of four numbers in a row.

13. Repeat Exercise 12 for four numbers in a column.

14. Repeat Exercise 12 for four numbers along any diagonal.

15. Make a new table by replacing each of the two-digit numbers in the given table, except those that are multiples of 10, by the fraction that is the ratio of the first digit to the second. For example, the third row becomes

$$\frac{3}{1} \quad \frac{3}{2} \quad \frac{3}{3} \quad \frac{3}{4} \quad \frac{3}{5} \quad \frac{3}{6} \quad \frac{3}{7} \quad \frac{3}{8} \quad \frac{3}{9}$$

Consider these fractions and test the following conjectures for any rectangular array of such fractions.

(**a**) The sums of the fractions in opposite corners are equal.

(**b**) The products of the fractions in opposite corners are equal.

16. Write any three-digit number, such as 123. Repeat the digits to form a six-digit number, such as 123,123. Now divide this number successively by 7, by 11, and by 13. Repeat this procedure for several other three-digit numbers and state a discovery that you find. Then try to show why this works. (*Hint*: Find the product $7 \times 11 \times 13$.)

17. Find the number of different ways that you can add four odd natural numbers to obtain a sum of 10. For this exercise, a sum such as $7 + 1 + 1 + 1$ is considered to be the same as the sum $1 + 1 + 7 + 1$. Use the strategy of making a list.

***18.** Find the smallest natural number such that when that number is divided by 2 the result is a perfect square and when it is divided by 3 the result is a perfect cube. (A *perfect square* is a number such as 1, 4, 9, 16, . . .; a *perfect cube* is a number such as 1, 8, 27, 64, ...) For this exercise use a judicious trial-and-error approach.

Perfect squares	Perfect cubes
$1 \times 1 = 1$	$1 \times 1 \times 1 = 1$
$2 \times 2 = 4$	$2 \times 2 \times 2 = 8$
$3 \times 3 = 9$	$3 \times 3 \times 3 = 27$
....	....

1	1	1
3	3	3
5	5	5
7	7	7

*19. In the array shown in the margin, circle three numbers whose sum is 22.

*20. Obtain a true statement by changing the position of one digit only: $102 + 1 = 101$. (No other symbols are to be moved.)

EXPLORATIONS

Consider the following table of multiplication facts from 1×1 through 9×9.

X	1	2	3	4	5	6	7	8	9
1	1	2	3	4	5	6	7	8	9
2	2	4	6	8	10	12	14	16	18
3	3	6	9	12	15	18	21	24	27
4	4	8	12	16	20	24	28	32	36
5	5	10	15	20	25	30	35	40	45
6	6	12	18	24	30	36	42	48	54
7	7	14	21	28	35	42	49	56	63
8	8	16	24	32	40	48	56	64	72
9	9	18	27	36	45	54	63	72	81

Try to find as many patterns as possible in the table. Here are several to get you started.

1. The entries in the diagonal from upper left to lower right are all perfect squares: 1, 4, 9, 16, 25, 36, 49, 64, 81. Explain why this is so.
2. The units digit for the multiples of 2 repeat: 2, 4, 6, 8, 0, 2, 4, 6, 8. Find other examples of such repetitions.
3. Note the sum of the digits for the multiples of 3: 3, 6, 9, 3, 6, 9, 3, 6, 9. Explore the sums of digits for other multiples in the table.
4. As in the two examples in the given figure select any square array of four numbers from the given table of multiplication facts. Find the cross products for the selected array. Explain why such cross products will always be equal.

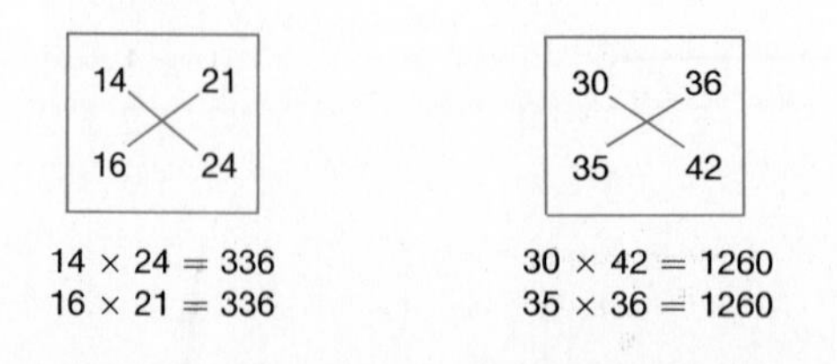

$14 \times 24 = 336$
$16 \times 21 = 336$

$30 \times 42 = 1260$
$35 \times 36 = 1260$

5. Extend the concept of Exploration 4 for the cross products of the corner entries in any rectangular array obtained from the given multiplication table.
6. For each column consider the respective entries in any two rows as the numerators and denominators of fractions. Then for rows 2 and 3 we obtain these fractions.

$$\frac{2}{3} \quad \frac{4}{6} \quad \frac{6}{9} \quad \frac{8}{12} \quad \frac{10}{15} \quad \frac{12}{18} \quad \frac{14}{21} \quad \frac{16}{24} \quad \frac{18}{27}$$

Show that for any two rows, not necessarily successive ones, a set of *equivalent fractions* is obtained. Explain why this is so.
7. Find the sum of all of the entries within the body of the table of multiplication facts. (*Hint*: Begin with several smaller tables first. Compare the sum of the entries within the table with the sum of the numbers across the top or side of the table, as in these two examples.)

This is another example of the problem-solving strategy of using a simpler but related problem.

X	1	2
1	1	2
2	2	4

Sum of entries = 9

Sum of numbers used as row (column) headings = 3

X	1	2	3
1	1	2	3
2	2	4	6
3	3	6	9

Sum of entries = 36

Sum of numbers used as row (column) headings = 6

1-3 Problem Solving with Logical Reasoning

Our English language and the structure it provides assists us in the communication, and understanding, of many problem situations. This aspect of problem solving deserves much more attention than it usually receives.

A sentence that can be identified as true or identified as false is often called a **statement**. Each of the following sentences is an example of a **simple statement**.

Commands such as "Stand up and be counted" and greetings such as "Hello" are neither true nor false and are not considered to be statements as we are using the word here.

I have made the last payment on my car.

I have recently purchased all of the books required for this semester.

A **compound statement** is formed by combining two or more simple statements, as in the following example.

I have made the last payment on my car *and* I have purchased all of the books required for this semester.

In this illustration the two simple statements are combined by the connective *and*. Other connectives could have been used. Consider the same simple statements used with the connective *or*.

> I have made the last payment on my car *or* I have purchased all of the books required for this semester.

We shall consider such compound statements and determine the conditions under which they are true or false, assuming that the simple statements are true or false. In doing this we use symbols to represent connectives and letters or variables to represent statements. The most common **connectives** are *and* ($\wedge$), *or* ($\vee$), and *not* ($\sim$). Then compound statements may be written in symbolic form and symbolic statements may be translated into words. For example, we may use p and q to represent these simple statements.

p: I have made the last payment on my car.
q: I have purchased all of the books required for this semester.

Connectives

$\wedge$: and
$\vee$: or
$\sim$: not

Then we may write the following compound statements.

$p \wedge q$: I have made the last payment on my car and I have purchased all of the books required for this semester.

$p \vee q$: I have made the last payment on my car or I have purchased all of the books required for this semester.

$\sim p$: I have not made the last payment on my car.

Example 1 Use p and q as in the preceding discussion and translate.
(a) $p \wedge (\sim q)$ **(b)** $\sim(\sim p)$

Solution
(a) I have made the last payment on my car and I have not purchased all of the books required for this semester.
(b) It is not true that I have not made the last payment on my car; that is, I have made the last payment on my car. ■

The statement $\sim(\sim p)$ is equivalent to, has the same meaning as, the statement p.

Example 2 Use p and q as in the preceding discussion and write each statement in symbolic form.
(a) I have not made the last payment on my car or I have not purchased all of the books for this semester.
(b) I have completed neither the last payment on my car nor the purchase of all the books required for the semester.

Solution **(a)** $(\sim p) \vee (\sim q)$ **(b)** $(\sim p) \wedge (\sim q)$ ■

p	$\sim p$
T	F
F	T

The truth values of compound statements depend on the truth values of the simple statements that are used.

$\sim p$ is true if p is false, false if p is true; that is, $\sim p$ and p are **contradictory statements**; each statement is the *negation* of the other. Such comparisons of truth values may be represented as in the adjacent array by a table of truth values, that is, a **truth table**.

$p \wedge q$ is true if *both* p and q are true, false in all other cases.

$p \vee q$ is true if *at least one of* the statements p, q is true; false if both p and q are false.

Note that the statement $p \vee q$ is true unless both p and q are false.

The last statement describes the *inclusive use* of *or* since if p or q or both p and q are true, then $p \vee q$ is true. The word *or* is widely used in everyday speech. Consider these statements.

At nine o'clock tomorrow morning I shall either be in a conference in New York or in my dentist's office in Boston.

Jane looks either very tired or ill.

The first of these two illustrations shows the *exclusive use* of *or*, since it is not possible for both statements to be true. The **exclusive or** (often denoted by $\underline{\vee}$, called *vel*) is true when, and only when, exactly one of the given statements is true. The second illustration shows the *inclusive use* of *or* since at least one, or possibly both, of the statements may be true. In mathematics, *or* is assumed to be the inclusive *or* unless otherwise specified.

p	q	$p \wedge q$	$p \vee q$
T	T	T	T
T	F	F	T
F	T	F	T
F	F	F	F

The statement $p \wedge q$ is the **conjunction** of p and q; the statement $p \vee q$ is the **disjunction** of p and q. Both conjunctions and disjunctions may be negated. A preface of "It is not true that" is sufficient, but other forms may be used. The truth values of conjunctions and disjunctions may be summarized in a truth table. Note that if a statement p is true, then a second statement q may be either true or false; also if p is false, then q may be true or false.

Example 3 Write two forms of the negation of the given statement.

(a) The weather is hot and I need to work outside.

(b) Today is a rainy day or it is Saturday.

Solution **(a)** The given statement has the form $p \wedge q$. The negation may be stated as

It is not true that the weather is hot and I need to work outside.

Note that the negation is true under any one of these three pairs of conditions.

The weather is hot and I do not need to work outside.
The weather is not hot and I need to work outside.
The weather is not hot and I do not need to work outside.

The negation of the given statement also may be stated as

Either the weather is not hot or I do not need to work outside.

(b) The given statement has the form $p \vee q$. The negation may be stated as

It is not true that today is a rainy day or it is not Saturday.

Note that the negation is true under only one pair of conditions and may also be stated as that pair of conditions.

Today is not a rainy day and it is Saturday. ■

Two statements that have the same truth values under all conditions are **equivalent statements**. The given statements and the second forms stated for their negations in Examples 3(a) and 3(b) provide illustrations for **De Morgan's laws for statements**.

Compare with De Morgan's laws for sets (Section 2-4.)

$\sim(p \wedge q)$	is equivalent to	$(\sim p) \vee (\sim q)$
$\sim(p \vee q)$	is equivalent to	$(\sim p) \wedge (\sim q)$

The British mathematician Augustus De Morgan (1806–1871) was an outstanding writer and teacher. A champion of religious and intellectual tolerance, he was one of the founders of the British Association for the Advancement of Science (1831). He is well known for his *Budget of Paradoxes*, a collection of many of his witticisms, that was edited after his death by his widow, Sophia Elizabeth De Morgan.

Many of the statements that we make in everyday conversation are based upon a condition. For example, consider the following statements.

If the telephone rings, then Jane will answer it.
If I have no homework, then I shall go bowling.

Each of these statements is expressed in the **if-then** form

If p, then q.

Any if-then statement can be expressed in symbols as $p \to q$, which is read "if p, then q." The **conditional symbol** $\to$ is another connective. It is used to form a compound statement $p \to q$, a **conditional statement.**

Our first task is to consider the various possibilities for p and q in order to define $p \to q$ for each of these cases. One way to do this would be to present a completed truth table and to accept this as our definition of $p \to q$. However, since a useful definition should fit the observed facts of our everyday world, let us attempt to justify the entries in such a table. Consider again the conditional statement

Note that definitions may be made to fit observed facts.

If the telephone rings, then Jane will answer it.

If the telephone rings and Jane answers it, then the given statement is obviously true. On the other hand, the statement is false if the telephone rings and Jane does not answer it. Assume now that the telephone does not ring. Then since there is no circumstance in which Jane fails to answer the telephone, there is no circumstance in which the given statement is false. Since the given statement cannot be false, we *define* it to be true.

p	q	$p \to q$
T	T	T
T	F	F
F	T	T
F	F	T

Consider the statement

If it rains, then I shall give you a ride home.

Have I lied to you

1. If it rains and I give you a ride home?
2. If it rains and I do not give you a ride home?
3. If it does not rain and I give you a ride home?
4. If it does not rain and I do not give you a ride home?

The given statement $p \rightarrow q$ is true unless the *premise* p is true and the *conclusion* q is false.

According to the accepted meanings of the words used, you have a right to feel that I lied to you only if it rains and I do not give you a ride home. In any conditional statement $p \rightarrow q$, the statement p is the **premise** and the statement q is the **conclusion**. Thus the conditional statement is true unless the premise is true and the conclusion is false.

Remember also that there need be no relationship between p and q in an if-then statement although we tend to use related statements in this way in everyday life.

If p is false, then the statement $p \rightarrow q$ is accepted as true regardless of the truth value of q. Under this definition each of the following statements is true.

If General George Washington, born in 1732, is alive today, then he is now the President of the United States of America.

If $2 + 3 = 7$, then the moon is made of green cheese.

If Wednesday is the day after Monday, then July is the month after June.

Example 4 Give the truth value of each statement.
(a) If $5 + 7 = 12$, then $6 + 7 = 13$.
(b) If $5 \times 7 = 35$, then $6 \times 7 = 36$.
(c) If $5 + 7 = 35$, then $6 + 7 = 13$.
(d) If $5 + 7 = 35$, then $6 \times 7 = 36$.

Solution Think of each statement in the form $p \rightarrow q$.
(a) For p true and q true, the statement $p \rightarrow q$ is *true*.
(b) For p true and q false, the statement $p \rightarrow q$ is *false*.
(c) For p false and q true, the statement $p \rightarrow q$ is *true*.
(d) For p false and q false, the statement $p \rightarrow q$ is *true*. ■

Any conditional *statement* such as

If I am hungry, then I eat

has three closely related conditional statements: the *inverse* of the given statement

If I am not hungry, then I do not eat,

the *converse* of the given statement

If I eat, then I am hungry,

and the *contrapositive* of the given statement

If I do not eat, then I am not hungry.

Statement:	$p \rightarrow q$	If p, then q.
Converse:	$q \rightarrow p$	If q, then p.
Inverse:	$(\sim p) \rightarrow (\sim q)$	If not p, then not q.
Contrapositive:	$(\sim q) \rightarrow (\sim p)$	If not q, then not p.

Each statement is false if its premise is true and its conclusion is false; otherwise the statement is true, as shown in the truth tables.

Note from these truth tables that:
Any conditional statement is equivalent to its contrapositive.
The converse of any conditional statement is equivalent to the inverse.

		STATEMENT			CONVERSE			INVERSE			CONTRAPOSITIVE		
p	q	$p \to q$			$q \to p$			$(\sim p) \to (\sim q)$			$(\sim q) \to (\sim p)$		
T	T	T	T	T	T	T	T	F	T	F	F	T	F
T	F	T	F	F	F	T	T	F	T	T	T	F	F
F	T	F	T	T	T	F	F	T	F	F	F	T	T
F	F	F	T	F	F	T	F	T	T	T	T	T	T

The following truth table shows that the negation of any conditional statement $p \to q$ is equivalent to $p \wedge (\sim q)$.

p	q	$p \to q$	$\sim(p \to q)$	$p \wedge (\sim q)$
T	T	T	F	F
T	F	F	T	T
F	T	T	F	F
F	F	T	F	F

The equivalence of any conditional statement $p \to q$ and $\sim[p \wedge (\sim q)]$ is often used to obtain the negation of a given conditional statement. For example the conditional statement

If Sue is wearing a coat, then the temperature is below freezing

is equivalent to

It is not true that Sue is wearing a coat and the temperature is not below freezing

and has as its negation

Sue is wearing a coat and the temperature is not below freezing.

In part (b) of Example 5 the contrapositive has the same truth values as the original statement. If you fail the course, then you have not worked hard, because if you had worked hard, then you would have passed the course.

Example 5 Write (**a**) the negation (**b**) the contrapositive of the following statement. (The forms of verbs may be changed to fit the situation.)

If I work hard, then I shall pass the course.

Solution (**a**) I work hard and I shall not pass the course.
(**b**) If I do not pass the course, then I have not worked hard. ■

EXERCISES *Consider the following statements.*

p: You pass the examination.
q: You pass the course.

Then translate each symbolic statement into an English sentence.

1. $p \rightarrow q$
2. $q \rightarrow p$
3. $(\sim p) \rightarrow (\sim q)$
4. $(\sim q) \rightarrow (\sim p)$

Give each of these statements in words. Use p and q as follows.

p: I like this book.
q: I like mathematics.

5. **(a)** $p \wedge q$ **(b)** $\sim q$
   (c) $\sim (\sim p)$ **(d)** $(\sim p) \wedge (\sim q)$
6. **(a)** $(\sim p) \wedge q$ **(b)** $p \vee q$
   (c) $\sim (p \wedge q)$ **(d)** $\sim [(\sim p) \wedge q]$
7. Assume that you like this book and that you like mathematics. Which of the statements in Exercises 5 and 6 are true for you?
8. Assume that you like this book but that you do not like mathematics. Which of the statements in Exercises 5 and 6 are true for you?

Think of "short" as "not tall" and use p and q as follows.

p: Jim is tall.
q: Bill is not tall.

9. Write each of these statements in symbolic form.
   (a) Jim is short and Bill is tall.
   (b) Neither Jim nor Bill is tall.
   (c) Jim is not tall and Bill is short.
   (d) It is not true that Jim and Bill are both tall.
   (e) Jim or Bill is tall.
10. Assume that Bill and Jim are both tall. Which of the statements in Exercise 9 are true?

Think of "sad" as "not happy" and use p and q as follows.

p: Joan is happy.
q: Mary is sad.

11. Write each of these statements in symbolic form.
   (a) Joan and Mary are both happy.
   (b) Joan is happy or Mary is happy.
   (c) Neither Joan nor Mary is happy.
   (d) It is not true that Joan and Mary are both sad.
   (e) It is not true that neither Joan nor Mary is happy.

12. Assume that Joan and Mary are both happy. Which of the statements in Exercise 11 are true?

Write two forms of the negation for the given statement.

13. Today is Monday.
14. My car is a Ford.
15. These two cars are not made by the same company.
16. These two lines do not intersect.
17. I am young and I am happy.
18. I worked hard and I did not pass.
19. Pedro has $50 and he has two tickets to the game.
20. Marcia will drive her car or she will fly.
21. The textbook is expensive or it is not used.
22. A vacation should be enjoyable or it should be profitable.

Complete the indicated statement so that it has the same meaning as the given statement.

23. There is no college that is not expensive.
    Every ....
24. There are courses with no required textbooks.
    Not all ....

For Exercises 25 and 26 consider these pictures of children with hats as shown.

(i) (ii) (iii)

Use the numbers (i), (ii), (iii) *to designate the pictures for which the given statement is true.*

25. **(a)** Every child has a hat.
    (b) No child has a hat.
    (c) No child is without a hat.
    (d) Every child is without a hat.
26. **(a)** Not every child has a hat.
    (b) Not every child is without a hat.
    (c) There is a child with a hat.
    (d) There is a child without a hat.
27. Give the truth value of each statement.
    (a) If $2 \times 3 = 5$, then $2 + 3 = 6$.
    (b) If $2 \times 3 = 5$, then $2 + 3 = 5$.
    (c) If $2 + 3 = 5$, then $2 \times 3 = 5$.
    (d) If $2 + 3 = 5$, then $2 \times 3 = 6$.

28. Give the truth value of each statement.
(a) If $5 \times 6 = 56$, then $5 - 6 = 11$.
(b) If $5 \times 6 = 42$, then $5 - 6 = 10$.
(c) If $5 \times 6 = 30$, then $5 + 6 = 10$.
(d) If $5 + 6 = 11$, then $5 \times 6 = 30$.

29. Assume that $2x = 6$, $x = 3$, and $x \neq 4$. Then give the truth value of each statement.
(a) If $2x = 6$, then $x = 3$.
(b) If $2x = 6$, then $x = 4$.
(c) If $3 = 4$, then $x = 4$.
(d) If $3 = 4$, then $x = 3$.

30. Assume that $a \times b = c$, $b \times c = d$, and $c \neq d$. Then give the truth value of each statement.
(a) If $a \times b = c$, then $b \times c = d$.
(b) If $a \times b = d$, then $b \times c = c$.
(c) If $a \times b = d$, then $b \times c = d$.
(d) If $a \times b = c$, then $b \times c = c$.

Write the negation, converse, inverse, and contrapositive of each statement.

31. If $x = 1$, then $x \neq 2$.
32. If $2x = 6$, then $x = 3$.
33. If we can afford it, then we shall buy a new car.
34. If we play tennis, then you will win the game.

EXPLORATIONS

There exist situations in which statements are neither true nor false and our usual rules of logic cannot be used.

1. Select a plain 3×5 card or similar piece of paper. On one side write

   The statement on the other side of this card is true.

   Then on the other side of the card write

   The statement on the other side of this card is false.

   Discuss the possible sets of truth values for the statements on the two sides of the card.

Exploration 2 is a famous paradox. Can you explain it?

2. There is reported to be a town in which the barber is a man who shaves all men who do not shave themselves. Who shaves the barber?
3. Discuss the truth values of the statement

   The sentence you are reading is false.

4. Start a collection of logical paradoxes; that is, sets of statements that do not seem to "make sense" under our usual laws of logic. For example, consider the statements given in Explorations 1, 2, and 3.

1-4 Solving Verbal Problems with Logical Reasoning

The richness of our spoken and written language in providing a wide variety of forms for statements makes it necessary to be very careful in the analysis of statements in problem solving. Note that a statement such as

All trees have green leaves

may be stated as

If the object is a tree, then it has green leaves.

Similarly, the statement

Any college student can do algebra

may be restated as

If a person is a college student, then the person can do algebra.

However, note also the statement

Not all trees have green leaves

does *not* have the same meaning as

All trees do not have green leaves.

The words *necessary* and *sufficient* are often used to express conditional statements in alternate forms. Consider the statement

Working hard is a sufficient condition for passing the course.

Suppose that you will get an A only if you work hard. Does this mean that you will get an A if you work hard? See Exploration 4.

Let us use p to mean *work hard* and q to represent *pass the course.* We need to decide whether the given statement means "if p, then q" or "if q, then p." The word sufficient can be interpreted to mean that working hard is adequate or enough, but possibly not always necessary, for passing. That is, there may be other ways to pass the course, but working hard will do it. Thus we interpret the statement to mean

If you work hard, then you will pass the course.

Then $p \rightarrow q$ may represent each of these statements.

If p, then q.
p is a sufficient condition for q.

Next consider this statement.

Working hard is a necessary condition for passing the course.

Here you are told that working hard is necessary or essential in order to pass. That is, regardless of what else you do, you must work hard if you wish to pass. However, there is no assurance that working hard alone will do the trick. It is necessary, but may not be sufficient. (You may also have to get good grades.) Therefore, we interpret the statement to mean

If you pass the course, then you have worked hard.

Then $q \rightarrow p$ may represent each of these statements.

If q, then p.
p is a necessary condition for q.

Still another form to consider is the statement "q, only if p." In terms of the example used in this section, we may write this as

You will pass the course only if you work hard.

This does *not* say that working hard will insure a passing grade. It does mean that if you have passed, then you have worked hard. That is, "q, only if p" is equivalent to the statement "if q, then p." We can also interpret this in another way. This statement "q, only if p" means "if not p, then not q." The contrapositive of this last statement, however, is "if q, then p." In terms of our illustration, this means that if you do not work hard, then you will not pass. Therefore, if you pass, then you have worked hard.

To summarize our discussion, each of the following statements represents a form of the conditional statement $p \rightarrow q$.

If p, then q.
q, if p.
p is a sufficient condition for q.
q is a necessary condition for p.
p, only if q.

The many distinct ways of expressing a conditional statement illustrate the difficulty of understanding the English language. We shall endeavor to reduce the confusion by expressing conditional statements in the form

If p, then q. Symbolically, $p \rightarrow q$.

Example 1 Write each statement in if-then form.

(a) All rainy days are cloudy.
(b) $x = 5$, only if $x \neq 0$.
(c) Any apple is a piece of fruit.
(d) Cats are mammals.

Solution

(a) If a day is rainy, then it is cloudy.
(b) If $x = 5$, then $x \neq 0$.
(c) If an object is an apple, then it is a piece of fruit.
(d) If an animal is a cat, then it is a mammal. ■

Example 2 Translate into symbolic form, using p and q as follows.

p: I shall work hard.
q: I shall get an A.

(a) I shall get an A only if I work hard.
(b) Working hard will be a sufficient condition for me to get an A.
(c) If I work hard, then I shall get an A, and if I get an A, then I shall have worked hard.

Solution **(a)** $q \rightarrow p$ **(b)** $p \rightarrow q$ **(c)** $(p \rightarrow q) \wedge (q \rightarrow p)$ ■

$p \to q$:

p is a sufficient condition for *q*.

$q \to p$:

p is a necessary condition for *q*.

$p \leftrightarrow q$:

p is a necessary and sufficient condition for *q*.

The statement $(p \to q) \wedge (q \to p)$ in Example 2(c) is one form of the **biconditional statement** $p \leftrightarrow q$ (read as "*p* if and only if *q*"); the symbol $\leftrightarrow$ is the **biconditional symbol.** Any biconditional statement $p \leftrightarrow q$ is a statement that *p* is a sufficient condition for *q* and also *p* is a necessary condition for *q*. We may condense this by saying that *p* is a **necessary and sufficient condition** for *q*. The biconditional statement may be stated in either of these forms.

p is a necessary and sufficient condition for *q*.
p if and only if *q*; that is, *p* **iff** *q*.

We may think of any biconditional statement $p \leftrightarrow q$ as a conjunction of conditional statements. Since $p \to q$ is true unless *p* is true and *q* is false and $q \to p$ is true unless *q* is true and *p* is false, both conditional statements and the biconditional statement $p \leftrightarrow q$ are true in the remaining two cases, that is, when *p* and *q* are both true or both false. Thus each of these biconditional statements is true.

$2 \times 2 = 4$ if and only if $7 - 5 = 2$ (both parts are true)
$2 \times 2 = 5$ if and only if $7 - 5 = 3$ (both parts are false)

Note that $p \leftrightarrow q$ is true if and only if *p* and *q* have the same truth values.

Each of the following biconditional statements is false because exactly one part of each statement is false.

$2 \times 2 = 4$ if and only if $7 - 5 = 3$
$2 \times 2 = 5$ if and only if $7 - 5 = 2$

How do you "prove" a statement to a friend? Undoubtedly there are several ways, including these three.

1. In a reference book or from a reliable authority find sufficient support for the statement so that your friend will accept it without further proof.
2. Prove to your friend that the statement is a necessary consequence of some statement that has already been accepted.
3. Prove to your friend that the statement cannot be false.

What does *proof* mean to you?

In mathematics there are also several ways of "proving" statements. In essence each proof is based upon statements that are accepted as true (assumed). Each *direct proof* consists of a sequence of statements (an argument) such that each statement is either assumed or is a *logical consequence* of the preceding statements, and the statement to be proved is included in the sequence. A statement may also be *proved indirectly* by proving that the statement cannot be false.

Any proof includes, at least informally, some given statements that are assumed to be true statements and one or more statements that are to be proved. The assumed statements are the "**given**" (often called the **premises**) of the proof. The statements that are to be proved are the **conclusions** of the proof. A correct mathematical proof is based on a *valid argument.* Specifically, any argument, mathematical or otherwise, is a **valid argument** if the conjunction of the premises implies the conclusions. In other words, *an argument is valid if, under the assumption of the premises, the conclusions cannot fail to be true.*

The law of detachment is sometimes called *modus ponens.*

One frequently used form of argument is the **law of detachment.**

If a statement of the form "If p, then q" is assumed to be true and if p is known to be true, then q must also be true.

Symbolically we write

Given:	$p \rightarrow q$	If p, then q; and
Given:	p	p
Conclusion:	q	imply q.

We may also write this argument as

$$[(p \rightarrow q) \wedge p] \rightarrow q$$

The truth table shows the validity of the law of detachment since the statement is true in all possible cases. The columns of the truth table should be completed in the indicated alphabetical order.

p	q	$[(p \rightarrow q)$	$\wedge$	$p]$	$\rightarrow$	q
T	T	T	T	T	T	T
T	F	F	F	T	T	F
F	T	T	F	F	T	T
F	F	T	F	F	T	F
		(a)	(c)	(b)	(e)	(d)

Example 3 Determine whether or not the following argument is valid.

Given: If Mary is a junior, then she is taking algebra.
Given: Mary is a junior.
Conclusion: Mary is taking algebra.

Solution Use

p: Mary is a junior.
q: Mary is taking algebra.

and think of the argument as

Given:	$p \rightarrow q$
Given:	p
Conclusion:	q

The argument has the form $[(p \rightarrow q) \wedge p] \rightarrow q$ and is therefore valid. ■

There are other forms of valid arguments. For example, any **argument by contraposition** is valid.

Any conditional statement is equivalent to its contrapositive, that is, $[p \to q]$ is equivalent to $[(\sim q) \to (\sim p)]$.

Given:	$p \to q$	If p, then q; and
Given:	$\sim q$	not q
Conclusion:	$\sim p$	imply not p.

This form of argument may be written as

$$[(p \to q) \wedge (\sim q)] \to (\sim p)$$

Example 4 Employ contraposition to give a valid argument using

p: You work hard.
q: You pass the course.

Solution If you work hard, then you pass the course. You do not pass the course. Therefore, you did not work hard.

$$\begin{array}{l} p \to q \\ \underline{\sim q} \\ \sim p \end{array}$$

■

Not all arguments are valid. Consider the following.

Given:	$p \to q$	If p, then q; and
Given:	q	q
Conclusion:	p	imply p.

The argument has the form

$$[(p \to q) \wedge q] \to p$$

If p is false and q is true, then the statement $[(p \to q) \wedge q]$ is true. The premise of the argument is true, and the conclusion p is false. Thus the argument is not valid. As an example of this type of reasoning consider the statement

Read this misleading advertisement and try to find others.

If you expect to be healthy, eat KORNIES.

The advertiser hopes that the consumer will assume, incorrectly, the converse statement

If you eat KORNIES, then you expect to be healthy.

The argument is not valid and is called a **fallacy**.

Here is another form of argument that is not valid and is a fallacy.

Given:	$p \to q$	If p, then q; and
Given:	$\sim p$	not p
Conclusion:	$\sim q$	imply $\sim q$.

Example 5 Determine whether or not the following argument is valid.

Given: If you worked hard, then you passed the course.
Given: You did not work hard.
Conclusion: You did not pass the course.

Solution For

p: You worked hard.
q: You passed the course.

the argument is not valid since it has the form

$$[(p \rightarrow q) \wedge (\sim p)] \rightarrow (\sim q) \blacksquare$$

Notice that anyone who uses the argument in Example 5 expects listeners to assume that the inverse $(\sim p) \rightarrow (\sim q)$ of any true statement $p \rightarrow q$ must be true. As in the case of a conditional statement and its converse, a statement does not necessarily imply its inverse. As another example of this type of reasoning, consider the advertisement

If you brush your teeth with SCRUB, then you will have no cavities.

The advertiser would like you to assume, incorrectly, the inverse statement

If you do not brush your teeth with SCRUB, then you will have cavities.

Another form of valid reasoning that we shall consider here is of the *chain-reaction* type.

Given:	$p \rightarrow q$	If p, then q; and
Given:	$q \rightarrow r$	if q, then r
Conclusion:	$\overline{p \rightarrow r}$	imply if p, then r.

The argument can be shown to be valid and has the form

$$[(p \rightarrow q) \wedge (q \rightarrow r)] \rightarrow (p \rightarrow r)$$

Here is an example of chain-reaction reasoning.

If you like this book, then you like mathematics.
If you like mathematics, then you are intelligent.
Therefore, if you like this book, then you are intelligent.

The first two statements are premises (assumed to be true) and the third statement is the conclusion. In this and in other cases in which the argument is valid, the conclusion is often called a **valid conclusion** of the premises.

Example 6

A valid conclusion may be either true or false. Thus in Example 6(c) the conclusion may be false but the argument is valid. If you accept the premises, then the conclusion must be accepted as a valid conclusion.

Determine whether each of the conclusions is or is not a valid conclusion.

Given: If you study mathematics, then you will be successful.
Given: If you are successful, then you will be rich.

Conclusions:

(a) If you study mathematics, then you will be rich.
(b) If you become rich, then you have studied mathematics.
(c) If you do not become rich, then you have not studied mathematics.

Solution (a) Valid: this argument is of the chain reaction type $[(p \to q) \wedge (q \to r)] \to (p \to r)$.

(b) Not valid; this argument uses the converse for its conclusion.

(c) Valid; this argument uses the contrapositive of part (a). ■

EXERCISES *Express each statement in if-then form.*

1. All apples are red.
2. All birds are beautiful.
3. All dogs are good watchdogs.
4. All ovals are round.
5. All squares are polygons.
6. All x's are y's.
7. Any two ball players are competitors.
8. Any large textbook is expensive.
9. Automobiles are expensive.
10. Calculators are useful.
11. You will like this book only if you like mathematics.
12. A necessary condition for liking this book is that you like mathematics.
13. To like this book it is sufficient that you like mathematics.
14. A sufficient condition for liking this book is that you like mathematics.
15. Liking this book is a necessary condition for liking mathematics.
16. A number is a natural number only if its square is a natural number.

Express each statement in if-then form and classify as true or false.

17. $12 - 4 = 7$ if $12 + 4 = 15$.
18. A necessary condition for $2 \times 2 \neq 4$ is $12 - 4 = 8$.
19. For $7 \times 4 = 20$ it is sufficient that $7 + 4 = 11$.
20. $7 \times 5 = 57$ is a sufficient condition for $7 + 5 = 13$.
21. $7 \times 5 = 75$ only if $15 \times 5 \neq 75$.
22. $7 \times 5 = 35$ only if $15 \times 5 \neq 75$.

Under what conditions is the given statement true?

23. I shall be happy if and only if I pass the test.
24. Studying hard is a necessary and sufficient condition for passing the test.
25. A necessary condition for passing the test is that you study hard.
26. A sufficient condition for passing the test is that you study hard.
27. I put on a sweater only if I feel chilly.
28. If I do not feel chilly, then I do not put on a sweater.

In each exercise assume that the premises are true and determine whether or not the argument is valid.

29. *Given*: If Elliot is a freshman, then Elliot takes mathematics.
    Given: Elliot is a freshman.
    Conclusion: Elliot takes mathematics.
30. *Given*: If you like dogs, then you will live to be 120 years old.
    Given: You like dogs.
    Conclusion: You will live to be 120 years old.
31. If the Braves win the game, then they win the pennant.
    They do not win the pennant.
    Therefore, they did not win the game.
32. If you like mathematics, then you like this book.
    You do not like mathematics.
    Therefore, you do not like this book.
33. If you work hard, then you are a success.
    You are not a success.
    Therefore, you do not work hard.
34. If you are reading this book, then you like mathematics.
    You like mathematics.
    Therefore, you are reading this book.
35. If you are reading this book, then you like mathematics.
    You are not reading this book.
    Therefore, you do not like mathematics.
36. If you work hard, then you will pass the course.
    If you pass the course, then your teacher will praise you.
    Therefore, if you work hard, then your teacher will praise you.
37. If you like this book, then you like mathematics.
    If you like mathematics, then you are intelligent.
    Therefore, if you are intelligent, then you like this book.

Exercise 38 is a famous nonstandard problem. Can you solve it?

*38. Suppose that you are a prisoner standing alone with the executioner in the execution chamber. You must choose one of two chairs and sit in it. One chair is an electrified chair that kills anyone who sits in it. The other chair is harmless.

You are allowed to ask the executioner just one question that he or she may answer by "yes" or "no." Furthermore, you know that the executioner either always tells the truth or always lies but you do not know whether he or she tells the truth or lies.

What question could you as the condemned prisoner ask and determine without any doubt which chair is the safe chair?

Suppose that you approach a fork in the road while traveling to the river in a strange country. What single question could you ask to find your way in each of these situations?

*39. You encounter a native who either always tells the truth or always lies but you do not know which.

*40. You encounter two natives, one of whom always tells the truth and one of whom always lies, but you do not know which is which.

EXPLORATIONS

1. Begin a collection of fallacies in reasoning that you find in newspapers or magazines, or that you hear on radio or television.
2. Examine a newspaper or magazine for if-then statements that are used in advertising. Then analyze the statement to see whether the converse or the inverse of the given statement is intended.
3. Give five valid arguments that could be used with elementary school children.
4. Give five arguments that could be used with elementary school children to illustrate arguments that are not valid.

Explorations 5, 6, and 7 are based upon sets of premises written by Charles Lutwidge Dodgson, who used the name of Lewis Carroll as author of *Alice's Adventures in Wonderland* and *Through the Looking Glass.* He was actually the mathematician C. L. Dodgson. His *Symbolic Logic* (fourth edition, 1886) and *The Game of Logic* (1886) were published as a single paperback volume in 1958 by Dover Publications, Inc. *Symbolic Logic* includes multitudes of intriguing illustrative examples, solutions, and an "appendix addressed to teachers."

Supply a conclusion so that each argument will be valid.

5. Babies are illogical.
   Nobody is despised who can manage a crocodile.
   Illogical persons are despised.
   Therefore, ...
6. No ducks waltz.
   No officers ever decline to waltz.
   All my poultry are ducks.
   Therefore, ...
7. No terriers wander among the signs of the zodiac.
   Nothing that does not wander among the signs of the zodiac is a comet.
   Nothing but a terrier has a curly tail.
   Therefore, ...

8. Discuss the validity of the arguments used in the following situation.

   A man approached the clerk at the checkout counter of a local market and asked the price of a box of blueberries.

   Clerk: "65 cents a box."

   Customer: "What! They're selling them for 55 cents a box across the street."

   Clerk: "Why don't you buy them there?"

   Customer: "Because they're sold out."

   Clerk: "Oh! If we were sold out, our price would be 45 cents a box."

Try to find other descriptions of situations, such as these, which involve logical concepts.

1-5 Problem Solving with Algebraic Patterns and Geometric Patterns

The use of algebra to generalize specific patterns is a powerful approach to problem solving. Consider, for example, the table of counting numbers that was introduced in Section 1-2. We can use algebra to show why the sums of the numbers in opposite corners of a rectangular array will always be equal. Consider any rectangular array in the table and use n to represent the first number shown. If we use an array of three rows and four columns, then the numbers in the first row can be represented as n, $n + 1$, $n + 2$, and $n + 3$. Also, each number in the second row will be 10 more than the corresponding number in the first row, and each number in the third row will be 20 more than the corresponding number in the first row.

n	$n+1$	$n+2$	$n+3$
$n+10$	$n+11$	$n+12$	$n+13$
$n+20$	$n+21$	$n+22$	$n+23$

Note that we have used algebraic facts here to establish that a particular pattern will always hold. That is, we have used algebra as a means of proof.

Now consider the sum of the numbers in opposite corners.

$$n + (n + 23) = 2n + 23 \qquad (n + 3) + (n + 20) = 2n + 23$$

Since these sums are equal, we have *proved* that the sums of numbers in opposite corners will *always* be equal for any rectangular array of three rows and four columns. Try to prove that such sums will be equal for any rectangular array of four rows and three columns.

Example 1 Consider a square array of nine numbers from the table of counting numbers in Section 1-2. Show that the positive difference of the products of the numbers in opposite corners is always constant, and state what that constant is.

Solution First we use a specific case to form a conjecture.

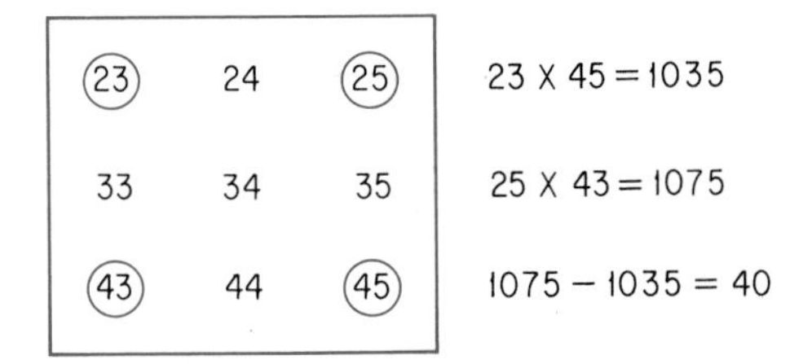

At this point we conjecture that the difference will always be 40. It is now important to test this conjecture on another sample; if the difference found is *not*

Note that a single *counterexample* is sufficient to show that a pattern does not hold, whereas any number of specific examples is insuffcent to prove that a particular pattern is always true.

40, then we have a *counterexample* showing our conjecture to be false and there is no need to test that conjecture further. The reader should show that the difference of products is 40 for another square array. Finally, we can prove our conjecture by generalizing and finding the difference of products for this general case.

(n)	$n+1$	$(n+2)$
$n+10$	$n+11$	$n+12$
$(n+20)$	$n+21$	$(n+22)$

$$n(n+22) = n^2 + 22n$$

$$(n+2)(n+20) = n^2 + 22n + 40$$

$$(n^2+22n+40) - (n^2+22n) = 40$$

Clearly the difference is 40 and will *always* be 40 for such square arrays of nine numbers. What will the difference be for a rectangular array of four rows and three columns? ■

Many magic tricks and "mind-reading" activities can be explained through the use of simple algebraic techniques. Consider as an example the "think of a number" type of mathematical trick.

Think of a number.
Add 3 to this number.
Multiply your answer by 2.
Subtract 4 from your answer.
Divide by 2.
Subtract the number with which you started.

If you follow these instructions carefully, your answer will always be 1, regardless of the number with which you start. We can explain why this trick works by using algebraic symbols or by drawing pictures, as shown below.

Think of a number:	n	□	(Number of coins in a box)
Add 3:	$n+3$	□ ○ ○ ○	(Number of original coins plus three)
Multiply by 2:	$2n+6$	□ ○ ○ ○ □ ○ ○ ○	(Two boxes of coins plus six)
Subtract 4:	$2n+2$	□ ○ □ ○	(Two bokes of coins plus two)
Divide by 2:	$n+1$	□ ○	(One box of coins plus one)
Subtract the original number, n:	$(n+1)-n=1$	○	(One coin is left)

Try to make up a similar trick of your own and use algebra to explain why the trick works.

Now let us try to "build" a trick together. We begin by forming a square array and placing any six numbers in the surrounding spaces as in the figure.

+	3	4	1
7			
2			
5			

The reader should try to reproduce this trick, using sets of numbers different from the ones shown.

The numbers 3, 4, 1, 7, 2, and 5 are chosen arbitrarily. Next find the sum of each pair of numbers as in a regular addition table.

+	3	4	1
7	10	11	8
2	5	6	3
5	8	9	6

Now we are ready to perform the trick. Have someone circle any one of the nine numerals in the box, say 10, and then cross out all the other numerals in the same row and column as 10.

+	3	4	1
7	(10)	~~11~~	~~8~~
2	~~5~~	6	3
5	~~8~~	9	6

Next circle one of the remaining numerals, say 3, and repeat the process. Circle the only remaining numeral, 9. The sum of the circled numbers is $10 + 3 + 9 = 22$, as shown in the figure.

+	3	4	1
7	(10)	~~11~~	~~8~~
2	~~5~~	~~6~~	(3)
5	~~8~~	(9)	~~6~~

The interesting item here is that the sum of the three circled numbers will always be equal to 22, regardless of where you start! Furthermore, note that 22 is the sum of the six numbers outside the square. Try to use algebra to explain why this trick works, and then build a table with 16 entries.

Geometric patterns also occur frequently in problem-solving settings. The ability to visualize problems from a geometric viewpoint is often a key feature in finding the solution to a problem.

Example 2 A basketball team consists of five members. Each member shakes hands with every other member of the team before the game starts. How many handshakes will there be in all?

Solution Consider a *pentagon* (a polygon of five sides) with each vertex representing a member of the team.

This problem is solved here through use of a diagram. It can also be solved by making a list. Thus the possible handshakes can be shown by listing all possible pairs of letters.

AB AC AD AE
BC BD BE
CD CE
DE

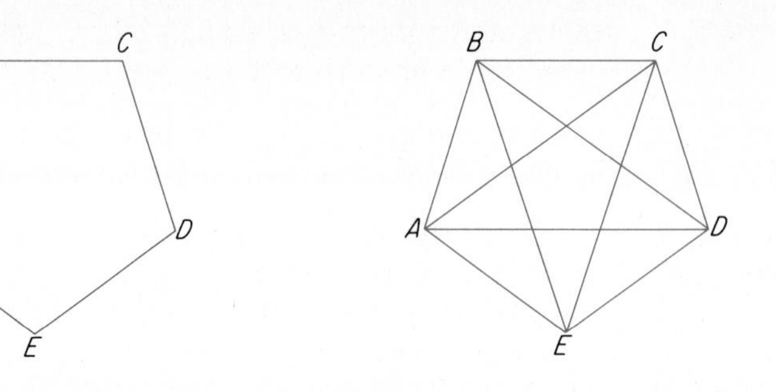

A handshake between two members of the team is represented by a line segment between two vertices, that is, either a diagonal or a side of the polygon. Thus there will be ten handshakes in all. ■

```
        .  1

        .
      . .  3

        .
      . .
    . . .  6

        .
      . .
    . . .
  . . . .  10

        .
      . .
    . . .
  . . . .
. . . . .  15
```

Triangular numbers provide another situation in which geometric reasoning appears in problem solving. The first five **triangular numbers** are 1, 3, 6, 10, and 15. The name *triangular numbers* is based upon the distinctive triangular patterns of the dot representations of these numbers.

The problem of identifying all numbers that are triangular numbers provides a challenge. However, the general solution may be obtained from a study of figures that are based on representations of the numbers. In particular consider the rectangular arrays obtained by placing a duplicate of the dot pattern for each triangular number above the original dot pattern, as in these figures.

```
                                        x x x x x
                              x x x x   x x x x .
                    x x x     x x x .   x x x . .
          x x       x x .     x x . .   x x . . .
x         x .       x . .     x . . .   x . . . .
.         . .       . . .     . . . .   . . . . .
```

Note that in each case the rectangular array has the same number of columns as the original triangular array; also, the rectangular array has one more row than the corresponding triangular array. The resulting *oblong* patterns have dimensions 1×2, 2×3, 3×4, 4×5, and 5×6. In general, the oblong array for the nth triangular number has n columns and $n + 1$ rows, contains $n(n + 1)$ dots, and contains twice as many dots as the nth triangular number. Hence, the nth triangular number is represented by $[n(n + 1)]/2$ dots. This means that if a number can be expressed as one-half of the product of two consecutive natural numbers, it is a triangular number. Also all triangular numbers can be expressed in this form.

Since the columns in the dot representation of a triangular number are formed by consecutive natural numbers of dots, that is, $1 + 2 + 3 + \cdots + n$, the total number of dots is the sum of the first n natural numbers. Therefore we know from our work with geometric patterns that

$$1 + 2 + 3 + \cdots + n = \frac{n(n + 1)}{2}$$

It is important to recognize that not all patterns lead to valid generalizations. There are times that an apparent pattern fails to continue beyond a certain point, as illustrated by the accompanying example.

Patterns offer an opportunity to make reasonable guesses, but these conjectures need to be proved before they can be accepted with certainty. Consider, for example, the maximum number of nonoverlapping regions into which a circular region can be separated by line segments joining given points of a circle in all possible ways.

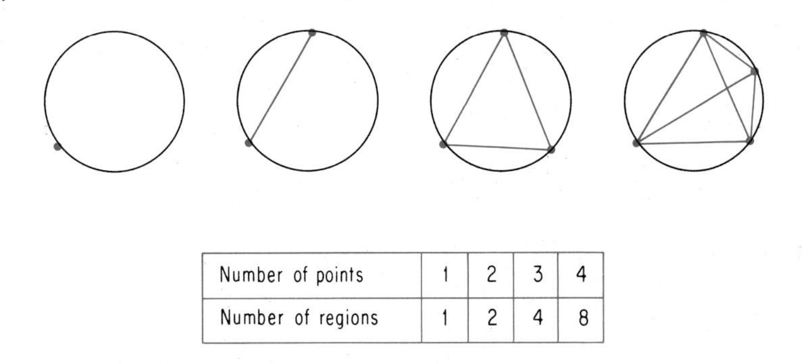

Number of points	1	2	3	4
Number of regions	1	2	4	8

Would you agree that a reasonable guess for the number of such regions derived from five points is 16? Draw a figure to confirm your conjecture. What is your guess for the maximum number of regions that can be derived from six points? Again, draw a figure to confirm your conjecture and count the number of regions formed. You may be in for a surprise!

EXERCISES

For Exercise 1 *through* 4 *use a table of natural numbers as in Section* 1-2. *Assume that the grid shown in each diagram is placed on top of the number table so that each box of the grid encloses exactly one number in the table.*

$n - 11$	$n - 10$	$n - 9$
$n - 1$	n	$n + 1$
$n + 9$	$n + 10$	$n + 11$

1. Select any square array of nine numbers from that table and prove that the sum of the numbers will be equal to nine times the center number. (*Hint*: Let the center number be n. Then the array of nine numbers will be as in the adjacent figure.)
2. For the arrays of numbers represented by the following diagrams, prove that the sum of all 11 numbers will be equal to 11 times the number n in the center.

(a) **(b)**

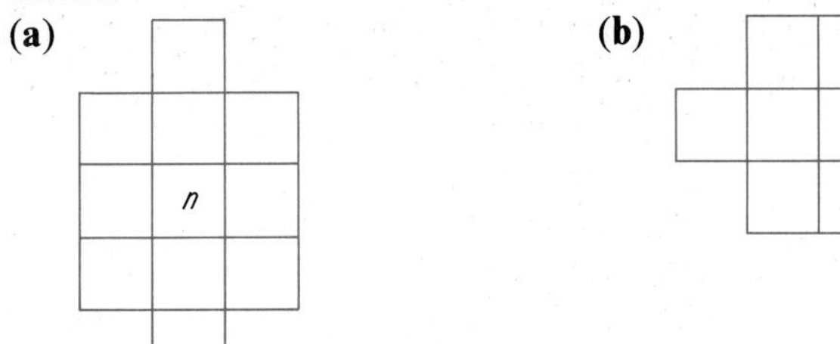

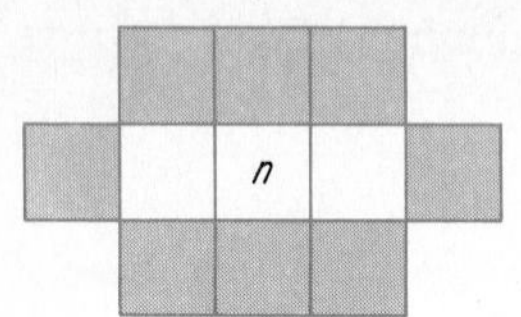

3. Prove that the sum of the numbers in the shaded regions will be equal to eight times the number n in the center for any such array in a number chart.
4. Consider the following magic trick. Have someone place an equal number of pennies (or other similar objects) in each of three positions, such as in spaces A, B, and C in the adjacent figure. For example, assume that 10 coins are placed in each position. Then provide these instructions.
   i. Take three coins from each of the positions A and C and add them to the coins in position B.
   ii. Count the number of coins in position A. Remove that number from position B.
   Start with a different number of coins and perform these same two steps. Show that the final number in position B again will be 9. Then prove that this number will always be 9. (*Hint*: Begin with n coins in each position.)

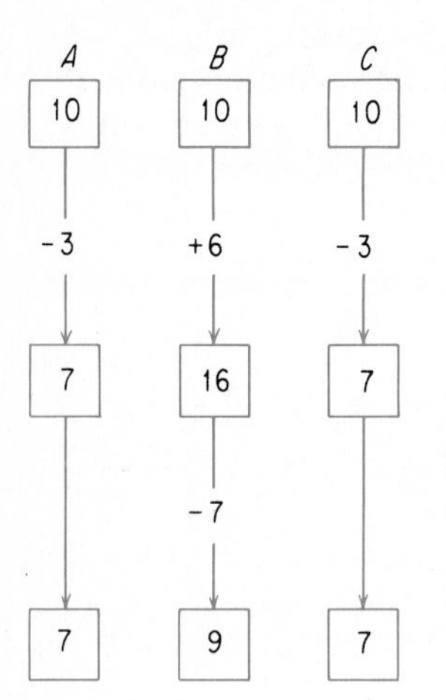

Use a calendar for any month of the year as a table of numbers.

5. Select any square array of nine numbers from the table and explain why the sums of the numbers in opposite corners must be equal.
6. Repeat Exercise 5 for a rectangular array of three rows and four columns.
7. Select any square array of nine numbers from the table, explain why each positive difference of the products of the numbers in opposite corners is the same constant, and identify that constant.
8. Repeat Exercise 7 for a rectangular array of three rows and four columns.

State the outcome of each of the following mathematical games. Use algebraic expressions to explain why each game works as it does.

9. Think of a number; add 7 to this number; multiply by 3; subtract 21; divide by 3; subtract the number with which you started.
10. Think of a number; multiply this number by 6; add 12; divide by 3; subtract 4; divide by the number with which you started.
11. Take a piece of paper and fold the paper in half as in the figure. Then fold it in half again and cut off a corner that does not involve an edge of the original piece of paper.

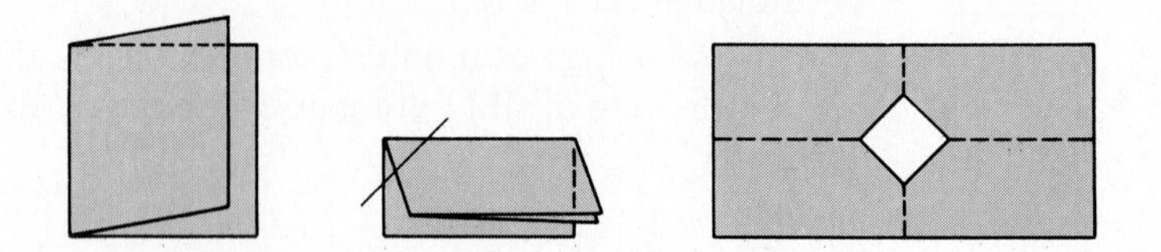

Your paper, when unfolded, should look like the preceding sketch. That is, with two folds we produced one hole. Repeat the same process but this time make three folds before cutting off an edge. Try to predict the number of holes that will be produced. How many holes will be produced with four folds? With n folds?

12. We wish to color each of the pyramids in the accompanying figure so that no two of the faces (sides and base) that have a common edge are of the same color.

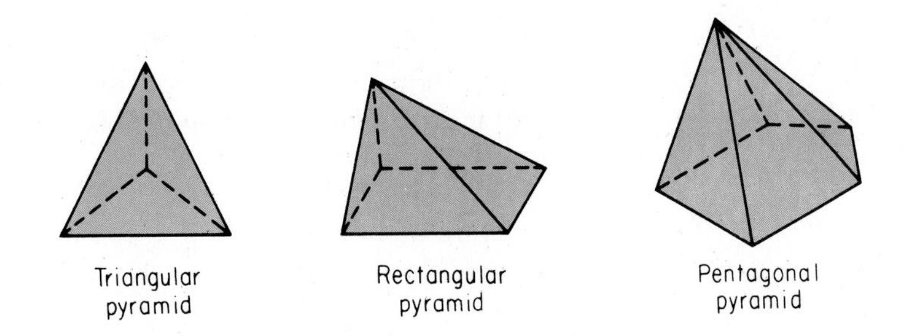

(a) What is the smallest number of colors required for each pyramid?

***(b)** What is the relationship between the smallest number of colors required and the number of faces of a pyramid?

13. Consider the following set of figures. In each figure we count the number V of vertices, the number A of arcs or line segments, and the number R of nonoverlapping regions into which the figure separates the plane. A square, for example, has four vertices, four arcs, and separates the plane into two regions (inside and outside the square). See if you can discover a relationship between V, R, and A that holds for each case. Confirm your generalization by testing it on several other figures.

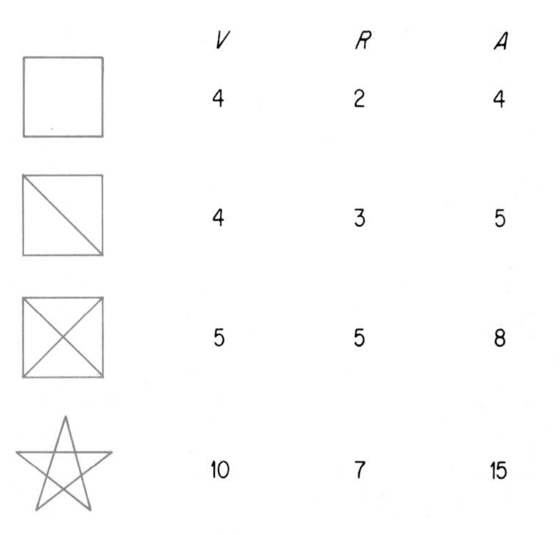

	V	R	A
	4	2	4
	4	3	5
	5	5	8
	10	7	15

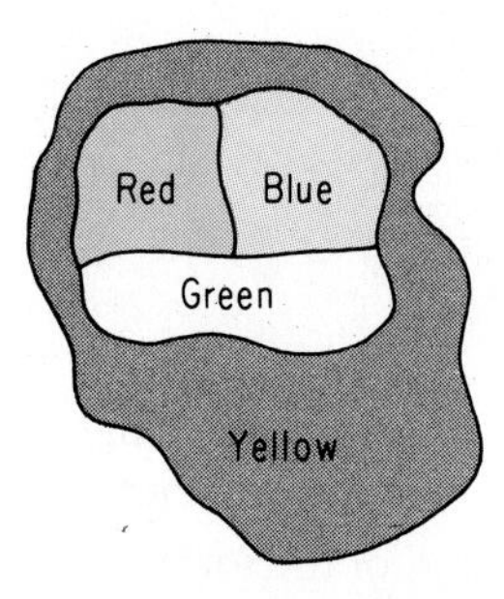

Four colors are required for this map

14. What is the largest number of pieces that can be obtained by making a cut along a plane through an orange once? Twice? Three times? Four times?

*15. The **four-color problem** was proposed in 1853 by a student at University College, London: Can every map in the plane be colored with four colors? Any two countries with common boundaries must have different colors. Two countries with only single points in common may have the same color. Mathematicians searched for a solution for this problem for more than 100 years. In 1976 two mathematicians finally used computer analyses of nearly 2000 cases and 10 billion logical alternatives to decide that four colors would always be sufficient.

(a) Draw a map of five countries that requires only three colors.

(b) Draw a map of five countries that requires only two colors.

*16. Begin with a large sheet of newspaper and fold it in half to form two layers. Then fold it in half again to form four layers. Repeat this process to form eight layers; and so on.

(a) Try this experiment with different sizes of paper and determine the maximum number of times it is possible to make such folds.

(b) Assume that you could continue this process 50 times and that the original paper was 0.003 inch thick. Approximately how thick (or high) would the final pile be? First estimate the answer. Then use a calculator, if available, to obtain a better approximation.

*17. You have an unlimited supply of cubes, a pail of blue paint, and a pail of red paint. Each face of each cube is to be painted either all blue or all red. How many distinguishable (different-appearing) cubes can be painted under these conditions?

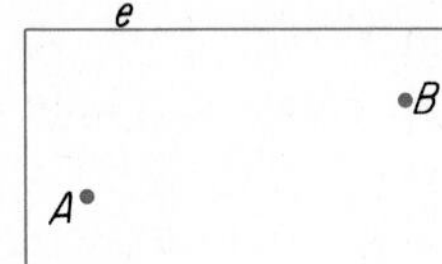

*18. One ball is on a pool table in position A of the adjacent figure, and another is in position B. Find approximately the point on the edge e of the table at which the ball at A must hit in order to rebound and hit the ball at B. To answer this question, note that the desired path from A to the edge of the table to B will be the shortest possible such path. Thus solve this problem by using a ruler and measuring different paths, such as those suggested by these figures.

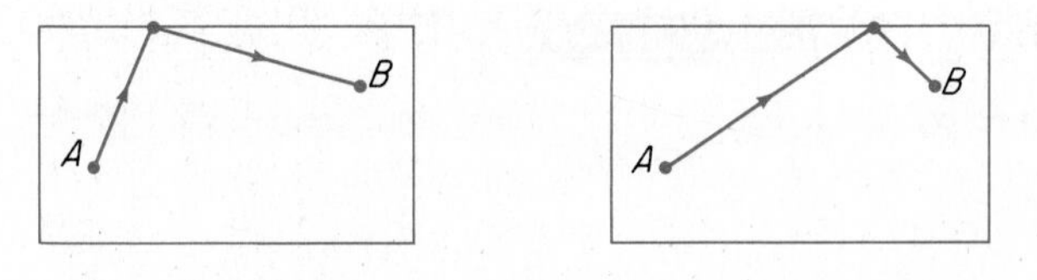

*19. Select any four-digit number, with each of its digits different, such as 3274. Form all possible *cyclic numbers* by keeping the digits in the same sequence. For example, consider the digits as on a clock and begin with a different digit each time as shown.

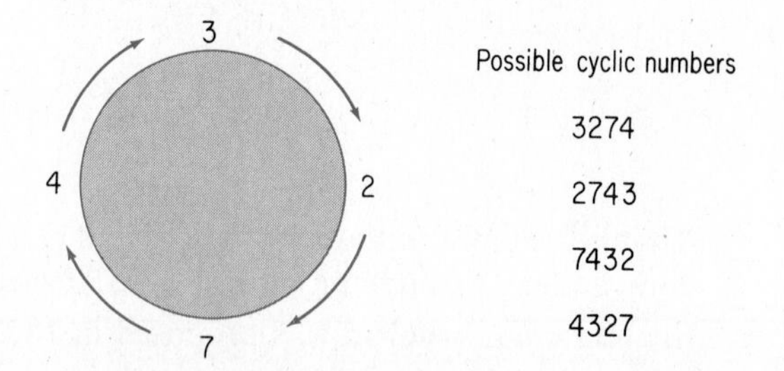

Find the sum of the four numbers and divide by the sum of the four digits. For this example, the sum is 17,776, the sum of the four digits is 16, and the quotient $17{,}776 \div 16 = 1111$. Repeat this procedure for a different four-digit number. Then prove that the final result will always be 1111. (*Hint*: Use $1000a + 100b + 10c + d$ to represent the given number.)

*20. Assume that you give a friend a book of 20 matches and that, without telling you any of the numbers, the friend follows these instructions:

Tear out and discard any number from 1 to 9 of matches.

Count the number N of matches that remain in the book.

Add the decimal digits of N. Let the sum be S.

Tear out and discard S more matches from the book.

Finally, burn, one at a time, any number from 1 to 9 of the matches that remain in the book.

You count the number of matches that are burned and state the number that are then left in the book of matches. Explain how you can tell the number of matches that are left in the book.

*21. Try the following card trick several times. Then explain mathematically why the trick always works.

First place a predetermined card, such as the ace of spades, as the 21st card from the top of a deck of cards. Then ask a friend to select any number n from 1 through 10 and to remove that number of cards from the top of the deck. Next you spread out the next 20 cards in a row face down. Finally, you ask your friend to count cards backward to n, the number originally selected by that person. You then announce, without seeing the card, that this card is the ace of spades, the predetermined card.

EXPLORATIONS

In this set of explorations we shall examine the ways in which any natural number (1, 2, 3, 4, 5, ...) can be represented as the difference of squares of whole numbers (0, 1, 2, 3, ...). Note, for example, the following.

$5 = 3^2 - 2^2 \qquad (5 = 9 - 4)$

$7 = 4^2 - 3^2 \qquad (7 = 16 - 9)$

Several interesting questions can be raised about such relationships.

(a) Can every natural number be represented as the difference of the squares of two whole numbers?

(b) Is there any pattern to those representations that exist?

The answers to these questions will be considered in the explorations that follow.

1. Represent each of the odd numbers from 1 through 15 as a difference of two squares. (Note that the answers for 5 and 7 are given above.) Study the pattern of answers and conjecture a manner in which any odd natural number can be written as a difference of squares.
2. An odd natural number can be represented by the form $2n + 1$ where n is a whole number. For example: when $n = 0$, $2n + 1 = 1$; when $n = 1$, $2n + 1 = 3$; when $n = 3$, $2n + 1 = 7$; and so on. Now consider the adjacent

n	$2n+1$	$(n+1)^2 - n^2$
0	1	$1^2 - 0^2$
1	3	$2^2 - 1^2$
2	5	$3^2 - 2^2$
3	7	$4^2 - 3^2$

partial table of odd numbers $2n + 1$ and their representations as differences of squares. It seems that for every whole number n, the odd number $2n + 1$ can be written as the difference $(n + 1)^2 - n^2$. Prove that this will be true in all such cases by simplifying the expression $(n + 1)^2 - n^2$.

3. Any natural number that is a multiple of 4 can be represented as $4n$ where n is a natural number. Copy and complete the adjacent table for the multiples of 4, 8, 12, and 16. Use several additional examples to test this conjecture: For every natural number n, the number $4n$ can be written as the difference $(n + 1)^2 - (n - 1)^2$.

n	$4n$	$(n+1)^2 - (n-1)^2$
1	4	$2^2 - 0^2$
2	8	
3	12	
4	16	

4. Prove the conjecture expressed in Exploration 3 by simplifying the algebraic expression given there.
5. Try to represent several of the remaining even numbers (2, 6, 10, 14, ...) as a difference of two squares. What do you conclude?
6. Find a magic card trick that you can demonstrate to the class. Explain the mathematical basis for the trick. For information on card tricks, see the many books by Martin Gardner, as well as books on tricks found in most magic shops.

Chapter Review

1. How many squares of all sizes are there on a standard checkerboard with eight unit squares on a side? Describe a problem-solving strategy that can be used to answer this question.
2. List all the possible sets of three natural numbers whose product is 72. Explain the method you used to obtain your answer.
3. Use a method similar to that of Gauss to find the sum of the first 150 natural numbers. Use a diagram to illustrate your procedure.
4. Repeat Exercise 3 for the sum of the even natural numbers from 2 through 200, that is,

$$2 + 4 + 6 + \cdots + 198 + 200$$

5. Consider a table of natural numbers in rows of ten (as described in Section 1-2) and describe five different patterns that can be observed in that table. Select one of these patterns and explain why it works.
6. Repeat Exercise 5 for a table of multiplication facts from 1×1 through 9×9.
7. Make a table of the natural numbers 1 through 60 in rows of twelve. Explain why the sums of the numbers in opposite corners of any rectangular array of four rows and five columns from the table must be equal.
8. Note the sums of these powers of 2.

$$2^0 + 2^1 + 2^2 = 1 + 2 + 4 = 7$$

$$2^0 + 2^1 + 2^2 + 2^3 = 1 + 2 + 4 + 8 = 15$$

$$2^0 + 2^1 + 2^2 + 2^3 + 2^4 = 1 + 2 + 4 + 8 + 16 = 31$$

(a) Describe a pattern for finding such sums without actual addition.
(b) Use this pattern to find the sum $1 + 2 + 4 + 8 + \cdots + 256$.

9. Find the total number of diagonals in an *octagon*, a polygon of eight sides.

10. At a committee meeting the eight members present each shake hands with every other member. Describe two different mathematical procedures for determining the total number of handshakes that take place.

11. Note the sums in this table of sums of natural numbers.

$$1 = 1$$
$$1 + 2 = 3$$
$$1 + 2 + 3 = 6$$
$$1 + 2 + 3 + 4 = 10$$

(a) Describe a pattern for the sums 1, 3, 6, 10.
(b) Use this pattern to predict the sums that would be obtained if the table were continued for three more rows.

Use

p: *Wendy is happy.*
q: *Jan is sad.*

Think of "sad" as "not happy" and write each of these statements in symbolic form.

12. (a) Wendy is happy and Jan is sad.
(b) Wendy is not happy but Jan is happy.

13. (a) Wendy or Jan is happy.
(b) It is not true that Wendy and Jan are both happy.

14. Assume that Wendy and Jan are both happy. Which of the statements in Exercises 12 and 13 are then true?

15. Write (a) the negation, and (b) the contrapositive, of the statement: If I work hard, then I shall pass the course.

16. Write the negation, the converse, the inverse, and the contrapositive of the given statement.

If apples are red, then they are ripe.

17. Give the truth value of each statement.
(a) If $3 \times 4 = 10$, then $4 \times 3 = 10$.
(b) If $3 \times 4 = 10$, then $4 \times 3 = 12$.
(c) If $3 \times 4 = 12$, then $4 \times 3 = 10$.
(d) If $3 \times 4 = 12$, then $4 \times 3 = 12$.

18. Write each statement in if-then form.
(a) All horses are quadrupeds.
(b) Knowing Judy is sufficient reason for liking her.

19. Write each statement in symbolic form. Use

p: I wear a coat.
q: Snow is falling.

(a) For me to wear a coat it is necessary that snow be falling.
(b) I wear a coat only if snow is falling.

20. Determine whether each of the conclusions is or is not a valid conclusion.

Given: If you enjoy this test, then you will get an A.
Given: If you get an A, then you will be happy.
Conclusions:
(a) If you enjoy this test, then you will be happy.
(b) If you do not enjoy this test, then you will not be happy.

COMPUTERS AND THE MATHEMATICS CLASSROOM

Microcomputers are used extensively in elementary school classrooms. Chapter 14 of this text is devoted to the development of programming skills in BASIC and in Logo for use in elementary school mathematics. To emphasize the many uses of computers, a section on computer applications is included at the end of each of the other chapters.

Computers, as we know them, were developed in the years of World War II to handle difficult computational problems involving resource allocation and engineering related problems. The first models employed huge vacuum tube arrays and miles of wiring. The preparation of one of these computers for solving a problem consisted of wiring the logical circuits needed to make the decision desired for that specific problem. This situation soon gave way to the development of coded input, called *machine language*, which consisted of sequences of 0's and 1's that gave the computer the directions for carrying out an involved computation. The size of these computers was drastically reduced through the use of transistors in place of the vacuum tubes.

This steady progress continues today with the application of microchip technology and the continued development of *programming languages*, which allow individuals to communicate their desires to computers without having to deal with machine language. The first of these programming languages to have a major influence on education was BASIC (Beginner's All-purpose Symbolic Instruction Code) developed by Professors John Kemeny and Thomas Kurtz at Dartmouth College in the early 1960's. This language was quickly adopted by many persons involved in educational computing due to the relative ease with which it could be used and the rapidity with which it could be taught and learned. BASIC will serve as the major language used in this text.

The second programming language to have a major influence on elementary school mathematics programs is Logo. Logo was developed in the 1970's by a group of computer scientists at Massachusetts Institute of Technology under the leadership of Professor Seymour Papert. This language also can be taught and learned in a short amount of time. In addition, it allows for experimentation with geometric situations through a feature known as "Turtle Graphics."

These advances allow for the use of the microcomputer in the mathematics classroom to achieve a wide range of goals. Some of these applications are:

- Provision of drill and practice on basic facts and skills.
- Provision of opportunities for additional tutoring on specific topics in the mathematics curriculum.
- Simulation of specific processes or situations involving mathematical processes or concepts.
- Problem-solving through the investigation of situations using computer programming or existing programs.

In the computer applications of each of the following chapters we explore the use of computer programs to examine major concepts and skills. Programs are provided and you have the opportunity to apply them in studying many of the mathematical concepts presented in the text.

2

Counting and Classifying—Sets and Reasoning

Georg Cantor
(1845–1918)

Georg Cantor was a German mathematician who provided us with the foundations of modern set theory and methods for working with the concept of infinity in mathematics. These breakthroughs were labeled by one of his contemporaries, David Hilbert, as "the most astonishing product of mathematical thought, one of the most beautiful realizations of human activity. . . ." While his contributions provided the basis for a great deal of twentieth-century progress in pure mathematics, he was virtually unrecognized in his own time.

2-1 Counting and Classifying

The study of mathematics involves many different skills. Two of the most primitive of these skills are the skills of counting and classifying. We use these skills daily as we answer questions such as "How many," "What kind," and "Do you have two number 2 pencils?" While we rarely think about it, we are constantly associating names with objects, developing systems for keeping track of the objects, and monitoring the number, or amount, of each object that we might have. These are basic mathematical skills, skills which we consider from a mathematical basis in this and following chapters. We first look at the various mathematical concepts that have been developed to assist us in counting and classifying objects.

The areas of mathematics most often associated with the development of counting and classifying skills are combinatorics and set theory. The former area deals with questions of "How many ways can...?" and the latter with questions concerning the relationships existing between two or more classes of objects.

Example 1 How many numbers are there in the set $\{35, 40, 45, \ldots, 270\}$?

Solution The first approach might be to note that all the numbers are multiples of 5, with the last number being 54 times 5. Then we can delete the first 6 multiples of 5—namely, 5, 10, 15, 20, 25, and 30—from that total, 54, to get 48 numbers in the set.

A second approach might be to realize that the first number 35 is $30 + (1 \times 5)$, the second is $30 + (2 \times 5)$, the third is $30 + (3 \times 5)$, and so on, until we get to 270, which is $30 + (48 \times 5)$. Hence we again see that there are 48 of these numbers, since 48 different multiples of 5 have been added to the original base of 30.

A third approach could be to subtract 30 from each of the numbers, giving the set $\{5, 10, 15, \ldots, 240\}$, and then to ask how many numbers are in this set. Each of these numbers could then be divided by 5, giving the numbers

$\{1, 2, 3, \ldots, 48\}$. As there are 48 numbers in the set 1 through 48, there must be 48 numbers in the original set.

A fourth approach to answering the question could be to write out (list) all of the multiples of 5 starting with 35 and ending with 270. Then count them. ■

The foregoing example shows that an elementary counting problem can be solved in a variety of ways. In the first approach, we changed the problem to finding the difference of the results of two simpler problems involving multiples of 5. In the second case, we looked at written expressions involving multiples of 5 and then used these expressions to find the answer in a different way. In the third approach, we again used the idea of multiples of 5 to reclassify the numbers. In the final case, we just resorted to brute-force counting.

The first three approaches made use of mathematics in considering the solution to the question at hand. The development of skills which allow us to cut down on the amount of work involved in solving problems will give us more leisure time, greater satisfaction, and, probably, higher grades in mathematics classes!

Example 2 Suppose that you have a cube, 4 units on an edge, made up of 64 small white unit cubes. Suppose that the outside of the original large cube is then sprayed with red paint. Each of the small cubes has either three, two, one, or no faces painted red. How many small cubes are there of each of these types?

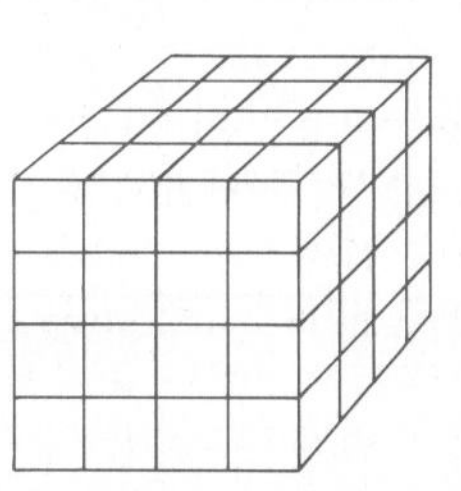

Solution The only small cubes having three faces red are the 8 corner cubes. The only small cubes having two faces red are the 2 cubes in the middle of each edge of the large cube. As there are 12 edges of the large cube, there must be 12×2, or 24, of the small cubes with two faces painted red. The 4 small cubes in the middle of each side of the large cube have only one face painted red. As there are 6 faces of the large cube, there must be 6×4, or 24, of the small cubes with one face painted red. The small cubes with no faces painted red are found in the interior of the large cube. Examining the structure of the large cube, we can see that there are 8 of the small cubes with no faces painted red. Finally, we check that the sum of the numbers of each of the types of small cubes is 64, the total number of small cubes. ■

The counting problem in Example 2 involving different types of cubes was solved by carefully classifying the small cubes, noting their positions, and carefully keeping track of the number of each type as the count took place.

Example 3 What is the smallest number of coins needed to make any sum of money less than $1?

Solution At most one half-dollar is needed. Similarly, at most one quarter, at most two dimes (since one quarter and a nickel can replace three dimes), at most one nickel, and at most four pennies can be needed. A supply of these nine coins enables us to make any total of money less than $1. However, note that even though one might need to use two dimes and one might need to use a dime and a nickel, one would never need to use two dimes and a nickel. In other words, at most eight of these nine coins are needed to make any specified sum of money less than $1. ■

Check several amounts of money less than $1 and observe that a supply of nine coins is sufficient and at most eight of the coins are needed for any given amount of money less than $1.

The solution to the counting problem in Example 3 was arrived at through a combination of logical reasoning, working forward to numbers of coins that should be available, and working backward from sample numbers to check on the number of coins needed. Such combinations of approaches are very common in counting problems.

Example 4 Suppose that a rectangular herb garden is three times as long as it is wide. The walk bordering the garden is made of black and white paving stones 1 foot on a side. The walk is not part of the garden. Each corner of this walk is made up of a square of white paving stones. All other stones are black. If there are 96 paving stones and 80 of them are black, what are the dimensions of the garden?

Solution Since there are 80 black stones, there must be 16 white stones. This means that there are 4 white stones in each of the four corners and each corner is a 2×2 square. Then the walk must be 2 stones wide. This implies that 40 of the black stones are on the inside row of the garden's border. Hence the perimeter of the garden, the sum of twice the length and twice the width, is 40 feet and the sum of the length and the width is 20 feet. Using the information that the garden is three times as long as it is wide, we are looking for two numbers such that their sum is 20 and one is three times the other. A quick guess and test tells us that the numbers are 15 and 5. Hence the herb garden must be 5 feet by 15 feet.

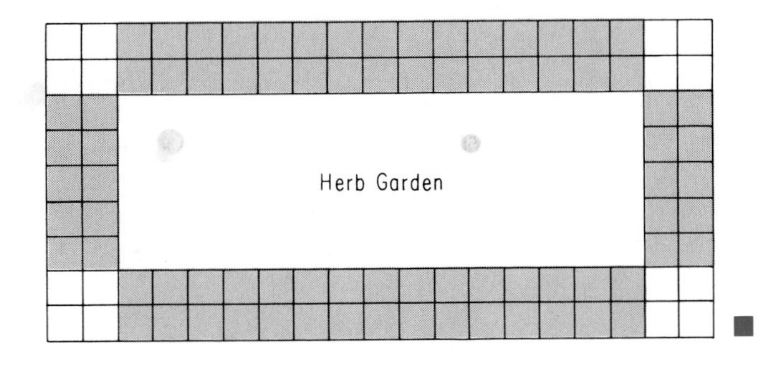

■

Example 4 was solved using the classification of the stones into two types. Then the number of white stones, the number of black stones, the number of stones along each side of the garden, and relations among these numbers were used. In the following exercises, explorations, and remaining sections of this chapter, you will see many different concepts and approaches to the solution of problems involving counting and classifications. Along the way, you will have a chance to develop your mathematical vocabulary and skills as you encounter new ideas and relationships.

EXERCISES

1. A hostess was planning the seating arrangement for a poolside party. She was planning to seat her guests at a set of card tables placed end to end and with only one guest at a side of a table. If she had invited 36 guests, how many tables would she need?
2. Mario observed the horses and chickens in his barnyard and counted 25 heads and 70 legs. How many horses and how many chickens were there?
3. There were nine daughters and no sons in the Smith family. Each of the daughters married and had three sons. How many grandsons did the elder Smiths have?
4. There were nine daughters in the Jones family. Each of the daughters had three brothers. How many children were there in the Jones family?
5. A man has three colors of socks in a drawer in his upstairs bedroom. How many socks must he select in the dark to insure that when he gets to the light he has at least one pair of the same color?
6. Suppose that there were 20 players in a tennis tournament and matches were played until there was a winner. How many matches were played? Check your answer to this problem by solving it in a second way.
7. Consider the accompanying triangular stacks of rows of blocks. How many blocks are needed to make the fifth such triangular pattern?

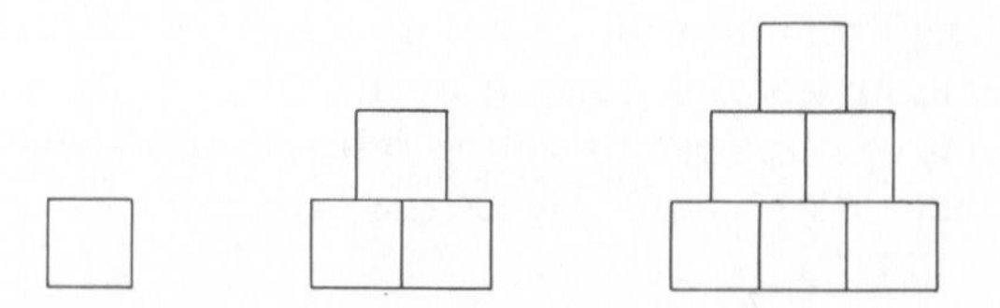

8. A triangle can be formed using three line segments; a square and its diagonals using six segments; a pentagon and its diagonals using ten segments. How many segments are needed for
   (a) an octagon (an eight-sided polygon) and its diagonals?
   *** (b)** An n-sided polygon and its diagonals?

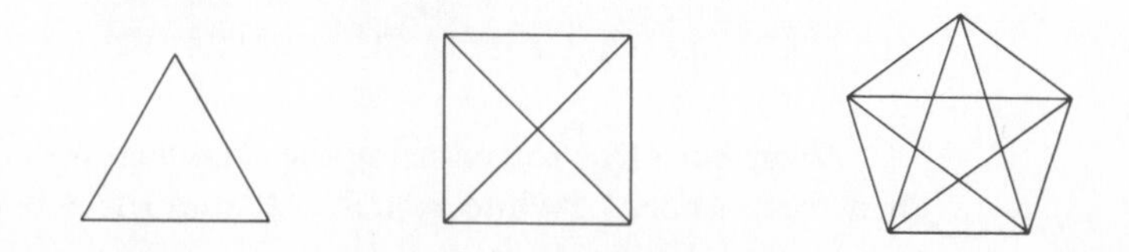

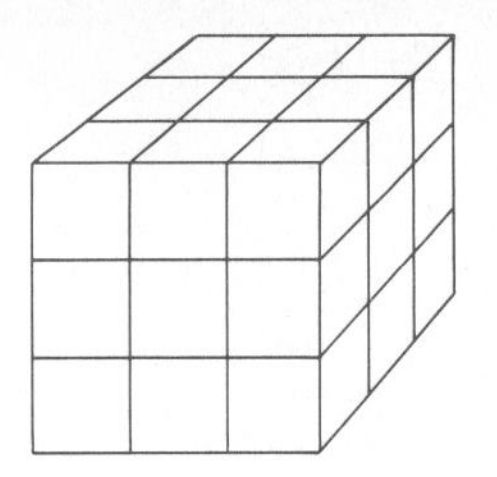

9. Suppose that you have a large cube, 3 units on an edge, made up of small white unit cubes. Suppose that you then spray the entire exterior of the large cube with blue paint. Find the number of small cubes that are now blue (**a**) on three faces, (**b**) on two faces, (**c**) on one face, and (**d**) on no faces.

10. Suppose that you have a similar situation to that described in Exercise 9, but the original cube has five units on its edge. Following the painting, find the number of small cubes that were painted blue (**a**) on four faces, (**b**) on three faces, (**c**) on two faces, (**d**) on one face, and (**e**) on no faces.

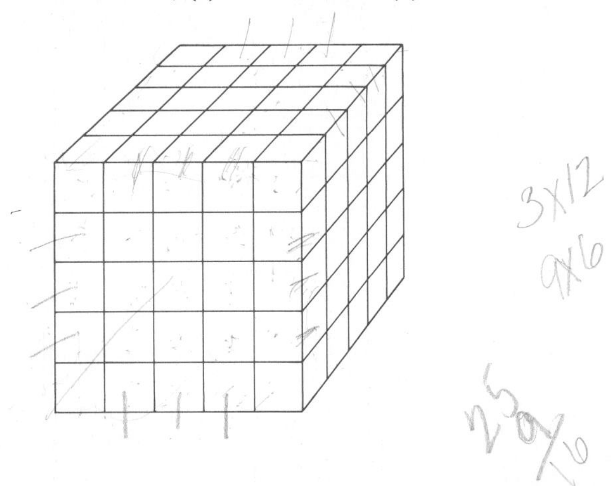

11. Suppose that you removed five small blocks to make a tunnel from the center of one face directly to the center of the opposite face of the large cube considered in Exercise 10. Furthermore, suppose that the sides of the inside of the tunnel are also painted blue. Find the number of small cubes that have (**a**) four faces painted blue, (**b**) three faces painted blue, (**c**) two faces painted blue, *(**d**) one face painted blue, and *(**e**) no faces painted blue.

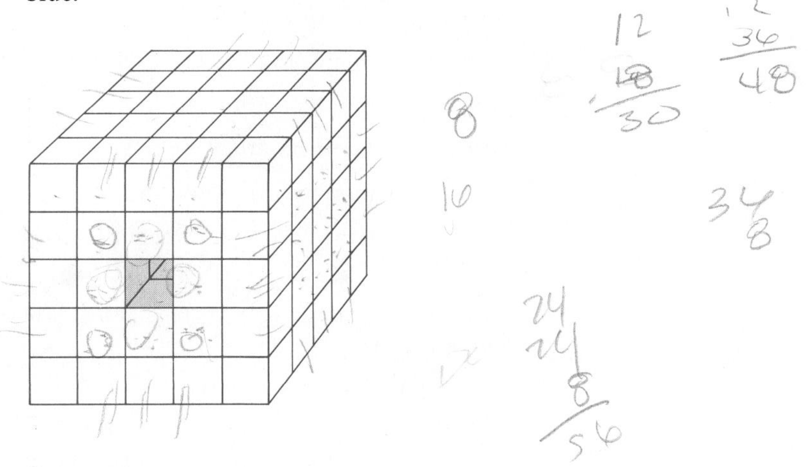

12. Suppose that the given map is a part of Everytown's street map. Further, suppose that we may travel from one point to another only by following

the streets. One path from point A to point B is illustrated. This path is 8 blocks long.

(a) What is the length of the shortest path from A to C?

(b) How many different paths are there from A to C of this minimal length?

Hint: Label each vertex with the number of paths from A to that vertex.

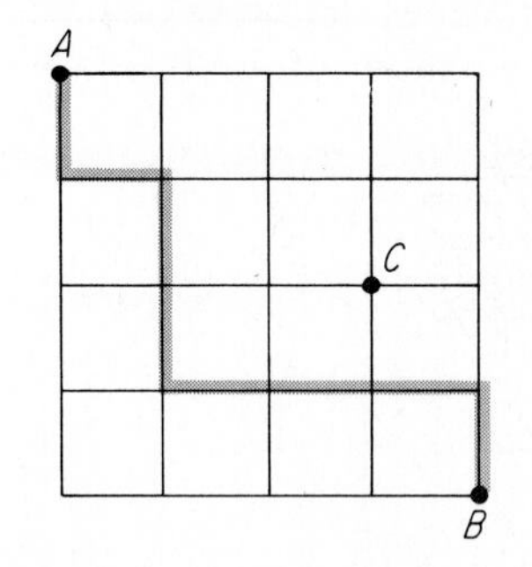

13. How many squares (any size) are shown in each figure?

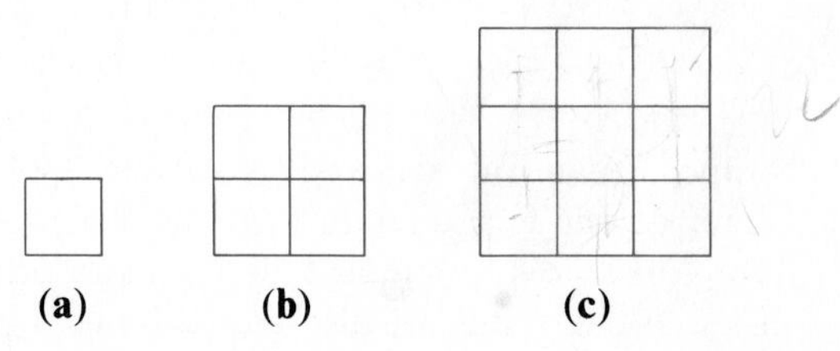

14. How many different shapes of rectangles (including squares) are shown in the figures for Exercise 13?

15. How many numbers are there in the set $\{15, 22, 29, 36, 43, \ldots, 78\}$?

16. How many numbers are there in the set $\{5, 12, 19, 26, 33, \ldots, 348\}$?

17. How many numbers are there in the set $\{2, 3, 5, 8, 12, \ldots, 93\}$?

18. How many numbers are there in the set $\{198, 184, 170, 156, 142, \ldots, 2\}$?

19. How many of the natural numbers 1 through 100 are multiples of 3 but not multiples of 5?

20. How many numbers between 1 and 100 are multiples of 2, 3, and 5 simultaneously?

21. How many numbers between 1 and 100 are even but are not multiples of 10?

*22. How many numbers are there in the set $\{0, 5, 25, 125, \ldots, 1{,}953{,}125\}$?

*23. How many numbers are there in the set $\{2, 4, 8, \ldots\}$ that are less than 5000?

EXPLORATIONS

1. Consider pyramids of blocks with one cube in the top layer, four cubes in the second layer, nine cubes in the third layer, and so forth. (a) How many cubes are in the nth layer? (b) How many cubes are there in a pyramid five layers high?

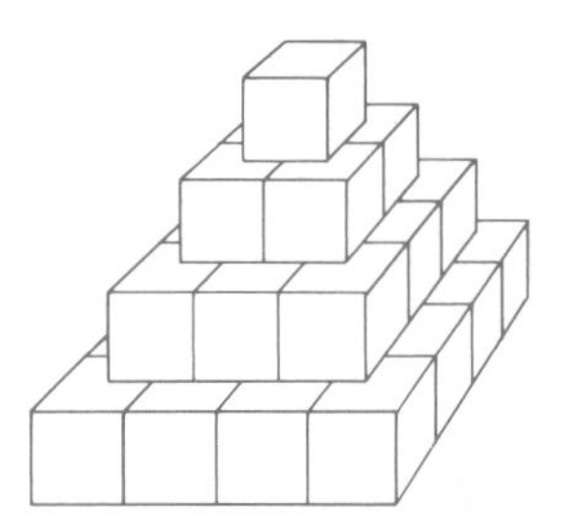

2. Repeat Exercise 9 for a large cube having n small cubes on an edge.
3. As in Exercise 13, give an expression for the number of squares (all sizes) shown in a square grid with n units on each side.

*4. In how many different ways can four checkers be arranged on a 4×4 checkerboard so that there is only one checker in each row and there is only one checker in each column? Arrangements, such as those shown, obtained from rotating the board are not counted as different arrangements.

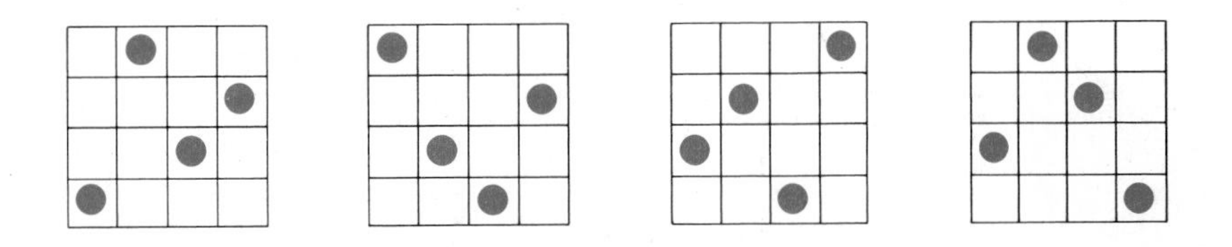

5. A man has 57¢ in change coming. How many different combinations of coins can be used to give the man his change, assuming that an adequate supply of each type of coin needed is available?

2-2 Sets and Set Notation

The authors welcome you to the *set* of readers of this book. You may be a *member* of a college or class for which this book is the assigned text. If so, the members of your class who read this page before it was discussed in class form a *subset* of the set of members of the class. Relative to the *universal set* of members of your class, the set of members who did not read this page before it was discussed in class is the *complement* of the set of members of the class who did. If each person in your class has exactly one copy of the textbook of his or her own, then there is a *one-to-one correspondence* between the set of members of your class and the set of their textbooks; that is, the set of members of your class and the set of their textbooks are *equivalent sets.*

The study of mathematics may be considered as the study of a new language. Mathematics has its own vocabulary and its own grammar (rules for combining its terms and symbols).

The terms set, member (or element), subset, universal set, complement, one-to-one correspondence, and equivalent sets, as used in the preceding paragraph, are part of the vocabulary of the language of mathematics. Each term has a very precise meaning. We shall use examples, counterexamples, exercises, and explorations to reinforce and "sharpen" your understanding of the terms considered.

There is a classic problem of which came first, the chicken or the egg. We have a similar problem in the language of mathematics—we cannot define everything. Thus we must assume that you already understand some terms from your previous experiences. For example, we assume that you know what is meant by a *set* (collection) of elements and also what is meant by a particular element *being a member of* (*belonging to*) a specified set.

Consider these two sets.

The set of letters of the English alphabet.

The set of states of the United States of America on January 1 of the bicentennial year 1976.

You can tell whether or not any specified element belongs to either of these given sets; that is, each set is a **well-defined set**.

These sets are not well-defined sets because there can be differences of opinion, possibly based on different standards, whether or not a particular person is a good tennis player or a successful country music singer.

Whenever there can be doubt as to the membership of an element in a set, the set is not a well-defined set. For example, the following sets are not well-defined sets.

The set of good tennis players.

The set of successful country music singers.

Mathematics is primarily concerned with well-defined sets. The elements of these sets may be numbers, points, geometric figures, blocks of various shapes (sizes, colors, materials, and so on), statements (equations, inequalities, and so on), people, or other identifiable entities.

The natural numbers are also called **counting numbers**.

Consider these sets of numbers.

$N = \{1, 2, 3, 4, \ldots\}$ the set of **natural numbers**
$W = \{0, 1, 2, 3, \ldots\}$ the set of **whole numbers**
$D = \{0, 1, 2, 3, \ldots, 9\}$ the set of **decimal digits**

We have used braces and indicated the sets N, W, and D in **set notation**. The three dots are used to indicate elements that are not explicitly listed. The list continues in the indicated pattern either indefinitely as in sets N and W or until the specified last element is reached as in set D. The sets N and W do not have a last element (they continue indefinitely) and are **infinite sets**. The set D has a last element and is a **finite set**.

In general, the order in which the elements of a set are listed is unimportant. However, when three dots are used to indicate elements that are not explicitly listed, it is necessary to state the elements in some order so that a pattern can be observed and used to identify the missing elements. Any such ordered set of elements is a ***sequence***.

The *elements* of a set are the members of the set. The membership symbol is $\in$. For example, we may write

$2 \in \{1, 2, 3, 4\}$; that is, 2 is an element of the set $\{1, 2, 3, 4\}$.
$7 \notin \{1, 2, 3, 4\}$; that is, 7 is not an element of the set $\{1, 2, 3, 4\}$.

The symbol "$\in$" reminds us of "E" and thus *element.* We read "$\in$" as "is an element of the set."

Frequently capital letters are used to name sets as we did for the set N of natural numbers. Sometimes it is convenient to seek verbal descriptions for given sets of elements that have already been listed.

Example 1 Give a verbal description for the set

$$Y = \{1, 3, 5, 7, 9\}$$

Solution There are several correct responses. Two of these are: "The set of odd numbers 1 through 9" and "The set of odd numbers between 0 and 11." Note that, as in the first response, the word "through" implies that the first and last numbers (1 and 9) are included; as in the second response the word "between" implies that the first and last numbers (0 and 11) are not included as members of the set. ■

The use of a *variable* x that can be replaced by any one of a set of elements is a basic concept of *algebra.*

The set of elements x such that $x \in \{1, 2, 3, 4, 5\}$ may be written in **set-builder notation** as

$$\{x \mid x \in \{1, 2, 3, 4, 5\}\}$$

and is precisely the set $\{1, 2, 3, 4, 5\}$. Note that the vertical line is read "such that." In some texts a colon is used in place of the vertical line.

Example 2 Use set-builder notation to describe each set.

(a) $\{1, 2, 3, 4, 5\}$ **(b)** $\{0, 2, 4, 6, 8\}$
(c) $\{0\}$ **(d)** $\{4, 7, 10, 13, \ldots, 100\}$

Solution
(a) $\{x \mid (x \in N) \text{ and } (x \leq 5)\}$
(b) $\{x \mid (x \in W) \text{ and } (x \leq 10) \text{ and } x \text{ is even}\}$
(c) $\{x \mid (x \in W) \text{ and } (x + a = a)\}$
(d) $\{x \mid x = 3n + 1 \text{ for } n \text{ a natural number less than } 34\}$ ■

Often several equivalent solutions are possible. For Example 2(d) consider also $\{x \mid x = 3n + 4 \text{ for } n \text{ a whole number less than } 33\}$

Two sets that have precisely the same elements are **equal sets**. We write

$$\{1, 2, 3\} = \{3, 2, 1\}$$

As sets of elements $\{1, 2, 1\} = \{1, 2\}$.

since order of listing does not affect membership. Furthermore, we shall not repeat elements within a listing since listing an element more than once does not increase the number of different members of the set. In general, we write $A = B$ to show that sets A and B have the same members; that is, A and B are *two names for the same set.* Similarly, $A \neq B$ indicates that sets A and B do not have the same members and thus are not equal sets.

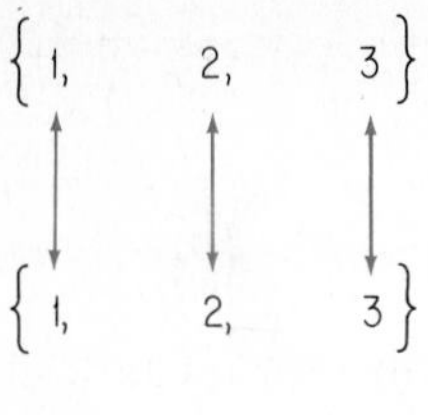

$\{1, 2, 3\}$
$\updownarrow \quad \updownarrow \quad \updownarrow$
$\{1, 2, 3\}$

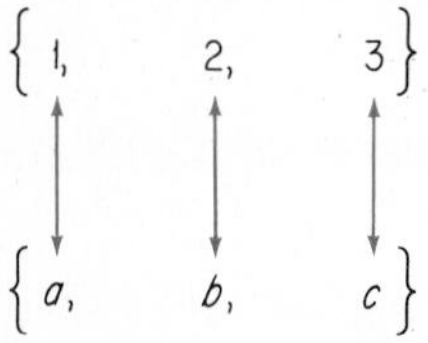

$\{1, 2, 3\}$
$\updownarrow \quad \updownarrow \quad \updownarrow$
$\{a, b, c\}$

Two sets $X = \{x_1, x_2, x_3, \ldots\}$ and $Y = \{y_1, y_2, y_3, \ldots\}$ are said to be in **one-to-one correspondence** if we can find a pairing of the x's and y's such that each x corresponds to one and only one y and each y corresponds to one and only one x. Any two equal sets may be placed in one-to-one correspondence since each element may be made to correspond to itself as in the adjacent illustration.

Also, it may be possible to place the members of two sets with different elements in one-to-one correspondence. Any two sets that can be placed in one-to-one correspondence are **equivalent sets** in the sense that they have the *same number of elements*. In general, any two equal sets are equivalent sets but, as in the accompanying illustration, two equivalent sets are not necessarily equal sets.

Example 3 (a) Is the set of letters in the word "nets" equivalent to the set of letters in the word "tens"?
(b) Are the two sets equal?

Solution (a) Yes, consider, for example, either of these correspondences.

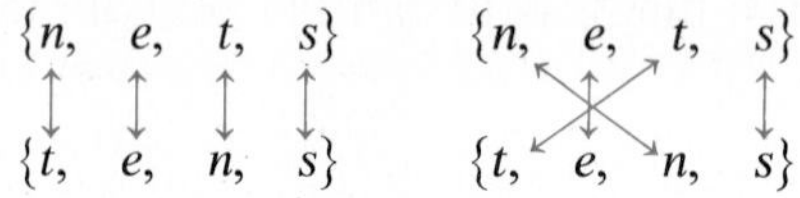

$$\begin{array}{cccc} \{n, & e, & t, & s\} \\ \updownarrow & \updownarrow & \updownarrow & \updownarrow \\ \{t, & e, & n, & s\} \end{array} \qquad \begin{array}{cccc} \{n, & e, & t, & s\} \\ & \times & & \updownarrow \\ \{t, & e, & n, & s\} \end{array}$$

(b) Yes, since the two sets have exactly the same members. ■

Any two equivalent sets have the same number of elements, that is, the same **cardinal number**. We denote the cardinal number of a set A by $n(A)$, read as "the cardinal number of A."

Example 4 For natural numbers N, whole numbers W, and decimal digits D, does the given equality hold?
(a) $n(\{2, 3, 4, 5\}) = n(\{r, e, s, t\})$
(b) $n(N) = n(W)$
(c) $n(D) = n(\{1, 2, 3, 4, 5, 6, 7, 8, 9, 0\})$

Solution (a) Yes, since the two sets can be placed in one-to-one correspondence by a pairing such as $2 \leftrightarrow r$, $3 \leftrightarrow e$, $4 \leftrightarrow s$, and $5 \leftrightarrow t$.
(b) Yes, since the two sets can be placed in one-to-one correspondence; for example, consider

$$\begin{array}{l} W = \{0, 1, 2, 3, 4, 5, 6, \ldots, \quad n, \ldots\} \\ \quad\;\; \updownarrow\; \updownarrow\; \updownarrow\; \updownarrow\; \updownarrow\; \updownarrow\; \updownarrow \qquad\quad \updownarrow \\ N = \{1, 2, 3, 4, 5, 6, 7, \ldots, n + 1, \ldots\} \end{array}$$

Note from their one-to-one correspondence that the sets N and W must have the same number of elements and that "number" is called a transfinite cardinal number. (See Section 4-1, Explorations 5 through 11.)

(c) Yes, since the two sets are equal, they are equivalent and therefore have the same cardinal number. ■

The set containing all of the elements for any particular discussion is called the **universal set** $\mathcal{U}$, and may vary for each discussion: the set of readers of this book, the set of decimal digits, the set of natural numbers, and so forth. The

$A' = \mathscr{U} - A$

complement A' (also written $\bar{A}$) of a given set A is the set of those elements of the universal set $\mathscr{U}$ that are not elements of A. This complement of A relative to $\mathscr{U}$ may also be denoted by $\mathscr{U} - A$, that is, $A' = \mathscr{U} - A$.

For any two sets A and B, the set of elements of B that are not elements of A is the **difference set** $B - A$. The complement of the universal set $\mathscr{U}$ is the **empty set (null set)** and may be denoted by either $\varnothing$ or $\{\ \}$. The empty set is the set that contains no elements.

Example 5 Let $\mathscr{U} = \{1, 2, 3, \ldots, 10\}$, $A = \{1, 2, 3, 4, 5, 7, 9\}$, and $B = \{2, 4, 6, 8\}$. Find
(a) A' **(b)** B' **(c)** $A - B$ **(d)** $B - B$.

Solution **(a)** $A' = \{6, 8, 10\}$ **(b)** $B' = \{1, 3, 5, 7, 9, 10\}$
(c) $A - B = \{1, 3, 5, 7, 9\}$ **(d)** $B - B = \varnothing$ ■

Consider the set B of natural numbers 1 through 9.

$$B = \{1, 2, 3, 4, 5, 6, 7, 8, 9\}$$

A set A is a **subset** of a set B, written $A \subseteq B$, if and only if every element of set A is also an element of set B. The given set B has many subsets. Here are a few of them.

$$A_1 = \{1, 2, 3\} \qquad A_2 = \{1, 5, 7, 8, 9\} \qquad A_3 = \{2\} \qquad A_4 = \{1, 2, 3, 4, 5, 6, 7, 8, 9\}$$

Note in particular that set A_4 contains each of the elements of B and is classified as a subset of B. Any set is said to be a subset of itself. Note also that if a set A is a subset of B, then A does not contain any elements that are not elements of B. This statement can be taken as a definition of *subset*. This interpretation is needed to explain why the empty set is a subset of every set. We could have used the relationship of being a subset to define *equality of sets*: For any sets A and B, $A = B$ if and only if $A \subseteq B$ and also $B \subseteq A$.

$A_1 \subset B$
$A_2 \subset B$
$A_3 \subset B$
$A_4 \subseteq B$

A *set* A is a **proper subset** of a set B if A is a subset of B and there is at least one element of B that is not an element of A. We write $A \subset B$ (read "A is a proper subset of B"). Intuitively, we speak of a proper subset as part of, but not all of a given set. The sets A_1, A_2, A_3, and A_4 are subsets of B; the sets A_1, A_2, and A_3 are proper subsets of B; and the set A_4 is not a proper subset of B.

Example 6 List three proper subsets of $\{1, 2, 3, 4\}$.

Solution Among others: $\{1, 2\}$ $\{1, 3, 4\}$ $\{1\}$. ■

All possible subsets of $\{1, 2, 3, 4\}$ may be found by considering the elements of the set none at a time (the empty set), one at a time, two at a time, three at a time, and four at a time (the given set). This procedure may be simplified by pairing the subsets with their complements. This pairing is done vertically in the following **array**, in this case a rectangular pattern of sets.

$\varnothing$	$\{1\}$	$\{2\}$	$\{3\}$	$\{4\}$	$\{1, 2\}$	$\{1, 3\}$	$\{1, 4\}$
$\{1, 2, 3, 4\}$	$\{2, 3, 4\}$	$\{1, 3, 4\}$	$\{1, 2, 4\}$	$\{1, 2, 3\}$	$\{3, 4\}$	$\{2, 4\}$	$\{2, 3\}$

In this array the selections of elements two at a time were made by considering the first element with each of the other elements. It was not necessary to continue the selections on the first row of the array since all other subsets of two or more elements had already been listed on the second row. Similar procedures may be used to find all subsets of any given set.

Example 7 List all possible subsets of $\{1, 2, 3\}$.

Solution

$\varnothing$	$\{1\}$	$\{2\}$	$\{3\}$
$\{1, 2, 3\}$	$\{2, 3\}$	$\{1, 3\}$	$\{1, 2\}$ ■

$A \cap A' = \varnothing$

For any sets A and B, the set of elements that are members of *both* A and B is the **intersection** of the sets A and B, written $A \cap B$. Thus

$$\boxed{A \cap B = \{x \mid (x \in A) \quad \text{and} \quad (x \in B)\}}$$

If $A = \{1, 2, 3, 4, 5\}$ and $B = \{4, 5, 6, 7\}$, then $A \cap B = \{4, 5\}$. Note that for all sets A and B, $A \cap B \subseteq A$ and $A \cap B \subseteq B$. If the intersection of two sets is the empty set, that is, if the sets have no elements in common, then the sets are **disjoint sets**. For example, $\{4, 5\}$ and $\{7, 8\}$ are disjoint sets.

For any sets A and B the set of elements that are members of *at least one* of the sets A and B is the **union** of A and B, written $A \cup B$. Thus

$$\boxed{A \cup B = \{x \mid (x \in A) \quad \text{or} \quad (x \in B)\}}$$

If $A = \{1, 2, 3, 4, 5\}$ and $B = \{4, 5, 6, 7\}$, then $A \cup B = \{1, 2, 3, 4, 5, 6, 7\}$. Note that for all sets A and B, $A \subseteq A \cup B$ and $B \subseteq A \cup B$. If A and A' are complementary sets, then

$$A \cap A' = \varnothing \quad \text{and} \quad A \cup A' = \mathscr{U}$$

that is, A and A' are disjoint sets whose union is the universal set.

Two ordered pairs are equal if and only if they have the same first element and the same second element.

If the elements of a pair are designated so that a particular element is the first element and the other is the second element, the pair is called an **ordered pair**. For any given sets A and B the set of all possible ordered pairs with an element of A as the first element and an element of B as the second element is the **Cartesian product** of A with B, written $A \times B$ and read "A cross B." Thus

$$\boxed{A \times B = \{(a, b) \mid (a \in A) \quad \text{and} \quad (b \in B)\}}$$

For $A = \{p, q, r\}$ and $B = \{1, 2\}$

$$A \times B = \{(p, 1), (p, 2), (q, 1), (q, 2), (r, 1), (r, 2)\}$$

$$B \times A = \{(1, p), (1, q), (1, r), (2, p), (2, q), (2, r)\}$$

In general, $A \times B \neq B \times A$.

The elements (a, b) of any Cartesian product may be represented in an array with the first element of each ordered pair serving as an *abscissa* to identify the column and the second element of each pair serving as an *ordinate* to identify the row of the array. For the given sets $A \times B$ may be represented as follows.

2	$(p, 2)$	$(q, 2)$	$(r, 2)$
1	$(p, 1)$	$(q, 1)$	$(r, 1)$
	p	q	r

Example 8 If $A = \{a, b, c\}$, $B = \{0, 1\}$, and $C = \{a, c\}$, find
(a) $A \times B$ **(b)** $B \times A$ **(c)** $C \times B$ **(d)** $A \times \varnothing$.

Solution **(a)** $A \times B = \{(a, 0), (a, 1), (b, 0), (b, 1), (c, 0), (c, 1)\}$
(b) $B \times A = \{(0, a), (0, b), (0, c), (1, a), (1, b), (1, c)\}$
(c) $C \times B = \{(a, 0), (a, 1), (c, 0), (c, 1)\}$
(d) $A \times \varnothing = \varnothing$ ■

The cross products obtained in Example 8 provide a basis for a discussion of relationships among cross products. As in Examples 8(a) and 8(b) we expected $A \times B \neq B \times A$. Would $A \times D = D \times A$ if $D = A$? Yes, but note from Example 8(d) that the cross products would also be equal if either A or D were the empty set. From Examples 8(b) and 8(c) note that $C \subseteq A$ and $C \times B \subseteq A \times B$. This relationship holds for any set B whenever C is a subset of A.

EXERCISES

Consider each of the sets **(a)** $\{s, e, n, d\}$ **(b)** $\{m, o, r, e\}$ **(c)** $\{m, o, n, e, y\}$.

1. Is n a member of the given set?
2. Is s a member of the given set?
3. Is it true that x is not a member of the given set?
4. Is it true that y is not a member of the given set?

Give a verbal description for each of the following sets.

5. $R = \{1, 2, 3, \ldots, 99\}$
6. $S = \{51, 52, 53, \ldots\}$
7. $M = \{5, 10, 15, 20, \ldots\}$
8. $K = \{10, 20, 30, \ldots, 150\}$
9. $T = \{1, 4, 9, 16, 25, 36\}$
10. $P = \{0, 2, 6, 12, 20, \ldots, 72, 90\}$

In Exercise 10 look for a pattern. Find the differences between successive elements of the set.

Describe each set using set-builder notation.

11. The set of whole numbers less than 10.
12. The set of even natural numbers.
13. $\{10, 20, 30, 40, \ldots, 190\}$
14. $\varnothing$

State whether or not the given sets are **(a)** *equivalent sets* **(b)** *equal sets.*

15. $\{t, o, n\}$, $\{n, o, t\}$
16. $\{c, a, r, t\}$, $\{r, a, c, k\}$
17. $\{d, o, n, t\}$, $\{d, o, n, e\}$
18. $\{h, o, t\}$, $\{h, e, a, t\}$

Find the cardinal number of the set given in the indicated exercise.

19. **(a)** Exercise 5 **(b)** Exercise 9
20. **(a)** Exercise 8 **(b)** Exercise 10

List all proper subsets of the given set.

21. **(a)** $\{p\}$ **(b)** $\{r, s, t, u\}$
22. **(a)** $\{p, q\}$ **(b)** $\{a, b, c\}$

For each of the following sets find **(a)** $A \cup B$ **(b)** $A \cap B$.

23. $A = \{1, 3, 4\}$, $B = \{1, 3, 5, 7\}$
24. $A = \{3, 4, 5\}$, $B = \{4, 5, 6, 7\}$
25. $A = \{2, 4, 6, 8\}$, $B = \{4, 6, 7, 8\}$
26. $A = \{1, 3, 5, \ldots\}$, $B = \{2, 4, 6, \ldots\}$
27. $A = \varnothing$, $B = \{1, 2, 3, \ldots\}$
28. $A = \{1, 2, 3, \ldots\}$, $B = \{1, 3, 5, \ldots\}$

For each of the given universal sets find **(a)** A' **(b)** B' **(c)** $A' \cup B'$ **(d)** $A' \cap B'$.

29. $\mathscr{U} = \{1, 2, 3, 4, 5\}$; $A = \{1, 2\}$, $B = \{1, 3, 5\}$
30. $\mathscr{U} = \{1, 2, 3, \ldots, 10\}$; $A = \{1, 3, 5, 7, 9\}$; $B = \{2, 4, 6, 8, 10\}$
31. $\mathscr{U} = \{1, 2, 3, 4, 5, 6, 7\}$; $A = \varnothing$, $B = \{1, 2, 3, 4, 5, 6, 7\}$
32. $\mathscr{U} = \{1, 2, 3\}$; $A = \{1\}$, $B = \{3\}$

Let $A = \{0, 1, 2\}$, $B = \{3, 4\}$, *and* $\mathscr{U} = \{0, 1, 2, 3, 4\}$, *and find the indicated cross product.*

33. $A \times B$
34. $A \times A$
35. $B \times B$
36. $A \times B'$
37. $A \times \varnothing$
38. $A' \times B'$

Consider the answers obtained in Exercises 33 through 38 and state **(a)** *whether the given statement is true for the given sets and* **(b)** *why you would or would not expect the given statement to be true in general.*

39. $A \times B$ is equivalent to $A \times B'$.
40. $A \times B$ is equivalent to $A' \times B'$.
41. $A \times \varnothing$ is a subset of $B \times B$.
42. $A \times B'$ is equal to $A \times A$.

Find the sets A and B used to form each of the given Cartesian products $A \times B$.

43. $\{(1, 2), (1, 3), (2, 2), (2, 3), (3, 2), (3, 3)\}$

44. $\{(a, 0), (b, 0), (c, 0), (a, 1), (b, 1), (c, 1)\}$

45. $\{(n, z), (a, z), (t, z), (e, z), (w, z), (p, z)\}$

46. $\{(a, a), (c, b), (c, a), (b, c), (a, c), (c, c), (b, b), (a, b), (b, a)\}$

State whether you would expect each of the following statements to be always true or not always true. If the statement is not always true, then give at least one counterexample.

47. (a) $(A \cap B) \subseteq A$ **(b)** $(A \cap B) \subset B$

48. (a) $B \subset (A \cup B)$ **(b)** $A \subseteq (A \cup B)$

49. (a) $(A - B) \subset A$ **(b)** $(A - B) \subseteq A$

Describe, if possible, the conditions on the sets A and B such that each statement will always be true.

*50. $(A - B) = A$

*51. $(A \cap B) = A$

*52. $(A \cup B) = A$

*53. $(A \cap B) = (B \cap A)$

*54. $(A \cup B) = (B \cup A)$

*55. $[A \cap (A \cup B)] = A$

*56. $[A \cap (A \cup B)] = B$

*57. $(A \cup B) = (A \cap B)$

*58. $A \times B = B \times A$

*59. $A \times (B \cup C) = (A \times B) \cup (A \times C)$

EXPLORATIONS

1. Name at least 30 words that indicate sets. For example, consider a *school* of fish, a *swarm* of bees, and a *squadron* of planes.

2. Select a newspaper or news magazine and list several of the references to sets and subsets that are made in that paper or magazine. Don't miss the committees and subcommittees.

Warning! Exploration 3 is tricky.

3. Give at least two verbal descriptions for the set $\{8, 5, 4, 9, 1, 7, 6, 3, 2, 0\}$. Include a verbal description that identifies the order in which the elements are listed but does not include a listing of them.
4. Use specific examples as necessary and complete the following table.

Number of elements	0	1	2	3	4	5	6	10	50
Number of subsets	1								
Number of proper subsets									

5. Use the results obtained in Exploration 4 to conjecture for a set consisting of n elements
   (**a**) a formula for the number N of subsets
   (**b**) a formula for the number P of proper subsets.

2-3 Venn Diagrams

Suppose that in a class of 32 students, 25 students did the odd-numbered problems, 20 did the even-numbered problems, and 15 did all the problems. Is it possible that everyone did either the odd- or the even-numbered problems? Try to solve this problem before reading the solution near the end of this section.

Relationships among sets are often represented by sets of points. In this text, we use a rectangular region to represent the universal set. Then a particular subset A is represented by a circular, or other convenient, region. In the second figure, the dashed line indicates that the points of the circle are elements of A and are *not* elements of A'.

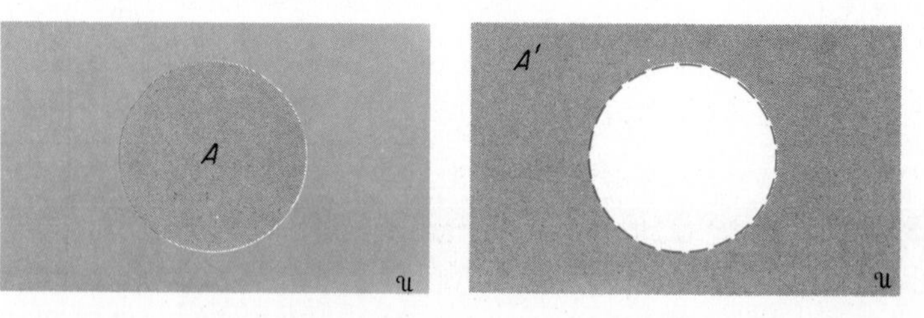

The diagrams are sometimes called *Euler circles* in honor of the Swiss mathematician Leonhard Euler (1707–1783), who first used circular regions in the discussion of principles of logic.

The shading is often omitted when the meaning is clear without the shading. Leonhard Euler used circular regions and arranged the regions so that each drawing by itself showed the relationships among the subsets of the universal set. For example, the intersection of two sets may be considered for sets that do not intersect (disjoint sets), for sets with one set a subset of the other, or for sets that have common elements but neither is a subset of the other.

Leonhard Euler

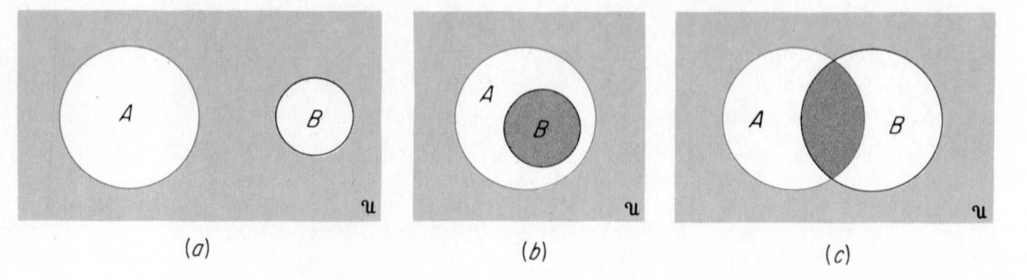

(a) (b) (c)

Note that in part (**a**) of the preceding figure, $A \cap B$ is the empty set; in part (**b**), $A \cap B = B$. We may also use such diagrams to show the union of any two sets. Note that $A \cup B = A$ in part (**b**) of the next figure.

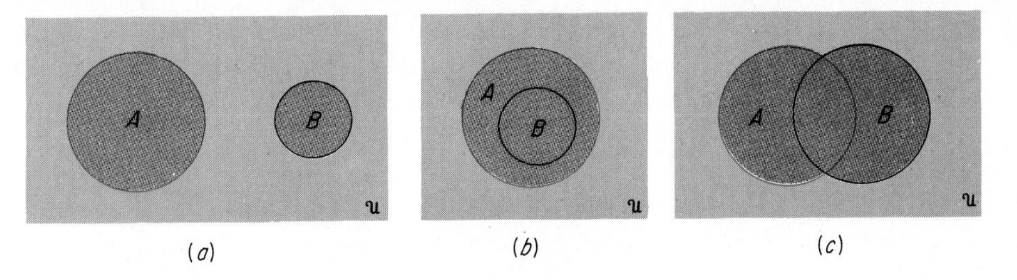

The figures for intersection and union may be used to illustrate the following properties of any two sets A and B.

$$(A \cap B) \subseteq A \qquad (A \cap B) \subseteq B$$

$$A \subseteq (A \cup B) \qquad B \subseteq (A \cup B)$$

We consider only well-defined sets (Section 2-2), and thus each element of the universal set $\mathscr{U}$ is a member of exactly one of the sets A and A'. Unless other relations are specified, we assume that any two given sets A and B determine four subsets of the universal set. Then we draw the diagram accordingly.

$$\begin{aligned} A \cap B;& \quad \text{that is,} \quad \{x \mid (x \in A) \text{ and } (x \in B)\} \\ A \cap B';& \quad \text{that is,} \quad \{x \mid (x \in A) \text{ and } (x \notin B)\} \\ A' \cap B;& \quad \text{that is,} \quad \{x \mid (x \notin A) \text{ and } (x \in B)\} \\ A' \cap B';& \quad \text{that is,} \quad \{x \mid (x \notin A) \text{ and } (x \notin B)\} \end{aligned}$$

Since these four sets include all elements of the universal set and no element belongs to more than one of the four sets, the universal set is said to be *partitioned* into four subsets by the sets A and B. This approach was used by John Venn who required that his drawings, now known as **Venn diagrams**, include a region for each logically possible case. Then the regions under special consideration can be identified by shading as in the solution for Example 1.

These diagrams are called Venn diagrams in honor of the English logician John Venn (1834–1923).

Example 1 Show by means of a Venn diagram that $(A \cup B)' = A' \cap B'$.

Solution We make separate Venn diagrams for $(A \cup B)'$ and $A' \cap B'$. The diagram for $(A \cup B)'$ is made by shading the region for A horizontally, shading the region for B vertically, identifying the region for $A \cup B$ as consisting of the points in regions that are shaded in any way (horizontally, vertically, or both horizontally and vertically), and identifying the region for $(A \cup B)'$ as consisting of the points in the region without horizontal or vertical shading.

The statements

$(A \cup B)' = A' \cap B'$
$(A \cap B)' = A' \cup B'$

are **De Morgan's laws for sets.**

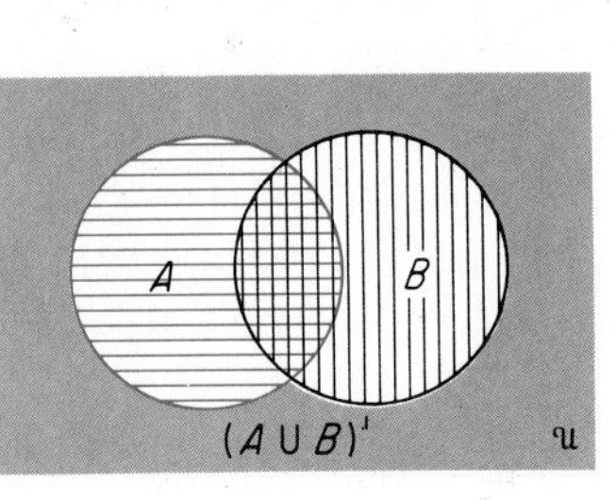

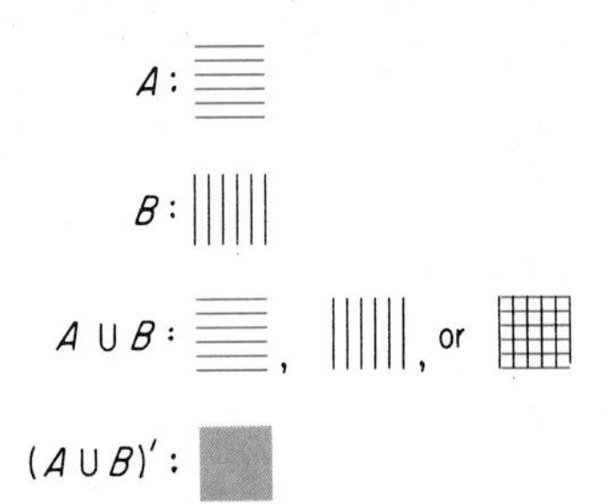

The diagram for $A' \cap B'$ is made by shading the region for A' horizontally, shading the region for B' vertically, and identifying the region for $A' \cap B'$ as consisting of all points in regions that are shaded both horizontally and vertically.

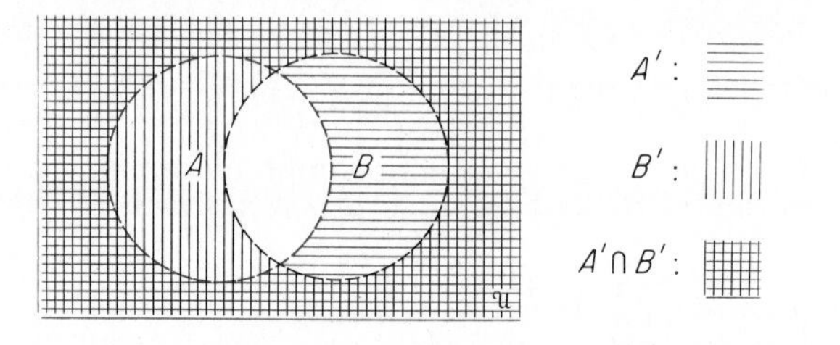

The solution is completed by observing that the region for $(A \cup B)'$ in the first Venn diagram is identical with the region for $A' \cap B'$ in the second Venn diagram; the two sets have the same elements and therefore are equal. ■

Euler used circles with a region for each case that was present in the particular problem under consideration. Venn used diagrams with a region for each case that was logically possible without regard to the particular problem under consideration.

The distinction between Euler circles and Venn diagrams is generally ignored in contemporary elementary mathematics and the term *Venn diagram* is often used to apply to both types of figures.

Suppose that $\mathcal{U}$ is the set of students who answered a specified question on a particular quiz. The statement

p: The student answered the question correctly

is true for a subset A of $\mathcal{U}$ and the statement

(not p): The student did not answer the question correctly

is then true for the subset A' of $\mathcal{U}$. To determine a second set B, consider the statement

q: The student is a boy.

Then the statement

(not q): The student is not a boy

is true for the subset B' of U. We can now identify each of the four subsets of U in terms of the statements p and q.

$A \cap B$ — p and q: The student answered the question correctly and is a boy.

$A \cap B'$ — p and not q: The student answered the question correctly and is not a boy.

$A' \cap B$ — (not p) and q: The student did not answer the question correctly but is a boy.

$A' \cap B'$ — (not p) and not q: The student did not answer the question correctly and is not a boy.

Similarly, we have

$A \cup B$ — p or q: The student answered the question correctly or the student is a boy.

Then DeMorgan's first law, $(A \cup B)' = A' \cap B'$, indicates that the statement

> It is not true that the student answered the question correctly or the student is a boy

has the same meaning as the statement

> The student did not answer the question correctly and is not a boy.

From the second of DeMorgan's laws $(A \cap B)' = A' \cup B'$, we observe that the statement

> It is not true that the student answered the question correctly and is a boy

has the same meaning as

> The student did not answer the question correctly or is not a boy.

Such interpretations of Venn diagrams illustrate very briefly how, in advanced courses, diagrams can be used to aid the understanding of statements. Often three or more statements are considered and three or more subsets of the universal set are needed.

In the most general situation for three sets A, B, and C the universal set is partitioned into eight subsets. We assume that this situation holds unless other relations are specified. The corresponding Venn diagram with eight regions is usually drawn as follows.

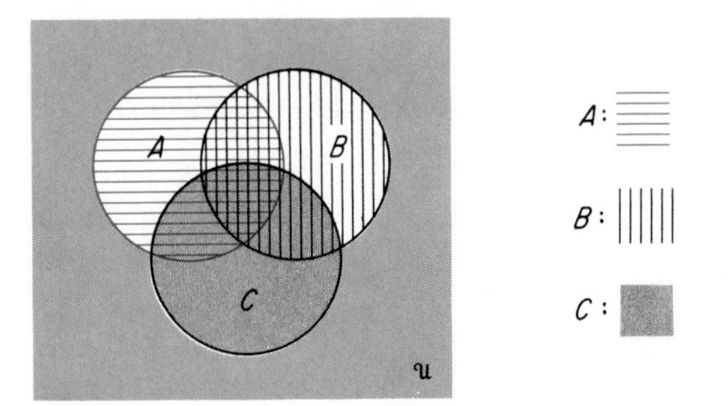

Example 2 Show that $A \cap (B \cup C) = (A \cap B) \cup (A \cap C)$.

Solution Set A is shaded with vertical lines; $B \cup C$ is shaded with horizontal lines. The intersection of these sets, $A \cap (B \cup C)$, is the region that has both vertical and horizontal shading.

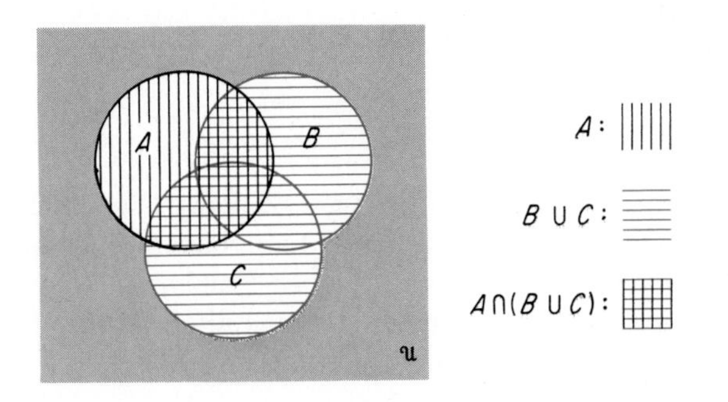

The set $A \cap B$ is shaded with horizontal lines; $A \cap C$ is shaded with vertical lines. The union of these sets is the subset of $\mathscr{U}$ that is shaded with lines in either or in both directions.

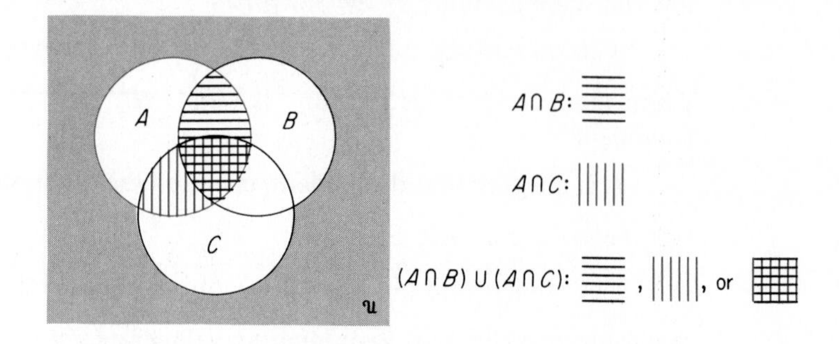

Note that the final results in the two diagrams designate the same regions, thus $A \cap (B \cup C)$ and $(A \cap B) \cup (A \cap C)$ are shown to be equal sets. ■

Recall that the *number of elements in a set S* is denoted by $n(S)$. For the set A in the figure for Example 3, we have

$$n(A) = 3 + 2 + 1 + 4 = 10$$

Example 3 For the given Venn diagram find **(a)** $n(A \cap B \cap C)$ **(b)** $n(A \cap B' \cap C)$.

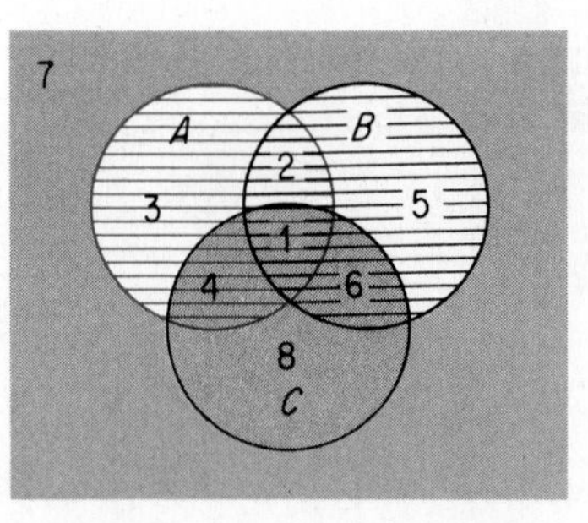

Solution **(a)** There is one element in the intersection of all three sets. Thus $n(A \cap B \cap C) = 1$.

(b) There are four elements that are in both sets A and C, but not in set B. Thus $n(A \cap B' \cap C) = 4$. ■

Example 4 In a group of 35 students, 15 are studying French, 22 are studying English, 14 are studying Spanish, 11 are studying both French and English, 8 are studying English and Spanish, 5 are studying French and Spanish, and 3 are studying all three subjects. How many are taking only English? How many of these students are not taking any of these subjects?

Challenge: Try to solve Example 4 before reading the solution.

Solution This problem can easily be solved by means of a Venn diagram with three circles to represent the set of students in each of the listed subject-matter areas. It is helpful to start with the information that there are 3 students taking all three subjects. We write the number 3 in the region that is the intersection of all three circles. Then we work backward: Since 5 are taking French and Spanish, and 3 of these have already been identified as also taking English, there must be exactly 2 taking only French and Spanish. That is, there must be 2 in the region representing French and Spanish but *not* English. Continuing in this manner, we enter the given data in the figure.

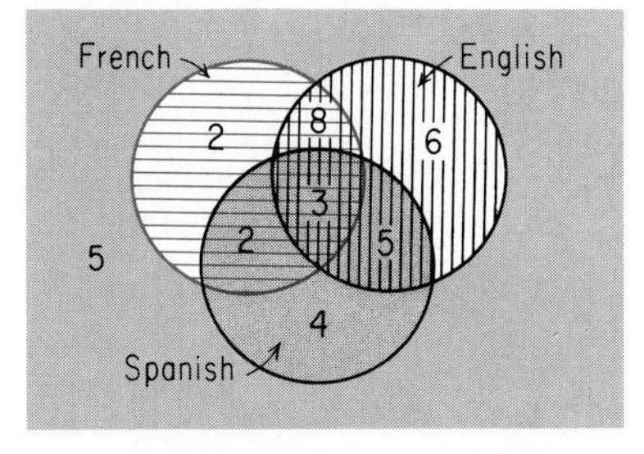

Reading directly from the figure, we find that there are 6 students taking only English. Since the total of the numbers in the various areas is 30, there must be 5 students not taking any of these specified subjects. ■

EXERCISES

Consider the given diagram and find each number.

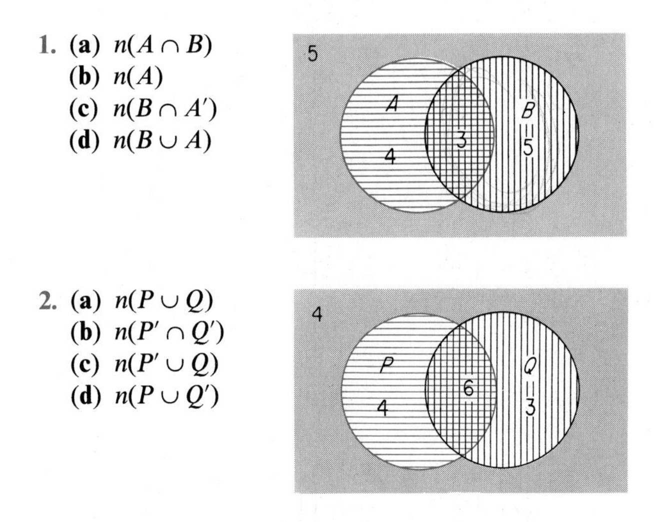

1. **(a)** $n(A \cap B)$
   (b) $n(A)$
   (c) $n(B \cap A')$
   (d) $n(B \cup A)$

2. **(a)** $n(P \cup Q)$
   (b) $n(P' \cap Q')$
   (c) $n(P' \cup Q)$
   (d) $n(P \cup Q')$

Represent each set by a Venn diagram.

3. $A' \cup B$ 4. $A' \cap B$ 5. $A \cap B'$ 6. $A \cup B'$

Show each relation by Venn diagrams.

7. $(A \cap B)' = A' \cup B'$

8. $A \cup B' = (A' \cap B)'$

Consider the given diagram and find each number.

9. **(a)** $n(A \cap B \cap C)$
   (b) $n(A \cap B \cap C')$
   (c) $n(A \cap B' \cap C')$
   (d) $n(A)$
   (e) $n(A \cup B)$
   (f) $n(B \cup C)$

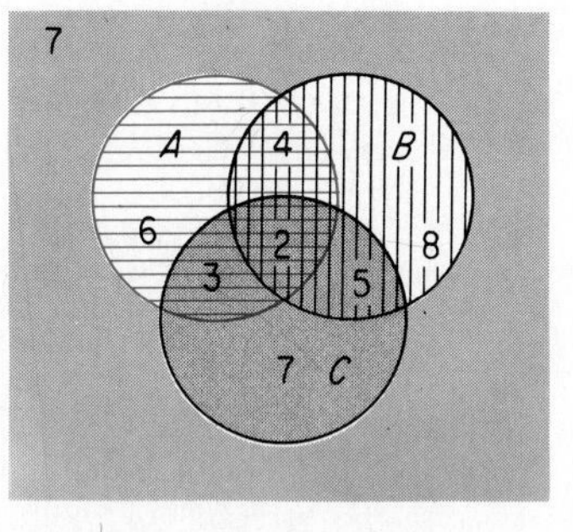

10. **(a)** $n(R' \cap S \cap T)$
    (b) $n(R')$
    (c) $n(R' \cup S)$
    (d) $n(S' \cup T')$
    (e) $n(R' \cup S' \cup T')$
    (f) $n(R \cup S' \cup T)$

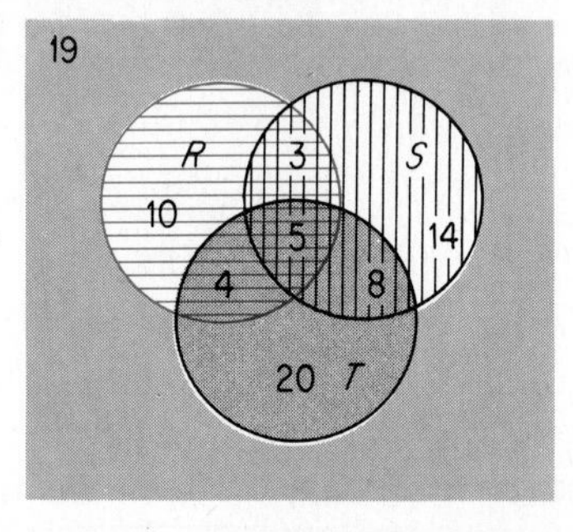

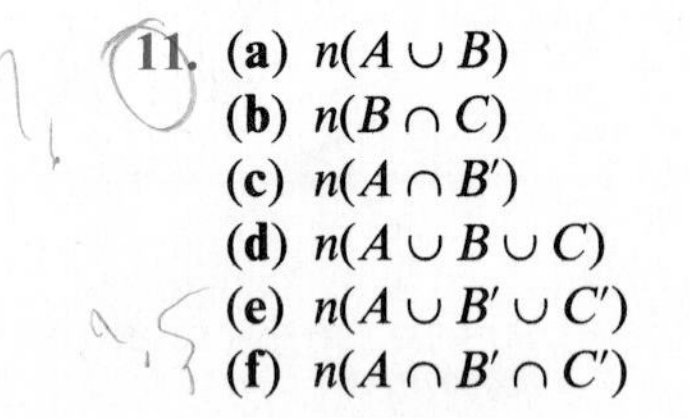

11. **(a)** $n(A \cup B)$
    (b) $n(B \cap C)$
    (c) $n(A \cap B')$
    (d) $n(A \cup B \cup C)$
    (e) $n(A \cup B' \cup C')$
    (f) $n(A \cap B' \cap C')$

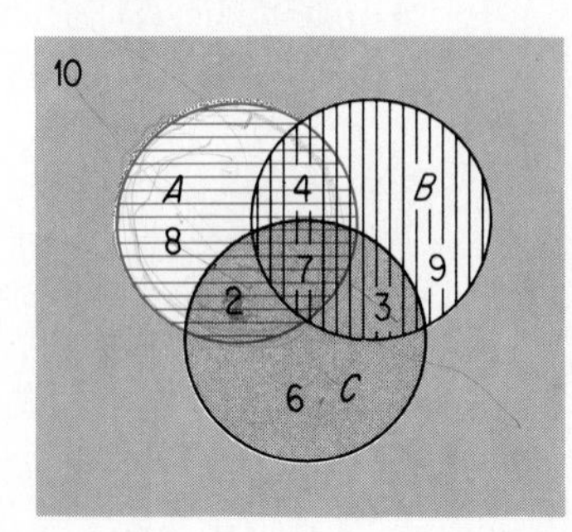

12. **(a)** $n(X \cup Y)$
    (b) $n(X \cap Z)$
    (c) $n(X')$
    (d) $n(X \cup Y')$
    (e) $n(X' \cap Y \cap Z)$
    (f) $n(X \cap Y' \cap Z')$

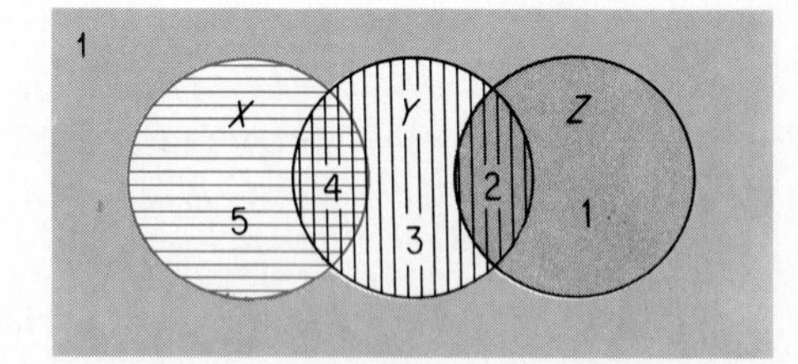

Represent each of the following by a Venn diagram.

13. **(a)** $A \cap B \cap C$ **(b)** $A \cap B \cap C'$
    (c) $A \cap B' \cap C$ **(d)** $A \cap B' \cap C'$
14. **(a)** $A' \cap B \cap C$ **(b)** $A' \cap B \cap C'$
    (c) $A' \cap B' \cap C$ **(d)** $A' \cap B' \cap C'$

Identify each gray region by set notation.

15\. 16. 17. 18.

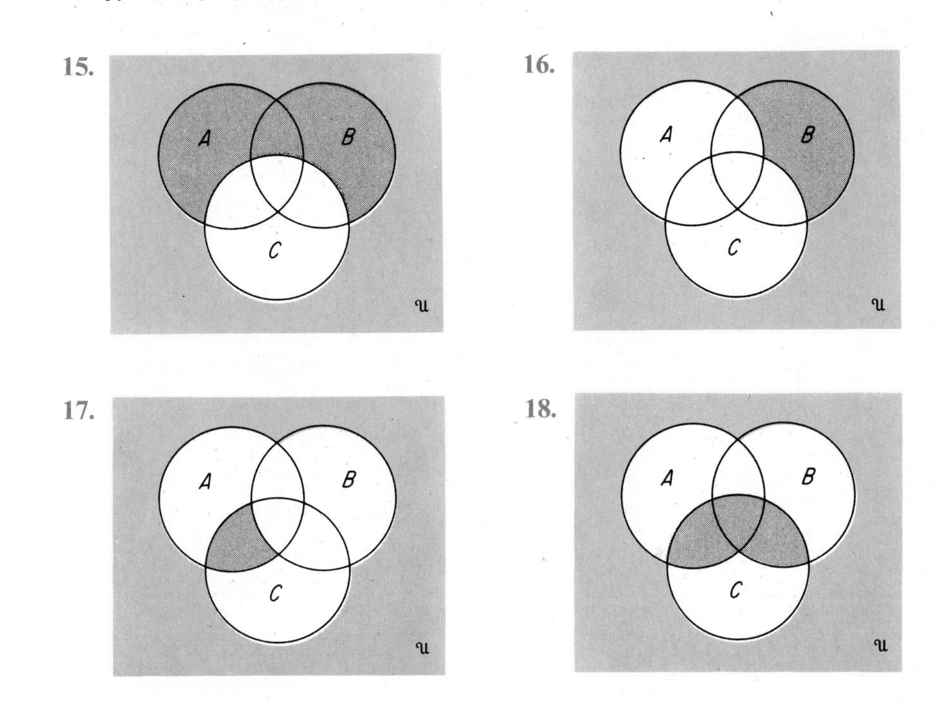

Use Venn diagrams to solve each problem.

Study Example 4 before trying Exercises 19 through 27.

19. Suppose that among 50 people, 16 have blond hair and 20 have blue eyes. Also, 9 of these people have both blond hair and blue eyes
    (a) How many have only blond hair?
    (b) How many have only blue eyes?
    (c) How many have neither blond hair nor blue eyes?
20. In a survey of 100 book readers, it was found that 44 liked fiction, 37 liked biography, and 23 liked both types.
    (a) How many of those surveyed did not like either type?
    (b) How many liked only one of the types?
    (c) How many liked only fiction?
    (d) How many liked only biography?
21. In a survey of 50 students, the following data were collected: There were 19 taking biology, 20 taking chemistry, 19 taking physics, 7 taking physics

and chemistry, 8 taking biology and chemistry, 9 taking biology and physics, 4 taking all three subjects.

(a) How many of the group were not taking any of the three subjects?
(b) How many were taking only chemistry?
(c) How many were taking physics and chemistry but not biology?

For the question in the margin near the beginning of this section consider the following Venn diagram.

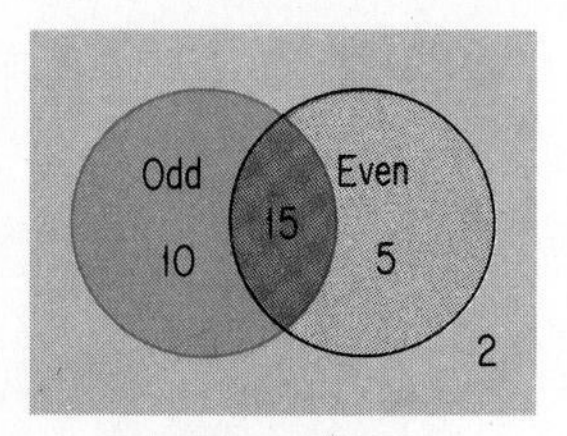

Two students failed to do either the odd- or even-numbered problems.

22. Fifty cars belong to students in a certain dormitory. Of these cars 42 have radios, 15 have tape decks, 10 have air conditioners, 2 have all three, 6 have a radio and an air conditioner, 5 have a tape deck and an air conditioner, and 10 have a radio and a tape deck.
(a) How many of these cars have an air conditioner but no radio?
(b) How many have no radio, no tape deck, and no air conditioner?
23. Repeat Exercise 22 for 4 cars having all three items and all other data as before.
24. Suppose that the student who collected the data for Exercise 22 stated that 5 of the cars had all three items and gave the other data as before.
(a) Did the student make a careful survey?
(b) Explain your answer to part (a).
25. A survey was taken of 30 students enrolled in three different clubs, A, B, and C. Show that the following data that were collected are inconsistent: 18 in A, 10 in B, 9 in C, 5 in B and C, 6 in A and B, 9 in A and C, 3 in A, B, and C.
26. In a survey at a community college, the following data emerged:

    244 students were taking English.
    208 were taking a foreign language.
    152 were taking computer science.
    72 were taking courses in English and language.
    46 were taking courses in English and computers.
    60 were taking courses in language and computers.
    24 were taking courses in all three areas.
    150 were not taking courses in any of the three areas.

    How many students were surveyed?
27. Using the data in Exercise 26, how many students were taking an English course but not taking a course in languages or computer science?

EXPLORATIONS

1. Select any sets A and B with $n(A) = 6$, $n(B) = 4$, and such that
(a) $n(A \cup B) = 7$ (b) $n(A \cup B) = 6$
(c) $n(A \cup B) = 9$ (d) $n(A \cap B) = 2$
2. Select sets A and B as in Exploration 1 and find
(a) $n(A \cap B)$ if $n(A \cup B) = 7$.
(b) $n(A \cap B)$ if $n(A \cup B) = 6$.
(c) $n(A \cap B)$ if $n(A \cup B) = 10$.
(d) $n(A \cup B)$ if $n(A \cap B) = 2$.
(e) $n(A \cup B)$ as an expression in terms of $n(A)$, $n(B)$, and $n(A \cap B)$.

3. Venn diagrams may be drawn for four sets W, X, Y, and Z. The region Z must intersect each of the eight regions of the Venn diagram for W, X, and Y. The new Venn diagram has 16 regions as lettered in the adjacent figure. Identify each of the 16 regions by its letter and as an intersection of four of the sets $W, X, Y, Z, W', X', Y', Z'$.

Use letters from the figure for Exploration 3 to identify each region. For example, the region $W \cap Y$ consists of the regions j, k, n, and o.

4. W
5. Z
6. X'
7. $(W \cup X) \cap Z$
8. $W' \cap (X \cup Z)$
9. $(W \cup Y) \cap (X' \cup Z')$

Use a Venn diagram for four sets to determine whether each statement is true or false.

10. $A \cap (B \cup C \cup D) = (A \cap B) \cup (A \cap C) \cup (A \cap D)$
11. $A \cup (B \cap C \cap D) = (A \cup B) \cap (A \cup C) \cap (A \cup D)$

12. Review *Symbolic Logic* by C. L. Dodgson, the author of *Alice's Adventures in Wonderland* (see Section 1-4, Explorations 5 through 7). Do not miss the introduction, "To Learners." Report on aspects of the book that you find most interesting. Dodgson claimed that he used all these materials with children at most 14 years old.

2-4 Properties of Set Operations

In the preceding sections we have developed the major concepts concerning sets and seen various ways of representing them. In this section we look at several properties of the various set operations, properties that will provide a basis for later discussions of operations with whole numbers.

In Section 2-2 we defined the concepts of union and intersection of sets. In each case a pair of sets was given and a new set created from them, following the appropriate method. Since each new resulting set is of the same type as the given sets, we think of the operations as being **closed**.

Closure Property of Set Union

The union of two sets A and B is a set, $A \cup B$.

Closure Property of Set Intersection

The intersection of any two sets A and B is a set, $A \cap B$.

A second property of the operations of set union and set intersection is that of **commutativity**, the fact that the order in which the sets are listed in forming the union or intersection has no effect on the resulting set. The accompanying Venn diagrams show that for any sets A and B, $A \cup B = B \cup A$ and $A \cap B = B \cap A$.

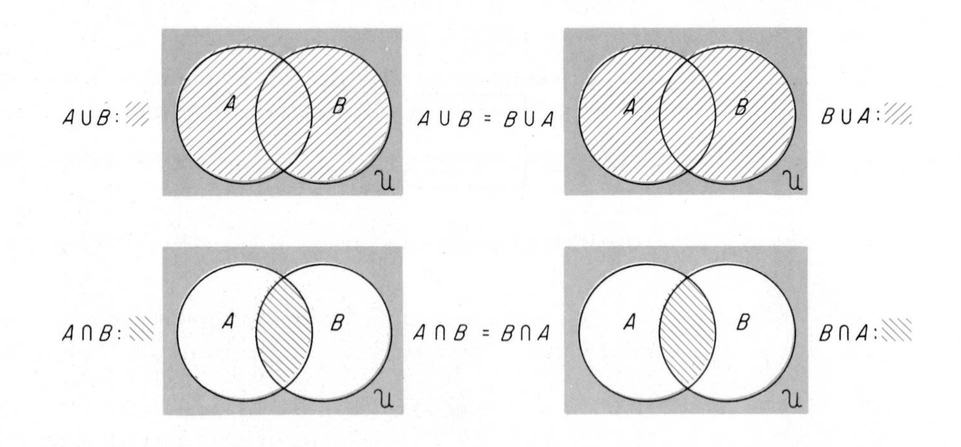

The equalities $A \cup B = B \cup A$ and $A \cap B = B \cap A$ can also be observed in the following examples.

Example 1 If $A = \{1, 2, 3, 4, 5\}$ and $B = \{3, 4, 5, 6, 7, 8\}$, find

(a) $A \cup B$ **(b)** $B \cup A$
(c) $A \cap B$ **(d)** $B \cap A$

Solution

If $x \in A \cup B$, then $x \in B \cup A$.
If $x \in B \cup A$, then $x \in A \cup B$.
If $x \in A \cap B$, then $x \in B \cap A$.
If $x \in B \cap A$, then $x \in A \cap B$.

(a) $A \cup B = \{1, 2, 3, 4, 5, 6, 7, 8\}$
(b) $B \cup A = \{1, 2, 3, 4, 5, 6, 7, 8\}$ and hence $A \cup B = B \cup A$.
(c) $A \cap B = \{3, 4, 5\}$
(d) $B \cap A = \{3, 4, 5\}$ and hence $A \cap B = B \cap A$. ■

An equality of sets may be proved by showing that the set on each side is a subset of the set on the other side. We leave such formal proofs for more advanced courses.

Commutative Property of Set Union

For all sets A and B, $A \cup B = B \cup A$.

Commutative Property of Set Intersection

For all sets A and B, $A \cap B = B \cap A$.

Another important property of the set operations of union and intersection is that of **associativity**. This property guarantees that in forming the union or the intersection of three sets, the second set may be associated with either the first set or the third set with no effect on the resulting set.

Associative Property of Set Union
For all sets A, B, and C, $A \cup (B \cup C) = (A \cup B) \cup C$.

Associative Property of Set Intersection
For all sets A, B, and C, $A \cap (B \cap C) = (A \cap B) \cap C$.

The associative property of set union can be easily illustrated by Venn diagrams.

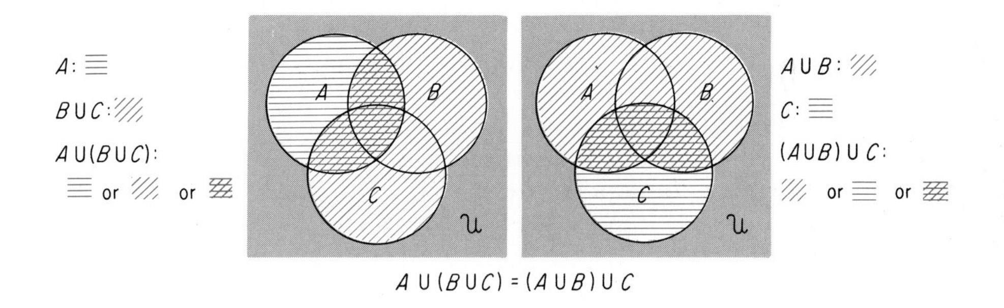

A similar illustration could be drawn for the associative property of set intersection.

The identity properties of set union and set intersection are statements of the conditions under which any set is unchanged, is identically itself, under union with another set and under intersection with another set. Intuitively, a set is unchanged under union if nothing is added, that is, $A \cup \varnothing = A$.

Identity Property of Set Union
For all sets A, $A \cup \varnothing = A$.

The intersection of a set A with another set is the set A only if each element of A is an element of the other set. The only set that contains all elements of any given set is the universal set $\mathcal{U}$; thus $A \cap \mathcal{U} = A$.

Identity Property of Set Intersection
For all sets A, $A \cap \mathcal{U} = A$.

Two additional properties of set operations that will be important for us in later chapters are the **distributive properties** of Cartesian product over set union.

Since the operation of Cartesian product is not commutative, we must state two separate forms of the distributive property.

Distributive Properties of Cartesian Product over Set Union

For all sets A, B, and C, the following two properties hold:

Left Distributive Property

$$A \times (B \cup C) = (A \times B) \cup (A \times C)$$

Right Distributive Property

$$(A \cup B) \times C = (A \times C) \cup (B \times C)$$

Example 2 If $A = \{1, 2\}$, $B = \{3, 4\}$, and $C = \{a, b\}$, find each of the following.

(a) $A \times (B \cup C)$ **(b)** $(A \times B) \cup (A \times C)$
(c) $(A \cup B) \times C$ **(d)** $(A \times C) \cup (B \times C)$

Solution

(a) $B \cup C = \{3, 4, a, b\}$ and $A \times \{3, 4, a, b\} = \{(1, 3), (1, 4), (1, a), (1, b), (2, 3), (2, 4), (2, a), (2, b)\}$.

(b) $A \times B = \{(1, 3), (1, 4), (2, 3), (2, 4)\}$, $A \times C = \{(1, a), (1, b), (2, a), (2, b)\}$, and $(A \times B) \cup (A \times C) = \{(1, 3), (1, 4), (1, a), (1, b), (2, 3), (2, 4), (2\ a), (2, b)\}$.

(c) $A \cup B = \{1, 2, 3, 4\}$ and $(A \cup B) \times C = \{1, 2, 3, 4\} \times C = \{(1, a), (1, b), (2, a), (2, b), (3, a), (3, b), (4, a), (4, b)\}$.

(d) $A \times C = \{(1, a), (1, b), (2, a), (2, b)\}$, $B \times C = \{(3, a), (3, b), (4, a), (4, b)\}$, and $(A \times C) \cup (B \times C) = \{(1, a), (1, b), (2, a), (2, b), (3, a), (3, b), (4, a), (4, b)\}$. ■

A comparison of the answers for parts (a) and (b) indicates that for these sets A, B, and C, the left distributive property of Cartesian product over set union holds.

A comparison of the answers for parts (c) and (d) indicates that for these sets A, B, and C, the right distributive property of Cartesian product over set union holds.

The relations $A = B$, $A \subseteq B$, and $A \subset B$ have been defined. We also define $A \supseteq B$, read "A contains B," to hold whenever $B \subseteq A$, and we define $A \supset B$, read "A properly contains B," to hold whenever $B \subset A$. The properties of set operations often enable us to simplify statements involving sets and to identify relations among them.

Example 3 Give the truth value, true or false, for the statement $A \cap B = B$ under the stated conditions.

(a) $A \subseteq B$ **(b)** $A \supset B$ **(c)** $B \subseteq A$
(d) $B \supset A$ **(e)** $A = B$

Solution

(a) False. **(b)** True. **(c)** True.
(d) False. **(e)** True. ■

We have studied only a few of the many properties that hold for sets under the operations of set union, intersection, difference, and Cartesian product. The study of these properties and the related theory constitutes a portion of the branch of mathematics known as **set theory**.

EXERCISES

1. Use a Venn diagram to illustrate the associative property of set intersection.
2. Use the sets $A = \{1, 2, 3,\}$, $B = \{3, 4\}$, and $C = \{1, 4, 5\}$ and illustrate
   (a) the commutative property of set union,
   (b) the associative property of set intersection.

Each of the following sets may be identified as A, A', $\mathcal{U}$, or $\varnothing$. Identify each set.

3. **(a)** $\mathcal{U} \cup A$ **(b)** $\varnothing \cap \mathcal{U}$
4. **(a)** $\mathcal{U} \cup \varnothing$ **(b)** $\mathcal{U}'$
5. **(a)** $\varnothing'$ **(b)** $A \cup \mathcal{U}'$
6. **(a)** $A \cap A'$ **(b)** $A \cup A'$
7. **(a)** $(A')'$ **(b)** $A - A'$
8. **(a)** $\varnothing' - A$ **(b)** $\mathcal{U}' \cap A$

Identify each statement as true for all sets A or false (not always true). If the statement is false give at least one counterexample.

9. **(a)** $A \cap A = A$ **(b)** $A \cap \varnothing = A$
10. **(a)** $A \cap \varnothing = \varnothing$ **(b)** $A \cup \varnothing = \varnothing$
11. **(a)** $A - \varnothing = A$ **(b)** $\mathcal{U} - A = A'$
12. $A \cap (A - B) = A \cap B'$
13. $A \cap (B \cup A') = A \cap B$
14. $A \cup (B \cup A') = A \cup A'$

From the sets A, A', B, B', $\mathcal{U}$, and $\varnothing$ find all possible replacements for the set Y such that the given statement is true.

15. **(a)** $A \cup Y = \mathcal{U}$ **(b)** $Y \cap A' = \varnothing$
16. **(a)** $A \cap Y = \varnothing$ **(b)** $A \cup Y = \mathcal{U}$
17. **(a)** $A \cup B = Y \cup B$ **(b)** $Y \cap \mathcal{U} = A'$
18. **(a)** $Y \cap \mathcal{U} = B$ **(b)** $A \cap Y = B' \cap A$

Give the truth values, true or false, for each statement under the stated conditions.

(a) $A \subseteq B$ **(b)** $A \subset B$ **(c)** $A \supseteq B$
(d) $A \supset B$ **(e)** $A = B$

19. $A \cup B = A$ 20. $A \cup B = B$
21. $A' \cap B = \varnothing$ 22. $A \cap B' = \varnothing$
23. $A - B' = A$ 24. $A' - B \subset A$

Draw a Venn diagram for each member of the given equation and identify the equation as true (always true) or false (not always true).

25. $(A \cup B) \cap C = A \cup (B \cap C)$
26. $A \cap (B \cup C) = (A \cap B) \cup C$
27. $A \cup (B \cap C) = (A \cup B) \cap (A \cup C)$
28. $A \cup (B' \cup C') = A \cup (B \cap C)'$

Identify each statement as true or false. If the statement is false give a counterexample.

29. $A \cup (B \times C) = (A \cup B) \times (A \cup C)$
30. $(A \cap B) \times C = (A \times C) \cap (B \times C)$

EXPLORATIONS

Logical statements are often represented by electric circuits. For a single statement q, think of an electric light cord from a wall outlet to an electric light bulb and with a switch in the middle of the cord.

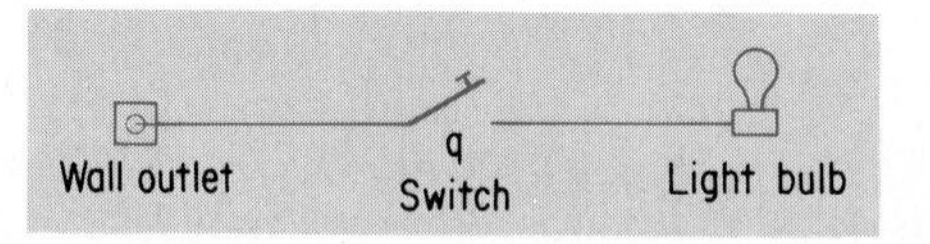

If the switch is closed the bulb is on; if the switch is open the bulb is off.

For two statements p and q, we use two switches. The following circuits are particularly useful and common.

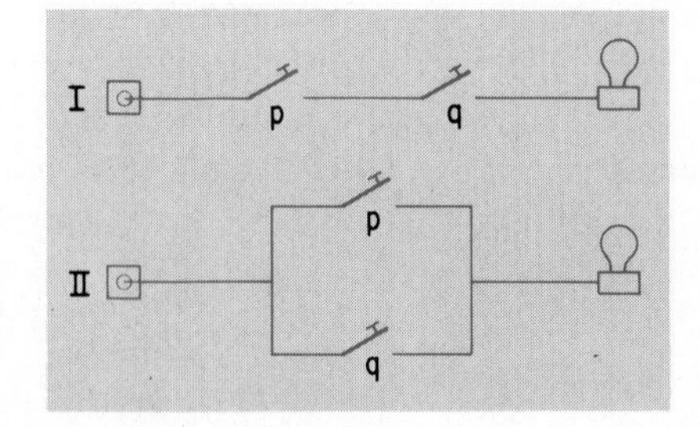

1. Assume that circuit I is in the specified condition and indicate whether the light is on or off.
   (a) p closed, q closed. **(b)** p closed, q open.
   (c) p open, q closed. **(d)** p open, q open.
2. Explain why it seems reasonable to call circuit I a **p and q circuit**.
3. Repeat Exploration 1 for circuit II.
4. Explain why it seems reasonable to call circuit II a **p or q circuit**.

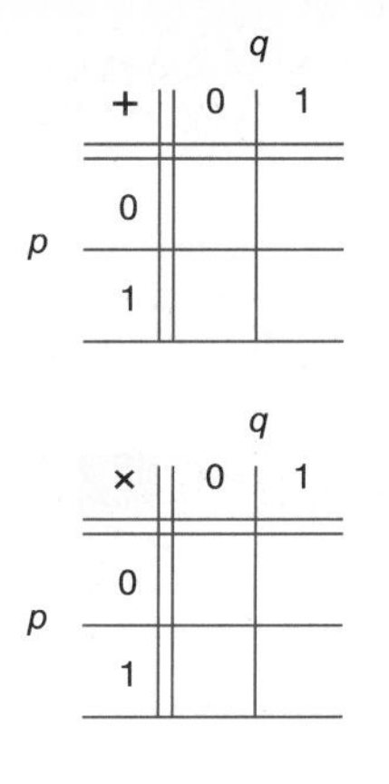

5. The p or q circuit provides a basis for defining addition of two elements:

   1 presence of current at light bulb or through a switch.
   0 absence of current at light bulb or through a switch.

   Use sketches of a p or q circuit as necessary and complete the adjacent *addition table.*

6. Use sketches of a p and q circuit as necessary and complete the adjacent *multiplication table.*

The algebra of the two elements 0 and 1 with addition and multiplication defined as in Explorations 5 and 6 is the **binary Boolean algebra**. It is named after the English mathematician George Boole (1815–1864). We compare this *algebra of the numbers* 0 and 1 with the *algebra of sets* $\varnothing$ and $\mathscr{U}$, where $\varnothing$ is the empty set and $\mathscr{U}$ is the universal set.

7. Interpret $+$ as $\cup$ and complete the following addition table.

$+$	$\varnothing$	$\mathscr{U}$
$\varnothing$		
$\mathscr{U}$		

8. Interpret $\times$ as $\cap$ and complete the following multiplication table.

$\times$	$\varnothing$	$\mathscr{U}$
$\varnothing$		
$\mathscr{U}$		

We shall make use of isomorphic sets when we identify natural numbers with positive integers and also when we identify integers with rational numbers represented by fractions with denominators of 1.

9. Two sets of elements, each with relations that may be interpreted as $+$ and $\times$, are said to be **isomorphic sets** if there is a one-to-one correspondence of the elements of the sets such that each sum corresponds to the sum of corresponding elements, and each product corresponds to the product of corresponding elements. Show that the set of elements of the binary Boolean algebra considered in Explorations 5 and 6 and the sets of the algebra of sets considered in Explorations 7 and 8 are isomorphic.

2-5 Use of Classifications in Reasoning

The ability to sort elements into categories and to use this sorting to structure and solve a problem situation is one of the powerful problem-solving skills. Earlier in this chapter we have seen the use of Venn diagrams to solve counting problems involving various sets of elements. In this section we examine another type of categorization system which is useful in problem solving where classifications are involved.

One type of classification problem involves the order in which objects are arranged in a row. Such problems are usually solved using a very simple representation—for example, a set of name holders on a line.

Example 1 Kevin is taller than Ron and younger than Cheryl. Cheryl is neither as tall as Ron nor as old as Ron. List the three siblings **(a)** by decreasing age **(b)** by increasing height.

Solution **(a)** Since Kevin is younger than Cheryl, we place his name to the right of Cheryl's in a listing by decreasing age.

Cheryl, Kevin.

Since Cheryl is not as old as Ron, we see that Ron is older than Cheryl. Thus we list his name to the left of Cheryl's and obtain the list of siblings by decreasing age:

Ron, Cheryl, Kevin.

(b) Note the statements

Kevin is taller than Ron.
Cheryl is not as tall as Ron.

When listed by increasing height, we have

Cheryl, Ron, Kevin. ■

The solutions of several other types of classification problems can be based upon the completion of a table, or grid.

Example 2 Douglas, Kim, and Rosha, not necessarily in that order, each play on the track, basketball, or football team. No two play on the same team. Douglas is not the track or basketball player. Rosha is not one of the track team members. On which team does each play?

Solution We have three players and three sports. Thus we use a 3 × 3 grid with a row for each player and a column for each sport. We use "Yes" in the "box" on a given row and column to indicate that the player (row) is on the team of the sport (column); we use an X to indicate that the player is not on that team.

	TRACK	BASKETBALL	FOOTBALL
DOUGLAS			
KIM			
ROSHA			

Then we *reread the problem*, note that the players and sports are properly identified on the table, and look for statements that can be represented in the table. The statement

Douglas is not the track or basketball player

enables us to eliminate those two sports for him. Then, since Douglas plays one of the sports, we know that

Douglas is the football player.

Also since no two play on the same team,

Kim does not play football.
Rosha does not play football.

Then since Rosha is not a track team member,

Rosha is not the track team player.
Rosha is the basketball player.
Kim is not the basketball player.
Kim is the track team member.

Thus Douglas is on the football team, Rosha is on the basketball team, and Kim is on the track team. The completed grid has the following form.

	TRACK	BASKETBALL	FOOTBALL
DOUGLAS	×	×	Yes
KIM	Yes	×	×
ROSHA	×	Yes	×

■

In dealing with classifications such as the one in Example 2, the basic strategy is to focus on the positions on one row (or column), to eliminate all but one of those possibilities in order to establish the remaining one, and thereby eliminate all others in its column (and row). Then the procedure is repeated for a grid with one less row and one less column and repeated again and again, until one position in each row and column has been established.

Example 2 can also be solved through the use of deductive reasoning as developed in Section 1-3. Since no two players are on the same team and Douglas is not the track or basketball player, Douglas is the football player and the others are the track and basketball players. Then since Rosha is not a track team member, Rosha is on the basketball team and Kim is on the track team.

Example 3 Determine the individual order of finish and each runner's colors in the race described by the following information.

Otto, Spence, and Vanden Eynden were the first three finishers in a 100-meter race. The colors of their uniforms were red, green, and yellow, but not necessarily in that order. The individual in the red uniform finished second. The runner in the green uniform finished ahead of Otto, who was not wearing red. Vanden Eynden finished first.

Solution Since Vanden Eynden finished first, the runner in red finished second, and Otto was not wearing red, we know the order of finish was Vanden Eynden, Spence, and Otto. Thus, Spence was wearing red. Since green finished ahead of Otto, Vanden Eynden must have worn green. This leaves the yellow uniform for Otto. ■

In the solution of Example 3 the available information was structured to obtain inescapable (valid) conclusions. However, not all such problems yield readily to such an analysis. Hence we provide an alternative approach using a table as in Example 2.

For Example 3 we have three runners, three positions, and three colors. The following table has a row for each runner, a set of three columns for the colors, and a set of three columns for the colors. We use "Yes" to indicate that a runner has a particular position or color and × to indicate that the runner does not have that position or color. We could have used either positions or colors for the rows and formed similar tables.

	FIRST	SECOND	THIRD	RED	GREEN	YELLOW
OTTO						
SPENCE						
VANDEN EYNDEN						

Then we *reread the problem*, note that the runners, positions, and colors have been properly identified in the table, and look for statements that can be represented in the table. The following statements are listed in one order that may be used to complete the table. Note that some of these statements are directly from the given information; others are based upon the entries in the table.

Vanden Eynden finished first.
Vanden Eynden did not finish second.
Vanden Eynden did not finish third.
Spence did not finish first.
Otto did not finish first.
Otto was not wearing green.
Otto was not wearing red.
Otto was wearing yellow.
Spence was not wearing yellow.
Vanden Eynden was not wearing yellow.
Otto did not finish second (was not wearing red).
Otto finished third.
Spence did not finish third.
Spence finished second.
Spence wore red.
Spence did not wear green.
Vanden Eynden did not wear red.
Vanden Eynden wore green.

	FIRST	SECOND	THIRD	RED	GREEN	YELLOW
OTTO	×	×	Yes	×	×	Yes
SPENCE	×	Yes	×	Yes	×	×
VANDEN EYNDEN	Yes	×	×	×	Yes	×

Many statements and patterns of reasoning are most easily understood when they are expressed in *if-then form.* In Example 1 we observed from the given data that

Kevin is younger than Cheryl.
Cheryl is not as old as Ron.

We reasoned

If Kevin is younger than Cheryl, then Cheryl is older than Kevin
If Cheryl is not as old as Ron, then Ron is older than Cheryl

to obtain the list of siblings by decreasing age:

Ron, Cheryl, Kevin.

Our reasoning in Example 2 may be summarized as follows, given that Douglas, Kim, and Rosha each play on one of the track, basketball, or football teams and no two play the same sport.

If Douglas is not a track or basketball team member, then Douglas is the football player.

If Douglas is the football player, then Rosha is on either the track team or the basketball team.

If Rosha is on either the track team or the basketball team and he is not on the track team, then he is on the basketball team.

If Douglas is on the football team and Rosha is on the basketball team, then Kim is on the track team.

The English language is both powerful and, at times, confusing because of the many equivalent forms of statements, that is, the many ways of saying the same thing. For example, the following statements are all equivalent.

If $x = -1$, then $x^2 = 1$.
$x^2 = 1$, if $x = -1$.
$x = -1$ is sufficient for $x^2 = 1$.
$x^2 = 1$ is necessary for $x = -1$.
$x = -1$ implies $x^2 = 1$.
$x = -1$ only if $x^2 = 1$.

The following statements are also equivalent.

$x^2 = 1$ if and only if $x = 1$ or $x = -1$.
$x^2 = 1$ is necessary and sufficient for $x = 1$ or $x = -1$.
$x = 1$ or $x = -1$ if and only if $x^2 = 1$.
$x = 1$ or $x = -1$ is necessary and sufficient for $x^2 = 1$.

An informal working knowledge of forms of equivalent statements is essential for all types of reasoning. Anyone who "explains" anything uses, or misuses, such statements. Teachers have special needs for clear explanations. Formal work on equivalent statements provides one basis for proofs and justifications in many different situations.

EXERCISES

1. Karen and Chad are the names of an educator and a gardener, but not necessarily in that order. Karen is not the gardener. What does each do?
2. Bruce was younger than Phil, and John was older than Phil. Was John younger than Bruce, the same age as Bruce, or older than Bruce?
3. Three mathematics students, Jim, John, and Dennis, were marching through graduation. As they marched, the oldest was first, the youngest next, and the other third. Jim was younger than Dennis. John was older than Jim but was younger than Dennis. What was the order in which they marched?
4. Douglas and Anne each have two degrees. The four degrees are in law, medicine, engineering, and mathematics. The engineer does not know how to stitch a patient. Douglas is not the mathematician and Anne is not the engineer. Identify the degrees for each.
5. Three professors are assigned to teach one section of mathematics each. The professors are named Edge, Ha, and Tipnis. The subjects to be taught are statistics, graph theory, and analysis. The statistics teacher is neither the oldest nor the youngest of the three professors. Ha is not the oldest. The oldest professor, who teaches analysis, had the youngest professor as a student several years ago. Tipnis teaches graph theory. Who teaches each of the other two courses?
6. Three special education teachers have a total of 14 students. The teacher's names are Sylvia, Ashley, and Martha. Sylvia has 2 girls and Ashley has 2 boys. Sylvia, who has 3 students, has at least as many children as Ashley. Martha has 6 more boys than girls and the same number of girls as Sylvia has boys. How many boys do Sylvia and Martha each have?
7. Anne was making boxes of fudge, macaroons, and chocolate chip cookies for the school social. Each box was correctly labeled for one of these items. Her son, Douglas, put some fudge in with the chocolate chip cookies and then changed the lids so that no box had the lid with its correct label. Her daughter, Elise, an excellent mathematics student, came home from school and asked what she was doing. Anne told her of the lid switch, gave her daughter the following statements as clues, and asked her to identify the contents of the box with each label.

   One box contained fudge.
   One box contained macaroons.
   One box contained both fudge and chocolate chip cookies.
   None of the boxes contains any of the items specified on its lid.

   Which boxes had which labels?
8. Three farmers, Kennell, Wettstein, and Remmert, have acres of corn and beans on their farms. Wettstein has 400 acres of beans and Kennell has 400

acres of corn. Kennell has 100 acres more planted than Wettstein, who has 500 acres. Remmert has 500 more acres of corn than beans and the same number of acres of beans as Wettstein has acres of corn. How many acres of corn do Wettstein and Remmert each have?

9. Rose, Sue, and Theresa each drive trucks. The trucks are a van, a panel truck, and a pickup truck. Rose does not drive the panel truck. Sue does not drive the pickup truck. Theresa does not drive either the panel truck or the pickup. Identify the driver of each type of truck.

10. Don, Roland, and Jack are pen pals. Each lives in a different country. The countries are the United States of America, Germany, and Australia. The Australian is younger than Roland. Jack is the youngest. The oldest lives in the United States. Roland lives in Germany. In which country does Jack live?

11. Andrea, Babette, and Chandra each entered a flower arrangement in the 4-H show. Their arrangements, which received the top three awards, were a daisy, an iris, and a rose arrangement. Use the following information to find each girl's arrangement and the award it received.

    This was the first time Andrea had entered her arrangement in a 4-H show.

    Babette and Chandra entered similar arrangements last year.

    Babette's arrangement was not the daisy arrangement.

    In last year's 4-H fair, the daisy arrangement won a better award than the iris arrangement.

    Andrea's arrangement received a higher award than Chandra's arrangement.

    Babette's arrangement received a higher award than Andrea's arrangement.

12. The first, second, third, and fourth positions in a race for handicapped children were awarded books as prizes. The other participants received other types of recognition. Albert, Bob, Charles, and Dana received books as prizes. Albert was neither first nor last. Bob was ahead of Charles and behind Dana. Charles was not second. Dana was followed by Bob. Find the order in which they finished the race.

EXPLORATIONS

1. Prepare a report on solving arithmetic problems. Include solutions of at least ten problems representing several types. The 1980 Yearbook of the National Council of Teachers of Mathematics, *Problem Solving in School Mathematics*, edited by Stephen Krulik, provides a general reference. Examples and solutions may be found in books such as *Mathematical Puzzles for Beginners and Enthusiasts* by Geoffrey Mott-Smith, Dover Publications, Inc., and *What is the Name of This Book? The Riddle of Dracula and other Logical Puzzles* by Raymond Smullyan, Prentice-Hall, Inc.

2. Repeat Exploration 1 for verbal problems.
3. Repeat Exploration 1 for problems involving geometric figures.

Chapter Review

1. Find the next five numbers in the sequence 1, 3, 7, 15, 31, ...
2. Suppose that you have a large cube having three units to an edge. Further suppose that this cube is made of small unit cubes. If the exterior of the large cube is painted red, how many of the unit cubes will have **(a)** one red face **(b)** two red faces **(c)** three red faces?
3. How many numbers are there in the set of multiples of 5 between 45 and 105?
4. How many multiples of 3 between 1 and 99 are also multiples of 2?
5. Give a verbal description for the set $Y = \{1, 3, 5, 7, 9\}$.
6. Show that $\{c, r, a, b\}$ and $\{f, i, s, h\}$ are
   (a) equivalent sets of letters **(b)** not equal sets.
7. List all proper subsets of $\{4, 5, 6\}$.

In Exercises 8 through 11 let $\mathcal{U} = \{1, 2, 3, \ldots, 10\}$, $A = \{1, 2, 3, 4, 5, 7, 9\}$, *and* $B = \{2, 4, 8\}$. *Find each set.*

8. A'
9. B'
10. $A - B$
11. $B - B$

12. If $X = \{1, 2, 3, 4, 5\}$ and $Y = \{4, 5, 6, 7\}$, find $X \cap Y$.
13. If the universal set is $\{1, 2, 3, \ldots, 10\}$, $A = \{1, 2, 3, 4, 5\}$, and $B = \{3, 4, 5, 6, 7\}$, find $A' \cap B'$.
14. If $A = \{1, 2, 3, 4, 5\}$ and $B = \{3, 5, 7, 9\}$, find $A \cup B$.
15. Let $A = \{1, 2, 3\}$ and $B = \{r, s\}$ and find the cross product $B \times A$.
16. Show by means of a Venn diagram that $(A' \cup B')' = A \cap B$.
17. Represent each of the following by a Venn diagram.
    (a) $A \cap (B \cup C)$ **(b)** $(A \cup B) \cap (A \cup C)$
18. Angela speaks three languages; Mario speaks two. There are four languages that at least one of them can speak. How many languages are spoken by both of them?
19. In a group of 35 students, 15 are studying French, 22 are studying English, 14 are studying Spanish, 11 are studying both French and English, 8 are studying English and Spanish, 5 are studying French and Spanish, and 4 are studying all three subjects.
    (a) How many are taking only English?
    (b) How many of these students are not taking any of these subjects?

In Exercises 20 through 23 identify as true or not always true (false).

20. The sets $\{b, a, t\}$ and $\{1, a, b\}$ are
    (a) equal sets **(b)** equivalent sets
21. **(a)** $(\mathcal{U} \cup B) \subseteq \mathcal{U}$ **(b)** $\mathcal{U} \cap \varnothing \subset \varnothing$
22. **(a)** $(A \cup B) \subset (B \cup A)$ **(b)** $(A \cap B) \subset (A \cup B)$
23. **(a)** $75 \in \{2, 4, 6, \ldots\}$ **(b)** $75 \in \{1, 3, 5, \ldots\}$

24. Alice, Betty, and Cathy took the first three places.

Alice was not first.
Betty was not third.
Cathy won a book.
The first prize was a cash prize.

Who was in first place?

25. Find the order of Bill, Frank, Andy, and John in increasing age if

Frank is younger than Bill but older than Andy.
Andy is younger than Frank but older than John.

PROGRAMMING IN BASIC

These computer application sections are intended to provide examples of computer programs which can be typed in and used with little, if any, prior microcomputer experience. Chapter 14 contains an introduction to the BASIC language and provides additional information on writing programs in that language.

The development of programs to solve problems in mathematics can vary greatly in difficulty. In order to help you get started in an easy fashion, we introduce BASIC programming commands and processes in a gradual fashion as we move through the textbook. We begin by providing you with a completed program and examining its components.

```
NEW
 5  HOME
1Ø  PRINT "N", "N + N", "N * N"
2Ø  FOR N = 1 TO 2Ø
3Ø  PRINT N, N + N, N * N
4Ø  NEXT N
5Ø  END
```

Note that the program consists of a set of numbered lines. Each of these lines presents a specific instruction to the computer. The **line numbers** in front of the lines indicate the order in which the computer will execute the commands, unless the program instructs the computer to do otherwise. That is, the computer will begin at the line with the smallest line number and proceed through the lines in increasing numerical order.

The command NEW is used at the beginning of a new program to indicate that any other commands in the computer's memory should be ignored. The command HOME on line 5 instructs the computer to clear the screen before the program is run. The command END on the last line of the program signals to the computer that there are no more instructions to be completed.

The BASIC language uses the symbol * to indicate multiplication, as the symbol × would be confused with the letter *x* by the computer. The symbol Ø is used for the numeral 0 to avoid confusion with the capital letter O. The commas in lines 1Ø and 3Ø instruct the computer to print the output in prespecified columns. Thus the given program instructs the computer

To print (line 1Ø) headings for three columns, and

On a row for each of the natural numbers 1 through 20 as values of N (line 2Ø)

To print (line 3Ø) the corresponding values of the expressions in the headings.

The command in line 2Ø FOR N = 1 TO 2Ø and the command in line 4Ø NEXT N work together to cycle the computer through calculating the double and square of each of the first 20 natural numbers. When the computer is executing the program, it first processes the number 1; prints it, its double, and its square; it then looks for and finds the next number, 2, prints it, its double, and its square; then it finds the next natural number, and so forth until the last specified number has been processed. When the computer finishes processing the natural number 20, the computer finds that there are no values of N to be considered when it tries to obey the command NEXT N on line 4Ø. Therefore the computer goes to the next numbered line, that is, to line 5Ø, which instructs the computer that the task has been completed.

If you have access to a microcomputer, type the program into your microcomputer, using a RETURN or ENTER at the end of each line of typing. When you have finished entering your program, type the single word RUN and touch RETURN or ENTER to signal the computer to execute the program. Did it give the following output?

```
RUN
N           N + N        N * N
1             2            1
2             4            4
3             6            9
4             8           16
5            1Ø           25
6            12           36
7            14           49
8            16           64
9            18           81
1Ø           2Ø          1ØØ
11           22          121
12           24          144
13           26          169
14           28          196
15           3Ø          225
16           32          256
17           34          289
18           36          324
19           38          361
2Ø           4Ø          4ØØ
```

Modify the program so that it will print out the square and cube (N * N * N) of each of the first 50 natural numbers. Type this program into your computer, run it, and check your output for correctness.

3

Numeration and Whole Numbers

Leonardo of Pisa
(ca. 1180–1250)

Leonardo of Pisa, better known as Fibonacci, was an Italian merchant of the thirteenth century. His travels and trade brought him into constant contact with the Arabic traders from North Africa. Through them he mastered numeration and computational expertise in the Hindu-Arabic numeration system. His text Liber Abaci, *published in 1202, provided an introduction to these concepts and methods for Europeans of his period.*

3-1 Base Ten Notation

We habitually take for granted the use of our system of numeration, as well as our computational procedures. However, these represent the creative work of human beings through the ages. We can gain a better appreciation of our system of numeration and methods of computation by examining other systems.

Four thousand years ago the early Egyptians used groupings of strokes so that they could recognize the number of strokes for each of the numbers 1 through 9.

|, ||, |||, ||||, ||| ||, ||| |||, |||| |||, |||| ||||, ||| ||| |||

1987 can be written using early Egyptian notation in this way:

𓆼 𓍢𓍢𓍢𓍢𓍢𓍢𓍢𓍢𓍢 ∩∩∩∩∩∩∩∩ |||||||

Probably because we have ten fingers, a new symbol is often introduced for ten. We write this symbol as 10 in our system of notation. The early Egyptians represented 10 by the symbol ∩ and introduced a new symbol for each power of ten.

Symbol	Name	Value
|	Vertical staff	1
∩	Heel-bone	10
𓍢	Scroll	100
𓆼	Lotus flower	1000
𓂭	Pointing finger	10,000

The scroll is often called a coil of rope and was sometimes "coiled" counterclockwise instead of clockwise.

Note these comparisons of decimal and early Egyptian number symbols.

3				
30	∩∩∩
300	999
25	∩∩|||||
142	9∩∩∩∩||
12,321	[illegible]999∩∩|

Recall that fingers are also called *digits*. Finger reckoning is still used extensively in some parts of the world.

The Egyptian system is said to have a **base** of ten because the symbols represent powers of ten. Our system of numeration is called a *decimal system* to emphasize the use of powers of ten for each *place value*. The Egyptian system has no place value. The absence of a place value means that the position of the symbol does not affect the number represented. For example, in our decimal system of numeration, 23 and 32 represent different numbers. In the Egyptian system ∩∩||| and |||∩∩ are different representations of the same numeral. Without the concept of place value the early Egyptians needed different symbols for different powers of ten.

The early Babylonians used the symbol < for ten. The Romans used X. Our **decimal system of numeration** makes use of the ten *decimal digits*.

0 1 2 3 4 5 6 7 8 9

This system has its roots in the works of Hindu scholars about A.D. 500. Originally there were nine digits. The tenth digit, 0, was added about A.D. 850. Arabic scholars adopted the Hindu system and introduced it in Europe as the Islamic Empire with its Arabic roots expanded. The many advantages of the *Hindu-Arabic system of numeration* over other systems were based upon its use of **place value**. The value represented by each digit in a numeral depends on the position that digit occupies in its particular sequence of digits. For example, 1987 is read as

One thousand nine hundred eighty-seven

$$1987 = (1 \times 1000) + (9 \times 100) + (8 \times 10) + (7 \times 1)$$

The digit 1 represents 1 thousand, the digit 9 represents 9 hundreds, the digit 8 represents 8 tens, and the digit 7 represents 7 ones in the numeral 1987. All symbols for numbers are **numerals**.

Note that zero plays an important part in place-value notation. For example, the numerals

43 403 430 400,300 4,003,000

each represent different numbers. By filling a place that is not otherwise occupied, the symbol zero allows us to use the same decimal digits for all powers of ten (place values).

1 10 100 1000 10000 ...

In contrast, early Egyptians needed a different symbol for each power of ten.

It is convenient to use exponents when representing the place values of digits. We define $10^0 = 1$ so that each place value may be written as a power of ten. For example,

$$1000 = 10^3 \qquad 100 = 10^2 \qquad 10 = 10^1 \qquad 1 = 10^0$$

Then we may write 1987 in **expanded notation** as

$$1987 = (1 \times 10^3) + (9 \times 10^2) + (8 \times 10^1) + (7 \times 10^0)$$

Example 1 Use exponents and write 2306 in expanded notation.

Solution $2306 = (2 \times 10^3) + (3 \times 10^2) + (0 \times 10^1) + (6 \times 10^0)$ ■

Example 2 Write in decimal notation.

$$(3 \times 10^5) + (2 \times 10^4) + (7 \times 10^3) + (0 \times 10^2) + (1 \times 10^1) + (3 \times 10^0)$$

Solution 327,013 ■

About 1650 B.C. an Egyptian scribe named Ahmes copied an earlier manuscript which he described as "the entrance into knowledge of all existing things and all obscure secrets." Ahmes' copy is often called the **Rhind Papyrus**, named after the Englishman A. Henry Rhind, who purchased the papyrus in Egypt. A copy of the papyrus was published in 1927. The original is now in the British Museum. The Rhind Papyrus provides us with information concerning the mathematics and methods of computation of the early Egyptians.

The development of a base ten system of numeration with place value was a major step in the development of mathematics. Yet its amazingly simple structure arises from just these two main ideas—groupings in sets of ten and place value. The early Egyptians used groupings in sets of ten without place value. Note that the symbols used to represent numbers affect the procedures (*algorithms*) used for computations. In the early Egyptian system computations are possible but tedious. For example, we may use these steps to add 27 and 35.

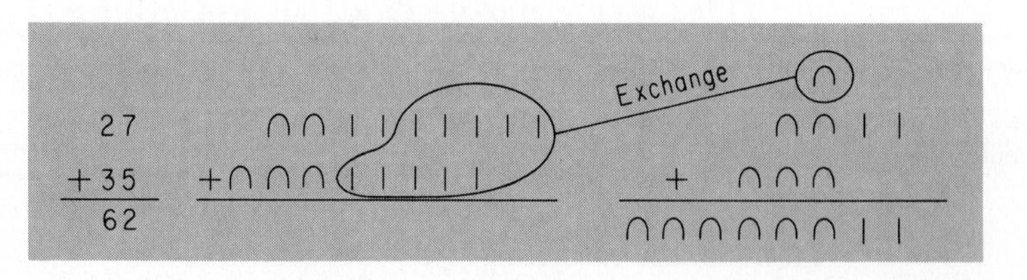

Observe that, in this Egyptian system, an indicated collection of ten ones was replaced by a symbol for ten before the final computation took place. In our decimal system we mentally perform a corresponding exchange of ten ones for a ten when we express $(7 + 5)$ as one ten and two ones. We exchange kinds of units in a similar manner in subtraction.

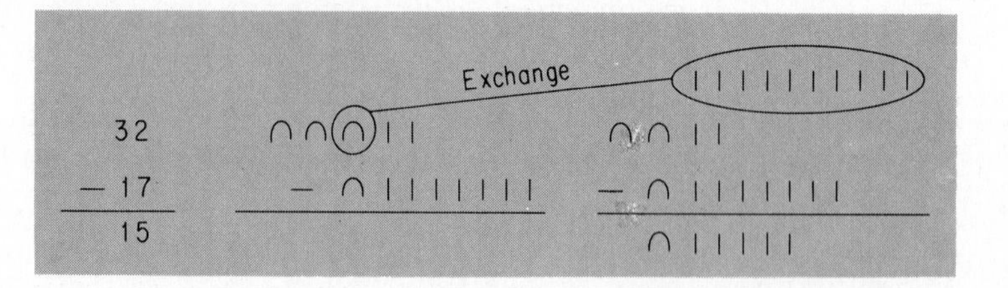

We often introduce groupings in sets of ten using models such as **base ten arithmetic blocks**.

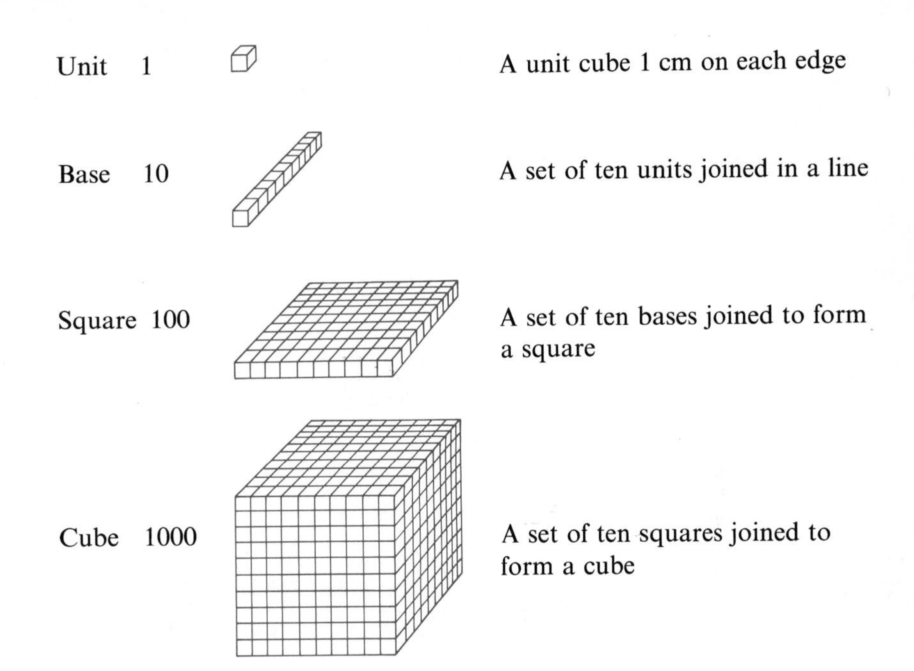

These blocks may be used to show the grouping of ten units into one base unit, the grouping of ten bases into one square of 10^2 units, and the groupings of ten squares into one cube of 10^3 units. The procedure for finding the base ten numeral for any given number of objects, considered as units represented by unit cubes, then consists of repetitions of a simple pattern.

> First group the units into as many sets of 10, bases, as possible; the remaining units represent the units digit of the numeral.
>
> Then group the bases into as many sets of 10, squares, as possible; the remaining bases represent the tens digit of the numeral.
>
> Then group the squares into as many sets of 10, cubes, as possible; the remaining squares represent the hundreds digit of the numeral.
>
> Introduce new representations for additional sets of tens as needed and continue to group in sets of ten and use the remaining elements to represent that place-value of the numeral.

Note that the procedure is precisely the same as the one used by the early Egyptians. Any natural number, such as 123, may be represented in this way:

123

Addition and subtraction may be visualized in terms of regrouping. Multiplication in terms of repeated additions and division in terms of repeated subtractions are very tedious. The early Egyptians multiplied by a process of doubling. This process is based on the fact that $1 = 2^0$ and that any number may be represented as a sum of powers of two. For example, $19 = 1 + 2 + 16$. To find the product 19×25 we first double 25; we then continue to double and keep a record of the indicated multiples of 25.

$$\textcircled{1} \times 25 = \textcircled{25}$$
$$\textcircled{2} \times 25 = \textcircled{50}$$
$$4 \times 25 = 100$$
$$8 \times 25 = 200$$
$$\textcircled{16} \times 25 = \textcircled{400}$$

Then we find the product 19×25 by adding the multiples of 25 that correspond to 1, 2, and 16.

$$19 = 1 + 2 + 16$$
$$19 \times 25 = (1 + 2 + 16) \times 25$$
$$= 25 + 50 + 400 = 475$$

Example 3 Use the Egyptian method of doubling to find the product 23×41.

Solution

$$\textcircled{1} \times 41 = \textcircled{41}$$
$$\textcircled{2} \times 41 = \textcircled{82}$$
$$\textcircled{4} \times 41 = \textcircled{164}$$
$$8 \times 41 = 328$$
$$\textcircled{16} \times 41 = \textcircled{656}$$

$$23 = 1 + 2 + 4 + 16$$
$$23 \times 41 = (1 + 2 + 4 + 16) \times 41$$
$$= 41 + 82 + 164 + 656 = 943$$

■

EXERCISES *Write in early Egyptian notation.*

1. 35
2. 246
3. 3417
4. 60
5. 12,307
6. 21,532

Write in decimal notation.

7. 𓎆𓎆𓏺𓏺
8. 𓍢𓍢𓎆𓏺
9. 𓆼𓍢𓏺𓏺
10. 𓆼𓆼𓍢𓍢𓎆𓏺
11. 𓆼𓍢𓍢𓎆𓏺𓏺
12. 𓆼𓍢𓍢𓍢𓎆𓎆𓏺𓏺𓏺𓏺

Write in expanded notation.

13. 257 14. 372 15. 3504 16. 5240

17. 235,100 18. 304,065 19. 500,200 20. 100,090

Write in decimal notation.

21. $(8 \times 10^3) + (1 \times 10^2) + (6 \times 10^1) + (5 \times 10^0)$
22. $(4 \times 10^5) + (3 \times 10^4) + (0 \times 10^3) + (4 \times 10^2) + (2 \times 10^1) + (8 \times 10^0)$
23. $(6 \times 10^5) + (9 \times 10^3) + (5 \times 10^2) + (2 \times 10^0)$
24. $(8 \times 10^6) + (6 \times 10^5) + (4 \times 10^4) + (3 \times 10^3)$
25. (6×10^7)
26. (8×10^9)
27. (7×10^8)
28. (5×10^6)
29. Thirty-seven thousand nineteen.
30. Three million five hundred five.
31. Three hundred thousand twenty-three.
32. Four billion three thousand eleven.

Read each decimal numeral.

33. 5370
34. 5730
35. 205,030
36. 2,250,300
37. 25,203,500
38. 502,320,035

Write the decimal numeral for each set of base ten arithmetic blocks.

39. 3 squares, 4 bases, 5 units.
40. 3 cubes, 2 bases, 1 unit.
41. 2 cubes, 3 squares.
42. 5 cubes, 6 squares, 3 bases, 7 units.

Sketch the outline of base ten arithmetic block representations for each number.

43. 32 44. 101 45. 1010

46. 324 47. 1023 48. 1234

Write in early Egyptian notation and perform the indicated operation in that system.

49. $\begin{array}{r} 82 \\ +32 \\ \hline \end{array}$

50. $\begin{array}{r} 123 \\ +\ 47 \\ \hline \end{array}$

51. $\begin{array}{r} 2358 \\ +\ 176 \\ \hline \end{array}$

52. $\begin{array}{r} 541 \\ -217 \\ \hline \end{array}$

53. $\begin{array}{r} 2651 \\ -\ 528 \\ \hline \end{array}$

54. $\begin{array}{r} 3023 \\ -\ 715 \\ \hline \end{array}$

Use the Egyptian method of doubling to find these products.

55. 19×45
56. 25×35
57. 17×55
58. 29×41
59. 43×29
60. 31×47

EXPLORATIONS Roman numerals have been widely used in the past and are not unusual in MCMLXXXVII, that is, 1987. These explorations are intended to help you recognize that it is relatively easy to represent whole numbers as Roman numerals. Some computations with Roman numerals are as easy as with our ordinary numerals, but other computations with Roman numerals are very awkward.

1. Use Roman numerals and list the natural numbers I through XXIX.
2. Use Roman numerals and list by fives the numbers V through L.
3. Describe the use of addition and subtraction in the representation of numbers by Roman numerals, such as XXIX.
4. Roman numerals are often used to state dates of construction of buildings on the cornerstones and in other places as well. Often the pages that precede the introduction to a book are given in Roman numerals. Find as many examples as you can of the use of such numerals, and then develop a short unit on Roman numerals for use at a specified elementary school grade level.
5. The early Egyptians often used their numerals to form patterns, or pictures. For example, they were able to write 25 in such ways as these.

   ||∩|∩|| ∩|||||∩ |∩|||∩|

   Why can't we do likewise with our numerals? What are the basic advantages of our decimal system of notation that make it desirable to give up such opportunities for "artistic effect"?
6. Sketch base ten arithmetic block representations for each number and then use these models to perform the operations indicated in Exercises 49 through 54.
7. Select an elementary school grade level and describe another early (before A.D. 400) system of numeration for students in that grade.

3-2 Whole Numbers

Natural numbers may be used in at least three different ways. They may be used for **identification** such as your social security number, your telephone number, and the number of your driver's license. Natural numbers may be used as **ordinal numbers** to assign an order (first, second, third, ...) to the elements of a finite set. Natural numbers may be used as *cardinal numbers* to specify the number of elements in a set.

Example 1 Tell whether each specified number in the following statement is used for identification, as an ordinal number, or as a cardinal number.

The *second* train through Peoria consisted of *thirty* cars pulled by engine number *534*.

Solution *Second* is used as an ordinal number, *thirty* is used as a cardinal number, and *534* is used for identification. ■

The set of *whole numbers* consists of zero and the natural numbers

$W = \{0, 1, 2, 3, \ldots\}$

Whole numbers are often used to answer questions such as

How many? How much? Which one?

The famous mathematician Leopold Kronecker (1823–1891) said "God created the natural numbers; everything else is man's handiwork."

These uses are based upon a variety of everyday events involving counts or measures. The events are often the result of joining sets of objects, removing a subset from a set of objects, or separating a set of objects in such a fashion that each of the subsets formed has the same number of elements. These procedures give rise to the concept of an *operation.* A **binary operation** is a procedure by which *two* elements are combined according to a specific rule to produce a unique third element, called the *result* of the operation. The basic binary operations in arithmetic are addition, subtraction, multiplication, and division.

The operation of **addition** can be represented by the *union* of two parts to form a whole—often to answer the question

How many objects are there all together?

For example, an elementary school model for $3 + 4 = 7$ might involve 3 geometric solids and 4 toys shown as sets of 3 objects and 4 objects before being joined and also shown as a set of 7 objects after being joined together as a single set.

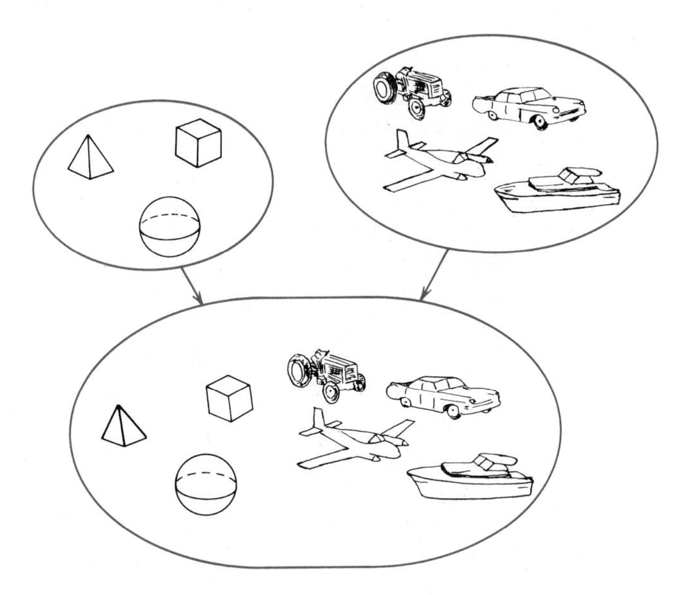

Note that this is an application of cardinal numbers of sets (Section 2-2). For any two sets A and B with 3 elements and 4 elements, respectively, and no elements in common,

$$n(A) = 3 \qquad n(B) = 4 \qquad n(A \cup B) = 7$$

Addition can be defined in general as follows:

> If A and B are any two sets such that $A \cap B = \varnothing$, then
>
> $$n(A) + n(B) = n(A \cup B).$$

Let $2 = n(\{a, b\})$ and $3 = n(\{c, d, e\})$. Then $2 + 3 = n(\{a, b\} \cup \{c, d, e\})$ $= n(\{a, b, c, d, e\}) = 5$

In a similar manner, the operation of **subtraction** can, when the difference is a whole number, be represented by the separation of a whole into subsets—the part removed and the part remaining as in $7 - 3 = 4$. Subtraction can also be thought of, as in the case of $7 - 3$, as asking what number must be added to 3 in order to get a sum of 7. This is sometimes called the **add-on interpretation of subtraction**.

The operation of **multiplication** can be represented by repeated additions of a set of fixed size. For example, if one were to join 4 basketball teams of 5 members each, the problem might be thought of as repeatedly adding 5 four times. This procedure could be modeled using a rectangular array in which teams are shown in columns and positions of individual players on rows.

	Celtics (C)	*Lakers (L)*	*Bucks (B)*	*Nets (N)*
Guard (RG)	×	×	×	×
Guard (LG)	×	×	×	×
Center (CE)	×	×	×	×
Forward (RF)	×	×	×	×
Forward (LF)	×	×	×	×

Such models are widely used for introducing the concept of multiplication to elementary school students. Note that the model is an application of the Cartesian product (Section 2-2)

$$\{C, L, B, N\} \times \{RG, LG, CE, RF, LF\}$$

The rectangular array provides a model for the statement $4 \times 5 = 20$. In general, multiplication can be defined as follows:

Let $1 = n(\{0\})$ and $2 = n(\{a, b\})$. Then $1 \times 2 = n(\{0\} \times \{a, b\})$ $= n(\{(0, a), (0, b)\}) = 2$

> If A and B are any two sets, then
>
> $$n(A) \times n(B) = n(A \times B)$$

The cardinal number of the empty set is zero, and for any whole number b

$$b + 0 = b \quad \text{since} \quad n(B) + n(\varnothing) = n(B \cup \varnothing) = n(B).$$

$$b \times 0 = 0 \quad \text{since} \quad n(B) \times n(\varnothing) = n(B \times \varnothing) = n(\varnothing).$$

However, since empty sets are not easy to visualize we turn to another representation.

We start by representing the set of whole numbers on a number line. To obtain a number line, draw any line, select any point of that line as the **origin** with *coordinate* 0 (zero), and select any other point of the line as the **unit point** with coordinate 1 (one). Usually the number line is considered in a horizontal position with the unit point on the right of the origin.

We can now graph different sets of numbers on a number line as in the following examples. We shall use the verb *graph* to mean *draw the graph of.*

The length of the line segment with the origin and the unit point as endpoints is the **unit distance,** or **unit of length**, for marking off a scale on the line. The points representing any given natural numbers may be obtained by marking off successive units to the right of the origin. The numbers are the **coordinates** of the points; the points are the **graphs** of the numbers.

Example 2 Graph on a number line the set of whole numbers less than 3.

Solution Draw a number line and place a solid dot at each of the points that correspond to 0, 1, and 2.

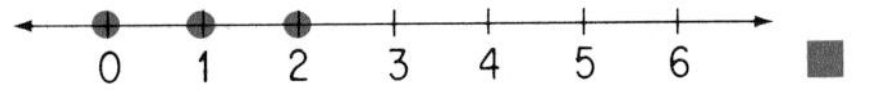

On a number line addition may be represented by successive moves or jumps. For example, to represent 3 + 4

Start at the origin.

Represent 3 by a move of 3 units to the right—arriving at the point with coordinate 3.

Then represent the addition of 4 by a move from the point with coordinate 3 of 4 more units to the right—arriving at the point with coordinate 7.

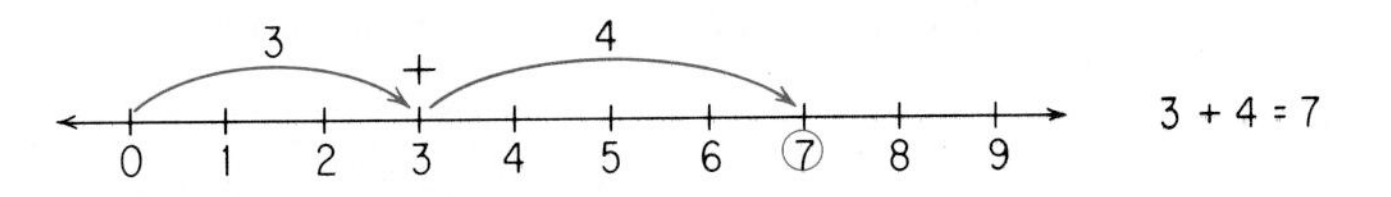

Since the addition of 0 units is represented by remaining in place, $3 + 0 = 3$, and in general for any whole number n we have the **addition property of zero.**

Zero, +: $n + 0 = n$

Then $0 + 0 = 0$ and repeated additions of 0 continue to have sum 0. Thus for any whole number n we have the **multiplication property of zero.**

Zero, $\times$: $n \times 0 = 0$

$$0 \quad 1 \quad 2 \quad 3 \quad 4 \quad 5 \quad 6 \quad 7 \quad 8$$

Try to illustrate each of these basic properties of whole numbers with at least two numerical examples. Use drawings, sketches, or other models as needed to convince yourself that each general statement is true.

On the number line the point with coordinate 6 is 2 units to the right of the point with coordinate 4, that is, $6 = 4 + 2$ and *6 is greater than 4* by 2. We write $6 > 4$ (read "6 is greater than 4") and $4 < 6$ (read "4 is less than 6"). In general, $a > b$ and $b < a$ if and only if there is a natural number c such that $a = b + c$.

The properties of the set of whole numbers under addition and multiplication are closely related to those of the set operations (Section 2-4). We now summarize these properties for any whole numbers a, b, and c. Each of the operations, addition and multiplication, associates one number with *two* given numbers. For example, $4 + 3 = 7$ and $2 \times 3 = 6$. In general, if $a + b = c$, then a and b are **addends** and c is their **sum**; if $r \times s = t$, then r and s are **factors** of their **product** t.

Since the sum of any two whole numbers is a unique whole number, the set of whole numbers is **closed under addition.**

Closure, $+$: There is one and only one whole number $a + b$.

Since the product of any two whole numbers is a unique whole number, the set of whole numbers is **closed under multiplication.**

Closure, $\times$: There is one and only one whole number $a \times b$.

The order of the addends does not affect the sum.

$$2 + 3 = 3 + 2$$

This property is the *commutative property of addition.*

Commutative, $+$: $a + b = b + a$

Similarly, the order of the factors does not affect the product.

$$2 \times 3 = 3 \times 2$$

Commutative, $\times$: $a \times b = b \times a$

When three or more numbers are to be added, we may write, for example, $2 + 3 + 4$ because the same answer is obtained whether we associate the second addend with the first $(2 + 3) + 4$ or associate the second addend with the third $2 + (3 + 4)$.

Associative, $+$: $(a + b) + c = a + (b + c)$

Similarly, whenever we have a product, such as $2 \times 3 \times 4$, of three numbers we may associate the second factor with either the first $(2 \times 3) \times 4$ or the third $2 \times (3 \times 4)$.

Associative, $\times$: $(a \times b) \times c = a \times (b \times c)$

If the whole number 0 is used as an addend, the sum is the same as, identical with, the other addend. Because of this addition property of zero the whole number 0 is the **additive identity element** and also is called the **identity element for addition**.

Identity, $+$: $a + 0 = 0 + a = a$

If the whole number 1 is used as a factor, the product is the same as, identical with, the other factor. For example,

$$1 \times 5 = 5 \qquad 7 \times 1 = 7$$

Because of this *multiplication property of one* the whole number 1 is the **multiplicative identity element** and also is called the **identity element for multiplication**.

Identity, $\times$: $a \times 1 = 1 \times a = a$

When a sum is multiplied by another factor we usually find the sum first. For example,

$$5 \times (7 + 11) = 5 \times 18 = 90$$

$5 \times (7 + 11)$ means the same thing as $5(7 + 11)$

However, note that if we first distribute the other factor with each of the addends, the same result is obtained.

$$5 \times (7 + 11) = (5 \times 7) + (5 \times 11) = 35 + 55 = 90$$

This is the **distributive property for multiplication with respect to addition**, or simply the *distributive property*.

Distributive property: $a \times (b + c) = (a \times b) + (a \times c)$

The distributive property allows us either to add first and then multiply, or to find the two products first and then add. Note that addition is *not* distributive with respect to multiplication since, according to convention, multiplication must be done before addition. For example,

$$3 + (5 \times 8) \neq (3 + 5) \times (3 + 8)$$

$3 + (5 \times 8) = 3 + 40 = 43$
$(3 + 5) \times (3 + 8)$
$= 8 \times 11 = 88$

that is,

$$3 + 40 \neq 8 \times 11$$

It is the distributive property that allows us, in algebra, to make such statements as

$$2(a + b) = 2a + 2b$$

$$3(x - y) = 3x - 3y$$

It is the distributive property that elementary school youngsters use in multiplication. Consider the problem 7×43. The distributive property is used by thinking of 7×43 as

$$7 \times (40 + 3) = (7 \times 40) + (7 \times 3) = 280 + 21 = 301$$

In our usual algorithm we have

$$\begin{array}{rl} 43 & \\ \times \quad 7 & \\ \hline 21 & = 7 \times 3 \\ 280 & = 7 \times 40 \\ \hline 301 & = (7 \times 3) + (7 \times 40) = (7 \times 40) + (7 \times 3) \end{array}$$

The distributive property can also be used in developing shortcuts in multiplication. Thus the product 8×99 can be found quickly as

$$8 \times 99 = 8(100 - 1) = 800 - 8 = 792$$

We are now ready to discuss in some detail the *algorithms* (procedures) for addition and subtraction of whole numbers. To do so we first explore several different approaches that may be used to justify the manner in which we add. One approach is to use expanded notation and the properties of the set of whole numbers. Explain each step shown in the following illustrative example, where the numbers 35 and 49 are called the *addends* in the sum $35 + 49$.

$$\begin{aligned} 35 + 49 &= [(3 \times 10) + (5 \times 1)] + [(4 \times 10) + (9 \times 1)] \\ &= [(3 \times 10) + (4 \times 10)] + [(5 \times 1) + (9 \times 1)] \\ &= (7 \times 10) + (14 \times 1) \\ &= (7 \times 10) + [(10 \times 1) + (4 \times 1)] \\ &= (7 \times 10) + [(1 \times 10) + (4 \times 1)] \\ &= [(7 + 10) + (1 \times 10)] + (4 \times 1) \\ &= (8 \times 10) + (4 \times 1) \\ &= 84 \end{aligned}$$

Of course we normally do not solve problems of addition in this cumbersome manner. Instead, our usual procedure looks something like this.

$$\begin{array}{r} 1 \\ 35 \\ +49 \\ \hline 84 \end{array}$$

Think: $5 + 9 = 14$; write the 4 in the ones place and exchange the other 10 units for 1 ten. Then $1 + 3 + 4 = 8$; write the 8 in the tens place.

This process may be shown in detail as follows.

$$\begin{array}{rl} 35 & (3 \times 10) + (5 \times 1) \\ +49 & (4 \times 10) + (9 \times 1) \\ \hline & (7 \times 10) + (14 \times 1) \end{array}$$

Exchange (14×1) for $(1 \times 10) + (4 \times 1)$:

$$\begin{array}{rl} & (1 \times 10) \\ 35 & (3 \times 10) + (5 \times 1) \\ +49 & (4 \times 10) + (9 \times 1) \\ \hline & (8 \times 10) + (4 \times 1) = 84 \end{array}$$

The following sequence of steps is yet another way to justify the usual addition algorithm and frequently helps clarify the procedures used.

$$\begin{array}{rl} 35 & 30 + 5 \\ +49 & 40 + 9 \\ \hline & 70 + 14 = 70 + (10 + 4) \\ & \qquad\quad = (70 + 10) + 4 \\ & \qquad\quad = 80 + 4 \\ & \qquad\quad = 84 \end{array}$$

Example 3 Express 387 and 259 in terms of hundreds, tens, and ones. Then find the sum $387 + 259$.

Solution First add the column of ones.

$$\begin{array}{rr} 387 & 300 + 80 + \ 7 \\ +259 & 200 + 50 + \ 9 \\ \hline & 16 \end{array}$$

Next rewrite 16 as $10 + 6$ and "carry" the 10 to the tens column. Add the column of tens.

Use or sketch base ten arithmetic blocks and model the sum $35 + 49$. Note that the visualization of steps is the same for all models (representations) but the selection of appropriate models is very important in developing the addition algorithm with children.

$$\begin{array}{rr} & 10 \\ 300 + & 80 + 7 \\ 200 + & 50 + 9 \\ \hline & 140 + 6 \end{array}$$

Now write 140 as $100 + 40$ and exchange, or "carry," the 100 to the hundreds column. Add the column of hundreds.

$$\begin{array}{l} 100 \\ 300 + 80 + 7 \\ 200 + 50 + 9 \\ \hline 600 + 40 + 6 = 646 \ ■ \end{array}$$

Note, in Example 3, that the usual method of adding columns and "carrying" is possible because of the place-value style of our decimal system of notation. In abbreviated form, this example would be completed in this way.

(i) Add the ones.	(ii) Carry 1 ten to the tens' column and add.	(iii) Carry 1 hundred to the hundreds' column and add.
	1	1
387	387	387
+ 259	+ 259	+ 259
16	6	6
	14	4
		6
		646

In reality, we are regrouping at each step and exchanging 10 ones for 1 ten and then 10 tens for 1 hundred. The following display should serve to further clarify this process.

$$\begin{array}{rl} 387 & 300 + \ \ 80 + \ \ 7 \\ +259 & 200 + \ \ 50 + \ \ 9 \\ \hline & 500 + 130 + 16 = 500 + 130 + (10 + 6) \\ & = 500 + (130 + 10) + 6 \\ & = 500 + 140 + 6 \\ & = 500 + (100 + 40) + 6 \\ & = (500 + 100) + 40 + 6 \\ & = 600 + 40 + 6 \\ & = 646 \end{array}$$

Another approach is to add the numbers of ones, tens, and hundreds separately, taking advantage of place value, and then add these partial sums to obtain the sum of the given numbers.

387	
+ 259	
16	(7 ones + 9 ones = 16 ones = 16)
130	(8 tens + 5 tens = 13 tens = 130)
500	(3 hundreds + 2 hundreds = 5 hundreds = 500)
646	

We noted earlier that sets could be used to represent the subtraction problem 7 − 3 by a "take-away" model or by an "add-on" model. Any subset of a given set may be removed and subtraction used to indicate the number of remaining elements. However, in the set of whole numbers subtraction is not always possible. For example, 6 − 9 does not represent a whole number. In other words, *the set of whole numbers is not closed under subtraction.* We limit our discussion here to the procedure for the subtraction of whole numbers in cases for which the difference is a whole number.

Use or sketch base ten arithmetic blocks to model the difference 73 − 28. Note that the visualization of the steps is the same for all models.

Consider 73 − 28. The usual algorithm for subtraction can be explained by first writing each term in this expanded form:

$$\begin{array}{rl} 73 & (7 \times 10) + (3 \times 1) \\ -\underline{28} & \underline{(2 \times 10) + (8 \times 1)} \end{array}$$

We need to subtract 8 ones from 3 ones. Since this is not possible with whole numbers, we "borrow" from the tens column. Rewrite 73 in this way:

$$\begin{aligned} 73 = (7 \times 10) + (3 \times 1) &= [(6 \times 10) + (1 \times 10)] + (3 \times 1) \\ &= [(6 \times 10) + (10 \times 1)] + (3 \times 1) \\ &= (6 \times 10) + [(10 \times 1) + (3 \times 1)] \\ &= (6 \times 10) + (13 \times 1) \end{aligned}$$

Our work now looks like this:

$$\begin{array}{rl} 73 & (6 \times 10) + (13 \times 1) \\ -\underline{28} & \underline{(2 \times 10) + (\ 8 \times 1)} \\ & (4 \times 10) + (\ 5 \times 1) = 45 \end{array}$$

$$\begin{array}{rr} 73 & \overset{6}{\not7}\,{}^{1}3 \\ -28 & 2\ 8 \\ \hline & 4\ 5 \end{array}$$

In actual practice, of course, we do not write all the steps shown. Instead we usually cross out the 7, and replace it by 6. This shows that we are borrowing 1 ten from the 7 tens, leaving 6 tens. We then show a numeral 1 alongside the ones column to denote the borrowed 10 as 10 ones.

Example 4 Express 237 and 805 in terms of hundreds, tens, and ones. Then find the difference 805 − 237.

Solution This problem may be done in three steps.

(i)

$$\begin{array}{rr} 805 & 800 + \ \ 0 + \ 5 \\ -\underline{237} & \underline{200 + 30 + \ 7} \end{array}$$

(ii) Next, rewrite 800 as 700 + 10 tens, or 100.

$$\begin{array}{r} 700 + 100 + \ 5 \\ \underline{200 + \ 30 + \ 7} \end{array}$$

(iii) Now write 100 as 90 + 10. Add 10 + 5. Then subtract.

$$\begin{array}{r} 700 + 90 + 15 \\ \underline{200 + 30 + \ 7} \\ 500 + 60 + \ 8 = 568 \ \blacksquare \end{array}$$

In actual practice, the work demonstrated in Example 4 usually is completed by the following corresponding steps.

$$\text{(i)}\ \begin{array}{r} 805 \\ \underline{-\ 237} \end{array} \qquad \text{(ii)}\ \begin{array}{r} \overset{7}{\not8}\,{}^{1}05 \\ \underline{2\ 37} \end{array} \qquad \text{(iii)}\ \begin{array}{r} \overset{7}{\not8}\,{}^{1}\overset{9}{\not0}\,{}^{1}5 \\ \underline{2\ 3\ 7} \\ 5\ 6\ 8 \end{array}$$

EXERCISES

Tell whether the number specified in the given statement is used for identification, as an ordinal number, or as a cardinal number.

1. There are *20* volumes in the set of encyclopedias?
2. Mathematics is discussed in the *12th* volume.
3. Dorothy is in the *fourth* row.
4. There are *35* students in the class.
5. I am listening to *104* on the FM dial.
6. It takes *9* players to field a baseball team.

In Exercises 7 through 12 find the cardinal number of each set.

7. $\{p\}$
8. $\{11, 12, \ldots, 18\}$
9. $\{100, 101, \ldots, 110\}$
10. $\varnothing$
11. $\{1, 3, 5, 7, \ldots, 19\}$
12. $\{2, 4, 6, \ldots, 102\}$
13. Find $A \times B$ for $A = \{1, 2\}$ and $B = \{1, 2, 3, 4\}$, thereby showing that $2 \times 4 = 8$.
14. Find $P \times Q$ for $P = \{1, 2, 3\}$ and $Q = \{1, 2, 3\}$, thereby showing that $3 \times 3 = 9$.

For each exercise draw a number line and graph the specified set.

15. $\{1, 2, 3, 4, 5\}$
16. $\{3, 7, 8\}$
17. $\{0, 2, 4, 6, 8\}$
18. $\{0, 1\}$
19. The set of natural numbers less than 6.
20. The set of whole numbers less than 6.
21. The set of whole numbers between 0 and 5.
22. The set of natural numbers between 1 and 3.
23. The set of natural numbers between 0 and 1.
24. The set of whole numbers greater than or equal to 5 and less than 9.

In Exercise 21 recall that the word **between** indicates that we are not to include the points for 0 and 5.

For each arithmetic statement, name the property, or properties, of whole numbers that is illustrated.

25. $5 + 0 = 5$
26. $3 \times 0 = 0$
27. $(17 + 3) \times 0 = 0$
28. $0 \times (28 + 19) = 0$
29. $0 + (28 + 19) = 28 + 19$
30. $2 + (3 \times 5) = 2 + (5 \times 3)$

For Exercises 31 and 32, think of a multiplication table. If necessary, make a multiplication table for the whole numbers 0 through 5. Describe the possible values for whole numbers a and b under the given condition.

31. **(a)** $a \times b = 2$ **(b)** $a \times b = 3$ **(c)** $a \times b = 4$
32. **(a)** $a \times b = 1$ **(b)** $a \times b = 0$

33. Show that subtraction of whole numbers is not commutative.
34. Show that division of whole numbers is not commutative.

35. Does $8 - (3 - 2) = (8 - 3) - 2$? Is subtraction of whole numbers associative?

36. Does $12 \div (6 \div 2) = (12 \div 6) \div 2$? Is division of whole numbers associative?

37. Show that addition of whole numbers is not distributive with respect to multiplication.

38. In ordinary arithmetic, is multiplication distributive with respect to multiplication? That is, does $a \times (b \times c) = (a \times b) \times (a \times c)$ for all possible replacements of a, b and c?

Use the distributive property to find the given product by means of a shortcut.

39. 7×79 40. 6×58 41. 8×92 42. 9×63

As in Examples 3 and 4 of this section, complete each problem by using ones, tens, hundreds, and thousands, as needed.

43. $\begin{array}{r} 45 \\ +38 \\ \hline \end{array}$ 44. $\begin{array}{r} 56 \\ +29 \\ \hline \end{array}$ 45. $\begin{array}{r} 375 \\ +287 \\ \hline \end{array}$

46. $\begin{array}{r} 509 \\ +238 \\ \hline \end{array}$ 47. $\begin{array}{r} 1309 \\ +2578 \\ \hline \end{array}$ 48. $\begin{array}{r} 4793 \\ +8147 \\ \hline \end{array}$

49. $\begin{array}{r} 95 \\ -32 \\ \hline \end{array}$ 50. $\begin{array}{r} 85 \\ -37 \\ \hline \end{array}$ 51. $\begin{array}{r} 304 \\ -128 \\ \hline \end{array}$

52. $\begin{array}{r} 350 \\ -179 \\ \hline \end{array}$ 53. $\begin{array}{r} 5023 \\ -2709 \\ \hline \end{array}$ 54. $\begin{array}{r} 8301 \\ -2076 \\ \hline \end{array}$

EXPLORATIONS

A *worksheet* is a practical tool for helping students to develop skills on a particular topic. Normally, a worksheet will consist of one or more illustrative examples, detailed solutions with models or diagrams, and several additional practice exercises for the student to solve. Often partial solutions are shown for some of the initial exercises.

1. Elementary school students use the distributive property of multiplication with respect to addition in many ways. For example:

$$30 + 40 = (3 \times 10) + (4 \times 10) = (3 + 4) \times 10 = 7 \times 10 = 70$$

$$(17 \times 3) + (13 \times 3) = (17 + 13) \times 3 = 30 \times 3 = 90$$

Make a worksheet of 20 problems to help elementary school students recognize the usefulness of this distributive property.

2. The properties of natural numbers are used to simplify arithmetic computations throughout the elementary school grades. For example, the number fact $7 \times 10 = 70$ is easily learned. However, the problem 10×7 often leads an elementary school student to computations 1×7, 2×7, $3 \times 7, \ldots, 10 \times 7$ and is rather hard *until* the student realizes that 7×10

and 10 × 7 are two ways of stating the same problem. Many educators feel that students *should not be told* but rather *should be helped to discover* for themselves such useful properties of numbers. In other words, many matched pairs of problems (7 × 10 and 10 × 7, 2 × 15 and 15 × 2, and so forth) should be given until the student observes the pattern of the commutativity of multiplication. Such experiences of discovery and the recognition of the laborsaving effect of the pattern that is discovered should be developed whenever possible. Make a worksheet of 20 problems to help elementary school students recognize and make effective use of the commutative property for multiplication.

3. Repeat Exploration 2 for the commutative property for addition.
4. Sketch base ten arithmetic blocks and model the problems in Exercises 43 through 46 and 49 through 52.
5. In the introduction to this section there appears a quotation from the German mathematician Kronecker. Start a collection of such famous sayings that might be of interest in mathematics classes. For example, Archimedes is supposed to have said, "Give me a place to stand and a lever long enough and I will move the earth."
6. Another puzzle is to represent the set of natural numbers using the digits of a *year in the order in which they appear*. Here are some examples, using the bicentennial year 1976.

$$1 = 1^{976} \qquad 2 = 1^9 + 7 - 6 \qquad 3 = 1 + \sqrt{9} - 7 + 6$$

   How many others can you find?

7. In a subtraction problem such as 85 − 37, 85 is sometimes called the *minuend*, and 37 is called the *subtrahend*. Explore an elementary textbook series to determine what attention if any, is given to these words.
8. Subtraction is often introduced as the inverse operation of addition. For example, since 5 + 7 = 12 and 7 + 5 = 12, it then follows that 12 − 5 = 7 and 12 − 7 = 5. Prepare a worksheet of problems for a specified elementary grade level that emphasizes these related number facts.
9. Elementary school teachers are frequently called upon to diagnose student errors and then to suggest remedial activities to correct the error patterns. Each of the following illustrations displays a type of error that some students make in addition and subtraction with whole numbers. Identify the error being made and suggest at least two teaching strategies that might be used to help correct the procedures being used.

$$\textbf{(a)}\quad \begin{array}{r} 78 \\ +25 \\ \hline \end{array} \qquad \begin{array}{r} 78 \\ +\ 25 \\ \hline 913 \end{array} \qquad\qquad \textbf{(b)}\quad \begin{array}{r} 573 \\ +846 \\ \hline \end{array} \qquad \begin{array}{r} 34 \\ 573 \\ +\quad 846 \\ \hline 1113 \end{array}$$

$$\textbf{(c)}\quad \begin{array}{r} 723 \\ -485 \\ \hline \end{array} \qquad \begin{array}{r} 723 \\ -485 \\ \hline 362 \end{array} \qquad\qquad \textbf{(d)}\quad \begin{array}{r} 587 \\ -132 \\ \hline \end{array} \qquad \begin{array}{r} 7\phantom{{}^{1}7} \\ 5\not{8}\,{}^{1}7 \\ -13\ 2 \\ \hline 4415 \end{array}$$

3-3 Early Procedures for Multiplication

There exist numerous examples of early algorithms for multiplication of numbers expressed in decimal notation. The following algorithm for multiplication appeared in one of the first published arithmetic texts in Italy, the *Treviso Arithmetic* (1478). The method was used by early Hindus and Chinese before being widely used by the Arabians, who passed it on to the Europeans during the Middle Ages. We shall refer to the method here as *galley multiplication,* although it was called "Gelosia" multiplication in the original text and is sometimes called "lattice" multiplication. Let us use this method to find the product of the two *factors* 457 and 382.

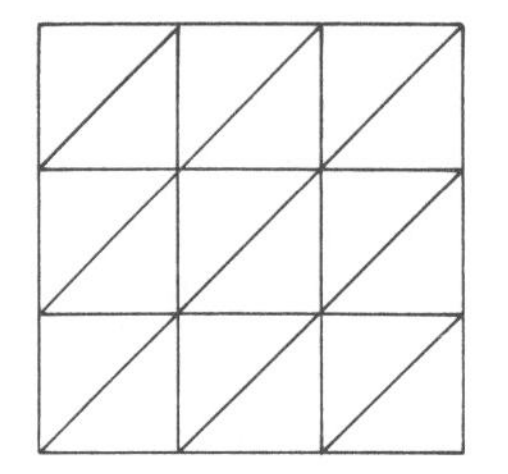

First prepare a "galley" with three rows for the digits of 382 and three columns for the digits of 457 and draw the diagonals, as in the adjacent figure. Place the digits 3, 8, and 2 of the factor 382 in order from top to bottom at the right of the rows. Place the digits 4, 5, and 7 in order from left to right at the tops of the columns. Then each product of a digit of 457 and a digit of 382 is called a partial product and is placed at the intersection of the column and row of the digits. The diagonal separates the digits of the partial product (tens digit above ones digit). For example, $3 \times 7 = 21$, and this partial product is placed in the upper right-hand corner of the galley; $5 \times 8 = 40$, and this partial product is placed in the center of the galley; $4 \times 2 = 8$, and this partial product is entered as 08 in the lower left-hand corner of the galley. See if you can justify each of the entries in the completed array.

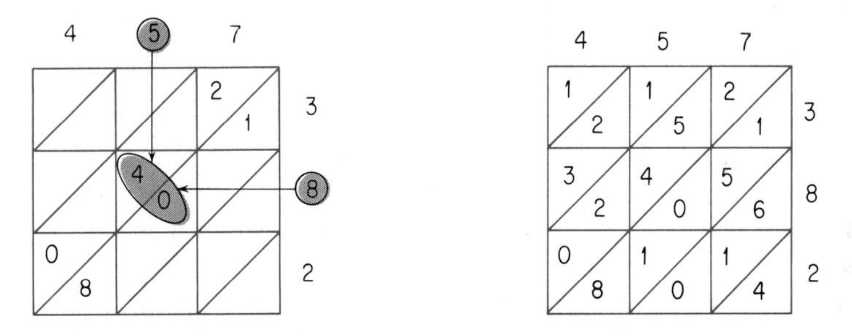

Check the product 457×382 using this galley.

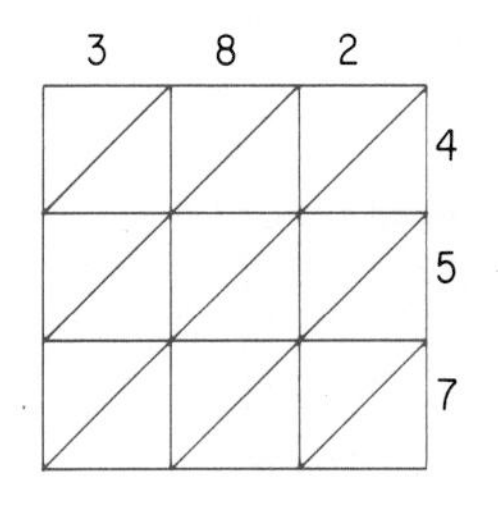

After all partial products have been entered in the galley, we add along diagonals, starting in the lower right-hand corner and carrying to the next diagonal sum where necessary. The next diagram indicates this pattern. The completed problem appears in the figure on the right. We read the final answer, as indicated by the curved arrow in the figure, as 174,574. Note that we read the digits in the opposite order to that in which they were obtained.

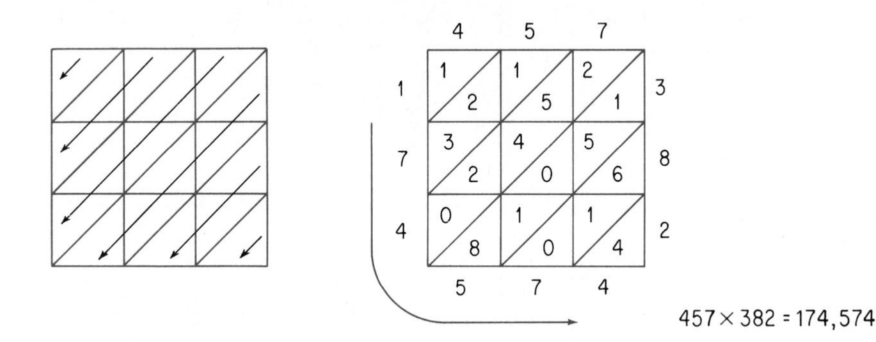

Example 1 Use galley multiplication and multiply 372 by 47.

Solution

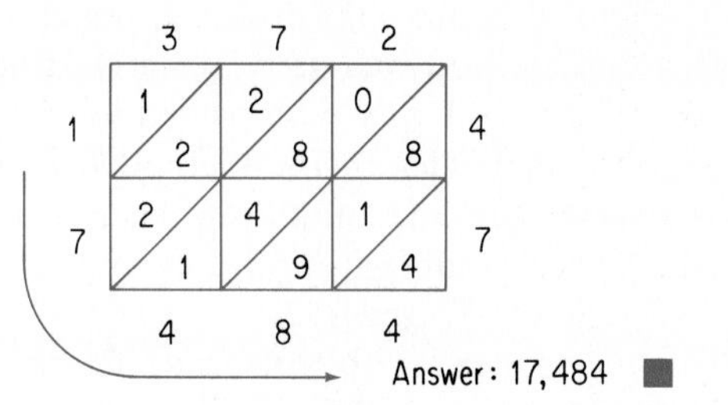

Galley multiplication works because we are really listing all partial products before we add. Compare the following two computations and note that the numerals along the diagonals correspond to those in the columns at the right.

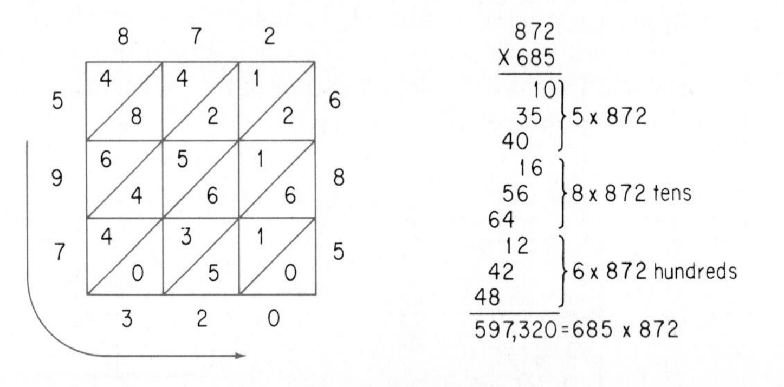

Nicaragua honored Napier on one of its stamps and, on the back of the stamp, described his work as follows: "With the invention of logarithms, Napier gave to the world a powerful arithmetic shorthand. It permitted people to do multiplication or division simply by adding or subtracting logarithms of numbers, and it meant that they could rapidly carry out these and more complicated operations for numbers containing many digits. The impact of logarithms on fields such as astronomy and navigation was enormous and comparable to the computer revolution of today."

The Scottish mathematician John Napier made use of galley multiplication as he developed what proved to be one of the forerunners of the modern computing machines. His device is referred to as **Napier's rods**, or **Napier's bones**, named after the material on which he had numerals printed. Napier (1550–1617) is often spoken of as the inventor of logarithms.

John Napier
1550-1617

Con la invención de logarítmos, Napier dió al mundo una taquigrafía poderosa de aritmética.
Permitió a los hombres hacer multiplicación ó división simplemente al sumar ó restar los logarítmos de números y significa que ellos podrían llevar a cabo éstos y más operaciones complicadas y rápidas de números conteniendo muchas cifras. El impacto del logarismo en campos como astronomía y navegación son enormes y comparables con la computadora revolucionaria de hoy.

To make a set of these rods, we need to prepare a collection of strips of paper, or other material, with multiples of each of the digits listed. Study the set of rods shown in part (a) of the next figure. Note, for example, that the rod headed by the numeral 9 lists the multiples of 9.

9 18 27 36 45 54 63 72 81

We can use these rods to multiply two numbers. To multiply 7 × 483, place the rods headed by numerals 4, 8, and 3 alongside the index, as shown in part (b) of the same figure.

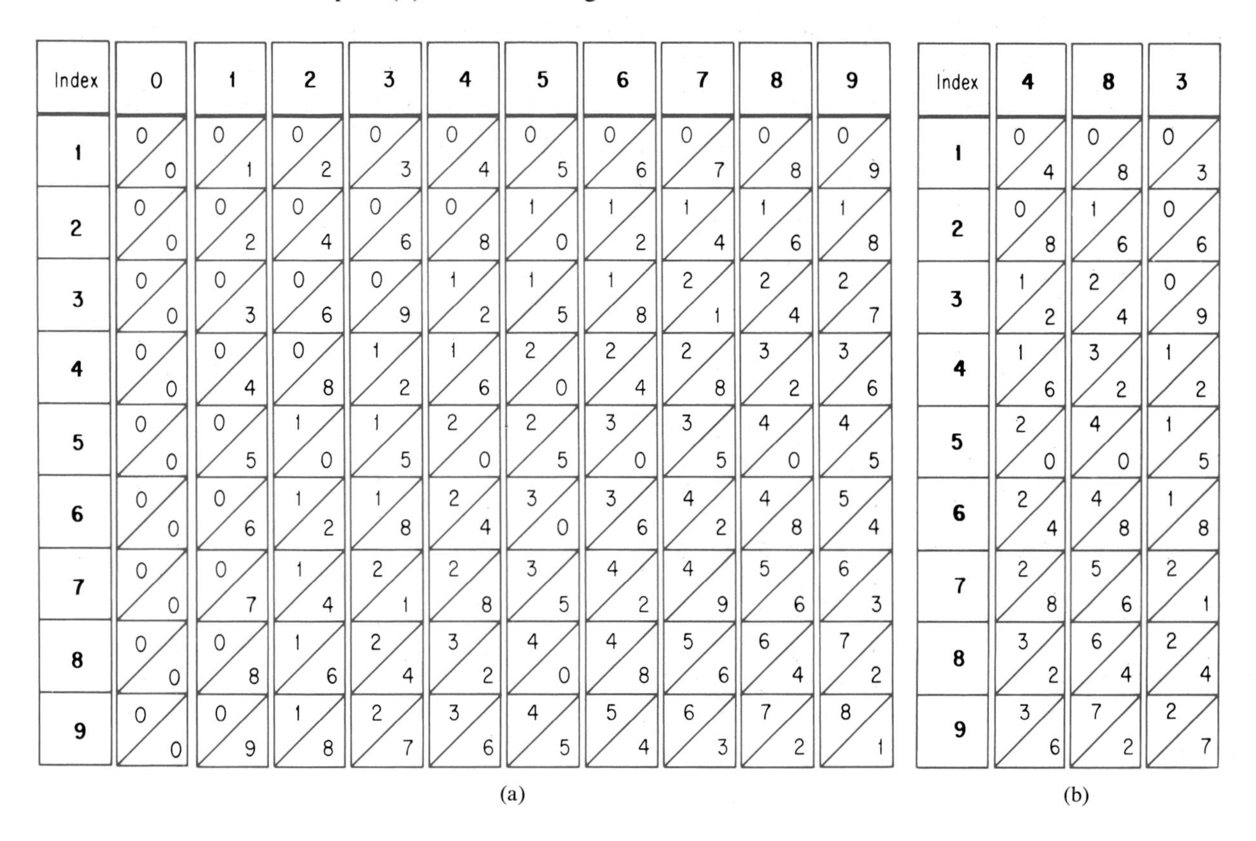

(a) (b)

Consider the row of numerals alongside the factor 7 on the index.

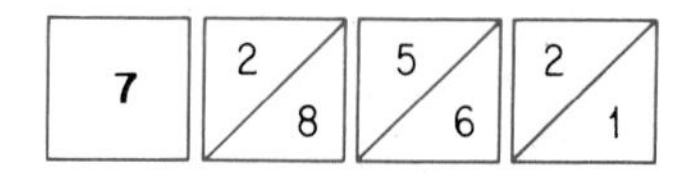

Add along the diagonals, as in galley multiplication.

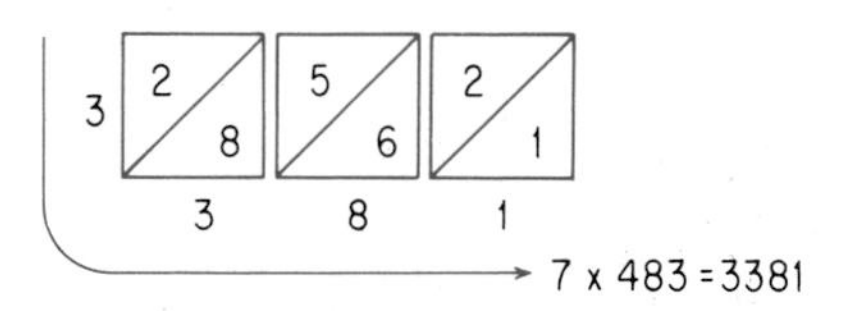

Example 2 The diagram shows a use of Napier's rods. Express the product in terms of the indicated factors.

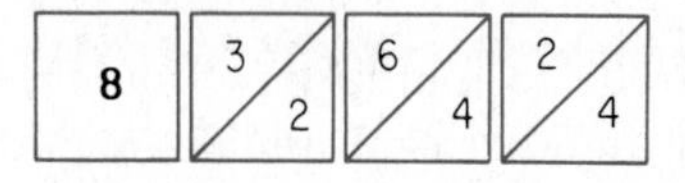

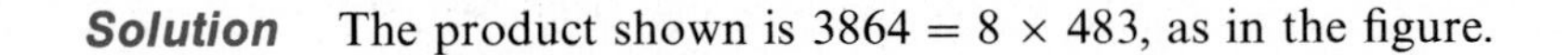

Solution The product shown is $3864 = 8 \times 483$, as in the figure.

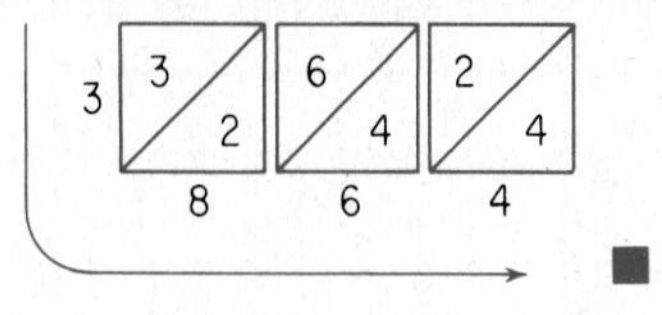

■

Note that we are able to read only products with a one-digit multiplier directly from the rods. The two products $7 \times 483 = 3381$ and $8 \times 483 = 3864$ can be used to find the product 87×483.

$$\begin{array}{r} 483 \\ \times\ \underline{87} \end{array} \qquad \begin{array}{rl} 7 \times 483 = & 3{,}381 \\ \underline{80 \times 483 =} & \underline{38{,}640} \\ 87 \times 483 = & 42{,}021 \end{array}$$

EXERCISES *Each diagram shows a use of Napier's rods. Express each product in terms of the indicated factors.*

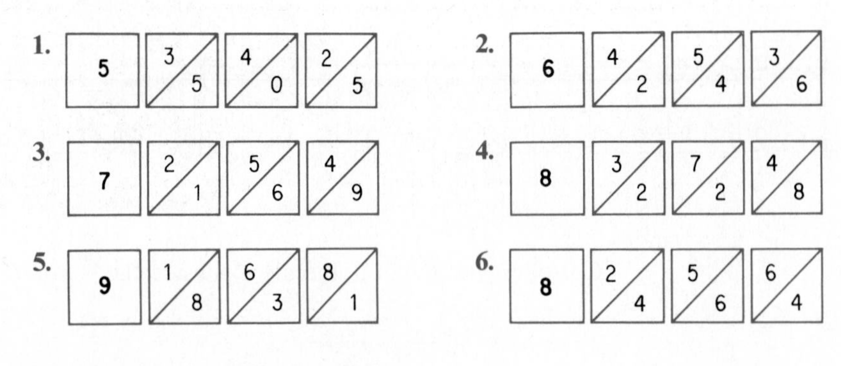

Multiply, using the galley method.

7. $\begin{array}{r} 942 \\ \times\ \underline{37} \end{array}$ 8. $\begin{array}{r} 586 \\ \times\ \underline{492} \end{array}$ 9. $\begin{array}{r} 234 \\ \times\ \underline{762} \end{array}$

10. $\begin{array}{r} 8764 \\ \times\ \underline{37} \end{array}$ 11. $\begin{array}{r} 8035 \\ \times\ \underline{289} \end{array}$ 12. $\begin{array}{r} 8501 \\ \times\ \underline{3726} \end{array}$

Construct a set of Napier's rods and use them to find each product.

13. $\begin{array}{r} 365 \\ \times\ \underline{7} \end{array}$ 14. $\begin{array}{r} 472 \\ \times\ \underline{9} \end{array}$ 15. $\begin{array}{r} 873 \\ \times\ \underline{5} \end{array}$

16. $\begin{array}{r} 259 \\ \times \underline{\quad 8} \end{array}$ 17. $\begin{array}{r} 3765 \\ \times \underline{\quad 6} \end{array}$ 18. $\begin{array}{r} 4239 \\ \times \underline{\quad 4} \end{array}$

19. $\begin{array}{r} 472 \\ \times \underline{\ 63} \end{array}$ 20. $\begin{array}{r} 587 \\ \times \underline{\ 78} \end{array}$ 21. $\begin{array}{r} 936 \\ \times \underline{\ 95} \end{array}$

22. $\begin{array}{r} 495 \\ \times \underline{187} \end{array}$ 23. $\begin{array}{r} 379 \\ \times \underline{258} \end{array}$ 24. $\begin{array}{r} 759 \\ \times \underline{567} \end{array}$

EXPLORATIONS

1. Write a *lesson plan* on galley multiplication for use in an elementary school class (specify the grade that you are considering). Your lesson plan should include objectives, proposed activities, teaching aids to be used, suggested assignments, and means of evaluation.
2. Repeat Exploration 1 for multiplication using Napier's rods.
3. Prepare a demonstration set of Napier's rods for use in front of a class or prepare a set for use on an overhead projector.
4. Find the product 39 × 756 by using the galley method and by using Napier's rods. Which process is easier to use? Which process do you think would be the better to use in an elementary mathematics class in order to show the comparison with our usual multiplication process?
5. Some educators believe that galley multiplication can be used effectively with slow learners as a "low-stress" approach to multiplication of whole numbers. For a theoretical class of seventh or eighth graders who are deficient in basic skills, prepare a plan for a 10-minute lesson using galley multiplication as a means of reviewing multiplication skills.

3-4 Multiplication and Division of Whole Numbers

In this section we examine some of the algorithms that are used for multiplication and division of whole numbers. Let us begin with a product such as 7 × 12. A computer might handle this problem by treating the operation of multiplication as repeated addition. The product 7 × 12 may be thought of as the sum obtained by using 7 as an addend 12 times, or by using 12 as an addend 7 times.

$$12 \times 7 = 7 + 7 + 7 + 7 + 7 + 7 + 7 + 7 + 7 + 7 + 7 + 7$$
$$7 \times 12 = 12 + 12 + 12 + 12 + 12 + 12 + 12$$

The distributive property may be used to show this product. Note the use of grouping by tens and ones.

$$\begin{aligned} 7 \times 12 &= 7 \times (10 + 2) \\ &= (7 \times 10) + (7 \times 2) \\ &= 70 + 14 \\ &= 70 + 10 + 4 \\ &= 80 + 4 \\ &= (8 \times 10) + (4 \times 1) \\ &= 84 \end{aligned}$$

Let us return once again to the product 7×12 and see how this can be shown using the traditional vertical arrangement.

$$\begin{array}{rl} 12 & \\ \times\ \underline{7} & \\ 14 & (7 \times 2) \\ \underline{70} & (7 \times 10) \\ 84 & (7 \times 12) \end{array}$$

In actual practice, we normally think of 7×2 as 14. We then write the 4 and "carry" the 1. Actually we are exchanging 10 ones for 1 ten. We indicate this by writing the numeral 1 in the tens column as in these two steps.

Step 1:

$$\begin{array}{r} 1 \\ 12 \\ \times\ \underline{7} \\ 4 \end{array}$$

7×2 (ones) $=$ 14 (ones); write 4 (ones) and carry 1 (ten).

Step 2:

$$\begin{array}{r} 1 \\ 12 \\ \times\ \underline{7} \\ 84 \end{array}$$

7×1 (ten) $=$ 7 (tens); plus the 1 (ten) carried equals 8 (tens).

Note that the regrouping corresponds to exchanging 10 ones for 1 ten and the steps are the same as in the numerical model.

The base ten arithmetic block model for 7×12 by repeated addition may be represented by a **graph-paper model** using individual colored squares of the grid to represent the tops of unit cube blocks. Start with 7 rows of 12, that is, with 7 rows with one base rod of ten and 2 units on each row. Then, as in the figure, regroup the 14 units into one set of ten and 4 units to obtain a total of 8 sets of ten and 4 units.

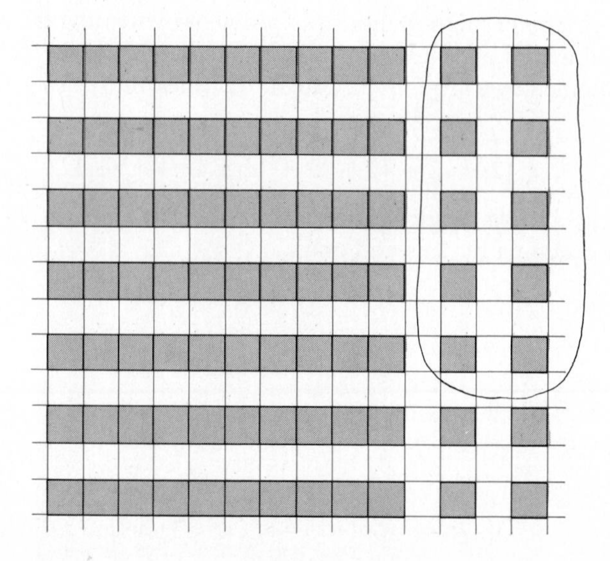

Example 1 Use the distributive property to find the product 7×235.

Solution

$$\begin{aligned} 7 \times 235 &= 7 \times (200 + 30 + 5) \\ &= (7 \times 200) + (7 \times 30) + (7 \times 5) \\ &= 1400 + 210 + 35 \\ &= 1400 + 200 + 10 + 30 + 5 \\ &= 1600 + 40 + 5 \\ &= (16 \times 100) + (4 \times 10) + (5 \times 1) \\ &= (1 \times 1000) + (6 \times 100) + (4 \times 10) + (5 \times 1) \\ &= 1645 \quad \blacksquare \end{aligned}$$

Let us explore a slightly more complex example, 23×34. One method of explaining the usual algorithm for finding such products is to use an expanded form of the distributive property as follows.

$$\begin{aligned} 23 \times 34 &= (20 + 3) \times (30 + 4) \\ &= [20 \times (30 + 4)] + [3 \times (30 + 4)] \\ &= [(20 \times 30) + (20 \times 4)] + [(3 \times 30) + (3 \times 4)] \\ &= 600 + 80 + 90 + 12 \\ &= 600 + 170 + 12 \qquad \text{(adding numbers of tens)} \\ &= 600 + (100 + 70) + (10 + 2) \\ &= (600 + 100) + (70 + 10) + 2 \\ &= 700 + 80 + 2 \\ &= 782 \end{aligned}$$

Note where each of the terms appears in this vertical arrangement.

$$\begin{array}{rl} 34 & \\ \times\ 23 & \\ \hline 12 & (3 \times 4) \\ 90 & (3 \times 30) \\ 80 & (20 \times 4) \\ 600 & (20 \times 30) \\ \hline 782 & \end{array}$$

In actual practice, these steps are condensed. Also, because of the place-value nature of our decimal system we need not insert all the 0's as shown in the preceding example. Instead, our work appears as follows.

$$\begin{array}{rl} 34 & \\ \times\ 23 & \\ \hline 102 & (3 \times 34) \\ 68\ & \longleftarrow \\ \hline 782 & \end{array}$$

This is really $20 \times 34 = 680$. The place-value position allows us to think of $2 \times 34 = 68$ instead of $20 \times 34 = 680$.

The graph-paper model for 23×34 consists of 23 rows with 3 base rods of ten and 4 units on each row. Graph paper with 10 × 10 squares accented is particularly useful for showing, as in the figure, the partial products that are obtained by using the distributive property as the 6 squares of 100, the 9 + 8 sets of ten, and the 12 units before further regrouping is considered. Such models for problems at this level provide reinforcement as students complete their shift to numerical models.

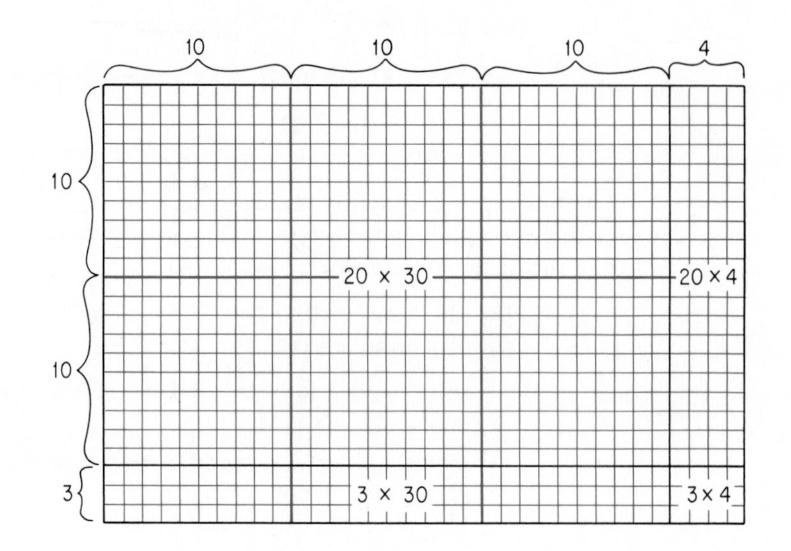

Division is related to multiplication in the same fashion that subtraction is related to addition. For example,

$3 + 4 = 7 \quad 4 + 3 = 7 \quad 7 - 3 = 4 \quad 7 - 4 = 3$

$3 \times 4 = 12 \quad 4 \times 3 = 12 \quad 12 \div 3 = 4 \quad 12 \div 4 = 3$

Students often learn their number facts by using such **families of related number facts**. As in the case of subtraction, division is not always possible in the set of whole numbers. For example, $12 \div 3 = 4$ but $7 \div 3$ does not represent a whole number. In other words, *the set of whole numbers is not closed under division.*

Multiplication can be performed by repeated addition. Division can be performed by repeated subtraction. Consider, for example, $7 \div 3$.

$7 - 3 = 4 \quad 4 - 3 = 1 \quad 7 = 2 \times 3 + 1$

Thus $7 \div 3$ is 2 with *remainder* 1. We write $7 \div 3 = 2\frac{1}{3}$ to indicate that the subtraction was completed twice and the remainder 1 has not yet been divided by 3.

There are two models for division of whole numbers—the measurement model and the partition model. In the **measurement model** for $7 \div 3$ we ask

How many sets of 3 are there in a set of 7?

Use base ten arithmetic blocks and show $45 \div 11$.

To answer the question we subtract sets of 3 from a set of 7 as many times as possible, that is, twice, and note the remainder of 1.

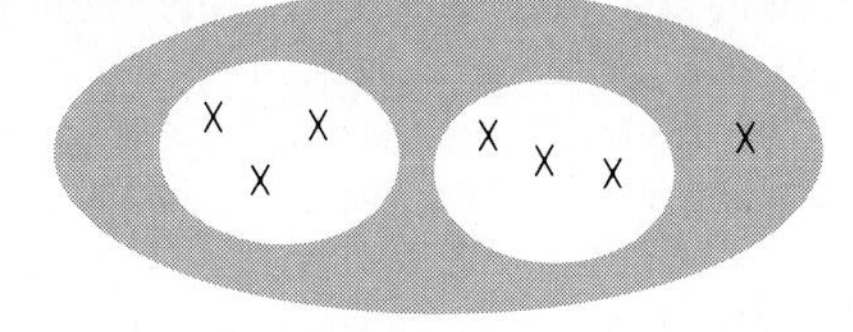

In the **partition model**, or sharing model, for $7 \div 3$ we ask

If 7 items are to be shared equally by 3 people, how many items does each person get?

The interpretation of the procedure is slightly different than before but the result continues to be 2 with a remaining one to be shared. In all cases $7 \div 3$ is 2 with remainder 1, that is $7 = 3 \times 2 + 1$. We speak of 7 as the **dividend**, 3 as the **divisor**, 2 as the **quotient**, and 1 as the **remainder**. In cases such as $12 \div 3 = 4$ in which the remainder is zero, the divisor is also a *factor* of the dividend.

The following examples illustrate the manner in which multiplication and division are *inverse operations*; each "undoes" the other.

$$12 \div 3 = 4 \quad \text{and} \quad 4 \times 3 = 12, \quad \text{that is,} \quad (12 \div 3) \times 3 = 12$$
$$2 \times 5 = 10 \quad \text{and} \quad 10 \div 5 = 2, \quad \text{that is,} \quad (2 \times 5) \div 5 = 2$$

Thus $12 \div 3$ is the number n such that $3 \times n = 12$. Such concepts are used by elementary school students when they check their division problems.

$$3\overline{)12}^{\,4} \qquad \text{Check:} \quad 3 \times 4 = 12$$

If the subtraction approach to division were tried for division by zero, the repeated subtractions of zero would not diminish the other number. Accordingly, *division by zero is not permitted*, and it is important that we understand why this is so. Consider the problem $7 \div 0$.

$$0\overline{)\;7}^{\,n} \qquad 7 \div 0 \quad \text{is the number } n \text{ such that} \quad 0 \times n = 7$$

For $n \neq 0$, $n \div 0$ is undefined.

But $0 \times n = 0$ for all n and can never be equal to 7. We could have used any number other than 7 with similar results. Therefore we say that division by 0 is not possible.

Example 2 Show that $0 \div 0$ does not represent a unique rational number.

Solution First we note that $0 \div 0$ is the number n such that $0 \times n = 0$. But $0 \times n = 0$ for *all n*. Therefore $0 \div 0$ has no unique solution and is *indeterminate.* ■

To arrive at the usual algorithm used for division, we consider the division problem $235 \div 5$. As one concrete representation of this problem, we may imagine that we have 235 pencils and need to package these in bundles of 5. We need to determine how many such bundles of 5 we will have. One way to arrive at the answer is to take successive guesses.

$$\begin{array}{r} 10 \\ 5\overline{)235} \\ \underline{50} \\ 185 \end{array}$$

Since $10 \times 5 = 50$, and 50 is less than 235, we certainly have at least 10 bundles of 5. By subtraction, we would then have 185 remaining pencils to package.

Repeat this procedure as often as possible, subtracting successive quantities of 50 each, until a known multiple of 5 is reached. This procedure makes use of the measurement interpretation of division.

$$\begin{array}{r} 10 \\ 5\overline{)185} \\ \underline{50} \\ 135 \end{array} \qquad \begin{array}{r} 10 \\ 5\overline{)135} \\ \underline{50} \\ 85 \end{array} \qquad \begin{array}{r} 10 \\ 5\overline{)85} \\ \underline{50} \\ 35 \end{array} \qquad \begin{array}{r} 7 \\ 5\overline{)35} \\ \underline{35} \\ 0 \end{array}$$

Next, add all of the quotients used to determine the final quotient.

$$10 + 10 + 10 + 10 + 7 = 47; \qquad 235 \div 5 = 47$$

$$\left.\begin{array}{r} 7 \\ 10 \\ 10 \\ 10 \\ 10 \end{array}\right\} 47$$
$$\begin{array}{r} 5\overline{)235} \\ \underline{50} \\ 185 \\ \underline{50} \\ 135 \\ \underline{50} \\ 85 \\ \underline{50} \\ 35 \\ \underline{35} \end{array}$$

These individual divisions and subtractions can be shown in a single vertical arrangement with each partial quotient placed above the line as shown in the margin.

With practice one can use this process more efficiently by determining the largest multiple of some power of 10 to use at each step. Thus, in the example under discussion, we note that $40 \times 5 = 200$, whereas $50 \times 5 = 250$. Thus our first partial quotient is 40 as shown.

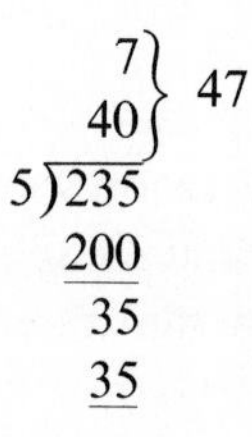

The following partition, or sharing, approach is probably closest to the interpretations used by most students. Consider $235 \div 5$ in terms of a graph-paper model of base ten arithmetic blocks. For a given model of 235 an effort is made to share the squares of 100. Then the squares that have not been shared are exchanged for base ten rods and the base rods are shared. Finally, the base rods that have not been shared are exchanged for units and the units are shared. The steps may be represented numerically as follows:

$$\begin{array}{r} \\ 5\overline{)235} \end{array} \qquad \begin{array}{r} 4 \\ 5\overline{)235} \\ \underline{20} \\ 3 \end{array} \qquad \begin{array}{r} 47 \\ 5\overline{)235} \\ \underline{20} \\ 35 \\ \underline{35} \\ 0 \end{array}$$

The last form represents our traditional *division algorithm* and may be obtained from each of the previous approaches.

Let us examine one more division problem, $4961 \div 23$, in detail. The explanation will be given in individual steps.

Step 1: Consider the largest multiple of 100×23 that can be used as a quotient. Since $2(100 \times 23) = 4600$, and $3(100 \times 23) = 6900$, we see that the first partial quotient is 200. (This is what is really meant when we divide 23 into 49, as shown in the abbreviated process on the right.)

$$\begin{array}{r@{\,}r} & 200 \\ 23 & \overline{)4961} \\ & \underline{4600} \\ & 361 \end{array} \qquad \begin{array}{r@{\,}l} & 2 \\ 23 & \overline{)4961} \\ & \underline{46} \\ & \;\;3 \end{array}$$

Step 2: Next consider multiples of 10×23 that can be divided into 361. Since $1(10 \times 23) = 230$ and $2(10 \times 23) = 460$, we find the next partial quotient to be 10. (In the abbreviated process we "bring down" 6, and divide 23 into 36.)

$$\begin{array}{r@{\,}r} & 10 \\ & 200 \\ 23 & \overline{)4961} \\ & \underline{4600} \\ & 361 \\ & \underline{230} \\ & 131 \end{array} \qquad \begin{array}{r@{\,}l} & 21 \\ 23 & \overline{)4961} \\ & \underline{46} \\ & \;\;36 \\ & \;\;\underline{23} \\ & \;\;13 \end{array}$$

Step 3: We now consider the largest multiple of 23 that we can divide into 131, and find this to be 5; $5 \times 23 = 115$. (In the abbreviated form we "bring down" 1, and divide 23 into 131.) Then add the partial quotients to obtain a quotient of 215, and a remainder of 16.

$$\begin{array}{r@{\,}r} & \left.\begin{array}{r} 5 \\ 10 \\ 200 \end{array}\right\} \; \begin{array}{l} 215 \\ \text{(quotient)} \end{array} \\ 23 & \overline{)4961} \\ & \underline{4600} \\ & 361 \\ & \underline{230} \\ & 131 \\ & \underline{115} \\ & 16 \;\; \text{(remainder)} \end{array} \qquad \begin{array}{r@{\,}l} & 215 \\ 23 & \overline{)4961} \\ & \underline{46} \\ & \;\;36 \\ & \;\;\underline{23} \\ & \;\;131 \\ & \;\;\underline{115} \\ & \;\;\;\;16 \end{array} \qquad \begin{array}{r} \textit{Check: } 215 \\ \underline{\times 23} \\ 645 \\ \underline{430\;} \\ 4945 \\ \underline{+16} \\ 4961 \end{array}$$

$$4961 = (215 \times 23) + 16$$

In general, for any natural numbers n and d, $d \neq 0$,

$n \div d = q$ with remainder r if and only if
$n = (d \times q) + r$ where $0 \leq r \leq d$.

This formulation of the division process involving the dividend n, the divisor d, the quotient q, and the remainder r is often called the **division algorithm.**

EXERCISES

Use repeated addition to find each product.

1. 5×20
2. 6×15
3. 8×5
4. 4×7
5. 7×4
6. 9×18

Use repeated subtraction to find each quotient.

7. $40 \div 5$
8. $40 \div 8$
9. $40 \div 4$
10. $120 \div 20$
11. $105 \div 15$
12. $144 \div 12$

Use the distributive property to find each product.

13. 8×15
14. 7×25
15. 6×45
16. 9×36
17. 5×435
18. 35×45
19. 12×15
20. 16×23
21. 42×57

Find each quotient by using multiples of 100 and 10 as in the illustrative problem of this section.

22. $7712 \div 32$
23. $13{,}803 \div 43$
24. $4255 \div 23$
25. $23{,}328 \div 54$
26. $8970 \div 26$
27. $9509 \div 37$
28. $5683 \div 27$
29. $7943 \div 18$
30. $6154 \div 32$

Check each division. If incorrect, find the correct quotient and remainder.

31. $1107 \div 23 = 48$, remainder 3
32. $1895 \div 29 = 68$, remainder 10
33. $3163 \div 37 = 83$, remainder 18
34. $11{,}828 \div 83 = 142$, remainder 42

EXPLORATIONS

1. Use base ten arithmetic blocks or graph-paper models to explain the exercises of this section.
2. Examine a recently published elementary textbook series and determine each of the following.
   (a) The grade level at which various multiplication facts are introduced.
   (b) The use made of the distributive property in explaining multiplication algorithms.
   (c) The manner in which the division algorithm is first introduced.
   (d) The attention given to multiplication and division as inverse operations.
3. A device to motivate drill on multiplication consists of starting with a "magic" number. For example, consider the number 12,345,679. If this number is multiplied by 18, that is, 2×9, the product will consist of 2's only. If multiplied by 27, that is, 3×9, the resulting product will have 3's only.
   (a) Verify this description by multiplying 12,345,679 by 45, that is 5×9 and by 63, that is, 7×9.
   (b) Use the number 15,873 and multiply by successive multiples of 7 to discover another "magic" number.

4. Identify the student errors being made in each of the following computations. Then suggest at least two teaching strategies that might be used to help correct the procedural error.

(a)
$$\begin{array}{r} 37 \\ \times\ 4 \\ \hline \end{array} \qquad \begin{array}{r} \boxed{2} \\ 37 \\ \times\ 4 \\ \hline 118 \end{array}$$

(b)
$$\begin{array}{r} 58 \\ \times\ 6 \\ \hline \end{array} \qquad \begin{array}{r} \boxed{4} \\ 58 \\ \times\ 6 \\ \hline 548 \end{array}$$

(c)
$$7\overline{)861} \qquad \begin{array}{r} 321 \\ 7\overline{)861} \\ \underline{7} \\ 16 \\ \underline{14} \\ 21 \\ \underline{21} \end{array}$$

(d)
$$9\overline{)5438} \qquad \begin{array}{r} 64 \text{ R } 2 \\ 9\overline{)5438}\phantom{\text{ R } 2} \\ \underline{54}\phantom{\text{ R } 2} \\ 38\phantom{\text{ R } 2} \\ \underline{36}\phantom{\text{ R } 2} \\ 2\phantom{\text{ R } 2} \end{array}$$

3-5 Other Systems of Numeration

In our decimal system objects are grouped and counted in tens and powers of ten. For example, the diagram shows how we might group and count 134 items.

$134 = (1 \times 10^2) + (3 \times 10) + (4 \times 1)$

We could, however, just as easily group sets of items in other ways. In the next figure we see 23 asterisks grouped in three different ways.

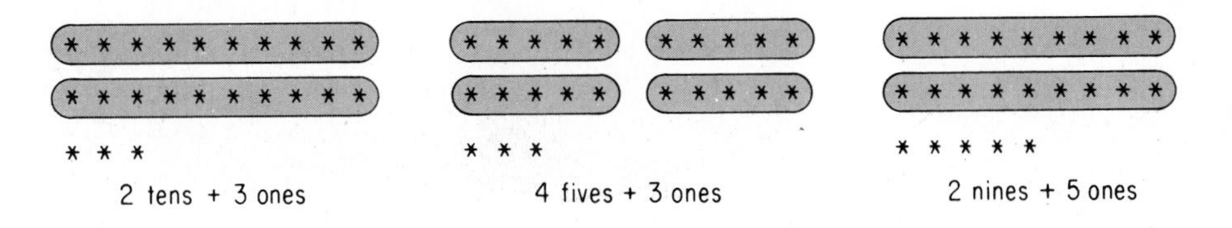

If we use a subscript to indicate our manner of grouping, we may write many different numerals (names) for the number of items in the same collection.

$$23_{\text{ten}} = 43_{\text{five}} = 25_{\text{nine}}$$
(2 *tens* + 3 ones) (4 *fives* + 3 ones) (2 *nines* + 5 ones)

43_{five} is read "four three, base five."
25_{nine} is read "two five, base nine."

Each of these numerals represents the number of asterisks in the same set of asterisks. Still another numeral for this number is 35_{six}.

$$35_{\text{six}} = 3 \textit{ sixes} + 5 \text{ ones} = 18 + 5 = 23$$

We call our decimal system of numeration a **base ten** system; when no subscript is used for a numeral with two or more digits, the numeral is understood to be expressed in base ten. Numerals with single digits do not depend upon the manner of grouping and do not require subscripts. When we group by fives, we have a **base five** system of numeration. In general, we name any system of numeration by the manner in which the grouping is accomplished.

Example 1 Draw a diagram for 18 objects and write the corresponding numeral **(a)** in base five notation **(b)** in base eight notation.

Solution **(a)**

(* * * * *)
(* * * * *)
(* * * * *)
* * *

33_{five}
(3 fives + 3 ones)

(b)

(* * * * * * * *)
(* * * * * * * *)
* *

22_{eight}
(2 eights + 2 ones) ■

Example 2 Write a numeral for 27 in base six notation.

Solution We note that $27 = 4$ sixes $+ 3$ ones; thus $27 = 43_{\text{six}}$. ■

The use of dollar bills and coins is a helpful aid in seeing the relationship between numbers expressed in base five and base ten. Consider, for example, a sum of 123¢.

Using a dollar, dimes, and pennies we may think of this sum as follows.

123¢ = 1 dollar + 2 dimes + 3 pennies

Compare this expression with the expanded notation for 123.

$$\begin{aligned} 123 &= (1 \times 10^2) + (2 \times 10^1) + (3 \times 10^0) \\ &= (1 \times 10^2) + (2 \times 10) + (3 \times 1) \end{aligned}$$

Now let us consider the base five numeral 123_{five}. A concrete representation of this numeral is to think of it in terms of quarters, nickels, and pennies.

$$123_{\text{five}} = 1 \text{ quarter} + 2 \text{ nickels} + 3 \text{ pennies}$$

For expanded base five notation we write numbers in terms of powers of five and use the digits 0, 1, 2, 3, and 4.

$$\begin{aligned} 123_{\text{five}} &= (1 \times 5^2) + (2 \times 5^1) + (3 \times 5^0) \\ &= (1 \times 5^2) + (2 \times 5) + (3 \times 1) \end{aligned}$$

Note that $5^0 = 1$. Using the monetary representation or the expanded base five notation, we see that $123_{\text{five}} = 38_{\text{ten}}$. We can illustrate this equality by one group of 25, two groups of five, and three ones, as in the following diagram.

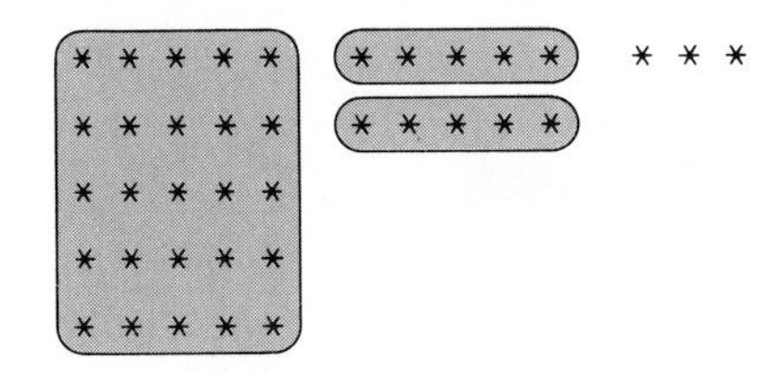

Arrays may be used in place of powers of the base to indicate place value for any base. The next two arrays may be used for 123 and 123_{five}.

10^2	10^1	10^0
1	2	3

$100 + 20 + 3 = 123$

5^2	5^1	5^0
1	2	3

$25 + 10 + 3 = 38$

Throughout the remainder of this section we shall work almost exclusively with base five notation, although the principles developed apply to all other bases as well. For convenience we shall write all numbers in base five notation using the numeral 5, rather than the word "five," as a subscript. Thus we shall write 123_5 although there is no numeral 5 in this system of notation. We begin with a table that contains the numbers 1 to 30 written in base five notation.

Base Ten	***Base Five***	***Base Ten***	***Base Five***	***Base Ten***	***Base Five***
1	1	11	21_5	21	41_5
2	2	12	22_5	22	42_5
3	3	13	23_5	23	43_5
4	4	14	24_5	24	44_5
5	10_5	15	30_5	25	100_5
6	11_5	16	31_5	26	101_5
7	12_5	17	32_5	27	102_5
8	13_5	18	33_5	28	103_5
9	14_5	19	34_5	29	104_5
10	20_5	20	40_5	30	110_5

To translate a number that is not listed in the table from base five notation to base ten notation, express the number in terms of powers of five and simplify.

Example 3 Write 432_5 in base ten notation.

Solution We may think of this expression in terms of quarters, nickels, and pennies.

$$432_5 = 4 \text{ quarters} + 3 \text{ nickels} + 2 \text{ pennies} = 117¢ = 117$$

Alternatively, we may use expanded notation.

$$\begin{aligned} 432_5 &= (4 \times 5^2) + (3 \times 5^1) + (2 \times 5^0) \\ &= (4 \times 25) + (3 \times 5) + (2 \times 1) \\ &= 117 \quad \blacksquare \end{aligned}$$

Example 4 Write 3214_5 in base ten notation.

Solution

$$\begin{aligned} 3214_5 &= (3 \times 5^3) + (2 \times 5^2) + (1 \times 5^1) + (4 \times 5^0) \\ &= (3 \times 125) + (2 \times 25) + (1 \times 5) + (4 \times 1) \\ &= 434 \quad \blacksquare \end{aligned}$$

To translate from base ten to base five, any one of several procedures may be adopted. Consider this problem.

$$339 = (\qquad)_5$$

When a number is expressed in base five, it is written in terms of powers of five.

$$5^0 = 1 \qquad 5^1 = 5 \qquad 5^2 = 25 \qquad 5^3 = 125 \qquad 5^4 = 625 \qquad \ldots$$

The highest power of five that is not greater than 339 is 5^3, that is, 125.

$$\begin{array}{r} 2 \\ 125\overline{)339} \\ \underline{250} \\ 89 \end{array} \qquad 339 = 2(125) + 89$$

The next lower power of five is 5^2, that is, 25.

$$\begin{array}{r} 3 \\ 25\overline{)89} \\ \underline{75} \\ 14 \end{array} \qquad \begin{aligned} 89 &= 3(25) + 14 \\ 339 &= 2(125) + 3(25) + 14 \end{aligned}$$

Then divide by 5.

$$\begin{array}{r} 2 \\ 5\overline{)14} \\ \underline{10} \\ 4 \end{array} \qquad 14 = 2(5) + 4$$

$$\begin{aligned} 339 &= 2(125) + 3(25) + 2(5) + 4 \\ &= (2 \times 5^3) + (3 \times 5^2) + (2 \times 5^1) + (4 \times 5^0) \\ &= 2324_5 \end{aligned}$$

$$\begin{array}{r l}
125\overline{)339} & 2 \\
\underline{250} & \\
89 & \\
25\overline{)89} & 3 \\
\underline{75} & \\
14 & \\
5\overline{)14} & 2 \\
\underline{10} & \\
4 & \\
1\overline{)4} & 4 \downarrow
\end{array}$$

The quotients may be written at the right and the divisions by 5^3, 5^2, 5, and 5^0 arranged, as in the adjacent figure, so that the digits of the numeral in base five are in a single column, which we read downward as 2324_5. Thus, a group of 339 elements can be considered as two groups of 125 elements, three groups of 25 elements, two groups of 5 elements, and 4 elements.

An alternative procedure for changing 339 to the base five depends upon successive division by 5.

$$339 = 67 \times 5 + 4$$

$$67 = 13 \times 5 + 2$$

$$13 = 2 \times 5 + 3$$

Next, substitute from the third equation into the second. Then substitute from the second equation into the first, and simplify.

$$13 = 2 \times 5 + 3$$

$$67 = 13 \times 5 + 2 = (2 \times 5 + 3) \times 5 + 2$$

$$\begin{aligned}
339 &= 67 \times 5 + 4 = [(2 \times 5 + 3) \times 5 + 2] \times 5 + 4 \\
&= [(2 \times 5^2) + (3 \times 5) + 2] \times 5 + 4 \\
&= (2 \times 5^3) + (3 \times 5^2) + (2 \times 5^1) + (4 \times 5^0) \\
&= 2324_5
\end{aligned}$$

The arithmetical steps, often called an *algorithm*, involved in these computations can be performed as shown in the following array.

$$\begin{array}{r l}
5\underline{)339} & \\
5\underline{)67} & \text{—}4 \uparrow \\
5\underline{)13} & \text{—}2 \\
5\underline{)2} & \text{—}3 \\
0 & \text{—}2
\end{array}$$

Read upward as 2324_5.

Note that the remainder is written after each division by 5. Then the remainders are read in reverse order to obtain the expression for the number in the base five. *This procedure works for integers only, not for fractional parts of a number.*

Example 5 Write 423 in base five notation.

Solution

$$\begin{array}{r l}
5\underline{)423} & \\
5\underline{)84} & \text{—}3 \uparrow \\
5\underline{)16} & \text{—}4 \\
5\underline{)3} & \text{—}1 \\
0 & \text{—}3
\end{array}$$

Answer: 3143_5.

Check:

$$\begin{aligned}
3143_5 &= (3 \times 5^3) + (1 \times 5^2) + (4 \times 5^1) + (3 \times 5^0) \\
&= 375 + 25 + 20 + 3 = 423 \ \blacksquare
\end{aligned}$$

The algorithm just developed will work for other bases as well. For example, let us write 339 in base eight notation. For base eight notation we divide by 8 until a quotient of 0 is obtained.

$$\begin{array}{r l}
8\overline{)339} & \\
8\overline{)42}\text{——}3 & \uparrow \\
8\overline{)5}\text{——}2 & \Big| \quad \text{Read upward as } 523_8. \\
0\text{——}5 & \Big|
\end{array}$$

Here is a check of this result.

$$\begin{aligned}
523_8 &= (5 \times 8^2) + (2 \times 8^1) + (3 \times 8^0) \\
&= (5 \times 8^2) + (2 \times 8) + (3 \times 1) \\
&= 320 + 16 + 3 = 339
\end{aligned}$$

EXERCISES *Write numerals for each of the following collections in the bases indicated by the manner of grouping.*

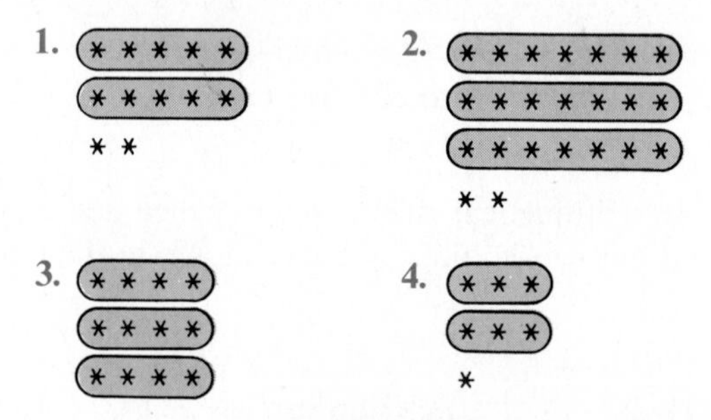

Draw a diagram to show the meaning of each of the following.

5. 34_{five} 6. 25_{six} 7. 25_{seven}
8. 13_{four} 9. 112_{four} 10. 122_{three}

Write each number in base ten notation.

11. 34_5 12. 25_7 13. 43_8
14. 54_6 15. 43_5 16. 21_4
17. 413_5 18. 330_5 19. 344_5
20. 3210_5 21. 1421_5 22. 2413_5
23. 4032_5 24. 4341_5 25. 1234_5

Write each number in base five notation.

26. 182 27. 493 28. 982
29. 596 30. 816 31. 337
32. 756 33. 607 34. 3125

Change to the stated base.

35. $339 = (\quad)_6$ 36. $339 = (\quad)_7$

37. $258 = (\quad)_4$ 38. $258 = (\quad)_8$

Extend the concepts of this section and write each number in base ten notation.

39. 337_8 40. 3013_4 41. 421_{12}

42. 4352_8 43. 314_{15} 44. 11011_2

EXPLORATIONS

1. For base five notation try to find rules for divisibility by
   (a) 10_5, that is, 5 **(b)** 4 **(c)** 2.
2. Prepare a brief introduction to base five notation for a specific elementary grade level. Use as many ways as possible to create interest in the topic.
3. Prepare a 10-minute talk, appropriate for a P.T.A. meeting, that explains the meaning of base five notation and that also attempts to convince the parents of the value of such a unit of study in the elementary curriculum.
4. Some first grade classes have been introduced to the basic idea of base five notation through the use of *hand numerals*. Thus 32_H may be used to represent 3 hands and 2 fingers, that is, 17. Prepare a table that shows how to represent the numbers from 1 through 24 using hand numerals.
5. Does the use of coins (pennies, nickels, quarters) for base five numerals illustrate a place-value system? Explain your answer and then discuss both the advantages and limitations of the use of these coins to teach the concepts of base five notation.
6. Try to find some other concrete representations for writing numbers in bases other than five and ten.
7. Explain why we cannot use 1 as a base in a system of numeration.

3-6 Computation in Other Bases

Techniques of computation with base five numerals will now be illustrated. The principles apply also for all other bases.

Coins can be used effectively to help develop an understanding of the principles involved in addition and subtraction in base five. Consider the problem $134_5 + 142_5$.

$$\begin{array}{r} 134_5 \\ +\ 142_5 \\ \hline \end{array}$$

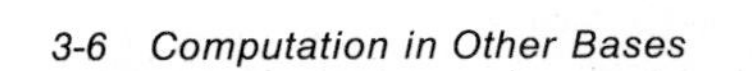

The 6 pennies are exchanged for 1 nickel and 1 penny. In the column of nickels we then have $1 + 3 + 4$ nickels. These 8 nickels are then exchanged for 1 quarter and 3 nickels. Finally in the column of quarters we have 3 quarters. Each of the two exchanges is the equivalent of "carrying" in place value notation.

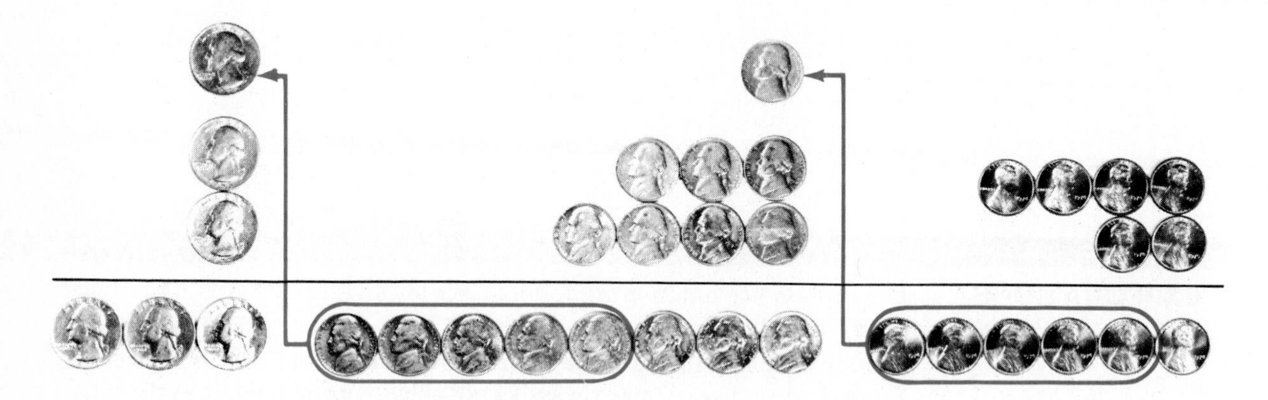

The final sum is 3 quarters, 3 nickels, and 1 penny. In base five notation this sum may be written as 331_5. To compute without such aids as coins we need to know the basic addition facts in base five notation. For example, we need to be able to find such sums as $4 + 3$. We can represent this sum as

$$**** + *** \quad \text{and regroup as} \quad (*****) + **$$

This grouping can be expressed as 12_5. That is, $4 + 3 = 12_5$. In base ten $4 + 3 = 7$, which is *one* group of five and *two* ones.

Here is a table of the number facts needed for addition problems in base five. (You should verify each entry.)

+	0	1	2	3	4
0	0	1	2	3	4
1	1	2	3	4	10_5
2	2	3	4	10_5	11_5
3	3	4	10_5	11_5	12_5
4	4	10_5	11_5	12_5	13_5

These addition facts may be used in finding sums of numbers, as illustrated in the following example.

Example 1 Find the sum of 432_5 and 243_5. Then check in base ten notation.

Solution

$$\begin{array}{r} 432_5 \\ +\ \underline{243_5} \\ 1230_5 \end{array}$$

Check:

$$\begin{array}{rl} 432_5 = & (4 \times 5^2) + (3 \times 5^1) + (2 \times 5^0) = 117 \\ +\ \underline{243_5} = & (2 \times 5^2) + (4 \times 5^1) + (3 \times 5^0) = \underline{\ 73} \\ 1230_5 = (1 \times 5^3) + & (2 \times 5^2) + (3 \times 5^1) + (0 \times 5^0) = 190 \end{array}$$ ■

Here are the steps used in Example 1. First add the column of ones.

$$\begin{array}{r} (4 \times 5^2) + (3 \times 5^1) + (2 \times 5^0) \\ \underline{(2 \times 5^2) + (4 \times 5^1) + (3 \times 5^0)} \\ 10_5 \end{array}$$

Write the sum 10_5 of the ones as 1×5^1 above the fives column and 0×5^0 below the ones column. Next add the column of fives.

$$\begin{array}{r} 1 \times 5^1 \qquad\qquad \\ \\ (4 \times 5^2) + (3 \times 5^1) + (2 \times 5^0) \\ \underline{(2 \times 5^2) + (4 \times 5^1) + (3 \times 5^0)} \\ (13_5 \times 5^1) + (0 \times 5^0) \end{array}$$

Write $13_5 \times 5^1$ as 1×5^2 above the 5^2 column and 3×5^1 below the fives column. Next add the column of 5^2 entries.

$$\begin{array}{r} 1 \times 5^2 \qquad\qquad\qquad\qquad\qquad \\ \\ (4 \times 5^2) + (3 \times 5^1) + (2 \times 5^0) \qquad\quad \\ \underline{(2 \times 5^2) + (4 \times 5^1) + (3 \times 5^0) \qquad\quad} \\ (12_5 \times 5^2) + (3 \times 5^1) + (0 \times 5^0) = 1230_5 \end{array}$$

Note that we "carry" groups of five and powers of five in base five computation, just as we "carry" groups of ten and powers of ten in decimal computation.

The next array provides a short form for these computations. The 1's marked by color indicate the groups that were "carried."

	5^3	5^2	5^1	5^0
	1	1	1	
		4	3	2
+		2	4	3
	1	2	3	0 $= 1230_5$

The table of addition facts in base five also may be used for subtraction if we recognize the relationship between addition and subtraction. As in the case of multiplication and division, the operations of addition and subtraction are *inverse operations*; each "undoes" the other.

$$17 - 5 = 12 \quad \text{and} \quad 12 + 5 = 17, \quad \text{that is,} \quad (17 - 5) + 5 = 17$$
$$8 + 3 = 11 \quad \text{and} \quad 11 - 3 = 8, \quad \text{that is,} \quad (8 + 3) - 3 = 8$$

Example 2 Subtract in base five and check in base ten $211_5 - 142_5$.

Solution First rewrite 211_5 in expanded notation and make the necessary exchanges.

$$\begin{aligned} 211_5 &= (2 \times 5^2) + (1 \times 5^1) + (1 \times 5^0) \\ &= (1 \times 5^2) + (11_5 \times 5^1) + (1 \times 5^0) \\ &= (1 \times 5^2) + (10_5 \times 5^1) + (11_5 \times 5^0) \end{aligned}$$

Then perform the subtraction.

$$\begin{array}{rr} 211_5 & (1 \times 5^2) + (10_5 \times 5^1) + (11_5 \times 5^0) \\ -\ 142_5 & (1 \times 5^2) + \quad (4 \times 5^1) + \quad (2 \times 5^0) \\ \hline 14_5 & (1 \times 5^1) + \quad (4 \times 5^0) \end{array}$$

Check:

$$\begin{array}{rl} 211_5 &= 56 \\ -\ 142_5 &= 47 \\ \hline 14_5 &= \ \ 9 \end{array} \blacksquare$$

Here are two alternative short forms for the steps of the solution of Example 2.

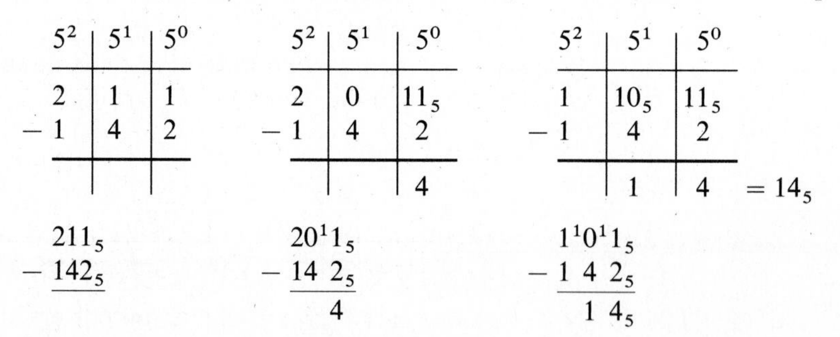

Next we consider multiplication. A table of multiplication facts is usually helpful (see Exercise 1).

Example 3 Find the product of 243_5 and 4.

Solution We need the following multiplication facts in base five.

$$4 \times 3 = 12_{10} = (2 \times 5^1) + (2 \times 5^0) = 22_5$$

$$4 \times 4 = 16_{10} = (3 \times 5^1) + (1 \times 5^0) = 31_5$$

$$4 \times 2 = \ 8_{10} = (1 \times 5^1) + (3 \times 5^0) = 13_5$$

The pattern for the computation when multiplying and "carrying" in base five notation is precisely the same as that used in base ten computation. A long form may be used to avoid "carrying" or a condensed form may be used. In the long form, zeros may be added when needed for each partial product, or indentation of partial products may be used to indicate the powers of five involved.

$$\begin{array}{rrl} & 243_5 & \\ \times & 4 & \\ \hline & 22_5 & = 4 \times 3 \\ & 310_5 & = 4 \times 40_5 \\ & 1300_5 & = 4 \times 200_5 \\ \hline & 2132_5 & \end{array} \qquad \begin{array}{rr} & 243_5 \\ \times & 4 \\ \hline & 22_5 \\ & 31_5 \\ & 13_5 \\ \hline & 2132_5 \end{array}$$

Condensed form:

$$\begin{array}{rr} & 243_5 \\ \times & 4 \\ \hline & 2132_5 \end{array} \quad \blacksquare$$

Multiplication by a multiplier with more than one digit is possible. The pattern for computation is similar to that used in base ten computation. Consider, for example, the product $34_5 \times 243_5$, that is, $(30_5 + 4) \times 243_5$.

$$\begin{array}{rr} & 243_5 \\ \times & 4 \\ \hline & 22_5 \\ & 31_5 \\ & 13_5 \\ \hline & 2132_5 \end{array} \qquad \begin{array}{rr} & 243_5 \\ \times & 30_5 \\ \hline & 0 \\ & 14_5 \\ & 22_5 \\ & 11_5 \\ \hline & 13340_5 \end{array} \qquad \begin{array}{rrl} & 243_5 & \\ \times & 34_5 & \\ \hline & 2132_5 & = 4 \times 243_5 \\ & 13340_5 & = 30_5 \times 243_5 \\ \hline & 21022_5 & = 34_5 \times 243_5 \end{array}$$

The pattern used for division in base ten can also be used for division in base five. In general, tables for both addition facts and multiplication facts (see Exercise 1) are useful in doing division problems. Consider, for example, this problem.

$$4\overline{)121_5}$$

The multiplication table includes these facts for 4.

$$1 \times 4 = 4 \qquad 2 \times 4 = 13_5 \qquad 3 \times 4 = 22_5 \qquad 4 \times 4 = 31_5$$

Since 12_5 is greater than 1×4 and less than 2×4, the first digit of the quotient is 1.

$$\begin{array}{r} 1 \\ 4\overline{)121_5} \\ 4 \\ \hline 31_5 \end{array}$$

As in base ten computation, we multiply 1×4 and subtract this from 12_5 in the dividend. The next digit is then brought down. Note that $12_5 = 7$ and thus $12_5 - 4 = 3$.

Next we divide 31_5 by 4. Note in the list of multiples of 4 that $4 \times 4 = 31_5$. Thus we can complete the division, and there is no remainder.

$$\begin{array}{r} 14_5 \\ 4\overline{)121_5} \\ 4 \\ \hline 31_5 \\ 31_5 \\ \hline \end{array} \qquad \textit{Check:} \quad \begin{array}{rr} & 14_5 \\ \times & 4 \\ \hline & 121_5 \end{array}$$

Example 4 Divide 232_5 by 3 and check by multiplication in base five.

Solution

$$1 \times 3 = 3 \qquad 2 \times 3 = 11_5 \qquad 3 \times 3 = 14_5 \qquad 4 \times 3 = 22_5$$

$$\begin{array}{r} 42_5 \\ 3\overline{)232_5} \\ \underline{22_5} \\ 12_5 \\ \underline{11_5} \\ 1 \end{array}$$

Check:

$$\begin{array}{rl} 42_5 & \text{(quotient)} \\ \times \quad 3 & \text{(divisor)} \\ \hline 231_5 & \\ + \quad 1 & \text{(remainder)} \\ \hline 232_5 & \text{(dividend)} \end{array}$$

The quotient is 42_5 and the remainder is 1. Note that in the check we multiply the quotient by the divisor, and add the remainder to obtain the dividend. ■

As indicated earlier, the principles that have been developed for computation in base five notation apply to other bases as well. As one illustration of this, consider the addition problem in Example 5.

Example 5 Find the sum of 543_8 and 376_8. Then check in base ten notation.

Solution

$$\begin{array}{r} 543_8 \\ + \ 376_8 \\ \hline 1141_8 \end{array}$$

Check:

$$\begin{array}{rl} 543_8 = & (5 \times 8^2) + (4 \times 8^1) + (3 \times 8^0) = 355 \\ + \ 376_8 = & (3 \times 8^2) + (7 \times 8^1) + (6 \times 8^0) = \underline{254} \\ 1141_8 = & (1 \times 8^3) + (1 \times 8^2) + (4 \times 8^1) + (1 \times 8^0) = 609 \end{array}$$ ■

EXERCISES

1. Complete a table showing the basic multiplication facts for base five.

Perform the indicated operation in base five and check in base ten.

2. $42_5 + 34_5$
3. $44_5 + 14_5$
4. $243_5 + 334_5$
5. $433_5 + 443_5$
6. $3343_5 + 3422_5$
7. $4124_5 + 4442_5$
8. $243_5 - 31_5$
9. $41_5 - 24_5$
10. $322_5 - 123_5$
11. $422_5 - 244_5$
12. $1021_5 - 403_5$
13. $3004_5 - 1312_5$
14. $400_5 - 133_5$
15. $2003_5 - 1014_5$
16. $1000_5 - 444_5$

17. $\begin{array}{r} 342_5 \\ \times \quad 3 \\ \hline \end{array}$

18. $\begin{array}{r} 243_5 \\ \times \quad 4 \\ \hline \end{array}$

19. $\begin{array}{r} 1424_5 \\ \times \quad 3 \\ \hline \end{array}$

20. $\begin{array}{r} 2104_5 \\ \times \quad 2 \\ \hline \end{array}$

21. $\begin{array}{r} 44_5 \\ \times\ 32_5 \\ \hline \end{array}$

22. $\begin{array}{r} 33_5 \\ \times\ 43_5 \\ \hline \end{array}$

23. $\begin{array}{r} 342_5 \\ \times\ 34_5 \\ \hline \end{array}$

24. $\begin{array}{r} 3244_5 \\ \times\ 324_5 \\ \hline \end{array}$

25. $\begin{array}{r} 3042_5 \\ \times\ 203_5 \\ \hline \end{array}$

26. $4\overline{)233_5}$

27. $3\overline{)231_5}$

28. $3\overline{)1243_5}$

29. $4\overline{)3014_5}$

30. $22_5\overline{)143_5}$

31. $32_5\overline{)4340_5}$

32. $23_5\overline{)1341_5}$

33. $13_5\overline{)4314_5}$

34. $41_5\overline{)33443_5}$

35. Complete the following tables of addition and multiplication facts for base four.

+	0	1	2	3
0				
1				
2				
3				

×	0	1	2	3
0				
1				
2				
3				

Perform the indicated operations in the stated base and check in base ten.

36. $\begin{array}{r} 325_6 \\ +\ 453_6 \\ \hline \end{array}$

37. $\begin{array}{r} 475_8 \\ +\ 535_8 \\ \hline \end{array}$

38. $\begin{array}{r} 1323_4 \\ +\ 332_4 \\ \hline \end{array}$

39. $\begin{array}{r} 372_8 \\ -\ 156_8 \\ \hline \end{array}$

40. $\begin{array}{r} 542_7 \\ -\ 256_7 \\ \hline \end{array}$

41. $\begin{array}{r} 1012_4 \\ -\ 233_4 \\ \hline \end{array}$

42. $\begin{array}{r} 356_8 \\ \times \quad 7 \\ \hline \end{array}$

43. $\begin{array}{r} 2354_6 \\ \times \quad 4 \\ \hline \end{array}$

*44. $\begin{array}{r} 312_4 \\ \times\ 23_4 \\ \hline \end{array}$

*45. $5\overline{)276_8}$

*46. $4\overline{)1132_6}$

*47. $23_4\overline{)10111_4}$

EXPLORATIONS

1. Prepare a worksheet suitable for a specific elementary grade level with problems translating from base five to base ten. Begin the sheet with an illustrative concrete example to aid the students in their work.
2. Repeat Exploration 1 for translation from base ten to base five.
3. Draw a set of pictures of hands (five fingers) and fingers that might be used for an early elementary class to explain the solution to these problems.

   (a) $\begin{array}{r} 24_5 \\ +\ 13_5 \\ \hline \end{array}$ **(b)** $\begin{array}{r} 41_5 \\ +\ 23_5 \\ \hline \end{array}$

4. Explore a set of recently published elementary school mathematics textbooks to determine what consideration, if any, is given to work on computation in other number bases.

Number bases greater than 10 are possible but require the introduction of new symbols. Consider, for example, a base twelve system of numeration. The numeral 10_{12} represents $(1 \times 12) + (0 \times 1)$; that is, 12. Thus new symbols are necessary for 10 and 11. Let us use t for 10 and e for 11 and count in base twelve.

$$1 \quad 2 \quad 3 \quad 4 \quad 5 \quad 6 \quad 7 \quad 8 \quad 9 \quad t \quad e \quad 10_{12} \quad 11_{12} \quad 12_{12} \quad \ldots$$

We can use expanded notation to change a base twelve numeral to base ten.

$$2te5_{12} = (2 \times 12^3) + (10 \times 12^2) + (11 \times 12^1) + (5 \times 12^0)$$
$$= 3456 + 1440 + 132 + 5 = 5033$$

5. Write $t37e_{12}$ as a base ten numeral.
6. Write 589 as a base twelve numeral.
7. Write a lesson plan and prepare to deliver a 15-minute presentation on base twelve numeration. Include both counting in base twelve and performing arithmetic operations. Use a comparison with the arithmetic on a 12-hour clock to explain the ones digit. For example, perform each operation (i) as on a 12-hour clock (ii) as in base twelve notation.
   (a) $8 + 8$ **(b)** $9 + 5$ **(c)** $11 + 11$
   (d) 4×6 **(e)** 7×7 **(f)** 8×9

3-7 Binary Notation

Here are the first 16 counting numbers written in binary notation.

Base Ten	*Base Two*
1	1
2	10_2
3	11_2
4	100_2
5	101_2
6	110_2
7	111_2
8	$1\,000_2$
9	$1\,001_2$
10	$1\,010_2$
11	$1\,011_2$
12	$1\,100_2$
13	$1\,101_2$
14	$1\,110_2$
15	$1\,111_2$
16	$10\,000_2$

The representation of numbers by using only two digits, usually 0 and 1, is of special interest because of its applications in modern electronic computers. One of the two digits is represented by the presence of an electric signal; the other digit is represented by the absence of an electric signal. The presence or absence of electric signals may be controlled by the presence or absence of holes in a punched card, the presence or absence of magnetic fields on a magnetic tape, and in other ways. For example, the pictures of the surface of the planet Mars taken by the Mariner spacecraft in 1976 were represented by dots on a coordinate plane. The location, color, and intensity of the dots were transmitted back to Earth in binary notation. Then computers were used to reassemble the pictures so that they could be displayed on television and in other media.

The place value numeration system in which only two digits are used is called **binary notation**. There is some evidence that the basic concepts of binary notation were known to the ancient Chinese about 2000 B.C. However, it is only in relatively recent times that binary notation has been widely applied in card-sorting operations and in computer mathematics. Here are some of the place values in the binary system of numeration.

2^7	2^6	2^5	2^4	2^3	2^2	2^1	2^0
128	64	32	16	8	4	2	1

Binary numerals often involve many digits. We set off these digits in sets of three, as in Example 1, to aid the reading of the numerals.

Example 1 Write $11\ 011\ 101_2$ in base ten notation.

Solution

$$\begin{aligned} 11\ 011\ 101_2 &= (1 \times 2^7) + (1 \times 2^6) + (0 \times 2^5) + (1 \times 2^4) \\ &\quad + (1 \times 2^3) + (1 \times 2^2) + (0 \times 2^1) + (1 \times 2^0) \\ &= 128 + 64 + 16 + 8 + 4 + 1 = 221 \ \blacksquare \end{aligned}$$

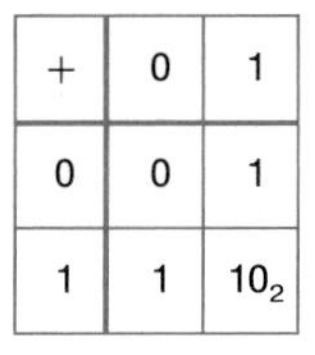

+	0	1
0	0	1
1	1	10_2

The adjacent tables are for addition and multiplication in binary notation.

Example 2 Multiply $101_2 \times 1101_2$.

Solution

×	0	1
0	0	0
1	0	1

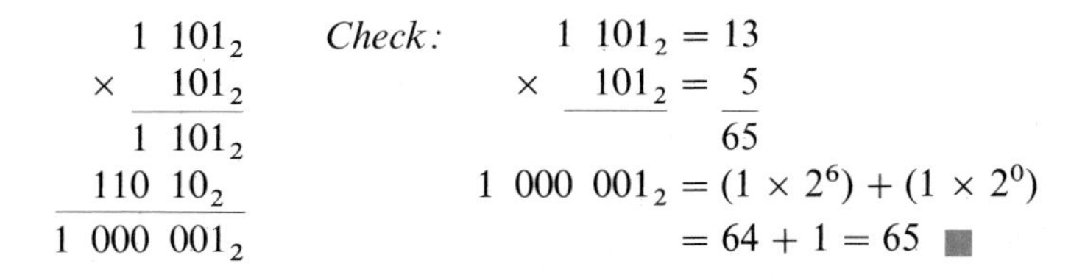

$$\begin{array}{r} 1\ 101_2 \\ \times \quad 101_2 \\ \hline 1\ 101_2 \\ 110\ 10_2 \\ \hline 1\ 000\ 001_2 \end{array}$$

Check:

$$\begin{array}{rl} 1\ 101_2 &= 13 \\ \times \quad 101_2 &= \ \ 5 \\ \hline & \ \ 65 \end{array}$$

$$\begin{aligned} 1\ 000\ 001_2 &= (1 \times 2^6) + (1 \times 2^0) \\ &= 64 + 1 = 65 \ \blacksquare \end{aligned}$$

Binary numerals can be shown by means of electric lights, as in Exploration 4. If a light is on, the digit 1 is represented; if the light is off, the digit 0 is represented. Similarly, the binary concept can be used in a card-sorting operation. An exploration of the processes involved in a card-sorting operation provides insight into the processes that occur in a computer. Prepare a set of 16 index cards with four holes punched in each and a corner cut off as in the adjacent figure. At this stage the 16 cards should be exactly alike.

Next represent the numbers 0 through 15 on these cards in binary notation. Cut out the space above each hole to represent 1; leave the hole untouched to represent 0. Several cards are shown in the next figure.

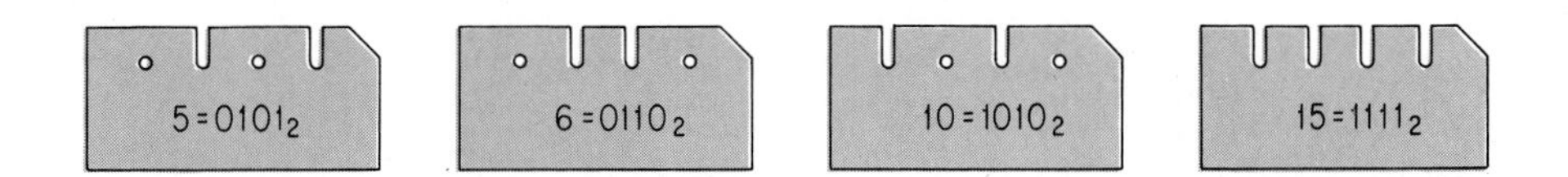

After all the cards have been completed in this manner, shuffle them thoroughly and align them, making certain that they remain "face up." (The position of the cut off corner indicates when a card is right side up.) Then, going from right to left, perform the following operation: Stick a pencil or other similar object through the first hole and lift up. Some of the cards will come up, namely, those in which the holes have not been cut through to the edge of the card (those cards representing numbers whose units digit in binary notation is 0).

Place the cards that have been lifted up in front of the other cards and repeat the same operation for the remaining holes in order from right to left. When you have finished, the cards should be in numerical order, 0 through 15.

Note that only four operations are needed to arrange the 16 cards. As the number of cards is doubled, only one additional operation will be needed each time to place them in order. That is, 32 cards may be placed in numerical order with five of the described card-sorting operations; 64 cards may be arranged with six operations; 128 cards with seven operations; and so forth. Thus a large number of cards may be arranged in order with a relatively small number of operations. For example, over one billion cards may be placed in numerical order with only 30 sortings.

The seven-digit numeral 1 000 001 is the name for the letter A when that letter is sent over teletype to a computer. The American Standard Code for Information Interchange (ASCII) was adopted in 1967 and is used to convert letters, numerals, and other symbols into binary notation.

Electronic computers are designed to have specified capacities both as to the number of digits that can be represented for a given number and the number of digits that are processed at one time in performing an operation. In the previous example of card sorting, only numbers that required at most four digits in binary notation were represented and the digits were processed one at a time. In "computer language" each binary digit is a *bit* and the *word length* for a particular computer is the number of bits that are processed simultaneously. Some early computers had a word length of three bits and essentially worked in octal notation (see Exercises 39 through 51). As computer technology has improved, the word lengths of computers have increased to 8, 16, 32, and even 64 or more bits.

EXERCISES

Write each number in binary notation.

1. 38	**2.** 35	**3.** 29
4. 75	**5.** 93	**6.** 129
7. 156	**8.** 173	**9.** 200
10. 425	**11.** 437	**12.** 511

Change each number to decimal notation.

13. 1110_2	**14.** $10\ 100_2$	**15.** $11\ 011_2$
16. $100\ 111_2$	**17.** $111\ 011_2$	**18.** $101\ 110_2$
19. $101\ 011_2$	**20.** $1\ 001\ 100_2$	**21.** $1\ 101\ 010_2$
22. $11\ 101\ 011_2$	**23.** $10\ 111\ 001_2$	**24.** $10\ 101\ 010_2$

Perform the indicated operation in binary notation; check in base ten.

25. $\begin{array}{r} 1111_2 \\ +\ 1011_2 \\ \hline \end{array}$ **26.** $\begin{array}{r} 10\ 001_2 \\ +\ 10\ 101_2 \\ \hline \end{array}$ **27.** $\begin{array}{r} 10\ 011_2 \\ +\ 10\ 101_2 \\ \hline \end{array}$

28. $\begin{array}{r} 100\ 101_2 \\ +\ 10\ 111_2 \\ \hline \end{array}$ **29.** $\begin{array}{r} 1111_2 \\ -\ 1011_2 \\ \hline \end{array}$ **30.** $\begin{array}{r} 11\ 001_2 \\ -\ 10\ 110_2 \\ \hline \end{array}$

31. $\begin{array}{r} 100\ 101_2 \\ -\ 10\ 111_2 \\ \hline \end{array}$ **32.** $\begin{array}{r} 100\ 011_2 \\ -\ 10\ 101_2 \\ \hline \end{array}$ **33.** $\begin{array}{r} 1\ 000\ 100_2 \\ -\ 111\ 111_2 \\ \hline \end{array}$

34. $\begin{array}{r} 1111_2 \\ \times\ 11_2 \\ \hline \end{array}$ **35.** $\begin{array}{r} 10\ 111_2 \\ \times\ 101_2 \\ \hline \end{array}$ **36.** $\begin{array}{r} 110\ 110_2 \\ \times\ 110_2 \\ \hline \end{array}$

37. In the ASCII code, the numbers for *B*, *D*, and *G* are 66, 68, and 71, respectively. Find the binary representation of **(a)** *B* **(b)** *D* **(c)** *G*.

38. In the ASCII code, letters A, B, C,... in alphabetical order are assigned numbers 65, 66, 67,.... What is the word transmitted by the code 1 010 010, 1 010 101, 1 001 110?

39. Write the number 234 in base eight and then in base two notation. Can you discover a relationship between these two notations?

Base eight notation is often called **octal notation**. *Use the relationship discovered in Exercise* 39 *and write in octal notation.*

40. $101\ 111\ 001\ 010_2$

41. $11\ 101\ 011\ 001_2$

42. $1\ 001\ 101\ 110\ 010_2$

43. $10\ 010\ 101\ 011\ 011_2$

44. $111\ 110\ 101\ 101\ 011\ 001_2$

45. $1\ 101\ 011\ 001\ 000\ 100\ 110_2$

Use the relationship discovered in Exercise 39 *and write in binary notation.*

46. 335_8

47. 5023_8

48. 4357_8

49. 4624_8

50. $42\ 345_8$

51. $36\ 543_8$

EXPLORATIONS

Many recreational items are based on the binary system of notation. Consider, for example, the boxes shown, within which the numbers 1 to 15 are placed according to the following scheme.

D	C	B	A
8	4	2	1
9	5	3	3
10	6	6	5
11	7	7	7
12	12	10	9
13	13	11	11
14	14	14	13
15	15	15	15

In box A place all numbers that have a 1 in the ones place when written in binary notation. In box B place those with a 1 in the second position from the right in binary notation. In C and D are those numbers with a 1 in the third and fourth positions, respectively.

Next ask someone to select a natural number less than 16 and tell you in which box or boxes it appears. You add the first number in each of the designated boxes and state the sum as the selected number. For example, if the number is 11, the designated boxes are A, B, and D. You then find the selected number as $1 + 2 + 8$.

1. Explain why the method given for finding a number after knowing the boxes in which it appears works as it does.
2. Extend the boxes to include all the numbers through 31. (A fifth box, E, will be necessary.) Then explain for the set of five boxes how to find a number if the boxes in which it appears are known.
3. You can use binary notation to identify any one of eight numbers by means of three questions that can be answered by *yes* or *no*. Consider the numbers 0 through 7 written in binary notation. The place values are identified by columns A, B, and C.

	C	B	A
0	0	0	0
1	0	0	1
2	0	1	0
3	0	1	1
4	1	0	0
5	1	0	1
6	1	1	0
7	1	1	1

The three questions to be asked are: Does the number have a 1 in position A? Does the number have a 1 in position B? Does the number have a 1 in position C? Suppose the answers are yes, no, yes. Then the number is identified as 101_2; that is, 5. Extend this process to show how you can find any selected number from 0 through 15.

4. Consider five light bulbs in a row on a panel. The bulbs may be labeled so that they can be used in the same manner as the columns in Exploration 3. In the adjacent figure three bulbs are lighted to represent 1110_2, that is, 14.

Draw a panel of five bulbs and show how they should be lighted to represent 21. Draw another panel of bulbs on which 57 is represented. How many bulbs would you need in a panel on which the whole numbers 1 through 127 are to be represented?

Chapter Review

1. Write 5280 in expanded notation.
2. Write $(4 \times 10^4) + (2 \times 10^2)$ in decimal notation.
3. Write four million thirty thousand twenty in decimal notation.
4. Use the Egyptian method of doubling to find the product 21×43.
5. Tell whether the number specified in the given statement is used for identification, as an ordinal number, or as a cardinal number.
   (a) Bob has *two* cars.
   (b) Mary sits in the *second* row.
6. Find the cardinal number of each set.
   (a) $\{1, 4, 7, 10, \ldots, 22\}$ **(b)** $\{3, 6, 9, 12, \ldots, 51\}$
7. Draw a number line and graph the specified set.
   (a) The whole numbers less than 5.
   (b) The natural numbers between 1 and 5.
8. Which of the four basic arithmetic operations (addition, subtraction, multiplication, and division) upon whole numbers are **(a)** commutative? **(b)** associative?
9. Use galley multiplication to find the product 273×47.
10. The following diagram shows a use of Napier's rods. Express the product in terms of the indicated factors.

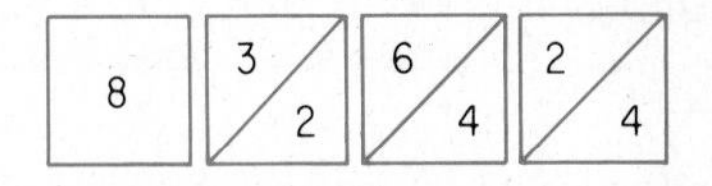

11. Use repeated subtraction to find the quotient $55 \div 17$.
12. Use the distributive law to find the product 7×208.
13. Write 342_5 in base ten notation.
14. Write 2341_5 in base ten notation.
15. Write 341 in base five notation.
16. Write 7894 in base five notation.
17. Use base five notation to find the sum of 234_5 and 123_5. Then check your result in base ten notation.
18. Subtract in base five and check in base ten: $213_5 - 142_5$.
19. Find the product of 324_5 and 4.
20. Divide 223_5 by 3 and check by multiplication in base five.

21. Write $10\ 101\ 111_2$ in base ten notation.
22. Write 77 in binary notation.
23. Write 241 in binary notation.

Perform the indicated operation in binary notation; check in base ten.

24. $10\ 101_2 - 1011_2$

25. $101_2 \times 1011_2$

LONG DIVISION AND THE COMPUTER

One of the first things that primary grade students notice in dealing with a hand calculator or computer is the form in which answers are given to long division problems. When they request the answer to $1 \div 3$ they get 0.33333333.... This is a puzzle to someone who has seen quotients expressed only in terms of a whole number plus a possible remainder.

This puzzle can be clarified on the computer through the use of the following program, which provides the answer to a long division problem in terms of the whole number quotient and the remainder. The development of the program requires the introduction of a few more concepts from the BASIC language. In addition, we use some other programming commands and techniques to make the results, or output, easily readable.

We begin the program with the command NEW, which tells the computer that a new program is being inserted. This is followed on line 5 by the command HOME. HOME instructs the computer to clear the TV monitor screen of any other printing that might be on it when the program begins to run. Another procedure used in the program is the PRINT statement standing alone on a line, as in lines 11, 13, 15, 19, and 2Ø. This causes the computer to leave a line blank at these positions on the screen or paper, as in double spacing when typing.

```
NEW
 5 HOME
1Ø PRINT "THIS PROGRAM WILL FIND THE QUOTIENT"
11 PRINT
12 PRINT "FOR A NATURAL NUMBER DIVISION PROBLEM."
13 PRINT
14 INPUT "ENTER THE DIVIDEND. ";N
15 PRINT
16 INPUT "ENTER THE DIVISOR. ";D
17 Q = INT(N/D)
18 R = N - (Q * D)
2Ø PRINT
21 PRINT N;"/";D;" IS ";Q;" WITH REMAINDER ";R;"."
22 END
```

A major new point in the program occurs in line 14 and again in line 16. The symbols N and D are variables standing for arbitrary natural numbers that are to be entered by the user of the program. These values are entered through the use of INPUT statements. An INPUT statement, like a PRINT statement, causes any specified statement to be printed, but then the computer waits for the user to type a value for the specified variable before proceding. Once a value is entered for the variable, the computer will retain that value and use it each time that variable occurs. In our program the variables occur in the computation in lines 17 and 18 for the values of N, the dividend, and D, the divisor.

Another new feature is the function INT, the greatest integer function, $[x]$, in line 17. This function is used often in higher mathematics to obtain the greatest integer that is less than or equal to a specified real number that may have fractional or decimal parts. For example, $[3] = 3$, $[4.5] = 4$, $[-3] = -3$, $[-3.5] = -4$, and $[0.0025] = 0$. For any real number x the integer $[x]$ may be viewed on a number line as representing the integral point that is at or immediately on the left of the point represented by x.

x

-6 -5 -4 -3 -2 -1 0 1 2 3 4 5 6 $[x] = 1$

The program for finding the quotient for a long division problem employs the greatest integer function to find the remainder. Consider the problem $7 \div 3$. We would normally think of the problem as

$$3\overline{)7}$$

and then write 2 above the 7, subtract $2 * 3$, and cite a remainder of 1. This is what is happening in line 18. The remainder R is found as the dividend, N, minus the number, INT(N/D), of groups of the divisor, D, contained in the dividend.

The use of the program is illustrated for the problem $138{,}745 \div 7246$. The output for this division problem follows.

```
RUN
THIS PROGRAM WILL FIND THE QUOTIENT

FOR A NATURAL NUMBER DIVISION PROBLEM.

ENTER THE DIVIDEND.  138745

ENTER THE DIVISOR.  7246

138745/7246 IS 19 WITH REMAINDER 1Ø71.
```

Type the program into your computer and use it to find the quotient and remainder for each of the following long division problems: $46{,}287 \div 37$, $343 \div 7$, and $98 \div 99$.

4

Integers

Carl Friedrich Gauss
(1777–1855)

Carl Friedrich Gauss is perhaps the greatest mathematician who ever lived. He began his career as a child prodigy and from the age of 19 he devoted himself to mathematics and applications of mathematics. He provided a solid core for the theory of numbers in his 1801 text Disquisitiones Arithmeticae. *In it he carefully developed the theory of modular arithmetic as well as several other topics related to number theory and algebra. In addition to his important contributions in mathematics, Gauss also made major contributions to astronomy, physics, and other sciences.*

4-1 The Set of Integers

With only the set of whole numbers at our disposal, we have no numbers to represent such quantities as a deficit of \$18, or a temperature of 10° below zero. Furthermore, with only the set of whole numbers at our disposal, we are unable to find replacements for n that make sentences such as these true.

$$5 + n = 2 \qquad n + 4 = 0 \qquad 2 - 7 = n$$

We need to extend our number system. We start by drawing a number line and then considering points to the left of the origin 0. The origin has coordinate 0. For any positive integer n, the point n units to the right of the origin has coordinate n and we may locate another point that is n units to the left of the origin. We call the coordinate of this new point **negative *n*,** written $-n$, and refer to the number that $-n$ represents as the **opposite** of n. The point 3 units to the right of the origin has coordinate 3. The point 3 units to the left of the origin has coordinate -3. The points are the graphs of the numbers.

The graph of the set of integers continues indefinitely both in the positive direction, to the right of the origin, and in the negative direction, to the left of the origin, as shown by the arrows on the number line.

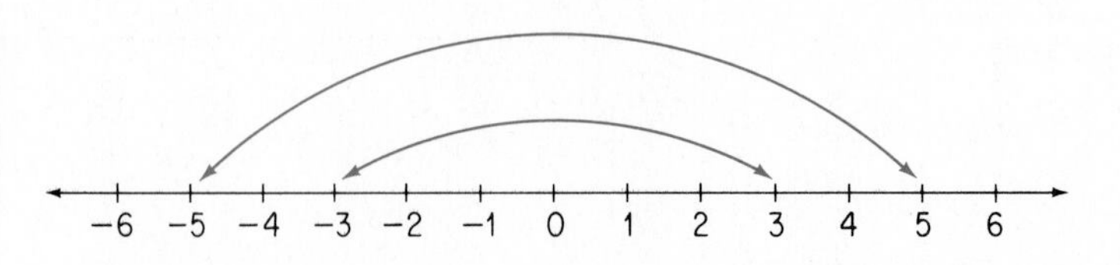

We say that the opposite of 3 is -3; also, the opposite of -3 is 3. Similarly, the point with coordinate -5 is located five units to the left of the origin, and we say that 5 and -5 are opposites of each other. Finally we agree that the opposite of 0 is 0; that is, $(-1) \times 0 = 0$.

We call the set of whole numbers together with the set of the opposites of the whole numbers, the set Z of **integers**.

$$Z = \{\ldots, -3, -2, -1, 0, 1, 2, 3, \ldots\}$$

If p is a positive integer, then p is a whole number and $p = 0 + p$. If n is a negative integer, then $-n$ is a whole number and $n = 0 - (-n)$.

The set of integers may be considered as the union of three sets of numbers.

Positive integers: $\{1, 2, 3, 4, 5, \ldots\}$

Zero: $\{0\}$

Negative integers: $\{\ldots, -5, -4, -3, -2, -1\}$

At times, especially for emphasis, plus signs are used for positive integers.

$\{+1, +2, +3, +4, +5, \ldots\}$

$|-3| = 3$

$|3| = 3$

$|0| = 0$

$-2 > -5$ since $-5 + 3 = -2$

We consider such expressions as $+4$, read "positive four," and 4 as two names for the same number, the number with its graph located at the point four units to the right of the origin on a number line. The positive integers and the negative integers are the **signed numbers**. Zero is neither a positive integer nor a negative integer and is not a signed number. When the signs of numbers are disregarded, the remaining value is called the **absolute value** and is indicated by placing a vertical bar on each side of the numeral. For example, we write $|+4| = 4$ and $|-4| = 4$. Order for integers is the same as for whole numbers. That is, $a < b$ and $b > a$ if and only if there is a natural number c such that $a + c = b$.

Example 1 Graph the set of integers between -3 and 2.

Solution

−6 −5 −4 −3 −2 −1 0 1 2 3 4 5 6 ■

Note that $-b$ may represent either a positive integer or a negative integer.
If $b = 5$, then $-b = -5$; if $b = -7$, then $-b = 7$.

The sum of any whole number n and its opposite $-n$ is 0; $n + (-n) = 0$. Consider the equation $n + (-n) = (-n) + n$ and extend the concept of opposites so that each of the numbers n and $-n$ is the opposite of the other. For example, -5 is the opposite of 5 and 7 is the opposite of -7. Then every integer n has an opposite $-n$ such that $n + (-n) = 0$. This property is sometimes called the *addition property of opposites.* Each of the numbers n and $-n$ is the **additive inverse** of the other.

Inverse, $+$: $n + (-n) = (-n) + n = 0$

The basic properties of the set of integers under addition and multiplication may be summarized as follows:

Closure, $+$	Closure, $\times$
Commutative, $+$	Commutative, $\times$
Associative, $+$	Associative, $\times$
Identity, $+$	Identity, $\times$
Inverse, $+$	

Distributive property

Example 2 Show that the subtraction of integers is not commutative.

Solution We can show this by a single counterexample. That is, we merely need to show one example where the property does *not* hold.

$$5 - 3 = 2 \qquad 3 - 5 = -2 \qquad 2 \neq -2$$

Then $5 - 3 \neq 3 - 5$ and subtraction of integers is not commutative. ■

A set of elements with the closure property, the associative property, an identity element, and for each of its elements an inverse element under a specified operation forms a **group** under that operation. A group with the commutative property is a **commutative group**. The set of integers forms a commutative group under addition. Since the set of integers does not, for example, contain an integer n such that $3 \times n = 1$, that is, a multiplicative inverse for 3, the set of integers does not form a group under multiplication. The concept of a group under an operation is widely used in advanced mathematics and its applications.

The need for additive inverses provided the basis for the extension of the set of whole numbers to the set of integers. The need for multiplicative inverses provides the basis for the extension, in Section 6-1, of the set of integers to the set of rational numbers.

We are now ready to explore a number of proofs that illustrate the type of reasoning that mathematicians use. First we need several definitions; then we shall consider several proofs and suggest others in the exercises. Some of these proofs will be needed in Chapter 7, where we prove that $\sqrt{2}$ cannot be expressed as a quotient of integers.

An integer is an **even integer** if it is a multiple of 2, that is, if it may be expressed as $2k$, where k stands for an integer. Then the set of even integers is

$$\{\ldots, -6, -4, -2, 0, 2, 4, 6, \ldots, 2k, \ldots\}$$

Even natural numbers: $\{2, 4, 6, \ldots, 2k, \ldots\}$

Odd natural numbers: $\{1, 3, 5, \ldots, 2k + 1, \ldots\}$ where $k \in N$.

An integer that is not even is said to be an **odd integer**. Each odd integer may be expressed in the form $2k + 1$, where k stands for an integer. Then the set of odd integers is

$$\{\ldots, -7, -5, -3, -1, 1, 3, 5, 7, \ldots, 2k + 1, \ldots\}$$

Example 3 Prove that the sum of any two even natural numbers is an even natural number.

Proof Any two even natural numbers m and n may be expressed as $2k$ and $2r$, where k and r stand for natural numbers. Then

$$m + n = 2k + 2r = 2(k + r)$$

where $k + r$ stands for a natural number, since the sum of any two natural numbers is a natural number. Therefore $m + n$ is an even natural number. ■

Example 4 Prove that the square of any even natural number is an even natural number.

Proof Any even natural number may be expressed as $2k$, where k stands for a natural number. Then the square of the natural number may be expressed as $(2k)^2$, where

$$(2k)^2 = (2k)(2k) = 2(2k^2)$$

Since k, k^2, and $2k^2$ all stand for natural numbers, $(2k)^2$ stands for an even natural number. ■

EXERCISES

1. Show that there is a one-to-one correspondence between the elements of the set of positive integers and the elements of the set of negative integers.
2. Show that the set of whole numbers does not form a group under addition.

Graph each set of numbers on a number line.

3. The set of integers between -3 and 5.
4. The set of integers -3 through 5.
5. The set of integers that are the opposites of the first six natural numbers.
6. The set of integers that are the opposites of the members of the set $M = \{-4, -3, -2\}$.

Classify each statement as true or false. If the statement is false, give a counterexample to justify your answer.

7. Every natural number is an integer.
8. Every whole number is an integer.
9. Every integer is a whole number.
10. Every integer is either positive or negative.
11. Every integer is the opposite of some integer.
12. The set of integers is the same as the set of the opposites of the integers.
13. The set of negative integers is the same as the set of the opposites of the whole numbers.
14. The set of integers is closed under multiplication.
15. The set of integers is closed under division.
16. The set of even integers forms a group under addition.
17. The set of odd integers forms a group under addition.

18. What is the intersection of the set of positive integers and the set of negative integers?
19. Is the union of the set of positive integers and the set of negative integers equal to the set of integers? Explain your answer.
20. Show that there is a one-to-one correspondence between the elements of the set of whole numbers and the elements of the set of integers.
21. Show that there is a one-to-one correspondence between the set of integers and the set of natural numbers.

Prove each statement.

22. The sum of any two odd integers is an even integer.
23. The product of any two even integers is an even integer.
24. The square of any odd integer is an odd integer.

*25. If the square of an integer is odd, the integer is odd; if the square of an integer is even, the integer is even.

EXPLORATIONS

1. Refer to a recently published elementary mathematics textbook series and determine the grade level at which negative integers are first introduced. Also report on the manner in which they are introduced.
2. List as many practical situations as you can think of that might involve the use of negative integers, and that would be meaningful to elementary school students.
3. Prepare a plan for a 20-minute lesson that introduces the concept of negative integers at a specific early elementary grade level. Include the use of at least one visual aid in your lesson.
4. The National Assessment of Educational Progress (NAEP) is funded by the Department of Education and attempts to gather information about the educational attainments of 9-, 13-, and 17-year-olds in various learning areas. In particular, surveys of mathematical achievement were completed in 1973, and 1978, and 1982. Prepare a report that summarizes their findings in 1982 and the changes in achievement between the tests. Comment in detail concerning their findings in the area of problem solving. For information write to NAEP at Suite 700, 1860 Lincoln Street, Denver, Colorado 80295. Also search through the 1980–84 issues of the *Arithmetic Teacher* and the *Mathematics Teacher* for interpretive articles concerning the results of the tests.

The cardinal number of the set W of whole numbers is the same as the cardinal number of the set N of natural numbers since there is a one-to-one correspondence between the elements of the two sets.

$$\begin{array}{llllllll} N = \{1, & 2, & 3, & 4, & \ldots, & n, & \ldots\} \\ \quad\;\updownarrow & \updownarrow & \updownarrow & \updownarrow & & \updownarrow & \\ W = \{0, & 1, & 2, & 3, & \ldots, & n-1, & \ldots\} \end{array}$$

Show that there is a one-to-one correspondence between the set of natural numbers and the given set.

5. $\{5, 6, 7, 8, 9, 10, 11, \ldots\}$
6. $\{5, 10, 15, 20, 25, 30, \ldots\}$
7. The set $\{2, 4, 6, 8, 10, \ldots\}$ of even positive integers.
8. The set $\{1, 3, 5, 7, 9, \ldots\}$ of odd positive integers.

The cardinal number of the set of natural numbers is not an integer but is the smallest of the **transfinite cardinal numbers**, $\aleph_0$, aleph-null. Since the number of whole numbers is one more than the number of natural numbers, the one-to-one correspondence between the set of natural numbers and the set of whole numbers shows that

$$\aleph_0 = \aleph_0 + 1$$

Use a one-to-one correspondence to illustrate each given statement.

9. $\aleph_0 = \aleph_0 + 4$ 10. $\aleph_0 = 5\aleph_0$ 11. $\aleph_0 = \aleph_0 + \aleph_0$

12. Read about the life of Evariste Galois, one of the mathematicians who is given credit for original work on group theory. A fascinating account of his life, and death at age 20, is given in the book by Leopold Infeld, *Whom the Gods Love: The Story of Evariste Galois*, published as a "Classic in Mathematics Education" by the National Council of Teachers of Mathematics.

4-2 Addition and Subtraction of Integers

The number line may be used to illustrate the addition of any two integers. Represent the first addend by an arrow that starts at 0, the origin. From the tip of the first arrow, draw a second arrow to show the second addend. Positive numbers are shown by arrows that go to the right, and negative numbers by arrows that go to the left. Finally, the sum is found as the coordinate of the point at the tip of the second arrow. Examples 1 and 2 illustrate this number-line method for addition of integers.

Example 1 Illustrate $(+3) + (-7)$ on a number line.

Solution Start at 0 and draw an arrow that goes three units to the right. Then show an arrow that goes seven units to the left. As in the figure, the tip of the second arrow is at -4. Thus, $(+3) + (-7) = -4$.

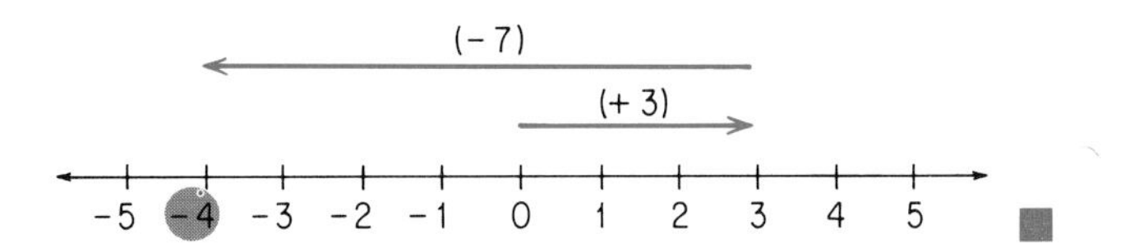

■

Example 2 Illustrate $(-2) + (-3)$ on a number line.

Solution

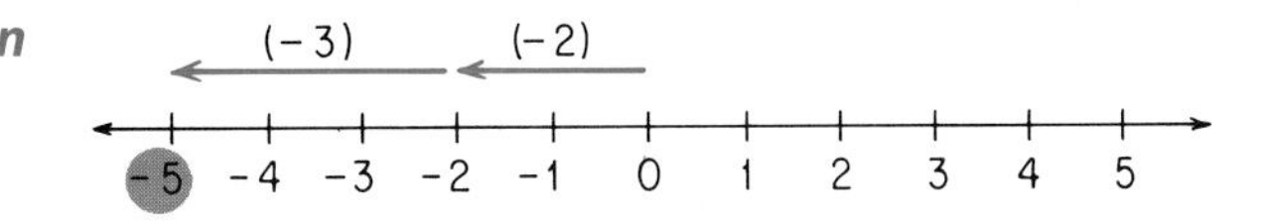

From the figure we see that $(-2) + (-3) = -5$. ■

Example 3 Find the sum $(-3) + (+3)$.

Solution Think of a number line. From the origin, move three units to the left to represent -3. From this point, move three units to the right, back to the origin. Thus, $(-3) + (+3) = 0$. One can also obtain the answer by applying the inverse property for addition. That is, $-(-3) = +3$, and the given sum is 0. ■

Example 4 Find the sum $(+3) + (-5) + (-2)$.

Solution Use the associative property and find the sum of two addends at a time. This grouping is essential because of the binary nature of addition; only two numbers can be added at a time.

This answer also can be obtained from the grouping $(+3) + [(-5) + (-2)]$ since $(+3) + (-7) = -4$.

$$\begin{aligned}(+3) + (-5) + (-2) &= [(+3) + (-5)] + (-2) \\ &= (-2) + (-2) \\ &= -4 \quad \blacksquare\end{aligned}$$

The relationship between addition and subtraction holds for integers as well as whole numbers. Thus any subtraction problem may be replaced by an equivalent addition problem. For example,

$$17 - 9 = n \quad \text{if and only if} \quad n + 9 = 17$$

In general, for any two integers a and b

$$a - b = n \quad \text{if and only if} \quad n + b = a$$

To find n in the equation $n + (+3) = -2$, we need to find the number n that must be added to $+3$ to obtain -2. On a number line we need to determine the number of units and direction to move from $+3$ to -2.

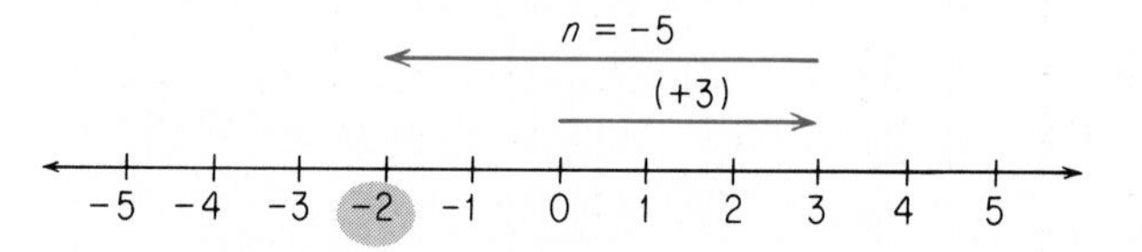

We see that the answer is five units in the negative direction, so that $n = -5$. Therefore, $(+3) + (-5) = -2$, $(-5) + (+3) = -2$, and $(-2) - (+3) = -5$.

Example 5 Find the difference $(-2) - (-8)$.

Solution Let $(-2) - (-8) = n$. Then $n + (-8) = -2$ and $(-8) + n = -2$. Thus we need to find a number that must be added to -8 to obtain -2. On a number line, we must move six units to the right to go from -8 to -2. Therefore $n = 6$, and $(-2) - (-8) = +6$. ■

The solution of Example 5 illustrates the correspondence of the subtraction problem $(-2) - (-8) = +6$ and the addition problem $(-2) + (+8) = +6$. We use this relationship to define the subtraction of integers a and b.

This is the common rule "to subtract a number, change its sign and add."

$$\boxed{a - b = a + (-b)}$$

Example 6 Subtract $(+5) - (-3)$.

Solution By the definition of subtraction, $a - b = a + (-b)$. Therefore

Or subtract: $\begin{array}{r} +5 \\ \underline{-3} \end{array}$

$$\begin{aligned}(+5) - (-3) &= (+5) + [-(-3)] \\ &= (+5) + (+3) = +8 \quad \blacksquare\end{aligned}$$

EXERCISES

Illustrate each sum on a number line.

1. $(+5) + (-7)$
2. $(-3) + (-4)$
3. $(+3) + (+4)$
4. $(-2) + (-5)$

Find each sum.

5. $(+8) + (-12)$
6. $(-3) + (-7)$
7. $(-8) + (+12)$
8. $(-23) + (+23)$
9. $(-15) + (-12)$
10. $(+13) + (-20)$
11. $(-5) + (-3) + (-7)$
12. $(+6) + (-8) + (-7)$
13. $(+12) + (-12) + (-9)$
14. $(-15) + (-7) + (+8)$
15. $(+5) + (-7) + (-6) + (+8)$
16. $(-7) + (-3) + (+12) + (-9)$

Add.

17. $\begin{array}{r} +12 \\ \underline{-15} \end{array}$
18. $\begin{array}{r} -15 \\ \underline{-10} \end{array}$
19. $\begin{array}{r} -13 \\ \underline{+25} \end{array}$
20. $\begin{array}{r} +16 \\ \underline{-24} \end{array}$

Find each difference.

21. $(+5) - (+2)$
22. $(-5) - (-2)$
23. $(+5) - (-2)$
24. $(-5) - (+2)$
25. $(-12) - (+15)$
26. $(+13) - (+7)$
27. $(+15) - (+25)$
28. $(-15) - (-25)$
29. $(+11) - (-5)$
30. $(+11) - (-20)$

Subtract.

31. $\begin{array}{r} -12 \\ \underline{-11} \end{array}$
32. $\begin{array}{r} +7 \\ \underline{-9} \end{array}$
33. $\begin{array}{r} -19 \\ \underline{+12} \end{array}$
34. $\begin{array}{r} +13 \\ \underline{+17} \end{array}$

Perform the indicated operations.

35. $[(-5) + (-3)] - (+3)$
36. $(-5) + [(-3) - (+3)]$
37. $[(-8) - (-2)] + (-5)$
38. $(-8) - [(-2) + (-5)]$
39. $[(+5) + (-8)] - [(-3) + (-7)]$
40. $[(-7) - (-3)] + [(+5) - (-8)]$

EXPLORATIONS

1. Another approach to addition of integers is to use the concept of positive and negative arrows of unit length. For example, consider the following: These two arrows together illustrate $+2$.

   ⟶ ⟶

   These three negative unit arrows together illustrate -3.

   ⟵ ⟵ ⟵

Positive and negative arrows are then combined, as indicated by the circles, to give zero pairs as follows.

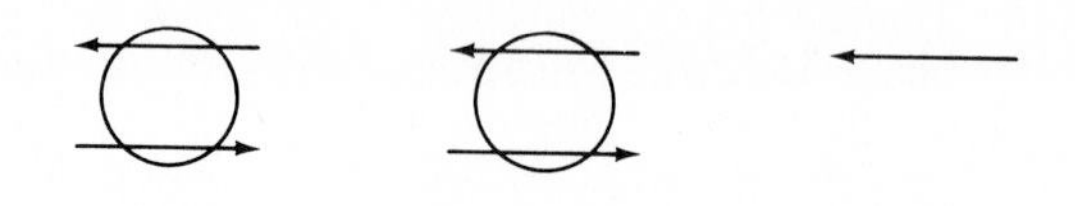

We have one negative arrow left. Thus $(-3) + (+2) = -1$.

Prepare a short unit introducing addition of integers in this way.

A **nomograph** is a device for performing computations in a simple manner. The following nomograph can be used to find the sums of integers. Just connect the point representing one addend on the A scale, with the corresponding point for the other addend on the B scale. The point where the line crosses the S scale will give the sum. The following figure shows the sum $(+4) + (-6) = -2$.

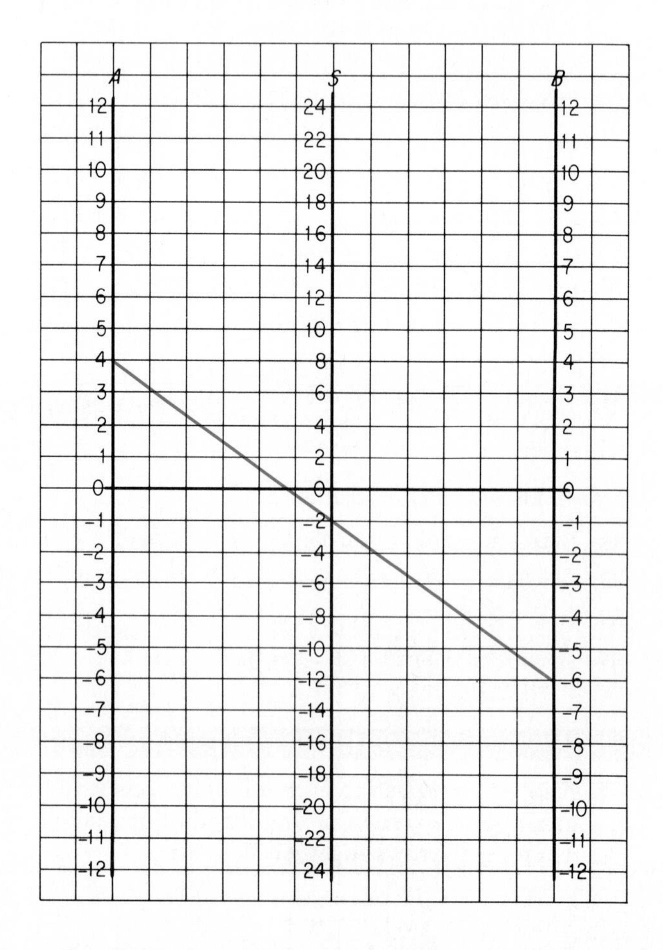

2. Construct your own nomograph on graph paper and use it to find various sums of integers.

3. See if you can use your knowledge of elementary geometry to explain why the nomograph shown works as it does.
4. Try to determine how the nomograph shown can be used for subtraction of integers.
5. Prepare a nomograph that will enable you to find $a + 2b$ directly for any two integers a and b. If possible, prepare a suitable transparency and demonstrate this nomograph on an overhead projector.

4-3 Multiplication and Division of Integers

There are a variety of ways in which we can consider the rules for multiplication of integers. Applications of numbers to physical situations can be helpful. Consider, for example, the situation of earning \$5 a day for two days. This can be expressed as the product of two positive integers, $(+2) \times (+5)$. We may think of multiplication by a positive integer as repeated addition.

$$(+2) \times (+5) = (+5) + (+5) = +10$$

Thus the result is a gain of \$10, a positive number. In general, the product of two positive integers is a positive integer.

To evaluate the product of a positive integer and a negative integer, consider the situation of losing \$5 each day for two days. This can be expressed as the product $(+2) \times (-5)$.

$$(+2) \times (-5) = (-5) + (-5) = -10$$

For any positive integer b, $(-1) \times b = -b$.

Thus the result is a loss of \$10, a negative number. In general, the product of a positive integer and a negative integer is a negative integer. Then, since multiplication is commutative, the product of a negative integer and a positive integer is also a negative integer. For example,

$$(-5) \times (+2) = (+2) \times (-5) = -10$$

Justifying the product of two negative numbers is somewhat more difficult. Informally, we might say that if you lose \$5 a day for two days, then two days *ago* you had \$10 more than you have today. Therefore $(-2) \times (-5) = +10$. However, such arguments are often unconvincing. Another approach is to consider patterns in a multiplication table such as the following. The entries in the lower right-hand corner represent the products of positive integers.

X	−3	−2	−1	0	+1	+2	+3
−3				0			
−2				0			
−1				0			
0	0	0	0	0	0	0	0
+1				0	+1	+2	+3
+2				0	+2	+4	+6
+3				0	+3	+6	+9

Next, note the pattern of entries as read from right to left, or from the bottom up. In the +3 row and column the entries are decreasing by 3's; in the +2 row and column, they are decreasing by 2's; and in the +1 row and column they are decreasing by 1's. Use these patterns to complete two more portions of the table, and note that the resulting table shows the product of a positive integer and a negative integer to be a negative integer, as shown in the next figure.

X	−3	−2	−1	0	+1	+2	+3
−3				0	−3	−6	−9
−2				0	−2	−4	−6
−1				0	−1	−2	−3
0	0	0	0	0	0	0	0
+1	−3	−2	−1	0	+1	+2	+3
+2	−6	−4	−2	0	+2	+4	+6
+3	−9	−6	−3	0	+3	+6	+9

To complete the table, note the pattern of entries once again. For example, consider the first row and read the entries from right to left.

— — — 0 −3 −6 −9

From right to left, the entries are increasing by 3's. Also, in the first column, the entries from the bottom up are increasing by 3's. Similar patterns for the other rows and columns allow us to complete the multiplication table as shown next. Note that the patterns show the product of two negative integers to be a positive integer.

X	−3	−2	−1	0	+1	+2	+3
−3	+9	+6	+3	0	−3	−6	−9
−2	+6	+4	+2	0	−2	−4	−6
−1	+3	+2	+1	0	−1	−2	−3
0	0	0	0	0	0	0	0
+1	−3	−2	−1	0	+1	+2	+3
+2	−6	−4	−2	0	+2	+4	+6
+3	−9	−6	−3	0	+3	+6	+9

Example 1 Find each product

(a) $(+4) \times (+7)$ **(b)** $(-3) \times (+6)$

(c) $(+5) \times (-8)$ **(d)** $(-7) \times (-8)$

Solution (a) The product of two positive integers is a positive integer; $(+4) \times (+7) = +28$.

(b) The product of a negative integer and a positive integer is a negative integer; $(-3) \times (+6) = -18$.

(c) The product of a positive integer and a negative integer is a negative integer; $(+5) \times (-8) = -40$.

(d) The product of two negative integers is a positive integer; $(-7) \times (-8) = +56$. ■

Example 2 Find the product $(-5) \times (-7) \times (-2)$.

Solution Use the associative property and then find the product of two factors at a time.

This answer also can be obtained from the grouping $(-5) \times [(-7) \times (-2)]$ since $(-5) \times (+14) = -70$.

$$\begin{aligned}(-5) \times (-7) \times (-2) &= [(-5) \times (-7)] \times (-2) \\ &= (+35) \times (-2) \\ &= -70 \quad ■\end{aligned}$$

Let us consider a mathematical justification for the product of two negative integers. We use the convention

$$(-2)(+5) = (-2) \times (+5) \qquad ab = a \times b$$

omitting the multiplication symbol for convenience and begin by exploring a specific example.

$$(-2)[(+5) + (-5)]$$

Because we wish to preserve the distributive property for the set of integers, we agree that there are two ways to obtain an answer to this problem. If we add first, within the brackets, and then multiply, we have

$$\begin{aligned}(-2)[(+5) + (-5)] &= (-2)(0) \\ &= 0\end{aligned}$$

Now we shall use the distributive property: multiply first, and then add. The final result must be the same, namely 0.

$$\begin{aligned}(-2)[(+5) + (-5)] &= [(-2)(+5)] + [(-2)(-5)] \\ &= (-10) + (?)\end{aligned}$$

The rules for multiplication of integers can be confirmed by mathematical procedures rather than by the somewhat intuitive approach we have just been using. To preserve the distributive property, the product of two negative integers must be a positive integer.

We have been forced to use a question mark for the product $(-2) \times (-5)$ because this is precisely the product that we are seeking. We do know that the final sum must be 0 if the distributive property is to hold. Furthermore, we know that $(-10) + (+10) = 0$. Therefore we conclude that $(-2) \times (-5)$ must be equal to $+10$; the product of two negative integers is a positive integer.

We can generalize this approach by considering any positive integers c and b, and the expression $(-c) \times [(b) + (-b)]$. We then evaluate this product in two different ways.

1. $$\begin{aligned}(-c) \times [(b) + (-b)] &= (-c) \times 0 \\ &= 0\end{aligned}$$

2. $$\begin{aligned}(-c) \times [(b) + (-b)] &= [(-c) \times (b)] + [(-c) \times (-b)] \\ &= -cb + (?)\end{aligned}$$

Since the final result must be 0, the question mark represents the opposite of $-cb$. Thus $(-c) \times (-b)$ must be the opposite of $-cb$ and $(-c) \times (-b) = cb$; the product of any two negative integers is a positive integer. Note that the relationship between opposites and multiplication by -1 has now been established.

For any integer b, $(-1) \times b$ is the opposite of b:

$(-1) \times 3 = -3$

$(-1) \times (-5) = 5$

$$\boxed{(-1) \times b = -b \qquad (-1) \times (-b) = b}$$

Any sum or product of integers is an integer since the set of integers is closed under addition and also under multiplication. Any difference of integers is an integer since the difference is equivalent to a sum. However, a quotient of integers $a \div b$ is an integer if and only if there is some integer n such that $b \times n = a$. For example, there is no integer n such that $3 \times n = 8$ and the quotient $8 \div 3$ cannot be named by an integer.

The set of integers is not closed under division.

The relationship between multiplication and division

$$a \div b = c \quad \text{if and only if} \quad a = bc \quad (b \neq 0)$$

provides us with rules for signs of quotients of integers. The quotient of a positive integer and a negative integer is negative. The quotient of two positive integers is positive. The quotient of two negative integers is positive.

The relationship between opposites and numbers expressed with + or − signs may be summarized as follows for any integers a and b:

$$-(a) = -a$$

$$-(-a) = a$$

$$a + (-b) = a - b$$

$$a - (-b) = a + b$$

$$a \times (-b) = -(ab)$$

$$(-a) \times (-b) = ab$$

Example 3 Find the quotient $(-12) \div (-3)$.

Solution Let $(-12) \div (-3) = n$. Then $(-3) \times n = (-12)$, and $n = +4$. Thus the quotient of these two negative integers is a positive integer. ■

In the division of rational numbers we think of b as $b/1$ and define the quotient $a \div b$ of two integers a and b, $b \neq 0$, as a product.

$$\boxed{a \div b = a \times \frac{1}{b}}$$

Then a quotient such as $(-12) \div (+3)$ can be found as follows.

$$(-12) \div (+3) = (-12) \times \left(+\frac{1}{3}\right) = -4$$

EXERCISES *Find each product.*

1. $(-5) \times (+9)$
2. $(-5) \times (-9)$
3. $(+5) \times (-9)$
4. $(+5) \times (+9)$
5. $(-8) \times (+12)$
6. $(+12) \times (-12)$
7. $(-25) \times (-25)$
8. $(-15) \times (+15)$
9. $(+10) \times (-17)$
10. $(-11) \times (-11)$
11. $(-3) \times (-7) \times (-5)$
12. $(-3) \times (+7) \times (-7)$
13. $(-8) \times (-17) \times 0$
14. $(-2) \times (-2) \times (-2)$
15. $(-1) \times (-2) \times (-3) \times (-4)$
16. $(+3) \times (-3) \times (-5) \times (-5)$

Find each quotient.

17. $(-24) \div (+3)$
18. $(+24) \div (+3)$
19. $(+24) \div (-3)$
20. $(-24) \div (-3)$
21. $(-36) \div (-18)$
22. $(+72) \div (-12)$
23. $(-144) \div (-12)$
24. $(+125) \div (-25)$
25. $(-100) \div (+100)$
26. $0 \div (-5)$
27. $[(-60) \div (-3)] \div (-4)$
28. $(-100) \div [(-50) \div (+10)]$
29. $[(-48) \div (+6)] \div (-1)$
30. $(+144) \div [(+24) \div (-3)]$
31. $[(+24) \div (-2)] \div [(-18) \div (-3)]$
32. $[(-150) \div (+5)] \div [(+225) \div (-15)]$

Perform the indicated operations.

33. $(-8) + [(-5) + (-7)]$
34. $[(-8) + (-5)] + (-7)$
35. $(-2) \times [4 \times (-5)]$
36. $[(-2) \times 4] \times (-5)$
37. $12 \div [6 \div (-2)]$
38. $(12 \div 6) \div (-2)$
39. Use the results of Exercises 37 and 38 to form a conjecture about the associativity of integers with respect to division. Then find another example to help confirm your conjecture.
40. Use a counterexample to show that division of integers is not a commutative operation.

EXPLORATIONS

1. Many elementary textbooks use raised negative signs to introduce the set Z^- of negative integers.

$$Z^- = \{\ldots, {}^-5, {}^-4, {}^-3, {}^-2, {}^-1\}$$

Discuss the pedagogical value of introducing the set in this manner, thus reserving the minus sign, $-$, for subtraction. Refer to a first-year algebra text to see how the transition is made from raised negative signs to lowered ones.

2. *Function machines* can be used to motivate practice with integers, as shown in the following illustrative example. The student is to use the rule

shown on the function machine to alter the input to the machine. Consider the example of introducing a $+2$ into a "$\times(-2)$" machine. The output is $(+2) \times (-2)$, that is, -4. The student is to write the correct output on the line with each input in the table.

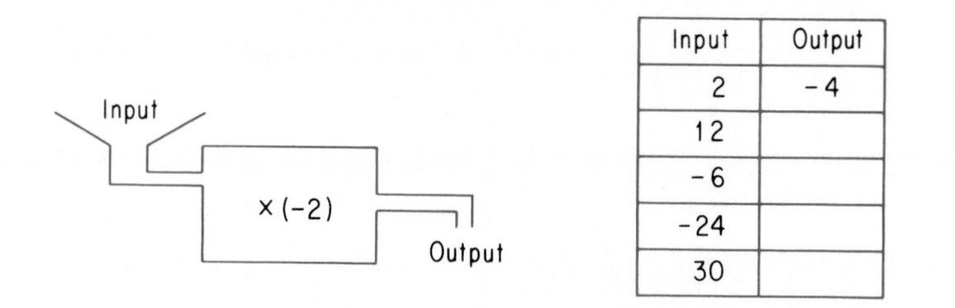

Input	Output
2	-4
12	
-6	
-24	
30	

When two machines are used sequentially, the output from the first is used as the input for the second.

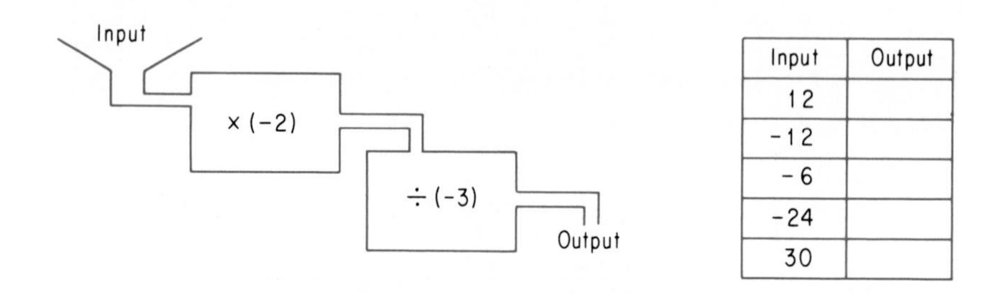

Input	Output
12	
-12	
-6	
-24	
30	

Use function machines and make a worksheet of practice problems on computations with integers.

3. The product of a positive integer and a negative integer can be shown on a number line as repeated addition. The figure that follows shows the product $5 \times (-2)$ as five "steps" of two units each, in the negative direction from the origin. That is, $5 \times (-2) = -10$.

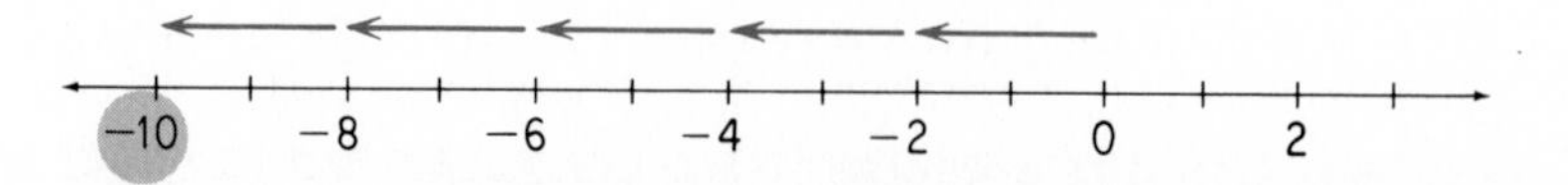

Prepare a plan for a 15- to 20-minute lesson that uses this number-line approach to show such products for a class at a specified elementary grade level. If possible, include the use of the overhead projector as an aid to your lesson.

4-4 Clock Arithmetic

A **mathematical system** includes a set of elements (such as the set of natural numbers), one or more operations (such as addition, subtraction, multiplication, and division), one or more relations (such as equality), and some axioms (rules) which the elements, operations, and relations satisfy. For example, we assume that every system includes the relation of equality and the axiom that $a = a$; that is, any quantity is equal to itself. Actually, you have been working with mathematical systems ever since you started school. We now study a specific mathematical system, and later will relate the properties of this system to those of familiar sets of numbers.

If it is now 9 P.M. as you begin to read this section, what time will it be in 5 hours? (We hope it won't take you that long to complete your reading!) Do you see that the statement

$$9 + 5 = 2$$

is a correct statement if we are talking about positions on a 12-hour clock? It represents one of the fundamental addition facts of **12-hour clock arithmetic**.

Consider the numerals 1 through 12 on a 12-hour clock as the elements of a set T and consider addition on this clock to be based upon counting in a clockwise direction. Thus to find the sum $9 + 5$ we start at 9 and count 5 units in a clockwise direction to obtain the result 2.

Verify that each of the following statements is correct.

$8 + 7 = 3$ (on a 12-hour clock)

$5 + 12 = 5$ (on a 12-hour clock)

$3 + 11 = 2$ (on a 12-hour clock)

Here is a table of addition facts on a 12-hour clock.

+	1	2	3	4	5	6	7	8	9	10	11	12
1	2	3	4	5	6	7	8	9	10	11	12	1
2	3	4	5	6	7	8	9	10	11	12	1	2
3	4	5	6	7	8	9	10	11	12	1	2	3
4	5	6	7	8	9	10	11	12	1	2	3	4
5	6	7	8	9	10	11	12	1	2	3	4	5
6	7	8	9	10	11	12	1	2	3	4	5	6
7	8	9	10	11	12	1	2	3	4	5	6	7
8	9	10	11	12	1	2	3	4	5	6	7	8
9	10	11	12	1	2	3	4	5	6	7	8	9
10	11	12	1	2	3	4	5	6	7	8	9	10
11	12	1	2	3	4	5	6	7	8	9	10	11
12	1	2	3	4	5	6	7	8	9	10	11	12

Note that regardless of where we start on the clock, we shall always be at the same place 12 hours later. Thus for any element t of set T we have

$$t + 12 = t \qquad \text{(on a 12-hour clock)}$$

Let us attempt to define several other operations for this arithmetic on the 12-hour clock. What does multiplication mean? Multiplication by a natural number may be considered as repeated addition. For example, 3×5 on the 12-hour clock is equivalent to $5 + 5 + 5$. Since $5 + 5 = 10$ and $10 + 5 = 3$, we know that $3 \times 5 = 3$ on a 12-hour clock.

Verify that each of the following statements is correct.

$$4 \times 5 = 8 \qquad \text{(on a 12-hour clock)}$$

$$3 \times 9 = 3 \qquad \text{(on a 12-hour clock)}$$

$$3 \times 7 = 9 \qquad \text{(on a 12-hour clock)}$$

Example 1 Solve the equation $t + 6 = 2$ for t, where t may be replaced by any one of the numerals on a 12-hour clock.

Solution Observe from row 6 of the table of addition facts that on a 12-hour clock $8 + 6 = 2$; therefore $t = 8$. Note that we may also solve for t by starting at 6 on the clock and then counting hours in a clockwise direction until we reach 2. The number of hours counted is the desired value of t since if $6 + t = 2$, then $t + 6 = 2$. ■

Example 2 Using the numerals on a 12-hour clock, find a replacement for t such that $9 \div 7 = t$.

Solution The statement "$9 \div 7 = t$" can be expressed as "$9/7 = t$" and thus as one of these four equivalent statements

$$9 = 7 \times t \qquad 9 = t \times 7 \qquad \frac{9}{7} = t \qquad \frac{9}{t} = 7$$

Then we use the sentence $9 = 7 \times t$ to solve the problem. How many groups of 7 must be added to obtain 9 on a 12-hour clock? We can count by 7's clockwise from 12 as many times as necessary to obtain 9.

$$7 \times 1 = 7$$

$$7 \times 2 = 14 = 2 \qquad \text{(on a 12-hour clock)}$$

$$7 \times 3 = 21 = 9 \qquad \text{(on a 12-hour clock)}$$

We can also use row 7 on a table of multiplication facts for a 12-hour clock (see Exercise 1), or trial and error, to find that $7 \times 3 = 9$ (on a 12-hour clock); thus $t = 3$. ■

Example 3 What is $3 - 7$ on a 12-hour clock?

Solution Let t represent a numeral on a 12-hour clock. The statement $3 - 7 = t$ is equivalent to $t + 7 = 3$. From the table of addition facts, $8 + 7 = 3$ (on a 12-hour clock); thus $t = 8$. ■

From a slightly different point of view, we may solve Example 3 by counting in a clockwise direction from 12 to 3, and then counting 7 units in a counterclockwise direction. We complete this process at 8. Thus $3 - 7 = 8$ (on a 12-hour clock).

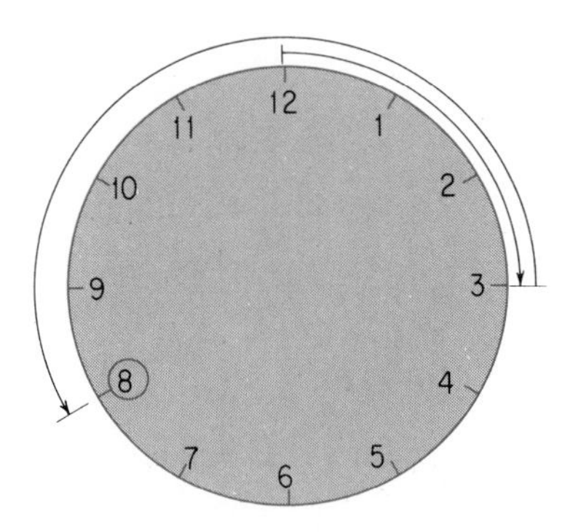

Note that clock arithmetic may be considered as an arithmetic of remainders. Multiples of 12 are discarded so that each answer may be expressed as a number 1 through 12. For example, consider the product $5 \times 8 = 40$.

$$5 \times 8 = 40 = (3 \times 12) + 4$$

In clock arithmetic (3×12) represents three rotations and is disregarded; this is analogous to dividing 40 by 12 and considering the remainder only.

$$5 \times 8 = 4 \qquad \text{(on a 12-hour clock)}$$

EXERCISES

1. Make a table of the multiplication facts for the numbers on a 12-hour clock.
2. Use the results of Exercise 1 to list as many special patterns as you can discover about the multiplication on a 12-hour clock.

Solve each problem as on a 12-hour clock.

3. $9 + 8$	**4.** $7 + 11$	**5.** $7 - 12$
6. $6 - 9$	**7.** 4×9	**8.** 7×7
9. $11 \div 5$	**10.** $1 \div 7$	**11.** 8×8
12. $6 + 10$	**13.** 3×8	**14.** 9×9
15. $2 - 10$	**16.** $5 - 11$	**17.** $4 \div 7$
18. $4 \div 5$	**19.** $12 \div 5$	***20.** $6 \div 10$

For each sentence find all possible replacements for t for which the sentence is a true statement for the numerals on a 12-hour clock.

21. $t + 8 = 5$ **22.** $t - 5 = 11$ **23.** $8 + t = 3$

24. $4 - t = 10$ **25.** $3 \times t = 3$ **26.** $7 \times t = 11$

27. $t \div 5 = 7$ **28.** $t \div 7 = 8$ ***29.** $2 \div t = 3$

***30.** $2 + t = 2 - t$ ***31.** $t + 12 = t$ ***32.** $3 - t = 5 + t$

EXPLORATIONS

1. Elementary school students are often taught to consider number facts by families of related facts. For example, here is a family of related number facts.

$$5 - 2 = 3 \qquad 5 = 2 + 3 \qquad 5 = 3 + 2 \qquad 5 - 3 = 2$$

Here is another family of related number facts.

$$6 \div 2 = 3 \qquad 6 = 2 \times 3 \qquad 6 = 3 \times 2 \qquad 6 \div 3 = 2$$

Any two members of a family of related number facts are equivalent statements. Any statement may be used to "solve" or to "check" another statement in the family.

Use families of number facts to explain your answers to Exercises 9 and 15 of this section.

2. Prepare a plan for a 20-minute lesson that introduces a unit on clock arithmetic at a specified elementary grade level. Make use of at least one visual aid in your lesson. If possible, present this lesson to a class and report on your results.

4-5 Modular Arithmetic

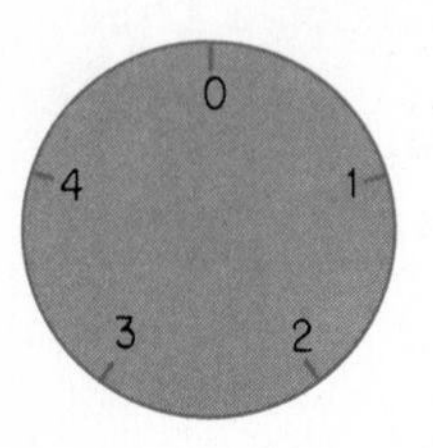

Let us next consider a mathematical system based on a clock with five elements numbered 0, 1, 2, 3, and 4, as in the accompanying figure. Addition on this clock may be performed by counting as on an ordinary clock. However, it seems easier to think of addition as rotations of a hand of a clock in a clockwise direction. Thus $3 + 4$ indicates that the hand starts at 0, moves 3 units, then moves 4 more units. The hand stops at 2; we say that $3 + 4 = 2$ on a 5-hour clock.

We interpret 0 to mean no rotation as well as to designate a position on the clock. Thus we have such facts as these:

$3 + 0 = 3$ (on a 5-hour clock)

$4 + 1 = 0$ (on a 5-hour clock)

Regardless of where we start on a 5-hour clock, we shall always be at the same place 5 hours later.

Multiplication on a 5-hour clock may be considered as repeated addition. For example, 3×4 on a 5-hour clock is equivalent to the sum $4 + 4 + 4$. Since $4 + 4 = 3$, and $3 + 4 = 2$, we see that $3 \times 4 = 2$ on a 5-hour clock.

Verify that each of the entries in these addition and multiplication tables on a 5-hour clock is correct.

+	0	1	2	3	4
0	0	1	2	3	4
1	1	2	3	4	0
2	2	3	4	0	1
3	3	4	0	1	2
4	4	0	1	2	3

×	0	1	2	3	4
0	0	0	0	0	0
1	0	1	2	3	4
2	0	2	4	1	3
3	0	3	1	4	2
4	0	4	3	2	1

We define subtraction and division using equivalent statements. Since the statements $a - b = x$ and $a = b + x$ are equivalent, we define *subtraction* by the statement

$$a - b = x \quad \textit{if and only if} \quad b + x = a$$

Since the statements $a \div b = x$ and $a = b \times x$ are equivalent, we define *division* by the statement

$$a \div b = x \quad \textit{if and only if} \quad b \times x = a$$

The arithmetic on a 5-hour clock is a **modular arithmetic**, in particular, arithmetic **modulo 5**. If any whole number is divided by 5, then the remainder must be 0, 1, 2, 3, or 4. Thus each whole number is congruent modulo 5 to exactly one element of the set F, where $F = \{0, 1, 2, 3, 4\}$. The elements of F are the elements of arithmetic modulo 5.

In general, two numbers are **congruent modulo 5** if and only if they differ by a multiple of 5. Thus 3, 8, 13, and 18 are all congruent to each other modulo 5. We may write, for example,

$18 \equiv 13 \pmod{5}$, read "18 is congruent to 13, modulo 5"

$8 \equiv 3 \pmod{5}$, read "8 is congruent to 3, modulo 5"

Example 1 Solve in arithmetic modulo 5. **(a)** $2 - 3$ **(b)** $2 \div 4$

Solution **(a)** $2 - 3 \equiv x \pmod{5}$ where $2 \equiv 3 + x \pmod{5}$. We use the addition table and in the row for 3 find 2 in the column headed by 4. Therefore we have $3 + 4 \equiv 2 \pmod{5}$ and $2 - 3 \equiv 4 \pmod{5}$.

(b) $2 \div 4 \equiv x \pmod{5}$ where $2 \equiv 4 \times x \pmod{5}$. We use the multiplication table and in the row for 4 find 2 in the column headed by 3. Therefore we have $4 \times 3 \equiv 2 \pmod{5}$ and $2 \div 4 \equiv 3 \pmod{5}$. ■

Example 2 Solve in arithmetic modulo 5. **(a)** $x + 2 \equiv 1 \pmod{5}$ **(b)** $2 - 3 \equiv x \pmod{5}$ **(c)** $4 \div 3 \equiv x \pmod{5}$

Solution (a) In the addition table in the row for 2 we find 1 in the column headed by 4. Therefore, $2 + 4 \equiv 1 \pmod{5}$, $4 + 2 \equiv 1 \pmod{5}$, and $x \equiv 4 \pmod{5}$.

(b) $2 - 3 \equiv x \pmod{5}$ if and only if $2 \equiv 3 + x \pmod{5}$. In the addition table in the row for 3 we find 2 in the column headed by 4. Therefore $3 + 4 \equiv 2 \pmod{5}$ and $x \equiv 4 \pmod{5}$.

(c) $4 \div 3 \equiv x \pmod{5}$ if and only if $4 \equiv 3 \times x \pmod{5}$. In the multiplication table in the row for 3 we find 4 in the column headed by 3. Therefore, $3 \times 3 \equiv 4 \pmod{5}$ and $x \equiv 3 \pmod{5}$. ■

The arithmetic on a 12-hour clock is a *modular arithmetic* with *modulus 12*. We may use the arithmetic on a 12-hour clock as a model for arithmetic modulo 12 with 0 represented by 12. In general, two integers are **congruent modulo** m if the integers differ by a multiple of m.

Example 3 Suppose that you leave at 9 A.M. on a 74-hour business flight around the world. What time will it be when you return?

Solution Since time is measured in 24-hour days, we think in terms of arithmetic modulo 24.

$$74 = 24 \times 3 + 2 \equiv 2 \pmod{24}$$

The time of your return will be two hours later than the time of your departure, that is, 11 A.M. ■

If New Year's Day is on Sunday this year, what day of the week is July 4, the 185th day of the year? The days of the week may be considered in arithmetic modulo 7.

$$185 = 7 \times 26 + 3 \qquad 185 \equiv 3 \pmod{7}$$

If the first day is Sunday, then the third day (and the 185th day) is Tuesday. The fact that Tuesday, July 4, is 26 weeks later is not significant. Only the day of the week was requested; only arithmetic modulo 7 is needed.

The theory of modular congruences was developed by Carl Friedrich Gauss. This theory was included in *Disquisitiones Arithmeticae*, which Gauss completed in 1801 at the age of 24.

The arithmetic of congruences is very similar to our usual arithmetic. Note that if $a \equiv b \pmod{m}$ and $c \equiv d \pmod{m}$, then for any integer k

$$a + c \equiv b + d \pmod{m}$$

$$ak \equiv bk \pmod{m}$$

$$ac \equiv bd \pmod{m}$$

We assume these properties and leave their proofs for more advanced courses.

EXERCISES *Solve in arithmetic modulo* 12.

1. $9 + 8$ **2.** $7 + 11$ **3.** $6 + 10$

4. 4×9 **5.** 3×8 **6.** 9×9

Solve in arithmetic modulo 5.

7. $3 + 4$	8. $2 + 3$	9. $4 + 4$	10. 2×3
11. 2×4	12. 4×3	13. 3×3	14. 4×4
15. $2 - 4$	16. $1 - 2$	17. $1 - 4$	18. $1 - 3$
19. $3 - 4$	20. $2 \div 3$	21. $3 \div 2$	22. $3 \div 4$
23. $4 \div 3$	24. $1 \div 3$		

In Exercises 25 through 42 solve in the indicated modular arithmetic.

25. $3 + x \equiv 1 \pmod 5$	26. $x + 4 \equiv 1 \pmod 5$
27. $4 \times x \equiv 2 \pmod 5$	28. $3 \times x \equiv 2 \pmod 5$
29. $1 \div 3 \equiv x \pmod 5$	30. $2 \div 3 \equiv x \pmod 5$
31. $x + 5 \equiv 0 \pmod 7$	32. $x - 3 \equiv 2 \pmod 4$
33. $3x \equiv 1 \pmod 7$	34. $x \times x \equiv 1 \pmod 8$
35. $x \div 4 \equiv 3 \pmod 9$	36. $2 \div x \equiv 3 \pmod 7$
37. $1 - x \equiv 4 \pmod 6$	38. $4 + x \equiv 1 \pmod 7$
39. $x + 5 \equiv 1 \pmod 8$	40. $2 - x \equiv 3 \pmod 6$
41. $2x \equiv 3 \pmod 6$	42. $3 \div x \equiv 3 \pmod 9$

43. Make a multiplication table for arithmetic modulo 12.
44. Use the multiplication table from Exercise 43 and solve in arithmetic modulo 12.

(a) $11 \div 5$	**(b)** $1 \div 7$	**(c)** $4 \div 5$
(d) $7 \div 11$	**(e)** $6 \div 7$	**(f)** $6 \div 11$

45. We describe the property illustrated by the equation $3 \times 4 \equiv 0$ in arithmetic modulo 12, where the product is zero but neither number is zero, by saying that 3 and 4 are **zero divisors**. Are there other zero divisors in arithmetic modulo 12? If so, list them.
46. Consider a system modulo 7 where the days of the week correspond to numbers as follows: Monday—0; Tuesday—1; Wednesday—2; Thursday—3; Friday—4; Saturday—5; Sunday—6. Memorial Day, May 30, is the 150th day of a certain year and falls on a Thursday. In that same year, on what day of the week does July 4, the 185th day of the year, fall? On what day does Christmas, the 359th day of the year, fall?

EXPLORATIONS

1. Note $10 \equiv 1 \pmod 9$ and $10^n \equiv 1 \pmod 9$. Show that any integer N expressed in decimal notation is congruent modulo 9 to the sum of its digits. Restate this fact as a rule for divisibility by 9.
2. Use $10^n \equiv 1 \pmod 9$ as in Exploration 1 and develop a check modulo 9 for addition, subtraction, multiplication, and division of integers. Look up the method of *casting out nines* and compare your procedure with this method. Note that since, for example, $1984 \equiv 1894 \pmod 9$, a check modulo 9 is not a "complete" check (see, for example, Section 1-2, Exercise 5).

3. Note that $10 \equiv -1 \pmod{11}$ and $10^n \equiv (-1)^n \pmod{11}$. As in Exploration 1 develop a check modulo 11 for addition, subtraction, and multiplication of integers. Note that since, for example $1984 \equiv 1489 \pmod 9$ and $1984 \equiv 1489 \pmod{11}$, answers that check both modulo 9 and modulo 11 may still not be correct.
4. Make a multiplication table for arithmetic modulo 4. One of the following equations has the empty set as its solution set on a 4-hour clock.

$$t = 3 \qquad 2t = 3 \qquad 3t = 3$$

Identify the given impossible equation, and give another impossible equation in the arithmetic on a 4-hour clock.
5. Give at least four impossible equations in the arithmetic on a 12-hour clock.
6. Explore the possible types of impossible equations in the arithmetic on a 12-hour clock.
7. Use congruences as in Exploration 1 and develop rules for the divisibility by 2, 3, 4, 5, and 8 of any natural number N in terms of its decimal digits.

Chapter Review

1. Graph the set of integers between -2 and 3.
2. State five properties of the set of integers under addition.

Classify each statement as true or false.

3. **(a)** The set of whole numbers is a subset of the set of integers.
   (b) The opposite of each negative integer is a whole number.
4. **(a)** The opposite of each whole number is a negative integer.
   (b) The opposite of each integer is an integer.

Illustrate each sum on a number line.

5. $(-3) + (+7)$
6. $(-2) + (-4)$

In Exercises 7 *through* 18 *perform the indicated operations.*

7. $(-20) + (+3)$
8. $(+7) + (-18)$
9. $(-3) - (-5)$
10. $(-2) - (+8)$
11. $(-5) \times (+8)$
12. $(-9) \times (-8)$
13. $(+36) \div (-4)$
14. $(-42) \div (-7)$
15. $(-3) + [(+2) + (-7)]$
16. $[(-9) - (-2)] - (+3)$
17. $(-5) \times [(-2) \times (-1)]$
18. $[(-40) \div (+5)] \div (-2)$

19. Show that the set of integers is not closed with respect to the operation of division.
20. Solve either as on a 12-hour clock or in arithmetic modulo 12.
    (a) $5 - 7$ **(b)** 5×7

In Exercises 21 and 22 solve in the indicated modular arithmetic.

21. $5 + x \equiv 3 \pmod{7}$

22. $4 \times x \equiv 3 \pmod{5}$

23. Solve in arithmetic modulo 12.
(**a**) $8 + 7$ (**b**) $3 - 7$ (**c**) 3×10

24. Solve in arithmetic modulo 5.
(**a**) $2 - 4$ (**b**) $3 \div 4$ (**c**) $2 \div 3 \equiv x \pmod{5}$

25. If New Year's Day is on Tuesday this year, what day of the week is July 4, the 185th day of the year?

ESTIMATION AND LONG DIVISION ON THE COMPUTER

One of the skills that is essential for students learning to use various forms of computational technology is that of estimation. Although they may not be actually carrying out the computations themselves, they need to develop the skills which will allow them to make reasonable estimates of the answers. This will allow them to determine if the answers given are reasonable. The following program provides a game to strengthen a student's estimation skills in dealing with expected quotients in long division problems.

In the program, the command of HOME clears the screen and allows printing to begin in the upper left corner of the screen. The command PRINT standing in a line by itself instructs the computer to leave the line empty. For this program the output will be double spaced to make reading the sentences on the screen easier. Type in the following program and then type RUN to use it.

```
5  HOME
1Ø PRINT "THIS PROGRAM PRESENTS A GAME TO TEST"
11 PRINT
12 PRINT "YOUR ABILITY TO ESTIMATE THE QUOTIENT"
13 PRINT
14 PRINT "IN A NATURAL NUMBER DIVISION PROBLEM."
15 PRINT
16 PRINT "THE COMPUTER WILL PRESENT YOU WITH A"
17 PRINT
18 PRINT "DIVISION PROBLEM FOR WHICH YOU MUST"
19 PRINT
2Ø PRINT "TYPE IN YOUR ESTIMATE OF THE QUOTIENT."
21 PRINT
22 PRINT "THE COMPUTER WILL THEN TEST TO SEE IF"
23 PRINT
24 PRINT "YOUR MARGIN OF ERROR IS LESS THAN 5%."
25 PRINT
3Ø N = INT(1ØØØ * (RND(1)))
35 D = INT(1ØØ * (RND(1)))
```

```
40  IF D> =N THEN 35
45  PRINT "WHAT IS THE QUOTIENT IN ";N;"/";D;"?"
46  INPUT Q
50  C=INT(N/D)
60  IF (0.95) * C< =Q AND Q< =1.05 * C
       THEN PRINT "YOU WIN!"
70   IF Q<(0.95) * C OR Q>1.05 * C THEN
         PRINT "YOU LOSE, THE CORRECT QUOTIENT
         WAS ";C;"."
80  PRINT
81  PRINT "IF YOU WANT TO TRY AGAIN, TYPE IN 'Y'."
82  PRINT "IF YOU DO NOT WANT TO TRY AGAIN, TYPE IN 'N'."
83  INPUT T$
84  IF T$="Y" THEN 30
85  END
```

This program makes use of three new BASIC concepts. The first of these is the function RND in line 30. This is a command to the computer to select a random decimal number between 0 and 1. This decimal is then multiplied by 1000 in line 30 to form the dividend in the decimal division problem. A second decimal number, between 0 and 1, is generated in line 35 and multiplied by 100 to get the divisor for the division problem.

The IF-THEN form used in line 40 asks the computer to make a decision in the execution of the program. In this case, the command asks the computer to determine if the divisor is greater than or equal to the dividend. If so, the command directs the computer to return to line 35 and generate a new value for the divisor, as only divisors less than the dividend are desired in this particular program. Here is an instance in which the actual execution of the lines of the program, by line number, might be altered by the program itself!

If the divisor is less than the dividend, the computer proceeds to the next line, line 45. The fraction N/D is now determined and you are asked on line 46 to enter your estimate Q of the largest integer C that is less than or equal to N/D. Then on line 50, the computer determines the value of C and on line 60 determines whether your estimate is within 5 percent of the correct answer C.

The third new concept is found in lines 82 and 83. Here we see the use of the symbol T$. This is an instance of a string variable. Here the computer is instructed to look for a variable value which comes from the alphabet, not from the set of digits. The computer will expect that the value of a variable will be numbers unless it is told differently. The addition of the dollar sign on the end of a variable name is the signal that a letter of the alphabet may be expected.

The given examples show the use of this game with two different runs of the program.

```
RUN
THIS PROGRAM PRESENTS A GAME TO TEST

YOUR ABILITY TO ESTIMATE THE QUOTIENT

IN A NATURAL NUMBER DIVISION PROBLEM.
```

```
THE COMPUTER WILL PRESENT YOU WITH A

DIVISION PROBLEM FOR WHICH YOU MUST

TYPE IN YOUR ESTIMATE OF THE QUOTIENT.

THE COMPUTER WILL THEN TEST TO SEE IF

YOUR MARGIN OF ERROR IS LESS THAN 5%.

WHAT IS THE QUOTIENT FOR 984/79?
?12
YOU WIN!

IF YOU WANT TO TRY AGAIN, TYPE IN 'Y'.
IF YOU DO NOT WANT TO TRY AGAIN, TYPE IN 'N'.
?Y
WHAT IS THE QUOTIENT FOR 289/71?
?5
YOU LOSE, THE CORRECT QUOTIENT WAS 4.

IF YOU WANT TO TRY AGAIN, TYPE IN 'Y'.
IF YOU DO NOT WANT TO TRY AGAIN, TYPE IN 'N'.
?N
```

Modify the program to make the dividends 10,000 times the value of the random digit computed and then play the game several times, keeping track of your success rate.

5

Elements of Number Theory

Julia Robinson
(1919–1985)

Julia Robinson was an outstanding mathematician on the faculty of the University of California at Berkeley. She gained fame by her work on the solution of Hilbert's tenth problem, one of the difficult problems stated by the famous German mathematician David Hilbert in 1900. This problem dealt with the solution of modular equations in advanced number theory. This and other major contributions led to her election to the National Academy of Sciences and as President of the American Mathematical Society.

5-1 Factors, Multiples, and Divisibility Rules

Numbers and relations among numbers have had a major role in the development of our culture, our concept of the world in which we live, the sciences, and philosophy. The early Pythagoreans believed that all events could be represented by numbers. For example, astronomy and music were mathematical subjects based upon natural numbers. Often numbers were assumed to have mystical qualities. Even today many people associate special qualities with numbers such as 7, 11, and 13. Relations among numbers were used to explain and to predict events.

One of the basic relationships among numbers is divisibility. The natural number 24 is divisible by the natural number 3 since there is another natural number 8 such that $24 = 3 \times 8$. Then 3 is a *divisor*, also called a *factor*, of 24 and 8 is the *other factor* when 24 is divided by 3.

> A natural number n is **divisible** by a natural number s if and only if there is a natural number k such that $n = s \times k$.

The natural number multiples of any number n may be obtained by repeated addition of n.

$$n + n = 2n \qquad 2n + n = 3n \qquad 3n + n = 4n \qquad \ldots$$

Multiples are easily visualized on a number line in terms of successive *jumps* starting from the origin. For $n = 3$ we have

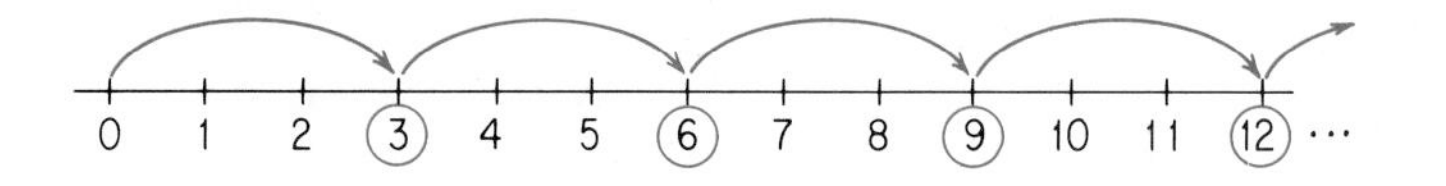

Since 24 is divisible by each of the numbers in the set $\{1, 2, 3, 4, 6, 8, 12, 24\}$, each of these numbers is a *divisor* of 24, each is a *factor* of 24, and 24 is a *multiple*

of each of them. On the other hand 24 is not divisible by 5 since there is no natural number b such that $24 = 5 \times b$. Therefore 5 is not a divisor of 24, 5 is not a factor of 24, and 24 is not a multiple of 5.

> If n is divisible by s, then s is a **divisor** of n, s is a **factor** of n, and n is a **multiple** of s.

The natural numbers were orginally associated with collections of objects such as small stones. The factors of a number n represented numbers of rows and columns in rectangular arrays of n elements. The arrays for 16 are

```
o o o o o o o o o o o o o o o o      1 × 16

o o o o o o o o
                    2 × 8
o o o o o o o o

o o o o
o o o o
            4 × 4
o o o o
o o o o
```

Since 16 elements can be arranged in a square array with 4 elements on each side, we speak of 16 as the *square* of the number 4.

Only natural numbers are considered as factors of other natural numbers. The factors of a given number may be found by listing pairs of numbers whose product is the given number. Consider these listings for 12, 16, and 18.

12	16	18
1×12	1×16	1×18
2×6	2×8	2×9
3×4	4×4	3×6
$\{1, 2, 3, 4, 6, 12\}$	$\{1, 2, 4, 8, 16\}$	$\{1, 2, 3, 6, 9, 18\}$

The listing of pairs of factors may be continued

1×12	1×16
2×6	2×8
3×4	4×4
4×3	8×2
6×2	16×1
12×1	

However all factors have already been found when the second factor becomes less than the first, that is, when the square of the first factor is greater than the number.

The natural numbers 1, 2, 3, . . . are tried in numerical order as possible factors. The process is terminated as soon as one of these two situations arises:

The two factors are alike, as in $4 \times 4 = 16$, or
The second factor is the next larger natural number that is a factor of the given number, as in $3 \times 4 = 12$ and $3 \times 6 = 18$.

Example 1 Find the set of factors of 19.

Solution The numbers 2, 3, and 4 are not factors of 19, and 5^2 is greater than 19. Therefore there is one and only one pair of natural numbers that have 19 as their product: $\{1, 19\}$. ■

Example 2 Show that (**a**) 245 is a multiple of 5 (**b**) 257 is not a multiple of 3.

Solution

Every natural number is a multiple of, and divisible by, each of its factors.

(**a**) Since $245 = 5 \times 49$, the natural number 245 is divisible by 5 and is a multiple of 5.

(**b**) There is no natural number k such that $257 = 3 \times k$. By division, $257 \div 3 = 85\frac{2}{3}$, which is not a natural number. ■

Example 3 List the first ten multiples of 5.

Solution 5, 10, 15, 20, 25, 30, 35, 40, 45, 50 ■

Because we are restricting our discussion in this section to the set of natural numbers, no mention has been made of the number 0. It should be noted that in some texts 0 is considered to be a multiple of every number. For example, 0 is a multiple of 9 since $0 \times 9 = 0$ and $0 \div 9 = 0$. However, because of problems with 0 that would be encountered in later sections, we shall continue to use only natural numbers as we consider multiples and factors of numbers.

Any natural number *is divisible by* another natural number if the division produces a zero remainder. Divisibility rules are obtained from patterns that arise when the numbers are expressed in decimal notation. Any natural number N less than 10,000 can be expressed in the form

Note that T represents the number of thousands while t represents the number of tens. Usually $T \neq t$.

$$N = 1000T + 100h + 10t + u$$

where T, h, t, and u are elements, not necessarily distinct, of the set $\{0, 1, 2, 3, 4, 5, 6, 7, 8, 9\}$. The letter u is used for the ones (units) digit to avoid confusion of o and 0.

$$1988 = 1000(1) + 100(9) + 10(8) + 8$$

The expanded notation can be used to establish rules for divisibility.

Divisibility by 2:

$$\frac{N}{2} = \frac{1000T}{2} + \frac{100h}{2} + \frac{10t}{2} + \frac{u}{2} = 500T + 50h + 5t + \frac{u}{2}$$

Thus $N/2$ is a natural number if and only if $u/2$ is a natural number; that is, *N is divisible by 2 if and only if u is divisible by 2.*

$$\frac{1988}{2} = 500(1) + 50(9) + 5(8) + \frac{8}{2}$$

Since the units digit, 8, is divisible by 2, the number 1988 is divisible by 2.

Divisibility by 3:

$$\frac{N}{3} = \frac{(999 + 1)T}{3} + \frac{(99 + 1)h}{3} + \frac{(9 + 1)t}{3} + \frac{u}{3}$$

$$= 333T + 33h + 3t + \frac{T + h + t + u}{3}$$

Thus $N/3$ is a natural number if and only if $(T + h + t + u)/3$ is a natural number; that is, *N is divisible by 3 if and only if the sum of its decimal digits is divisible by 3.*

$$\frac{1988}{3} = 333(1) + 33(9) + 3(8) + \frac{1 + 9 + 8 + 8}{3}$$

Since the sum of the digits, 26, is not divisible by 3, the number 1988 is not divisible by 3.

Divisibility by 4:

$$\frac{N}{4} = \frac{1000T}{4} + \frac{100h}{4} + \frac{10t}{4} + \frac{u}{4} = 250T + 25h + \frac{10t + u}{4}$$

Thus $N/4$ is a natural number if and only if $(10t + u)/4$ is a natural number; that is, *N is divisible by 4 if and only if the number represented by its tens and ones digits is divisible by 4.*

$$\frac{1988}{4} = 250(1) + 25(9) + \frac{10(8) + 8}{4}$$

Other rules for divisibility are considered in the exercises.

Since the number represented by its tens and ones digits, 88, is divisible by 4, the number 1988 is divisible by 4.

EXERCISES

List the first five multiples of the given number.

1. 3
2. 4
3. 7
4. 8
5. 11
6. 15
7. 25
8. 40

Find the factors of the given number.

9. 20
10. 32
11. 36
12. 60
13. 64
14. 72
15. 80
16. 84
17. 92
18. 95
19. 111
20. 450

For decimal notation find a rule for divisibility by the given number.

21. 5
22. 6
23. 8
24. 9

Test each number for divisibility by **(a)** 2 **(b)** 3 **(c)** 4 **(d)** 5 **(e)** 6 **(f)** 8 **(g)** 9.

25. 5280
26. 225
27. 1728
28. 16,275
29. 17,540
30. 19,678
31. 36,000
32. 27,600
33. 45,460
34. 80,124
35. 100,200
36. 100,100

For natural numbers n, s, and t give at least three numerical examples of these general rules of divisibility.

37. If ns divides t, then n divides t.
38. If n divides s and s divides t, then n divides t.

39. If n divides s and n divides t, then n divides $s + t$ and, assuming that s is greater than t, $s - t$.
40. If n divides s and n divides t, then n divides $js + kt$ for any natural numbers j and k.

In Exercises 41 *through* 43 *give an example for which the given statement is true and at least two counterexamples to show why the statement cannot be a general rule for natural numbers n, s, and t.*

41. If n divides st, then n divides t.
42. If n divides t and s divides t, then ns divides t.
43. If n does not divide s and n does not divide t, then n does not divide st.

EXPLORATIONS

1. The numbers 84 and 108 are each divisible by 12.
   (a) List the natural numbers that divide both 84 and 108.
   (b) List the factors of 12.
   (c) Compare your answers for parts (a) and (b) and conjecture a rule for identifying at least five natural numbers less than 12 that must be factors of any number that is divisible by 12.
   (d) Conjecture a similar rule for factors of any number that is divisible by 30 and find seven numbers less than 30 that must be factors of any number that is divisible by 30.
   (e) Conjecture a general rule for factors of any number b that is divisible by a natural number s.

Consider as many numerical examples as necessary and explain why each statement must be true for all natural numbers.

2. The sum of any two consecutive natural numbers, in general n and $n + 1$, is an odd number.
3. The product of any two consecutive natural numbers is an even number.
4. The product of any three consecutive natural numbers is divisible by 3 and by 6.
5. The product of any four consecutive natural numbers is divisible by 24.
6. A natural number N has an odd number of natural numbers as factors if and only if N is the square of a natural number.

4/18/73

5-2 Prime Numbers

Study the set of factors for each of the following numbers.

2: {1, 2} 3: {1, 3} 5: {1, 5} 7: {1, 7}

$n = 1 \times n$

In each case the number shown has exactly two distinct natural numbers as factors, the number itself and 1, and is said to be a *prime number*.

A **prime number** is a natural number that has exactly two distinct factors.

The set of prime numbers is an infinite set, and to this date the search continues for larger and larger prime numbers. The advent of the computer has made such searches possible, and the discovery of a new prime number invariably finds its way as a news article into the daily papers.

As shown, 2, 3, 5, and 7 are examples of prime numbers. On the other hand, 6 is not a prime number because it can be factored as 1×6 and 2×3. That is, 6 has more than two distinct factors. The set of factors of 6 is $\{1, 2, 3, 6\}$.

The natural numbers that are greater than 1 and are not prime are **composite numbers**. Note that every natural number greater than 1 is either prime or composite. The number 1 is neither prime nor composite but is a **unit** since every natural number is a multiple of 1. Thus we may classify any natural number as belonging to one of these three sets.

The set whose only element is the unit 1.
The set of prime numbers.
The set of composite numbers.

Example 1 Classify as prime or composite. **(a)** 24 **(b)** 31

Solution

(a) Composite. The number 24 can be factored as 1×24, 2×12, 3×8, or 4×6 and has more than two factors.

(b) Prime. The number 31 can be factored only as 1×31 and has exactly two distinct factors. ■

Eratosthenes (ca. 276–194 B.C.) was a gifted mathematician, astronomer, geographer, historian, philosopher, poet, and athlete. Ptolemy III invited him to Alexandria, Egypt to tutor his son and take charge of the library of the University.

The following method for identifying prime numbers was discovered over two thousand years ago by a Greek mathematician named Eratosthenes. This method, known as the **sieve of Eratosthenes**, is illustrated for the set consisting of the natural numbers through 100.

First prepare a table as shown, and cross out the unit 1. Draw a circle around 2, the smallest prime number. Then cross out every multiple of 2 that follows, since each one is divisible by 2 and thus is not prime. That is, cross out the numbers in the set $\{4, 6, 8, \ldots, 100\}$.

~~1~~	(2)	3	~~4~~	5	~~6~~	7	~~8~~	9	~~10~~
11	~~12~~	13	~~14~~	15	~~16~~	17	~~18~~	19	~~20~~
21	~~22~~	23	~~24~~	25	~~26~~	27	~~28~~	29	~~30~~
31	~~32~~	33	~~34~~	35	~~36~~	37	~~38~~	39	~~40~~
41	~~42~~	43	~~44~~	45	~~46~~	47	~~48~~	49	~~50~~
51	~~52~~	53	~~54~~	55	~~56~~	57	~~58~~	59	~~60~~
61	~~62~~	63	~~64~~	65	~~66~~	67	~~68~~	69	~~70~~
71	~~72~~	73	~~74~~	75	~~76~~	77	~~78~~	79	~~80~~
81	~~82~~	83	~~84~~	85	~~86~~	87	~~88~~	89	~~90~~
91	~~92~~	93	~~94~~	95	~~96~~	97	~~98~~	99	~~100~~

Think of the numbers in the table as if they were separated by *unit spaces* on a number line. Cross out the unit 1. Circle the next number after the number 1—it is 2, the smallest prime number p. Then cross out the numbers obtained by repeated jumps of 2 spaces. These two steps—circle the next prime number p and cross out the numbers obtained by repeated jumps of p spaces—are repeated over and over until for some number p all of the numbers obtained by jumps of p spaces have already been crossed out.

Draw a circle around 3, the next prime number in the list. Then cross out each succeeding multiple of 3. Some of these numbers, such as 6 and 12, will already have been crossed out because they are also multiples of 2.

Draw a circle around 5, the next prime number. Then exclude each fifth number after 5. The next prime number is 7; exclude each seventh number after 7. The next prime number is 11. For this table of the first 100 natural numbers we find that all other multiples of 11 have already been crossed out. This implies that all of the remaining numbers that have not been crossed out are prime numbers and may be circled.

~~1~~	2	3	~~4~~	5	~~6~~	7	~~8~~	~~9~~	~~10~~
11	~~12~~	13	~~14~~	~~15~~	~~16~~	17	~~18~~	19	~~20~~
~~21~~	~~22~~	23	~~24~~	~~25~~	~~26~~	~~27~~	~~28~~	29	~~30~~
31	~~32~~	~~33~~	~~34~~	~~35~~	~~36~~	37	~~38~~	~~39~~	~~40~~
41	~~42~~	43	~~44~~	~~45~~	~~46~~	47	~~48~~	~~49~~	~~50~~
~~51~~	~~52~~	53	~~54~~	~~55~~	~~56~~	~~57~~	~~58~~	59	~~60~~
61	~~62~~	~~63~~	~~64~~	~~65~~	~~66~~	67	~~68~~	~~69~~	~~70~~
71	~~72~~	73	~~74~~	~~75~~	~~76~~	~~77~~	~~78~~	79	~~80~~
~~81~~	~~82~~	83	~~84~~	~~85~~	~~86~~	~~87~~	~~88~~	89	~~90~~
~~91~~	~~92~~	~~93~~	~~94~~	~~95~~	~~96~~	97	~~98~~	~~99~~	~~100~~

To determine whether or not a natural number N is a prime number try the prime numbers p such that $p^2 \leq N$ as factors of N. For example, to check 217 try 2, 3, 5, 7, 11, and 13.

Notice that 49 is the first number that is divisible by 7 and is not also divisible by a prime number less than 7. In other words, each composite number less than 7^2 has at least one of its factors less than 7. Similarly, each composite number less than 5^2 has at least one factor less than 5. In general, *for any prime number p each composite number less than p^2 has a prime number less than p as a factor*. A natural number N is a prime number if and only if N is not divisible by any prime number p where p^2 is less than or equal to N.

We use this property to tell us when we have excluded all composite numbers from a set. In the set $\{1, 2, \ldots, 100\}$ we have considered the prime numbers 2, 3, 5, and 7. The next prime number is 11 and 11^2 is greater than 100. Thus we have already excluded all composite numbers and identified all prime numbers that are less than or equal to 100.

Example 2 List the set of prime numbers less than 70.

Solution From the table, this set is

$$\{2, 3, 5, 7, 11, 13, 17, 19, 23, 29, 31, 37, 41, 43, 47, 53, 59, 61, 67\} \blacksquare$$

EXERCISES

Classify the given number as prime or composite.

1. 42	**2.** 51	**3.** 71	**4.** 92
5. 89	**6.** 101	**7.** 147	**8.** 203
9. 257	**10.** 707	**11.** 727	**12.** 729

Express the given number as the product of as many different pairs of natural numbers as possible.

13. 16	**14.** 17	**15.** 21	**16.** 24
17. 31	**18.** 51	**19.** 54	**20.** 60
21. 100	**22.** 120	**23.** 125	**24.** 150

List the prime numbers between the given numbers.

25. 1 and 20	**26.** 30 and 40	**27.** 40 and 50
28. 60 and 70	**29.** 80 and 100	**30.** 100 and 120

List the composite numbers between the given numbers.

31. 1 and 10	**32.** 20 and 30	**33.** 30 and 50
34. 50 and 70	**35.** 80 and 90	**36.** 110 and 120

37. Is every odd natural number a prime number? Is every prime number an odd number?

38. Exhibit a pair of prime numbers that differ by 1 and show that there is only one such pair possible.

Christian Goldbach (1690–1764) was a German mathematician who left his mathematical work for a career in the Russian civil service. His conjecture has been viewed as a wild guess but no one has been able to prove that it is wrong.

39. Here is a famous theorem that has not yet been proved: Every even number greater than 2 is expressible as the sum of two prime numbers, not necessarily distinct. (This theorem is often called **Goldbach's conjecture.**) Express each of the even numbers 4 through 40 as a sum of two prime numbers.

40. Here is another famous theorem. Two prime numbers such as 17 and 19 that differ by 2 are called **twin primes**. It is believed but has not yet been proved that there are infinitely many twin primes. Find a pair of twin primes between (**a**) 35 and 45 (**b**) 65 and 75 (**c**) 95 and 105.

41. A set of three prime numbers that differ by 2 is called a **prime triplet**. Exhibit a prime triplet and explain why it is the only possible triplet of prime numbers.

Note that the conjecture stated in Exercise 42 is an extension of Goldbach's conjecture to three addends.

42. It has been conjectured but not proved that every odd number greater than 5 is expressible as the sum of three prime numbers, not necessarily distinct. Verify this for the numbers 7, 9, 11, 13, and 15.

43. What is the largest prime number that you need to consider to be sure that you have excluded all composite numbers less than or equal to (**a**) 200 (**b**) 500 (**c**) 1000?

***44.** Extend the sieve of Eratosthenes to find the prime numbers less than or equal to 200.

EXPLORATIONS

1. No one has ever been able to find a formula that will produce only prime numbers. At one time it was thought that the expression $n^2 - n + 41$ would give only prime numbers for the set of natural numbers as replacements for n. Show that prime numbers are obtained when n is replaced by 1, 2, 3, 4, and 5. Show that the formula fails for $n = 41$.

Pierre de Fermat (1601–1665) was a lawyer who spent his spare time studying mathematics. He corresponded with many of the leading mathematicians of his time, helped spread the discoveries of others, and made major contributions of his own. He was the founder of modern number theory.

2. A theorem about prime numbers first stated in 1640 by Pierre de Fermat (1601–1665) states that if p is a prime number, then for every integer a, $a^p - a$ is divisible by p. For example, if $p = 2$, then $a^2 - a$ is divisible by 2 for all integers a. Test that this is so for at least five different values of a. Then let $p = 3$ and test again. Then let $p = 4$, not a prime, and show by a single counterexample that the theorem does not hold for composite numbers.

3. For $n = 0, 1, 2, 3, \ldots$, numbers of the form

$$2^{(2^n)} + 1$$

are known as **Fermat numbers**. Fermat conjectured, incorrectly, that all such numbers would be prime. For $n = 0$, we have

$$2^{(2^0)} + 1 = 2^1 + 1 = 3$$

a prime number. Show that prime numbers are obtained for $n = 1$, $n = 2$, and $n = 3$.

For $n = 4$ the Fermat number 65,537 is prime. For $n = 5$ the number $4{,}294{,}967{,}297 = 641 \times 6{,}700{,}417$. No one has found a prime Fermat number for n greater than 4 and it is not known whether or not any exist.

4. It can be proved that every natural number greater than 11 is the sum of two composite numbers. Show, by example, that this is true for the natural numbers 12 through 25.

5. Euclid proved that the set of prime numbers is infinite. Refer to a textbook or a book on the history of mathematics and study this simple, yet elegant proof.

6. On October 30, 1978, two 18-year-old students discovered the largest prime number known at that time. On February 9, 1979, one of these students found a larger prime number. Try to find out what is the largest prime number known today and by whom it was discovered.

The conjecture stated in Exploration 7 has been proved to be true.

*7. Does the following conjecture, **Bertrand's conjecture**, appear to be true? Explain your reasoning.

   For any natural number n greater than 1 there exists at least one prime number p such that $n < p < 2n$.

5-3 Fundamental Theorem of Arithmetic

We have seen that every natural number greater than 1 is either a prime number or a composite number. Now we shall find that, except for the order of the factors, every composite number greater than 1 can be expressed as the product of one and only one set of powers of prime numbers.

Consider these factorizations of 24.

In this sense the prime numbers are the "building blocks" of the natural numbers.

$24 = 1 \times 24$

$24 = 2 \times 12$

$24 = ? \times 8$

$24 = 4 \times 6$

$24 = 2 \times 2 \times 6$

$24 = 2 \times 3 \times 4$

$24 = 2 \times 2 \times 2 \times 3 = 2^3 \times 3$

In the last factorization the set $\{2^3, 3\}$ of powers of primes is unique. The actual factorization in terms of prime numbers can be written in other ways such as $2 \times 3 \times 2^2$. However, these ways are equivalent, since the order of the factors does not affect the product. Thus 24 can be expressed as the product of a unique set of powers of its prime factors.

One of the easiest ways to find the prime factors of a number is to consider the prime numbers

2, 3, 5, 7, 11, 13, 17, 19, 23, 29, 31, ...

in order and use each one as a factor as many times as possible. Then for 24 we would have

$$\begin{aligned} 24 &= 2 \times 12 \\ &= 2 \times 2 \times 6 \\ &= 2 \times 2 \times 2 \times 3 \end{aligned}$$

These steps may be performed by successive division.

$$\begin{array}{r} 2\overline{)24} \\ 12 \end{array} \qquad \begin{array}{r} 2\overline{)24} \\ 2\overline{)12} \\ 6 \end{array} \qquad \begin{array}{r} 2\overline{)24} \\ 2\overline{)12} \\ 2\overline{)6} \\ 3 \end{array} \qquad \begin{array}{r} 2\overline{)24} \\ 2\overline{)12} \\ 2\overline{)6} \\ 3\overline{)3} \\ 1 \end{array}$$

Since 3 is a prime number, no further steps are needed and $24 = 2^3 \times 3$.

Example 1 Express 3850 in terms of its prime factors.

Solution

$$\begin{array}{r} 2\overline{)3850} \\ 5\overline{)1925} \\ 5\overline{)385} \\ 7\overline{)77} \\ 11\overline{)11} \\ 1 \end{array} \qquad 3850 = 2 \times 5^2 \times 7 \times 11 \quad \blacksquare$$

The prime factors of a composite number may also be shown as numbers at the ends of branches of a **prime factor tree**.

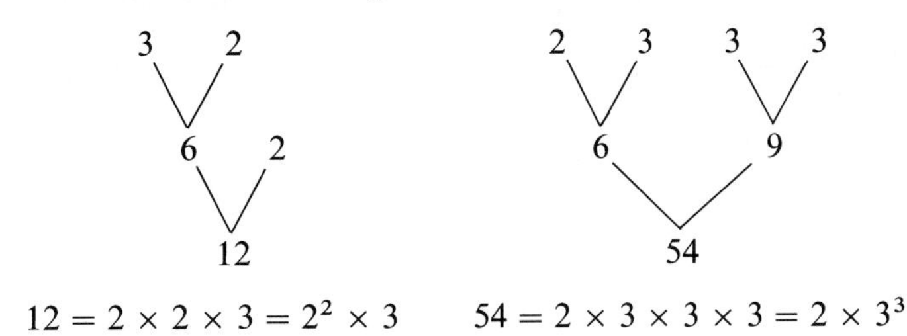

$12 = 2 \times 2 \times 3 = 2^2 \times 3$ $\qquad$ $54 = 2 \times 3 \times 3 \times 3 = 2 \times 3^3$

Example 2 Use a prime factor tree and express 120 in terms of its prime factors.

Solution Among others:

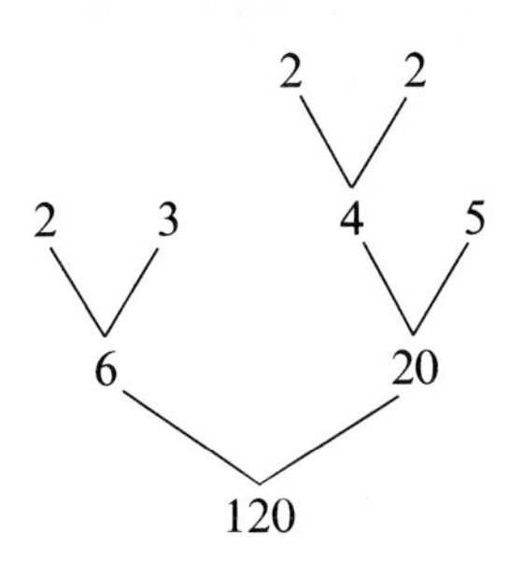

$120 = 2^3 \times 3 \times 5$ ■

It is customary to write the prime factors in increasing order as in Examples 1 and 2.

Note that in the solution of Example 2 several different appearing factor trees may be found but the resulting factors are always the same. For example,

The expression of any composite number n as a product of its prime factors is called the **prime factorization** of n. In general, if a natural number n is greater than 1, then n has a prime number p_1 as a factor. Suppose that

$$n = p_1 n_1$$

Then if n is a prime number, $n = p_1$ and $n_1 = 1$. If n is not a prime number, then n_1 is a natural number greater than 1. In this case n_1 is either a prime number or a composite number. Suppose that

$$n_1 = p_2 n_2 \quad \text{and thus} \quad n = p_1 p_2 n_2$$

where p_2 is a prime number. As before, if $n_2 \neq 1$, then

$$n_2 = p_3 n_3 \quad \text{and thus} \quad n = p_1 p_2 p_3 n_3$$

where p_3 is a prime number, and so forth. We may continue this process until some $n_k = 1$, since there are only a finite number of natural numbers less than n and

$$n > n_1 > n_2 > n_3 > \cdots > n_k = 1$$

Then we have an expression for n as a product of prime numbers,

$$n = p_1 p_2 p_3 \cdots p_k$$

The uniqueness of the prime factorization can be observed for any given number such as 120, but in general this property is a theorem—the **fundamental theorem of arithmetic**.

> Each composite number has, except for the order of the factors, one and only one prime factorization.

There are two common methods for finding the prime factorization of a given natural number n greater than 1. These methods were illustrated in Examples 1 and 2 and may also be used to determine whether or not the given number n is itself a prime number. In Example 2 the given number is first expressed as a product of any two factors greater than 1 and the prime factors of these smaller numbers are then sought. In Example 1 the prime numbers 2, 3, 5, 7,... are tried in order with each used as many times as possible. This process cannot continue indefinitely since each quotient is smaller than the previous number and prime numbers greater than n do not need to be considered. Indeed, if $n = bc$, then b and c are not both greater than $\sqrt{n}$. In other words, to find all prime factors of a natural number n, it is sufficient to test for divisibility by prime numbers whose squares are not greater than n. If no prime factor is found in this way and n is greater than 1, then n is a prime number. For example, 101 is not divisible by 2, 3, 5, or 7 and therefore is a prime number.

In the sections that follow, we shall find that the prime factorization of a number can be used to develop techniques for simplifying fractions, as well as for computing with fractions. The prime factorization of a number also may be used to determine the set of all factors of that number. Consider, for example, the prime factorization of 30.

$$30 = 2 \times 3 \times 5$$

Select the prime factors of 30 one at a time, two at a time, and three at a time as in the array.

One at a Time	*Two at a Time*	*Three at a Time*
2	2×3	$2 \times 3 \times 5$
3	2×5	
5	3×5	

The number 1, which is a factor of every number, and the numbers shown in the array form the set $\{1, 2, 3, 5, 6, 10, 15, 30\}$ of factors of 30.

EXERCISES

Find the number represented by each prime factorization.

1. $3 \times 5^2 \times 11$
2. $2 \times 3 \times 5^2$
3. $2^5 \times 3^3$
4. $2 \times 3^2 \times 5 \times 7$
5. $2^5 \times 3 \times 5^2$
6. $11 \times 13 \times 17^2$
7. $7^3 \times 13 \times 17$
8. $3^2 \times 13 \times 17^2 \times 19$

Identify each number as a prime number or as a composite number in terms of its prime factorization.

9. 75
10. 76
11. 77
12. 78
13. 79
14. 103
15. 257
16. 357
17. 618
18. 938
19. 1001
20. 2425
21. 3000
22. 4800
23. 4895
24. 5780

Use the prime factorization shown to write the set of factors of each number.

25. $2 \times 5 \times 7$
26. $3 \times 7 \times 11$
27. $2^2 \times 5$
28. $3^2 \times 7$
29. $2 \times 3^2 \times 5$
30. $2^2 \times 3 \times 5$

Find the given number N, where each of the other letters represents a prime number.

31. $N = 2 \times a^2 \times 5 = b \times 3^2 \times c$
32. $N = d^3 \times 3 \times e = 2^3 \times f \times 5$
33. $N = g^2 \times h \times 7 = 2^2 \times 5 \times j$
34. $N = k^2 \times 5^2 \times m \times 13 = 2^2 \times p^2 \times 11 \times q$

EXPLORATIONS

Here is a trick that depends upon the prime factorization of a number for its explanation. Begin with a box of colored objects such as beads, blocks, or marbles. Assign different prime numerical values to the colors. For example, suppose that a box contains red beads, green beads, and blue beads and suppose that the beads are assigned these values:

Red—2 Green—3 Blue—5

Next ask someone to select any number of beads from the box and to find the product of their values. Suppose the person selects three red, two green, and four blue beads. Then the product of their values is

$$\underbrace{2 \times 2 \times 2}_{\text{red}} \times \underbrace{3 \times 3}_{\text{green}} \times \underbrace{5 \times 5 \times 5 \times 5}_{\text{blue}} = 45{,}000$$

At this point you ask for the product of the values of the selected beads. From this product, you are able to determine the number of beads of each color that have been chosen. The trick is to write the prime factorization of the product. Since

$$45{,}000 = 2^3 \times 3^2 \times 5^4$$

the beads chosen are three red (2^3), two green (3^2), and four blue (5^4). Note that the exponent of the associated prime number value indicates the number of beads that have been chosen of each color.

In Explorations 1 through 3 consider the given number as the product of the values of the beads in a selection and find the number of beads that have been chosen of each color.

1. 2400 2. 10,000 3. 6750
4. Find the set of natural numbers less than 100 that have **(a)** only one divisor **(b)** exactly two divisors **(c)** an odd number of divisors **(d)** exactly eight divisors **(e)** exactly ten divisors **(f)** as many divisors as possible.
5. **(a)** Note that $1 \times 2 \times 3 \times 4 \times 5 + b$ is divisible by 2, 3, or 5 if b is divisible by 2, 3, or 5. Use this observation to find at least four consecutive composite numbers.
   (b) Describe a procedure for finding ten consecutive composite numbers and, in general, k consecutive composite numbers for any natural number k.

5-4 Greatest Common Divisor

Consider the sets of divisors of 12 and of 18. The circled numbers are those that are divisors of both 12 and 18 and are called the **common divisors** of the two numbers.

Divisors of 12: {①, ②, ③, 4, ⑥, 12}

Divisors of 18: {①, ②, ③, ⑥, 9, 18}

Common divisors of 12 and 18: {1, 2, 3, 6}

The largest member of the set of common divisors of two numbers is called their **greatest common divisor** (GCD) or their *greatest common factor*.

GCD(12, 18) = 6

We may use the prime factorization of two natural numbers to find their greatest common divisor. First represent each number by its prime factorization. Then consider the prime numbers that are divisors of both of the given numbers and take the product of those prime numbers with each raised to the highest power that is a divisor of both of the given numbers. For example $12 = 2^2 \times 3$, $18 = 2 \times 3^2$, and GCD(12, 18) $= 2 \times 3$, that is, 6. To visualize this process we may use a different card for each factor, possibly cards of different colors for different given numbers. We line up the cards for each given number and match them for the two numbers as much as possible.

12 = 2 × 2 × 3 [2] [2] [3]

18 = 2 × 3 × 3 [2] [3] [3]

GCD (12, 18) = 2 × 3, that is, 6.

The product of the numbers on the matched cards (one number for each matched pair) is the greatest common divisor of the given numbers.

Example 1 Find the greatest common divisor of 60 and 5280.

Solution

$$60 = 2^2 \times 3 \times 5$$
$$5280 = 2^5 \times 3 \times 5 \times 11$$
$$\text{GCD}(60, 5280) = 2^2 \times 3 \times 5, \quad \text{that is,} \quad 60. \blacksquare$$

Example 2 Find GCD(3850, 5280).

Solution

$$3850 = 2 \times 5^2 \times 7 \times 11$$
$$5280 = 2^5 \times 3 \times 5 \times 11$$
$$\text{GCD}(3850, 5280) = 2 \times 5 \times 11, \quad \text{that is,} \quad 110. \blacksquare$$

Example 3 Find the greatest common divisor of 12, 36, and 60.

Solution

$$12 = 2^2 \times 3 \qquad 36 = 2^2 \times 3^2 \qquad 60 = 2^2 \times 3 \times 5$$
$$\text{GCD}(12, 36, 60) = 2^2 \times 3, \quad \text{that is,} \quad 12.$$

Alternate solution Make an array with the given numbers on a line. As in the prime factorization try the prime numbers that are factors of at least one of the numbers and list each factor used in the column of divisors D at the left of the array. If the number is a factor of all three of the numbers, list it also in the column A at the right. On each row write in a column below a given number either the quotient or, if division was impossible, the number that was on the previous row. Continue the process until a row of 1's is obtained.

D	12	36	60	A
2	6	18	30	2
2	3	9	15	2
3	1	3	5	3
3	1	1	5	
5	1	1	1	

Then GCD(12, 36, 60) $= 2 \times 2 \times 3 = 12$, the product of the numbers in the column A. Also the *least common multiple* of 12, 36, and 60 (Section 5-5, Example 3) is $2 \times 2 \times 3 \times 3 \times 5 = 180$, the product of the numbers in the column D. ■

The greatest common divisor can be used to *simplify* (reduce) a fraction.

We next use greatest common divisors to simplify fractions. For any natural number n we know that $n/n = 1$. For any natural numbers a, b, and c the fraction a/b has *numerator* a and *denominator* b. Also

$$\frac{ac}{bc} = \frac{a}{b} \times \frac{c}{c} = \frac{a}{b} \times 1 = \frac{a}{b}$$

If c is the greatest common divisor of ac and bc, then the fraction ac/bc when expressed in the form a/b is expressed in *simplest form*. For example,

$$\frac{60}{4880} = \frac{2^2 \times 3 \times 5}{2^4 \times 5 \times 61} \qquad \text{GCD}(60, 4880) = 2^2 \times 5$$

$$\frac{60}{4880} = \frac{(2^2 \times 5) \times 3}{(2^2 \times 5) \times (2^2 \times 61)} = \frac{3}{2^2 \times 61} = \frac{3}{244}$$

Since 3 is the only prime factor of the numerator and 3 is not a factor of the denominator, the fraction 3/244 is expressed in simplest form. That is, the numerator and denominator do not have any common prime factors. Any two natural numbers that do not have a common prime factor are said to be **relatively prime.** Note that two numbers, such as 15 and 16, may be relatively prime without themselves being prime numbers.

Example 4 Simplify 60/168.

Solution

$$60 = 2^2 \times 3 \times 5$$
$$168 = 2^3 \times 3 \times 7 \qquad \text{GCD}(60, 168) = 2^2 \times 3$$

$$\frac{60}{168} = \frac{(2^2 \times 3) \times 5}{(2^2 \times 3) \times (2 \times 7)} = \frac{5}{14} \blacksquare$$

EXERCISES *Find the set of common divisors for the given pair of numbers.*

1. 8 and 10 **2.** 12 and 30 **3.** 16 and 40
4. 9 and 30 **5.** 5 and 7 **6.** 17 and 24
7. 10 and 30 **8.** 12 and 60 **9.** 25 and 75
10. 93 and 155 **11.** 202 and 606 **12.** 103 and 600

Write the prime factorizations and find the greatest common divisor of the given numbers.

13. 42 and 60 **14.** 68 and 96 **15.** 123 and 287
16. 96 and 1425 **17.** 123 and 615 **18.** 285 and 1425
19. 68 and 112 **20.** 112 and 480 **21.** 600 and 800
22. 1850 and 7400 **23.** 2450 and 3500 **24.** 2025 and 5400
25. 12, 18, and 21 **26.** 18, 24, and 45 **27.** 15, 25, and 40
28. 12, 30, and 48 **29.** 10, 20, and 35 **30.** 24, 40, and 64

Simplify.

31. $\frac{8}{20}$ **32.** $\frac{12}{18}$ **33.** $\frac{18}{45}$

34. $\frac{30}{45}$ **35.** $\frac{60}{72}$ **36.** $\frac{54}{90}$

37. $\frac{68}{112}$

38. $\frac{112}{480}$

39. $\frac{128}{320}$

40. $\frac{378}{405}$

41. $\frac{2450}{3500}$

42. $\frac{2025}{5400}$

In Exercises 43 *through* 47 *classify each statement as true or false. If false, give a specific counterexample to justify your answer.*

43. If two numbers are relatively prime, then each of the numbers must be prime numbers.
44. If two numbers are relatively prime, they must both be odd numbers.
45. If two numbers are relatively prime, then one number must be even and the other number must be odd.
46. Every two natural numbers have at least one common factor.
47. The greatest common divisor of any two prime numbers is 1.

In 1866 Nicolo Paganini, a 16-year-old Italian boy, discovered that 1184 and 1210 are amicable numbers. More than 1000 pairs of amicable numbers have been identified. For example, 2,844,637,606,234,215 and 2,884,708,168,019,865 are amicable numbers.

*48. The factors of a given number N that are less than N are often called **proper divisors** of the given number. Two numbers are called **amicable,** or *friendly,* **numbers** if each is the sum of the proper divisors of the other. Show that 220 and 284 are amicable numbers.

EXPLORATIONS

Extend the first three columns of this table for natural numbers 1 through 17.

Natural number	*Factors*	*Number of Factors*	*Sum of Factors*
1	1	1	1
2	1, 2	2	3
• • •	• • •	• • •	• • •

1. Can any natural number greater that 1 have only one factor?
2. Can the number of its factors be used to identify any given number as a prime number?
3. Can the number of its factors be used to identify any given number as a composite number?

4. Give a rule for determining whether or not a natural number is a prime number if the number of its factors is known.

5. Find a way to distinguish in terms of the number of factors the natural numbers that are squares of natural numbers.

Additional examples may be found by extending the table.

6. A number is a **deficient number** if the sum of its factors is less than twice the number. List the first five deficient numbers.

7. A number is a **perfect number** if the sum of its factors is equal to twice the number. Find at least one perfect number.

8. A number is an **abundant number** if the sum of its factors is more than twice the number. Find at least one abundant number.

9. Consider the mathematical system that consists of the set of natural numbers and the operation # defined to mean "find the greatest common divisor of the two numbers."
   (a) Find 8 # 15 and 12 # 20.
   (b) Is the set of natural numbers closed with respect to #?
   (c) For natural numbers m and n, does $m \# n = n \# m$?
   (d) For natural numbers m, n, and p, does $m \# (n \# p) = (m \# n) \# p$?
   (e) Is there an identity element with respect to #? If so, what is it?

Graphs can be used to explore patterns formed by the greatest common divisors of numbers.

10. The following graph of the greatest common divisors of 6 and the natural numbers 1 through 12 can be extended as needed to identify the continuing pattern.

The greatest common divisor of

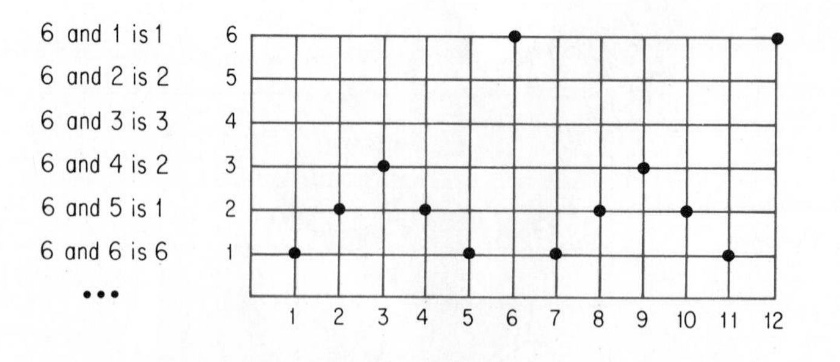

Copy and extend the graph for the natural numbers 1 through 18. From the pattern identify **(a)** the greatest common divisor of 6 and any odd number that is not divisible by 3; **(b)** the greatest common divisor of 6 and any even number that is not divisible by 6.

11. As in Exploration 10 draw a graph to show the pattern formed by the greatest common divisors of 4 and the natural numbers 1 through 12.
   (a) Identify the greatest common divisor of 4 and any odd natural number.
   (b) Identify the patterns formed by the greatest common divisors of 4 and the even numbers.

5-5 Least Common Multiple

Let us now turn our attention to the concept of a multiple of a number. Consider for example the multiples of 5 and 6 as shown on the number line.

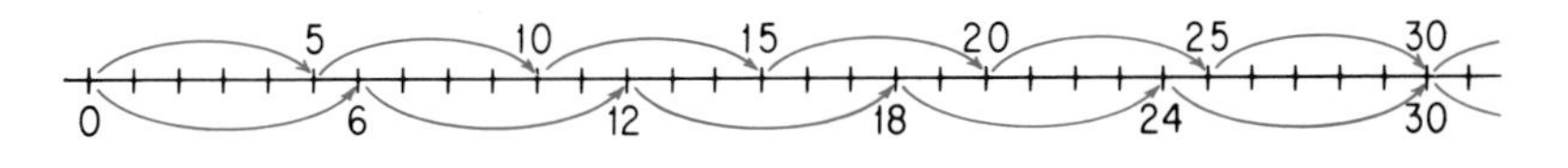

The smallest natural number that is a multiple of 5 and also a multiple of 6 is 30, the *least common multiple* of 5 and 6. Next consider the sets of multiples of 12 and of 18. The circled numbers are those that are multiples of both 12 and 18 and are called the **common multiples** of the two numbers.

Multiples of 12: {12, 24, (36), 48, 60, (72), 84, 96, (108), 120, ...}

Multiples of 18: {18, (36), 54, (72), 90, (108), 126, ...}

Common multiples of 12 and 18: {36, 72, 108, ...}

The set of common multiples is an infinite set. There is no greatest common multiple. The smallest member of the set of common multiples is their **least common multiple** (LCM).

LCM(12, 18) = 36

To visualize the process of finding the least common multiple of 12 and 18, we may use cards for the prime factors of each number and match them as was done for GCD(12, 18) above Example 1 in Section 5-4. The product of the numbers on the unmatched cards and one number for each matched pair of cards is the least common multiple.

$\text{LCM}(12, 18) = 2 \times 2 \times 3 \times 3$, that is, 36.

Example 1 List the first three common multiples of 6 and 9. Then find LCM(6, 9).

Solution

Multiples of 6: {6, 12, (18), 24, 30, (36), 42, 48, (54), 60, ...}

Multiples of 9: {9, (18), 27, (36), 45, (54), 63, 72, ...}

Common multiples of 6 and 9: {18, 36, 54, ...}
LCM(6, 9) = 18 ■

We may use the prime factorizations of two natural numbers to find their least common multiple. For example, $24 = 2^3 \times 3$ and $90 = 2 \times 3^2 \times 5$. Then LCM(24, 90) must be divisible by 2^3 and by 2. Since any number that is divisible by 2^3 must be divisible by 2, it is sufficient to select 2^3, the highest power of 2 in the prime factorizations of 24 and 90, as a factor of LCM(24, 90). Since we seek the least common multiple, we do not want a greater power of 2. Similarly, we select 3^2, the highest power of 3, and 5, the highest power of 5.

$\text{LCM}(24, 90) = 2^3 \times 3^2 \times 5$, that is, 360.

In general, to find the least common multiple of two natural numbers consider the prime numbers that are factors of either of the given numbers and take the product of these prime numbers with each raised to the highest power that occurs in either of the prime factorizations of the given numbers.

Example 2 Find LCM(3850, 5280).

Solution

$$3850 = 2 \times 5^2 \times 7 \times 11$$

$$5280 = 2^5 \times 3 \times 5 \times 11$$

$$\text{LCM}(3850, 5280) = 2^5 \times 3 \times 5^2 \times 7 \times 11, \quad \text{that is,} \quad 184{,}800. \blacksquare$$

Example 3 Find LCM(12, 36, 60).

Solution

$$12 = 2^2 \times 3 \qquad 36 = 2^2 \times 3^2 \qquad 60 = 2^2 \times 3 \times 5$$

$$\text{LCM}(12, 36, 60) = 2^2 \times 3^2 \times 5, \quad \text{that is,} \quad 180. \blacksquare$$

For an alternate method see the alternate solution for Example 3 in Section 5-4.

We use the least common multiple of the denominators of any two fractions first to express the fractions as *like fractions* and then to add or subtract the numbers represented by the fractions. Consider for instance 7/12 and 5/18. Recall that LCM(12, 18) = 36.

See the matching of cards for the factors of 12 and 18 in Section 5-4. Note that to obtain like fractions, the fraction with denominator 12 is multiplied by n/n, where n is the product of the numbers on cards for factors of 18 that were not matched. Similarly, the fraction with denominator 18 is multiplied by k/k, where k is the product of the numbers on cards for factors of 12 that were not matched.

$$\frac{7}{12} = \frac{7}{12} \times \frac{3}{3} = \frac{21}{36}$$

$$\frac{5}{18} = \frac{5}{18} \times \frac{2}{2} = \frac{10}{36}$$

$$\frac{7}{12} + \frac{5}{18} = \frac{21}{36} + \frac{10}{36} = \frac{31}{36}$$

$$\frac{7}{12} - \frac{5}{18} = \frac{21}{36} - \frac{10}{36} = \frac{11}{36}$$

The least common denominator of two or more fractions is the least common multiple of their denominators.

The answers 31/36 and 11/36 are in simplest form since 31 and 36 are relatively prime and 11 and 36 are relatively prime. In general, the least common multiple of the denominators of two or more fractions is the **least common denominator** (LCD) of the fractions. For any two like fractions a/c and b/c, the least common denominator is c.

$$\frac{a}{c} + \frac{b}{c} = \frac{a+b}{c} \qquad \frac{a}{c} - \frac{b}{c} = \frac{a-b}{c}$$

In adding or subtracting fractions it is customary to use the least common denominator. However, any common multiple of the denominators may be used.

For any two numbers represented by fractions a/b and c/d

$$\frac{a}{b} = \frac{a}{b} \times \frac{d}{d} = \frac{ad}{bd} \qquad \frac{c}{d} = \frac{b}{b} \times \frac{c}{d} = \frac{bc}{bd}$$

$$\frac{a}{b} + \frac{c}{d} = \frac{ad}{bd} + \frac{bc}{bd} = \frac{ad + bc}{bd}$$

$$\frac{a}{b} - \frac{c}{d} = \frac{ad}{bd} - \frac{bc}{bd} = \frac{ad - bc}{bd}$$

Example 4 Subtract $\frac{3}{4} - \frac{2}{5}$.

Solution LCM(4, 5) = 20. To subtract the given numbers first express the fractions as like fractions with common denominator 20.

$$\frac{3}{4} - \frac{2}{5} = \frac{3}{4} \times \frac{5}{5} - \frac{2}{5} \times \frac{4}{4}$$

$$= \frac{3 \times 5}{4 \times 5} - \frac{2 \times 4}{5 \times 4}$$

$$= \frac{15}{20} - \frac{8}{20} = \frac{7}{20} \quad \blacksquare$$

Example 5 Simplify $\frac{7}{12} - \frac{11}{20}$.

Solution The least common multiple of $12 = 2^2 \times 3$ and $20 = 2^2 \times 5$ is $60 = 2^2 \times 3 \times 5$.

Recall that the instruction "simplify" is used to mean "perform the indicated operations and express the answer in simplest form."

$$\frac{7}{12} - \frac{11}{20} = \frac{7}{2^2 \times 3} \times \frac{5}{5} - \frac{11}{2^2 \times 5} \times \frac{3}{3}$$

$$= \frac{35}{60} - \frac{33}{60} = \frac{2}{60} = \frac{1}{30} \quad \blacksquare$$

EXERCISES *List the first three common multiples of each pair of numbers.*

1. 3 and 4
2. 4 and 5
3. 3 and 5
4. 5 and 7
5. 3 and 9
6. 4 and 9
7. 10 and 30
8. 12 and 20
9. 14 and 20
10. 2 and 17
11. 32 and 48
12. 48 and 60

Write the prime factorization and find the least common multiple of the given numbers.

13. 14 and 40
14. 68 and 96
15. 123 and 287
16. 96 and 1425
17. 123 and 615
18. 285 and 1425
19. 68 and 112
20. 112 and 480
21. 600 and 800
22. 1850 and 7400
23. 2450 and 3500
24. 2025 and 5400
25. 12, 18, and 21
26. 18, 24, and 45
27. 15, 25, and 40
28. 12, 30, and 48
29. 10, 20, and 35
30. 24, 40, and 64

Simplify.

31. $\frac{5}{8} + \frac{7}{12}$
32. $\frac{5}{12} + \frac{9}{16}$
33. $\frac{9}{10} + \frac{1}{15}$
34. $\frac{11}{12} + \frac{7}{18}$
35. $\frac{11}{12} - \frac{7}{15}$
36. $\frac{7}{10} - \frac{5}{18}$
37. $\frac{11}{14} - \frac{3}{40}$
38. $\frac{7}{68} + \frac{13}{96}$
39. $\frac{9}{123} - \frac{7}{615}$
40. $\frac{5}{68} + \frac{7}{112}$
41. $\frac{17}{800} - \frac{13}{600}$
42. $\frac{7}{285} + \frac{2}{1425}$

Classify each statement as true or false. If false, give a specific counterexample to justify your answer.

43. The least common multiple of two prime numbers is their product.
44. If a number m is a multiple of another number n, then their least common multiple is m.
45. There is no greatest common multiple for two counting numbers.
46. The least common multiple of a prime number and composite number is their product.

State necessary conditions upon m and n such that each statement is true.

47. $\text{LCM}(m, n) = m$
48. $\text{LCM}(m, n) = mn$
49. $\text{GCD}(m, n) = m$

EXPLORATIONS

Solve each problem. Identify a possible use of common factors, greatest common factors, common multiples, or least common multiples in solving problems of each type.

1. An interior decorator would like a wallpaper pattern that would fit exactly in a room with walls 8 feet high and also fit exactly under window sills that are 30 inches from the floor. What is the height of the largest pattern that can be considered?

2. Describe the sizes of square tiles that could be used for the floor of a room 10 feet by 15 feet without cutting any tiles.
3. Describe some of the shapes of rooms whose floors can be tiled completely using tiles 9 inches square without cutting any tiles.
4. A manufacturer ships one product in packages 4 by 6 by 6 inches and another in packages 2 by 8 by 9 inches. Cartons are to be made that can be used without extra space for one dozen of either product. Find the dimensions of at least three shapes of cartons that are at most 2 feet on their longest side.

5. Venn diagrams with the prime factors, including repeated factors, as elements may be used to represent the procedures for finding the greatest common factor and the least common multiple of any two or more given natural numbers. Consider for example

$12 = 2 \times 2 \times 3$ $\qquad$ $18 = 2 \times 3 \times 3$

12: (2 2 3) $\qquad$ 18: (2 3 3) $\qquad$ 12, 18: (2 (2 3) 3)

$\text{GCD} = 2 \times 3$

$\text{LCM} = 2 \times 2 \times 3 \times 3$

Explore this use of Venn diagrams for finding the greatest common divisor and the least common multiple of the given numbers.
(a) 8 and 18 **(b)** 27 and 36 **(c)** 12 and 35
(d) 9, 12, and 15 **(e)** 18, 24, and 60 **(f)** 15, 24, and 50

6. Explain why for any two natural numbers m and n the product of LCM(m, n) and GCD(m, n) must be equal to mn, the product of the given numbers. Then describe a method for finding LCM(m, n) whenever m, n, and GCD(m, n) are known.
7. Consider the mathematical system that consists of the set of natural numbers and the operation $*$ defined to mean "find the least common multiple of the two numbers."
(a) Find $7 * 9$ and $12 * 5$.
(b) Is the set of natural numbers closed with respect to $*$?
(c) For natural numbers m and n, does $m * n = n * m$?
(d) For natural numbers m, n, and p, does $m * (n * p) = (m * n) * p$?
(e) Is there an identity element with respect to $*$? If so, what is it?

5-6 Euclidean Algorithm and Applications

The distributive property of multiplication

$$ca + cb = c(a + b) \qquad ca - cb = c(a - b)$$

was used in Section 5-1 to show that if c is a divisor of both ac and bc, then c is a divisor of $ac + bc$ and of $ac - bc$. Euclid showed that this idea can be used to find the greatest common divisor of any two natural numbers. For Example 2 of Section 5-4 Euclid would have used these steps.

$$5280 = 1 \times 3850 + 1430$$
$$3850 = 2 \times 1430 + 990$$
$$1430 = 1 \times 990 + 440$$
$$990 = 2 \times 440 + 110$$
$$440 = 4 \times 110 + 0$$

Note the repeated shifts indicated by the arrows.

Then considering the steps in reverse order

110 is a divisor of 440.

110 is a common divisor of 110 and 440 and is therefore a divisor of 990.

110 is a common divisor of 440 and 990 and is therefore a divisor of 1430.

110 is a common divisor of 990 and 1430 and is therefore a divisor of 3850.

110 is a common divisor of 1430 and 3850 and is therefore a divisor of 5280.

Therefore 110 is a common divisor of the given numbers 5280 and 3850. Also the equations at each step may be rewritten as

$$1430 = 5280 - 1 \times 3850$$
$$990 = 3850 - 2 \times 1430$$
$$440 = 1430 - 1 \times 990$$
$$110 = 990 - 2 \times 440$$

to show that

Any common divisor of 5280 and 3850 is a divisor of 1430.

Any common divisor of 3850 and 1430 is a divisor of 990.

Any common divisor of 1430 and 990 is a divisor of 440.

Any common divisor of 990 and 440 is a divisor of 110.

Therefore 110 is a common divisor of 5280 and 3850 and is divisible by all common divisors, that is, 110 is the greatest common divisor of 5280 and 3850.

This procedure for finding the greatest common divisor is called the **Euclidean algorithm**. The procedure is particularly useful because it can be performed using only repeated subtractions without considering factors, divisibility, multiplication, or the usual form of division. Considering 5280 and 3850 once more, note that

3850 could be subtracted from 5280 once and the remainder was 1430.

1430 could be subtracted from 3850 twice and the remainder was 990.

990 could be subtracted from 1430 once and the remainder was 440.

440 could be subtracted from 990 twice and the remainder was 110.

110 could be subtracted from 440 four times and the remainder was 0.

The last nonzero remainder in the Euclidean algorithm procedure is the greatest common divisor of the two given numbers.

Example 1 Use the Euclidean algorithm to find GCD(19, 23).

Solution

$$23 = 1 \times 19 + 4$$
$$19 = 4 \times 4 + 3$$
$$4 = 1 \times 3 + 1$$
$$3 = 3 \times 1 + 0 \qquad \text{GCD}(19, 23) = 1 \quad \blacksquare$$

The Euclidean algorithm and the relation

$$\text{LCM}(m, n) = \frac{mn}{\text{GCD}(m, n)}$$

(see Exploration 6(b) of Section 5-5) may be used to find the least common multiple of any two natural numbers. Note that since 23 and 19 are both prime numbers, LCM(23, 19) = 23 × 19, that is, 437.

The steps of the Euclidean algorithm can also be used to express the greatest common divisor of any two natural numbers as a difference of multiples of the two numbers. For 23 and 19 consider the steps in reverse order.

$$1 = 4 - 1 \times 3$$
$$3 = 19 - 4 \times 4$$
$$4 = 23 - 1 \times 19$$

Substitute for the remainder 3 from the second equation into the first.

$$1 = 4 - 1 \times (19 - 4 \times 4)$$
$$1 = -1 \times 19 + [1 - (-4)] \times 4 = -1 \times 19 + 5 \times 4$$

Then substitute for the remainder 4 from the third equation.

$$1 = -1 \times 19 + 5 \times (23 - 1 \times 19)$$
$$1 = 5 \times 23 - 6 \times 19$$

Note that care must be taken not to perform operations too soon since the resulting equality at each step, 1 = 1, is not informative.

The relation $1 = 2 \times 3 - 1 \times 5$ from Example 2 has historical significance since the equation

$$\frac{1}{15} = 2 \times \frac{3}{15} - 1 \times \frac{5}{15}$$
$$\frac{1}{15} = 2 \times \frac{1}{5} - 1 \times \frac{1}{3}$$

obtained by dividing both members of the previous equation by 15 enabled Carl Friedrich Gauss to inscribe a regular polygon of 15 sides in a circle. The desired central angle was simply the difference of two central angles of a regular pentagon and one of an equilateral triangle.

Example 2 Express GCD(5, 3) as a difference of multiples of 5 and 3.

Solution By the Euclidean algorithm

$$5 = 1 \times 3 + 2$$
$$3 = 1 \times 2 + 1$$
$$2 = 2 \times 1 + 0 \qquad \text{GCD}(5, 3) = 1$$

Then

$$1 = 3 - 1 \times 2$$

$$1 = 3 - 1 \times (5 - 1 \times 3)$$

$$1 = 2 \times 3 - 1 \times 5 \quad \blacksquare$$

Example 3 Express GCD(147, 64) as a difference of multiples of the given numbers.

Solution

$$147 = 2 \times 64 + 19$$

$$64 = 3 \times 19 + 7$$

$$19 = 2 \times 7 + 5$$

$$7 = 1 \times 5 + 2$$

$$5 = 2 \times 2 + 1$$

$$2 = 2 \times 1 + 0 \qquad \text{GCD}(147, 64) = 1$$

Then

$$1 = 5 - 2 \times 2 = 5 - 2 \times (7 - 1 \times 5) = -2 \times 7 + 3 \times 5$$

$$1 = -2 \times 7 + 3 \times (19 - 2 \times 7) = 3 \times 19 - 8 \times 7$$

$$1 = 3 \times 19 - 8 \times (64 - 3 \times 19) = -8 \times 64 + 27 \times 19$$

$$1 = -8 \times 64 + 27 \times (147 - 2 \times 64) = 27 \times 147 - 62 \times 64$$

Check: $1 = 3969 - 3968$ ■

EXERCISES *Classify each statement as true or false. In each case give three numerical examples or counterexamples to explain your answer.*

1. If $\text{GCD}(m, n) = 1$, then m and n are relatively prime and $\text{LCM}(m, n) = mn$.
2. If $\text{LCM}(m, n) = mn$, then m and n are relatively prime and $\text{GCD}(m, n) = 1$.
3. If $\text{GCD}(m, n) = m$, then $\text{LCM}(m, n) = n$.
4. If $\text{LCM}(m, n) = m$, then $\text{GCD}(m, n) = \text{n}$.

Use the Euclidean algorithm to find the greatest common divisor and the least common multiple of the given numbers.

5. 66 and 96
6. 125 and 450
7. 154 and 462
8. 468 and 546
9. 1236 and 1545
10. 1506 and 1530

For Exercises 11 through 16 express the greatest common divisor of the given numbers as a difference of multiples of the numbers.

11. 15 and 40
12. 111 and 370
13. 64 and 108
14. 64 and 147
15. 952 and 3961
16. 1275 and 4680

17. Cal is home for lunch every third day; Doris is home every fourth day. How often are they able to have lunch at home together?
18. A pan of brownies is to be cut so that each member of the family can have the same number. How many brownies should there be if four members of the family will definitely be present and they may or may not be joined by **(a)** one other member? **(b)** two other members?
19. Bob and Jean are both off from work today. Bob has every sixth day off; Jean has every fifth day off. How long will it be before the next day they are both off?
20. Two bus drivers, Virginia and Ralph, both arrived at the railway station at 8 A.M. Virginia arrives every 30 minutes. Find the time at which they will next both arrive at the station if Ralph arrives **(a)** every 35 minutes **(b)** every 40 minutes **(c)** every 42 minutes **(d)** every 45 minutes.

EXPLORATIONS

Each exploration is concerned with the division of a long strip of paper into pieces of equal length.

1. John knows how to divide any strip of paper into two pieces of equal length.
   (a) Explain how John can divide any given strip of paper into 2, 4, 8, or 16 pieces of equal length.
   (b) Give a general expression for the number of pieces of equal length that can be obtained in this way.
2. Betty knows how to divide a strip of paper only into either two pieces of equal length or three pieces of equal length. List the numbers n less than 20 such that Betty can divide any given strip of paper into n pieces of equal length.
3. Repeat Exploration 2 if procedures are known only for dividing a given strip of paper into 2, 3, or 5 pieces of equal length.
4. Extend the list in Exploration 2 for n less than 50 and give a general expression for these numbers.
5. Repeat Exploration 4 for the list in Exploration 3.

Chapter Review

1. Find the factors of **(a)** 42 **(b)** 71.
2. Use the general divisibility rules to test 2358 for divisibility by **(a)** 3 **(b)** 4 **(c)** 6 **(d)** 9.
3. Express 98 as a product of as many different pairs of natural numbers as possible.
4. List the prime numbers between 70 and 80.
5. What is the largest prime number that you need to consider to be sure that you have excluded all composite numbers less than 600?

Find the number represented by the given prime factorization.

6. $2^3 \times 3^2 \times 5^3$

7. $2 \times 5^2 \times 7 \times 11$

Find the prime factorization of the given number.

8. 60

9. 1500

Find the set of common divisors for the given pair of numbers.

10. 9 and 15

11. 24 and 40

Find the greatest common divisor of the given numbers.

12. 120 and 140

13. 24, 30, and 42

In Exercises 14 *and* 15 *simplify.*

14. $\dfrac{24}{108}$

15. $\dfrac{130}{195}$

16. List the first three common multiples for 5 and 9.

Find the least common multiple of the given numbers.

17. 90 and 1500

18. 12, 20, and 24

In Exercises 19 *through* 21, *simplify.*

19. $\dfrac{7}{12} + \dfrac{5}{16}$

20. $\dfrac{11}{123} - \dfrac{7}{615}$

21. $\dfrac{13}{2^2 \times 3} - \dfrac{17}{2 \times 3^2}$

22. Use the Euclidean algorithm to find the greatest common divisor and the least common multiple of 48 and 120.

23. Express the greatest common divisor of 48 and 120 as a difference of multiples of the numbers.

Under what conditions, if any, upon m and n is the given statement true?

24. $\text{LCM}(m, n) = \text{GCD}(m, n)$

25. $\text{LCM}(m, n) < \text{GCD}(m, n)$

THE EUCLIDEAN ALGORITHM AND THE COMPUTER

The following algorithm, written in BASIC, makes use of several advanced programming techniques to calculate the greatest common divisor of two natural numbers using the Euclidean Algorithm. Several of the commands are beyond the scope of this book, but you can still type the program into a computer and use it. Note the use of arbitrary line numbers to number the lines of the program. These numbers can be any natural numbers less than 64,000. The only requirement is that their order matches the order in which the steps of the program are to be executed. The missing numbers in the sequence of line numbers allow one to insert additional lines at a later time in the program's development.

```
  10  HOME
  20  PRINT "THIS PROGRAM USES THE EUCLIDEAN"
  21  PRINT
  22  PRINT "ALGORITHM TO FIND THE GCD OF TWO"
  23  PRINT
  24  PRINT "NATURAL NUMBERS."
  30  PRINT
  40  PRINT
  50  PRINT "ENTER THE TWO NATURAL NUMBERS A AND B."
  52  INPUT A,B
  55  IF A < = 0 OR B < = 0 THEN PRINT "BOTH NUMBERS
         MUST BE NATURAL NUMBERS!": GOTO 50
  56  IF A  < >  INT(A) OR B  < >  INT(B) THEN PRINT
         "BOTH NUMBERS MUST BE NATURAL NUMBERS!" GOTO 50
  60  X = A
  61  Y = B
  70  DIM A(25),B(25),Q(25),R(25)
  80  IF X > = Y THEN A(1) = X
  90  IF X > = Y THEN B(1) = Y
 100  IF X < Y THEN A(1) = Y
 110  IF X < Y THEN B(1) = X
 120  C = 1
 130  Q = (A(C) / B(C))
 140  Q(C) = INT(Q)
 150  R(C) = A(C) - Q(C) * B(C)
 160  IF R(C) = 0 THEN GOTO 1000
 170  C = C + 1
 180  A(C) = B(C - 1)
 190  B(C) = R(C - 1)
 200  GOTO 130
1000  HOME
1020  PRINT "GCD ( ";X;", ";Y;" ) = ";R(C - 1)
1030  PRINT
1035  PRINT
1040  PRINT "A = Q * B + R"
2000  C = 1
2010  IF A(C) = 0 THEN END
2020  PRINT "C = ";C;" ";A(C);" = ";Q(C);" * ";B(C);" + ";R(C)
2030  PRINT
2040  C = C + 1
2050  GOTO 2010
```

This program for performing the Euclidean algorithm for two natural numbers A and B requires the program to check in lines 80 through 110 to see which of the two numbers A and B is larger, if either. Lines 120 through 200 then instruct the computer to carry out the repeated division process which is at the heart of the Euclidean algorithm. Lines 1000 through 2050 contain the instructions on how to print out the results of the process.

The use of this program to find the greatest common divisor of 23,456 and 71 is shown by the following printout:

```
RUN

THIS PROGRAM USES THE EUCLIDEAN

ALGORITHM TO FIND THE GCD OF TWO

NATURAL NUMBERS.


ENTER THE TWO NATURAL NUMBERS A AND B.
? 23456, 71
GCD ( 23456, 71 ) = 1

A = Q * B + R

C = 1    23456 = 330 * 71 + 26

C = 2    71 = 2 * 26 + 19

C = 3    26 = 1 * 19 + 7

C = 4    19 = 2 * 7 + 5

C = 5    7 = 1 * 5 + 2

C = 6    5 = 2 * 2 + 1

C = 7    2 = 2 * 1 + 0
```

The notations C = 1 through C = 7 at the left of the steps used in determining the greatest common divisor, 1, indicate the number of iterations, or repetitions, of the division process in finding the desired result.

If possible, use the program on a computer to find the greatest common divisor of two numbers in several cases. Then see if you can predict the maximum number of iterations needed to find the greatest common divisor of two given large natural numbers. You might wish to look up a theorem called *Lamé's theorem* in an advanced number theory book as a help.

6

Rational Numbers

Simon Stevin
(1548–1620)

Simon Stevin was a Flemish engineer and mathematician who lived in Brugge, Belgium. His 1585 text De Thiende *(The Tenth) provided the first popular explanation of the use of decimal fractions for Europeans. While the concept had been recommended earlier by a number of other mathematicians, Stevin was the first to provide a systematic introduction to them. He also popularized the use of double-entry bookkeeping.*

6-1 Concept of Rational Number

There are many practical problems that cannot be solved when only the integers are available.

Not one of these problems has a solution if we are restricted to the use of integers only. Think of several examples of your own need for fractions.

1. A cook cannot use a recipe that calls for 3 cups of flour if only one-half the quantity is desired. The problem $3 \div 2$ has no solution in the system of integers.
2. A child cannot find an integer to represent one-half of a candy bar; $1 \div 2$ is not an integer.
3. A driver does not have an integer to represent the average number of miles per gallon of gas for a car that travels 200 miles using 7 gallons; $200 \div 7$ has no solution in the system of integers.
4. When grapefruit are three for a dollar, a different price must be charged for one grapefruit; $100 \div 3$ has no solution in the system of integers.

Furthermore, with only the set of integers at our disposal, we are unable to find replacements for n that make these sentences true.

$$2 \times n = 7 \qquad n + \frac{1}{4} = \frac{7}{8} \qquad 5 \div 3 = n$$

The solution set for each of the preceding sentences is the empty set if only integers may be used as possible replacements for n. To have numbers as solutions of such sentences, as well as to make division always possible (except division by zero), we must extend our set of numbers. In this section we make such an extension and call this new set of numbers the set of rational numbers.

> A **rational number** is a number that can be expressed in the form
>
> $$\frac{a}{b}$$
>
> (often written a/b), where a and b are integers and $b \neq 0$.

Each of the following fractions represents a rational number.

$$\frac{2}{3} \quad \frac{1}{2} \quad \frac{-7}{3} \quad \frac{0}{1} \quad \frac{215}{524}$$

Note that $b \neq 0$ for any fraction a/b, since division by zero is not permitted (see Section 3-4).

From the definition every rational number can be represented by a quotient a/b. Thus $a \div b$ is also an expression for a rational number. The expression a/b is a **fraction** with *numerator* a and *denominator* b. With the set of rational numbers at our disposal each of the problems and sentences considered in the first paragraph of this section has a unique solution.

If a rectangular region is divided equally into six parts and one part is shaded, we say that one-sixth of the rectangular region is shaded and five-sixths are not shaded. For each of the fractions 1/6 and 5/6 the *denominator* 6 indicates the portion (*denomination* 1/6) of the basic unit (given rectangular region) that is used as a subunit to measure the regions. The *numerator* indicates the *number* of the subunits equivalent to the measured region.

$a \div b = \frac{a}{b} = a \times \frac{1}{b}$

$$\frac{1}{6} = 1 \times \frac{1}{6} \qquad \frac{5}{6} = 5 \times \frac{1}{6} \qquad \frac{a}{b} = a \times \frac{1}{b}$$

Fractions indicate comparisons.

The fractions 1/6 and 5/6 indicate comparisons of the regions (shaded and unshaded) being measured with the basic unit (given rectangular region). Any fraction a/b may be considered as a *ratio* of a to b indicating a comparison of a measured quantity to a basic unit.

$1 = \frac{k}{k}$ for any natural number k.

Any two fractions that represent the same number are **equivalent fractions**. Each division of a rectangular region into k parts provides a basis for recognizing 1 in the form k/k for a natural number k. Thus the fractions k/k are equivalent and we write

$b \times \frac{1}{b} = \frac{b}{b} = 1$ for any natural number b.

$$\frac{1}{1} = \frac{2}{2} = \frac{3}{3} = \frac{4}{4} = \frac{5}{5} = \cdots$$

In the figure that follows, the rectangular region has been divided into fourths. Three of these parts have been shaded to denote 3/4 of the original rectangular region.

This rectangular region may be divided into halves so that each of the fourths is also divided into halves.

$$\frac{3}{4} = \frac{3 \times 2}{4 \times 2} = \frac{6}{8}$$

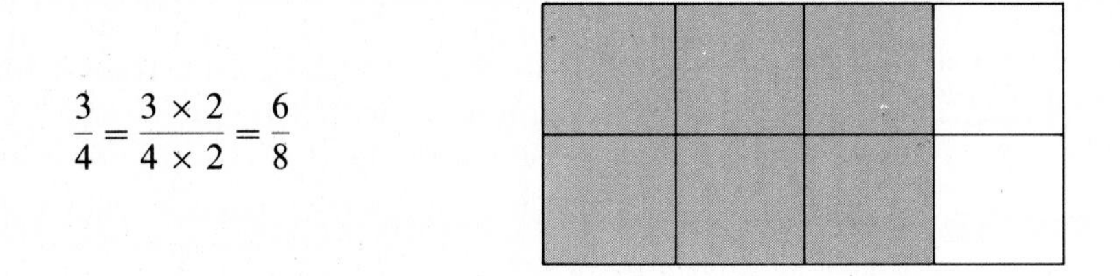

The original rectangular region also may be divided into thirds so that each of the fourths is also divided into thirds.

$$\frac{3}{4} = \frac{3 \times 3}{4 \times 3} = \frac{9}{12}$$

The fractions 3/4, 6/8, and 9/12 represent the same region in terms of different subunits. Relative to the basic unit they represent the same quantity and thus the same number.

$$\frac{3}{4} = \frac{6}{8} = \frac{9}{12}$$

In general we have the familiar rule that is sometimes called the **fundamental law of fractions.**

$$\frac{a \times k}{b \times k} = \frac{a}{b}$$

For any rational number a/b and any integer $k \neq 0$

$$\frac{a}{b} \times \frac{k}{k} = \frac{ak}{bk} = \frac{a}{b}$$

The numerator and the denominator of any fraction may be multiplied or divided by any integer $k \neq 0$ without changing the rational number represented by the fraction. In Section 5-4 we used this form of the fundamental law of fractions to express fractions in simplest form.

Example 1 Express each fraction in simplest form.

(a) $\frac{6}{10}$ **(b)** $\frac{44}{66}$ **(c)** $\frac{72}{100}$

Solution Use the greatest common divisor of the numerator and the denominator.

(a) $\frac{6}{10} = \frac{3 \times 2}{5 \times 2} = \frac{3}{5}$

(b) $\frac{44}{64} = \frac{11 \times 4}{16 \times 4} = \frac{11}{16}$

(c) $\frac{72}{100} = \frac{18 \times 4}{25 \times 4} = \frac{18}{25}$ ■

The fundamental law of fractions enables us to represent any rational number by as many different-appearing fractions as we wish. For example,

$$\frac{2}{3} = \frac{4}{6} = \frac{6}{9} = \frac{8}{12} = \cdots = \frac{-2}{-3} = \frac{-4}{-6} = \frac{-6}{-9} = \frac{-8}{-12} = \cdots$$

In general, two given fractions a/b and c/d represent the same rational number if and only if $ad = bc$.

$$\boxed{\frac{a}{b} = \frac{c}{d} \quad \text{if and only if} \quad ad = bc}$$

Example 2 Find replacements for n to make the given statement a true statement.

(a) $\frac{2}{5} = \frac{6}{n}$ (b) $\frac{2}{n} = \frac{6}{9}$ (c) $\frac{5}{3} = \frac{n}{6}$

Solution We use the definition $a/b = c/d$ if and only if $ad = bc$.
(a) $2 \times n = 6 \times 5, \quad 2n = 30, \quad n = 15$
(b) $2 \times 9 = 6 \times n, \quad 18 = 6n, \quad n = 3$
(c) $5 \times 6 = 3 \times n, \quad 30 = 3n, \quad n = 10$ ■

If a given fraction has a negative denominator, an equivalent fraction can be found with a positive denominator. For example,

$$\frac{4}{-5} = \frac{4 \times (-1)}{(-5) \times (-1)} = \frac{-4}{5}$$

A fraction m/n is a **proper fraction** if m and n are natural numbers and m is less than n, that is, $0 < m/n < 1$. Any fraction that is not equivalent to a proper fraction may be called an **improper fraction**. Thus

$$\frac{7}{2}, \quad \frac{40}{10}, \quad \frac{5}{-8}, \quad \frac{-3}{4}, \quad \frac{6}{6}$$

are improper fractions. Numbers represented by improper fractions, such as 7/2, can be expressed in **mixed form** as the sum of a natural number and a number represented by a proper fraction.

$$\frac{7}{2} = 7 \div 2 = 3 + \frac{1}{2} = 3\frac{1}{2}$$

Example 3 Represent each expression by a single fraction with a natural number as denominator.

(a) $2 \div 7$ (b) $1\frac{3}{8}$ (c) $3/(-5)$

Solution (a) $2 \div 7 = \dfrac{2}{7}$

(b) $1\dfrac{3}{8} = 1 + \dfrac{3}{8} = \dfrac{8}{8} + \dfrac{3}{8} = \dfrac{11}{8}$

(c) $3/(-5) = \dfrac{3}{-5} = \dfrac{3 \times (-1)}{(-5) \times (-1)} = \dfrac{-3}{5}$ ■

$m = \dfrac{m}{1}$

Rational numbers on a number line.

Every integer is a rational number since $m = m/1$. Every rational number is the coordinate of exactly one point on a number line. Consider a number line and locate a point midway between the points with coordinates 0 and 1. The coordinate of the midpoint is called $\frac{1}{2}$. Then locate all points that correspond to "halves," such as $\frac{3}{2} = 1\frac{1}{2}$, $\frac{5}{2} = 2\frac{1}{2}$, $\frac{7}{2} = 3\frac{1}{2}$, and so forth. Also locate points with coordinates $-\frac{1}{2}$, $-1\frac{1}{2}$, $-2\frac{1}{2}$, $-3\frac{1}{2}$, and so forth. The graph of any rational number $n/2$, where n is an integer, may be found in this way.

$-3 \quad -2\frac{1}{2} \quad -2 \quad -1\frac{1}{2} \quad -1 \quad -\frac{1}{2} \quad 0 \quad \frac{1}{2} \quad 1 \quad 1\frac{1}{2} \quad 2 \quad 2\frac{1}{2} \quad 3$

$0 = \frac{0}{3} \quad \frac{1}{3} \quad \frac{2}{3} \quad \frac{3}{3} = 1$

Rational numbers $n/3$ may be graphed on a number line. The graphs of $\frac{1}{3}$ and $\frac{2}{3}$ may be found by dividing the unit interval from 0 to 1 into three congruent parts. Then the points corresponding to 4/3, 5/3, 6/3, 7/3, ... and also the points corresponding to $-1/3$, $-2/3$, $-3/3$, $-4/3$, ... may be located. Similarly, the points corresponding to multiples of 1/4, 1/5, 1/6, and so forth, may be located on a number line.

The set of rational numbers can be classified as being positive (greater than 0), negative (less than 0), or 0. The number 0 is a rational number, but it is neither positive nor negative. As in the case of integers, positive rational numbers are the coordinates of points on the right of the origin and negative rational numbers are coordinates of points on the left of the origin.

Any two rational numbers can be represented by fractions in simplest form with natural numbers as denominators. The least common multiple of these denominators may be found as in Section 5-5 and used as the common denominator of *like fractions* to represent the given rational numbers. Such representations are particularly useful since numbers represented by like fractions can be easily added by using the distributive law for multiplication with respect to addition.

$\dfrac{a}{b} + \dfrac{c}{b} = \dfrac{a+c}{b}$

$$\frac{a}{b} + \frac{c}{b} = \left(a \times \frac{1}{b}\right) + \left(c \times \frac{1}{b}\right) = (a + c) \times \frac{1}{b} = \frac{a+c}{b}$$

Example 4 Represent each pair of fractions by like fractions.

(**a**) $\frac{1}{2}$ and $\frac{1}{3}$ (**b**) $\frac{5}{6}$ and $\frac{3}{4}$ (**c**) $\frac{7}{8}$ and $\frac{3}{4}$

Solution (**a**) LCM(2, 3) = 6

$$\frac{1}{2} = \frac{1 \times 3}{2 \times 3} = \frac{3}{6} \qquad \frac{1}{3} = \frac{1 \times 2}{3 \times 2} = \frac{2}{6}$$

(**b**) LCM(6, 4) = 12

$$\frac{5}{6} = \frac{5 \times 2}{6 \times 2} = \frac{10}{12} \qquad \frac{3}{4} = \frac{3 \times 3}{4 \times 3} = \frac{9}{12}$$

(**c**) LCM(8, 4) = 8

$$\frac{7}{8} = \frac{7}{8} \qquad \frac{3}{4} = \frac{3 \times 2}{4 \times 2} = \frac{6}{8}$$ ■

We have already shown that rational numbers such as 4/(−5) can be expressed in the form (−4)/5. We can now show that 4/5 and (−4)/5 are opposites of one another, that is, their sum is 0.

$$\frac{4}{5} + \frac{-4}{5} = [4 + (-4)] \times \frac{1}{5} = 0 \times \frac{1}{5} = 0$$

In general, there are three signs associated with any fraction a/b.

The sign of the fraction, as in $-\frac{c}{d}$.

The sign of the numerator, as in $\frac{-c}{d}$.

The sign of the denominator, as in $\frac{c}{-d}$.

Thus any two of the three signs associated with a fraction may be changed without changing the value of (rational number represented by) the fraction.

$$\boxed{-\frac{c}{d} = \frac{-c}{d} = \frac{c}{-d}}$$

Example 5 Simplify $\frac{6}{-8}$.

Solution The denominator should be a counting number.

This result is usually written as − 3/4.

$$\frac{6}{-8} = \frac{-6}{8} = \frac{-(3 \times 2)}{4 \times 2} = \frac{-3}{4}$$ ■

2 II $\frac{1}{2}$ ⌾II

3 III $\frac{1}{3}$ ⌾III

About four thousand years ago the early Egyptians annotated their symbols for counting numbers to obtain symbols for the **unit fractions** 1/2, 1/3, 1/4, 1/5, ... Each unit fraction $1/n$ was used to denote the measure of one of n parts when a whole quantity was divided equally into n parts. The early Egyptians had numerals only for 2/3, 3/4, and the unit fractions. They could indicate 5/6 either as 2/3 and 1/6 or as 1/2 and 1/3. Indeed, they could represent any positive rational number as an integer, a sum of numbers represented by unit fractions, or a sum of an integer and numbers represented by unit fractions.

Example 6 Express each rational number as an integer, a sum of numbers represented by unit fractions, or a sum of an integer and numbers represented by unit fractions.

(a) $\frac{5}{2}$ **(b)** $\frac{40}{10}$ **(c)** $\frac{5}{7}$

Solution Among others:

(a) $\frac{5}{2}=\frac{2+2+1}{2}=\frac{2}{2}+\frac{2}{2}+\frac{1}{2}=1+1+\frac{1}{2}=2+\frac{1}{2}=2\frac{1}{2}$

(b) $\frac{40}{10}=\frac{10+10+10+10}{10}=\frac{10}{10}+\frac{10}{10}+\frac{10}{10}+\frac{10}{10}=1+1+1+1=4$

(c) $\frac{5}{7}=\frac{10}{14}=\frac{7+2+1}{14}=\frac{7}{14}+\frac{2}{14}+\frac{1}{14}=\frac{1}{2}+\frac{1}{7}+\frac{1}{14}$ ■

EXERCISES *Each fraction m/n has a numerator m and a denominator n.*

1. Identify the numerator of **(a)** $\frac{2}{3}$ **(b)** $\frac{1}{5}$ **(c)** $\frac{2}{7}$ **(d)** $\frac{0}{7}$.
2. Identify the denominator of each fraction in Exercise 1.

Express as a fraction $\frac{a}{b}$.

3. $7\frac{1}{8}$
4. $2\frac{7}{8}$
5. $-4\frac{2}{3}$
6. $-1\frac{9}{10}$

Express in mixed form.

7. $\frac{13}{4}$
8. $\frac{7}{2}$
9. $-\frac{15}{7}$
10. $-\frac{19}{3}$

Express each fraction as an integer, a sum of numbers represented by unit fractions, or a sum of an integer and numbers represented by unit fractions.

11. $\frac{7}{3}$
12. $\frac{15}{4}$
13. $\frac{5}{8}$
14. $\frac{7}{12}$

15. $\frac{4}{9}$ 16. $\frac{7}{8}$ 17. $\frac{25}{4}$ 18. $\frac{26}{5}$

19. $\frac{30}{6}$ 20. $\frac{67}{10}$ 21. $\frac{73}{10}$ 22. $\frac{78}{15}$

Simplify.

23. $\frac{18}{60}$ 24. $\frac{45}{70}$ 25. $\frac{48}{-64}$ 26. $\frac{32}{-36}$

Find a replacement for n to make the given statement a true statement.

27. $\frac{2}{3} = \frac{n}{12}$

28. $\frac{3}{4} = \frac{9}{n}$

29. $\frac{7}{n} = \frac{21}{30}$

30. $\frac{18}{n} = \frac{9}{24}$

31. $\frac{7}{12} = \frac{n}{60}$

32. $\frac{8}{15} = \frac{n}{120}$

33. $\frac{5}{6} = \frac{n}{30}$

34. $\frac{7}{8} = \frac{28}{n}$

35. $\frac{9}{n} = \frac{81}{72}$

36. $\frac{n}{7} = \frac{42}{49}$

37. $\frac{100}{n} = \frac{10}{3}$

38. $\frac{144}{12} = \frac{n}{1}$

Represent each expression by a single fraction with a natural number as denominator.

39. $1\frac{3}{4}$ 40. $2\frac{5}{8}$ 41. $3 \div 5$

42. $11 \div 7$ 43. $\frac{6}{-7}$ 44. $\frac{5}{-12}$

45. $4 \div (-3)$ 46. $9 \div (-5)$

Represent each pair of fractions by like fractions.

47. $\frac{2}{3}$ and $\frac{3}{4}$

48. $\frac{2}{5}$ and $\frac{3}{4}$

49. $\frac{3}{8}$ and $\frac{5}{16}$

50. $\frac{4}{5}$ and $\frac{7}{10}$

51. $\frac{5}{12}$ and $\frac{3}{8}$

52. $\frac{7}{8}$ and $\frac{3}{10}$

EXPLORATIONS

1. Rational numbers that are not integers are sometimes called **fractional numbers**. One way to introduce the need for fractional numbers is to present a class with a set of questions and ask whether a fractional number or a whole number is needed to answer the question. For example, consider such questions as these.

   How many students are in this class?

   A pound of butter is usually packed as four bars. Each bar is what part of a pound?

   Prepare a set of five similar questions that illustrate this difference between natural numbers and fractional numbers.

2. Make a collection of different ways that rational numbers are used in practical situations. For example, stock market reports are quoted in fractional terms and generally end with a final column that indicates the daily rise or fall of particular stocks in terms of positive or negative rational numbers.

3. Prepare a worksheet of ten examples based on a stock market report from a daily newspaper and suitable for upper elementary or junior high school students.

4. Take a small strip of paper about 12 inches long and label it 0 at the left end and 1 at the right end. Then fold the paper once to obtain the midpoint of the strip. Label this point as 1/2. Continue with this paper-folding experiment to construct points representing the fourths and points representing the eighths. The corresponding number line can be very helpful in giving young students an intuitive concept of order.

   $0 \quad \frac{1}{8} \quad \frac{1}{4} \quad \frac{3}{8} \quad \frac{1}{2} \quad \frac{5}{8} \quad \frac{3}{4} \quad \frac{7}{8} \quad 1$

   1/2 is to the left of 7/8; therefore $1/2 < 7/8$.
   1/2 is to the right of 1/8; therefore $1/2 > 1/8$.

   Prepare a 10-minute lesson that illustrates order of fractional numbers through the use of a number-line approach.

5. Any number represented by a unit fraction may be expressed as a sum of numbers represented by unit fractions. For example,

   $$\frac{1}{2} = \frac{1}{3} + \frac{1}{6} \qquad \frac{1}{7} = \frac{1}{8} + \frac{1}{56}$$

   (a) Express 1/3, 1/4, 1/5, 1/10, and 1/100 as sums of numbers represented by unit fractions.
   (b) Give or describe a general rule for expressing any number represented by a unit fraction as a sum of numbers represented by unit fractions.
   (c) Assume that any number represented by a proper fraction can be expressed as a sum of numbers represented by unit fractions in at least

one way and show that any number represented by a proper fraction can be expressed as a sum of numbers represented by unit fractions in infinitely many ways.

6. Explore a recently published elementary textbook series and determine the grade level at which different fractional concepts are introduced, as well as the manner in which any visual aids are used to introduce fractions.

6-2 Equivalence and Order Relations

We have been using the equality symbol (=) to show that two symbols or expressions are names for the same number. The **equality relation** (*is equal to*) is only one of many relations that we study in mathematics. In geometry we study such relations as

$\parallel$ is parallel to
$\perp$ is perpendicular to
$\cong$ is congruent to

We may also consider such **order relations** as

$<$ is less than
$\not<$ is not less than
$\leq$ is less than or equal to
$>$ is greater than
$\not>$ is not greater than
$\geq$ is greater than or equal to

In everyday language we consider many relations such as

is a brother of
is a friend of
is more expensive than
is lighter than
is further than

In each case we are concerned with the *correspondence* of the first element of an ordered pair of elements with the second element of that pair. For *is less than* we could consider the correspondence of integers with integers; for *is a brother of*, people with people; for *is a citizen of*, people with countries; and so forth. The *relation* may be considered as the set of ordered pairs for which the statement is true, that is, the set of ordered pairs for which the first element has the stated relationship with the second element. In this sense a **relation** is a subset of a Cartesian product of two sets or of a set with itself.

Recall (Section 2-2) that two sets of elements are *equivalent sets* if their elements can be placed in one-to-one correspondence, that is, if the sets have the same number of elements. For example, $n(A) = n(B)$. We use equivalent sets to state the definitions of three general properties of relations.

For any sets A, B, and C

Note that the sets A, B, and C are not necessarily distinct.

Reflexive property: $n(A) = n(A)$.
Symmetric property: If $n(A) = n(B)$, then $n(B) = n(A)$.
Transitive property: If $n(A) = n(B)$ and $n(B) = n(C)$, then $n(A) = n(C)$.

Each of these properties can be verified for given sets using one-to-one correspondences. The relation *is equivalent to* for sets is said to be reflexive, symmetric, and transitive, that is, to have these three properties. In general, any relation that is reflexive, symmetric, and transitive is an **equivalence relation**.

Equality of natural numbers is an equivalence relation since equal numbers can be used as cardinal numbers (Section 2-2) of equivalent sets, that is, for $n(A) = a$ and $n(B) = b$ we have $a = b$ if and only if $n(A) = n(B)$. Then, since $a/b = c/d$ if and only if $ad = bc$, *equality* of fractions is also an equivalence relation.

Reflexive: $\frac{a}{b} = \frac{a}{b}$ for any fraction $\frac{a}{b}$.

Symmetric: If $\frac{a}{b} = \frac{c}{d}$, then $\frac{c}{d} = \frac{a}{b}$ for any fractions $\frac{a}{b}$ and $\frac{c}{d}$.

Transitive: If $\frac{a}{b} = \frac{c}{d}$ and $\frac{c}{d} = \frac{e}{f}$, then $\frac{a}{b} = \frac{e}{f}$ for any fractions $\frac{a}{b}$, $\frac{c}{d}$, and $\frac{e}{f}$.

Accordingly, any two fractions that represent the same number are *equivalent fractions*.

Example 1 Give two numerical examples to illustrate the specified property of fractions. **(a)** Reflexive **(b)** Symmetric **(c)** Transitive.

Solution Among others:

(a) $\frac{2}{3} = \frac{2}{3}$ since $6 = 6$.

$\frac{5}{8} = \frac{5}{8}$ since $40 = 40$.

(b) Since $24 = 24$, $\frac{3}{4} = \frac{6}{8}$ and $\frac{6}{8} = \frac{3}{4}$.

Since $36 = 36$, $\frac{2}{3} = \frac{12}{18}$ and $\frac{12}{18} = \frac{2}{3}$.

(c) $\frac{2}{5} = \frac{4}{10}$ and $\frac{4}{10} = \frac{40}{100}$; therefore $\frac{2}{5} = \frac{40}{100}$.

$\frac{3}{4} = \frac{15}{20}$ and $\frac{15}{20} = \frac{75}{100}$; therefore $\frac{3}{4} = \frac{75}{100}$. ■

Example 2 Determine whether or not the specified relation is an equivalence relation.

(a) *Is the same age as* for people.
(b) *Is a brother of* for people.
(c) *Is greater than* for integers.
(d) *Is less than or equal to* for integers.

Solution **(a)** For people *is the same age as* is reflexive (a person's age is the same as that person's age), is symmetric (if Mary is the same age as John, then John is the same age as Mary), is transitive (if Mary and John are the same age and John and Bob are the same age, then Mary and Bob are the same age), and therefore is an equivalence relation.

(b) For people *is a brother of* is not reflexive, is symmetric, but is not transitive (if Bob is a brother of Jim and Jim is a brother of Bob, then it is not true that Bob is a brother of Bob).

(c) Recall from Section 4-1 that for any integers a and b we have $a > b$ and $b < a$ if and only if there is a natural number c such that $a = b + c$. For integers *is greater than* is not reflexive ($2 \not> 2$), is not symmetric ($3 > 2$ but $2 \not> 3$), is transitive ($10 > 7$ and $7 > 3$, then $10 > 3$), and therefore is not an equivalence relation.

(d) For integers *is less than or equal to* is reflexive ($5 \leq 5$), is not symmetric ($2 \leq 5$ but $5 \not\leq 2$), is transitive ($2 \leq 5$ and $5 \leq 6$, then $2 \leq 6$), and therefore is not an equivalence relation. ■

Any rational number can be represented by a fraction m/n, where $n \neq 0$ and n is a natural number. *Order relations* among rational numbers are defined in terms of such fractions.

For any two rational numbers a/b and c/d, where b and d are natural numbers,

$$\frac{a}{b} < \frac{c}{d} \quad \text{if and only if} \quad ad < bc.$$

$$\frac{a}{b} > \frac{c}{d} \quad \text{if and only if} \quad ad > bc.$$

$$\frac{a}{b} \leq \frac{c}{d} \quad \text{if and only if} \quad ad \leq bc.$$

$$\frac{a}{b} \geq \frac{c}{d} \quad \text{if and only if} \quad ad \geq bc.$$

If m and n are any two rational numbers, then either m is less than n, or m is equal to n, or m is greater than n.

The equality and order relations for integers are themselves related by the **trichotomy law**: For any two integers m and n exactly one of these three relations must hold

$$m < n, \qquad m = n, \qquad m > n$$

In advanced courses it is proved that since ad and bc are integers in the definitions for order relations among rational numbers, the trichotomy law also holds for rational numbers.

Example 3 Which relation symbol, $<$, $=$, or $>$, can be used to complete the given statement as a true statement?

(a) $\frac{2}{3} \square \frac{13}{20}$ (b) $\frac{7}{4} \square \frac{175}{100}$ (c) $\frac{5}{8} \square \frac{1}{-2}$

Solution (a) $>$, since $40 > 39$.

(b) $=$, since $700 = 700$.

(c) $>$, that is, $\frac{5}{8} > \frac{-1}{2}$, since $10 > -8$. ■

Every rational number has a point on the number line as its graph. However, not every point on the number line has a *rational number* as its coordinate.

Our number line has now become dense with points and "resembles" a complete line. The word "resembles" is used because we shall find later that there are still points on the number line that do not have rational numbers as their coordinates. However, the set of rational numbers is said to be **dense**, since between any two elements of the set, there is always another element of the set.

Example 4 Name a rational number between $\frac{17}{19}$ and $\frac{18}{19}$.

Solution Select a multiple of 19 that is greater than 19, such as 38. Express each given rational number as a fraction with a denominator of 38.

$$\frac{17}{19} \times \frac{2}{2} = \frac{34}{38} \qquad \frac{18}{19} \times \frac{2}{2} = \frac{36}{38}$$

Clearly, 35/38 lies between the two given rational numbers. Note also that

$$\frac{35}{38} = \frac{1}{2}\left(\frac{17}{19} + \frac{18}{19}\right) \quad ■$$

Example 5 Name a rational number between $\frac{17}{19}$ and $\frac{35}{38}$.

Solution Select a multiple of 19 and 38 that is greater than 38, such as 76. Express each given rational number as a fraction with a denominator of 76.

$$\frac{17}{19} \times \frac{4}{4} = \frac{68}{76} \qquad \frac{35}{38} \times \frac{2}{2} = \frac{70}{76}$$

Note that 69/76 lies between the two given rational numbers. Also,

$$\frac{69}{76} = \frac{1}{2}\left(\frac{17}{19} + \frac{35}{38}\right) \quad ■$$

We can summarize the results of the two preceding examples in the following manner.

$$\frac{17}{19} < \frac{35}{38} < \frac{18}{19}$$

$$\frac{17}{19} < \frac{69}{76} < \frac{35}{38}$$

This process can be extended indefinitely. For example, to locate a rational number between 17/19 and 69/76, express each number as a fraction with a denominator of 152. Then find one-half the sum of the two numbers. This existence of a rational number between any two given rational numbers is what is meant by the *density property.*

In this section we have found that every rational number is the coordinate of some point on the number line, but *not* every point on the number line can be named by a rational number. In Chapter 7, we extend our number system to establish a one-to-one correspondence between the points of the number line and a new set of numbers, thereby completing the number line.

EXERCISES *Determine whether or not the specified relation is* **(a)** *reflexive* **(b)** *symmetric* **(c)** *transitive* **(d)** *an equivalence relation.*

1. For people, *is a daughter of.*
2. For numbers, *is not equal to,* $\neq$.
3. For numbers, *is greater than or equal to,* $\geq$.
4. For books, *is heavier than.*
5. For people, *lives within a mile of.*
6. For lines in a plane, *is parallel to,* $\|$, if a line is defined to be parallel to itself.
7. For lines in a plane, *is parallel to,* $\|$, if a line is not parallel to itself.
8. For whole numbers, *is not divisible by.*

Which relation symbol, $<$, $=$, or $>$, *can be used to complete the given statement as a true statement?*

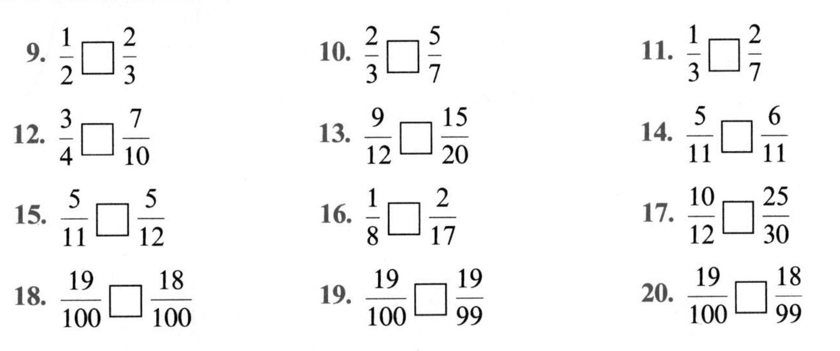

9. $\frac{1}{2} \square \frac{2}{3}$
10. $\frac{2}{3} \square \frac{5}{7}$
11. $\frac{1}{3} \square \frac{2}{7}$
12. $\frac{3}{4} \square \frac{7}{10}$
13. $\frac{9}{12} \square \frac{15}{20}$
14. $\frac{5}{11} \square \frac{6}{11}$
15. $\frac{5}{11} \square \frac{5}{12}$
16. $\frac{1}{8} \square \frac{2}{17}$
17. $\frac{10}{12} \square \frac{25}{30}$
18. $\frac{19}{100} \square \frac{18}{100}$
19. $\frac{19}{100} \square \frac{19}{99}$
20. $\frac{19}{100} \square \frac{18}{99}$

Assume that $a/b < c/d$, b and d are natural numbers, $0 < a/b < 1$, $k > 0$, and $n < 0$. Then complete the given statements as in Exercises 9 through 20.

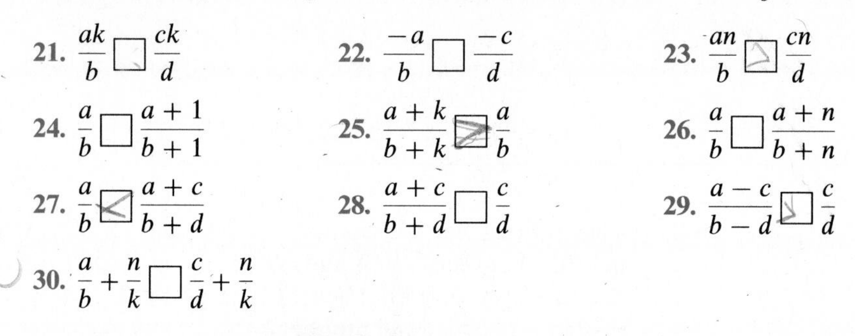

21. $\frac{ak}{b} \square \frac{ck}{d}$
22. $\frac{-a}{b} \square \frac{-c}{d}$
23. $\frac{an}{b} \square \frac{cn}{d}$
24. $\frac{a}{b} \square \frac{a+1}{b+1}$
25. $\frac{a+k}{b+k} \square \frac{a}{b}$
26. $\frac{a}{b} \square \frac{a+n}{b+n}$
27. $\frac{a}{b} \square \frac{a+c}{b+d}$
28. $\frac{a+c}{b+d} \square \frac{c}{d}$
29. $\frac{a-c}{b-d} \square \frac{c}{d}$
30. $\frac{a}{b} + \frac{n}{k} \square \frac{c}{d} + \frac{n}{k}$

Identify each person's statement as true or false and explain your answer.

31. Mary drives 8 miles to class. Susan drives 12 miles.
    (a) Mary said, "I drive one-third less."
    (b) Susan said, "I drive one-half more."
32. Jack and John both had summer jobs. Jack earned \$4.00 an hour. John earned \$5.00 an hour.
    (a) Jack said, "I earned one-fifth less."
    (b) John said, "I earned one-fourth more."

Assume that each letter represents a natural number and classify each statement as true or false. If false, give a specific counterexample to justify your answer.

33. $n^2 \geq n$
34. $s^2 > s$
35. If $ac < bc$, then $a < b$.
*36. If $p^k > p^n$ for a prime number p, then $k > n$.
*37. If $s^k = s^n$, then $k = n$.
*38. If $p^k < q^k$, then $p < q$.

Recall that a single counterexample is sufficient to show that a statement is not true in general.

EXPLORATIONS

Arrow diagrams such as the following may be used to show relations.

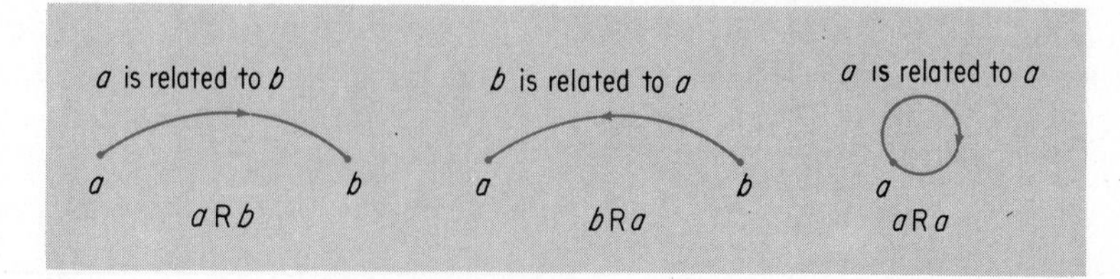

If a relation is reflexive, there is a loop at every element of the given set under discussion. If a relation is symmetric, then all arrows that go from an element a to an element b must also have an arrow that goes from b to a. A relation is transitive if whenever there is an arrow from a to b, and one from b to c, then there is also an arrow from a to c.

1. Here is a diagram for the relation *divides* for the set $M = \{2, 3, 6, 12\}$. From the diagram, show that the relation is reflexive and transitive but not symmetric.

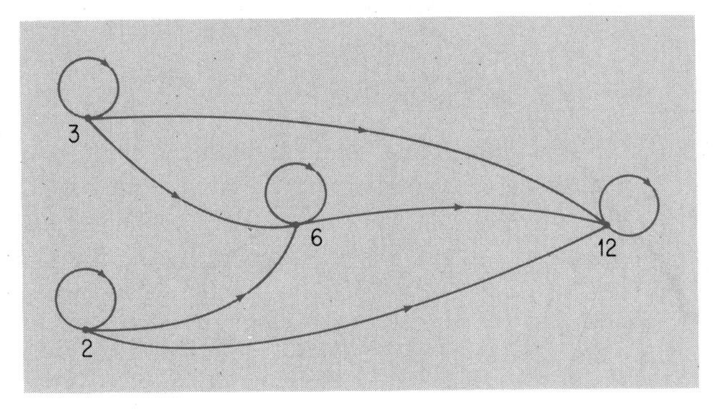

For Explorations 2 through 4 draw arrow diagrams for each of the specified relations on the indicated sets.

2. *Is greater than* for $A = \{2, 5, 6, 9, 10\}$.
3. *Is greater than or equal to* for $B = \{3, 8, 9, 12\}$.
4. *Is within three kilograms of* for the set of weights $C = \{130, 132, 134\}$.

5. For natural numbers a and b consider several numerical examples of the statement

$$\frac{a}{b} < \frac{a+1}{b+1}$$

(a) Find three examples that are true statements.
(b) Find three examples that are false statements.
(c) Find a condition on a/b that enables you to determine in advance whether or not the statement will be a true statement.

6. The set of all possible fractions p/q in lowest terms where $0 \leq p \leq q \leq n$ when arranged in order forms a **Farey sequence** of order n. The Farey sequence of order 3 is

$$\frac{0}{1} \quad \frac{1}{3} \quad \frac{1}{2} \quad \frac{2}{3} \quad \frac{1}{1}$$

Find the Farey sequence of order **(a)** 2, **(b)** 4, **(c)** 5.

7. Explain how the density property enables us to use rational numbers as measures of line segments.

8. The Greek philosopher Zeno, in the fifth century B.C., proposed several paradoxes that were based on the density property. Prepare a report on these paradoxes using a history of mathematics book as your reference. This report should be appropriate for upper elementary and junior high school students.

6-3 Addition and Subtraction of Rational Numbers

The addition of two rational numbers that are represented by fractions with like denominators is accomplished by adding the numerators and using the common denominators. Consider the problem

$$\frac{3}{8} + \frac{2}{8} = \frac{3+2}{8} = \frac{5}{8}$$

This problem may be illustrated by using a rectangular region divided into eighths, as in the given figure. Three of the parts are shaded with horizontal lines to represent 3/8. Two of the parts are shaded with vertical lines to represent 2/8. The sum 3/8 + 2/8 is represented by the five shaded parts: 3/8 + 2/8 = 5/8.

This sum may also be illustrated on a number line that is divided equally into eighths, as in the next figure.

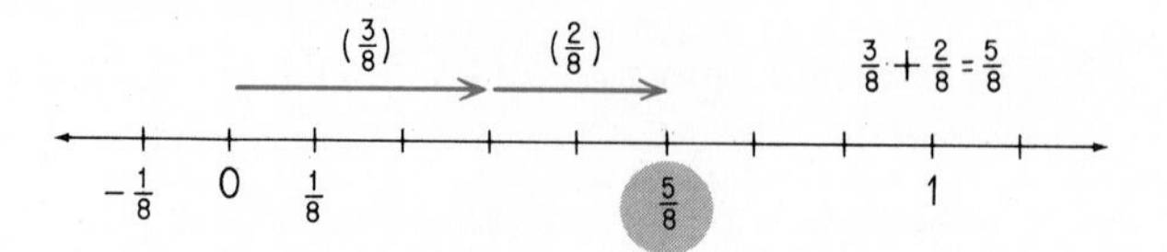

Rectangular regions may also be used to illustrate subtraction of rational numbers represented by fractions with like denominators. The next figure is divided equally into eight parts, with five of the parts shaded with horizontal lines to represent 5/8. Three of these five parts are then shaded with vertical lines to represent 3/8. The difference 5/8 − 3/8 is represented by the remaining parts that are shaded with horizontal lines only.

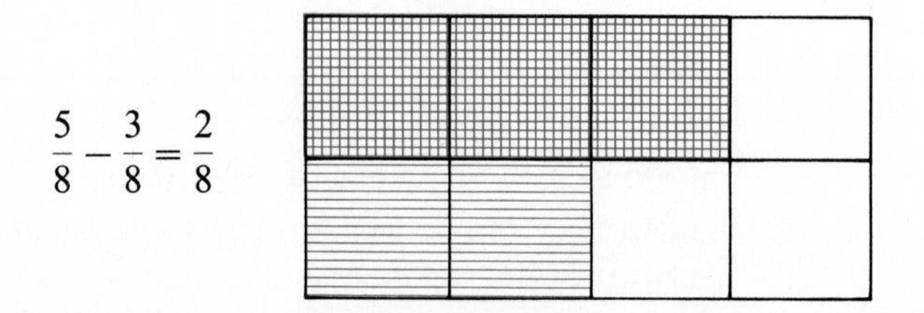

Example 1 Illustrate on a number line $\left(-1\frac{1}{2}\right)+\left(-1\frac{1}{4}\right)$.

Solution

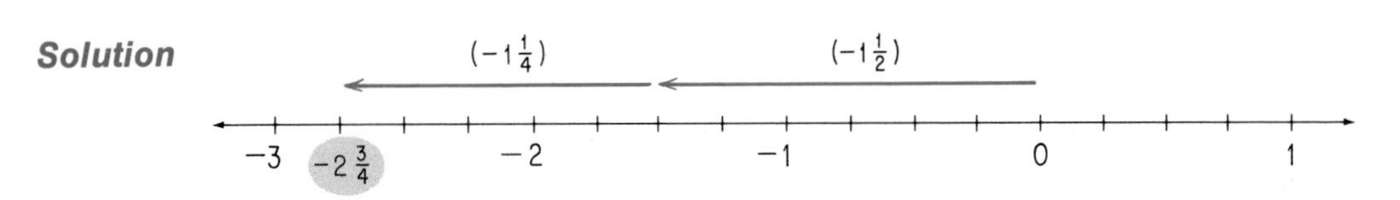

From the figure we see that $(-1\frac{1}{2}) + (-1\frac{1}{4}) = -2\frac{3}{4}$. Note that the scale on the number line used was divided into fourths because of the units used in the given problem. ■

The opposite of any rational number is a rational number. Thus any difference of rational numbers may be expressed as a sum. For example,

$$\left(-1\frac{1}{2}\right)-\left(+1\frac{1}{4}\right)=\left(-1\frac{1}{2}\right)+\left(-1\frac{1}{4}\right)=-2\frac{3}{4}$$

as in Example 1.

The procedures for adding and subtracting rational numbers represented by fractions with like denominators may be stated as in these equations.

$$\frac{a}{c}+\frac{b}{c}=\frac{a+b}{c} \qquad \frac{a}{c}-\frac{b}{c}=\frac{a-b}{c}$$

Example 2 Find the difference $\frac{2}{8}-\frac{7}{8}$.

Solution

$$\frac{2}{8}-\frac{7}{8}=\frac{2-7}{8}=\frac{-5}{8}=-\frac{5}{8} \quad ■$$

The sum of any two rational numbers is a rational number. Any difference of rational numbers is a rational number. The set of rational numbers is closed under addition and under subtraction.

The general procedures for the addition and subtraction of rational numbers were considered in the discussion of the least common multiple in Section 5-5.

For any rational numbers a/b and c/d,

$$\frac{a}{b}+\frac{c}{d}=\frac{ad+bc}{bd} \qquad \frac{a}{b}-\frac{c}{d}=\frac{ad-bc}{bd}$$

Example 3 Subtract $\frac{3}{4}-\frac{2}{5}$.

Solution LCM(4, 5) = 20; first express the fractions as like fractions with denominator 20.

$$\frac{3}{4} - \frac{2}{5} = \frac{3}{4} \times \frac{5}{5} - \frac{2}{5} \times \frac{4}{4}$$

$$= \frac{3 \times 5}{4 \times 5} - \frac{2 \times 4}{5 \times 4}$$

$$= \frac{15}{20} - \frac{8}{20}$$

$$= \frac{7}{20} \quad \blacksquare$$

Problems that involve addition and subtraction with fractions are often written in vertical form, as in the following display for the difference 3/4 − 1/5.

$$\begin{array}{rl} \frac{3}{4} & \frac{3}{4} \times \frac{5}{5} = \frac{15}{20} \\ -\frac{1}{5} & \frac{1}{5} \times \frac{4}{4} = \frac{4}{20} \\ \hline & \qquad\quad \frac{11}{20} \qquad \left(\frac{3}{4} - \frac{1}{5} = \frac{(3 \times 5) - (4 \times 1)}{4 \times 5} = \frac{11}{20}\right) \end{array}$$

Any number in mixed form can be expressed as a single fraction. Addition and subtraction of numbers in mixed form can be performed either directly or with the equivalent fractions.

Example 4 Add $\left(-2\frac{2}{3}\right) + \left(-3\frac{1}{2}\right)$.

Solution

$$-2\frac{2}{3} = -\frac{8}{3} \qquad -3\frac{1}{2} = -\frac{7}{2}$$

There are several possible ways to complete this addition. Write each addend in fractional form.

$$\left(-2\frac{2}{3}\right) + \left(-3\frac{1}{2}\right) = \left(-\frac{8}{3}\right) + \left(-\frac{7}{2}\right)$$

$$= \left(-\frac{16}{6}\right) + \left(-\frac{21}{6}\right)$$

$$= -\frac{37}{6}$$

$$= -6\frac{1}{6}$$

Alternate solution In vertical form, this addition appears as follows.

The alternate solution is generally the preferred approach when adding or subtracting numbers in mixed form.

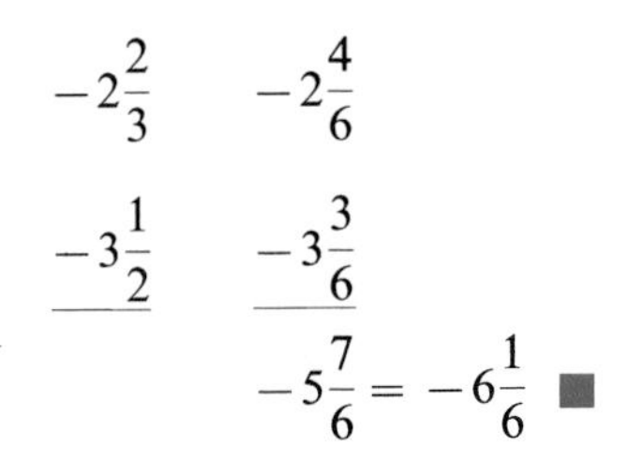

$$\begin{array}{rr} -2\frac{2}{3} & -2\frac{4}{6} \\ -3\frac{1}{2} & -3\frac{3}{6} \\ \hline & -5\frac{7}{6} = -6\frac{1}{6} \end{array} \quad \blacksquare$$

At times it may become necessary to make appropriate exchanges during a subtraction problem, as we shall illustrate for the example $8\frac{1}{2} - 4\frac{2}{3}$.

$$8\frac{1}{2} = 8\frac{3}{6}$$

$$4\frac{2}{3} = 4\frac{4}{6}$$

After the fractional parts have been expressed as like fractions, we find that we must subtract 4/6 from 3/6. To do this we think of $8\frac{3}{6}$ as follows.

$$8\frac{3}{6} = 7 + 1 + \frac{3}{6} = 7 + \frac{6}{6} + \frac{3}{6} = 7 + \frac{9}{6} = 7\frac{9}{6}$$

Of course, most of this work is done mentally but it is important to understand the procedure. In this example we exchange the number 1 for 6/6 and write our work in this way.

$$\begin{array}{rr} 8\frac{1}{2} & 8\frac{3}{6} = 7\frac{9}{6} \\ -4\frac{2}{3} & 4\frac{4}{6} = 4\frac{4}{6} \\ \hline & 3\frac{5}{6} \end{array} \quad \left(8\frac{1}{2} - 4\frac{2}{3} = 3\frac{5}{6}\right)$$

Note that both the integral and the fractional parts of the mixed form of a number must be considered in identifying the opposite of a number in mixed form. For example,

$$-1\frac{1}{4} = (-1) \times \left(1 + \frac{1}{4}\right) = -1 + \left(-\frac{1}{4}\right) = -\frac{5}{4}$$

$$-1\frac{1}{4} \neq (-1) + \frac{1}{4} = -\frac{3}{4}$$

EXERCISES

Illustrate each sum or difference on a number line.

1. $\frac{3}{5} + \frac{1}{5}$
2. $\frac{3}{10} + \frac{4}{10}$
3. $\frac{5}{8} + \frac{3}{8}$
4. $\frac{4}{5} - \frac{1}{5}$
5. $\frac{9}{10} - \frac{2}{10}$
6. $\frac{7}{8} - \frac{2}{8}$

Simplify.

7. $\frac{2}{3} + \frac{1}{4}$

8. $\frac{1}{6} + \frac{3}{4}$

9. $\frac{8}{9} + \frac{2}{3}$

10. $\frac{-2}{7} + \frac{-3}{7}$

11. $\frac{-3}{11} + \frac{-7}{11}$

12. $\frac{-8}{15} + \frac{-3}{15}$

13. $\frac{5}{12} + \frac{7}{8}$

14. $\frac{3}{8} + \frac{2}{5}$

15. $\frac{9}{10} + \frac{7}{12}$

16. $\frac{5}{6} - \frac{2}{3}$

17. $\frac{3}{5} - \frac{1}{4}$

18. $\frac{13}{8} - \frac{3}{4}$

19. $\frac{3}{10} - \frac{7}{10}$

20. $\frac{5}{12} - \frac{10}{12}$

21. $\frac{7}{8} - \frac{10}{8}$

22. $4\frac{1}{8} + 3\frac{3}{4}$

23. $5\frac{1}{6} + 4\frac{2}{3}$

24. $13\frac{1}{6} - 8\frac{2}{3}$

25. $9\frac{2}{3} + 5\frac{7}{8}$

26. $5\frac{1}{4} + 7\frac{3}{5}$

27. $12\frac{7}{9} + 9\frac{2}{3}$

28. $13\frac{4}{5} - 7$

29. $25 - 12\frac{7}{10}$

30. $15 - 7\frac{3}{5}$

31. $\left(-\frac{5}{8}\right) + \left(-\frac{7}{8}\right)$

32. $\left(-\frac{2}{3}\right) + \left(-\frac{1}{5}\right)$

33. $\left(-\frac{1}{4}\right) + \left(+\frac{1}{3}\right)$

34. $\frac{2}{3} + \left(\frac{3}{4} + \frac{7}{8}\right)$

35. $\left(\frac{2}{3} + \frac{3}{4}\right) + \frac{7}{8}$

36. $\frac{7}{8} - \left(\frac{3}{4} - \frac{1}{2}\right)$

37. $\left(\frac{7}{8} - \frac{3}{4}\right) - \frac{1}{2}$

38. $\left(-1\frac{1}{2}\right) + \left(-1\frac{1}{3}\right)$

39. $\left(-2\frac{2}{3}\right) + \left(-1\frac{1}{4}\right)$

40. $\left(-3\frac{2}{5}\right) + \left(-1\frac{3}{4}\right)$

41. $\left(2\frac{1}{2} + 3\frac{3}{4}\right) - \left(1\frac{1}{8} + 2\frac{1}{4}\right)$

42. $\left(5\frac{1}{3} - 1\frac{1}{2}\right) + \left(7\frac{1}{4} - 2\frac{2}{3}\right)$

For Exercises 43 through 46 represent each expression by a single fraction.

*43. $\frac{a}{b} + \left(\frac{c}{d} + \frac{e}{f}\right)$

*44. $\frac{a}{b} - \left(\frac{c}{d} + \frac{e}{f}\right)$

*45. $\left(\frac{a}{b} + \frac{c}{d}\right) + \frac{e}{f}$

*46. $\left(\frac{a}{b} - \frac{c}{d}\right) - \frac{e}{f}$

47. Find
   (a) $\frac{1}{1} + \frac{1}{2} + \frac{1}{3} + \frac{1}{6}$
   (b) $\frac{1}{1} + \frac{1}{2} + \frac{1}{4} + \frac{1}{7} + \frac{1}{14} + \frac{1}{28}$
   *(c) The sum of the reciprocals of the factors of any perfect number n (see Section 5-4, Exploration 7).

*48. Give a counterexample or explain each statement.
   (a) The sum of the reciprocals of the factors of any prime number p is greater than 1 and at most 1.5.
   (b) The sum of the reciprocals of the factors of any abundant number (see Section 5-4, Exploration 8) is greater than 2.
   (c) The sum of the reciprocals of the factors of any deficient number (see Section 5-4, Exploration 6) is greater than 1 and less than 2.

*49. List a sequence of steps that may be used for a hand calculator with at least two memory cells to obtain the numerator and the denominator of the single fraction equal to the sum $(a/b) + (c/d)$ of any two fractions. Check that your procedure is correct for $1/2 + 1/3$ and for $3/7 + 5/2$.

*50. Repeat Exercise 49 for the difference $(a/b) - (c/d)$ of any two fractions. Check that your procedure is correct for $1/2 - 1/3$ and for $5/2 - 3/7$.

EXPLORATIONS

1. Assume that you find some students in an elementary mathematics class committing errors such as these.
   (a) $\frac{1}{2} + \frac{1}{3} = \frac{1}{5}$
   (b) $\frac{3}{10} - \frac{1}{5} = \frac{2}{5}$
   (c) $\frac{1\not{6}}{\not{6}4} = \frac{1}{4}$
   (d) $\frac{1}{2} \times \frac{1}{3} = \frac{2}{6}$

   Prepare a 10-minute lesson that might be used to help correct and avoid such errors.
2. Prepare a collection of ten word problems suitable for a specified elementary grade level, and whose solutions involve addition and subtraction with fractions.
3. Examine a recently published elementary mathematics textbook series and determine the grade level at which addition and subtraction with fractions is introduced and developed.
4. Try to gather information on student achievement with fractions as found by the National Assessment of Educational Progress (NAEP). (See Exploration 4 of Section 4-1.) In particular, when asked to estimate

   $$\frac{12}{13} + \frac{7}{8}$$

only 24% of the 13-year-olds and 37% of the 17-year-olds selected the correct answer. Prepare a worksheet that will provide students in a specified upper elmentary class with practice in estimations using fractions.

5. On another NAEP question students were asked to express $1\frac{1}{5}$ as an improper fraction. Only 68% of the eighth graders selected the correct answer. Suggest some of the difficulties of the 32% who selected the wrong answer. Describe some procedures for helping these students.

6-4 Multiplication and Division of Rational Numbers

The multiplication of rational numbers is defined in terms of fractions. The fundamental law of fractions (Section 6-1) indicates that any fraction can be multiplied by 1 in the form k/k, where $k \neq 0$.

$$\frac{a}{b} = \frac{a}{b} \times \frac{k}{k} = \frac{a \times k}{b \times k} = \frac{ak}{bk}$$

The numerator and denominator of any fraction may be multiplied by any nonzero rational number k without changing the value of the fraction.

The general procedure for multiplying fractions may be introduced for two proper fractions using a rectangular region. Consider, for example, $(2/3) \times (3/4)$. We first draw a rectangular region as in the figure, use vertical lines to divide the region equally into four parts, and shade three of these parts to represent 3/4.

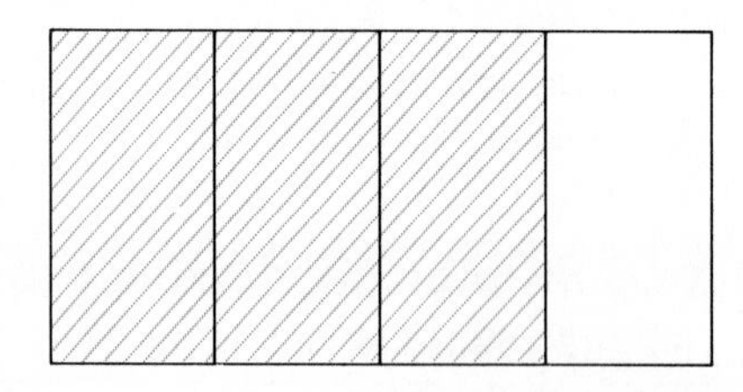

Then we use horizontal lines to divide the original rectangular region and each of its fourths equally into three parts and we shade two of these thirds in a direction different from the previous shading to represent 2/3.

The final picture shows that the rectangular region has been divided into 12 congruent parts and that six of these parts are shaded in both directions. Thus

$$\frac{2}{3} \times \frac{3}{4} = \frac{2 \times 3}{3 \times 4} = \frac{6}{12}$$

Such diagrams are used to justify the usual rule for multiplication of rational numbers written in fractional form.

For any rational numbers a/b and c/d

$$\frac{a}{b} \times \frac{c}{d} = \frac{ac}{bd}$$

The set of rational numbers is closed under multiplication.

The product of any two rational numbers represented by fractions can be represented by a fraction and thus is a rational number. The numerator of the product fraction is the product of the numerators of the factors. The denominator of the product fraction is the product of the denominators of the factors.

Example 1 Multiply $\frac{2}{3} \times \frac{3}{4} \times \left(-\frac{1}{8}\right)$.

Solution

Some of the steps shown are often performed mentally and not written. Reasoning such as

$$\frac{\not{2} \times \not{3} \times (-1)}{\not{3} \times \not{2} \times 2 \times 8} = \frac{-1}{16}$$

is useful but note that only common factors of the numerator and denominator may be crossed out. For example,

$$\frac{12}{12} = \frac{(2 \times 3) + 6}{2 \times 6} \neq 3$$

$$\begin{aligned}
\frac{2}{3} \times \frac{3}{4} \times \left(-\frac{1}{8}\right) &= \left(\frac{2}{3} \times \frac{3}{4}\right) \times \left(-\frac{1}{8}\right) \\
&= \frac{2 \times 3}{3 \times 4} \times \left(-\frac{1}{8}\right) \\
&= \frac{2 \times 3}{2 \times (2 \times 3)} \times \left(-\frac{1}{8}\right) \\
&= \frac{1}{2} \times \left(-\frac{1}{8}\right) \\
&= -\frac{1}{16} \quad \blacksquare
\end{aligned}$$

By expanding the number system from the set of integers to the set of rational numbers, we have included a **multiplicative inverse**, or **reciprocal**, for each rational number different from zero. For example, the multiplicative inverse of 2 is 1/2, of 2/3 is 3/2, and of $-3/4$ is $4/-3$. In each case, the product of the number and its multiplicative inverse is 1, the *identity element for multiplication.*

$$2 \times \frac{1}{2} = 1 \qquad \frac{2}{3} \times \frac{3}{2} = 1 \qquad \left(\frac{-3}{4}\right) \times \left(\frac{4}{-3}\right) = 1$$

The restriction that a/b is not equal to zero is needed because $0 \times n = 0$ for any rational number n, and thus zero does not have a reciprocal in the set of rational numbers.

Inverse ($\neq 0$), $\times$: If $\frac{a}{b} \neq 0$, $\frac{a}{b} \times \frac{b}{a} = 1$.

For any integers a and b where $b \neq 0$ we observed in Section 6-1 that

$$a \div b = a \times \frac{1}{b} = \frac{a}{b}$$

where $1/b$ is the multiplicative inverse of b. We use a similar approach for rational numbers represented by fractions and seek methods for rewriting the new fraction so that its denominator becomes 1. For example,

$$\frac{2}{3} \div \frac{5}{7} = \frac{\frac{2}{3}}{\frac{5}{7}}$$ (since these are just two different ways of expressing a quotient)

$$= \frac{\frac{2}{3}}{\frac{5}{7}} \times 1$$ (by the multiplication property of 1)

$$= \frac{\frac{2}{3}}{\frac{5}{7}} \times \frac{\frac{7}{5}}{\frac{7}{5}}$$ $\left(\text{since } \frac{7}{5} \div \frac{7}{5} = 1\right)$

$$= \frac{\frac{2}{3} \times \frac{7}{5}}{\frac{5}{7} \times \frac{7}{5}}$$ (by the rule for multiplication of rational numbers)

$$= \frac{\frac{2}{3} \times \frac{7}{5}}{1}$$ $\left(\text{since } \frac{5}{7} \times \frac{7}{5} = \frac{35}{35} = 1\right)$

$$= \frac{2}{3} \times \frac{7}{5}$$ (since any number divided by 1 is equal to that number)

Note that we began with a division problem, and showed that this could be written as an equivalent multiplication problem.

This is the common rule "to divide, invert and multiply," that is, "to divide by a fraction, multiply by the reciprocal of that fraction."

For any rational numbers a/b and c/d where $(c/d) \neq 0$

$$\frac{a}{b} \div \frac{c}{d} = \frac{a}{b} \times \frac{d}{c}$$

To divide by a rational number different from zero, we multiply by its reciprocal. This procedure may be justified in general as follows.

$$\frac{\frac{a}{b}}{\frac{c}{d}} = \frac{\frac{a}{b} \times \frac{d}{c}}{\frac{c}{d} \times \frac{d}{c}} = \frac{\frac{a}{b} \times \frac{d}{c}}{1} = \frac{a}{b} \times \frac{d}{c} = \frac{a \times d}{b \times c}$$

A division problem such as (2/3) ÷ (5/7) expressed in vertical form may be simplified by multiplying by 1 in a form such that the numerator and denominator become integers. For example, note that the least common multiple of the denominators of the fractions 2/3 and 5/7 is 21. Then

$$\frac{\frac{2}{3}}{\frac{5}{7}} \times \frac{21}{21} = \frac{14}{15}$$

This example illustrates the following general procedure.

$$\frac{a}{b} \div \frac{c}{d} = \frac{\frac{a}{b}}{\frac{c}{d}} = \frac{\frac{a}{b}}{\frac{c}{d}} \times \frac{b \times d}{b \times d} = \frac{a \times d}{b \times c} = \frac{a}{b} \times \frac{d}{c}$$

Note that the form obtained is the same as that obtained by the invert and multiply procedure.

Example 2 Simplify

$$\frac{\frac{2}{5}}{\frac{3}{4}}$$

The word *simplify* is used here and elsewhere to indicate that all operations are to be performed and the resulting expression written in simplest form.

Solution We consider two methods of solution for this example.

$$\frac{\frac{2}{5}}{\frac{3}{4}} = \frac{2}{5} \div \frac{3}{4} = \frac{2}{5} \times \frac{4}{3} = \frac{8}{15} \quad \text{(invert and multiply)}$$

$$\frac{\frac{2}{5}}{\frac{3}{4}} = \frac{\frac{2}{5}}{\frac{3}{4}} \times \frac{20}{20} = \frac{8}{15} \quad (\text{LCM}(5, 4) = 20) \blacksquare$$

Multiplication of numbers in mixed form can be completed in a variety of ways. Consider the product $2\frac{1}{2} \times 3\frac{1}{4}$. We may write each of these numbers as the sum of a whole number and a number represented by a proper fraction and then use the distributive property to find the product.

$$\begin{aligned}
2\frac{1}{2} \times 3\frac{1}{4} &= \left(2 + \frac{1}{2}\right) \times \left(3 + \frac{1}{4}\right) \\
&= 2\left(3 + \frac{1}{4}\right) + \frac{1}{2}\left(3 + \frac{1}{4}\right) \\
&= (2)(3) + (2)\left(\frac{1}{4}\right) + \left(\frac{1}{2}\right)(3) + \left(\frac{1}{2}\right)\left(\frac{1}{4}\right) \\
&= 6 + \frac{2}{4} + \frac{3}{2} + \frac{1}{8} \\
&= 6 + \frac{1}{2} + 1\frac{1}{2} + \frac{1}{8} \\
&= 8\frac{1}{8}
\end{aligned}$$

In actual practice if numbers are given in mixed form, it is usually easier to express each number first as an improper fraction and then multiply.

$$\begin{aligned}
2\frac{1}{2} \times 3\frac{1}{4} &= \frac{5}{2} \times \frac{13}{4} \\
&= \frac{65}{8} \\
&= 8\frac{1}{8}
\end{aligned}$$

Example 3 Divide $4\frac{1}{2} \div 1\frac{1}{3}$.

Solution

$$\begin{aligned}
4\frac{1}{2} \div 1\frac{1}{3} &= \frac{9}{2} \div \frac{4}{3} \\
&= \frac{9}{2} \times \frac{3}{4} \\
&= \frac{27}{8} \\
&= 3\frac{3}{8} \quad \blacksquare
\end{aligned}$$

Write each number as a fraction and then divide.

$\frac{a}{b} \div \frac{c}{d} = \frac{a}{b} \times \frac{d}{c}$

In Section 6-2 we observed that the set of rational numbers is *dense*, that is, there is at least one rational number between any two given rational numbers.

The other properties of the set of rational numbers may be summarized as follows.

Closure, +	Closure, ×
Commutative, +	Commutative, ×
Associative, +	Associative, ×
Identity, +	Identity, ×
Inverse, +	Inverse ($\neq 0$), ×

Distributive property

Mathematicians recognize the presence of these eleven properties by saying that the set of rational numbers with the operations of addition and multiplication forms a **field**.

EXERCISES *Classify the given statement as true or false. If the statement is false, give a counterexample.*

1. The set of rational numbers is closed under multiplication.
2. The set of rational numbers is closed under division.
3. Every rational number different from zero has a rational number as its reciprocal.
4. Every rational number has a rational number as its opposite.

Use divisions of a rectangular region equally into parts to simplify and explain each product.

5. $\frac{1}{2} \times \frac{1}{3}$
6. $\frac{1}{6} \times \frac{3}{4}$
7. $\frac{2}{5} \times \frac{3}{4}$
8. $\frac{5}{6} \times \frac{2}{3}$

Simplify.

9. $\frac{11}{12} \times \frac{3}{7}$
10. $\frac{9}{10} \times \frac{2}{3}$
11. $\frac{12}{5} \times \frac{15}{8}$
12. $\frac{7}{9} \times \frac{5}{11}$
13. $\frac{9}{14} \times \frac{21}{45}$
14. $\frac{11}{12} \times \frac{30}{44}$
15. $\frac{3}{5} \div \frac{3}{5}$
16. $\frac{3}{5} \div \frac{5}{3}$
17. $\frac{12}{7} \div \frac{9}{10}$
18. $\frac{8}{9} \div \frac{4}{3}$
19. $\frac{-3}{5} \times \frac{2}{5}$
20. $\frac{5}{-2} \times \frac{-3}{5}$
21. $\left(-\frac{3}{4}\right) \times \frac{2}{5}$
22. $\left(-\frac{7}{8}\right) \times \left(-\frac{4}{9}\right)$
23. $\frac{8}{9} \div \frac{-2}{3}$
24. $\frac{-5}{12} \div \frac{10}{-3}$
25. $\left(-\frac{5}{6}\right) \div \frac{1}{3}$
26. $\left(-\frac{4}{5}\right) \div \left(-\frac{2}{5}\right)$
27. $2\frac{2}{3} \times 12$
28. $1\frac{3}{4} \times 20$
29. $\frac{3}{5} \times 2\frac{1}{2}$

30. $1\frac{3}{8} \times 2\frac{2}{3}$

31. $1\frac{1}{5} \times 1\frac{1}{5}$

32. $\left(-3\frac{1}{8}\right) \times \left(-2\frac{3}{5}\right)$

33. $12 \div 1\frac{1}{3}$

34. $2\frac{2}{5} \div \frac{3}{5}$

35. $4\frac{2}{3} \div 7$

36. $1\frac{1}{2} \div 2\frac{1}{3}$

37. $3\frac{2}{3} \div 2\frac{1}{3}$

38. $\left(-4\frac{1}{4}\right) \div \left(-1\frac{1}{2}\right)$

39. $\frac{4}{5} \times \frac{3}{4} \times \frac{2}{3}$

40. $\frac{7}{8} \times \frac{4}{9} \times \left(-\frac{3}{2}\right)$

41. $\frac{5}{9} \div \left(\frac{2}{3} \div \frac{1}{2}\right)$

42. $\left(\frac{5}{9} \div \frac{2}{3}\right) \div \frac{1}{2}$

43. $\left(\frac{7}{9} \times \frac{3}{4}\right) \div \frac{1}{2}$

44. $\frac{7}{9} \times \left(\frac{3}{4} \div \frac{1}{2}\right)$

45. $\dfrac{\frac{1}{2} + \frac{2}{3}}{\frac{3}{4}}$

46. $\dfrac{\frac{3}{4}}{\frac{1}{2} + \frac{2}{3}}$

47. $\dfrac{\frac{1}{3} + \frac{3}{4}}{\frac{1}{2} + \frac{2}{3}}$

48. $\dfrac{\frac{3}{4} - \frac{1}{3}}{\frac{2}{3} - \frac{1}{2}}$

For Exercises 49 through 54 simplify by multiplying numerator and denominator by the least common multiple of the denominators of the fractions involved.

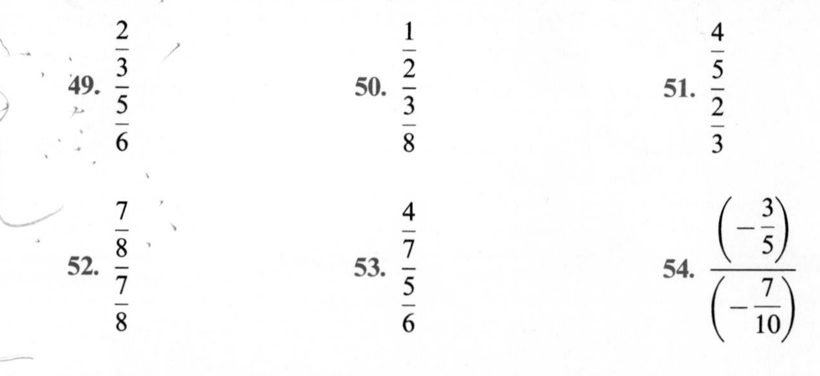

49. $\dfrac{\frac{2}{3}}{\frac{5}{6}}$

50. $\dfrac{\frac{1}{2}}{\frac{3}{8}}$

51. $\dfrac{\frac{4}{5}}{\frac{2}{3}}$

52. $\dfrac{\frac{7}{8}}{\frac{7}{8}}$

53. $\dfrac{\frac{4}{7}}{\frac{5}{6}}$

54. $\dfrac{\left(-\frac{3}{5}\right)}{\left(-\frac{7}{10}\right)}$

55. Bob bought a bag of marbles, gave one-third to his friend John, gave one-half of the remaining marbles to his friend Don, and had 12 marbles left.
   (**a**) After Bob gave one-third of the marbles to John, what part of the original bag of marbles was left?
   (**b**) Did Bob give the same number of marbles to John and to Don?
   (**c**) How many marbles did Bob buy?

56. A mother of three children made a batch of cookies for them as she had promised to do and then went shopping. Jane came home, took one-third of the cookies, and went to visit a friend. Mary came home without knowing that Jane had been home, took one-third of the remaining cookies, and went to her room to read. Jonathan came home without knowing that either of his sisters had been there, took one-third of the remaining cookies, and left the remaining 16 cookies on the table for his sisters.
   (**a**) How many cookies did the mother leave for the children?
   (**b**) How many cookies did each child take?

57. The results obtained for Exercises 41 and 42 provide a counterexample to show that the set of rational numbers is *not* associative for division. Provide another counterexample of your own to show this fact.

*58. Study the results for Exercises 43 and 44. Then show that for all natural numbers a, b, c, d, e, and f

$$\left(\frac{a}{b} \times \frac{c}{d}\right) \div \frac{e}{f} = \frac{a}{b} \times \left(\frac{c}{d} \div \frac{e}{f}\right)$$

*59. First use your own specific example, and then determine, in general, whether or not the following relationship is true for all natural numbers a, b, c, d, e, and f.

$$\frac{a}{b} \div \left(\frac{c}{d} \times \frac{e}{f}\right) = \left(\frac{a}{b} \div \frac{c}{d}\right) \times \frac{e}{f}$$

Copy the following table. Use "$\surd$" to show that the set of elements named at the top of the column has the property listed at the side. Use "$\times$" if the set does not have the property.

	PROPERTY	NATURAL NUMBERS	WHOLE NUMBERS	INTEGERS	POSITIVE RATIONALS	RATIONAL NUMBERS
60.	Closure, +					
61.	Associative, +					
62.	Identity, +					
63.	Inverse, +					
64.	Commutative, +					
65.	Commutative group, +					
66.	Closure, ×					
67.	Associative, ×					
68.	Identity, ×					
69.	Inverse, ($\neq 0$), ×					
70.	Commutative, ×					
71.	After excluding 0, commutative group, ×					
72.	Distributive					

EXPLORATIONS

1. Make up numerical examples and then explain why anyone wishing to share a collection of items with three other people needs to give one-fourth to the first person, one-third of the remaining items to the second person, one-half of the items then remaining to the third person, and keep the rest.
2. Extend the reasoning used in Exploration 1 and explain the corresponding general procedure for dividing a collection of items equally among n people for any natural number n.
3. Any ordinary fraction such as 39/11 can be expressed in the following form, called a **continued fraction**, where all numerators are 1's.

$$\frac{39}{11} = 3 + \frac{6}{11} = 3 + \frac{1}{\frac{11}{6}} = 3 + \frac{1}{1 + \frac{5}{6}} = 3 + \frac{1}{1 + \frac{1}{\frac{6}{5}}} = 3 + \frac{1}{1 + \frac{1}{1 + \frac{1}{\frac{1}{5}}}}$$

Express each fraction as a continued fraction.

(a) $\frac{17}{12}$ (b) $\frac{29}{8}$ (c) $\frac{127}{10}$

4. Express each continued fraction as an ordinary fraction.

(a) $2 + \cfrac{1}{3 + \cfrac{1}{5 + \cfrac{1}{2}}}$ (b) $1 + \cfrac{1}{2 + \cfrac{1}{3 + \cfrac{1}{4 + \cfrac{1}{5}}}}$

5. Prepare a collection of visual aids that can be used to present basic concepts of fractional numbers to elementary school children.

6-5 Decimal Fractions

The first six units of decimal fractions are summarized in the following chart.

Tenths	$\frac{1}{10}$	$\frac{1}{10}$	0.1
Hundredths	$\frac{1}{100}$	$\frac{1}{10^2}$	0.01
Thousandths	$\frac{1}{1000}$	$\frac{1}{10^3}$	0.001
Ten-thousandths	$\frac{1}{10{,}000}$	$\frac{1}{10^4}$	0.0001
Hundred-thousandths	$\frac{1}{100{,}000}$	$\frac{1}{10^5}$	0.00001
Millionths	$\frac{1}{1{,}000{,}000}$	$\frac{1}{10^6}$	0.000001

Initial zeros as in 0.1, 0.01, ... are included before the decimal points to help readers avoid overlooking the decimal points.

We may use multiples of 10 and 1/10 to write a decimal numeral as follows.

$$3256.78 = (3 \times 1000) + (2 \times 100) + (5 \times 10) + (6 \times 1) + \left(7 \times \frac{1}{10}\right) + \left(8 \times \frac{1}{100}\right)$$

This decimal is read "three thousand two hundred fifty-six and seventy-eight hundredths." We use the word *and* to represent the decimal point.

Integral powers of 10 are defined so that $10^0 = 1$ and $10^{-n} = 1/10^n$.

$$\ldots \quad 1000 = 10^3 \qquad 100 = 10^2 \qquad 10 = 10^1 \qquad 1 = 10^0 \qquad \frac{1}{10} = 10^{-1} \qquad \frac{1}{100} = 10^{-2} \quad \ldots$$

Then any given decimal numeral may be written in **expanded notation.**

$$3256.78 = (3 \times 10^3) + (2 \times 10^2) + (5 \times 10^1) + (6 \times 10^0) + (7 \times 10^{-1}) + (8 \times 10^{-2})$$

Example 1 Write in decimal notation

$$(5 \times 10^0) + (3 \times 10^{-1}) + (2 \times 10^{-2}) + (7 \times 10^{-3})$$

Solution

$$(5 \times 1) + \left(3 \times \frac{1}{10}\right) + \left(2 \times \frac{1}{100}\right) + \left(7 \times \frac{1}{1000}\right) = 5 + 0.3 + 0.02 + 0.007$$
$$= 5.327 \quad ■$$

Since $5.327 = 5\frac{327}{1000}$, this answer is read "five and three hundred twenty-seven thousandths."

Addition with decimals becomes a relatively easy task once an algorithm for the addition of whole numbers has been established. For example, consider the addition problem 0.35 + 0.49. Arrange the addends in this form.

$$\begin{array}{r} 0.35 \\ +\ 0.49 \\ \hline \end{array}$$

Because of the place value of our decimal system, we merely add in the hundredths column, carry as needed, and add in the tenths column.

A fraction expressed in decimal notation is often called a **decimal fraction.** By comparison a fraction expressed in the form a/b is called an **ordinary fraction** or a **common fraction.** Years ago the proper term for an ordinary fraction was a **vulgar fraction.**

$$\begin{array}{r} 1\ \ \\ 0.35 \\ +\ 0.49 \\ \hline 0.84 \end{array}$$

Notice that we actually line up place values by "lining up the decimal points" and then adding as with whole numbers.

The addition problem 0.35 + 0.49 may also be done through application of the distributive property $ac + bc = (a + b)c$.

$$0.35 + 0.49 = 35(0.01) + 49(0.01)$$
$$= (35 + 49)(0.01) = 84(0.01) = 0.84$$

This use of the distributive property is equivalent to doing the problem in hundredths; that is, 0.35 + 0.49 is (35 + 49) hundredths. Finally, note that this same problem may be stated and solved in terms of fractions.

$$0.35 + 0.49 = \frac{35}{100} + \frac{49}{100} = \frac{84}{100} = 0.84$$

Subtraction with decimals is also based upon the procedures that are used for whole numbers. We may line up the decimal points and then proceed to subtract as with whole numbers. Example 2 illustrates this method of subtraction, as well as other procedures for subtraction with decimals.

Example 2 Subtract $0.73 - 0.28$.

Solution Three methods of solution are given. Using fractions we have

$$\frac{73}{100} - \frac{28}{100} = \frac{73 - 28}{100} = \frac{45}{100} = 0.45$$

Lining up the decimal points we have

$$\begin{array}{r} 0.73 \\ -\,0.28 \\ \hline 0.45 \end{array}$$

Using the distributive property we have

$$\begin{aligned} 0.73 - 0.28 &= 73(0.01) - 28(0.01) \\ &= 73(0.01) + (-28)(0.01) \\ &= [73 + (-28)](0.01) = 45(0.01) = 0.45 \end{aligned}$$ ■

To explain the usual algorithm for multiplication with decimals, we examine several examples done as fractions and attempt to discover a pattern.

Example 3 Multiply **(a)** 0.7×0.8 **(b)** 0.7×0.08 **(c)** 0.07×0.08.

Solution **(a)** $0.7 \times 0.8 = \frac{7}{10} \times \frac{8}{10} = \frac{56}{100} = 0.56$

Note that a number of *tenths* multiplied by a number of *tenths* gives a number of *hundredths*.

(b) $0.7 \times 0.08 = \frac{7}{10} \times \frac{8}{100} = \frac{56}{1000} = 0.056$

Note that a number of *tenths* multiplied by a number of *hundredths* gives a number of *thousandths*.

(c) $0.07 \times 0.08 = \frac{7}{100} \times \frac{8}{100} = \frac{56}{10{,}000} = 0.0056$

Note that a number of *hundredths* multiplied by a number of *hundredths* gives a number of *ten-thousandths.* ■

Example 3 illustrates the rule for multiplying two rational numbers that are written as decimals. First, ignore the decimal points and multiply as with whole numbers. Then *the number of decimal places in the final product will be equal to the sum of the numbers of decimal places in the two factors that are multiplied.*

Example 4 Place the decimal point in the correct position in the product 0.93×5.4 if the sequence of digits in the answer is 5022.

Solution There are two decimal places in 0.93 and one decimal place in 5.4. The product should then have three decimal places: $0.93 \times 5.4 = 5.022$. The decimal point can also be located by estimation. Since 0.93 is approximately equal to 1 and 5.4 is approximately equal to 5, the product should be about 1×5, that is, 5. Therefore the decimal point must be located to show the product as 5.022. ■

To explain the usual algorithm for division with decimals we again use fractional equivalents. Consider the division problem $0.46 \div 2$. This problem can be expressed and solved in terms of fractions.

$$\frac{46}{100} \div 2 = \frac{46}{100} \times \frac{1}{2} = \frac{46 \times 1}{100 \times 2} = \frac{23}{100} = 0.23$$

We may think of any division problem in the form

$$\begin{array}{r} \text{quotient} \\ \text{divisor}\overline{)\text{dividend}} \end{array}$$

Thus the problem under consideration may be written as

$$\begin{array}{r} 0.23 \\ 2\overline{)0.46} \end{array}$$

Any division problem with ordinary decimals (decimals with a finite number of digits) may be solved by finding an equivalent problem in which the divisor is a natural number and dividing as with natural numbers.

Notice that the divisor is a natural number. Whenever a number represented by a decimal is divided by a natural number, we divide as with natural numbers and place the decimal point in the quotient directly over the decimal point in the dividend.

Consider the problem

$1.75 \div 2.5$

We first write the problem in the form

$$\frac{1.75}{2.5}$$

and multiply by 1 in the form 10/10 to obtain a fraction with a natural number as denominator.

$$\frac{1.75}{2.5} = \frac{1.75}{2.5} \times \frac{10}{10} = \frac{17.5}{25}$$

This rewriting of the fraction

$$\frac{1.75}{2.5} \quad \text{in the form} \quad \frac{17.5}{25}$$

corresponds to rewriting the division problem

$$2.5\overline{)1.75} \quad \text{in the form} \quad 25\overline{)17.5}$$

In other words, multiplying both the dividend and the divisor by 10 is equivalent to multiplying the numerator and the denominator of the corresponding fraction by 10. In each case the problem is unchanged.

Most people think of the above process as "moving" the decimal point one unit to the right in both the dividend and the divisor. In general, since the numerator and the denominator of the corresponding fraction may be multiplied by any power of 10, *the decimal point may be moved as many places as necessary to make the divisor a natural number and moved the same number of places in the dividend.*

Example 5 Find $3.726 \div 0.23$.

Solution Move the decimal point two places to the right in the divisor to make the divisor a natural number. Then move the decimal point two places to the right in the dividend so that both the dividend and the divisor are multiplied by 100. Then divide as for natural numbers and place the decimal point in the quotient directly over the new location of the decimal point in the dividend.

$$\begin{array}{r} 16.2 \\ 0.23\overline{)3.72.6} \\ 23 \\ \hline 142 \\ 138 \\ \hline 46 \\ 46 \\ \hline \end{array}$$

Thus $3.726 \div 0.23 = 372.6 \div 23 = 16.2$.

The decimal point can also be located by estimation. Note that 3.762 is approximately equal to 4 and 0.23 is approximately equal to 1/4. The quotient should be about $4 \div 1/4$, that is, 16. The decimal point must be placed in the quotient to show 16.2 as the result (rather than 0.162, 1.62, or 162). ■

Finally we explore the relationship between the fractional form of a rational number and its decimal form. Every rational number can be expressed in the form a/b. If the denominator b is a power of 10, then the given rational number may be written as a decimal directly, as in these cases.

$$\frac{3}{10} = 0.3 \qquad \frac{13}{100} = 0.13 \qquad \frac{123}{1000} = 0.123$$

The denominator may also be a factor of some power of 10. If so, we can multiply numerator and denominator by a power of 2 or a power of 5 in order to write an equivalent fraction with denominator that is a power of 10.

Example 6 Write as a decimal **(a)** $\frac{3}{5}$ **(b)** $\frac{5}{8}$.

Solution **(a)** $\frac{3}{5} = \frac{3}{5} \times \frac{2}{2} = \frac{6}{10} = 0.6$

(b) $\frac{5}{8} = \frac{5}{8} \times \frac{125}{125} = \frac{625}{1000} = 0.625$ ■

Decimal equivalents can also be found by division. For example, 5/8 is the indicated division $5 \div 8$.

$$\begin{array}{r} 0.625 \\ 8\overline{)5.000} \\ \underline{48} \\ 20 \\ \underline{16} \\ 40 \\ \underline{40} \end{array} \qquad \frac{5}{8} = 0.625$$

$$\begin{array}{r} 0.3333\ldots \\ 3\overline{)1.0000\ldots} \\ \underline{9} \\ 10 \\ \underline{9} \\ 10 \\ \underline{9} \\ 10 \\ \underline{9} \\ 1 \end{array}$$

For any fraction such as 1/3 whose denominator is not a factor of any power of 10, there is no natural number by which we can multiply numerator and denominator to obtain a fraction with a power of 10 as its denominator. If we resort to long division for 1/3 we obtain

$$\frac{1}{3} = 0.333\ldots = 0.\overline{3}$$

The bar is used over the first repeating digit or set of repeating digits on the right of the decimal point and indicates that this sequence of digits repeats endlessly. Such a set of digits is called the **repetend**.

Example 7 Write $\frac{3}{11}$ as a repeating decimal.

Solution By division we find that the repetend is 27.

$$\frac{3}{11} = 0.272727\ldots = 0.\overline{27} \quad ■$$

Decimals such as $0.\overline{3}$ and $0.\overline{27}$ with sequences of digits that repeat endlessly are called **nonterminating repeating decimals**. Decimals such as 0.6 and 0.625 in Example 6 that are expressed using a finite number of decimal places are called **terminating decimals**. Note that zeros may be added to express any terminating decimal as a nonterminating repeating decimal.

$$0.6 = 0.6000\ldots = 0.6\overline{0} \qquad 0.625 = 0.625000\ldots = 0.625\overline{0}$$

See Exploration 1 in Section 7-2.

It is easy to show that every rational number in fractional form can be represented as a repeating (or terminating) decimal. Consider, for example, any rational number such as 12/7. When we divide 12 by 7, the possible remainders are 0, 1, 2, 3, 4, 5, 6. If the remainder is 0, the division is exact; if any remainder occurs a second time, the terms after it will repeat also. Since there are only seven possible remainders when you divide by 7, the remainders must repeat or be exact by the seventh decimal place. Consider the determination of the decimal value of 12/7 by long division.

```
     1.714285
  7)12.000000
     7
     ⑤0
     49
      10
       7
      30
      28
       20
       14
        60
        56
         40
         35
          ⑤
```

The fact that the remainder 5 occurred again implies that the same steps will be used again in the long division process and the digits 714285 will be repeated over and over; that is, $12/7 = 1.\overline{714285}$. Similarly, any rational number p/q can be expressed as a terminating or repeating decimal, and at most q decimal places will be needed to identify it.

We can also show that any terminating or repeating decimal represents a rational number in the form a/b.

Any terminating decimal can be expressed as a fraction, a quotient of integers, with a power of ten as its denominator. For example, if $n = 0.75\overline{0}$, then $100n = 75$, and $n = 75/100$, which reduces to 3/4. Conversely, if a fraction can be expressed as a terminating decimal, its denominator must be a factor of a power of ten.

Any repeating decimal can be expressed as a quotient of integers. For example, if a decimal n repeats one digit, we can find $10n - n$. Suppose $n = 3.2\overline{4}$; then $10n = 32.\overline{4}$ and we have these equivalent differences.

$$\begin{array}{rr} 10n & 32.4\overline{4} \\ -\ \ n & -\ \ 3.2\overline{4} \\ \hline 9n & 29.2\overline{0} \end{array}$$

$$n = \frac{29.2}{9} = \frac{292}{90} = \frac{146}{45}$$

If a decimal n repeats two digits, we find $10^2n - n$; if it repeats three digits, we find $10^3n - n$; and so forth. We can also avoid the use of decimals as follows.

$$\begin{array}{rr} 100n & 324.\overline{4} \\ -\ \ 10n & -\ \ 32.\overline{4} \\ \hline 90n & 292.\overline{0} \end{array}$$

$$n = \frac{292}{90} = \frac{146}{45}$$

Example 8 Express as a quotient of integers.
(**a**) $0.\overline{36}$ (**b**) $0.78\overline{346}$

Solution (**a**) Let $n = 0.\overline{36}$; then $100n = 36.\overline{36}$.

$$\begin{array}{rr} 100n & 36.\overline{36} \\ -\quad n & -\ 0.\overline{36} \\ \hline 99n & 36.\overline{0} \end{array}$$

$$n = \frac{36}{99} = \frac{4}{11}$$

(**b**) Let $n = 0.78\overline{346}$ and $100n = 78.\overline{346}$ so that only the repeating digits occur on the right of the decimal point. Then multiply by 1000.

$$\begin{array}{rr} 100{,}000n & 78{,}346.\overline{346} \\ -\quad 100n & -\quad 78.\overline{346} \\ \hline 99{,}900n & 78{,}268.\overline{000} \end{array}$$

$$n = \frac{78{,}268}{99{,}900} = \frac{19{,}567}{24{,}975} \quad \blacksquare$$

We may now summarize the discussion of this section.

Every rational number can be represented either by a terminating decimal or by a repeating decimal.

Every terminating or repeating decimal represents a rational number.

EXERCISES

Write in expanded notation and include all zero digits.

1. 49.6 **2.** 3402.07 **3.** 5.0103
4. 0.0002 **5.** 3000.03 **6.** 2.0202

Write in decimal notation.

7. $(9 \times 10^1) + (0 \times 10^0) + (0 \times 10^{-1}) + (5 \times 10^{-2}) \times (3 \times 10^{-3})$
8. $(4 \times 10^{-2}) + (8 \times 10^{-3}) + (2 \times 10^{-4})$
9. $(6 \times 10^3) + (9 \times 10^2) + (0 \times 10^1) + (1 \times 10^0) + (9 \times 10^{-1})$
10. $(8 \times 10^0) + (0 \times 10^{-1}) + (5 \times 10^{-2}) + (3 \times 10^{-3})$

Find each sum or difference by using decimals and check by using fractions.

11. $0.14 + 0.16$ **12.** $0.527 + 0.174$ **13.** $0.925 + 0.386$
14. $0.28 - 0.09$ **15.** $0.423 - 0.187$ **16.** $2.317 - 0.789$

Find each product or quotient by using decimals and check by using fractions.

17. 0.08×0.05 **18.** 0.5×0.14 **19.** 0.03×0.005
20. $0.45 \div 0.9$ **21.** $2.55 \div 0.015$ **22.** $6.75 \div 0.026$

Add zeros as needed and place the decimal point in the correct position in the sequence of digits given for each product or quotient.

23. **(a)** 3.45×2.87; 99015 **(b)** 27.3×0.367; 100191

24. **(a)** 0.257×3.65; 93805 **(b)** 72.09×308; 2220372

25. **(a)** $244.72 \div 5.6$; 437 **(b)** $3.393 \div 8.7$; 39

26. **(a)** $0.7905 \div 0.93$; 85 **(b)** $1469.63 \div 0.281$; 523

Perform the indicated operations.

27. $3.75 + (2.87 + 9.56)$ **28.** $12.76 - (9.05 - 7.87)$

29. $5.01 \times (2.3 \times 0.68)$ **30.** $(22.14 \div 0.27) \div 0.2$

31. $3.75 \times (2.74 + 8.35)$ **32.** $5.95 \times (5.73 - 1.28)$

33. $(8.56 + 7.93) - (2.47 + 3.09)$ **34.** $(3.78 - 1.99) + (5.03 - 2.78)$

Select the best answer.

35. 0.89×0.9 is about **(a)** 0.8 **(b)** 8 **(c)** 0.08 **(d)** 0.9.

36. 79.6×9.8 is about **(a)** 8 **(b)** 80 **(c)** 800 **(d)** 8000.

37. 9.01×6.99 is about **(a)** 0.63 **(b)** 6.3 **(c)** 63 **(d)** 630.

38. 0.09×10.01 is about **(a)** 0.009 **(b)** 0.09 **(c)** 0.9 **(d)** 9.

39. $19.79 \div 3.98$ is about **(a)** 5 **(b)** 0.5 **(c)** 0.05 **(d)** 50.

40. $0.079 \div 0.02$ is about **(a)** 4 **(b)** 40 **(c)** 0.4 **(d)** 0.04.

41. $7.926 \div 1.9$ is about **(a)** 4 **(b)** 40 **(c)** 0.4 **(d)** 0.04.

42. $123.9 \div 0.2$ is about **(a)** 62 **(b)** 620 **(c)** 6.2 **(d)** 0.62.

Express each fraction as a decimal.

43. **(a)** $\frac{13}{20}$ **(b)** $\frac{17}{200}$ **44.** **(a)** $\frac{27}{15}$ **(b)** $\frac{1}{6}$

45. **(a)** $\frac{5}{7}$ **(b)** $\frac{13}{7}$ **46.** **(a)** $\frac{1}{13}$ **(b)** $\frac{9}{13}$

Express each decimal as a fraction.

47. $0.\overline{72}$ **48.** $0.\overline{81}$ **49.** $0.\overline{423}$ **50.** $0.\overline{414}$

51. $0.\overline{531}$ **52.** $0.\overline{9}$ **53.** $6.\overline{1}$ **54.** $8.\overline{2}$

55. $3.\overline{14}$ **56.** $3.\overline{25}$ **57.** $0.\overline{321}$ **58.** $65.\overline{268}$

59. Write the first ten digits to the right of the decimal point.
(a) $0.\overline{45}$ **(b)** $0.3\overline{572}$ **(c)** $0.8\overline{0}$ **(d)** $0.42\overline{89}$.

60. Write the digit in the fifteenth position to the right of the decimal point.
(a) $0.52\overline{0}$ **(b)** $0.\overline{78}$ **(c)** $0.\overline{342}$ **(d)** $0.7\overline{168}$.

61. Arrange in order, from smallest to largest.

\$2.59 \$2.89 \$2.37 \$2.65 \$2.08

62. Write each decimal to two decimal places and arrange in order, from smallest to largest.

4.37   4.42   4.4   4.39   4.51   4.3

63. Write each decimal to at least three decimal places and arrange in order, from smallest to largest.

$1.78 \quad 1.\overline{7} \quad 1.\overline{8} \quad 1.7\overline{8} \quad 1.\overline{78}$

64. Write each decimal to at least six decimal places and arrange in order, from smallest to largest.

$0.234 \quad 0.\overline{234} \quad 0.\overline{23} \quad 0.\overline{24} \quad 0.24$

Name a rational number that lies between each of these pairs of rational numbers.

*65. 0.234 and 0.235

*66. $0.\overline{234}$ and $0.\overline{235}$

*67. $0.23\overline{0}$ and $0.24\overline{0}$

*68. 0.234 and $0.\overline{234}$

EXPLORATIONS

For sevenths consider the circular array.

1. Write the decimal representation for each of the rational numbers 1/7, 2/7, 3/7, 4/7, 5/7, and 6/7. See if you can find a pattern that describes the manner in which the digits in each representation are related.
2. Repeat Exploration 1 for the multiples of 1/13 from 1/13 through 12/13.
3. Show that $0.\overline{9} = 1$. Then try to prepare an explanation of this fact to satisfy a seventh or eighth grader. Consider these possibilities.

(a)
$$\frac{1}{3} = 0.\overline{3}$$
$$+\frac{2}{3} = 0.\overline{6}$$
$$1 = 0.\overline{9}$$

(b)
$$\frac{1}{3} = 0.\overline{3}$$
$$3 \times \frac{1}{3} = 3 \times 0.\overline{3}$$
$$1 = 0.\overline{9}$$

4. If the denominator of a fraction can be expressed as the product of powers of 2 and 5 only, then the fraction can be written as a terminating decimal.
   (a) Show that this is so for several examples, such as 3/50 and 33/60.
   (b) Explain why
   $$\frac{n}{2^q 5^p}$$
   for any whole numbers n, p, and q can be represented by a terminating decimal.
   (c) Explain why a fraction that cannot be expressed in the form given in part (b) also cannot be expressed as a terminating decimal.
5. Review a recently published elementary mathematics textbook series. Determine the manner in which decimals are first introduced in the program, as well as the grade placement for the various operations with decimals.

6. Prepare a 30-minute test that evaluates an elementary student's knowledge of the fundamental skills of addition, subtraction, multiplication, and division with decimals.
7. Prepare a bulletin board display that shows the use of decimals in daily life activities.

6-6 Ratio and Proportion

Ratios are used in a wide variety of situations to show comparisons. For example, the ratio of girls to boys may be 18 to 15 in a school classroom, 4 to 4 in a choral octet, 0 to 20 on the boy's soccer squad, 25 to 0 on the girl's soccer squad, or 300 to 310 in an elementary school. Note that a ratio such as 25 to 0 has meaning even though the corresponding fraction is undefined. Note also that if the ratio of girls to boys is 18 to 15, then the ratio of boys to girls is 15 to 18.

Any fraction a/b may be interpreted as the *ratio* of a to b but note that ratios such as 25 to 0 cannot be represented by fractions.

Frequently we wish to state the equality of two ratios. For example, if the ratio of girls to boys in the fourth grade classes is 30 to 31 and the ratio in the elementary school is 300 to 310, we may write

$$\frac{30}{31} = \frac{300}{310}$$

In such circumstances the ratio of girls to boys in the fourth grade classes is said to be **proportional to** the ratio in the school.

The proportion property was recognized by the early Hindus as an arithmetic rule. In the seventh century it was called the **rule of three** and stated in words in the style of the times. Merchants regarded the rule highly and used it widely as a mechanical procedure without explanation. Prior to the nineteenth century the ability to use the rule of three was a mark of mathematical literacy.

Any statement of the equality of two rational numbers in fractional form $a/b = c/d$ is a **proportion**. For example,

$$\frac{3}{4} = \frac{75}{100} \qquad \frac{2}{3} = \frac{12}{n} \qquad \frac{5}{8} = \frac{n}{200}$$

The definition $a/b = c/d$ of equality for rational numbers in fractional form is often called the **proportion property** and is equivalent to the equality of the *cross products* of the proportions, that is, $ad = bc$.

$$3 \times 100 = 4 \times 75 \qquad 2n = 3 \times 12 \qquad 5 \times 200 = 8n$$

The proportion property may be used to find any one of the four terms of a proportion when the other three are known.

Example 1 Find the unknown term in each proportion.

(a) $\frac{a}{5} = \frac{8}{10}$ **(b)** $\frac{2}{b} = \frac{60}{100}$ **(c)** $\frac{5}{8} = \frac{c}{100}$ **(d)** $\frac{3}{8} = \frac{15}{d}$

Solution
(a) $10a = 40, \quad a = 4$
(b) $200 = 60b, \quad b = 3\frac{1}{3}$
(c) $500 = 8c, \quad c = 62.5$
(d) $3d = 120, \quad d = 40$ ■

Comparisons between ratios correspond to the comparisons between the fractions that correspond to the ratios. For example, if in a particular company the ratio of females to males at the executive level is 1 to 2 and that ratio at the employee level is 52 to 75; then in that company the ratio of female to male executives is *lower* (less) than that for employees. At the same time the ratio of female to male employees is *higher* (greater) than that for executives. The ratio 1 to 2 of female to male executives may be used to illustrate a difference that is often made between a *ratio* and a *fractional part* even though both may be represented by a fraction. A ratio may be used to compare disjoint sets; a fractional part may be used to compare a subset with the entire set. If a rectangular region is divided equally into three parts and one of those parts is shaded, then

> The ratio of the shaded part to the unshaded parts is 1 to 2, but the shaded part is one-third of the original rectangular region.

Similarly, for the executives of the company under consideration

> The ratio of the female to male executives is 1 to 2, that is, 1/2, and the female executives are one-third (1/3) of the entire set of executives.

In general, if the ratio of one set to another disjoint set is a to b, then the fractional part that the first set is of the entire set is $a/(a + b)$.

Example 2 Suppose that the only animals owned by the Jones family are cats and dogs and that the ratio of cats to dogs is 3 to 2. Find each of the following.

(a) The ratio of dogs to cats.
(b) The cats as a fractional part of the animals.
(c) The dogs as a fractional part of the animals.

Solution **(a)** The ratio of dogs to cats is 2 to 3.
(b) Three-fifths of the animals are cats.
(c) Two-fifths of the animals are dogs. ■

In Example 2 does the Jones family have 3 cats and 2 dogs? Six cats and 4 dogs? Nine cats? Ten cats? Twelve cats? We do not know the number of cats. They do not have 10 cats but the number of cats may be any multiple of 3. Ratios and fractional parts do not enable us to determine the sizes of sets without further information.

Suppose that in a class of 30 girls and 8 boys one-third of the girls and one-half of the boys had A's on their first hour examination. What fraction of the class had A's on the examination? If we add 1/3 and 1/2 it would appear that only one-sixth of the class failed to get A's, which certainly cannot be correct. To find the fractional part that had A's we must consider the number of students who had A's and the total number of students in the class.

$$\frac{\frac{1}{3} \times 30 + \frac{1}{2} \times 8}{30 + 8} = \frac{10 + 4}{38} = \frac{14}{38} = \frac{7}{19}$$

Thus seven-nineteenths of the class had A's. As illustrated by this example we cannot, in general, add ratios or fractional parts.

Example 3 There are 120 residents in the Star condominiums. The ratio of adults to children is 3 to 2. The ratio of females to males is 5 to 1 for adults and 1 to 1 for children. Find the number of residents who are **(a)** adults **(b)** adult males **(c)** female children. Find the fractional part of the residents who are **(d)** adult females **(e)** female children **(f)** females.

Solution **(a)** $\frac{3}{3+2} \times 120 = \frac{3}{5} \times 120 = 72$ adults

(b) $\frac{1}{5+1} \times 72 = \frac{1}{6} \times 72 = 12$ adult males

(c) $\frac{1}{1+1} \times \left(\frac{2}{2+3} \times 120\right) = \frac{1}{2} \times \left(\frac{2}{5} \times 120\right) = 24$ female children

(d) $\left(\frac{5}{5+1} \times 72\right) \div 120 = \left(\frac{5}{6} \times 72\right) \div 120 = \frac{60}{120} = \frac{1}{2}$

(e) $\frac{24}{120} = \frac{1}{5}$

(f) $\frac{24 + \left(\frac{5}{6} \times 72\right)}{120} = \frac{24+60}{120} = \frac{84}{120} = \frac{7}{10}$ ■

Constant ratios arise in many everyday situations. If pencils cost 20¢ each, then for any number n of pencils purchased, the ratio of n to the cost c in cents is 1 to 20.

$$\frac{n}{c} = \frac{1}{20} \qquad 20n = c \qquad n = \frac{1}{20}c$$

If Maria drives steadily at 55 miles per hour, then for any time t in hours the ratio of the distance d driven in miles to t is 55 to 1.

$$\frac{d}{t} = \frac{55}{1} \qquad d = 55t \qquad t = \frac{1}{55}d$$

In each case the constant ratio may be used in a proportion and each of the two quantities involved is said to be *directly proportional* to the other.

The cost of the pencils is directly proportional to the number purchased.
The number of pencils purchased is directly proportional to the cost.
The distance driven is directly proportional to the time.
The time for the trip is directly proportional to the distance driven.

In general, if x is **directly proportional to** y, then $x = ky$ for some constant k and k is called the **constant of proportionality**. For the cost of pencils at 20 cents each and the distance driven at 55 miles per hour

$c = 20n$ with constant of proportionality 20

$d = 55t$ with constant of proportionality 55 .

Note that in each case it is the constant of proportionality that keeps the ratios the same. For the example $d = 55t$ consider distances of 110 and 165 miles and times of 5 and 6 hours. The corresponding values of d and t may be represented in a table.

d	110	165	275	330
t	2	3	5	6

For each pair of values of d and t the ratio d/t is 55, the constant of proportionality.

EXERCISES *In Exercises* 1 *through* 12 *find the unknown term in each proportion.*

1. $\frac{n}{3} = \frac{8}{12}$
2. $\frac{n}{5} = \frac{40}{100}$
3. $\frac{n}{4} = \frac{5}{12}$
4. $\frac{4}{n} = \frac{12}{60}$
5. $\frac{3}{n} = \frac{6}{10}$
6. $\frac{10}{n} = \frac{80}{1000}$
7. $\frac{3}{5} = \frac{n}{20}$
8. $\frac{5}{8} = \frac{n}{24}$
9. $\frac{3}{4} = \frac{n}{10}$
10. $\frac{2}{7} = \frac{6}{n}$
11. $\frac{7}{10} = \frac{35}{n}$
12. $\frac{5}{11} = \frac{20}{n}$

13. The Brodeurs have a herd of 60 cows. One-third of the herd are Jerseys. The rest are Holsteins.
    (a) What fractional part of the herd are Holsteins?
    (b) What is the ratio of the Jerseys to the Holsteins?
    (c) How many of the cows are Holsteins?
14. Jane drives 25 miles to class, including 20 miles on an interstate highway.
    (a) What fractional part of her drive is on an interstate highway?
    (b) What fractional part of her drive is not on an interstate highway?
    (c) Describe two ways for obtaining the answer for part (b).
15. Assume that the ratio of female students to male students in a certain college class is f to m and find an expression for
    (a) The ratio of male students to female students.
    (b) The fractional part p of the students who are female.
    (c) In terms of f and m, the fractional part of the students who are male.
    (d) In terms of the answer p for part (b), the fractional part of the students who are male.

16. Suppose that grapefruit cost 40¢ each. Provide this same information using a statement involving
   (a) The ratio of the cost in cents to the number of grapefruit purchased.
   (b) A proportion.

Use ratios and/or proportions in solving the following problems if and only if such an approach is needed.

17. Assume that your car averages 24 miles per gallon of gasoline.
   (a) How many gallons of gasoline are needed for a trip of 600 miles?
   (b) How far can you drive on 5 gallons of gasoline?
18. Assume that apples cost 40¢ each.
   (a) How many apples can be bought for $2.00?
   (b) What is the cost of a dozen apples?
19. Assume that you have a brother who spent one year in each elementary school grade and was 12 years old in grade 6.
   (a) How old was he in grade 3?
   (b) How old was he in grade 1?
20. Assume that the cost of 3 tangerines is the same as the cost of 2 oranges.
   (a) If you bought a dozen oranges, how many tangerines could you have bought for the same price?
   (b) If you paid $1.20 for tangerines, how much would the same number of oranges cost?
   (c) If you paid $1.20 for oranges, how much would the same number of tangerines cost?
21. Assume that the air pressure at sea level is 1 atmosphere (14.7 pounds per square inch), the additional pressure on a deep sea diver is proportional to the depth below the surface, and the total pressure at a depth of 33 feet is 2 atmospheres (29.4 pounds per square inch).
   (a) At what depth is the total pressure 3 atmospheres, that is, 2 atmospheres more than on the surface?
   (b) At what depth is the total pressure 4 atmospheres?
   (c) What is the pressure in atmospheres at a depth of 165 feet?
   (d) What is the pressure to the nearest pound per square inch at a depth of 100 feet?
22. Assume that the cost of fuel oil for heating a house is proportional to the number of gallons used.
   (a) If your neighbor's heating bill was $600, your own bill was $500, and you used 480 gallons of fuel, how many gallons of fuel did your neighbor use?
   (b) If your bill last year was $500, a colder winter requiring one-quarter more fuel is expected this year, and the price of fuel oil has increased by 10%, how much should you expect to pay to heat your house this year?

*23. Assume that each cut of a log takes the same amount of time. If you can cut a log into two pieces in 6 minutes, how long will it take to cut the log into (a) four pieces (b) five pieces (c) k pieces?

EXPLORATIONS

Two quantities are *directly proportional* if their ratio is a constant. Two quantities are **inversely proportional** if their product is a constant. Consider the formula $d = rt$ for the *distance* in miles, the *rate* in miles per hour, and the *time* in hours. Note that

For a constant rate the distance is directly proportional to the time.

For a constant time the distance is directly proportional to the rate.

For a constant distance the time is inversely proportional to the rate.

In each case the constant is called the *constant of proportionality*.

In Explorations 1 through 5 use numerical examples and discuss briefly each of the following in terms of ratios, proportions, proportionality, and constants of proportionality.

1. Actual distances and distances on a map with a specified scale.
2. The heights of objects and the lengths of their shadows on level ground at a specified time.
3. The simple interest $I = Prt$ on a principal of P dollars at a rate r per year for t years.
4. The lengths of corresponding sides of similar triangles.
5. Areas and volumes of common geometric figures in terms of other measures of the figures.

6. Obtain a spring and a collection of washers or other objects of the same size and weight that can be hung on the spring. Suspend the spring from a hook at a convenient height and make a table of the distances that the spring stretches with 0, 1, 2, 3, 4, ... washers attached. Look for a relationship between the distances d and the numbers w of washers. Within the elastic limit of the spring, the relationship is **Hooke's law**.

6-7 Percent and Applications

The word *percent* comes from the latin *per centum* for per hundred and can be interpreted to mean *hundredths*. Thus we may write a percent as a fraction with denominator 100 or as a decimal.

$$75\% = \frac{75}{100} = 0.75$$

$$9\% = \frac{9}{100} = 0.09$$

$$125\% = \frac{125}{100} = 1.25$$

$$0.5\% = \frac{0.5}{100} = 0.005$$

$$100\% = \frac{100}{100} = 1$$

Many of the everyday applications of mathematics involve the use of percent. What is the sale price of an \$80 jacket that has been reduced 25%? What is the amount of a 5% tax on a \$4.20 purchase? How much interest can be expected the first year on \$500 in a savings account at 6% compounded quarterly?

These equivalents are used often and are worth memorizing for future use.

In computations with percents, it is often helpful to express the percent as a fraction in simplest form. Here are some of the most commonly used equivalents.

$$25\% = \frac{25}{100} = \frac{1}{4} \qquad 20\% = \frac{20}{100} = \frac{1}{5} \qquad 33\frac{1}{3}\% = \frac{33\frac{1}{3}}{100} = \frac{1}{3}$$

$$50\% = \frac{50}{100} = \frac{1}{2} \qquad 40\% = \frac{40}{100} = \frac{2}{5}$$

$$75\% = \frac{75}{100} = \frac{3}{4} \qquad 60\% = \frac{60}{100} = \frac{3}{5} \qquad 66\frac{2}{3}\% = \frac{66\frac{2}{3}}{100} = \frac{2}{3}$$

$$100\% = \frac{100}{100} = 1 \qquad 80\% = \frac{80}{100} = \frac{4}{5}$$

The preceding illustrations indicate the procedure for expressing a percent as a fraction. Write the percent as a number of hundredths and then reduce the fraction, if possible.

Example 1 Write as a fraction in simplest form.

(a) 70% **(b)** 45%

Solution **(a)** $70\% = \frac{70}{100} = \frac{7}{10}$ **(b)** $45\% = \frac{45}{100} = \frac{9}{20}$ ■

Example 2 Write as a percent.

(a) 0.32 **(b)** $\frac{7}{10}$

Solution **(a)** $0.32 = \frac{32}{100} = 32\%$ **(b)** $\frac{7}{10} = \frac{70}{100} = 70\%$ ■

One method of expressing a fraction as a percent is shown in Example 2. This method causes no difficulty if the given fraction can easily be rewritten as an equivalent fraction with denominator 100. Otherwise we may make use of a proportion (see Section 6-6).

Example 8 A television set costs \$245. For one week only it is offered on sale at 22% off. How much can be saved by buying the set during the sale week?

Solution We need to find 22% of \$245. Since $0.22 \times 245 = 53.90$, the saving during the sale is \$53.90. ■

Very often the consumer is not necessarily interested in an exact answer, but rather needs to make a quick estimate to determine whether or not a sale is worthwhile. Thus, in Example 8, you might think of the television set as costing *about* \$250, and the discount as *approximately* 20%. Since $20\% = 1/5$, the amount saved is approximately 1/5 of \$250. This computation can be done mentally to obtain an estimate of \$50 for the amount saved by buying on sale.

Example 9 The regular price of a radio is \$80. The sale price is \$64. What percent discount is being offered during the sale?

Solution The amount of the saving is \$80 – \$64, that is, \$16. The question may be restated as

16 is what percent of 80?

$$\frac{16}{80} = \frac{n}{100} \qquad 1600 = 80n \qquad n = 20$$

The radio is being offered at a 20% discount. ■

In Example 9 it is important to note that we compared the saving with the original, or regular, price. If you compare the saving with the sale price, you will find that 16 is 25% of \$64. However, it is *incorrect* to claim that this sale offers a 25% discount.

The solution for Example 9 can be checked mentally. Since 20% = 1/5, we merely need to find 1/5 of \$80.

$$\frac{1}{5} \times 80 = 16$$

This checks since the discount was found to be \$16.

There are many times in daily life when it is helpful to be able to compute a percent mentally.

A tip in a restaurant is almost universally computed at 15% of the cost of the food consumed. (That is, you need not pay 15% of any tax that may be added to the bill.) The tip can often be computed mentally by thinking of 15% as the sum of 10% and 5%. We can find 10% of a number because 10% = 1/10. To find 1/10 of a number, just divide by 10; that is, move the decimal point one decimal place to the left. Here are some examples.

$$10\% \text{ of } \$240 = \frac{1}{10} \times \$240 = \$24$$

$$10\% \text{ of } \$8.20 = \frac{1}{10} \times \$8.20 = \$0.82$$

Now suppose that the total for a bill is \$8.20 and you wish to compute a tip of 15%. First find 10% of that amount. Then find 5% of the amount, which will be one-half of 10%, and add.

$$\begin{array}{rl} 10\% \text{ of } \$8.20 = & \$0.82 \\ + \quad 5\% \text{ of } \$8.20 = & \$0.41 \qquad (0.82 \div 2 = 0.41) \\ \hline 15\% \text{ of } \$8.20 = & \$1.23 \end{array}$$

Hence, one would probably round the amount to \$1.25 and leave this amount for the tip.

EXERCISES

Write each percent as a decimal.

1. 57%
2. 89%
3. 3%
4. 225%
5. 0.95%
6. 1%
7. 250%
8. 125%
9. 100%
10. 0.5%

Write each percent as a fraction in simplest form.

11. 50%
12. 40%
13. 95%
14. 65%
15. 99%
16. 130%
17. 150%
18. 125%
19. 8%
20. 0.8%

Write each expression as a percent.

21. (a) 0.35 (b) 1.01 (c) 0.002
22. (a) 1.35 (b) 0.9 (c) 0.008
23. (a) 1.45 (b) 0.006 (c) 1.00
24. (a) $\frac{9}{100}$ (b) $\frac{92}{100}$ (c) $\frac{150}{100}$
25. (a) $\frac{450}{100}$ (b) $\frac{16}{20}$ (c) $\frac{38}{500}$
26. (a) $\frac{9}{1000}$ (b) $\frac{17}{10}$ (c) $\frac{42}{20}$

Copy and complete each table.

	Fraction	Decimal	Percent
27.	$\frac{3}{10}$		
28.		0.89	
29.			55%
30.		0.7	
31.	$\frac{17}{20}$		
32.			175%

	Fraction	Decimal	Percent
33.			5%
34.		0.17	
35.	$\frac{11}{10}$		
36.		0.01	
37.	$\frac{143}{200}$		
38.			120%

Use an equivalent fraction to find each percent.

39. Find 75% of 120.
40. Find 40% of 240.
41. Find 125% of 48.
42. Find $33\frac{1}{3}\%$ of 210.
43. Find $66\frac{2}{3}\%$ of 72.
44. Find 80% of 320.

Write and solve a proportion.

45. Find 40% of 60.
46. Find 80% of 140.
47. Find 35% of 80.
48. Find 120% of 75.
49. 20 is what percent of 160?
50. 15 is what percent of 120?
51. 120 is what percent of 160?
52. 160 is what percent of 120?
53. 25 is 20% of what number?
54. 40 is 75% of what number?
55. 80 is 125% of what number?
56. 80 is 20% of what number?

The ability to estimate is an important skill used often in daily life. Use Exercises 57 through 64 to practice this skill. Thereafter try to estimate your answers before you actually compute.

Select the best estimate mentally.

	(a)	(b)	(c)
57. 25% of $198	**(a)** $25	**(b)** $50	**(c)** $100
58. 34% of $241	**(a)** $8	**(b)** $60	**(c)** $80
59. 19% of $352	**(a)** $35	**(b)** $70	**(c)** $90
60. 49% of $81	**(a)** $20	**(b)** $42	**(c)** $50
61. 1% of $120	**(a)** $1	**(b)** $12	**(c)** $120
62. 9% of $230	**(a)** $2.30	**(b)** $23	**(c)** $2300
63. 26% of $81	**(a)** $2	**(b)** $20	**(c)** $25
64. 5% of $320	**(a)** $1.60	**(b)** $16	**(c)** $32

In Exercises 65 through 70 find the selling price of an item for each given regular price and discount.

65. $240, less 25%
66. $320, less 10%
67. $180, less 20%
68. $48, less $33\frac{1}{3}$%
69. $90, less 40%
70. $60, less 15%

71. A storekeeper advertises a sale offering a discount of $33\frac{1}{3}$% on all merchandise. A coat on display is marked with regular price $120 and sale price $90. Is this the correct selling price? If not, what should it be and how can you account for the merchant's error?

72. Wendy bought a coat on sale for $35.55. The regular price of the coat was $45.00.
(a) What percent discount did she receive?
(b) The sale price is what percent of the regular price of the coat?

73. The school population in a certain elementary school rose from 700 to 800 in a recent year. What was the percent of increase? That is, the increase was what percent of the original enrollment?

74. Roberto went on a diet and reduced his weight from 150 pounds to 135 pounds. What was the percent of decrease in his weight? That is, his loss of weight was what percent of his original weight?

***75.** A television set that regularly costs $250 is advertised on sale at a 15% discount. A week later it is further reduced by 10% of the discounted price. Find a single discount that is equivalent to these two successive discounts.

***76.** A storekeeper pays $80 for a coat and prices it at a markup of 15%. Later it is marked up an additional 10% of that price. Find a single markup equivalent to these two successive markups.

***77.** What single action on a given initial salary would be equivalent to a 20% cut followed by a 25% increase? Answer the same question for a 25% increase followed by a 20% cut.

EXPLORATIONS

1. The following figure illustrates a **percent chart**. Here is how this chart may be used to find 50% of 60. First a line is drawn from 0 on the percent scale to 60 on the base scale. Next 50 is located on the percent scale. Then the answer is found by reading across to the line and down to 30 on the base scale. Thus 50% of 60 is 30.

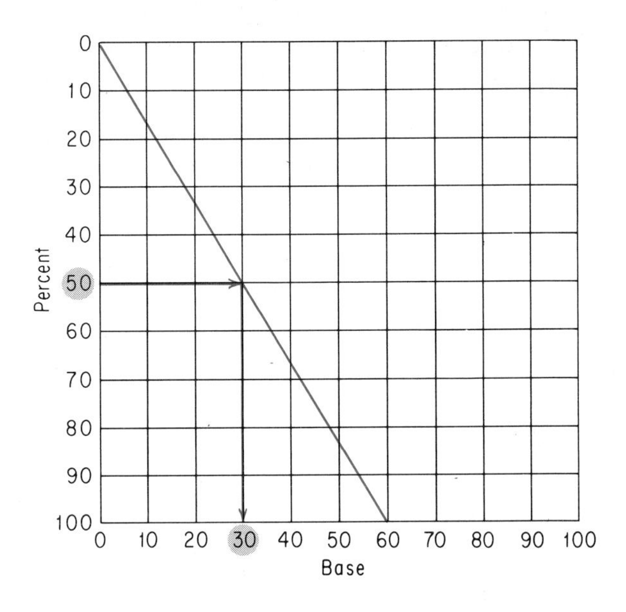

Construct a percent chart to illustrate each of these problems.

(**a**) Find 25% of 80.
(**b**) What percent of 50 is 40?
(**c**) 20 is 40% of what number?

2. Percent problems are often solved by using the formula $p = r \times b$, where p is the percentage, r is the rate, and b is the base. For example, consider the statement 25% of 80 is 20. Here 25% is the rate, 80 is the base, and 20 is the percentage. Now use this formula to solve Examples 4, 5, and 6 of this section.
3. Many traditional junior high school texts solve percent problems by considering three different "cases." Find out what these three types of problems are, and then identify the examples of this section according to this classification.
4. You purchase \$50 worth of books in a bookstore that offers a 20% discount. However, there is also a 5% state sales tax on all such purchases. Is it to your advantage to have the store give you the discount first and then add the sales tax, or add the sales tax first and then give you the discount? First guess which you think would be better for you, and then compute the cost using each approach.
5. Repeat Exploration 4 to determine which approach would be best for the storekeeper and which would be best for the state.

6. Investigate the procedure used by your bookstore in the event that they offer a discount and also must add a state sales tax to the purchase.
7. Begin a collection of as many applications of percent as you can find in newspapers and other periodicals.

Chapter Review

Classify each statement as true or false.

1. **(a)** Between any two rational numbers there is always another rational number.
   (b) There is a one-to-one correspondence between the set of rational numbers and the points of the number line.
2. **(a)** The number 0 does not have a multiplicative inverse.
   (b) Every rational number has an additive inverse.

In Exercises 3 *and* 4 *determine for natural numbers whether or not the given relation is* **(a)** *reflexive* **(b)** *symmetric* **(c)** *transitive* **(d)** *an equivalence relation.*

3. Is greater than or equal to, $\geq$.
4. Is a factor of.

5. Illustrate this sum on a number line: $(2\frac{1}{4}) + (-5\frac{1}{2})$.

In Exercises 6 through 13 simplify.

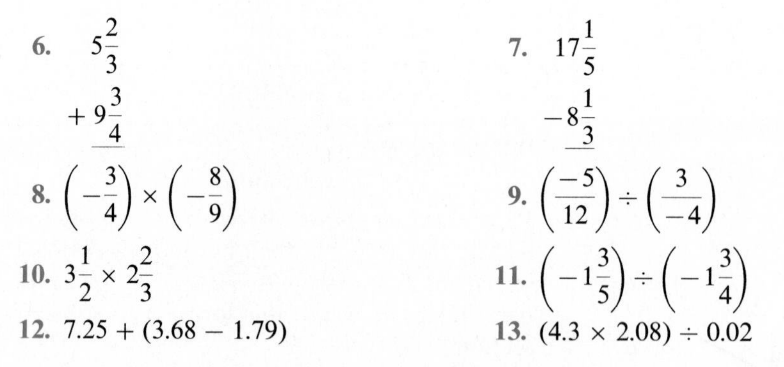

6. $5\frac{2}{3} + 9\frac{3}{4}$
7. $17\frac{1}{5} - 8\frac{1}{3}$
8. $\left(-\frac{3}{4}\right) \times \left(-\frac{8}{9}\right)$
9. $\left(\frac{-5}{12}\right) \div \left(\frac{3}{-4}\right)$
10. $3\frac{1}{2} \times 2\frac{2}{3}$
11. $\left(-1\frac{3}{5}\right) \div \left(-1\frac{3}{4}\right)$
12. $7.25 + (3.68 - 1.79)$
13. $(4.3 \times 2.08) \div 0.02$

14. Represent each expression by a single fraction.
    (a) $\frac{a}{b} + \frac{c}{d}$ **(b)** $\frac{a}{b} \div \frac{c}{d}$
15. Write 32.157 in expanded notation.
16. Name the property illustrated by each statement.
    (a) $(a + b) + c = a + (b + c)$ **(b)** $a(b + c) = ab + ac$

In Exercises 17 *and* 18 *find the unknown term in each proportion.*

17. $\frac{n}{4} = \frac{15}{12}$
18. $\frac{2}{n} = \frac{80}{100}$

19. Express each decimal as a fraction.
(a) $0.\overline{63}$ (b) $0.\overline{612}$

20. Express each fraction as a repeating decimal.
(a) $\frac{7}{12}$ (b) $\frac{7}{13}$

21. Jack teaches a class in which the ratio of girls to boys is 2 to 3. What fractional part of the class is boys?

In Exercises 22 and 23 write and solve a proportion.

22. 45 is what percent of 75? 23. 18 is 20% of what number?

24. The regular price of a certain item is \$120. The discount is 15%. Find the selling price.

25. The regular price is \$72. The sale price is \$60. Find the percent discount being offered.

CALCULATING DECIMAL EXPANSIONS

Calculating decimal equivalences of rational numbers is often difficult even when one uses a hand calculator. When the denominator is larger than eight, the possibility of a repeating pattern showing up on the display decreases. Hence a program which displays an arbitrary number of decimal places of the expansion of a given rational number is a handy tool. The program in this section displays 30 decimal places for a given rational number's expansion.

```
5  HOME
1Ø PRINT "THIS PROGRAM FINDS THE FIRST 3Ø DIGITS"
15 PRINT
2Ø PRINT "OF THE DECIMAL EXPANSION OF A FRACTION."
25 PRINT
26 PRINT
3Ø PRINT "ENTER THE FRACTION YOU WISH EXPANDED."
35 PRINT
4Ø PRINT "TYPE IN THE NUMERATOR AND DENOMINATOR,"
41 PRINT
42 PRINT "SEPARATED BY A COMMA."
43 PRINT
44 INPUT N,D
45 PRINT
46 PRINT N;"/";D;" = ";INT(N/D);".";
5Ø FOR C = 1 TO 3Ø
55 N = (N − D * INT(N/D)) * 1Ø
6Ø PRINT INT(N/D);
65 NEXT C
7Ø END
```

The semicolons at the ends of the PRINT statements on lines 46 and 6∅ instruct the computer to stay on the same line for the next PRINT instructions. The sample display of this program shows the expansion of the rational number 4/31. Here we can see the repeating pattern of 129032258064516, a repetend of 15 digits.

```
RUN
THIS PROGRAM FINDS THE FIRST 3∅ DIGITS

OF THE DECIMAL EXPANSION OF A FRACTION.


ENTER THE FRACTION YOU WISH EXPANDED.

TYPE IN THE NUMERATOR AND DENOMINATOR,

SEPARATED BY A COMMA.

4,31

4/31 = ∅.129∅32258∅645161290∅32258064516
```

The above program makes use of the related pair of BASIC commands FOR...TO... and NEXT... in lines 5∅ and 65. These commands are used to tell the computer to repeat the procedure described on the intervening lines 30 times. The FOR 1 TO 3∅ command tells the computer to start with C = 1 and then complete the commands between that line and the line containing the command NEXT C. At this time the value of C is set at 2 and the process is repeated. This continues until the value of C reaches 30. This time, when the program reaches the command NEXT C, it has used all of the values of C allowed in the FOR...TO... command, so the computer continues on to the next line in the program.

If possible run the program to investigate some of the repeating patterns for several rational numbers having denominators in the range from 10 to 30. Then change the program to allow printouts of repeating patterns up to 50 digits in length.

7

Real Numbers

Pythagoras
(ca. 580–500 B.C.)

Pythagoras was one of the first people to conceive of mathematics as a logical structure based on a set of common assumptions and properties which are logically developed from these assumptions. He, and his followers, formed a theory of numbers based on positive integers and their ratios. When one of his followers discovered an irrational number, legend has it that the follower was put to death. Much of the early work with irrational numbers arose out of applications of geometry related to work with applications of the Pythagorean theorem in theoretical and applied settings.

7-1 Concept of Real Number

A line of points with integral coordinates does not fit our concept of a line in the physical universe since there are obvious gaps in a line of integral points. A line of points with rational numbers as coordinates is dense and 3000 years ago appeared to fit the concept of a line in the physical universe.

There is a legend that the discovery of the need for numbers that are not rational numbers was made by a Pythagorean who was brutally punished for his unorthodox discovery. The Pythagoreans were a secret society of scholars about 2500 years ago. Their concept of a number was as a number of units—a counting number. Numbers such as $\frac{5}{2}$ were considered as five halves; that is, as five units where each unit was $\frac{1}{2}$ instead of 1. Today we still use this concept when we add two fractions with the same number as denominator, that is, *like fractions.* For example,

$$\frac{3}{2} + \frac{5}{2} = 3\left(\frac{1}{2}\right) + 5\left(\frac{1}{2}\right) = (3 + 5)\left(\frac{1}{2}\right) = 8\left(\frac{1}{2}\right) = 4$$

As a mystical sect the Pythagoreans believed that all events and all elements of the universe depended upon numbers (natural numbers). In particular, they believed that the lengths of any two line segments should be expressible as positive integral multiples of some common unit of length. This belief was shattered when it was discovered that there could not exist a common unit for a diagonal d and a side s of a square.

As indicated in the adjacent figures the *square* has four *sides*, each of length s, and four *angles*, each a *right angle*. The two *diagonals* are shown in the second figure. The Pythagoreans discovered that for any given square, there cannot exist a unit segment such that the lengths of both a diagonal and a side of the square are positive integral multiples of the length of that unit segment. We consider the case of a square with sides 1 unit long.

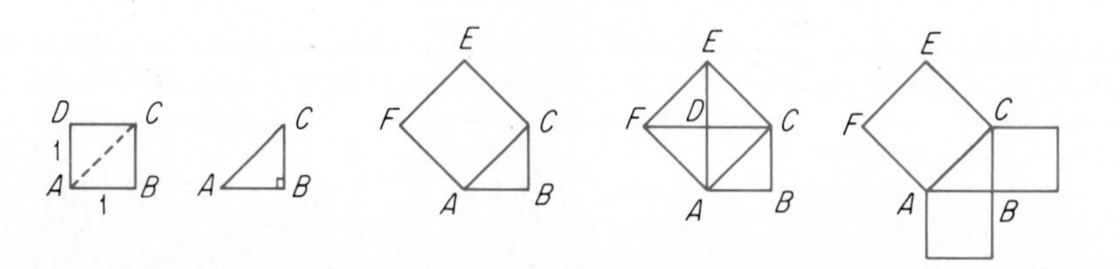

The unit of area is the area of a square that has a side of unit length.

The area of a square on a line segment of length s is s^2, which we read as "s squared."

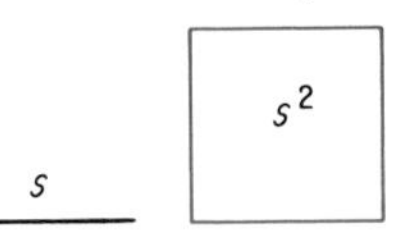

Let $ABCD$ be a square with sides 1 unit long. The area of this **unit square** is, by definition, 1 square unit. Draw the diagonal AC and consider the **right triangle** (a triangle with a right angle) ABC with sides AB and BC, each 1 unit long. The right angle of the triangle is at B, as indicated by the small square. The side AC is the **hypotenuse** (the side opposite the right angle). Next copy triangle ABC and, as in the figure, draw a square $ACEF$ with side AC. Draw the diagonals AE and CF of this square and label their intersection D, since $ABCD$ in this figure is a copy of the original square $ABCD$. The diagonal AC separates the square $ABCD$ into two *congruent* (same size and shape) right triangles ABC and ADC, each with area $\frac{1}{2}$ square unit. The two diagonals AE and CF separate the square $ACEF$ into four congruent right triangles, each with area $\frac{1}{2}$ square unit. Thus the square $ACEF$

has area 2 square units

has twice the area of the original square $ABCD$

has area equal to the sum of the areas of the squares on the other two sides of the right triangle ABC since each of those two squares is equal in area to the original square $ABCD$.

In modern notation we write $a^2 + b^2 = c^2$

c b a

We have demonstrated a special case of the famous **Pythagorean theorem**.

The sum of the areas of the squares on the *legs* (two short sides) of any right triangle is equal to the area of the square on the *hypotenuse*.

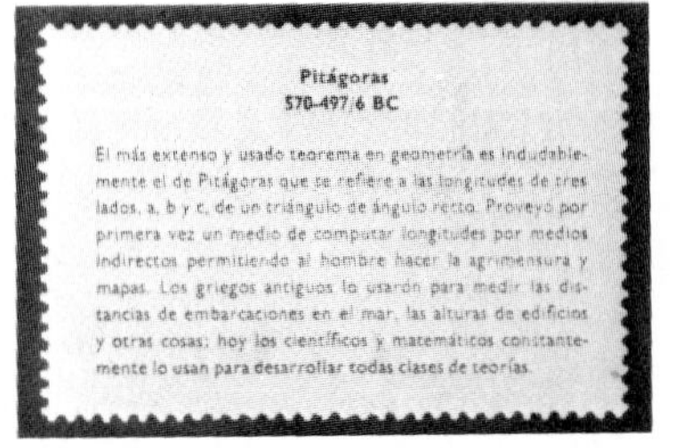

Pitágoras
570-497/6 BC

El más extenso y usado teorema en geometría es indudablemente el de Pitágoras que se refiere a las longitudes de tres lados, a, b y c, de un triángulo de ángulo recto. Proveyó por primera vez un medio de computar longitudes por medios indirectos permitiendo al hombre hacer la agrimensura y mapas. Los griegos antiguos lo usarón para medir las distancias de embarcaciones en el mar, las alturas de edificios y otras cosas; hoy los científicos y matemáticos constantemente lo usan para desarrollar todas clases de teorías.

Nicaragua honored Pythagoras on one of its stamps and, on the back of the stamp, described the Pythagorean theorem as follows: "The most widely used theorem in geometry is undoubtedly the Pythagorean theorem that refers to the lengths of three sides, a, b, and c, of a right triangle. It provided for the first time a way of computing lengths by indirect methods, thus permitting people to make surveys and maps. The ancient Greeks used it for measuring the distances between ships at sea, the heights of buildings, and other things. Today scientists and mathematicians constantly use it in developing all kinds of theories."

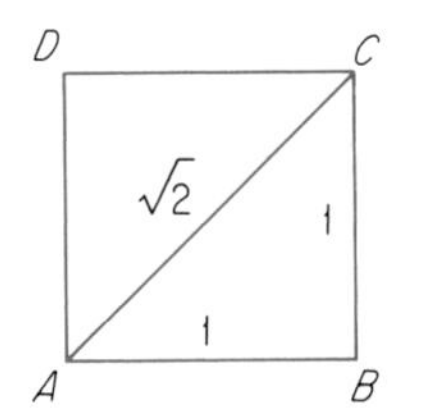

don't worry about proof √2

The diagonal AC of the unit square $ABCD$ has length d such that $d^2 = 2$, that is, $d = \sqrt{2}$, the *square root* of 2, the positive number whose square is 2. Since $\sqrt{2}$ represents the length of a line segment, $\sqrt{2}$ should be accepted as a number.

However, we know that $\sqrt{2}$ is not an integer since

$$1^2 < (\sqrt{2})^2 < 2^2 \quad \text{and therefore} \quad 1 < \sqrt{2} < 2$$

Could $\sqrt{2}$ be a rational number?

The proof that $\sqrt{2}$ is not a rational number depends on these properties of natural numbers (Section 4-1):

A natural number n is an even number if and only if $n = 2k$, where k is a natural number.

The square of an even natural number is an even number.

The square of an odd natural number is an odd number.

If the square of a natural number is even, then the number is an even number.

The proof is an *indirect proof*, that is, we assume that the opposite of what we are trying to prove is true and we show that this assumption leads to an *inconsistency*. (An inconsistency occurs whenever two statements cannot both be true.) Since inconsistencies are not acceptable, the assumption is impossible and its opposite, the statement to be proved, must be true.

Suppose that $\sqrt{2}$ is a rational number p/q. We may simplify p/q so that $p/q = a/b$, where a and b are not both even natural numbers.

The Pythagoreans used this argument more than 2000 years ago to show that it is impossible for $\sqrt{2}$ to be a rational number.

$$\sqrt{2} = \frac{a}{b} \qquad \text{(assumed)}$$

$$2 = \frac{a^2}{b^2} \qquad \text{(square both sides)}$$

$$2b^2 = a^2 \qquad \text{(multiply by } b^2\text{)}$$

Since $a^2 = 2b^2$ implies that a^2 is even, then a is even and $a = 2k$ for some natural number k.

$$2b^2 = (2k)^2 \qquad \text{(substitute } 2k \text{ for } a\text{)}$$

$$2b^2 = 4k^2 \qquad \text{(multiplication)}$$

$$b^2 = 2k^2 \qquad \text{(divide by 2)}$$

Since $b^2 = 2k^2$ implies that b^2 is even, then b is even. We have shown that if $\sqrt{2} = a/b$, then both a and b are even. However, this is contrary to our assumption that a and b are *not* both even. Thus our assumption that $\sqrt{2}$ is a rational number must be incorrect. We need to extend our concept of number if we are to have a number to represent the length of the diagonal of a unit square.

We assume that any length of a line segment is a **real number**. Then $\sqrt{2}$ is a real number that is not a rational number; in other words, $\sqrt{2}$ is an **irrational number**, a real number that cannot be represented as the quotient a/b of two integers.

Example 1 Use an indirect proof to prove

(a) $1 + \sqrt{2}$ is an irrational number.

(b) $a + b\sqrt{2}$ is an irrational number for any rational numbers a and b, where $b \neq 0$.

Solution **(a)** If $1 + \sqrt{2} = p/q$, a rational number, then

$$\sqrt{2} = \frac{p}{q} - 1 = \frac{p - q}{q}$$

where q and $p - q$ are both integers. In other words, if $1 + \sqrt{2}$ were a rational number, then $\sqrt{2}$ would also be a rational number. Since $\sqrt{2}$ is not a rational number, $1 + \sqrt{2}$ cannot be a rational number; that is, $1 + \sqrt{2}$ is an irrational number.

(b) If $a + b\sqrt{2} = p/q$, a rational number, then

$$\sqrt{2} = \frac{1}{b}\left(\frac{p}{q} - a\right) = \frac{p - aq}{bq}$$

where bq and $p - aq$ are both integers. That is, $\sqrt{2}$ is a rational number, which is impossible. Therefore, $a + b\sqrt{2}$ is an irrational number. ■

Real numbers (both rational and irrational) are needed so that all positive numbers have square roots.

Mathematicians have proved, and we assume, that

If a positive integer n is not the square of a positive integer, then $\sqrt{n}$ is an irrational number.

Then, for example, $\sqrt{3}$, $\sqrt{5}$, $\sqrt{6}$, and $\sqrt{15}$ are irrational numbers because 3, 5, 6, and 15 are not squares of positive integers; $\sqrt{4}$ and $\sqrt{9}$ are integers and are therefore rational numbers. Similarly, since $b \times b \times b = b^3$ (read as "b cubed"), and $(\sqrt[3]{b})^3 = b$, it can be shown that if a positive integer n is not the cube of a positive integer, then $\sqrt[3]{n}$ is an irrational number. Then, for example $\sqrt[3]{2}$, $\sqrt[3]{5}$, and $\sqrt[3]{75}$ are irrational numbers; $\sqrt[3]{8}$ is the rational number 2.

Example 2 Identify each number as a rational number or an irrational number.

(a) $\sqrt{19}$	**(b)** $2 + \sqrt[3]{7}$	**(c)** $3 + \sqrt{9}$
(d) $5\sqrt{11}$	**(e)** $\sqrt{13} - 2$	**(f)** $5 - \sqrt{15}$

Solution **(a)** Irrational number, since 19 is not the square of an integer.

(b) Irrational number, since 7 is not the cube of an integer.

(c) Rational number, since 9 is the square of an integer.

(d), **(e)**, **(f)** Irrational numbers since 11, 13, and 15 are not squares of integers. ■

The multitudes of irrational numbers become evident as we recognize some of the numerous infinite sets of irrational numbers.

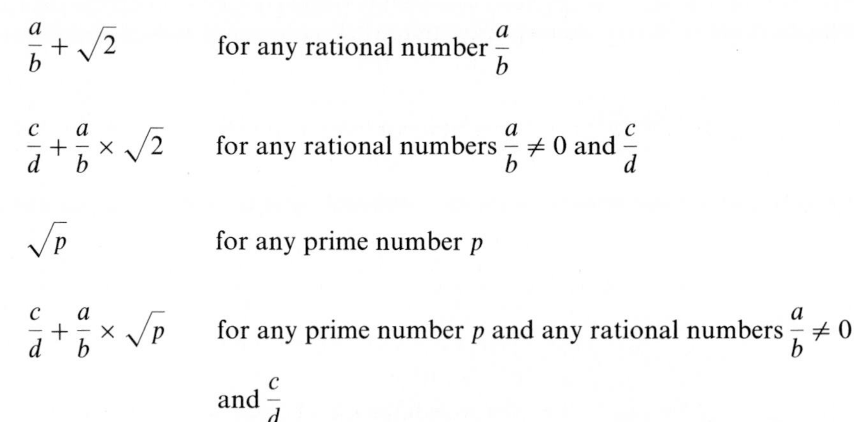

$\frac{a}{b} + \sqrt{2}$	for any rational number $\frac{a}{b}$
$\frac{c}{d} + \frac{a}{b} \times \sqrt{2}$	for any rational numbers $\frac{a}{b} \neq 0$ and $\frac{c}{d}$
$\sqrt{p}$	for any prime number p
$\frac{c}{d} + \frac{a}{b} \times \sqrt{p}$	for any prime number p and any rational numbers $\frac{a}{b} \neq 0$ and $\frac{c}{d}$

There are also infinitely many irrational numbers that cannot be expressed using the radical symbol; for example, $n\pi$ where n is a natural number and π is the ratio of the circumference to the diameter of a circle.

Real numbers are needed as lengths of line segments.

The assumption that each line segment has a real number as its length is equivalent to an assumption that each point on a number line has a real number as its coordinate. For example, every line segment AB of length d has the same length as the line segment OD, where O is the origin and D is the point with coordinate d on a number line.

There is a one-to-one correspondence between the elements of the set of real numbers and the set of points on the number line. Indeed, this is the distinguishing feature between the set of real numbers and the set of rational numbers. Thus every real number is the **coordinate** of a point on the number line, and every point on the number line is the **graph** of a real number. Accordingly, we refer to the number line as the **real number line** and the set of real numbers is said to be **complete**.

Graphs on a number line may be used to compare the sets of natural numbers, whole numbers, integers, and real numbers. Consider these graphs of $x \leq 3$.

Natural numbers: -5 -4 -3 -2 -1 0 1 2 3 4 5 $\{1, 2, 3\}$

Whole numbers: -5 -4 -3 -2 -1 0 1 2 3 4 5 $\{0, 1, 2, 3\}$

Integers: -5 -4 -3 -2 -1 0 1 2 3 4 5 $\{\ldots, -3, -2, -1, 0, 1, 2, 3\}$

Real numbers: -5 -4 -3 -2 -1 0 1 2 3 4 5 $\{x \mid x \leq 3\}$

The real numbers enable us to describe geometric figures on a line.

In the graph for integers the shaded arrowhead on the left of the line indicates that the set includes all negative integers.

Example 3 Graph each set of real numbers.

(a) 2 through 5, that is, $2 \leq x \leq 5$.
(b) Between 2 and 5, that is, $2 < x < 5$.
(c) Greater than or equal to 2, that is, $x \geq 2$.
(d) Less than 1, that is, $x < 1$.

Solution **(a)** The graph is a *line segment* with endpoints indicated by the heavy dots at the points with coordinates 2 and 5.

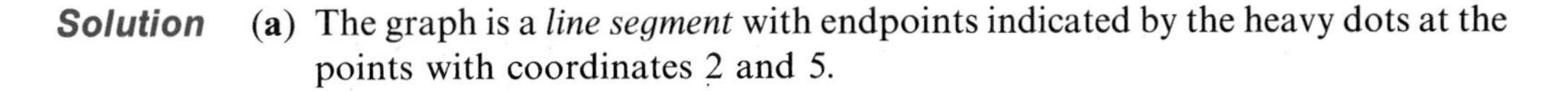

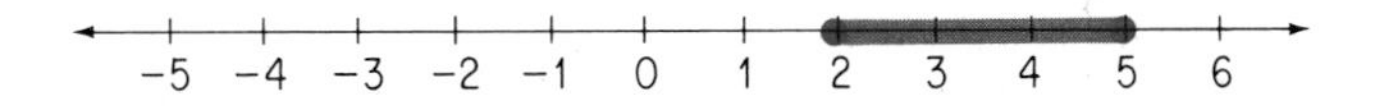

(b) This graph may be obtained from the graph in part (a) by excluding the points with coordinates 2 and 5. Use hollow dots to show that the points with coordinates 2 and 5 are not included in the graph. The graph is an *open line segment*.

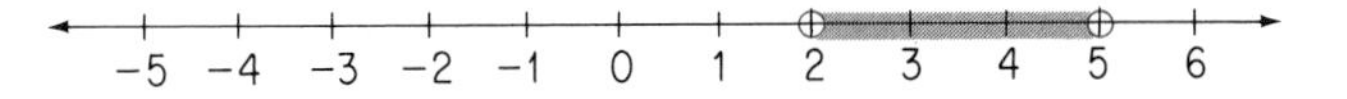

(c) A solid dot is placed at the point with coordinate 2 to indicate that this point is included in the graph of the set. Then a heavily shaded arrow is drawn to indicate that all points with coordinates greater than 2 are to be included. The graph is a *ray*.

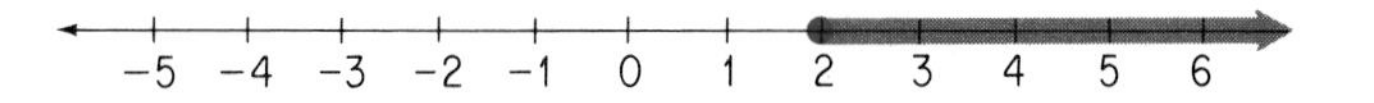

(d) The graph is a *half-line*.

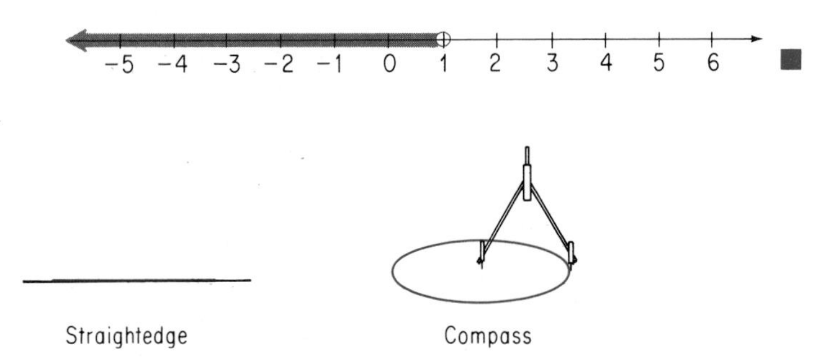

The Pythagoreans used line segments to represent real numbers because they had numerals only for natural numbers. They constructed number scales with a **straightedge** (unmarked ruler) and **compass**. The compass was used for drawing **circles**, that is, sets of points at a given distance (**radius**, plural **radii**)

from a given point (**center**), and **arcs** (connected subsets) of circles. For any setting of the compass (radius of the circle) as the unit distance, a scale can be marked off on a line using successive units from an initial point A.

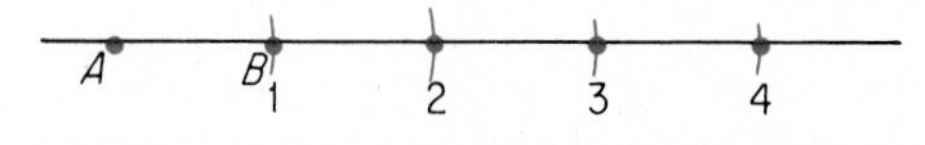

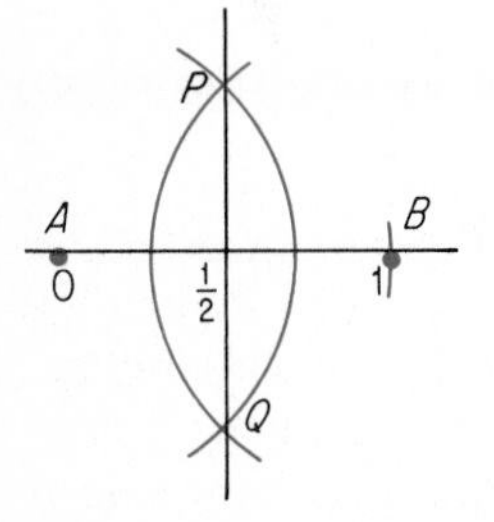

The point with coordinate $\frac{1}{2}$ can be constructed. Let B be the point with coordinate 1. With centers A and B, draw circles with equal radii selected such that the circles intersect in two points P and Q. Draw the line PQ; its intersection with the line AB is the **midpoint** of the line segment AB, that is, the point with coordinate $\frac{1}{2}$. The line PQ is the **perpendicular bisector** of the line segment AB.

To construct a line perpendicular to a given line AB at a given point B, first mark off the point S on AB such that B is the midpoint of AS (draw an arc of the circle with center B and radius AB as in the figure). Then construct the perpendicular bisector of the line segment AS and obtain the line that is perpendicular to the line AB at the point B.

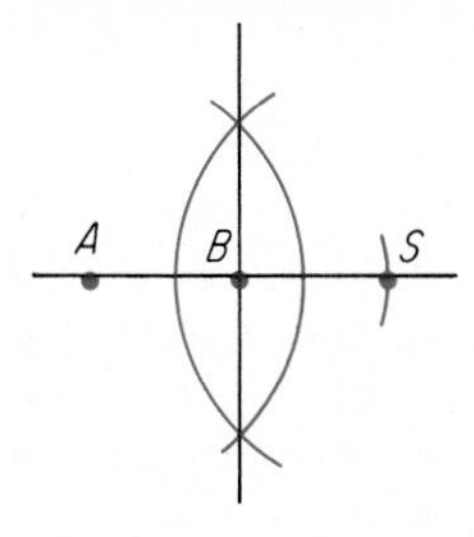

Example 4 Construct the graph on a number line for (**a**) $\sqrt{2}$ (**b**) $3\sqrt{2}$ (**c**) $1 + \sqrt{2}$.

Solution (**a**) On a number line let A be the point with coordinate 0, B the point with coordinate 1, and E the point with coordinate 2. Construct a line BF perpendicular to the number line at B. Mark off BC of length 1 on line BF. Draw the line segment AC. Then

$$1^2 + 1^2 = (AC)^2 \qquad (AC)^2 = 2$$

and line segment AC has length $\sqrt{2}$. Therefore the circle with center A and radius AC intersects the number line at the point with coordinate $\sqrt{2}$.

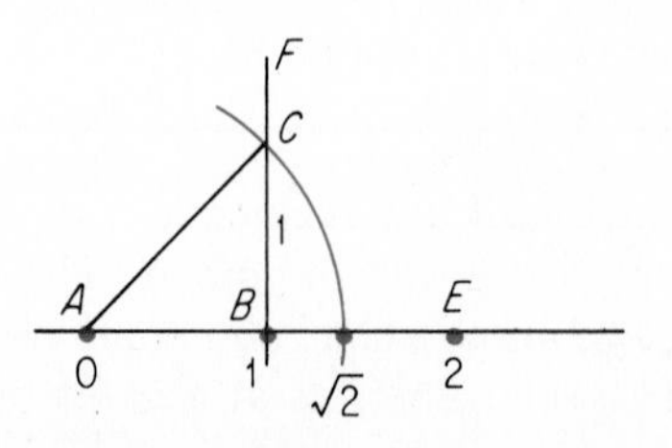

(b) Let P be the point with coordinate $\sqrt{2}$ in part (a) and mark off a scale with AP as its unit length.

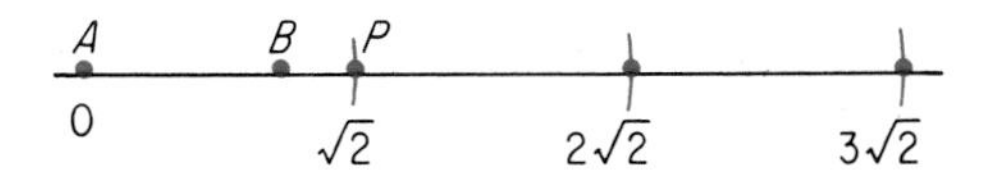

(c) The same construction as in part (a) may be used with A as the point with coordinate 1 and E as the point with coordinate 3.

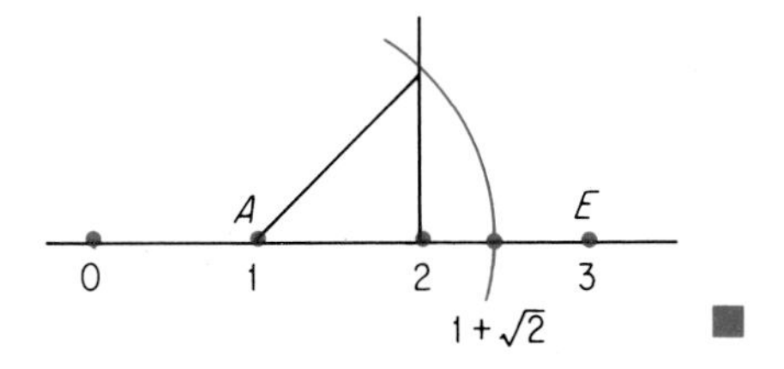

■

In Example 4(c) the point could also have been graphed by marking off one unit from the point P in Example 4(b), that is, $1 + \sqrt{2} = \sqrt{2} + 1$. Such constructions may be used to graph on a number line any number of the form $b\sqrt{2}$ or $a + b\sqrt{2}$, where a and b are integers. The method may be extended, although we shall not do so, to include any rational numbers a and b.

All decimals represent real numbers.

The set of real numbers may be obtained either by assuming that all line segments have real numbers as lengths or by assuming that all decimals represent real numbers. Any integer can be represented by a decimal. Any rational number a/b can be represented by a terminating decimal (division is exact at some stage) or by a nonterminating repeating decimal.

$$\frac{3}{8} = 3 \div 8 = 0.375 = 0.375000\ldots = 0.375\overline{0}$$

$$\frac{80}{11} = 80 \div 11 = 7.272727\ldots = 7.\overline{27}$$

Real numbers that are irrational numbers are represented by **nonterminating, nonrepeating decimals**, such as

$$\sqrt{2} = 1.414214\ldots \qquad \pi = 3.1415926\ldots$$

We consider these decimals rather than numbers such as π, the ratio of the circumference to the diameter of a circle, because advanced mathematical procedures are needed to determine whether π and other such numbers are rational numbers or irrational numbers.

In each of these examples the digits do not exhibit a fixed repeating pattern no matter how far they are extended. Some irrational numbers can be represented in decimal notation by a sequence of digits that has a pattern but does not repeat any particular sequence of digits. For example, each of the following decimals names an irrational number.

$0.20220222022220222220\ldots$

$0.305300530005300005\ldots$

$0.40400400040000400004\ldots$

Example 5 Write the digit in the sixteenth position to the right of the decimal point in **(a)** 0.272272227... **(b)** 12.301230012....

Solution **(a)** 0.2722722272222722...; the digit is 2.
(b) 12.3012300123000123...; the digit is 3. ■

Example 6 Name two irrational numbers between 0.47 and 0.48.

Solution It is easier to see the solution if the given numbers are written with several decimal places.

$$0.47 = 0.470000\overline{0}$$

$$0.48 = 0.480000\overline{0}$$

We need to write nonterminating, nonrepeating decimals for irrational numbers that are greater than 0.47 and less than 0.48. Here are two of the many possible answers.

0.4724722472224722222472...

0.475050050005000050000...

The patterns allow us to write as many digits of each decimal as we wish. Such a decimal will never terminate and will never consist of repetitions of any particular sequence of digits. ■

We have defined real numbers that are not rational numbers to be irrational numbers. But are irrational numbers useful in any way? Suppose that a and b, where $b \neq 0$, are integers. Then

$$a + b \qquad a - b \qquad a \times b \qquad a \div b$$

are rational numbers. The same is true of rational numbers a and b. Also any number that is obtained as a measure, obtained from a calculator, or obtained from a digital computer, is a rational number. Thus irrational numbers *are not needed* for the four basic operations of arithmetic, for measuring, for use with calculators, or for use with digital computers. Irrational numbers *are needed* as lengths of line segments, as coordinates of points on a number line, as square roots of most positive numbers, and for performing many operations in advanced mathematics. Any real number may be classified as

positive, zero, or negative, or as
a rational number or an irrational number.

At each step of the development of the set of real numbers we may think of the original set of numbers as a proper subset of the new set.

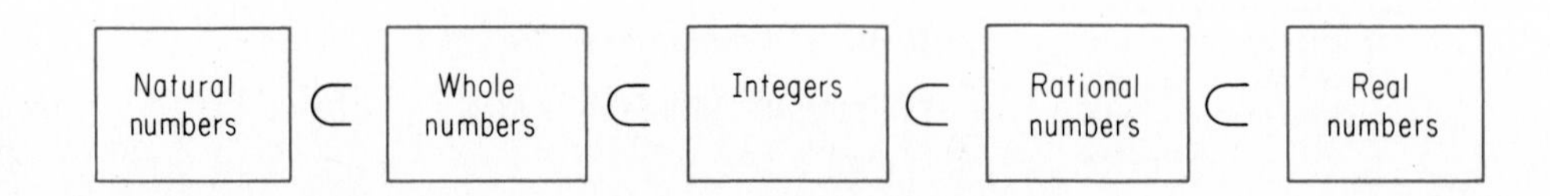

The numbers of each set are *ordered* in the same manner as their graphs on a number line, that is, for the number line in its usual position, $a < b$ and $b > a$ if and only if the point with coordinate a is on the left of the point with coordinate b. The set of real numbers with the operations addition and multiplication forms a **complete linearly ordered field**.

EXERCISES

Classify each statement as true or false. Give a counterexample for each false statement.

1. Every rational number is a real number.
2. Every real number is a rational number.
3. Every real number is either a rational number or an irrational number.
4. Every point on a number line has a real number as its coordinate.
5. Every real number has a point of the number line as its graph.
6. For a given unit of length every positive real number may be represented by the length of a line segment.
7. Every irrational number is either positive or negative.
8. The sum of any two irrational numbers is an irrational number.
9. The additive inverse of an irrational number is an irrational number.
10. The multiplicative inverse of an irrational number is an irrational number.

Identify the given number as a rational number or an irrational number.

11. **(a)** $\sqrt{11}$ **(b)** $5\sqrt{11}$
12. **(a)** $-4 + \sqrt{16}$ **(b)** $11 - \sqrt{63}$
13. **(a)** $7\sqrt{36}$ **(b)** $5\sqrt{37}$
14. **(a)** $\sqrt{20}$ **(b)** $\sqrt{63} + 1$
15. **(a)** $\sqrt{2}/\sqrt{98}$ **(b)** $-2\sqrt{25}$
16. **(a)** $\sqrt{25} - \sqrt{9}$ **(b)** $\sqrt{20} + \sqrt{45}$
17. **(a)** $2\sqrt{2} + (5 - \sqrt{8})$ **(b)** $2\sqrt{5} + (3 - \sqrt{20})$
18. **(a)** $(\sqrt{12} + \sqrt{27})/\sqrt{3}$ **(b)** $\sqrt{7}(\sqrt{28} + \sqrt{63})$
19. **(a)** $0.5\overline{67}$ **(b)** $0.567677677767\ldots$
20. **(a)** $0.\overline{325}$ **(b)** $0.325225222522\ldots$

Graph each set of numbers on a real number line.

21. The set of whole numbers less than 5.
22. The set of integers -6 through -1.
23. The set of real numbers between -4 and 5, that is, $-4 < x < 5$.
24. The set of real numbers greater than -2, that is, $x > -2$.
25. The set of real numbers less than or equal to 1, that is, $x \leq 1$.
26. The set of real numbers greater than or equal to -4, that is, $x \geq -4$.

Sketch the construction of the graph of each number on a number line.

27. $5\sqrt{2}$ **28.** $3 + \sqrt{2}$ **29.** $\sqrt{2} + \frac{1}{2}$

30. $\sqrt{2} - 1$ **31.** $5 - 2\sqrt{2}$ **32.** $\frac{5}{2} + \sqrt{2}$

Tell whether the given number can be represented by a terminating decimal, a nonterminating, repeating decimal, or a nonterminating, nonrepeating decimal. Give the decimal representation of each number that can be expressed as a terminating or repeating decimal.

33. **(a)** $\frac{3}{8}$ **(b)** $\frac{5}{12}$ **(c)** $\sqrt{8}$

34. **(a)** $\sqrt{3}$ **(b)** $\sqrt{100}$ **(c)** $\frac{13}{16}$

Identify the digit in the sixteenth position to the right of the decimal point in the given decimal.

35. **(a)** $0.\overline{23}$ **(b)** 0.131331333...

36. **(a)** $0.\overline{421}$ **(b)** 0.205200520005...

37. **(a)** 0.969969996... **(b)** 0.614611461114...

38. **(a)** 0.989819811981... **(b)** 0.272732733273...

In Exercises 39 *through* 41 *list the numbers of each set in order from smallest to largest.*

39. 0.45, 0.454554555..., 0.45455, $0.\overline{45}$, $0.4\overline{5}$

40. 2.525, 2.5252, $2.5\overline{2}$, 2.525225222..., 2.5

41. 0.067, $0.06\overline{7}$, 0.067677677767..., 0.06, $0.0\overline{6}$

42. State whether or not the indicated number is a rational number between 0.37 and 0.38.

(a) 0.375 **(b)** $0.3\overline{7}$ **(c)** 0.373773777... **(d)** $0.37\overline{8}$

43. State whether or not the indicated number is an irrational number between 0.234 and 0.235.

(a) 0.2345 **(b)** $0.\overline{234}$

(c) 0.234040040004... **(d)** 0.235545545554...

44. Name two rational numbers between 0.524244244424... and 0.525255255525....

45. Name two irrational numbers between 0.48 and 0.49.

46. Name two irrational numbers between $0.\overline{78}$ and $0.\overline{79}$.

Give at least three examples that satisfy the given condition.

47. A sum of irrational numbers that is 0.

48. A sum of irrational numbers that is 2.

49. A difference of irrational numbers that is 1.

50. A product of irrational numbers that is a rational number.
*51. A product of irrational numbers that is an irrational number.
*52. A quotient of irrational numbers that is a rational number.
*53. A quotient of irrational numbers that is an irrational number.

Prove that the number given in the indicated exercise is an irrational number.

*54. Exercise 27. *55. Exercise 28. *56. Exercise 29.
*57. Exercise 30. *58. Exercise 31. *59. Exercise 32.

Assume that $\sqrt{10}$ and $\sqrt{11}$ are irrational numbers. Prove that for any rational numbers p and q, where $q \neq 0$, the given number is an irrational number.

*60. $p + q\sqrt{10}$ *61. $p + q\sqrt{11}$

EXPLORATIONS

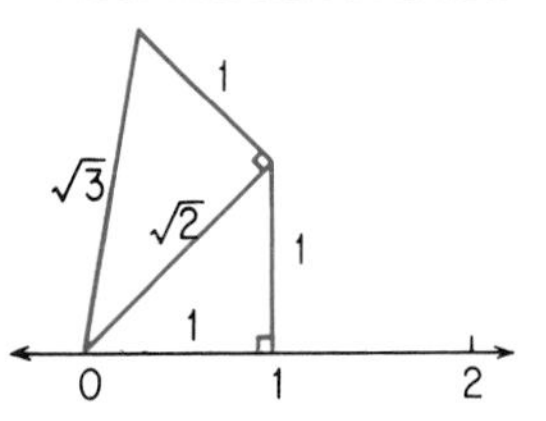

1. The construction for $\sqrt{2}$ as the length of a diagonal of a unit square can be extended to find segments whose measures are $\sqrt{3}$, $\sqrt{4}$, $\sqrt{5}$, and so on. Merely continue to construct right triangles, using the hypotenuse of the preceding triangle as one leg and a segment of one unit as the other leg. The adjacent figure shows the construction for $\sqrt{3}$. Continue this method of construction to find three more line segments with lengths that are irrational numbers.
2. Construct any right triangle and designate the lengths of its sides as a, b, and c, where c is the hypotenuse. Make three copies of this triangle to obtain four congruent right triangles. Cut out the four triangles. Arrange the four triangles as in the adjacent figure to represent

$$(4 \text{ triangles}) + c^2 = (a + b)^2$$

Arrange the four triangles as in the next adjacent figure to represent

$$(4 \text{ triangles}) + a^2 + b^2 = (a + b)^2$$

Explain the sense in which the Pythagorean theorem has been demonstrated.
3. (a) Write a nonterminating, nonrepeating decimal for a number between 5 and 6.
   (b) Explain why your answer for part (a) is considered to be nonterminating.
   (c) Explain why your answer for part (a) is considered to be nonrepeating.
4. Consider the irrational number 0.131131113...
   (a) Find the number of 1's between the 20th and 21st 3.
   (b) Find the total number of 1's preceding the 50th 3.
5. Prepare a worksheet for introducing the construction of line segments of lengths $a + b\sqrt{n}$ for any rational number a and any natural numbers b and n.

6. Look through recently published mathematics textbooks for grades 4 through 8 and prepare a summary of the ways in which irrational numbers (not necessarily by name) are introduced and used.

7-2 Decimal Representations

Integers can be represented in expanded decimal notation. Any rational number can, by definition, be represented by a fraction. If the denominator of the equivalent fraction in simplest form does not have any prime factors except possibly 2 and 5, there is an equivalent fraction with a power of 10 as its denominator and the fraction can be expressed as a terminating decimal. All fractions can, by long division, be expressed as nonterminating, repeating decimals. How can we represent irrational numbers as decimals?

Any decimal that represents an irrational number must be a nonterminating, nonrepeating decimal. We may write such decimals, for example, as

$$1.23233233323\ldots \qquad 9.8768877668887776668\ldots$$

Often irrational numbers are represented by special symbols such as $\sqrt{2}$, $\sqrt[3]{5}$, and π, the ratio of the *circumference* (distance around) of a circle to the *diameter* (greatest distance across) of the circle. If we wish to find the decimal representation of a particular irrational number, such as $\sqrt{2}$, the ratio of a diagonal to a side of a square, we may use a method of successive approximation.

On the number line the point with coordinate $\sqrt{2}$ is being identified as on a sequence of intervals with each interval a proper subset and less than one-half (in this case one-tenth) the length of its predecessor. Such intervals are called *nested intervals*, or Cantor's nested intervals in recognition of the work of Georg Cantor in establishing our concept of real numbers.

We know that

$$1 < \sqrt{2} < 2$$

since $1^2 = 1$, $(\sqrt{2})^2 = 2$, and $2^2 = 4$. We may verify that $(1.4)^2 = 1.96$ and $(1.5)^2 = 2.25$. Therefore

$$1.4 < \sqrt{2} < 1.5$$

Continuing in this manner we may verify each of these statements.

$$\left.\begin{aligned}(1.41)^2 &= 1.9881\\(1.42)^2 &= 2.0164\end{aligned}\right\}\quad 1.41 < \sqrt{2} < 1.42$$

$$\left.\begin{aligned}(1.414)^2 &= 1.999396\\(1.415)^2 &= 2.002225\end{aligned}\right\}\quad 1.414 < \sqrt{2} < 1.415$$

$$\left.\begin{aligned}(1.4142)^2 &= 1.99961644\\(1.4143)^2 &= 2.00034449\end{aligned}\right\}\quad 1.4142 < \sqrt{2} < 1.4143$$

$$\left.\begin{aligned}(1.41421)^2 &= 1.9999899241\\(1.41422)^2 &= 2.0000182084\end{aligned}\right\}\quad 1.41421 < \sqrt{2} < 1.41422$$

A calculator can be very helpful in this *squeezing process* for approximating square roots.

We are now caught up in a never-ending process. The decimal representation for $\sqrt{2}$ does not terminate, and it does not repeat. Given enough time and patience, we may express $\sqrt{2}$ correct to as many decimal places as we wish, but at each step we merely have a rational number approximation to this irrational number.

The early Babylonians used a procedure of *guess*, then *divide and average* to approximate irrational square roots. For $\sqrt{2}$ we would first need to make a guess such as 1, 2, or 1.5. Let us try 1. For approximations we use the **approximation symbol** " $\approx$ ", read "is approximately equal to."

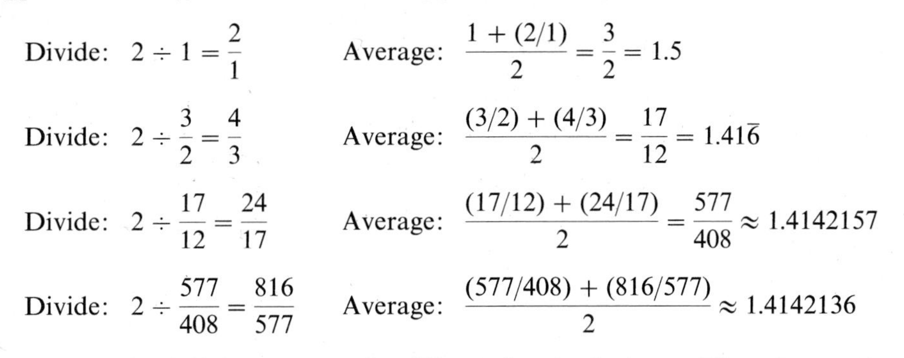

$$\text{Divide: } 2 \div 1 = \frac{2}{1} \qquad \text{Average: } \frac{1 + (2/1)}{2} = \frac{3}{2} = 1.5$$

$$\text{Divide: } 2 \div \frac{3}{2} = \frac{4}{3} \qquad \text{Average: } \frac{(3/2) + (4/3)}{2} = \frac{17}{12} = 1.41\overline{6}$$

$$\text{Divide: } 2 \div \frac{17}{12} = \frac{24}{17} \qquad \text{Average: } \frac{(17/12) + (24/17)}{2} = \frac{577}{408} \approx 1.4142157$$

$$\text{Divide: } 2 \div \frac{577}{408} = \frac{816}{577} \qquad \text{Average: } \frac{(577/408) + (816/577)}{2} \approx 1.4142136$$

The value found by the Babylonians for $\sqrt{2}$ was $1 + \frac{24}{60} + \frac{51}{3600} + \frac{10}{216{,}000}$ which is approximately 1.414213 and is accurate to six decimal places. The Babylonian facility with computations provided accuracy that was not improved upon for 3000 years.

The last decimal digits may vary for different hand calculators. The process can be continued, but note how rapidly a very good approximation can be obtained. The hundredths digit was known to be correct after the third step; the hundred-thousandths digit was known to be correct after the fourth step. Note also that

$$1^2 = 1$$
$$(1.5)^2 = 2.25$$
$$(1.41\overline{6})^2 \approx 2.006944$$
$$(1.4142156)^2 \approx 2.0000058$$
$$(1.4142136)^2 \approx 2.0000001$$

We shall refer to the Babylonian algorithm for approximating irrational square roots as the **divide and average algorithm**. It is often called **Newton's method** since he also used it.

The Babylonians made extensive use of tables. Square root tables may become obsolete as hand calculators are used more extensively. However, other tables, such as the tax tables that must be used to complete one's annual income tax return, continue to the widely used. Thus the ability to read a table remains an essential skill.

n	$\sqrt{n}$
2	1.414 214
3	1.732 051
4	2.000 000
5	2.236 068
6	2.449 490
7	2.645 751
8	2.828 427
9	3.000 000
10	3.162 278

Consider the adjacent brief table of square roots. The approximation for each irrational square root is given as a rational number with seven-digit accuracy. Suppose that we are only interested in three-digit accuracy. Is it desirable to *truncate* (cut off) the last four digits?

1.41   1.73   2.23   2.44   2.64   2.82   3.16

To obtain three-digit accuracy we need approximations that are correct to the nearest hundredth.

1.41   1.73   2.24   2.45   2.65   2.83   3.16

Thus it is not desirable to truncate and we must **round off** each decimal. Note that we *round down* (leave the hundredths digit unchanged) if the next (thousandths) digit is less than 5; we *round up* (increase the hundredths digit by 1 where $9 + 1 = 0$ with 1 carried) if the next digit is greater than 5 or the digit 5 with at least one nonzero digit after it. If the next digit is a 5 alone or a 5 followed by zeros, we have a choice. In most financial transactions it is customary to round up. In later work in mathematics, especially in statistics, a bias in the data is introduced by this procedure. The usual practice in such situations is to round so that the preceding numeral is even whenever a 5, alone or followed only by zeros, is the first digit to be eliminated. In other words, rounded to the nearest 10, we would round 25 down to 20, but we would round 35 up to 40. Then about half of the time we round down and about half of the time we round up. We shall use this convention at all times.

Example 1 Round each decimal to the nearest tenth.

(a) 24.36 **(b)** 34.52 **(c)** 3.15 **(d)** 8.25

Solution **(a)** 24.4 **(b)** 34.5 **(c)** 3.2 **(d)** 8.2 ■

Approximate values for the square root of any positive integer or any proper decimal fraction may also be found by considering the decimal digits in pairs from the decimal point. For $n = 250{,}000$ consider

$$n = 25 \times 100 \times 100 = 25\,00\,00$$

$$\sqrt{n} = \ \ 5 \times \ \ 10 \times \ \ 10 = \ \ 5 \ \ 0 \ \ 0 = 500$$

Similarly, if $n = 0.00000016$,

$$n = 0.00\,00\,00\,16$$

$$\sqrt{n} = 0.\ \ 0 \ \ 0 \ \ 0 \ \ 4 = 0.0004$$

The two roots just found may be checked by multiplication since $\sqrt{n} \times \sqrt{n} = n$. This procedure may also be used to obtain approximations. For example, if $n = 100{,}000$, then

$$n = 10\,00\,00 = 10^5$$

$$\sqrt{n} \approx \ \ 3 \ \ 0 \ \ 0$$

This approach is also helpful when we use approximate values for square roots from a table. For example, if

$$n = 10^5 = 10^2 \times 10^2 \times 10$$

$$\sqrt{n} = 10 \times 10 \times \sqrt{10} \approx 316.2278$$

Thus we may simply use the digits from the table.

$$n = 10\,00\,00$$

$$\sqrt{n} \approx \ \ 3 \ \ 1 \ \ 6.2278$$

Example 2 Use the value for $\sqrt{7}$ from the table and find $\sqrt{0.0007}$ to the nearest hundred thousandth.

Solution

$$n = 0.00\,07$$

$$\sqrt{n} \approx 0.\ 0\ \ 2645751 \approx 0.02646 \ \blacksquare$$

Our knowledge of squares of numbers usually enables us to estimate square roots with one-digit accuracy. The approximations

$$\sqrt{100{,}000} \approx 300 \quad \text{and} \quad \sqrt{0.0017} \approx 0.04$$

have *one significant digit* (see Section 9-2) since the zeros serve only to locate the decimal point.

Example 3 Estimate the square root of each number to one significant digit.
(a) 7800 **(b)** 0.08 **(c)** 0.0024 **(d)** 1,050,900

Solution
(a) $\sqrt{7800} \approx 90$
(b) $\sqrt{0.08} \approx 0.3$
(c) $\sqrt{0.0024} \approx 0.05$
(d) $\sqrt{1{,}050{,}900} = \sqrt{1\,05\,09\,00} \approx 1000 \ \blacksquare$

EXERCISES

1. Assume that $\pi = 3.141\ 592\ 653\ 589\ 793\ 238\ 462\ 643\ 383\ 279\ 50\ldots$ Round off the value of π to the nearest
   (a) tenth **(b)** thousandth **(c)** ten millionth.
2. As in Exercise 1 state the value of π with the specified accuracy.
   (a) 4 digit **(b)** 6 digit **(c)** 10 digit.
3. As in Exercise 1 round π to the indicated number of places to the right of the decimal point.
   (a) 4 **(b)** 6 **(c)** 13.
4. Assume that $1/\pi = 0.318\ 309\ 886\ 183\ 790\ 671\ 537\ 767\ 526\ 745\ 02\ldots$ Round off the value of $1/\pi$ to the nearest
   (a) thousandth **(b)** millionth **(c)** billionth.
5. About 240 B.C. Archimedes asserted that the value of π was between 223/71 and 22/7. Determine whether or not his assertion was correct.
6. About A.D. 480 an early Chinese writer used 355/113 as an approximation for π. To how many decimal places is this approximation correct?

Estimate the square root of the given number to one significant digit.

7. 15
8. 40
9. 800
10. 5000
11. 25,000
12. 10,000,000
13. 0.05
14. 0.06
15. 0.15
16. 0.015
17. 0.021
18. 0.0026

Use the squeezing method to find the indicated square root, correct to the nearest tenth.

19. $\sqrt{11}$ **20.** $\sqrt{17}$
21. $\sqrt{45}$ **22.** $\sqrt{82}$
23. $\sqrt{93}$ **24.** $\sqrt{107}$

Start with the natural number whose square is nearest the square of the given number. Then list this approximation with the next three approximations obtained by the divide and average method for finding the indicated square root.

25. $\sqrt{3}$ **26.** $\sqrt{5}$
27. $\sqrt{7}$ **28.** $\sqrt{11}$
29. $\sqrt{17}$ **30.** $\sqrt{45}$
31. $\sqrt{82}$ **32.** $\sqrt{93}$
33. $\sqrt{107}$ **34.** $\sqrt{127}$
35. $\sqrt{405}$ **36.** $\sqrt{911}$

For Exercises 37 through 42 make an estimate and list it with the next three approximations obtained by the divide and average method for finding the indicated square root.

37. $\sqrt{0.15}$ **38.** $\sqrt{0.23}$
39. $\sqrt{0.015}$ **40.** $\sqrt{0.023}$
41. $\sqrt{0.00024}$ **42.** $\sqrt{0.00041}$

43. Assume that $\sqrt{13} \approx 3.605$ and $\sqrt{130} \approx 11.40$. Find the values of the indicated square root with four-digit accuracy.

(a) $\sqrt{1.3}$ **(b)** $\sqrt{1300}$ **(c)** $\sqrt{13{,}000}$ **(d)** $\sqrt{0.013}$

44. Assume that $\sqrt{15} \approx 3.8730$ and $\sqrt{150} \approx 12.247$. Find the value of the indicated square root with five-digit accuracy.

(a) $\sqrt{1500}$ **(b)** $\sqrt{15{,}000}$ **(c)** $\sqrt{0.015}$ **(d)** $\sqrt{0.00015}$

45. To the nearest foot find the length of one side of a square house lot with an area of 1.00 acres (1 acre = 43,560 square feet).

46. To the nearest meter find the length of a side of a square field with an area 10.0 hectares (1 hectare is a square hectometer; 1 hectometer = 100 meters).

47. The "radio horizon" in kilometers for a certain radio station is $\sqrt{3h}$, where h is the height of the antenna in meters. To the nearest kilometer find the distance to the radio horizon for an antenna with height

(a) 50 meters **(b)** 100 meters.

48. In a vacuum a freely falling body falls s feet in $t = \sqrt{2s/g}$ seconds, where $g \approx 32$ feet per second per second. Find the time for a body to fall
(a) 100 feet **(b)** 400 feet **(c)** 40,000 feet.

EXPLORATIONS

Operations on a hand calculator are based upon rational numbers in decimal form with a specified number, depending on the calculator, of digits. This sometimes creates difficulty in identifying the set of repeating digits.

1. From advanced mathematics it is known that the number of digits in the set of repeating digits for any fraction with denominator d must be a factor of $d - 1$. Explain why the number of repeating digits is at most $d - 1$. (See Section 6-5)
2. Consider all proper fractions with denominator 17 and use the expected pattern to identify the first twenty decimal places of
(a) 2/17 **(b)** 13/17.
3. Repeat Exploration 2 for denominator 23 and find the decimal expansion for 7/23.

7-3 Integers as Exponents

Natural numbers have been used as exponents to tell how many times another number, called the base, is used as a factor in a product. In general,

$$a^m = \overbrace{a \times a \times a \times a \times a \times a \times \cdots \times a}^{m \text{ factors}}$$

where m is a natural number. This is read as "a to the mth power" where a is the **base**, m is the **exponent** and a^m is the **power**. For example, a^5 is read as "a to the fifth power" and means $a \times a \times a \times a \times a$.

Through the use of specific examples, we can develop rules for multiplying and dividing. Thus, consider the product $a^3 \times a^2$.

$$a^3 = a \times a \times a \qquad a^2 = a \times a$$

$$a^3 \times a^2 = (a \times a \times a) \times (a \times a) = a \times a \times a \times a \times a = a^5$$

In general, for all real numbers a and for all natural numbers m and n,

$$\mathbf{a^m \times a^n = a^{m+n}}$$

Note that in multiplication, the bases must be the same. Thus we may simplify $a^3 \times a^2$ by adding exponents to obtain a^5. However, an expression such as $a^3 \times b^2$ cannot be further simplified and would normally be written as a^3b^2.

Example 1 Simplify $x^3 \times x^4$.

Solution

$$x^3 \times x^4 = x^{3+4} = x^7$$

Although an intermediate step is shown here for illustrative purposes, it is usually omitted in actual practice. ■

Example 2 Simplify $(2a^2b^3)(3a^4b^5)$.

Solution As in Example 1, it is possible to proceed directly to the solution, but an intermediate step is shown to indicate groupings of like terms using the commutative and associative properties of multiplication.

$$(2a^2b^3)(3a^4b^5) = (2)(3)(a^2)(a^4)(b^3)(b^5) = 6a^6b^8 \quad ■$$

Next, let us consider the quotient $a^6 \div a^2$, $a \neq 0$. By writing the factors of each term we have the following.

$$\begin{aligned}\frac{a^6}{a^2} &= \frac{a \times a \times a \times a \times a \times a}{a \times a} \\ &= \frac{a}{a} \times \frac{a}{a} \times a \times a \times a \times a \\ &= 1 \times 1 \times a \times a \times a \times a \\ &= 1 \times a^4 \\ &= a^4\end{aligned}$$

From this example it seems clear that to divide powers of the same base we merely subtract the exponents, as in Example 3.

$$\mathbf{a}^m \div \mathbf{a}^n = \mathbf{a}^{m-n}, \mathbf{a} \neq \mathbf{0}$$

Example 3 Simplify **(a)** $\dfrac{x^{12}}{x^3}$ **(b)** $\dfrac{12a^6b^5}{3a^2b^3}$.

Solution **(a)** $\dfrac{x^{12}}{x^3} = x^{12-3} = x^9$

(b) $\dfrac{12a^6b^5}{3a^2b^3} = \left(\dfrac{12}{3}\right)\left(\dfrac{a^6}{a^2}\right)\left(\dfrac{b^5}{b^3}\right) = 4a^4b^2$ ■

Zero and negative integers have been used for powers of 10 in the expanded decimal notation for rational numbers. We now extend the use of exponents for all real numbers to the set of integers. First we give meaning to the use of zero as an exponent.

Consider the quotient $a^4 \div a^4$, $a \neq 0$. We know that if a number different from zero is divided by itself, the result is 1. However, if we wish to find such quotients by subtracting exponents, then we have the following:

$$\frac{a^4}{a^4} = a^{4-4} = a^0$$

Since we want to have a unique result, regardless of the procedure used, we are forced to adopt the following definition.

For all a, a $\neq$ 0, $a^0 = 1$.

The restriction $a \neq 0$ is needed since 0^0 is undefined. For example, 0^0 might arise from $0^2/0^2$, but division by 0 is not permitted.

Next we shall attempt to give meaning to the use of negative integers as exponents. For example, what do we mean by a^{-4}? It hardly makes sense to think of using the base a as a factor -4 times. Instead we approach the problem by the consideration of a specific example, $a^2 \div a^6$.

$$\frac{a^2}{a^6} = \frac{a \times a}{a \times a \times a \times a \times a \times a}$$

$$= \frac{a}{a} \times \frac{a}{a} \times \frac{1}{a \times a \times a \times a}$$

$$= 1 \times 1 \times \frac{1}{a \times a \times a \times a}$$

$$= 1 \times \frac{1}{a^4}$$

$$= \frac{1}{a^4}$$

Suppose we were to solve this same problem by subtracting exponents. We would then have the following result.

$$\frac{a^2}{a^6} = a^{2-6} = a^{-4}$$

Again, it is essential that we obtain a unique result. We therefore define a^{-4} to mean $1/a^4$. In general, we now define negative exponents in this way.

Note also that $a^m = \dfrac{1}{a^{-m}}$.

For all integers m, and for a $\neq$ 0, $a^{-m} = \dfrac{1}{a^m}$

Example 4 Simplify (**a**) 5^{-2} (**b**) 10^{-3} (**c**) $\left(\frac{1}{2}\right)^{-3}$.

Solution (**a**) $5^{-2} = \dfrac{1}{5^2} = \dfrac{1}{25}$ (**b**) $10^{-3} = \dfrac{1}{10^3} = \dfrac{1}{1000}$

(**c**) $\left(\dfrac{1}{2}\right)^{-3} = \dfrac{1}{\left(\dfrac{1}{2}\right)^3} = \dfrac{1}{\frac{1}{8}} = 8.$ ■

Having defined the meaning of zero and of negative integers as exponents, we may now remove the restrictions originally stated and assert that these rules hold for all integers used as exponents. To summarize:

For all a, and for all integers m and n	
$a^m \times a^n = a^{m+n}$	
$a^m \div a^n = a^{m-n}$	$(a \neq 0)$
$a^0 = 1$	$(0^0$ undefined$)$
$a^{-m} = \dfrac{1}{a^m}$	$(a \neq 0)$

The following example illustrates further applications of these rules. The word *simplify* is used here to imply that all possible operations are to be performed, and that the final result should be free of negative exponents.

Example 5 Simplify (**a**) $(xy)^{-3}$ (**b**) $(x^2)^3$ (**c**) $\left(\dfrac{x}{y^2}\right)^3$.

Solution (**a**) By definition of a negative exponent, $(xy)^{-3} = 1/(xy)^3$.

$$(xy)^{-3} = \frac{1}{(xy)^3} = \frac{1}{(xy)(xy)(xy)} = \frac{1}{x^3y^3}$$

(**b**) By the definition of an exponent, we may think of this example as $x^2 \times x^2 \times x^2 = x^{2+2+2} = x^6$.

(**c**) $\left(\dfrac{x}{y^2}\right)^3 = \dfrac{x}{y^2} \times \dfrac{x}{y^2} \times \dfrac{x}{y^2} = \dfrac{x^3}{y^6}$ ■

The results found in Example 5 lead to three additional rules for integral exponents.

For all $a \neq 0$, $b \neq 0$, m, and n
$(a^m)^n = a^{mn}$
$(ab)^m = a^m b^m$
$\left(\dfrac{a}{b}\right)^n = \dfrac{a^n}{b^n}$

The rules for exponents have a very useful application in our operations with numbers in decimal notation. They are particularly helpful when very large numbers or very small numbers are involved. Each number may be considered as in expanded notation with the terms for the zero digits omitted.

$$5{,}000{,}000 \times 8{,}000{,}000 = (5 \times 10^6) \times (8 \times 10^6) = 40 \times 10^{12} = 4 \times 10^{13}$$

$$0.006 \times 0.0007 = (6 \times 10^{-3}) \times (7 \times 10^{-4})$$
$$= 42 \times 10^{-7} = 4.2 \times 10^{-6}$$

$$8000 \div 0.02 = (8 \times 10^3) \div (2 \times 10^{-2}) = 4 \times 10^5$$

$$0.0006 \div 300 = (6 \times 10^{-4}) \div (3 \times 10^2) = 2 \times 10^{-6}$$

These procedures are often extended for any numbers by writing each number as the product of a number greater than or equal to 1 but less than 10 and some power of 10, that is, by expressing each number in **scientific notation**. For example,

$$5280 = 5.28 \times 10^3$$

$$0.0048 = 4.8 \times 10^{-4}$$

$$5280 \div 0.0048 = (5.28 \times 10^3) \div (4.8 \times 10^{-4}) = 1.1 \times 10^7$$

Unless otherwise instructed, we assume that in ordinary decimal notation, zeros that are not otherwise needed are used only to locate the decimal point.

EXERCISES *Perform the indicated operations and simplify where possible.*

1. **(a)** $2^3 \times 2^2$ **(b)** $2^5 \times 2^{-3}$ **(c)** $2^2 \div 2^{-3}$

2. **(a)** $3^2 \times 3^3$ **(b)** $3^2 \div 3^4$ **(c)** $3^4 \times 3^{-3}$

3. **(a)** $10^4 \times 10^3$ **(b)** $10^5 \div 10^2$ **(c)** $10^2 \div 10^{-3}$

4. **(a)** $\frac{10^7}{10^3}$ **(b)** $\frac{10^4}{10^6}$ **(c)** $\frac{10^{-2}}{10^3}$

5. **(a)** $(5x^5)(4x^4)$ **(b)** $(3x^7)(2x^5)$ **(c)** $(2a^3b^2)(3a^4b^5)$

6. **(a)** $(6x^5) \div (3x^3)$ **(b)** $(7a^2b^3)(2a^3b^{-2})$ **(c)** $(3x^2y^2) \div (x^{-1}y^{-3})$

7. **(a)** $\frac{24a^4b^5}{6a^2b^2}$ **(b)** $\frac{14m^2n^8}{2m^2n^4}$ **(c)** $\frac{8ab^5c^{10}}{2ab^2c^{10}}$

8. **(a)** $\frac{12x^{12}y^{15}}{3x^{10}y^5}$ **(b)** $\left(\frac{x}{2}\right)^3$ **(c)** $\left(\frac{a^3}{5}\right)^2$

9. **(a)** $(2x^3)^4$ **(b)** $(5x^3)^2$ **(c)** $(x^2/2)^2$

10. **(a)** $(3a^3)^2$ **(b)** $(5b^{-5})^2$ **(c)** $(3x^4)^0$

11. **(a)** $(5x^{-3})^0$ **(b)** $(-2x^3)^2$ **(c)** $(x^{-2}y)^{-3}$

12. **(a)** $(x^2)^3 + (y^3)^2$ **(b)** $(ab^2)^{-3}$ **(c)** $(a^4b^{-2})^{-2}$

Write each number in scientific notation.

13. **(a)** 90,000 **(b)** 7,500,000 **(c)** 450,000

14. **(a)** 750,000 **(b)** 235.4 **(c)** 32.5×10^3

15. **(a)** 12.35 **(b)** 451.6×10^2 **(c)** 0.004

16. **(a)** 0.0005 **(b)** 0.000 12 **(c)** 0.000 567

17. **(a)** 0.000 0982 **(b)** 25×10^{-4} **(c)** 26.5×10^{-3}

18. **(a)** 0.000 0675 **(b)** 350×10^{-5} **(c)** 79.3×10^5

19. 0.000 000 0315 centimeters, the diameter of a molecule of the chemical element nitrogen.

20. 0.000 000 000 000 000 000 000 001 6617 grams, the mass of a hydrogen atom.

21. One hundred-millionth of a centimeter, 1 *angstrom*, the smallest common unit of distance, a unit used in expressing lengths of light waves.
22. 19.2 quintillion miles, 1 *megaparsec*, the longest common unit of distance.

Write each number in ordinary decimal notation.

23. 2.4×10^5 miles, the distance from the earth to the moon.
24. 9.3×10^7 miles, the distance from the earth to the sun.
25. 24.5×10^{12} miles, the distance from the earth to the nearest star.
26. 2.5×10^4 miles per hour, the least speed with which an object can be shot upward if it is to escape from the earth.
27. 5.88×10^{12} miles, 1 light year, the distance light travels in a year.
28. 1.602×10^{-19} coulomb, the charge on an electron.

Use scientific notation and compute each result. Express the answer in scientific notation.

29. **(a)** $200{,}000 \times 30{,}000$
    (b) $12{,}000{,}000 \div 40{,}000$
30. **(a)** $250{,}000{,}000 \times 4000$
    (b) $720{,}000 \div 6000$
31. **(a)** $0.000\,003 \times 0.000\,004$
    (b) $0.000\,000\,45 \div 0.000\,03$
32. **(a)** $0.000\,012 \times 0.0001$
    (b) $2{,}000{,}000 \times 0.000\,015$
33. **(a)** $0.000\,45 \div 150$
    (b) $0.000\,025 \times 300{,}000$
34. **(a)** $0.000\,050 \div 250$
    (b) $2{,}000{,}000 \div 0.000\,005$

In Exercises 35 through 37 express each quantity in scientific notation.

35. The number of kilograms in a ton (one ton is 2000 pounds, 1 pound is 0.453 5924 kilograms).
36. The distance that light travels in 3.20 seconds if light travels at 1.86×10^5 miles per second. (Recall that $d = rt$.)
37. The distance that sound travels in one microsecond (10^{-6} second) it sound travels 1087 feet per second.

*38. If radio waves are sent out, reflected from an airplane, and return to the same radar in 600×10^{-6} seconds, how far is the airplane from the radar? Radio waves travel at the speed of light, about 328×10^6 yards per second.

EXPLORATIONS

1. Does $2^4 = 4^2$? Does this show that raising to a power is a commutative operation? Explore other possibilities and explain your answer.
2. Does $(2^2)^3 = 2^{(2^3)}$? Explain the need for the definition $a^{b^c} = (a^b)^c$. Determine the number of zeros in the decimal expansion of
   (a) 10^k **(b)** $(10^2)^k$ **(c)** $10^{(2^k)}$.
3. Make a collection of very large numbers and very small numbers that are used in science, expressing each in scientific notation.
4. Explore a recently published elementary mathematics textbook series to determine what use, if any, is made of scientific notation.

7-4 Simple and Compound Interest

The determination of amounts of interest to be paid or received provides one of the most common applications of integral exponents. In order to understand the usual procedures for determining interest, we first need to understand *simple interest.*

Suppose that you borrow \$2000 from a friend at the rate of 12% per year simple interest. The interest on the loan each year is

$$0.12 \times 2000, \quad \text{that is,} \quad \$240$$

The interest rates used here are for illustrative purposes only. In actual practice they vary greatly, especially in periods of rapid inflation.

The original amount of the loan is called the **principal** (P). The interest (I) is a percent (r) of the principal each year. The duration (time, t) of the loan is stated in years. The amount needed to pay off a loan depends on the duration of the loan. The amount needed in any one of the first five years to pay off a loan of \$2000 at 12% per year simple interest is shown in the following table.

At the End of	*Principal*	*Accumulated Interest*	*Amount Due*
1 year	\$2000	\$240	\$2240
2 years	2000	480	2480
3 years	2000	720	2720
4 years	2000	960	2960
5 years	2000	1200	3200

The **simple interest** (I) on a principal (P) at a rate (r) per year for t years is

$$I = Prt$$

Example 1 Find the simple interest on a loan of \$2500 at 6% for 3 years.

Solution $P = \$2500, \quad r = 6\%, \quad t = 3$:

$$I = Prt = \$2500 \times 0.06 \times 3 = \$450$$

The interest is \$450; at the end of 3 years the amount due is \$2500 + \$450, that is, \$2950. ■

In general practice, the interest charged on loans and paid on savings accounts is **compound interest**, that is, interest is charged (or paid) on the previously due (or credited) interest. On credit card balances, the interest is compounded and added monthly. On bank deposits the interest is usually added quarterly, but often is compounded daily. The effect of compounding may be seen in the following development where we shall consider a sum of money that is deposited in a bank where interest is compounded. To simplify the discussion, we begin by considering interest compounded yearly. The computations can be extended to interest compounded more frequently.

Suppose that $2500 is deposited at 6% interest compounded annually. Then, as in the next table, the interest is added to the account at the end of each year and the interest for the following year is computed on this new amount.

	***Principal* (at the start of the year)**	***Interest***	***Amount* (at the end of the year)**
First year	$2500	$2500 × 0.06 × 1 = $150	$2500 + $150 = $2650
Second year	2650	2650 × 0.06 × 1 = 159	2650 + 159 = 2809
Third year	2809	2809 × 0.06 × 1 = 168.54	2809 + 168.54 = 2977.54

Each year the procedure is the same. The **amount** (A) at the end of the year is equal to the sum of the principal (P) and the interest (Pr) for that year.

$$A = P + Pr = P(1 + r)$$

First year $\quad A = 2500(1 + 0.06) = 2650$

Second year $\quad A = 2650(1 + 0.06) = [2500(1 + 0.06)](1 + 0.06)$

$$= 2500(1 + 0.06)^2 = 2809$$

Third year $\quad A = 2809(1 + 0.06) = 2500(1 + 0.06)^3 = 2977.54$

The pattern continues.

For ten years $\quad A = 2500(1 + 0.06)^{10}$

For n years $\quad A = 2500(1 + 0.06)^n$

Since interest is paid (charged) on interest, compound interest is always more than simple interest. For example, at 6% the simple interest on $2500 is $150 per year and $450 for 3 years, as in Example 1. Thus after 3 years

$2977.54	(amount at compound interest)
− 2950.00	(amount at simple interest)
$ 27.54	(difference)

Although this difference over a 3-year period may seem inconsequential, the accumulation for larger sums over a longer period of time is quite substantial.

Furthermore, interest is usually compounded more frequently than once a year. Thus interest may be compounded twice a year (semiannually), or four times a year (quarterly), or more often. Indeed, as noted earlier, some funds compound money on a daily basis. In general, the more often money is compounded the more interest is earned up to a certain limit.

Example 2 Find the amount on deposit in a bank at the end of 1 year if $2000 is deposited at 5% compounded semiannually.

Solution A rate of 5% per year compounded semiannually is equivalent to a rate of $2\frac{1}{2}\%$ per half year.

	Principal	*Interest*	*Amount*
First half year	$2000	$\$2000 \times 0.05 \times \frac{1}{2} = \$50$	$\$2000 + \$50 = \$2050$
Second half year	$2050	$\$2050 \times 0.05 \times \frac{1}{2} = \$51.25$	$\$2050 + \$51.25 = \$2101.25$

The amount on deposit at the end of 1 year is $2101.25. ■

In Example 2 the total interest earned in one year on $2000 at 5% compounded semiannually is $101.25, which is 5.0625% of the original investment. This equivalent rate of 5.0625% simple interest is called the **effective annual rate** of interest, also called the **annual percentage rate.**

> The amount A, when interest is compounded annually for n years, and no further deposits or withdrawals are made, is
>
> $$A = P(1 + r)^n$$

We can modify this general formula to account for cases when interest is compounded more frequently than once a year. Thus in Example 2 the **period** (interval for compounding) is 6 months. There are two periods per year, the rate is $r/2$ per period, and the number of periods in n years is $2n$. Thus, when interest is compounded semiannually for n years, the amount A is

$$A = P\left(1 + \frac{r}{2}\right)^{2n}$$

Similarly, if interest is compounded quarterly, the rate per period is $r/4$, there are $4n$ periods in n years, and

$$A = P\left(1 + \frac{r}{4}\right)^{4n}$$

This pattern may be extended for compounding k times per year to obtain

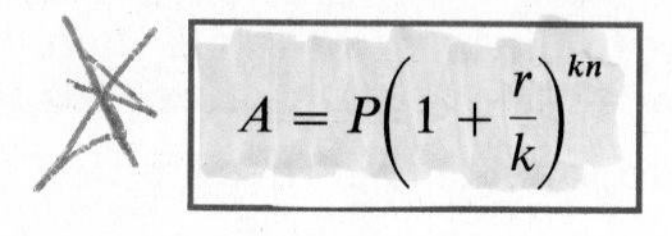

$$A = P\left(1 + \frac{r}{k}\right)^{kn}$$

Example 3 Give an expression for the amount on deposit in each case.

(a) \$1000 is deposited at 5% interest compounded annually for 15 years.
(b) \$2000 is deposited at 6% interest compounded semiannually for 20 years.
(c) \$5000 is deposited at 8% interest compounded quarterly for 6 years.

Computations of these amounts can be very tedious. Fortunately, it is a relatively simple matter with an appropriate calculator or a personal computer.

Solution **(a)** $A = 1000(1 + 0.05)^{15}$, that is, $\$1000(1.05)^{15}$

(b) $A = 2000\left(1 + \frac{0.06}{2}\right)^{2 \times 20}$, that is, $\$2000(1.03)^{40}$

(c) $A = 5000\left(1 + \frac{0.08}{4}\right)^{4 \times 6}$, that is, $\$5000(1.02)^{24}$ ■

A compound interest table may also be used to find amounts for different rates and periods. Thus Table 1 is a table of values of $(1 + i)^n$ where i is the interest rate per period and n is the number of periods. Let us use the table to evaluate the amount in Example 3(a), $\$1000(1.05)^{15}$.

Because of rounding procedures used, the answer obtained by use of the table may differ slightly from that obtained with a computer.

Use $i = 5\%$ and $n = 15$.
From Table 1, $(1 + i)^n = 2.0789$.
$A = \$1000(1.05)^{15} = \$1000(2.0789) = \$2078.90$

Note that your money has more than doubled. Compare this with the amount that you would have had at 5% simple interest for 15 years.

Example 4 Use Table 1 to find the amount for a deposit of \$400

(a) at 6% compounded semiannually for 4 years
(b) at 12% compounded quarterly for 5 years.

Solution **(a)** $P = \$400, \quad i = 3\%, \quad n = 8$:

$$A = 400(1.03)^8 = 400(1.2668) = 506.72$$

The amount is \$506.72.

(b) $P = \$400, \quad i = 3\%, \quad n = 20$:

$$A = 400(1.03)^{20} = 400(1.8061) = 722.44$$

The amount is \$722.44. ■

TABLE 1
Compound Interest $(1 + i)^n$

n	1%	$1\frac{1}{2}$%	2%	$2\frac{1}{2}$%	3%	4%	5%	6%	7%	8%
1	1.0100	1.0150	1.0200	1.0250	1.0300	1.0400	1.0500	1.0600	1.0700	1.0800
2	1.0201	1.0302	1.0404	1.0506	1.0609	1.0816	1.1025	1.1236	1.1449	1.1664
3	1.0303	1.0457	1.0612	1.0769	1.0927	1.1249	1.1576	1.1910	1.2250	1.2597
4	1.0406	1.0614	1.0824	1.1038	1.1255	1.1699	1.2155	1.2625	1.3108	1.3605
5	1.0510	1.0773	1.1041	1.1314	1.1593	1.2167	1.2763	1.3382	1.4026	1.4693
6	1.0615	1.0934	1.1262	1.1597	1.1941	1.2653	1.3401	1.4185	1.5007	1.5869
7	1.0721	1.1098	1.1487	1.1887	1.2299	1.3159	1.4071	1.5036	1.6058	1.7138
8	1.0829	1.1265	1.1717	1.2184	1.2668	1.3686	1.4775	1.5938	1.7182	1.8509
9	1.0937	1.1434	1.1951	1.2489	1.3048	1.4233	1.5513	1.6895	1.8385	1.9990
10	1.1046	1.1605	1.2190	1.2801	1.3439	1.4802	1.6289	1.7908	1.9672	2.1589
11	1.1157	1.1779	1.2434	1.3121	1.3842	1.5395	1.7103	1.8983	2.1049	2.3316
12	1.1268	1.1956	1.2682	1.3449	1.4258	1.6010	1.7959	2.0122	2.2522	2.5182
13	1.1381	1.2136	1.2936	1.3785	1.4685	1.6651	1.8856	2.1329	2.4098	2.7196
14	1.1495	1.2318	1.3195	1.4130	1.5126	1.7317	1.9799	2.2609	2.5785	2.9372
15	1.1610	1.2502	1.3459	1.4483	1.5580	1.8009	2.0789	2.3966	2.7590	3.1722
16	1.1726	1.2690	1.3728	1.4845	1.6047	1.8730	2.1829	2.5404	2.9522	3.4259
17	1.1843	1.2880	1.4002	1.5216	1.6528	1.9479	2.2920	2.6928	3.1588	3.7000
18	1.1961	1.3073	1.4282	1.5597	1.7024	2.0258	2.4066	2.8543	3.3799	3.9660
19	1.2081	1.3270	1.4568	1.5987	1.7535	2.1068	2.5270	3.0256	3.6165	4.3157
20	1.2202	1.3469	1.4859	1.6386	1.8061	2.1911	2.6533	3.2071	3.8697	4.6610
21	1.2324	1.3671	1.5157	1.6796	1.8603	2.2788	2.7860	3.3996	4.1406	5.0338
22	1.2447	1.3876	1.5460	1.7216	1.9161	2.3699	2.9253	3.6035	4.4304	5.4365
23	1.2572	1.4084	1.5769	1.7646	1.9736	2.4647	3.0715	3.8197	4.7405	5.8715
24	1.2697	1.4295	1.6084	1.8087	2.0328	2.5633	3.2251	4.0489	5.0724	6.3412
25	1.2824	1.4509	1.6406	1.8539	2.0938	2.6658	3.3864	4.2919	5.4274	6.8485
26	1.2953	1.4727	1.6734	1.9003	2.1566	2.7725	3.5557	4.5494	5.8074	7.3964
27	1.3082	1.4948	1.7069	1.9478	2.2213	2.8834	3.7335	4.8223	6.2139	7.9881
28	1.3213	1.5172	1.7410	1.9965	2.2879	2.9987	3.9201	5.1117	6.6488	8.6271
29	1.3345	1.5400	1.7758	2.0464	2.3566	3.1187	4.1161	5.4184	7.1143	9.3173
30	1.3478	1.5631	1.8114	2.0976	2.4273	3.2434	4.3219	5.7435	7.6123	10.0627
31	1.3613	1.5865	1.8476	2.1500	2.5001	3.3731	4.5380	6.0881	8.1451	10.8677
32	1.3749	1.6103	1.8845	2.2038	2.5751	3.5081	4.7649	6.4534	8.7153	11.7371
33	1.3887	1.6345	1.9222	2.2589	2.6523	3.6484	5.0032	6.8406	9.3253	12.6760
34	1.4026	1.6590	1.9607	2.3153	2.7319	3.7943	5.2533	7.2510	9.9781	13.6901
35	1.4166	1.6839	1.9999	2.3732	2.8139	3.9461	5.5160	7.6861	10.6766	14.7853
36	1.4308	1.7091	2.0399	2.4325	2.8983	4.1039	5.7918	8.1473	11.4239	15.9682
37	1.4451	1.7348	2.0807	2.4933	2.9852	4.2681	6.0814	8.6361	12.2236	17.2456
38	1.4595	1.7608	2.1223	2.5557	3.0748	4.4388	6.3855	9.1543	13.0793	18.6253
39	1.4741	1.7872	2.1647	2.6196	3.1670	4.6164	6.7048	9.7035	13.9948	20.1153
40	1.4889	1.8140	2.2080	2.6851	3.2620	4.8010	7.0400	10.2857	14.9745	21.7245
41	1.5038	1.8412	2.2522	2.7522	3.3599	4.9931	7.3920	10.9029	16.0227	23.4625
42	1.5188	1.8688	2.2972	2.8210	3.4607	5.1928	7.7616	11.5570	17.1443	25.3395
43	1.5340	1.8969	2.3432	2.8915	3.5645	5.4005	8.1497	12.2505	18.3444	27.3666
44	1.5493	1.9253	2.3901	2.9638	3.6715	5.6165	8.5572	12.9855	19.6285	29.5560
45	1.5648	1.9542	2.4379	3.0379	3.7816	5.8412	8.9850	13.7646	21.0025	31.9204

EXERCISES *Find the amount for each of these deposits at simple interest.*

1. \$450, 5%, 8 years
2. \$600, 6%, 5 years
3. \$1250, 4%, 3 years
4. \$2050, 5%, 6 years
5. \$3500, $4\frac{1}{2}$%, 4 years
6. \$4000, $5\frac{1}{4}$%, 2 years
7. \$2000, 9%, $3\frac{1}{2}$ years
8. \$2750, 8%, $4\frac{1}{2}$ years

Find the amount for each of these deposits at compound interest.

For Exercises 9 through 16 it is suggested that you find the answer using Table 1 and check your results using a calculator. The results by the two methods may differ slightly due to approximations arising from the number of places in the table or the capacity of the calculator.

9. \$1500, 6% annually, 4 years
10. \$3200, 5% annually, 5 years
11. \$750, 6% semiannually, 5 years
12. \$850, 5% semiannually, 3 years
13. \$150, 8% quarterly, 8 years
14. \$800, 8% quarterly, 10 years
15. \$3250, 12% quarterly, 6 years
16. \$7500, 10% quarterly, 4 years

Use Table 1 to answer each of the following.

17. Approximately how many years does it take for money to double in amount if the money is left on deposit at 6% interest compounded annually?
18. Repeat Exercise 17 for (**a**) 5% (**b**) 8%.
19. Find the effective annual rate of interest if \$100 is deposited for one year at 8% interest compounded
(**a**) semiannually (**b**) quarterly.
20. Repeat Exercise 19 for an initial rate of 10%.
21. A deposit of \$2000 earns 12% interest compounded annually. How much additional interest would be earned the first year if the interest were compounded
(**a**) semiannually (**b**) quarterly?
22. Repeat Exercise 21 for interest compounded monthly. Use a calculator to compute the interest since Table 1 does not provide the necessary information.
23. Assume that you earn \$12,000 per year and that the rate of inflation is 10% per year. How much must you earn 5 years from now just to keep up with inflation? (*Hint:* Compute $12{,}000(1.10)^5$.)
24. Repeat Exercise 23 for each of these annual inflation rates that have been experienced in other countries.
(**a**) 30% (**b**) 50% (**c**) 100% (**d**) 150%

*25. Let us explore one further use of the compound interest table. Suppose you determine that you must have \$15,000 in 10 years in order to finance a child's education. Furthermore, suppose that your bank offers 6% interest compounded quarterly. To the nearest dollar, how much should you

The answer to Exercise 25 is called the **present value** of \$15,000 under the specified conditions.

deposit today to have \$15,000 available when it is needed? In this problem we know the value for A; and we need to find the value of P in the compound interest formula $A = P(1 + i)^n$ where

$$i = \frac{0.06}{4} = 1.5\% \qquad n = 4 \times 10 = 40$$

and from the table

$$(1.015)^{40} = 1.8140$$

Therefore

$$15{,}000 = P(1.8140)$$

Solve for P by completing this division.

$$P = \frac{15{,}000}{1.8140}$$

***26.** Repeat Exercise 25 assuming that your money can earn 12% compounded quarterly.

***27.** You wish to have \$10,000 in 10 years and are making a single deposit for that purpose.
- **(a)** How much money would you need if you deposit your money at 6% interest compounded quarterly?
- **(b)** How much would you need if you deposit your money at 8% interest compounded semiannually?
- **(c)** What is the difference between the amounts needed in parts (a) and (b)?

EXPLORATIONS

1. Visit a local bank and obtain information concerning the different types of accounts offered and the interest rates used. Try to find out what method is used for compounding interest, that is, the frequency with which interest is compounded. Then compare your findings with that for other banks in your community.
2. At one time the United States sold Treasury bills in denominations of \$10,000 for 3 months at a discount rate of 12.35% per year. This means that you are credited with the interest when you purchase the bill and receive the face value of \$10,000 when it matures in 3 months. How much interest can you earn on such a purchase? How much must you invest to purchase a \$10,000 Treasury bill? What is the equivalent annual rate of simple interest on your investment?
3. Obtain details on Christmas savings clubs at one or more banks and compare this approach with other methods of saving.
4. Find out what is meant by a bank's *prime rate*.
5. In recent years interest rates have fluctuated widely from month to month. Watch the financial pages of a newspaper and keep a record of interest rates offered by one of your local banks for six-month *certificates of deposit*.

7-5 Powers and Roots

Irrational numbers often arise as roots of numbers such as $\sqrt{2}$ and $\sqrt[3]{5}$. If we assume that the rules for integral exponents apply also for fractional exponents, then roots may be expressed either as fractional powers or as radicals.

$$(2^{1/2})^2 = 2 \qquad (\sqrt{2})^2 = 2 \qquad 2^{1/2} = \sqrt{2}$$

$$(5^{1/3})^3 = 5 \qquad (\sqrt[3]{5})^3 = 5 \qquad 5^{1/3} = \sqrt[3]{5}$$

In general, the relationship between exponents and radicals may be stated as

$$\boxed{a^{1/n} = \sqrt[n]{a}}$$

The rules for integral powers (Section 7-3) are often called the *laws of exponents*. Fractional and other powers are defined so that these laws hold for any positive real number as base and any real numbers as exponents. Proofs of these properties are left for the mathematical specialists. Special problems may arise when roots of negative numbers are concerned, as for $\sqrt{-2}$ and $\sqrt{-3}$, which do not represent real numbers. However, there are some powers and roots of negative numbers that can be readily obtained from the corresponding powers and roots of positive numbers. For example, an integral power of a negative number is a positive number if the exponent is an even number, a negative number if the exponent is an odd number.

$$(-2)^6 = 64 \qquad (-2)^5 = -32$$

$$(-2)^{-5} = -\frac{1}{32} \qquad (-3)^{-4} = \frac{1}{81}$$

Also for any odd number n, the nth root of a negative number is the negative of the nth root of the corresponding positive number as in these examples.

$$\sqrt[5]{-32} = -\sqrt[5]{32} = -2 \qquad \sqrt[3]{-8} = -\sqrt[3]{8} = -2$$

We assume that such operations with negative numbers can be easily understood as soon as operations with positive numbers are understood. Thus we assume that the rules for integral powers hold for any *positive* numbers as bases and any *real* numbers as exponents.

An equation such as $x^2 = 4$ has both 2 and -2 as members of its solution set. However, both the radical symbol $\sqrt{}$ and the exponent 1/2 are used to represent exactly one number, the positive square root.

$$\sqrt{4} = 4^{1/2} = 2$$

This usage is consistent with the introduction of square roots as lengths of line segments, since lengths are always nonnegative. The fractional exponent 1/2 and the radical symbol may be used interchangeably at any time. Thus for any positive real numbers a and b, the statements

$$(ab)^{1/2} = a^{1/2}b^{1/2} \qquad \left(\frac{a}{b}\right)^{1/2} = \frac{a^{1/2}}{b^{1/2}}$$

may be represented by radicals as

$$\sqrt{ab} = \sqrt{a}\sqrt{b} \qquad \sqrt{\frac{a}{b}} = \frac{\sqrt{a}}{\sqrt{b}}$$

Example 1 Simplify **(a)** $\sqrt{18}$ **(b)** $\sqrt{9/2}$ **(c)** $\sqrt{18} + 4\sqrt{50}$.

Solution **(a)** $\sqrt{18} = \sqrt{9 \times 2} = \sqrt{9}\sqrt{2} = 3\sqrt{2}$

(b) $\sqrt{\frac{9}{2}} = \sqrt{\frac{9}{2} \times \frac{2}{2}} = \frac{\sqrt{9 \times 2}}{\sqrt{2 \times 2}} = \frac{3\sqrt{2}}{2}$

(c) $\sqrt{18} + 4\sqrt{50} = \sqrt{9 \times 2} + 4\sqrt{25 \times 2}$
$= 3\sqrt{2} + (4 \times 5)\sqrt{2} = 23\sqrt{2}$ ■

To simplify the square root of an integer, the prime factors are considered and grouped to obtain squares of integers where possible. To simplify the square root of a fraction, write the fraction in reduced form, multiply numerator and denominator by a number that makes the denominator the square of an integer, and then simplify the square root of the numerator. To add or subtract radicals, first simplify each radical. The most common difficulties in working with radicals arise from the use of incorrect rules (see Example 2(a)) and from the failure to simplify and take roots whenever possible before taking powers (see Example 2(b)).

Example 2 Simplify **(a)** $\sqrt{9 + 16}$ **(b)** $\left(\frac{88}{11}\right)^{2/3}$.

Solution **(a)** $\sqrt{9 + 16} = \sqrt{25} = 5$

(Note that $\sqrt{9 + 16} \neq \sqrt{9} + \sqrt{16} = 3 + 4 = 7$.)

(b) $\left(\frac{88}{11}\right)^{2/3} = 8^{2/3} = (8^{1/3})^2 = 2^2 = 4$ ■

Powers and roots, expressed either in terms of radicals or in terms of fractional exponents, are treated the same as other numbers in computations. For example,

$$(2 \times 11^3) + (3 \times 11^3) = 5 \times 11^3 \qquad 6\sqrt{17} - 2\sqrt{17} = 4\sqrt{17}$$

$$\frac{15\sqrt{6}}{3\sqrt{2}} = \frac{5\sqrt{2}\sqrt{3}}{\sqrt{2}} = 5\sqrt{3} \qquad 12^{1/2} \times 3^{1/2} = 36^{1/2} = 6$$

$V = e^3$

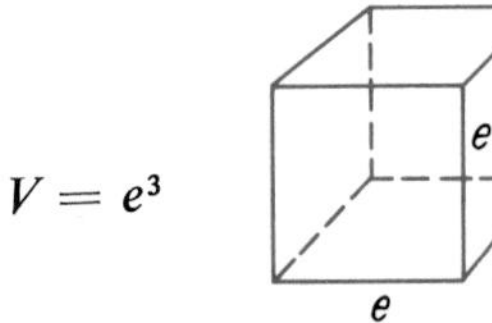

Cube roots occur when volumes of cubes are considered. A solid cube with e unit cubes along each edge contains e^3 unit cubes. If we use V for the volume of the cube in cubic units and e for the length of the edge in linear units, then a cube with each edge 2 centimeters long has volume 8 cubic centimeters, where

$8 = 2^3$. Next suppose that we wanted the volume of a cube to be 2 cubic centimeters. Then

$$2 = e^3 \qquad e = \sqrt[3]{2} \qquad e = 2^{1/3}$$

Cube roots are treated the same as other numbers in computations. There is one major difference between our use of cube roots and our use of square roots—negative numbers have real numbers as cube roots but do not have real numbers as square roots.

$$(-8)^{1/3} = -2 \quad \text{but} \quad (-9)^{1/2} \quad \text{does not represent a real number.}$$

When we simplify an expression involving powers and roots, we seek a form that can be used or approximated as easily as possible. This conventionally means that in the simplified form

Operations have been performed whenever possible.
There are no negative exponents.
Roots have been taken wherever possible.
There are no radicals or fractional exponents in any denominator.

Negative exponents in the numerator are equivalent to the corresponding positive exponents in the denominator. The process of eliminating radicals and fractional exponents from the denominator is called **rationalizing the denominator**, that is, making the denominator a rational number. If two given expressions each involve radicals and their product is free of radicals, then each of the given expressions is called a **rationalizing factor** of the other. For example, $\sqrt{2}$ is a rationalizing factor of $\sqrt{8}$ since

$$\sqrt{2} \times \sqrt{8} = \sqrt{16} = 4$$

Also $2 - 3\sqrt{5}$ is a rationalizing factor of $2 + 3\sqrt{5}$ since

$$(2 - 3\sqrt{5})(2 + 3\sqrt{5}) = 4 - 45 = -41$$

Example 3 Simplify **(a)** $\dfrac{6}{5\sqrt{3}}$ **(b)** $3 \times 5^{-1/2}$ **(c)** $\sqrt{\dfrac{6}{5}}$.

Solution **(a)** $\dfrac{6}{5\sqrt{3}} = \dfrac{6}{5\sqrt{3}} \times \dfrac{\sqrt{3}}{\sqrt{3}} = \dfrac{6\sqrt{3}}{5 \times 3} = \dfrac{2}{5}\sqrt{3}$

(b) $3 \times 5^{-1/2} = \dfrac{3}{5^{1/2}} \times \dfrac{5^{1/2}}{5^{1/2}} = \dfrac{3}{5}\sqrt{5}$

(c) $\sqrt{\dfrac{6}{5}} = \sqrt{\dfrac{6}{5} \times \dfrac{5}{5}} = \dfrac{\sqrt{30}}{\sqrt{25}} = \dfrac{1}{5}\sqrt{30}$ ■

In Example 3 the solutions could have been expressed entirely in terms of radicals or entirely in terms of fractional exponents. Our mixture of the two notations in part (b) was unorthodox but purposeful for pedagogical reasons. The recognition of the equivalence of fractional exponents and radicals is one of our primary objectives. This equivalence enables us to use the laws of exponents in simplifying expressions that involve radicals.

$$\boxed{\sqrt[m]{\mathbf{a}^n} = (\mathbf{a}^n)^{1/m} = \mathbf{a}^{n/m}}$$

Approximate values of square roots may be found from a table of square roots, a calculator, or arithmetic computation. Suppose that an approximate value is needed for

$$\frac{3}{5\sqrt{2}}$$

that no calculator is available, and that

$$\sqrt{2} \approx 1.414$$

Then $5\sqrt{2} \approx 7.070$, but division by such a number would be tedious. Accordingly, we first simplify the given expression and then complete the approximation.

$$\frac{3}{5\sqrt{2}} = \frac{3}{5\sqrt{2}} \times \frac{\sqrt{2}}{\sqrt{2}} = \frac{3\sqrt{2}}{10} \approx \frac{3 \times 1.414}{10} = \frac{4.242}{10} = 0.4242$$

To find an approximate value for an expression that involves irrational numbers, first simplify the expression and then substitute the approximate values of the irrational numbers.

Example 4 Assume that $\sqrt{2} \approx 1.41$ and find $3 + \sqrt{\frac{25}{8}}$ to the nearest hundredth.

Solution

$$3 + \sqrt{\frac{25}{8}} = 3 + \frac{\sqrt{25}}{\sqrt{8}} = 3 + \frac{5}{\sqrt{8}} \times \frac{\sqrt{2}}{\sqrt{2}}$$

$$= 3 + \frac{5}{4}\sqrt{2} \approx 3 + 1.76 = 4.76 \quad \blacksquare$$

EXERCISES *Express in terms of fractional exponents.*

1. $\sqrt{11}$
2. $\sqrt[3]{6}$
3. $\sqrt[3]{5^2}$
4. $\sqrt{3 \times 5}$
5. $2 + \sqrt[3]{7}$
6. $\sqrt{1 + \sqrt{2}}$

Express in terms of radicals.

7. $19^{1/2}$
8. $21^{1/3}$
9. $7^{2/3}$
10. $1 + 5^{1/2}$
11. $6 - 3^{2/3}$
12. $(2 + 6^{1/2})^{1/3}$

Give at least three numerical examples for each statement.

13. All integral powers of rational numbers are rational numbers.
14. An integral power of an irrational number may be an irrational number.
15. An integral power of an irrational number may be a rational number.
16. A root of a rational number may be a rational number.
17. A root of rational number may be an irrational number.
18. All roots of irrational numbers are irrational numbers.

Simplify.

19. $\sqrt{8}$
20. $\sqrt{75}$
21. $\sqrt{200}$
22. $\sqrt{1800}$
23. $\sqrt{4500}$
24. $\sqrt{25{,}000}$
25. $\sqrt[3]{16}$
26. $\sqrt[3]{54}$
27. $\sqrt[3]{200}$
28. $\sqrt{\frac{3}{4}}$
29. $\sqrt{\frac{8}{9}}$
30. $\sqrt{\frac{5}{9}}$
31. $\sqrt{\frac{5}{2}}$
32. $\sqrt{\frac{2}{3}}$
33. $\sqrt{\frac{21}{24}}$
34. $\sqrt{\frac{7}{18}}$
35. $\sqrt[3]{\frac{7}{8}}$
36. $\sqrt[3]{\frac{5}{9}}$
37. $\sqrt[3]{\frac{-112}{14}}$
38. $\sqrt[3]{\frac{-5}{7}}$
39. $\sqrt[3]{\frac{-24}{5}}$
40. $\sqrt{10 + 6}$
41. $\sqrt{81 + 144}$
42. $(25 + 144)^{1/2}$
43. $(100 - 36)^{1/2}$
44. $\sqrt{12} + \sqrt{75}$
45. $2\sqrt{8} - \sqrt{18}$
46. $\sqrt{72} - \sqrt{50}$
47. $\sqrt{\frac{1}{2}} - \sqrt{\frac{1}{8}}$
48. $\sqrt{\frac{3}{2}} - \sqrt{\frac{2}{3}}$

Find the smallest rationalizing factor for each expression.

49. $\sqrt{12}$
50. $\sqrt{20}$
51. $\sqrt[3]{9}$
52. $\sqrt[3]{16}$
53. $\sqrt[3]{3}$
54. $\sqrt[3]{54}$

In Exercise 55 through 66 use $\sqrt{2} \approx 1.414$ and $\sqrt{3} \approx 1.732$ and evaluate to the nearest hundredth.

55. $5 - \sqrt{3}$
56. $6 - \sqrt{2}$
57. $50^{1/2}$
58. $75^{1/2}$
59. $2 + 8^{1/2}$
60. $4 + \sqrt{12}$

61. $3 + \sqrt{\frac{1}{8}}$

62. $7 - \sqrt{\frac{1}{9}}$

63. $9 - \left(\frac{11}{22}\right)^{1/2}$

64. $6 + \left(\frac{7}{21}\right)^{1/2}$

*65. $2 + \sqrt{\frac{3}{8}}$

*66. $5 - \sqrt{6}$

*67. As a language evolves the fractional part p of words that remain basically unchanged after n millenia (thousands of years) is approximately $(0.864)^n$. Find the value of p to three decimal places after **(a)** 2000 years **(b)** 3000 years.

EXPLORATIONS

1. Suppose that you can find square roots from a table or using a calculator. Explain how the taking of square roots may be extended to find, or obtain approximate values for, the given expression.

   (a) $\sqrt[4]{16}$ **(b)** $\sqrt[4]{625}$ **(c)** $\sqrt[4]{17}$

   (d) $\sqrt[8]{256}$ **(e)** $\sqrt[8]{10}$ **(f)** $\sqrt[16]{1000}$

The relationship $(a + b)(a - b) = a^2 - b^2$ may be used to find rationalizing factors for many expressions. For example, $a\sqrt{x} + b\sqrt{y}$ is a rationalizing factor for $a\sqrt{x} - b\sqrt{y}$ for any rational numbers a, b, x, and y.

2. Find a rationalizing factor for each expression and check that the product is free of radicals.

   (a) $1 - \sqrt{2}$ **(b)** $2 + \sqrt{3}$ **(c)** $\sqrt{3} - \sqrt{2}$

3. Use the results obtained in Exploration 2 and simplify.

   (a) $\frac{5}{1 - \sqrt{2}}$ **(b)** $\frac{\sqrt{3}}{2 + \sqrt{3}}$ **(c)** $\frac{\sqrt{2}}{\sqrt{3} - \sqrt{2}}$

4. Simplify.

   (a) $2\sqrt{2} + 3\sqrt{8}$ **(b)** $5 - 2\sqrt{2} + 3\sqrt{18}$

   (c) $2\sqrt{6} + \frac{2\sqrt{3}}{\sqrt{3} + \sqrt{2}}$ **(d)** $\frac{1 - \sqrt{2}}{1 + \sqrt{2}} + \frac{1 + \sqrt{2}}{1 - \sqrt{2}}$

7-6 Complex Numbers

The development of our number system may be considered in terms of operations. We started our number system with the *natural numbers*. The set of natural numbers is closed under addition and multiplication, that is, for any natural numbers a and b

$$a + b \quad \text{and} \quad a \times b$$

are natural numbers. All *integers* are needed if subtraction is to be included. For any integers a and b

$$a + b \quad a \times b \quad \text{and} \quad a - b$$

are integers. All *rational numbers* are needed if division, except by zero, is included. For any rational numbers a/b and c/d

$$\frac{a}{b} + \frac{c}{d} \qquad \frac{a}{b} \times \frac{c}{d} \qquad \frac{a}{b} - \frac{c}{d} \quad \text{and for} \quad \frac{c}{d} \neq 0 \qquad \frac{a}{b} \div \frac{c}{d}$$

are rational numbers.

Integral powers are equivalent to repeated multiplication, or division, and do not require an extension of the set of numbers under consideration. Fractional powers and roots require the introduction of new numbers. The *real numbers* are needed as lengths of line segments, as coordinates of points on a number line, and as numbers associated with nonterminating, nonrepeating decimals. The real numbers enable us to find roots of positive numbers. A further extension to include *complex numbers* is needed for roots of all real numbers.

Our number system may also be introduced using equations. If a and b are natural numbers, then all equations of the form

$$x = a + b \qquad x = a \times b$$

have solutions in the set of natural numbers. Integers are needed to obtain solutions for all equations of the form

$$x = a - b \qquad x + b = a$$

and rational numbers are needed to obtain for $b \neq 0$ solutions for all equations of the form

$$x = a \div b \qquad bx = a$$

The rational numbers are sufficient when only the four *fundamental operations* of arithmetic (addition, subtraction, multiplication, and division, except division by zero) are used. For $a/b > 0$ the real numbers are needed to obtain solutions for all equations of the form

$$x^2 = \frac{a}{b} \qquad bx^2 = a$$

Complex numbers are needed if we wish to drop the restriction $a/b > 0$.

The square of any real number (positive, negative, or zero) is a positive number or zero. Thus we must extend our concept of number if we are to have numbers for symbols of the form $\sqrt{-b}$, $b > 0$. We define $\sqrt{-1}$, the **imaginary unit**, to be a new kind of number, usually represented by i, such that

$$i^2 = (\sqrt{-1})^2 = -1$$

Then

$$\sqrt{-3} = \sqrt{(-1) \times 3} = \sqrt{i^2 \times 3} = (\sqrt{3})i$$

$$\sqrt{-4} = \sqrt{(-1) \times 4} = \sqrt{i^2 \times 2^2} = i \times 2 = 2i$$

and in general

The symbol i for $\sqrt{-1}$ was introduced in 1777 by Leonard Euler, who also introduced many other symbols including e, the base of natural logarithms; $e = 2.71828\ldots$. The numbers i, e, π, 0, and 1 are related by the unusual equation $e^{\pi i} + 1 = 0$.

$$\sqrt{-n} = \sqrt{(-1) \times n} = \sqrt{i^2 \times n} = (\sqrt{n})i$$

The definition of $i = \sqrt{-1}$ enables us to express the square root of any negative number as the product of a real number and the number i. Numbers such as

$$2i \qquad (-5)i \qquad (\sqrt{2})i$$

of the form $\pm bi$ where b is a real number different from zero are called **pure imaginary numbers**. Any sum of a real number and a pure imaginary number is a **complex number**. For example,

$$3 + 2i \qquad 5\sqrt{2} + \sqrt{-3} \qquad 7 + 0\sqrt{-1}$$

and, in general, $a + bi$ for any real numbers a and b are complex numbers.

Unless otherwise requested, answers should be left in the form $a + bi$ or as real numbers

Each complex number $a + bi$ has a as its **real part** and bi as its **imaginary part**. If $a = 0$, the complex number is a *pure imaginary number*. If $b = 0$, the complex number is a *real number*. If $b \neq 0$, the complex number is called an **imaginary number**.

Example 1 Write in the form $a + bi$.

(a) $2 + \sqrt{-9}$ **(b)** 5 **(c)** $\sqrt{-3}$
(d) i^2 **(e)** $(-i)^2$

Solution
(a) $2 + \sqrt{-9} = 2 + \sqrt{9}\sqrt{-1} = 2 + 3i$
(b) $5 = 5 + 0i$
(c) $\sqrt{-3} = \sqrt{3}\sqrt{-1} = (\sqrt{3})i = 0 + (\sqrt{3})i$
(d) $i^2 = \sqrt{-1}\sqrt{-1} = -1 = -1 + 0i$
(e) $(-i)^2 = (-1)^2 i^2 = 1(-1) = -1 = -1 + 0i$ ■

As an extension of the last two parts of Example 1 we may note that just as the equation $x^2 - 4 = 0$ has two solutions 2 and -2, the equation $x^2 + 4 = 0$ has two solutions $2i$ and $-2i$.

The properties of the numbers in our number system have also evolved as we extended our concept of number. We started with the set of natural numbers and with addition and multiplication defined so that we had these eight properties.

Closure, +	Closure, ×
Commutative, +	Commutative, ×
Associative, +	Associative, ×
	Identity, ×

Distributive property

The first extension of our concept of a number was to the set of whole numbers. This extension enabled us to keep the eight properties of natural numbers and to gain one additional property.

IDENTITY, +: There is an element 0 in the set such that $0 + a = a + 0 = a$.

Then we extended our concept of a number to the set of integers. The nine properties of the whole numbers were retained and one additional property was gained.

INVERSE, +: For each element a there is an element $(-a)$ in the set such that $a + (-a) = 0$.

In Chapter 6 we again extended our concept of a number to include all rational numbers, retained all previously mentioned properties, and gained one more property.

INVERSE ($\neq 0$), $\times$: For each element a different from zero, the additive identity, there is an element $1/a$ in the set such that $a \times 1/a = 1$.

We summarized these eleven properties by saying that the set of rational numbers with the operations of addition and multiplication forms a *field*. We also noted that the rational numbers are *dense*, that is, between any two rational numbers there is at least one other rational number.

In the present chapter we have discovered that even with the density property, the rational numbers are not sufficient to serve as coordinates of all points on a line in our ordinary geometry. Accordingly, we introduced the real numbers in three equivalent ways.

As lengths of line segments.
As coordinates of points on a number line.
As decimals.

The set of coordinates for points on a line is now complete and the set of real numbers is said to be **complete.** There is a real number as the coordinate of each point of the line and each point of the line has a real number as its coordinate. The set of real numbers has the properties that we have considered for the set of rational numbers. Furthermore, a one-to-one correspondence can be established between the set of points on an ordinary line and the set of real numbers.

In this section we have extended our concept of number to include all complex numbers, that is, all numbers of the form $a + bi$, where a and b are real numbers. The set of complex numbers with the operations of addition and multiplication has the first eleven properties considered for real numbers and forms a *field*. However, the complex numbers are graphed on an ordinary plane rather than an ordinary line and accordingly are *not* linearly ordered. Thus statements such as $2i < 1 + i$ are not meaningful.

Properties of numbers seem a bit remote without a consideration of the uses of sets of numbers. Frequently these uses are related to the types of sentences that can be solved. The importance of recognizing the set of possible replacements being used is shown by the solutions of the equations in the following array. The word "none" implies that for the set of numbers named at the top of the column there is no replacement for n that will make the sentence at the left of the row a true statement.

SENTENCE	REPLACEMENT SET					
	NATURAL NUMBERS	WHOLE NUMBERS	INTEGERS	RATIONAL NUMBERS	REAL NUMBERS	COMPLEX NUMBERS
$n + 1 = 1$	None	0	0	0	0	0
$n + 1 = 0$	None	None	-1	-1	-1	-1
$2n = 1$	None	None	None	1/2	1/2	1/2
$n^2 - 1 = 0$	1	1	$1, -1$	$1, -1$	$1, -1$	$1, -1$
$n^2 - 2 = 0$	None	None	None	None	$\sqrt{2}, -\sqrt{2}$	$\sqrt{2}, -\sqrt{2}$
$n^2 + 1 = 0$	None	None	None	None	None	$i, -i$

The rules used for operations with real numbers may also be used for operations with complex numbers. The set of complex numbers with these same operations forms a field.

Example 2 Simplify.

(a) $(2 + 3i) + (4 - 2i)$ **(b)** $5(3 - 2i)$ **(c)** $(3i)(1 + 2i)$

Solution **(a)** $(2 + 3i) + (4 - 2i) = (2 + 4) + (3 - 2)i = 6 + i$

(b) $5(3 - 2i) = 5(3) - 5(2i) = 15 - 10i$

(c) $(3i)(1 + 2i) = 3i + (3i)(2i) = 3i + 6i^2 = -6 + 3i$ ■

EXERCISES *Give at least three numerical examples for each statement.*

1. A sum of imaginary numbers may be a real number.
2. A sum of imaginary numbers may be a pure imaginary number.
3. The additive inverse (opposite) of an imaginary number is an imaginary number.
4. The multiplicative inverse (reciprocal) of a pure imaginary number is a pure imaginary number.
5. The square of any pure imaginary number is a real number.
6. The square of an imaginary number may be an imaginary number.

Complete the array using "$\surd$" to show that the number at the left of the row is a member of the set listed at the top of the column and "$\times$" to show that the number is not a member.

	NUMBERS	SET					
		NATURAL NUMBERS	RATIONAL NUMBERS	IRRATIONAL NUMBERS	REAL NUMBERS	IMAGINARY NUMBERS	COMPLEX NUMBERS
7.	1						
8.	$\sqrt{1/4}$						
9.	$\sqrt{121}$						
10.	$\sqrt{50}$						
11.	$-3\sqrt{36}$						
12.	$\sqrt{12}/\sqrt{3}$						
13.	$\sqrt[3]{-8}$						
14.	$2\sqrt{2}-\sqrt{8}$						
15.	$1-\sqrt{8}$						
16.	$1+\sqrt{-8}$						

Simplify and express each number in the form $a + bi$.

17. $\sqrt{20}$ **18.** $\sqrt{-25}$ **19.** $3\sqrt{12}-\sqrt{-3}$
20. $\sqrt{9-25}$ **21.** $\sqrt{9}+\sqrt{-25}$ **22.** $\sqrt{9}-\sqrt{25}$
23. $2+i^2$ **24.** $4+i^4$ **25.** $3+i^3$
26. $5+i^5$ **27.** $6-i^6$ **28.** $7-i^7$

Find all possible replacements for n that will make the given sentence true where n must be a member of the set at the top of the column. If there is no such replacement, write "none."

	SENTENCE	REPLACEMENT SET					
		NATURAL NUMBERS	WHOLE NUMBERS	INTEGERS	RATIONAL NUMBERS	REAL NUMBERS	COMPLEX NUMBERS
29.	$n+5=0$						
30.	$n+3=5$						
31.	$3n-5=0$						
32.	$n^2-25=0$						
33.	$n^2-5=0$						
34.	$n^2+5=0$						

Use imaginary numbers and give at least three numerical examples of the indicated property of complex numbers.

35. Closure, $+$.	**36.** Closure, $\times$.	**37.** Associative, $\times$.
38. Associative, $+$.	**39.** Commutative, $+$.	**40.** Commutative, $\times$.
41. Distributive.	**42.** Inverse, $+$.	**43.** Inverse ($\neq 0$), $\times$.

EXPLORATIONS

The expression $\sqrt{-1}$ does not have any meaning in the set of real numbers, that is, the expression does not represent a real number.

1. For $n > 2$, give five expressions of the form $b^{1/n}$ for integers b and n such that the expressions
   (a) have meaning in the set of real numbers
   (b) do not have meaning in the set of real numbers.
2. Use as many examples as necessary and develop a general rule for determining whether or not an expression of the form $b^{1/n}$ has meaning in the set of real numbers for integers b and n.
3. Repeat Exploration 2 for the expression $b^{p/q}$ for integers b, p, and q.
4. Explore a recently published series of elementary mathematics textbooks to determine when and how **(a)** irrational numbers **(b)** imaginary numbers, are avoided, recognized, or introduced.

Chapter Review

In Exercises 1 and 2 classify the given statement as true or false.

1. **(a)** Every irrational number is not an integer.
   (b) There is a one-to-one correspondence between the set of rational numbers and the set of points on a number line.
2. **(a)** The set of irrational numbers is a subset of the set of real numbers.
   (b) Every irrational number is the coordinate of some point on a number line.

3. Sketch the construction of the graph of $3 + \sqrt{2}$ on a number line.
4. Prove that $3 + \sqrt{2}$ is an irrational number.
5. Write the first fifteen digits to the right of the decimal point in the numeral 0.232232223. . . .
6. List the given numbers in order from smallest to largest.

   $$2.\overline{56} \qquad 2.5\overline{6} \qquad 2.56 \qquad 2.566 \qquad 2.565$$

7. Name two irrational numbers between $0.\overline{56}$ and $0.5\overline{6}$.

Find the indicated square root.

8. **(a)** $\sqrt{1600}$ **(b)** $\sqrt{0.0025}$
9. **(a)** $\sqrt{400{,}000{,}000}$ **(b)** $\sqrt{0.0144}$

In Exercises 10 through 17 simplify.

10. $(3x^2)(5x^4)$

11. $(2a^3b^3)(3a^0b^2)(4ab)$

12. $\dfrac{6x^4y^{12}}{2xy^3}$

13. $3^0x^{-3}y^5$

14. $\sqrt{\dfrac{3}{4}}$

15. $(36 + 64)^{1/2}$

16. $(1 + 4^{3/2})^{1/2}$

17. $\sqrt{20} + \sqrt{45}$

18. Classify each statement as true or false. Give a counterexample for each false statement.
(a) The sum of any two irrational numbers is an irrational number.
(b) The product of any two irrational numbers is an irrational number.

19. Write each number in scientific notation.
(a) 2,530,000 **(b)** 0.000 0768

20. Find the amount for a deposit of \$1600 at 6% simple interest for 2 years.

21. Find the amount for a deposit of \$1200 at 8% interest compounded semi-annually for 5 years.

Find all possible replacements for n that will make the given sentence true where n must be a member of the set named at the top of the column. If there is no such replacement, write "none."

		REPLACEMENT SET			
	SENTENCE	INTEGERS	RATIONAL NUMBERS	IRRATIONAL NUMBERS	REAL NUMBERS
22.	$n + 3 = 3$				
23.	$5n = 8$				
24.	$n^2 = 3$				
25.	$n^2 + 9 = 0$				

COMPUTER APPROXIMATION OF SQUARE ROOTS

In Section 7-2 the algorithm for calculating the square root of a positive real number via guessing, dividing, and averaging was presented. This algorithm is well suited to adaptation for use on a computer. The following program illustrates the BASIC version of this process.

```
 5  HOME
1Ø  PRINT "THIS PROGRAM CALCULATES THE SQUARE"
15  PRINT
2Ø  PRINT "ROOT OF A POSITIVE REAL NUMBER."
25  PRINT
26  PRINT
3Ø  PRINT "ENTER THE POSITIVE REAL NUMBER WHOSE"
31  PRINT
32  PRINT "SQUARE ROOT YOU WANT TO SEE."
33  PRINT
34  INPUT R
35  IF R < = Ø THEN GOTO 3Ø
36  PRINT
38  PRINT "ENTER YOUR ESTIMATE OF THE SQUARE ROOT OF ";R;"."
39  PRINT
4Ø  INPUT G
41  S = R/G
42  PRINT
44  IF ABS (G - S) < Ø.ØØØØØ5 THEN PRINT
         "THE SQUARE ROOT OF ";R;" TO 6 PLACES IS ";G;".": END
45  G = (S + G)/2
46  PRINT G
5Ø  GOTO 41
```

This particular version of the algorithm uses one new facet of BASIC we have not encountered in BASIC programs presented earlier. That is the BASIC function ABS(X) in line 44. Here the program is checking to see if the absolute value of the difference between the last calculated guess for the square root and the latest average calculated is less than five-millionths. The absolute value of a number was introduced in Section 4-1.

The input requirements of the program call for the entry of the real number whose square root you desire and an estimate of the square root. The program then takes over and moves to find a good approximation of that square root. The sample printout shows the use of the program in finding an approximation of the square root of 34567 using an initial estimate of 34 for the square root. The actual square root is clearly between 100 and 200 and closer to 200. The "wild guess" of 34 is used to emphasize that one does not have to struggle to be "close," the computer does the work.

```
RUN
THIS PROGRAM CALCULATES THE SQUARE

ROOT OF A POSITIVE REAL NUMBER.

ENTER THE POSITIVE REAL NUMBER WHOSE

SQUARE ROOT YOU WANT TO SEE.

? 34567

ENTER YOUR ESTIMATE OF THE SQUARE ROOT OF 34567.

34

525.338236

295.568874

206.25981

186.9247Ø7

185.924716

185.922Ø27

THE SQUARE ROOT OF 34567 TO 6 PLACES IS 185.922Ø27.
```

The lines of output before the listing of the calculated approximation to the square root indicate the successive values calculated by the algorithm as it approaches the value of the square root of 34567.

If possible use the program to calculate the square roots of several real numbers. In particular, examine the difference in the number of steps required as one makes relatively poor and relatively good initial estimates of the magnitude of the square root.

8

An Introduction to Geometry

Hypatia
(ca. 370–415)

Hypatia was the first recorded female professor of mathematics. She taught mathematics in the ancient Egyptian city of Alexandria. The mathematical topics in which she specialized were the geometry of the conic sections (parabola, ellipse, and hyperbola) and finding integral solutions to equations of the form $ax + by = c$. Such equations are now known as Diophantine equations and play an important role in modern applications of mathematics in business and social science settings. Hypatia was murdered by a fanatical mob of religious zealots because she kept her beliefs in the Greek gods.

8-1 Geometry and Arithmetic

Mathematics has been described as the study of patterns. In this sense we think of patterns of specific numbers

$$2 \times 3 = 6 \qquad 1 + 3 + 5 = 9$$

as *arithmetic*, patterns of general numbers

$$a + b = b + a \qquad (a + b)(a - b) = a^2 - b^2$$

as *algebra*, and visual patterns

as *geometry*. Our concept of *number* probably started from efforts to distinguish between or identify the "number" (how many) property of different sets of elements (animals, people, nights, trees, and so forth). Our concept of *visual patterns* probably started from efforts to distinguish between or identify the "shapes" of different observed objects (cats, deer, the full moon, clusters of stars, edible plants, and so forth). Scratches (numerical tallies?) in sets of five made about 30,000 years ago on the leg bone of a wolf have been found in Czechoslovakia. Cave paintings with recognizable hunters, bows, arrows, and animals made about 15,000 years ago have been found in France and Spain.

The counting of objects did not start with large numbers such as the number of hairs on a person's head, the number of leaves on a full-grown tree, or the number of grains of sand on a beach. Small sets of elements were counted and large sets of elements were considered to have *many* elements. "Many" might mean more than three or it might mean much larger numbers such as thousands, depending on the extent of the development of the numerical concepts held by a specific group of people. Similarly, the identification of visual patterns did not start with the hexagonal cells of the honeycomb of a bee or a map of the known world. The shapes of simple figures were named; and attractive, detailed, or complicated figures, such as the reflection of a scene in a still lake, were considered simply to be *beautiful*.

We start counting small sets and make special use of the numbers of common sets such as

Two hands.
Five fingers on one hand.
Ten fingers.

In geometry we start with simple shapes and make special use of the shapes of common figures such as a

Circle: the apparent shape of the sun and full moon.

Line: the apparent path of an arrow.

Plane: the apparent shape of the surface of a calm lake or large level valley floor.

Unity, 1, was recognized as the basic element for counting. Gradually points and line segments were recognized as the basic elements of geometric figures.

The objects that were counted or whose shapes were considered were space figures (*solids*). We think of the earth as a sphere even though its surface is not smooth and its shape is not exactly spherical. Consider the set of elements consisting of the sun, the moon, and the earth. When we say that the set has three elements, we are abstracting a number property. When we say that the elements are spherical in shape, we are abstracting a geometric property. We feel confident of the arithmetic count since we can match the elements of the set with the numbers 1, 2, 3. The geometric property of the elements of the set as spheres seems a bit more abstract and in a sense an idealized model for the elements. Indeed, geometric figures are abstract idealizations of observed shapes. The abstractness of geometric figures enables us to use them in a wide variety of situations just as the abstractness of numbers enables us to use the same numbers for all sorts of elements. How difficult it would be if we used one number system for counting days, another for pencils, another for sheets of paper, and so forth. The different units and groupings of units are sufficiently confusing without using different sets of numbers. Consider

Days: week, fortnight, month, year, decade, century
Pencils: dozen, gross
Sheets: quire, ream

The Earth from 23,000 miles in space over Brazil.

Anyone making a ball from snow, clay, or other material tends to make the ball spherical as the material is tightly packed. A picture of the sphere is often a *circle* or a *circular disk* (region). These circles and circular disks are plane figures. Given any point O and any distance r, the set of points of the plane that are at a distance r from the point O forms a **circle** with **center** O and **radius** r. The set of points of the plane at a distance less than or equal to r from the point O forms a **circular disk (circular region)** with **center** O and **radius** r. The points of the circular disk that are not points of the circle are **interior points** of the circle. The points of the plane that are not points of the circular disk are **exterior points** of the circle.

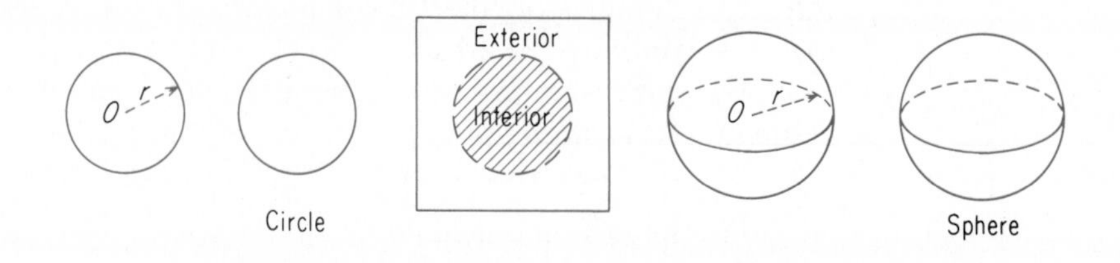

If we remove the restriction to the points of a plane, we have the definition of a sphere. Given any point O and any distance r, the set of points in space that are at a distance r from the point O forms a **sphere** with **center** O and **radius** r. Such precise definitions are needed for the study of geometry and to make sure that we are communicating with each other when these words are used. Such definitions are not needed to enjoy the visual patterns of geometric figures.

Numbers are used in many ways in geometry. For example, we have already noted that real numbers may be used to identify lengths of line segments, that is, as distances, and to identify points on a number line, that is, as coordinates. For a given unit of length the radius of any circle, circular disk, or sphere may be represented by a number.

We often recognize colors (such as red), people, a friend's car, and other objects without being able to describe in words the precise pattern that makes such recognition possible. For some people the same situation holds for many geometric figures, even for triangles and triangular regions, squares and square regions, lines that intersect each other at right angles, lines that intersect a plane at right angles (are perpendicular to the plane), rays (such as rays of light), cubes, cylinders, and many other common plane and space figures.

The recognition of such common three-dimensional figures (**solids**) as spheres, cubes, triangular pyramids, and right circular cylinders preceded the art of writing. Five thousand years ago in the great river valleys of the area that became Iraq and Iran, it was a common business practice to use a set of solid

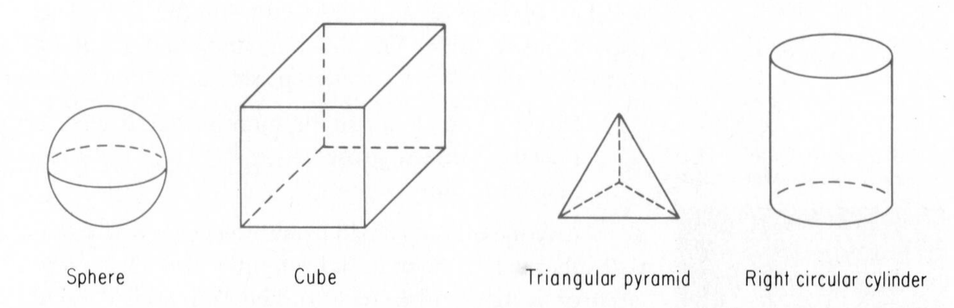

figures as an invoice with a shipment of goods. Clay was abundant, could be molded easily when wet, and hardened well in the sun. Sets of solid clay figures were made in easily recognized geometric shapes. A messenger conveying a shipment of sheep, loaves of bread, or other objects would also carry an invoice.

The invoice was a hollow clay ball about the size of a golf ball. Inside the ball would be a set of tiny geometric solids packed in straw. The number of these solids would be the same as the number of objects shipped. The person who received the shipment could check that all items were present by breaking the ball and matching the geometric solids, one-to-one, with the objects received. Geometric solids were used so that they would not be confused with the fragments of the broken hollow ball. The shipment could be checked without a knowledge of counting, without words for numbers, and without a written language. Over many centuries, pictures on the balls to indicate the kind of objects shipped and the number of objects shipped gradually evolved into the earliest known writing of words. At one time the shape of the solid and marks on the solid indicated the kind of object shipped.

The geometric solids were simply "recognized." There were no formal definitions. The shapes of the solids were identifiable "attributes" of the figures. Even today the initial recognition of a solid figure is "sensed"—often by both visual and tactile (touch, handling) senses. In order to identify a solid figure visually, the figure must be viewed from several directions (front, top, side, and so on). Suppose that you are shopping for a spherical lamp shade and that there are two types at the store. Each type has a small hole at the top for the lamp cord. One type is otherwise spherical; the other has a portion cut away so that the lamp provides extra light directly below it.

If the two lamp shades were on a shelf and on their sides so that you were looking directly at the small holes for the lamp cord, the lamp shades would look alike. But if they were turned a quarter turn, their differences would be apparent; one would appear to have a "flat side," as in the view of them hanging ready for use.

Our two eyes receive slightly different views of objects. We use this difference in our recognition of objects and also in determining the distances at which objects are located (depth perception). In the case of the visual appearance of objects, we often gain the use of a third direction by the lighting of the object. The slight shadows and different intensities of the light on different parts of an object help us identify those parts and thus the object.

The use of two directions comes naturally to us as we use two ears in hearing sounds. Frequently we can point toward the source of the sound because of the slightly different reception of the sound by our two ears. Similarly, our two eyes receive slightly different views of objects.

The complete identification of the shape of an unknown object requires that the object be viewed from all possible directions. Suppose that a bright moon were directly overhead and that shadows of an otherwise invisible Martian spaceship were visible on level ground. In order to identify the shape of the spaceship from its shadows, we would need to observe the changes in its shadow as the spaceship rolled and several silhouettes were viewed. Even then the identification would be difficult.

To illustrate the process of identifying space figures by their shadows, we shall consider a related game. An object is placed behind a screen, a bright light (such as the sun) is in the distant background, and on the screen only the shadow of the object is visible. To simplify the game we assume that the unknown object is one of the four solids mentioned at the beginning of this section, that is, a sphere, a cube, a triangular pyramid, or a right circular cylinder. The solid figures may be located in any position so that the shadow may be a top view, a side view, an end view, or a view from some other position. Opposite views, such as the top view and the bottom view, have the same shadow and therefore are considered to be the same, that is, *not different*, views. The light source is assumed to be very far away and the shadows are all assumed to be on planes that are perpendicular to the rays of light. The same effect can be achieved by placing an object on an overhead projector and observing the shadow on the screen. The shadows obtained using perpendicular rays are like the shadows obtained when the sun is directly overhead. The long shadows of the early morning or late evening provide excellent geometric illustrations, but of a different geometry.

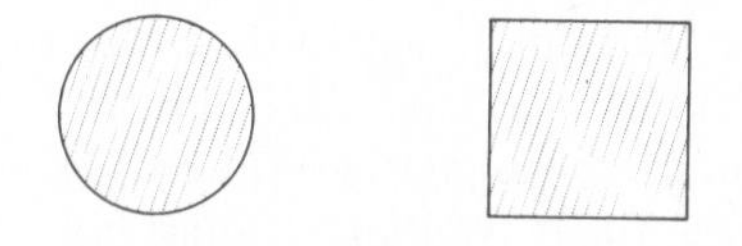

Suppose that the object has a circular region (*disk*) as its shadow. Can the object be

the sphere? the cube? the pyramid? the cylinder?

Suppose that the object has a square region as its shadow. Can the object be

the sphere? the cube? the pyramid? the cylinder?

Before the last question can be answered we need to know whether the height of the cylinder is equal to its diameter. For future discussion of these particular figures, we make four assumptions.

The sphere has diameter 1 meter.
The cube has edge 1 meter.
Each edge of the pyramid is 1 meter.
The cylinder has diameter 1 meter and height 125 centimeters.

Now the previous questions can be answered. Only the cube can have a square region as its shadow.

Which of the four solid figures can have a rectangular region that is not a square region as its shadow? Which of the solid figures can have a triangular region as its shadow? Both the cube and the cylinder can have shadows that are rectangular regions but not square regions. Only the pyramid can have a shadow that is a triangular region.

Example 1 A sphere with radius 10 centimeters is rolled around on a plane. What changes, if any, occur among the shadows of the sphere from light rays that intersect the plane of the shadows at right angles?

Solution All shadows are circular regions with radius 10 centimeters.

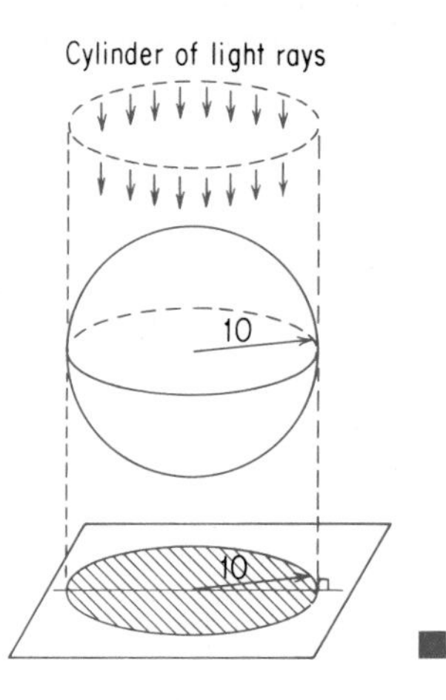

■

Example 2 A cube with edge 4 centimeters is rolled with an edge or face on the plane at all times. What changes, if any, occur among the shadows of the cube from light rays that intersect the plane at right angles?

Solution Each shadow is a square or a rectangular region with one side 4 centimeters long. The other side of the rectangular shadow may have any length from 4 centimeters (when the cube has a face on the plane) to $4\sqrt{2}$ centimeters (when the cube is balanced on an edge, that is, has a diagonal of a face parallel to the plane).

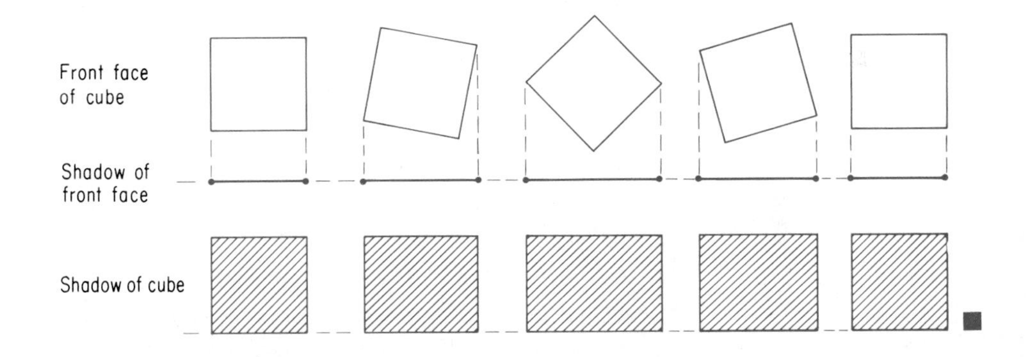

■

EXERCISES *Identify at least three objects that you see frequently and that have approximately the shape of the specified geometric figure.*

1. Sphere.
2. Circular disk.
3. Circular cylinder.
4. Circular cone.
5. Rectangular box.
6. Rectangular region.
7. Triangular pyramid.
8. Triangular region.
9. Cube.
10. Square region.
11. Circle.
12. Line segment.

In Exercises 13 through 32 all shadows are assumed to be formed by light rays that intersect the plane of the shadow at right angles. In Exercises 13 through 18 assume that the unknown figure is one of the four discussed in this section (sphere, cube, triangular pyramid, cylinder). Specify which of these figures is then identified by the listed shadows of two different views of the figure.

13. Two square regions.
14. A circular region and a rectangular region.
15. A square region and a rectangular region.
16. Two circular regions.
17. Two triangular regions.
18. A triangular region and the region of a quadrilateral that is not a rectangle.

Identify the given statement as true or false.

19. Every shadow of a sphere is a circular region.
20. Every shadow of a cube is a square region.
21. Every shadow of a triangular pyramid is a triangular region.
22. Every shadow of a cube is either a square region or a rectangular region.

The given figure represents a triangular wedge. The bases are on parallel planes and are isosceles triangles. One face is a square 10 *centimeters on each edge. Each of the edges of the wedge is either* 10 *centimeters or* 20 *centimeters long.*

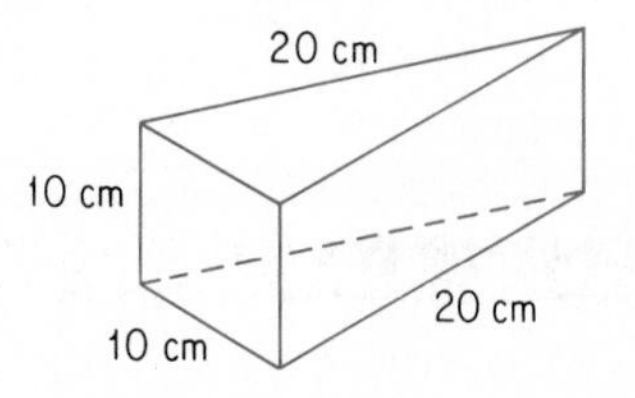

23. If the wedge can have a square shadow, what is (are) the possible lengths of the edges of the shadow?
24. Repeat Exercise 23 for rectangular shadows.
25. Repeat Exercise 23 for triangular shadows.
26. If the wedge can have a shadow that is not one of those considered in Exercises 23 through 25, sketch another possible type of shadow.

The given figure represents a cylindrical log with diameter 1 *meter and length* 3 *meters.*

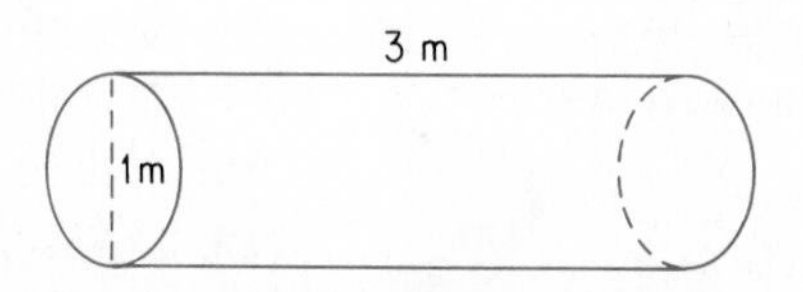

27. Describe the shadows that would be obtained if the log were
   (a) rolled on its cylindrical side, along the plane of the shadows
   (b) stood on end.

28. If for example, the log is in the process of being stood on end, can it have a shadow of a different shape than those considered in Exercise 27? If so, sketch one such shadow.

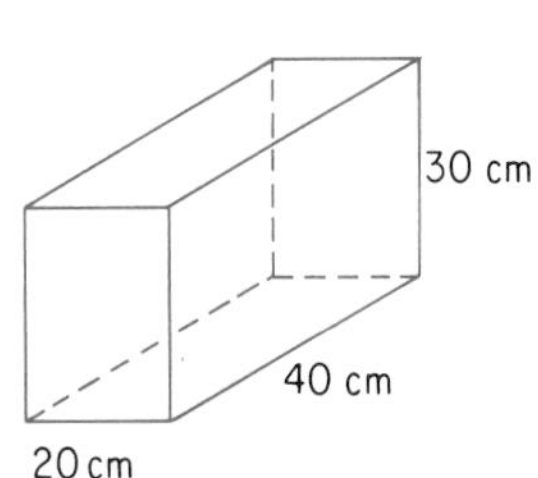

The adjacent figure represents a rectangular box with edges 20 *centimeters,* 30 *centimeters, and* 40 *centimeters long.*

*29. Can the box have a square shadow? If so, what is the length of the edge of the smallest possible square shadow?

*30. Think of the box as rolling end over end about its edges of length 20 centimeters in the plane of the shadow.
   (a) What sorts of figures are all the possible shadows as the box turns?
   (b) What are the dimensions of the smallest such shadow?
   (c) What are the dimensions of the largest such shadow?

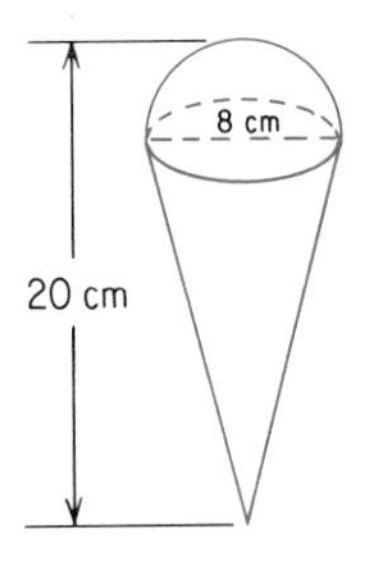

The adjacent figure represents an ice cream cone with a hemispherical cap of ice cream. The diameter of the cap is 8 *centimeters; the total height is* 20 *centimeters.*

31. Describe two possible shadows of the figure.

*32. (a) Describe the shadow with the smallest possible area.
   (b) Describe the shadow with the largest possible area.
   (c) Sketch the shape of a shadow that is neither the smallest nor the largest.

EXPLORATIONS

1. Graph paper is an excellent aid for sketching space figures on a plane. Consider the three steps shown below for sketching a cube. In Step 1 we draw the top face, with opposite sides parallel and equal in length. In Step 2 draw four edges of the cube as shown, parallel and of equal length. Note that the "hidden" edge is shown as a dashed line. The figure is completed in Step 3, again showing "hidden" edges with dashed line segments.

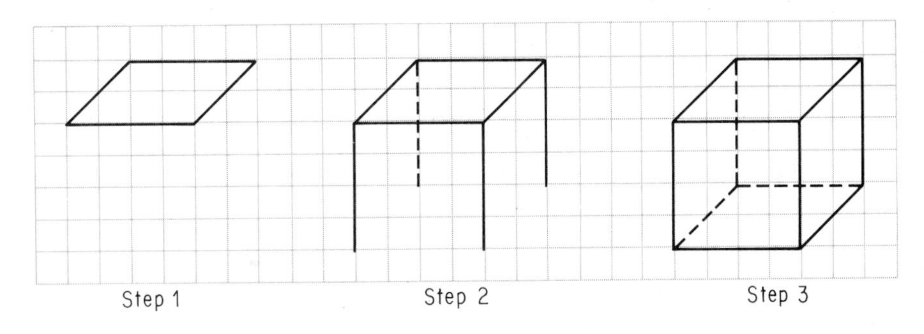

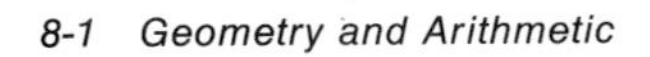

Use graph paper, follow these suggested steps, and draw a number of sketches of rectangular boxes, triangular prisms, and other prisms and pyramids.

2. The "pictures" (sketches, drawings) of space figures on a sheet of paper never completely represent the figure and often are hard to visualize. Consider these patterns and make a model of
   (**a**) a tetrahedron (triangular pyramid)
   (**b**) a cube.

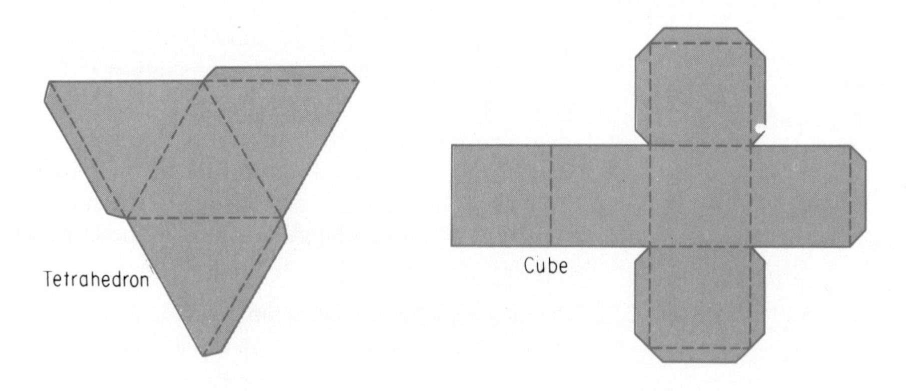

3. Use the patterns in Exploration 2 as a guide and design patterns for
   (**a**) a rectangular box with edges of given lengths
   (**b**) a triangular prism with edges of given lengths
   (**c**) a right circular cylinder of given radius and height.
4. A single square region, a **monomino**, can be arranged in only one way. Two square regions joined along their edges, a **domino**, also can be arranged to form only one figure, since the position of the 2×1 rectangle does not affect the arrangement. In general, the figures that can be formed by arranging square regions joined along their edges are **polyominoes**. Any two figures that have the same shape and size have the same arrangement of square regions.
   (**a**) Sketch the two possible distinct arrangements for three square regions, that is, **trominoes**.
   (**b**) Sketch the five possible distinct arrangements for four square regions, that is, **tetrominoes**.
   (**c**) Identify the distinct arrangements of five square regions, **pentominoes**, that can be cut out and folded to form a box (cube) without a top.

Many solid figures can be generated by revolving plane regions about a line (the **axis of revolution** for the figure). For example, if a circular region is spun (like a top) around a diameter, the solid figure generated is a sphere with the same radius as the circular region. If a circle is revolved about a line that is in the plane of the circle but does not intersect the circle, the solid figure generated is a **torus** (doughnut).

Try to generate some familiar solid figures and some other solid figures by revolving the specified figure about an axis parallel to a side. For each generated figure, describe the position of the axis of revolution relative to the given figure and describe, or sketch, the generated figure.

5. Square region.
6. Rectangular region that is not a square.
7. The region of a 3-4-5 right triangle.

8-2 Points and Lines on a Plane

Look at the plane figures that occur on solid figures around you.

The earliest geometry was concerned with space figures (*solids*) that could be recognized by their shapes, reproduced from clay or other materials, handled, admired, and used. In our three-dimensional world, experiences with geometry naturally started with solids. However, our descriptions of solid figures make use of the names of *plane figures*.

The annual flooding of the Nile River in Egypt made it necessary to relocate the boundaries of fields each year. This early surveying provided the basis for the word *geometry* (geo-metry) from the Greek words for *earth measure*. About 300 B.C. Euclid wrote these definitions.

> A *point* is that which has no part.
>
> A *line* is breadthless length.
>
> A *straight line* is a line which lies evenly with the points on itself.
>
> A *surface* is that which has length and breadth only.
>
> A *plane surface* is a surface which lies evenly with the straight lines on itself.

These descriptions no longer provide satisfactory definitions since we now recognize that some *undefined terms* are needed to provide words for use in defining other terms.

We take **point** and **line** (straight line) as undefined terms. We assume (*postulate*) these statements.

> Two points determine one and only one line.
>
> A line and a point that is not on the line determine one and only one plane.
>
> If two points of a line are on a plane, then every point of the line is on the plane.

A line is a set of points. Two lines that have exactly one point in common are **intersecting lines**. A plane is also a set of points. We think of a plane as a *flat surface*.

Any line may be taken as a *real number line*. Then each point of the line has a unique real number as its *coordinate* and each real number has a unique point of the line as its *graph*. Properties of real numbers may be used as in Section 7-1 to define half-line, opposite half-lines, ray, opposite rays, endpoint of a ray, line segment, and endpoints of a line segment. In the present section we use paper folding to explore a few geometric figures. Lines are represented by creases (folds) in the paper. The word *line* is used for both the abstract line and the crease, just as the word *number* is used for both the abstract number and its numeral.

Many statements in geometry can be demonstrated by paper folding. However, paper folding should be a "hands on" experience. The examples will be most beneficial to readers who do their own paper folding as they read this material.

Almost any kind of paper may be used for **paper folding**; however, the creases are particularly visible when waxed paper is used. A point is represented by the intersection of two lines, often by short creases "pinched" into the paper. Points and lines may be labeled for reference as on sketches and other representations of geometric figures.

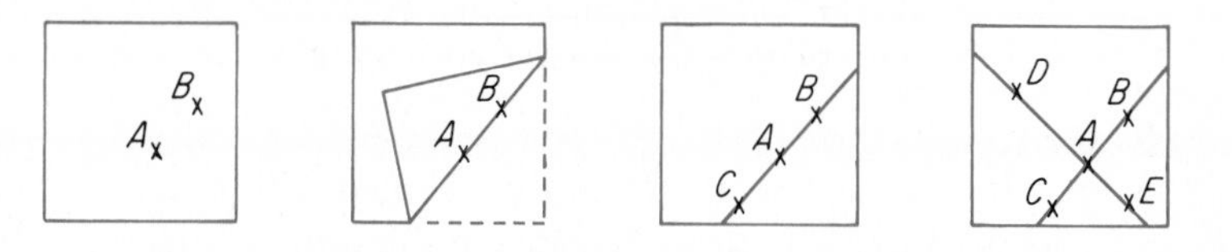

Shown in the first three figures are points A and B on a piece of paper, a fold through A and B, and a line AB. Since one and only one fold can be made through two given points, we have demonstrated that

There is one and only one line through two given points.

Points are usually named by capital letters. A line may be named by any two of its points or by a single lower case letter. For example, the line AB containing the point C also may be named AC or BC, or by a letter such as t or m. The *line* AB is often written as $\overleftrightarrow{AB}$, the *line segment* AB as $\overline{AB}$, and the *ray* AB as $\overrightarrow{AB}$.

Let C be any point of the ray opposite ray AB. Then A, B, and C are points of the same line, that is, **collinear points**. If ray AB is folded onto its opposite ray AC, the resulting crease can be formed in one and only one position. On this crease, label the opposite rays with endpoint A as AD and AE. The lines AB and AD appear to have a special relationship to each other. Before describing this relationship we define angles and right angles.

Any two rays PQ and PR with a common endpoint form an **angle**. The common endpoint P is the **vertex** of the angle; the rays PQ and PR are the **sides** of the angle. The angle may be named as $\angle QPR$ or as $\angle RPQ$ with the letter for the vertex as the middle letter, or if there is only one angle with vertex P, the angle may be named $\angle P$.

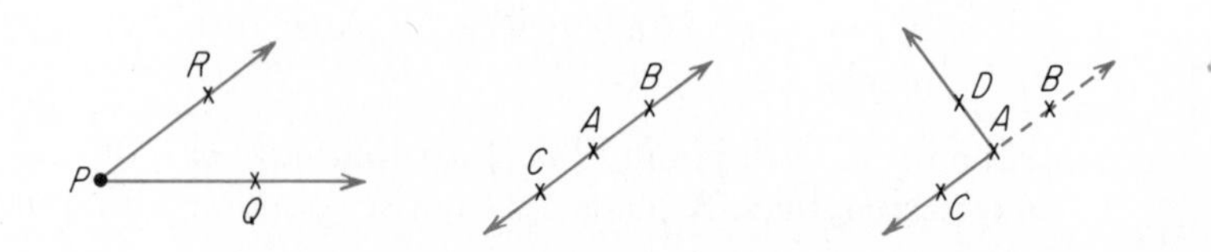

An angle with sides that are opposite rays, such as $\angle CAB$, is often referred to as a **straight angle**. In the paper folding of ray AB onto its opposite ray AC, line AD was formed and $\angle DAB$ was folded onto (**superimposed upon**) $\angle DAC$. We assume that if one figure may be superimposed upon another figure, the two figures are **congruent** (have the same measures). Then

$$\angle DAB \cong \angle DAC \qquad \text{(angle } DAB \textit{ is congruent to } \text{angle } DAC\text{)}$$

and each angle is one-half of a straight angle; that is, each angle is a **right angle**.

Any two intersecting lines that form right angles are **perpendicular lines**. Thus the special relationship between lines AB and AD in the paper-folding example may be described by identifying the lines as perpendicular at the point A. The procedure used demonstrates the statement

> On a plane there is one and only one line that is perpendicular to a given line at a given point of the line.

Statements in geometry may be **assumptions** (postulates and definitions) or **theorems** (statements proved from assumed statements). Whether a particular statement is an assumption or a theorem often depends upon the assumptions that have been selected.

Figures in geometry may also be represented by sketches. The sketches may be done freehand, using straightedge and compass, or using other instruments.

Example Sketch and label a figure consisting of three collinear points A, B, C such that

(a) C is on the line segment AB
(b) C is not on the line segment AB
(c) C is not on the line AB
(d) A is between B and C.

Solution Among others:

(a) A C B

(b) A B C

(c) Impossible since "collinear" means "on the same line" and there is one and only one line AB.

(d)

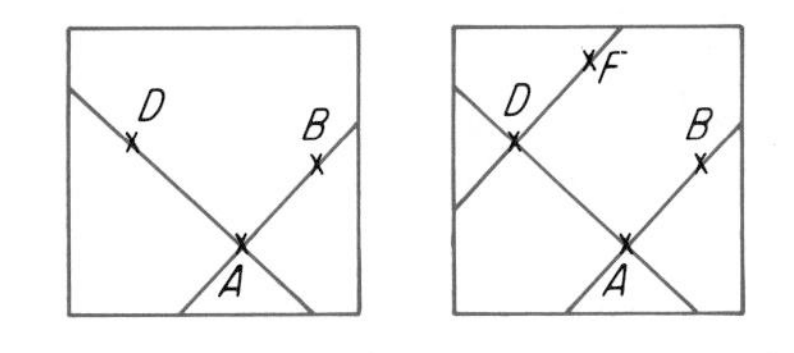

Let us return to the paper-folding example with perpendicular lines AB and AD. A line DF perpendicular to the line DA at D may be obtained by folding the ray DA onto its opposite ray. The lines DF and AB do not intersect on the paper and do not appear to intersect even if extended indefinitely. Since the lines are on the same plane, they are **coplanar lines**. Also as coplanar lines that do not intersect, lines AB and DF are **parallel lines**, which is written $AB \| DF$ (AB is parallel to DF). We have used paper folding to demonstrate the following theorem.

> Any two coplanar lines that are perpendicular to the same line are parallel lines.

Several definitions are included in the exercises. A few others are desirable for effective communication.

Two line segments are parallel if they are on parallel lines; two line segments are perpendicular if they are on perpendicular lines.

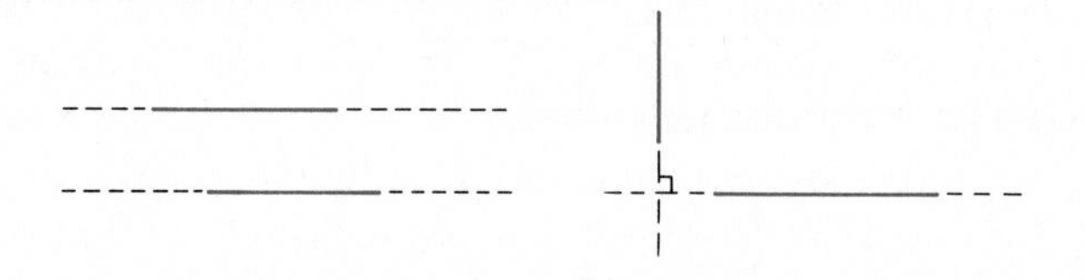

In preparation for the exercises on the development of common figures and the demonstrations of theorems by paper folding, we need to define a half-plane and provide another illustration of the use of paper folding.

A B

The definition of a half-plane is very similar to that of a half-line. A line may be determined by two points A and B. As shown in the figure above, the point A separates the line into three parts,

Your "line of vision" as you look in any specified direction represents a half-line.

> The point A.
> The **half-line** that contains B and has endpoint A.
> The opposite half-line.

The point A is often called the **endpoint** of each half-line even though it is not a point of either half-line.

A plane may be determined by a line t and a point D that is not a point of the line t. As shown in the figure, the line t separates the plane into three parts,

> The line t.
> The **half-plane** D-t that contains D and has edge t.
> The **opposite half-plane**.

When the horizon appears to be a line, the sky represents a half-plane and the scenery is on the opposite half-plane.

The line t is often called the **edge** of each half-plane even though it is not a part of either half-plane.

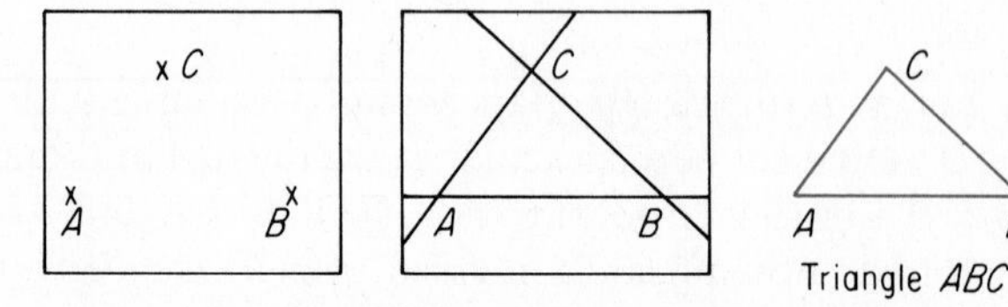

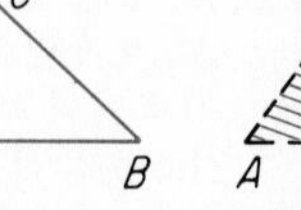

Triangle ABC

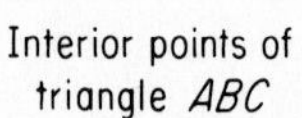

Interior points of triangle ABC

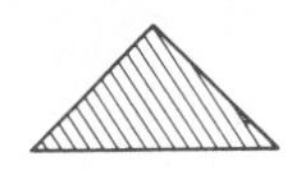
Triangular region

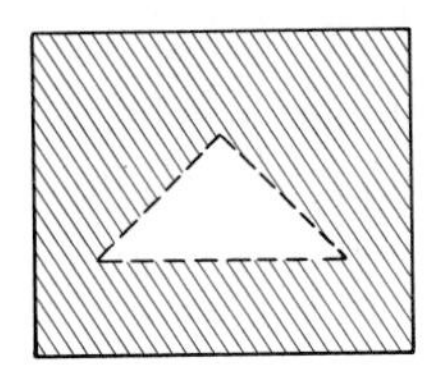
Exterior points of a triangular

Label three noncollinear points A, B, and C on a piece of paper. Make folds for the lines AB, BC, and AC. The union of the line segments AB, BC, and AC is the **triangle** ABC, written $\triangle ABC$. The three line segments are the **sides** of the triangle. The points A, B, and C are the **vertices** (singular, *vertex*) of the triangle. The **interior points** of the triangle are the points of intersection of

The half-plane that has edge AB and contains C.
The half-plane that has edge BC and contains A.
The half-plane that has edge AC and contains B.

The union of the triangle and its interior points is a **triangular region**. The points that are on the plane of the triangle but are not points of the triangular region are **exterior points** of the triangle.

EXERCISES

Sketch and label.

1. A line AB, a point C that is not on the line AB, a line CE that is perpendicular to AB at E, and a line CD that is parallel to AB.
2. A line FG, a half-plane $H\text{-}FG$, a point I of the half-plane $H\text{-}FG$, a point J of the half-plane opposite $H\text{-}FG$, a point K such that the line segment KI intersects the line FG, and a point L such that the line segment LI does not intersect the line FG.
3. A triangle PQR with an interior point S, an exterior point T, a point U such that the line segment SU does not intersect the triangle, and a point W that is not a point of the triangle but such that the line segment SW intersects the triangle.
4. A line ABC, a ray BD that is perpendicular to the line AC, a second ray BE that is perpendicular to the line AC, and a point G such that the line DG is parallel to AB and the line AG is parallel to BD.

Sketch as many figures as needed and identify each statement as true or false for points on a plane.

5. Let $C\text{-}AB$ be the half-plane that contains C and has edge AB. A point D is in the half-plane $C\text{-}AB$ if the line segment CD does not intersect
   (a) the line segment AB **(b)** the line AB.
6. A point E is in the half-plane opposite $C\text{-}AB$ if the line segment CE intersects
   (a) the line segment AB **(b)** the line AB.
7. Let C and D be any points that are not on the line AB.
   (a) If the line segment CD intersects the line AB, then $C\text{-}AB$ and $D\text{-}AB$ are opposite half-planes.
   (b) If the line segment CD does not intersect the line AB, then $C\text{-}AB$ and $D\text{-}AB$ are two names for the same half-plane.
8. Let D be any interior point of a triangle ABC and let E and F be points that are not points of the triangle.
   (a) If the line segment DE intersects the triangle, then E is an exterior point of the triangle.
   (b) If the line segment DF does not intersect the triangle, then F is an exterior point of the triangle.

9. Assume that all points are on the same plane and let AB be any line.
   (a) If the lines AC and AD are both perpendicular to AB, then AC and AD are two names for the same line.
   (b) If the lines AC and BE are both perpendicular to AB, then AC and BE are parallel lines.
10. Assume that all points are on the same plane, let AB be any line, and let the rays AC and AD each be perpendicular to AB.
   (a) The rays AC and AD must be either opposite rays or the same ray.
   (b) The rays AC and AD are opposite rays if the points C and D are not in the same half-plane with edge AB.

Use paper folding to illustrate the given theorem. In many cases steps, labels, and definitions are suggested in the exercises.

OK

11. *Theorem* There are infinitely many points on any given line.

    To illustrate this theorem it is only necessary to show that for any given number of points on a line, another point can be added.
   (a) Represent a line t containing points A and B.
   (b) Represent additional points C, D, E, F, G, and H on line t.

OK

12. *Theorem* There are infinitely many lines through any given point.

    To illustrate this theorem it is only necessary to show that for any given number of lines through a point, another line can be added.
   (a) Represent lines a and b through a point P.
   (b) Represent additional lines c, d, e, f, g, and h through the point P.

OK

13. *Theorems* Any line segment has one and only one midpoint.
    Any line segment has one and only one perpendicular bisector.

    The point M of a line segment AB such that the line segments AM and BM are congruent is the **midpoint** of the line segment AB. The line t that is perpendicular to line AB at M is the **perpendicular bisector** of line segment AB.
   (a) Represent a line containing points A and B.
   (b) Fold the paper so that the point B is superimposed upon the point A. Label the line obtained t and its intersection with AB as M.
   (c) Explain why M is the one and only midpoint of the line segment AB.
   (d) Explain why t is the one and only perpendicular bisector of AB.

OK

14. *Theorem* Any two lines that are respectively perpendicular to intersecting lines are themselves intersecting lines.
   (a) Represent an angle KLM such that the lines KL and LM are intersecting lines.
   (b) Represent the line k that is perpendicular to KL at K and the line m that is perpendicular to the line ML at M.

OK

15. *Theorem* The base angles of an isosceles triangle are congruent.

    Any triangle with at least two sides congruent is an **isosceles triangle**. The third side is the **base** of the isosceles triangle. The endpoints of the base are the vertices of the **base angles** of the isosceles triangle.
   (a) Represent a line segment RS with perpendicular bisector p and midpoint M.

(**b**) Select a point T different from M on the line p and make folds for the line segments TR and TS.

(**c**) Show that the line segments RT and ST are congruent and thus that the line segment RS is the base of an isosceles triangle RST with base angles RST and SRT.

(**d**) Show that the angles RST and SRT are congruent.

16. *Theorems* There are infinitely many isosceles triangles with a given line segment AB as a base.

There are infinitely many scalene triangles with a given line segment AB as a side.

There are infinitely many right triangles with a given line segment AB as a side and a right angle at A.

Any triangle that is not an isosceles triangle is a **scalene triangle**. Any triangle ABC with perpendicular sides AB and AC is a *right triangle* with the right angle at A and the side BC as *hypotenuse*.

17. *Theorem* There is one and only one line that is perpendicular to a given line and contains a given point that is not a point of the given line.

(**a**) Represent a line AB and a point C that is not a point of AB.

(**b**) Represent a line t that contains C and is perpendicular to AB.

18. *Theorem* The sum of the measures of the angles of a triangle is equal to the measure of a straight angle.

(**a**) Represent a triangle PQR with the vertices labeled so that there is a line PS perpendicular to the line QR and with S an interior point of the line segment QR.

(**b**) After cutting out the triangular region PQR, superimpose all three points P, Q, and R on the point S.

19. *Theorem* Any angle ABC such that the points A, B, and C are not collinear points has a unique angle bisector.

(**a**) Represent an angle ABC such that the points A, B, and C are not collinear points.

(**b**) Represent points D, E, F, G, and H that are on the intersection of the half-plane that has edge BA and contains C with the half-plane that has edge BC and contains A, that is, points that are **interior points** of $\angle ABC$.

(**c**) Represent the ray BK where K is an interior point of $\angle ABC$ and the ray BK is on the line obtained when the ray BA is superimposed on the ray BC.

(**d**) Explain why angles ABK and CBK are congruent, that is, ray BK is the **angle bisector** of $\angle ABC$.

20. *Theorem* The angle bisectors of any triangle are concurrent.

Three or more lines (or other figures) that contain a given point are **concurrent** at that point.

21. *Theorem* The perpendicular bisectors of the sides of any triangle meet at a single point, that is, are concurrent.

22. *Theorem* The medians of any triangle are concurrent.

A line segment with a vertex of a triangle as one endpoint and the midpoint of the *opposite side* (the side determined by the other two vertices) of the triangle as the other endpoint is a **median** of the triangle.

23. *Theorem* The altitudes of any triangle are on concurrent lines.

An **altitude** of a triangle is a line segment that has a vertex of the triangle as an endpoint and is perpendicular to the opposite side of the triangle. The intersection of the altitude with the line determined by the opposite side of the triangle is the other endpoint, often called the *foot*, of the altitude. The foot may be an interior point, an endpoint, or an exterior point of the opposite side.

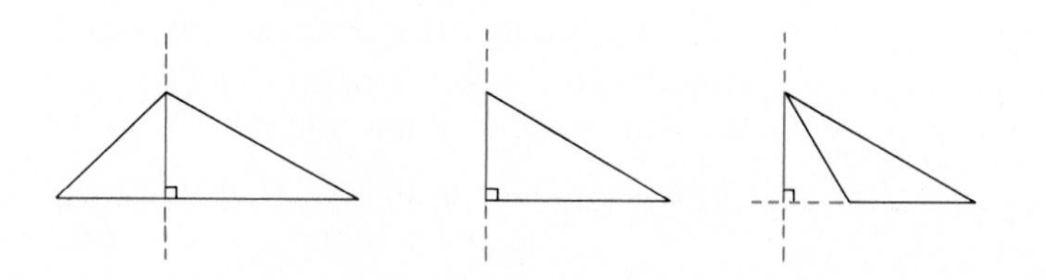

*24. *Theorem* There is one and only one line that is parallel to a given line m and contains a given point that is not on the line m.

*25. *Theorem* The midpoint of the hypotenuse of a right triangle is equidistant from the vertices of the triangle.

*26. *Theorem* There are infinitely many right triangles with a given line segment AB as hypotenuse.

*27. *Theorem* For a given line segment AB with midpoint M, it is possible to use paper folding to construct as many points as desired of the circle with center M and radius MA.

EXPLORATIONS

1. Four lines are to be drawn on a sheet of paper. Can the lines be drawn so that there are no points of intersection (even if the drawings were extended)? One point of intersection? Two points? Three points? Four points? Five points? Six points? Only one of these seven cases is impossible. Two cases can occur in different appearing ways. Sketch figures for all possible situations.
2. The early Greeks considered sets of points that formed geometric figures. Find a formula for T_n, the nth *triangular number*, where T_n is the sum of the integers 1 through n.

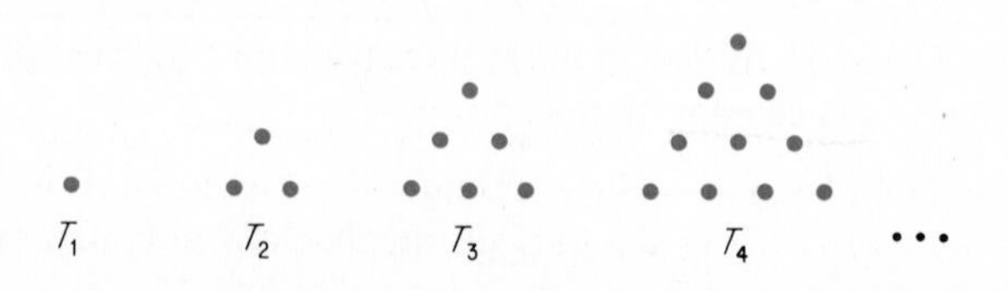

Hint: Compare R_n and T_n (see Exploration 2).

3. One line separates a plane into two regions. Two lines separate the plane into at most four regions. Three lines separate a plane into at most seven regions. Find a formula for the greatest possible number R_n of regions into which n lines can separate a plane.

When doing paper folding, designs and patterns may be formed by folding a given point P to each of many points of a given line, circle, or other figure. Frequently the creases represent the lines that are tangent to a familiar curve. For Explorations 4 through 9 explore the possibilities of such an approach using the indicated figure and point. A point S is an interior point of a circle with center O and radius r if $OS < r$; a point T is an exterior point if $OT > r$.

4. A circle and its center O.

5. A circle and a point Q that is a point of the circle.

6. A line m and a point P that is not a point of m.

7. A line m and a point P that is a point of m.

8. A circle and a point S that is an interior point but not the center of the circle.

9. A circle and a point T that is an exterior point of the circle.

10. Read and prepare a report on *Flatland: A Romance in Many Dimensions* by E. A. Abbott, fifth revised edition, Barnes and Noble, 1963.

11. Prepare a report on the *four-color problem.* For example, see "Snarks, Boojums, and Other Conjectures Related to the Four-Color-Map Theorem" by Martin Gardner on pages 126 through 130 of the April 1976 issue of *Scientific American* and "The Solution of the Four-Color-Map Problem" by Kenneth Appel and Wolfgang Haken on pages 108 through 121 of the October 1977 issue of *Scientific American.*

8-3 Constructions

The classical constructions of geometry were made with *straightedge* and *compass.* As in Section 7-1, the straightedge is an unmarked ruler and is used for drawing lines through two given points; the compass is used for drawing circles with a given center and radius. Among other figures, any line segment, angle, or polygon can be *copied* using straightedge and compass, that is, a figure congruent to the given figure can be constructed.

To copy a given line segment AB, draw a line CD. With $C = A'$ as center, intersect the line with an arc of a circle with radius AB and label the intersection B'. Then the distances AB and $A'B'$ are equal ($AB = A'B'$) and the line segments AB and $A'B'$ are congruent ($AB \cong A'B'$).

An explicit comparison of the meanings of "$\cong$" and "$=$" seems desirable since both were used in the previous sentence.

Congruence is used to indicate that two geometric figures *have the same measure(s).*

For example,

Line segments of the same length	$AB \cong A'B'$
Angles of the same measure	$\angle ABC \cong \angle PQR$
Triangles of the same measures	$\triangle ABC \cong \triangle DEF$

Equality is used to indicate that two expressions are *names for the same number or geometric figure* (set of points).

For example,

Same number	$3 + 2 = 5$
Same number	$x + 2 = 5$
Same angle	$\angle ABC = \angle CBA$
Same triangle	$\triangle ABC = \triangle BCA$
Same line segment	$AB = BA$

Note that with this distinction between "$\cong$" and "$=$," the context indicates the meaning of a symbol such as AB.

$AB = 3$	refers to a measure
$AB \cong A'B'$	refers to line segments

The construction of B' in copying line segment AB could have been done on either ray CD or its opposite ray. However, there is one and only one point B' on ray CD such that $CB' \cong AB$. As in Section 7-1, copying line segments can be used to construct number scales. Copying line segments can also be used to copy any given triangle.

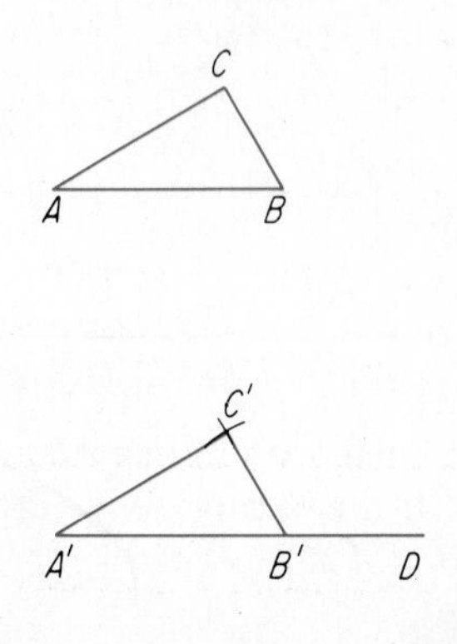

Use the following steps to copy a given triangle ABC.

Copy the line segment AB as $A'B'$ on a ray $A'D$.

Draw an arc of the circle with B' as center and BC as radius.

Draw an arc of the circle with A' as center and AC as radius.

Extend the arcs, if necessary, to obtain a point of intersection and label that point C'.

Draw line segments $A'C'$ and $B'C'$.

The point C' can be selected in either of the two half-planes with the line $A'D$ as edge, that is, either above or below the line in the figure. With B' on a given ray and C' in a given half-plane, there is one and only one triangle $A'B'C'$ such that $AB \cong A'B'$, $BC \cong B'C'$, and $AC \cong A'C'$. If the triangular region $A'B'C'$ were cut out, it could be superimposed upon the triangular region ABC. By superposition or by assumption (postulate) the two triangles are congruent. That is,

$$\triangle ABC \cong \triangle A'B'C'$$

Note that in performing the superposition and in the identification of the sides that are respectively congruent, the following condition must hold.

A corresponds to A'.
B corresponds to B'.
C corresponds to C'.

This correspondence is indicated in the statement

$$\triangle ABC \cong \triangle A'B'C'$$

where on each side of the statement of congruence, the first letters indicate corresponding vertices, the second letters indicate corresponding vertices, and the third letters indicate corresponding vertices.

We assume that

All corresponding parts of congruent figures are congruent.

Then a statement such as

$$\triangle OPQ \cong \triangle RST$$

of the congruence of $\triangle OPQ$ and $\triangle RST$ means that

$$OP \cong RS \qquad PQ \cong ST \qquad OQ \cong RT$$

$$\angle OPQ \cong \angle RST \qquad \angle PQO \cong \angle STR \qquad \angle QOP \cong \angle TRS$$

If two triangles are congruent, then there is a correspondence of their vertices such that corresponding sides are congruent and corresponding angles are congruent. We have used constructions to demonstrate that

If three sides of one triangle are congruent respectively to three sides of another triangle, then the triangles are congruent **(s.s.s.)**

Here, "**s.s.s.**" stands for *side-side-side* and is used to refer to the respective congruences of the three sides of the triangles.

The construction for copying a triangle ABC using the lengths of the sides provides a demonstration for this theorem. Except for its position on the plane, a triangle may be constructed and is uniquely determined by the lengths of its sides. Other constructions of triangles and demonstrations of the congruence of triangles are considered in the exercises.

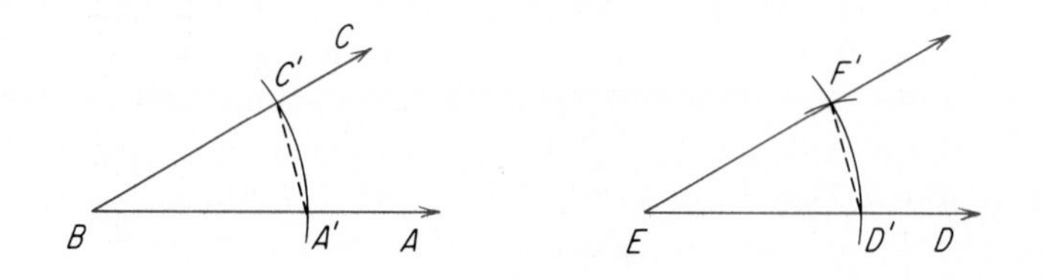

When we copy a given angle ABC, we may select any ray ED to correspond to ray BA and either half-plane with line ED as edge for the point corresponding to C. Use the following steps to copy a given angle ABC.

Select a point $A' \neq B$ on the ray BA.

Draw an arc of a circle with center B and radius BA' and label its intersection with the ray BC as C'.

Draw an arc of the circle with center E and radius BA' and label its intersection with the ray ED as D'.

Draw an arc of the circle with center D' and radius $A'C'$ and label its intersection with the arc with center E as F'.

Draw ray EF'.

The angles ABC and DEF' are central angles of circles with equal radii and intercept congruent arcs. The congruent arcs were constructed by making the line segments $A'C'$ and $D'F'$ (**chords**, line segments with points of the circle as endpoints) congruent. If the chords were drawn, then the triangles $A'BC'$ and $D'EF'$ would be congruent. (Why?) Thus $\angle ABC \cong \angle DEF'$.

Copying angles is used, for example, to copy triangles by *angle-side-angle* (Exercise 22) and by *side-angle-side* (Exercise 23). The construction of a right angle requires the bisection of a straight angle (Section 8-2).

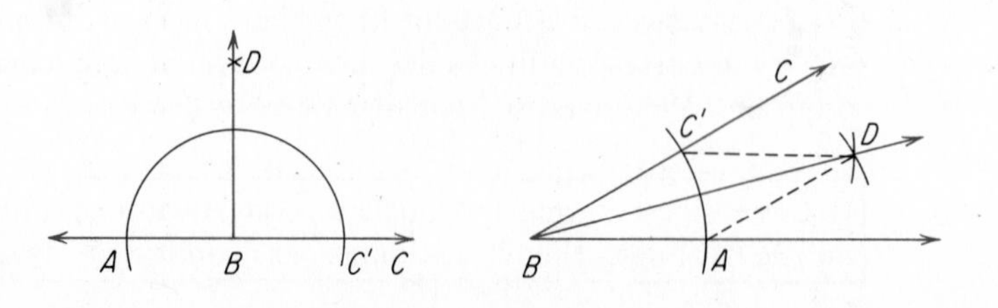

Use the following steps to bisect any angle ABC.

Draw an arc of the circle with center B and radius BA and label its intersection with the ray BC as C'.

Select a radius r so that the arcs will intersect, draw arcs of circles with centers A and C' and radius r, and label the intersection of these arcs as D.

Draw ray BD.

If chords AD and $C'D$ were drawn, then the triangles ABD and $C'BD$ would be congruent (**s.s.s.**). Thus $\angle ABD \cong \angle C'BD$ and ray BD is the **angle bisector** of $\angle ABC$. If rays BA and BC are opposite rays, then $\angle ABD$ is a right angle. Right angles are used, for example, in the construction of squares and rectangles.

The perpendicular bisector of a line segment, the midpoint of a line segment, and the line perpendicular to a given line at a given point of that line were constructed in Section 7-1. A construction may be given for the line that is perpendicular to a given line t and contains a given point P that is not a point of t. Draw a circle that has center P and intersects the line t in two points A and B. Then construct the perpendicular bisector m of the line segment AB. The line m will contain the given point P and be perpendicular to the given line t. Constructions for several other figures are considered in the exercises.

EXERCISES *Use straightedge and compass for all constructions.*

1. Draw a line segment PQ about 3 centimeters long and copy it onto a given ray RS.
2. Draw a triangle ABC with sides approximately 5 centimeters, 3 centimeters, and 4 centimeters long. Use the lengths of the sides and make a copy, triangle DEF, of triangle ABC.
3. Draw an obtuse angle HIJ and make a copy, angle KLM, of angle HIJ.
4. Draw an obtuse angle PQR and construct its angle bisector QT.

Construct an angle with the given measurement.

5. $45°$ 6. $135°$ 7. $22\frac{1}{2}°$
8. $11\frac{1}{4}°$ 9. $67\frac{1}{2}°$ 10. $157\frac{1}{2}°$

Construct the indicated figures for a given line AB and a point C that is not a point of AB.

11. The line that is perpendicular to the line AB and contains the point C.
12. A square with the line segment AB as side.

Draw and label a line segment TW approximately 3 *centimeters long, a line segment XY approximately* 4 *centimeters long, and an angle PQR of approximately* 30°. *Then construct a right triangle ABC with the given congruences and right angle ABC.*

13. $AB \cong TW$ and $BC \cong TW$
14. $AB \cong TW$ and $BC \cong XY$
15. $AB \cong TW$ and $AC \cong XY$

*16. $AC \cong XY$ and $\angle BCA \cong \angle PQR$

Explain the sense in which the construction in the specified exercise provides a basis for a demonstration of the given statement.

17. *Exercise 15.* If the hypotenuse and a leg of one right triangle are congruent respectively to the hypotenuse and a leg of another right triangle, then the triangles are congruent.

***18.** *Exercise 16.* If the hypotenuse and an acute angle of one right triangle are congruent respectively to the hypotenuse and an acute angle of another right triangle, then the triangles are congruent.

Draw line segments PQ, RS, and TU with lengths approximately 2 centimeters, 3 centimeters, and 4 centimeters, respectively. Draw angles MNO and JKL with measurements approximately $40°$ and $70°$, respectively.

19. Construct a line segment $AB \cong PQ$ and a point C such that $AC \cong PQ$ and $BC \cong PQ$. Draw line segments AC and BC to obtain **equilateral triangle** ABC, that is, a triangle with all sides congruent.

20. Construct $\triangle ABC$ with $AB \cong RS$, $AC \cong PQ$, and $BC \cong PQ$; that is, construct **isosceles triangle** ABC with **base** AB and congruent sides AC and BC.

21. Construct $\triangle ABC$ with $AB \cong PQ$, $BC \cong RS$, and $AC \cong TU$.

22. Construct $\triangle ABC$ with $AB \cong TU$, $\angle ABC \cong \angle MNO$, and $\angle BAC \cong \angle JKL$.

23. Construct $\triangle ABC$ with $AB \cong TU$, $\angle ABC \cong \angle MNO$, and $BC \cong RS$.

24. In each case how many triangles ABC of different shapes can be constructed for the given set of conditions?

(a) $AB \cong TU$, $\angle ABC \cong \angle JKL$, $AC \cong PQ$

(b) $AB \cong TU$, $\angle ABC \cong \angle MNO$, $AC \cong RS$

(c) $AB \cong RS$, $\angle ABC \cong \angle MNO$, $AC \cong TU$

Use a construction and demonstrate each theorem.

25. *Theorem* An equilateral triangle may be constructed with its side of any given length.

Select any line segment as a side.

26. *Theorem* If two angles and the included side of one triangle are congruent respectively to two angles and the included side of another triangle, then the triangles are congruent (**a.s.a.**).

Take one triangle as given, construct the other, and check the congruence by superposition.

27. *Theorem* If two sides and the included angle of one triangle are congruent respectively to two sides and the included angle of another triangle, then the triangles are congruent (**s.a.s.**).

EXPLORATIONS

1. Use straightedge and compass constructions to copy each figure.

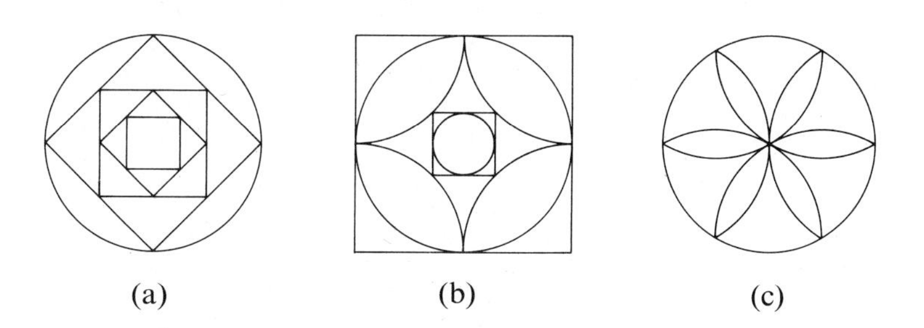

(a) (b) (c)

2. Create several designs, such as those in Exploration 1, that can be constructed using straightedge and compass.
3. Construct a polygon with 24 congruent sides and all of its diagonals.
4. Select line segments PQ and RS of different lengths. Try to construct an isosceles triangle ABC with $AB \cong PQ$, $AC \cong PQ$, and $BC \cong RS$. Repeat for different selections of the lengths until a necessary and sufficient condition for the possibility of the construction can be stated.
5. Repeat Exploration 4 for three line segments and the construction of a triangle with sides congruent to the given line segments.
6. Repeat Exploration 4 for two angles and a line segment that are to be used (if possible) to construct a triangle with the line segment as the included side between the two angles.
7. Select two angles A' and B'. Try to construct triangles ABC with $A \cong A'$ and $B \cong B'$. Repeat for several selections of sizes for the angles A' and B' until a necessary and sufficient condition for such a construction can be conjectured.
8. As in Exploration 7 construct triangles ABC for given angles A' and B' and several selections of the length of line segment AB. Compare the triangles obtained; they are **similar triangles**. Conjecture a theorem regarding
   (a) the corresponding angles of similar triangles;
   (b) the corresponding sides of similar triangles.
9. Compare at least two K-8 series of mathematics textbooks relative to their treatments of similar triangles. Then prepare a lesson plan on similar triangles for use at a specified elementary school grade level.

8-4 Polygons

Any set of points on a plane may be called a **plane figure**. Thus lines, rays, angles, and triangular regions are examples of plane figures. In the following set of four diagrams, each figure is a union of line segments. Each figure also appears to have three separate parts. Identify those parts and notice that any two distinct parts are not "connected"; that is, you could not draw both parts without removing the pencil from the paper to move from one part to the other. We assume that the word *connected* is understood even though it has not been defined.

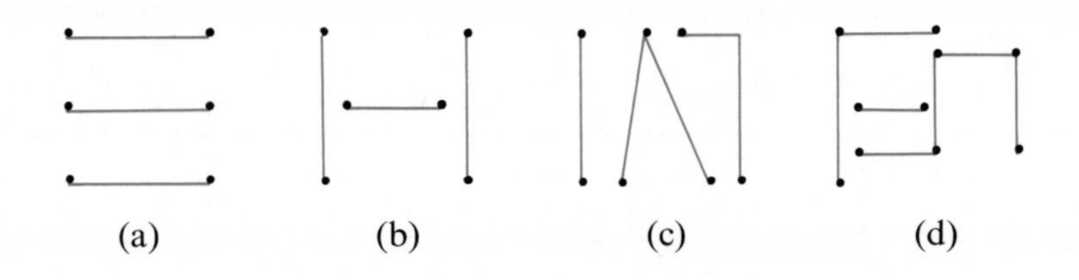

Each of the three figures in the next diagram is a union of line segments, and each figure is connected. Each may be drawn by starting at one of the points A or B, and ending at the other.

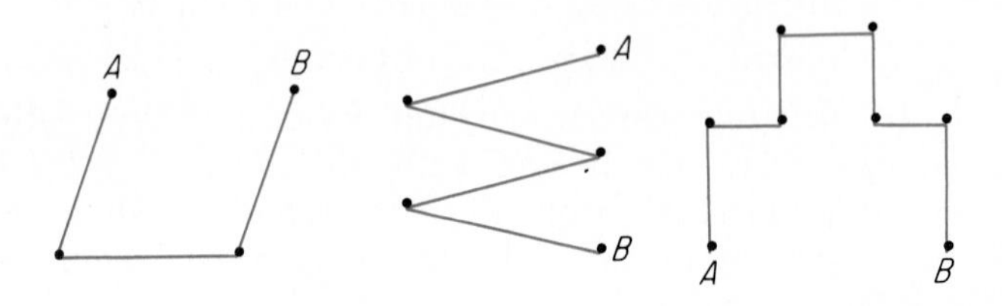

Each of the five figures in the next diagram is a union of line segments, each is connected, each may be drawn as a continuous line by starting at one of the points R or S and ending at the other, and each differs from the figures involving A and B in the preceding set. See if you can identify that difference.

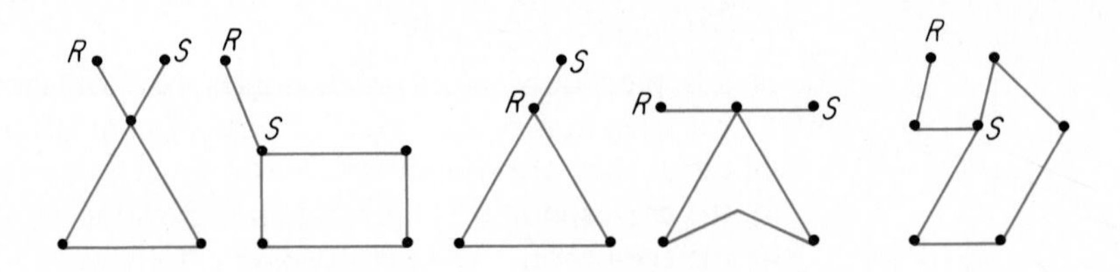

Any connected union of line segments is a **broken line**. Compare the three broken lines AB with the five broken lines RS in the preceding sets of figures. Each broken line AB may be drawn starting at A and ending at B without retracing any line segments or arriving at any point a second time. Each broken line RS may be drawn starting at R and ending at S without retracing any line segments, but for each broken line RS some point, possibly the starting point, must be used twice in the drawing. Briefly, we say that each broken line RS intersects itself. That is, each broken line RS contains at least one point that can be approached along at least three "paths." Such a point may appear, for example, as in any of the figures at the top of the next page.

The word "simple" is used to distinguish between figures that do not intersect themselves and figures that do intersect themselves. A figure such as one of the broken lines AB, which does not intersect itself, is a **simple figure**. Figures such as triangles, squares, circles, and rectangles are also simple figures. A figure such as one of the broken lines RS that does intersect itself is *not a simple figure.*

The broken lines AB and RS differ from the first two of the next three figures. Each of these two figures is simple; each is connected; each is a union of line segments. Each differs from the previous figures in that when it is drawn one starts at a point and returns to that point. We say that such a figure is **closed**.

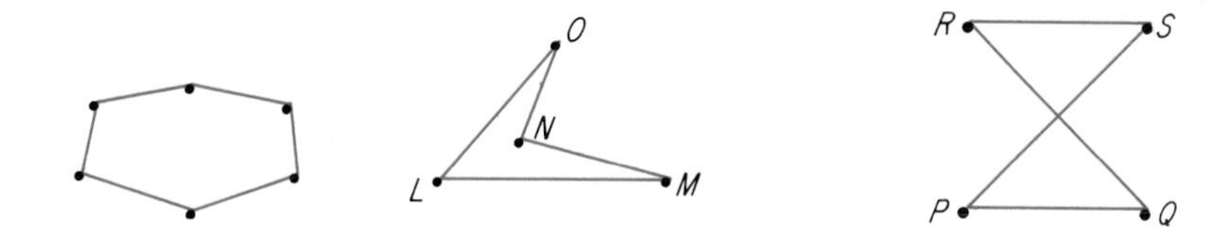

Figures may be either simple or not simple; they may also be either closed or not closed. For example, the figure $LMNO$ is simple and closed; the figure $PQRS$ is closed and not simple; the broken lines AB are simple and not closed; the broken lines RS are neither simple nor closed.

As examples of polygons consider the figures formed by the exterior walls of most buildings.

Any plane figure that is a simple, closed, connected, union of line segments is a **polygon**. The line segments are **sides** of the polygon; the endpoints of the line segments are **vertices** of the polygon. A polygon is a **convex polygon** if every line segment PQ with points of the polygon as endpoints is either a subset of a side of the polygon or has only the points P and Q in common with the polygon. A polygon that is not a convex polygon is a **concave polygon**.

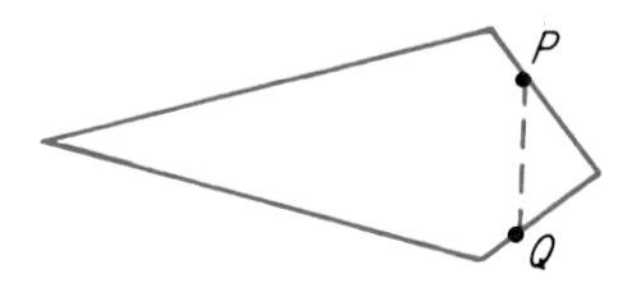

Convex polygon

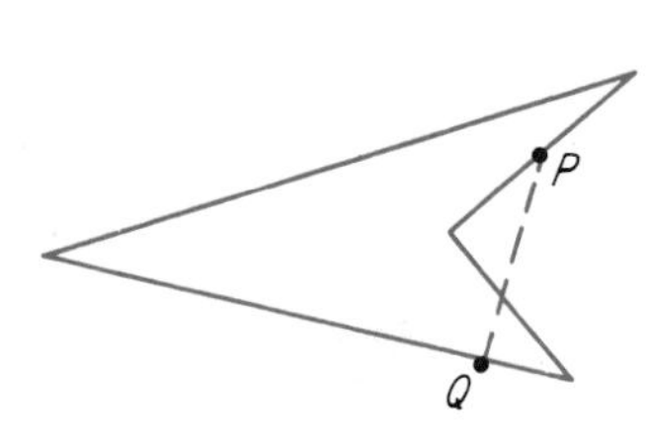

Concave polygon

If P and Q are points of a convex polygon and the line segment PQ is not a subset of a side of the polygon, then the points of the open line segment PQ are **interior points** of the convex polygon. The union of the points of a convex polygon and its interior points is a **polygonal region**. The points of the plane of a convex polygon that are not points of the polygonal region are **exterior points** of the polygon.

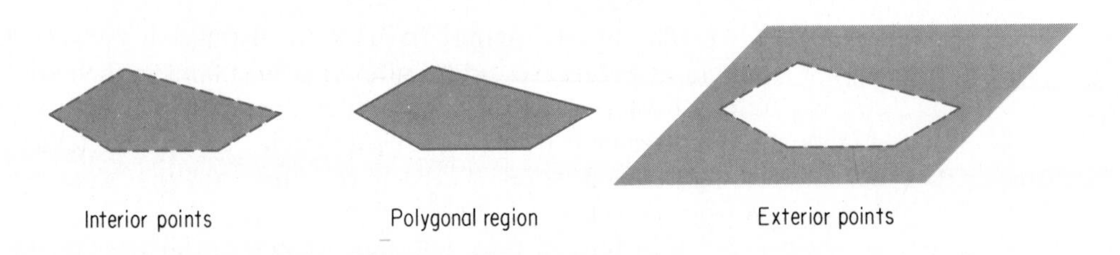

Fold a sheet of paper to represent each type of polygon. Note that a polygonal region must have at least three sides.

Plane polygons are classified according to the number of sides that each has. Most of the common names are shown in the next array.

Number of Sides	*Name of Polygon*
3	triangle
4	quadrilateral
5	pentagon
6	hexagon
7	heptagon
8	octagon
9	nonagon
10	decagon
12	dodecagon
n	n-gon

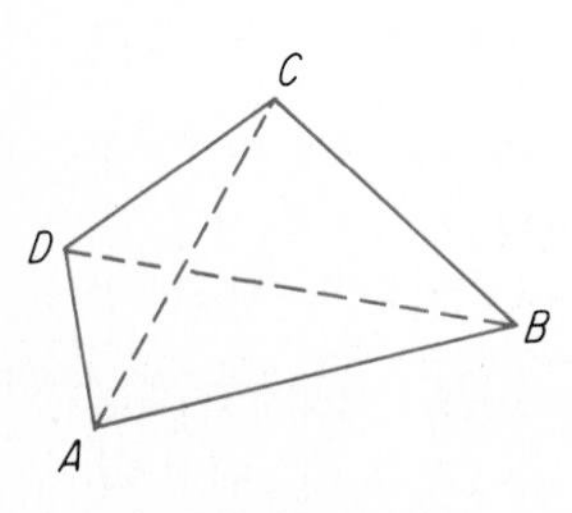

Any quadrilateral may be named by listing its vertices as if you were walking around the figure.

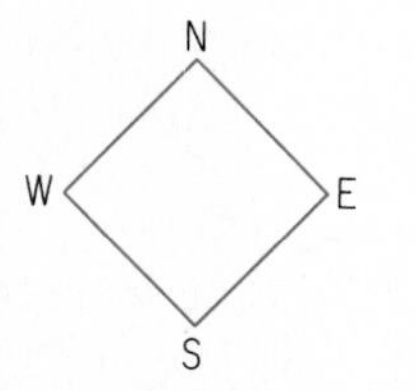

This square has two names starting with S. They are SENW and SWNE. You can start at any one of the four vertices and proceed from that vertex in either of two directions. State the other six ways of using its vertices to name the given square.

The quadrilateral has the points A, B, C, and D as vertices. Any two vertices, such as A and B, that are endpoints of a common side are **consecutive vertices**. The quadrilateral has the line segments AB, BC, CD, and DA as sides and the line segments AC and BD as **diagonals**. Any two sides, such as AB and BC, that have a common endpoint are **adjacent sides**. Any two sides, such as AB and CD, that are not adjacent sides are **opposite sides**. The quadrilateral may be named by listing the points A, B, C, and D in any order of consecutive vertices. For example, $CBAD$ and $CDAB$ are names for the given quadrilateral but $CADB$ and $CBDA$ are not names for the given quadrilateral. The four vertices of the quadrilateral determine six line segments; the four sides and the two diagonals AC and BD. In the name of any quadrilateral, consecutive letters must identify a side, and not a diagonal, of the quadrilateral.

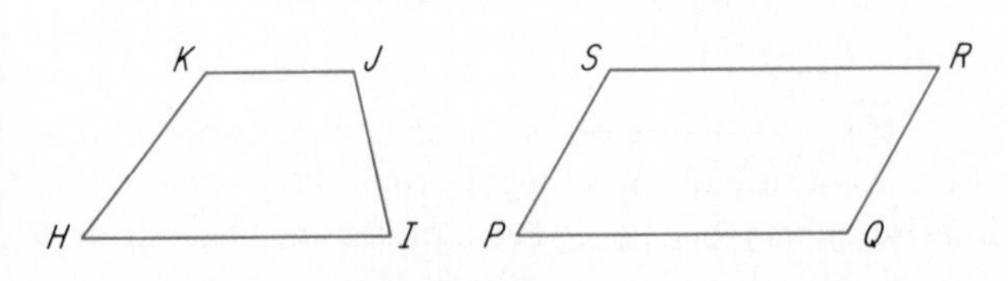

Parallelism and perpendicularity (right angles) may be used to classify quadrilaterals. A quadrilateral is a **trapezoid** if at least one pair of opposite sides are parallel; a quadrilateral is a **parallelogram** if both pairs of opposite sides are parallel. In the figures *HIJK* and *PQRS* both appear to be trapezoids; only *PQRS* appears to be a parallelogram. A parallelogram with at least one (actually all four) of its pairs of adjacent sides perpendicular is a **rectangle**. A parallelogram with at least one (actually all four) of its pairs of adjacent sides congruent is a **rhombus**. A parallelogram that is both rhombus and a rectangle is a **square**.

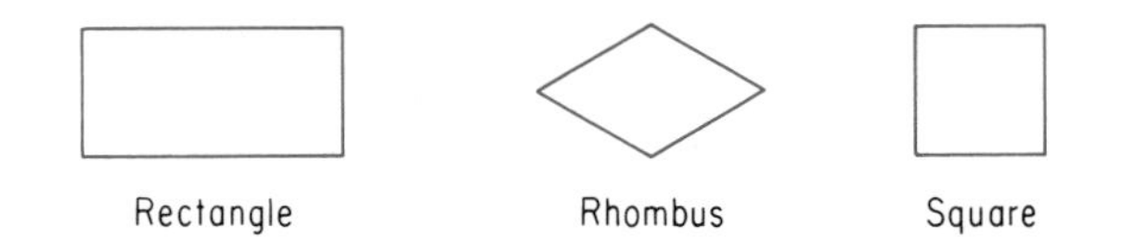

Mathematics has been described as the study of patterns—number patterns, geometric patterns, and all sorts of regularities that provide insights into a wide variety of concepts. Some of the apparent patterns can be misleading. However, the process of the discovery and testing of apparent patterns can provide some of the most interesting and fruitful experiences of mathematics. A conjecture is a suggested theorem; no proof is implied. Any statement that appears to be true is a reasonable conjecture. Often a careful tabulation or arrangement of known facts or observations suggests a pattern that might be conjectured as being true in general. We illustrate this process by an example.

Let *P* be a vertex of a polygon of *n* sides, that is, an ***n*-gon**. For polygons of different numbers of sides, does there appear to be a pattern that would support a conjecture regarding the number of diagonals with a particular vertex *P* as endpoint? Does there appear to be a basis for a conjecture regarding the number of triangles formed by these diagonals and the side of the polygon? If so, state the conjectures.

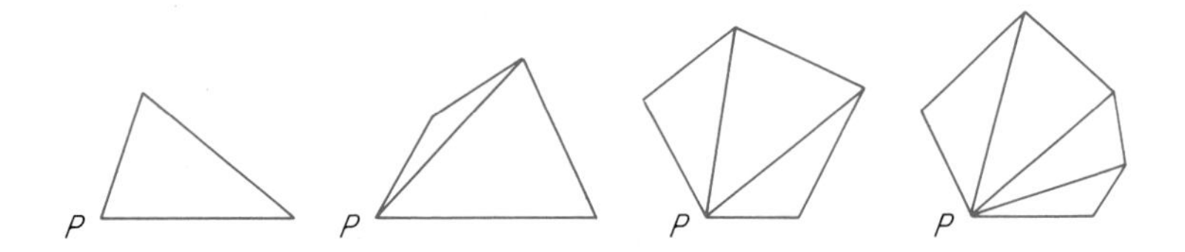

The simplest polygon is a triangle with three sides; there are no diagonals and only one triangle. Next polygons with 4, 5, and 6 sides are considered in order. Figures are drawn with the vertex *P* labeled and diagonals drawn. The related data include the number of sides of the polygon, the number of diagonals with the vertex *P* as an endpoint, and the number of triangles formed.

Number of sides	3	4	5	6	...	n
Number of diagonals from P	0	1	2	3	...	?
Number of triangles	1	2	3	4	...	?

The pattern for the number of sides is the sequence of positive integers 3, 4, 5, 6, ... Similar patterns can be observed in the other two rows of the tabulation. These patterns may be used to conjecture elements in the second and third rows for any given number in the first row.

The sequence of numbers of diagonals with the point P as an endpoint appears to be the sequence of whole numbers. If such is the case, then for $n = 7$ the number of diagonals should be 4; for $n = 8$ the number of diagonals should be 5; for $n = 9$ the number of diagonals should be 6; for $n = 23$ the number of diagonals should be 20; and, in general, for any n-gon the number of diagonals should be $n - 3$. The conjecture is

If P is a vertex of an n-gon, then there are $n - 3$ diagonals with the point P as an endpoint.

The conjecture is reasonable since there are $n - 1$ vertices that are different from the point P, two of these are endpoints of sides with one endpoint at P, and the remaining $n - 3$ are endpoints of diagonals with one endpoint at P.

The sequence of numbers of triangles appears to be the sequence of counting numbers. For any polygon the number of triangles appears to be two less than the number of sides. The conjecture is

If P is a vertex of an n-gon, then the n-gon and the diagonals with an endpoint at P form $n - 2$ triangles.

The conjecture seems reasonable since, after excluding the two sides with an endpoint at P, a triangle may be associated with each of the remaining $n - 2$ sides.

These identifications of patterns and formulations of conjectures illustrate the following general procedure.

Consider the cases in order of increasing difficulty.
For each case, draw figures or make models as needed.
Tabulate the related data.
Consider additional cases until a pattern is found.
Use the pattern to formulate a conjecture for the general case.
Check the conjecture as a reasonable general statement.

Conjectures have provided the bases for many, possibly most, mathematical proofs. If a statement appears to be true, the efforts to prove it may lead to this statement, or a different statement, as a theorem. For example, a person looking at a list of prime numbers greater than 100

101, 103, 107, 109, 113, 127, 131, 137,...

might conjecture that all prime numbers are odd numbers. The conjecture would be wrong, but the general procedure for testing the conjecture would lead to the following theorem.

The only prime number that is also an even number is 2.

Comparable situations exist in geometry. An instance in which a conjecture is false is a *counterexample*. Relative to the conjecture that all prime numbers were odd numbers, 2 was a counterexample. A single counterexample is sufficient to prove a conjecture false. It is often more difficult to prove a conjecture true. In many cases no proof may be known and no counterexample found; that is, the problem of proving the conjecture is unsolved. In other cases the counterexamples prove that the conjectures are false without suggesting any other possible conjectures (see Exercise 27).

The use of patterns, conjectures, and counterexamples may appear unusual to readers who have thought of mathematics as a set of long-established rules. However, in an informal and realistic sense these experiences are much closer to the actual nature of mathematics than routine computations and proofs.

EXERCISES

Sketch a union of line segments that satisfies the given condition.

1. **(a)** Connected. **(b)** Not connected.

2. **(a)** Simple. **(b)** Not simple.

3. **(a)** Closed. **(b)** Not closed.

Sketch a connected union of line segments (broken line) that satisfies the given condition.

4. **(a)** Simple but not closed. **(b)** Simple and closed.

5. **(a)** Closed but not simple. **(b)** Neither closed nor simple.

6. **(a)** A convex pentagon. **(b)** A concave hexagon.

Sketch and label the indicated figure.

7. A rectangle *ABCD* that is not a square.

8. A trapezoid *EFGH* that is not a parallelogram.

9. A parallelogram *JKLM* that is not a rectangle.

10. A quadrilateral *PQRS* that is not a trapezoid.

11. A quadrilateral with diagonals *AP* and *RS*.

12. A rectangle with diagonals *DH* and *BM*.

Describe a method for constructing a figure of the specified type by paper folding.

13. A rectangle.

14. A parallelogram that is not a rectangle.

15. A trapezoid that is not a parallelogram.

16. A convex pentagon.

Use paper folding for each exercise.

17. **(a)** Make folds for the sides of a triangle ABC and the altitude AD (see Section 8-2, Exercise 23).

(b) Fold the point A onto the point D and label the crease EF, where E is on the line AB and F is on AC.

(c) Identify the type of quadrilateral represented by $BCFE$.

18. **(a)** Repeat Exercise 17 for an isosceles triangle ABC with A on the perpendicular bisector of the side BC.

(b) Explain why line segments BE and CF are congruent, that is, why the trapezoid $BCFE$ is an **isosceles trapezoid**.

19. For any given triangle ABC describe a construction by paper folding for a parallelogram $ABCD$ and a different parallelogram $ABFC$.

20. **(a)** Make folds for any line segment AB with lines AM perpendicular to AB at A and BK perpendicular to AB at B.

(b) Make a fold for the bisector of $\angle BAM$ and label the intersection C of this angle bisector and line BK.

(c) Make a fold for the line that is perpendicular to line BC at C and label its intersection D with line AM.

(d) Explain why quadrilateral $ABCD$ is a rectangle.

(e) Make creases for the angle bisectors and show that the sides of the rectangle are all congruent; that is, the rectangle is a *square*.

Use paper folding to demonstrate each theorem.

21. For any rectangle:

(a) The opposite sides of the rectangle are congruent.

(b) The diagonals separate the rectangular region into two pairs of congruent triangular regions.

(c) The triangles are isosceles triangles.

(d) The diagonals of a rectangle are congruent line segments and bisect each other.

22. The diagonals of any parallelogram bisect each other.

Conjecture a general statement for each specified situation.

23. A single point on a line separates the line into two nonintersecting (*disjoint*) parts, besides the point itself. Into how many disjoint parts, besides the points themselves, do n distinct points of a line separate the line?

24. A single point of a circle does not separate a circle; thus there remains one part besides the point. Into how many disjoint parts, besides the points themselves, do n distinct points of a circle separate the circle?

25. Consider a map composed of a single central region and n neighboring regions as in the figure, where $n = 4$. If $n = 1$, the map can be colored with two colors. In general, regions that have a common edge (boundary) should have different colors. For this special type of map, how many

In 1852 Francis Guthrie mentioned in a letter to his brother Frederick that it seemed that every map drawn on a sheet of paper could be colored with only four colors in such a way that countries sharing a common boundary have different colors. For 124 years the problem provided a challenge for both amateur and professional mathematicians. In 1976 Kenneth Appel and Wolfgang Haken proved Francis Guthrie's conjecture by making extensive use of computers. Searches continue for a proof that does not depend on computers.

colors are needed (what is the smallest possible number of colors) for the central region and its n neighbors?

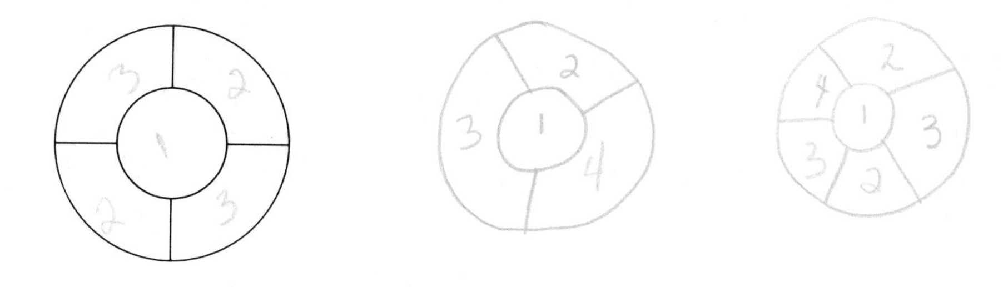

26. **(a)** Make a tabulation of the number of sides, the number of diagonals on each vertex, and the total number of diagonals for a polygon of n sides where $n = 4$, 5, and 6.
    (b) Use additional figures as needed and conjecture an expression in terms of n for the total number of diagonals.

*27. Two points on a circle determine one chord, which separates the interior of the circle into two regions; three points on a circle determine three chords, which separate the interior of the circle into four regions.

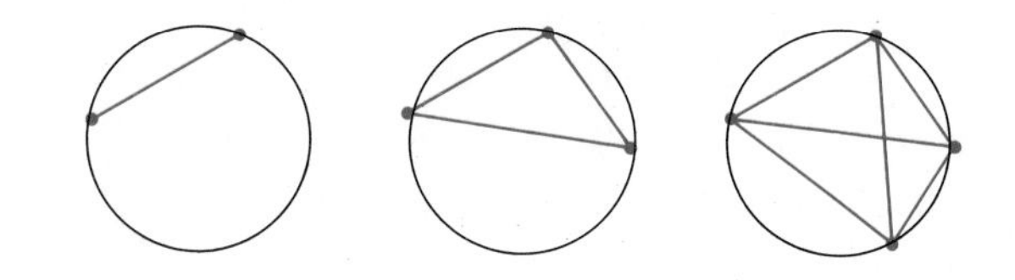

Conjecture: On a circle n points determine chords that separate the interior of the circle into 2^{n-1} regions.

Test this conjecture until a counterexample is found.

EXPLORATIONS

Activities such as the following can provide elementary school children with an opportunity to make explorations on their own as well as to study various geometric figures.

1. For any given triangle two line segments are to be drawn with points of the triangle as endpoints. Three or four nonoverlapping polygons are formed and the sum of the numbers of sides of these polygons determined. Here are two examples.

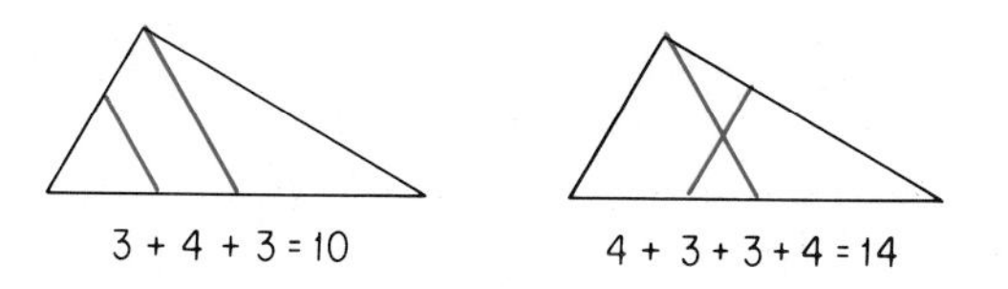

Sketch such figures for as many different sums as possible and list those sums from smallest to largest.

Games may be used to present many mathematical concepts very effectively. Frequently slight modifications of the rules for a particular game lead to another very different game. The following games are described in terms of points and line segments on a piece of paper. However other procedures such as pegs and elastics on a peg board may be used instead.

2. Make several dots for points on a piece of paper as in the figure. Two players take turns drawing line segments with the given dots as endpoints. No line segment may cross another or contain more than two of the given points. The last person able to play is the winner.

3. Proceed as in Exploration 2, but with the added restriction that no point may be the endpoint of more than two line segments.
4. Proceed as in Exploration 2, but with the added restriction that no point may be the endpoint of more than three line segments.
5. In Exploration 2, can you win if you begin with three dots and your opponent has the first move?

8-5 Polyhedra

Identify a few objects that have approximately the shapes of common space figures such as spheres, rectangular boxes, cubes, pyramids, or cones.

On a plane a simple, closed, connected union of line segments is a **polygon**. In space a simple, closed, connected union of polygonal regions is a **polyhedron** (plural **polyhedra**). Although slight extensions of the terms *simple* and *closed* have been required, their meanings should be clear from our knowledge of triangular pyramids, rectangular boxes, and cubes.

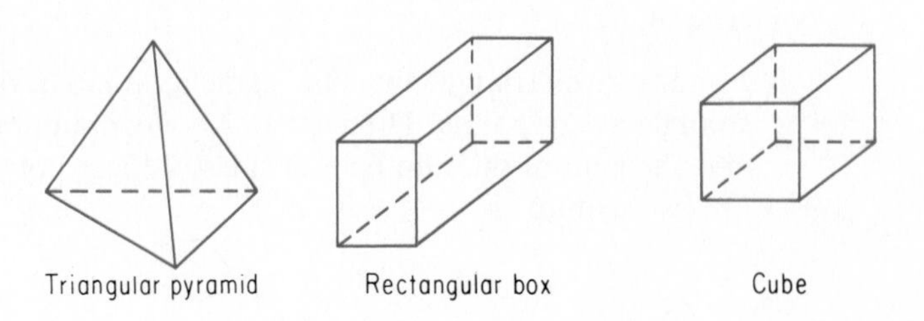

The polygonal regions that are used to form a polyhedron are **faces** of the polyhedron; the sides of the polygonal regions are **edges** of the polyhedron; the vertices of the polygonal regions are **vertices** of the polyhedron. A polyhedron with four triangular regions as faces is a **triangular pyramid**, also called a **tetrahedron**. A polyhedron with rectangular faces is a **rectangular box**, which is

Note that a triangular pyramid has:

6 edges

4 faces

4 vertices

also called a **rectangular parallelepiped** since the opposite faces are on **parallel planes**, that is, planes that have no points in common. A **cube** is a rectangular box with square faces.

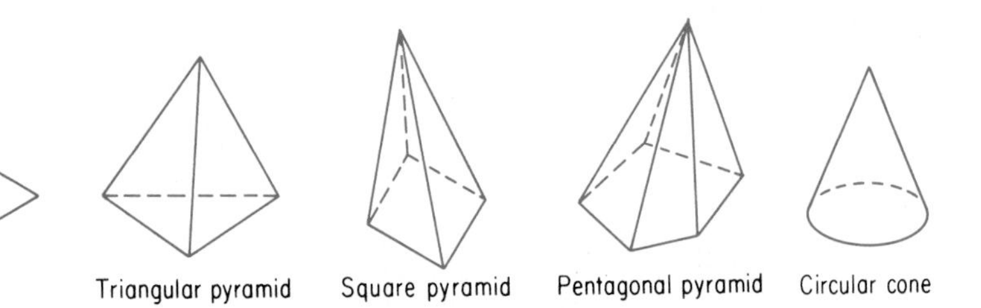

A polyhedron with all but one of its vertices on a plane is a **pyramid**. The vertices on a plane are vertices of the **base** of the pyramid. Pyramids are often classified as triangular, rectangular, square, and so forth, by the shapes of their bases. Since a circle is not a polygon (a circle does not have a finite number of sides), a circular cone is not a pyramid or a polyhedron. However, a **cone** may be considered to be obtained from pyramids by increasing the number of sides of the base indefinitely.

Example 1 The pyramid $ABCDE$ is a rectangular pyramid. Name its (**a**) vertices (**b**) edges (**c**) faces.

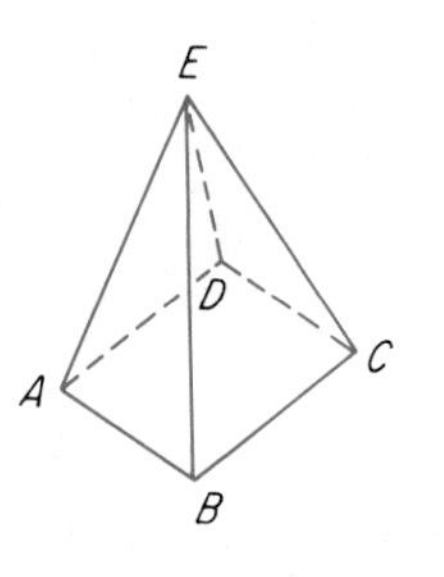

Solution (**a**) Points A, B, C, D, and E.
(**b**) Line segments AB, BC, CD, DA, AE, BE, CE, and DE.
(**c**) Rectangular region $ABCD$ and triangular regions ABE, BCE, CDE, and DAE. ■

We use the edges and faces of a rectangular box to explore some relationships among lines and planes in space before considering other polyhedra. Suppose that the edges of the box shown in the figure at the top of the next page are located so that AB is an east-west line, AE is a north-south line, and AD is a vertical line. (If you are in a rectangular room, its walls, floor, and ceiling may provide a better example than the box.) The lines AB, EF, HG, and DC are all east-west lines and therefore parallel lines. The fact that the lines AB

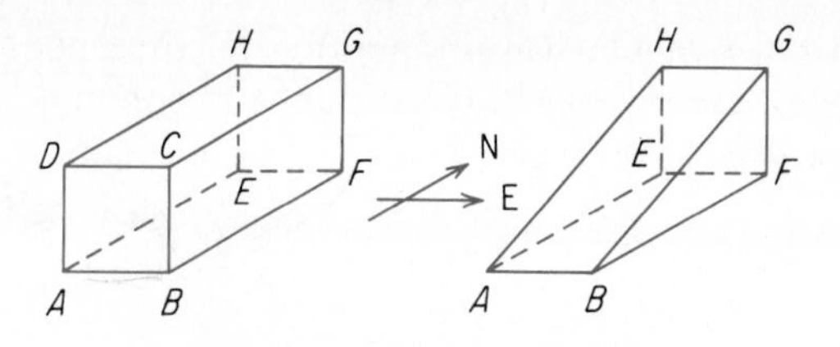

and GH are parallel implies that they are on a plane $ABGH$, as shown in the second figure. In general, two lines in space are

Intersecting lines if they have exactly one point in common.
Parallel lines if they are coplanar and do not intersect.
Skew lines if they are not coplanar.

For example, the lines AB and CG are skew lines.

The vertical line BC is perpendicular to the line BA and also to the line BF. If a line intersects a plane so that the line is perpendicular to at least two lines on the plane, then the line is perpendicular to the plane. Thus the line BC is perpendicular to the plane ABF. This implies that the line BC is perpendicular to every line that is on the plane ABF and contains the point B.

Example 2 The figure represents a rectangular box. Name the edges that
(**a**) are parallel to the line SZ
(**b**) intersect SZ
(**c**) form skew lines with SZ
(**d**) are perpendicular to the plane WXY.

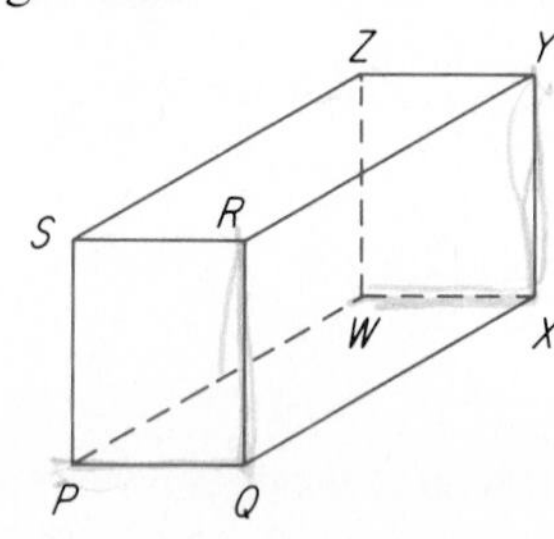

Solution (**a**) PW, QX, and RY
(**b**) SP, SR, ZW, and ZY
(**c**) PQ, RQ, WX, and XY
(**d**) SZ, RY, PW, and QZ ■

Any two planes in space are either parallel (no points in common) or intersect in a line. (It is not possible in three-dimensional space for two planes to have one and only one point in common.) If a polyhedron has all of its vertices in two parallel planes, then the edges in those planes are the edges of the two bases of the polyhedron. The edges that are not edges of a base are lateral edges. If the lateral edges are on parallel lines, then the polyhedron is a prism. Prisms are often classified as triangular, square, rectangular, and so forth according to the shapes of their bases.

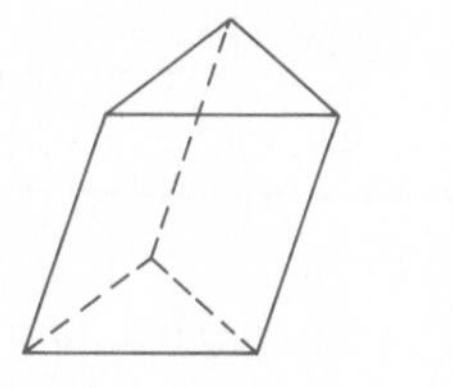

Triangular prism

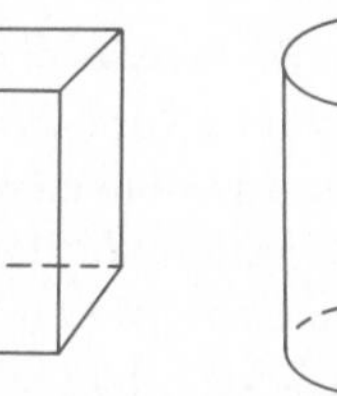

Square prism

Right circular cylinder

A prism with a lateral edge perpendicular to the plane of a base is a **right prism**. The square prism in the figure is a right prism; the triangular prism is not a right prism. A **cylinder** is not a prism (or polyhedron) but may be considered to be obtained by increasing the number of sides of the base of a prism indefinitely. A circular cylinder with the line through the centers of its bases perpendicular to the plane of a base is a **right circular cylinder**.

Example 3 The figure represents a portion of a square pyramid viewed from above. Identify the given statement as true or false.

(a) Edges *AD* and *BC* intersect.
(b) Lines *AD* and *BC* intersect.
(c) Line *AD* intersects plane *BCE*.
(d) Line segment *AD* intersects plane *BCE*.
(e) Face *ADGH* intersects plane *BCE*.
(f) Plane *ADG* intersects plane *BCE*.

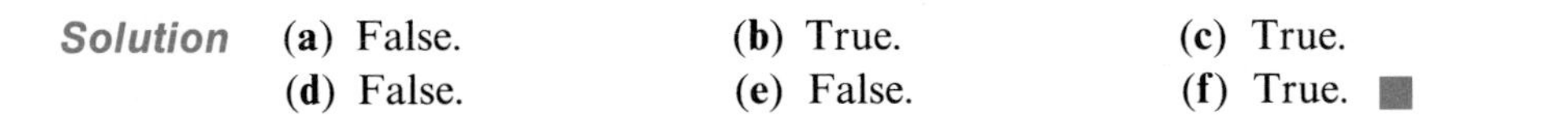

Solution **(a)** False. **(b)** True. **(c)** True.
(d) False. **(e)** False. **(f)** True. ■

EXERCISES

1. The tetrahedron *MNOP* is a triangular pyramid. Name its
   (a) vertices **(b)** edges **(c)** faces.

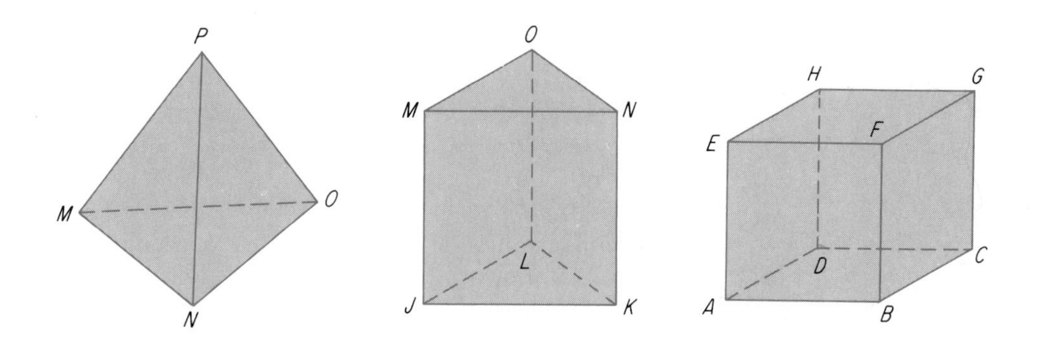

2. Repeat Exercise 1 for the triangular prism *JKLOMN*.
3. Repeat Exercise 1 for the cube shown in the figure.
4. Think of the lines along the edges of the cube in Exercise 3 and identify the lines that appear to be
   (a) parallel to *AB*
   (b) skew to *AB*
   (c) parallel to the plane *ABFE*
   (d) perpendicular to the plane *ABFE*.

From the given figures, identify those of the specified type.

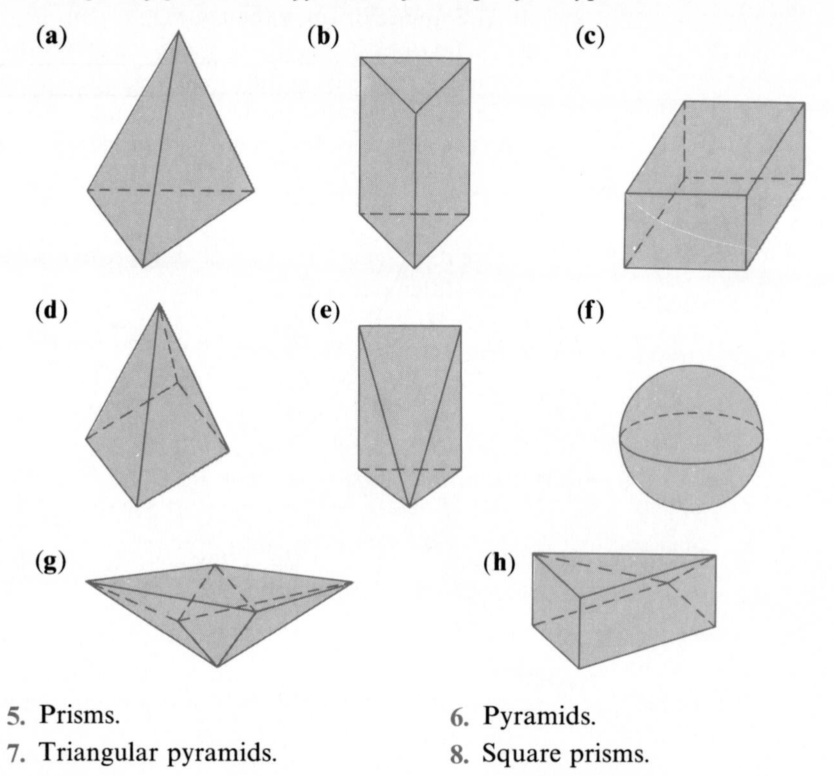

5. Prisms.
6. Pyramids.
7. Triangular pyramids.
8. Square prisms.

Sketch.

9. A triangular pyramid.
10. A triangular prism.
11. A square pyramid.
12. A square prism.
13. The intersection of a plane and a square pyramid when that intersection is
    (**a**) a square
    (**b**) a triangle
    (**c**) a trapezoid that is not a square.
14. The intersection of a plane and a cube when that intersection is
    (**a**) a square
    (**b**) a triangle
    (**c**) a rectangle that is not a square.

*15. A polyhedron with six triangular regions as its faces.

*16. A polyhedron with eight triangular regions as its faces.

Use pencils, table tops, and so on to represent lines and planes and determine whether each statement appears to be true or false in Euclidean geometry.

17. Given any line m and any point P that is not a point of m, there is exactly one line t that is parallel to m and contains P.
18. Given any line m and any point P that is not a point of m, there is exactly one line q such that q contains P and m and q are intersecting lines.

19. Given any line m and any point P that is not a point of m, there is exactly one line s such that s contains P and s and m are skew lines.
20. Given any line m and any point P that is not a point of m, there is exactly one plane that is parallel to m and contains P.
21. Given any plane ABC and any point P that is not a point of ABC, there is exactly one plane that contains P and intersects ABC.
22. Given any plane ABC and any point P that is not a point of ABC, there is exactly one plane that contains P and is parallel to ABC.
23. Given any plane ABC and any point P that is not a point of ABC, there is exactly one line that contains P and intersects ABC.
24. Given any plane ABC and any point P that is not a point of ABC, there is exactly one line that contains P and is parallel to ABC.
25. Given any plane ABC and any line m that is parallel to ABC, there is exactly one plane that contains m and intersects ABC.
26. Given any plane ABC and any line m that is parallel to ABC, there is exactly one plane that contains m and is parallel to ABC.
27. Given any plane ABC and any line m that is parallel to ABC, there is for any given point P that is not on m and not on ABC exactly one line that is parallel to m and also parallel to ABC.

A triangular pyramid KEFJ has been removed from a cube as indicated in the figure. Identify the given statement as true or false.

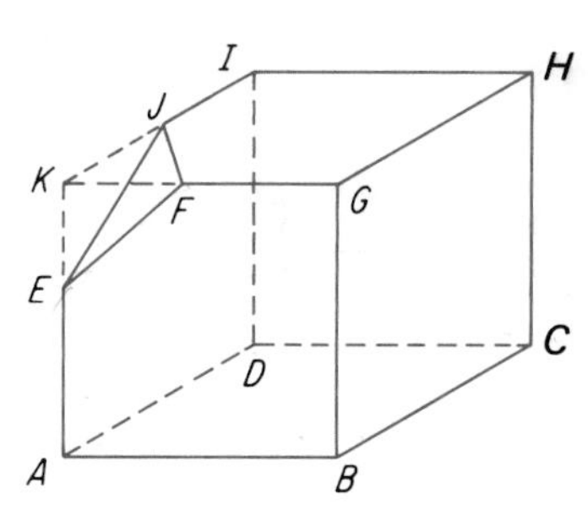

28. Line segment EF intersects plane
    (a) GHI
    (b) ABC
    (c) DCH.
29. Line EF intersects plane
    (a) GHI
    (b) ABC
    (c) DCH.
30. Triangular region EFJ intersects plane
    (a) GHI
    (b) ABC
    (c) DCH.
31. Plane EFJ intersects plane
    (a) GHI
    (b) ABC
    (c) DCH.

Conjecture a general statement for each specified situation.

32. Pyramids are considered with bases that are polygonal regions with 3, 4, 5, . . . , n sides. Each face, including the base, is to be colored. Two faces that have a common edge are to have different colors. What is the smallest possible number of colors for a pyramid with an n-sided base?

odd-4

33. (a) Make a table of the number F of faces, the number V of vertices, and the number E of edges for a triangular pyramid, a square pyramid, a pyramid with a five-sided base, and a cube.
(b) Use additional polyhedra as needed and conjecture a relation among F, V, and E. (You should obtain the famous *Euler formula for polyhedra.*)

EXPLORATIONS

The representations of space figures on a plane is both an art and a science. One of the common procedures for representing space figures, **orthographic projection**, makes use of views from three mutually perpendicular directions (top view, front view, and side view) and includes dashed lines for hidden edges. The pattern for presenting the views, *orthographic projections*, is always the same; the side view is conventionally the right side.

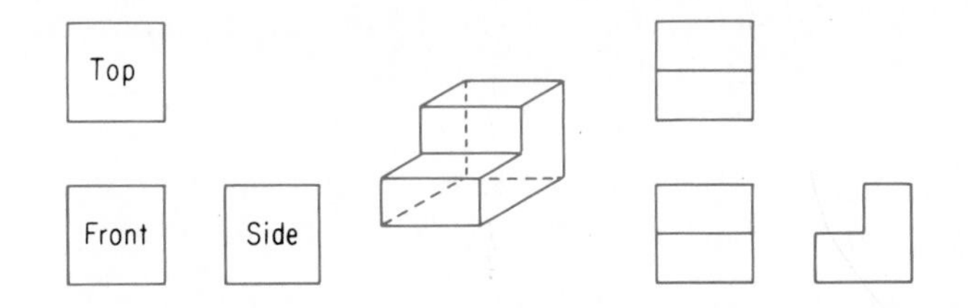

Suppose that one quarter of a cube were removed as in the figure; then the three views would be as shown. Each of the following figures has the same top and front views as the previous example but the side view would be different as shown with the figure.

Sketch the indicated figure and its three orthographic projections.

1. A cube.
2. A sphere.
3. A cylinder with $h = 2r$.
4. A square pyramid.

Sketch a space figure that has the three given orthographic projections.

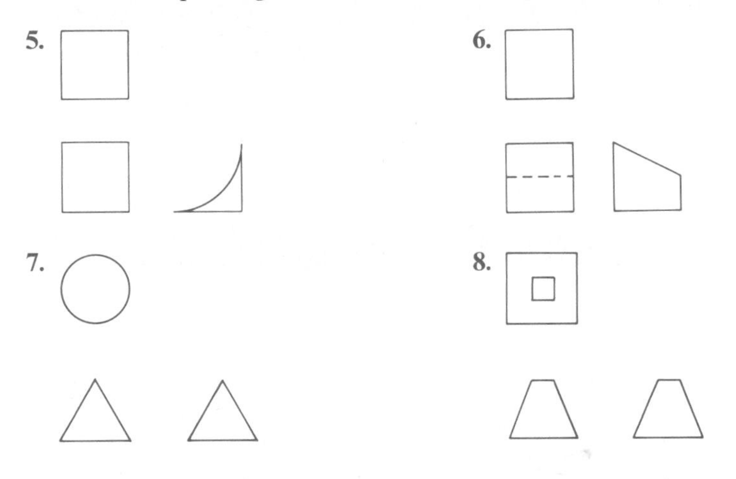

For Explorations 9 through 12 sketch at least three different figures and their side views such that all figures have the given top and front views.

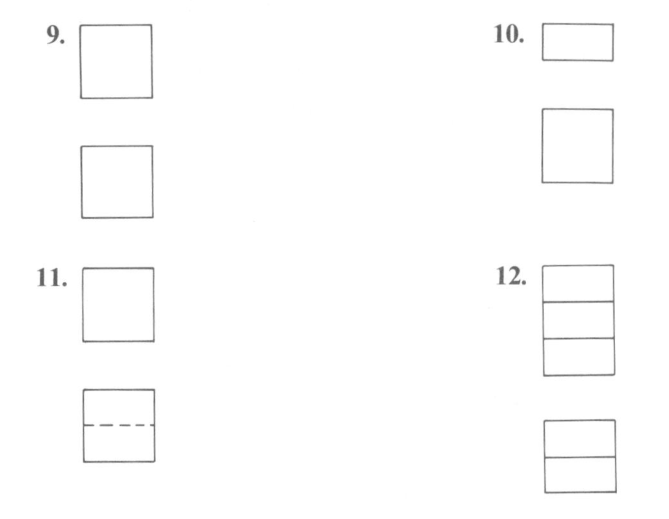

13. Sketch at least five other space figures and their orthographic projections.

8-6 Networks and Graphs

Geometry and arithmetic (also algebra) supplement each other by providing different approaches to mathematics. Some mathematical problems can be easily visualized and solved using geometry. Some problems can be solved using only arithmetic. Many problems can be solved most easily by using both geometry and arithmetic. The use of geometry in solving problems provides an important basis for extending our concept of geometry, just as solving problems stimulated the expansion of our concept of number. We consider two of the many topics in geometry that have evolved from efforts to solve problems.

The people of Königsberg in Prussia (now Kaliningrad in the Soviet Union) used to enjoy walking over the bridges of the Pregel River. There were two islands in the river and seven bridges. The locations of the bridges are indicated in the figure.

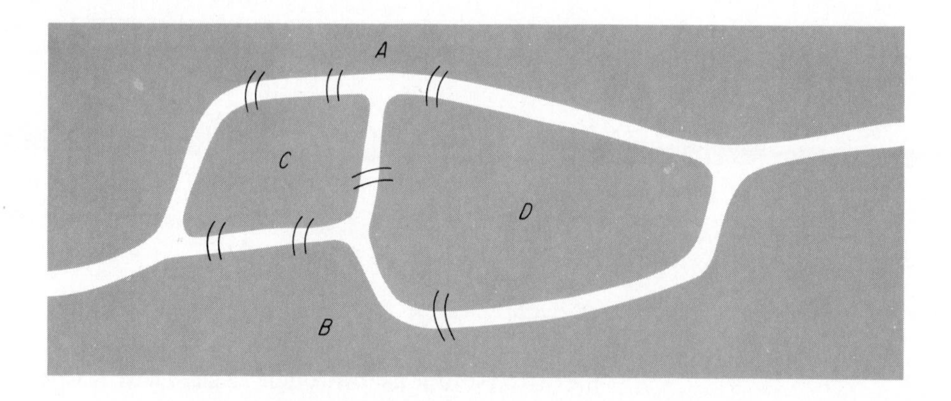

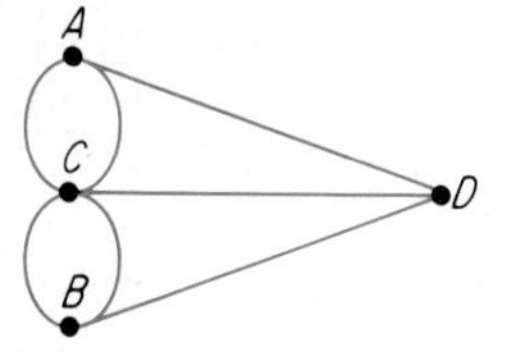

The walkers tried to find a route that would take them across each bridge exactly once. Gradually people observed that the basic problem was concerned with paths between the two sides A and B of the river and the two islands C and D as in the adjacent figure.

Any set of line segments (or arcs) forms a **network**. If the network can be drawn by tracing each line segment exactly once without removing the point of the pencil from the paper, then the network is **traversable**. The representation of the **Königsberg bridge problem** by a network made it unnecessary to walk across the bridges when trying to solve the problem. Instead, the traversability of the network could be studied.

It is possible to walk completely around an ordinary city block, and it is not necessary to start at any particular point in order to do so. In general, it is possible to trace any simple closed broken line with one uninterrupted movement of the pencil. Next consider walking around two blocks and down the street BE that separates them. This problem is a bit more interesting in that it is necessary to start at B or E. Furthermore, a walk that is begun at B terminates at E, and conversely, as shown in the second figure.

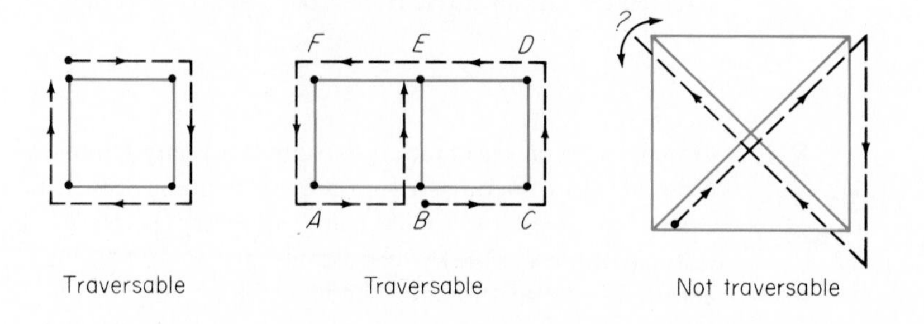

Note that it is permissible to pass through a vertex several times, but a line segment, or arc, may be traversed only once. The peculiar property of the vertices B and E is that each of these vertices is an endpoint of three line segments. Each of the other vertices is an endpoint of two line segments. A similar observation led the famous mathematician Leonhard Euler (1707–1783) to devise a theory for the traversability of networks.

Euler classified the vertices of a network as *odd* or *even*. For example, in the adjacent figure the vertex A is an endpoint of three arcs AB, AC, and AD and thus is an odd vertex; B is an endpoint of two arcs BA and BC and is an even vertex; C is an endpoint of four arcs CB, CA, CD, and CE and is an even vertex; D is an endpoint of two arcs DA and DC and is an even vertex; E is an endpoint of one arc EC and is an odd vertex. Thus the figure has two odd vertices A and E and three even vertices B, C, and D. For any network a vertex that is an endpoint of an odd number of arcs is an **odd vertex**; a vertex that is an endpoint of an even number of arcs is an **even vertex**. Since each arc has two endpoints, there must be an even number of odd vertices in any network.

Any network, such as a square, that has only even vertices is traversable, and any vertex may be the starting point. That same vertex will be the terminal point. If a network has exactly two odd vertices, the network is traversable, but one of the odd vertices must be the starting point and the other odd vertex will be the terminal point. If a network has more than two odd vertices, that network is not traversable in a single trip. In general, a network with $2k$ odd vertices may be traversed in k distinct trips. The network for the Königsberg bridge problem has four odd vertices and thus is not traversable in a single trip. Notice that the Königsberg bridge problem is independent of the size and shape of the river, bridges, or islands.

Our second topic may be introduced using a theoretical problem for plane figures. Note that this is not a problem in three-dimensional space. Suppose that three houses are each to be connected to three utilities—water, electric power, and sewage disposal. The question as to whether the connections can be made on a plane and with no two of the lines (arcs) crossing is the classical **problem of the three houses and three utilities**. If you have not previously tried to solve the problem, copy the figure, and try to find suitable arcs before reading further.

Electric power

Water

Sewage disposal

The solution of the problem depends upon the **Jordan curve theorem**. Any simple closed plane curve separates the plane into three sets of points:

The points of the curve.
The interior points of the curve.
The exterior points of the curve.

If A is an interior point of a simple closed plane curve and B is an exterior point of the curve, then on the plane every arc AB must intersect the curve.

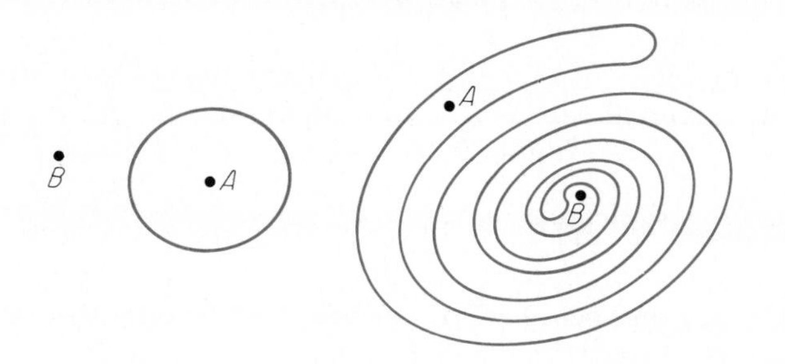

The classic utility problem may be represented using points P, Q, and R for houses and points X, Y, and Z for utilities. As in the next figure, each of the houses may be joined to the utilities X and Z and two of the houses may be joined to the third utility. In the figure the house R and the utility Y are in different regions of the simple closed curve $PXQZP$. Thus the "solution" of the utility problem is that the three houses cannot each be connected on a plane to the three utilities in the prescribed manner. In practice we build our utility lines in three-space, rather than a plane, so that the lines may "cross" without intersecting.

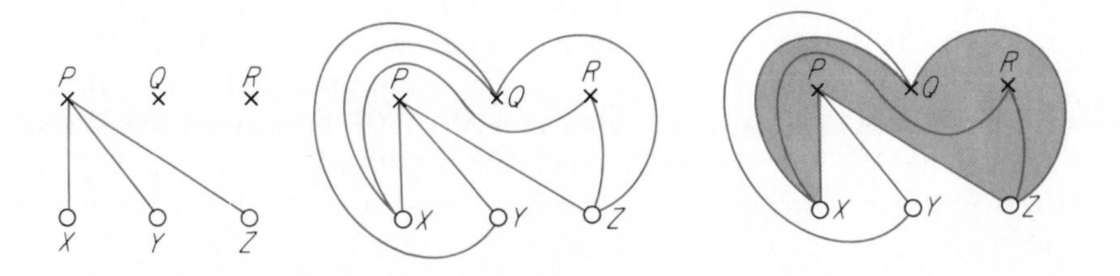

A **graph** is a set of points (**vertices**) and a subset of the line segments or arcs (**edges**) with the vertices as endpoints. Any network is a special type of graph and the study of networks is a special case of graph theory. If each edge of a graph may be traversed exactly once, the graph is traversable and the path is an **Euler path**. Thus, the highway inspector in Exercise 23 seeks an Euler path. An Euler path that returns to its starting point is an **Euler circuit**.

Example 1 State whether or not the given graph has (**a**) an Euler path (**b**) an Euler circuit.

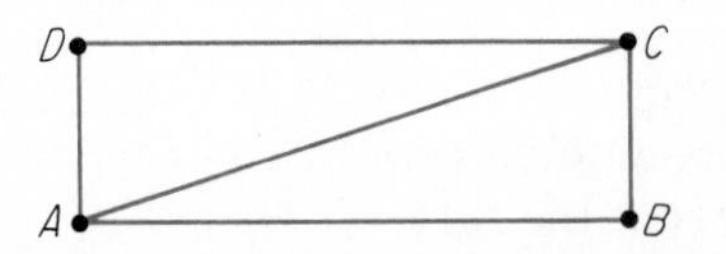

Solution (**a**) The graph has an Euler path such as *ABCDAC*.
(**b**) The graph does not have an Euler circuit since each Euler path must start at one of the vertices *A* or *C* and stop at the other. ■

If each vertex of a graph is passed through exactly once, the path is a **Hamiltonian path**. Thus the salesperson in Exercise 26 seeks a Hamiltonian path. A Hamiltonian path that returns to its starting point is a **Hamiltonian circuit**. For the figure for Example 1 the paths *ABCD* and *BACD* are among the Hamiltonian paths; the path *ABCDA* is among the Hamiltonian circuits.

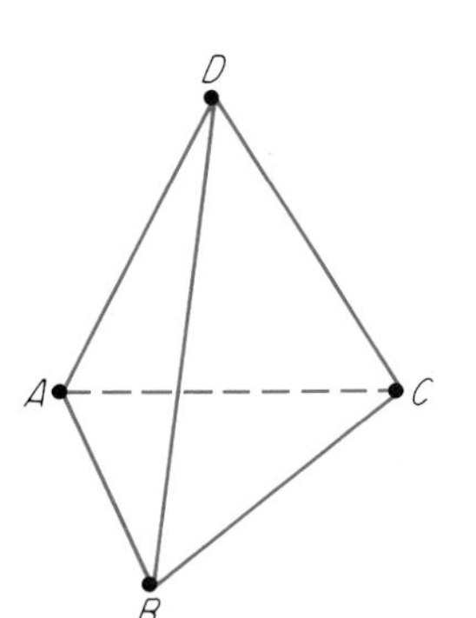

The adjacent figure is a picture of a triangular pyramid. Thus we know that the line segments *AC* and *BD* do not intersect. The edges and vertices of the pyramid form a graph. However, the graph is represented by a picture which is drawn on a plane. As in this picture, we assume that edges of a graph intersect only at their endpoints. We also recognize that a graph is not necessarily a plane figure.

Suppose that the points *A*, *B*, *C*, *D*, and *E* represent five baseball teams that have formed a league. Each edge represents a game that has been played by the two teams whose points are the endpoints of that edge. Before the season opens no games have been played. Then the graph consists only of a set of points with no edges. Such a graph is a **null graph**. If the season's schedule consists of each team playing each of the other teams exactly once, the season's graph is a **complete graph**, that is, a graph with each pair of points the endpoints of exactly one edge.

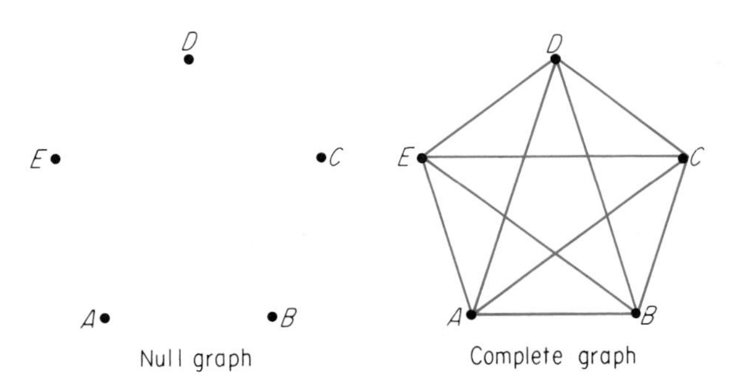

In the complete graph the edges *AD* and *BE* have no vertices of the graph in common and are thought of as passing each other, possibly in space, without intersecting. Whatever interpretation is used, the apparent intersections that are not vertices of the graph do not represent teams and are not considered to be points (endpoints of edges) of the graph.

In a graph of the games of the five baseball teams, each edge represents a game. If there is a winner for each game, then each edge can be directed from the point of the loser to the point of the winner. A graph with all of its edges directed is a **directed graph**.

A graph with some but not all of its edges directed is a **mixed graph**. A map of a city with some but not all of its streets designated as one-way streets is an example of a mixed graph.

A directed graph in which each directed line segment RS denotes that S is a child of R is a **genetic graph**. Not all parents or descendants need to be represented and not all directed graphs can be interpreted as genetic graphs since the descendants of an individual cannot include that individual and no person can have more than two parents.

Example 2 Interpret, if possible, as a genetic graph.

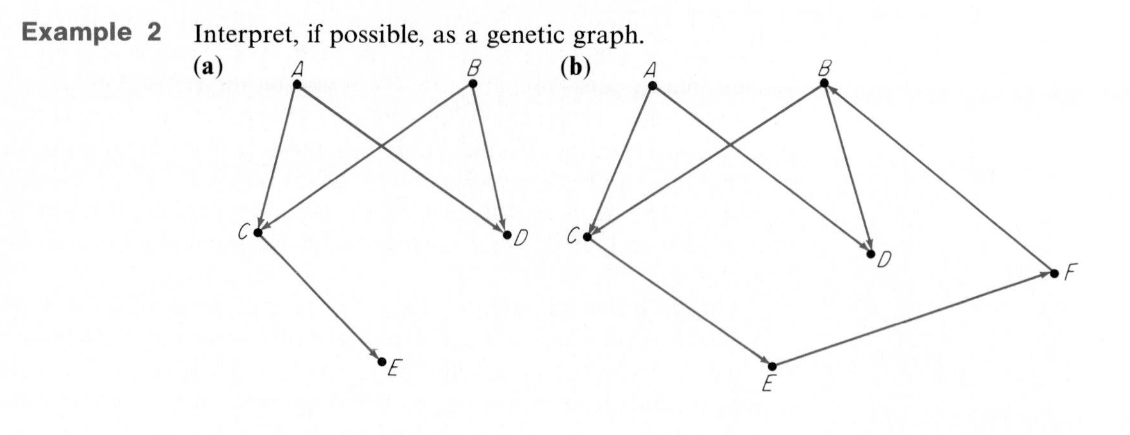

Solution **(a)** The graph can be interpreted as a genetic graph with C and D both children of A and B; E a child of C and a descendant of A and B.

(b) If we attempt to interpret the graph as a genetic graph, we find that C is a child of B, E is a child of C, F is a child of E, and B is a child of F. The directed edges would imply that B is a descendant of F, E, C, and B which, of course, is impossible. Thus the given graph cannot be interpreted as a genetic graph. ■

The study of simple closed curves, networks, and graphs is a part of **topology**, a very general type of geometry. Each of the geometric topics considered in this text, including the Möbius strips in the explorations, is a special case of topology.

EXERCISES *Identify* **(a)** *the number of even vertices* **(b)** *the number of odd vertices* **(c)** *whether or not the network is traversable and, if it is traversable, the vertices that are possible starting points.*

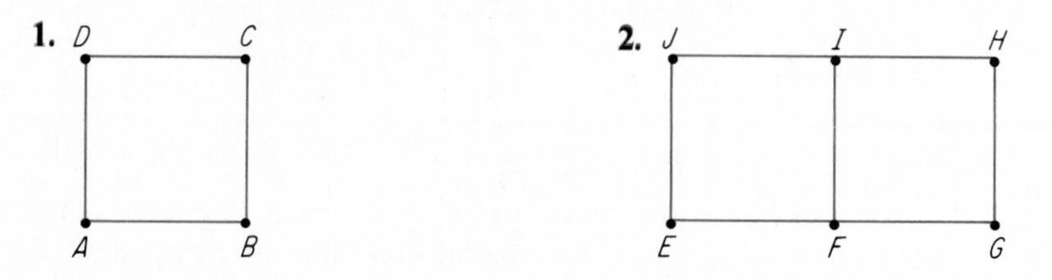

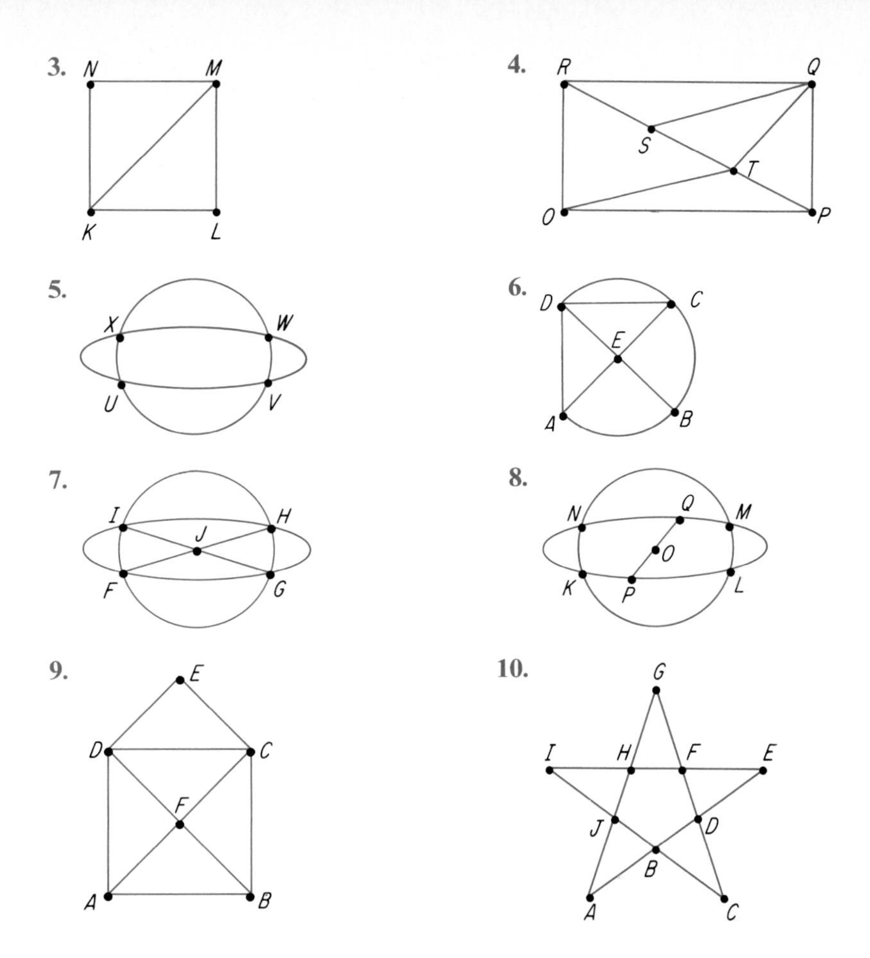

11. The network formed by the edges of a triangular pyramid.
12. The network formed by the edges of a cube.

In Exercises 13 *through* 16, *state whether or not the given graph has* **(a)** *an Euler path* **(b)** *an Euler circuit. If such a path or circuit exists, identify at least one.*

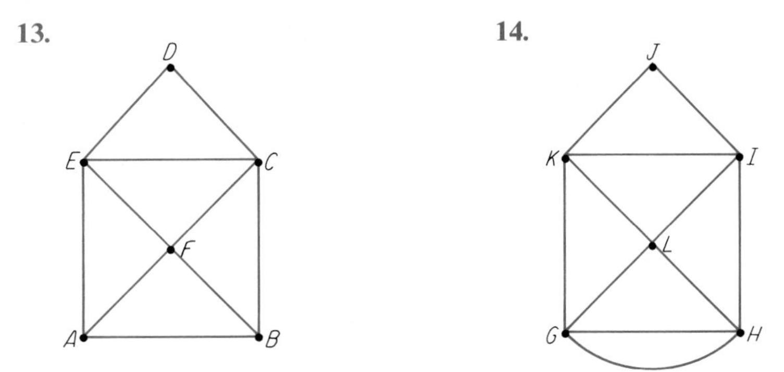

15.

16.

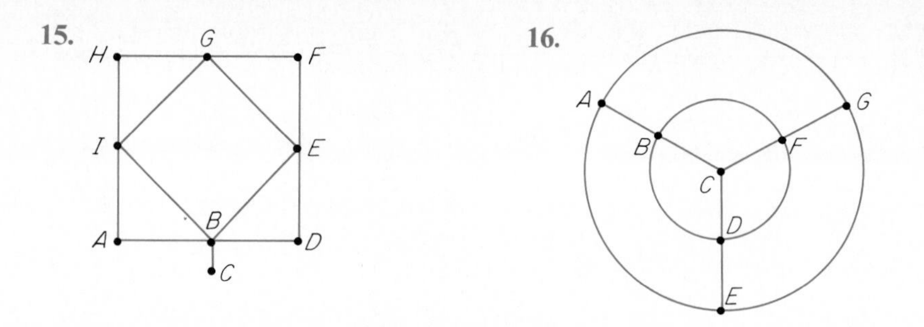

17. Identify the possible numbers of odd vertices for a network that has an Euler path.

18. Repeat Exercise 17 for an Euler circuit.

In Exercises 19 *through* 22 *state whether or not the graph in the indicated exercise has* **(a)** *a Hamiltonian path* **(b)** *a Hamiltonian circuit. If such a path or circuit exists, identify at least one.*

19. Exercise 13.

20. Exercise 14.

21. Exercise 15.

22. Exercise 16.

23. Explain how a county highway inspector can use a network to determine whether or not each section of a highway can be inspected without traversing any section twice.

24. Use a related network to obtain and explain your answer to the following question: Is it possible to take a trip through a house with the floor plan indicated in the figure and pass through each doorway exactly once?

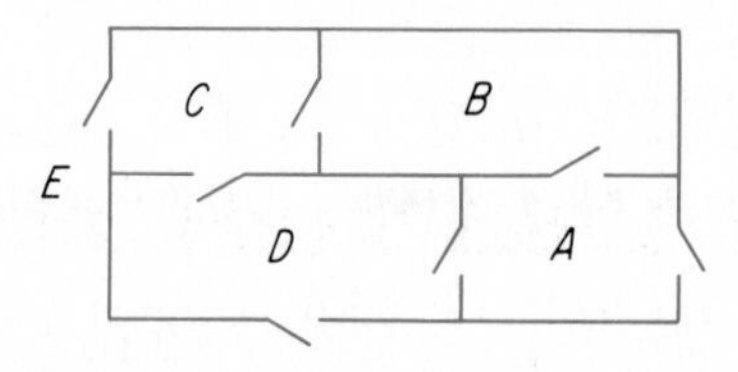

25. Explain whether or not it is possible to draw a simple connected broken line cutting each line segment of the given figure exactly once.

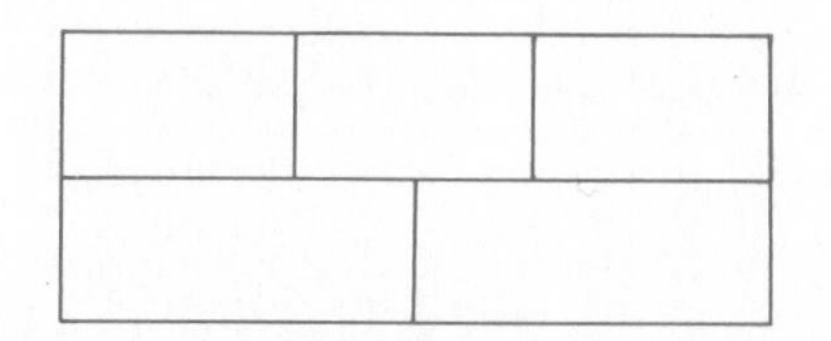

26. Consider the use of highway networks by a salesperson who wants to visit each town exactly once.
    (a) Does the salesperson always need to travel each highway at least once?
    (b) Consider each of the networks in Exercises 1 through 4 and indicate whether or not the salesperson can visit each town exactly once without retracing any highway.
    (c) On the basis of your answer for part (b), does traversability of the network appear to be of interest to the salesperson?

In Exercises 27 through 29 draw a genetic graph and label it such that A and B have the indicated relationship.

27. A and B are children of C and D.
28. A and B are first cousins.
29. A is the niece (or nephew) of B.

30. Describe two modifications of the seven bridges in Königsberg such that a traversable network is obtained in each case.
31. The caliph's problem. (According to tradition, part (b) of this problem was posed by a Persian caliph to select a suitor for his daughter.) Draw arcs that have as endpoints the points with the same numbers and so that the arcs do not otherwise intersect the given figure or each other.

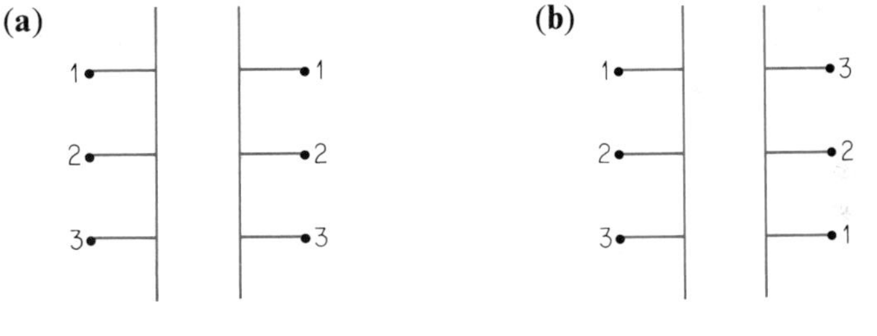

EXPLORATIONS

1. Design an efficient route (traverse each street at least once, keep total distance as short as possible, and return to starting point A) for a street-sprinkler truck and the indicated street plan.

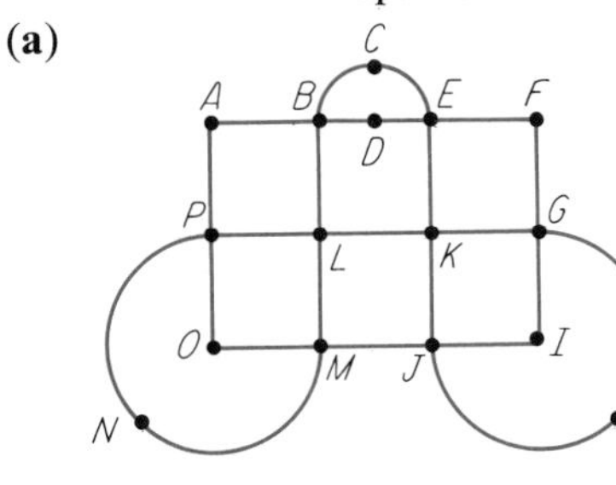

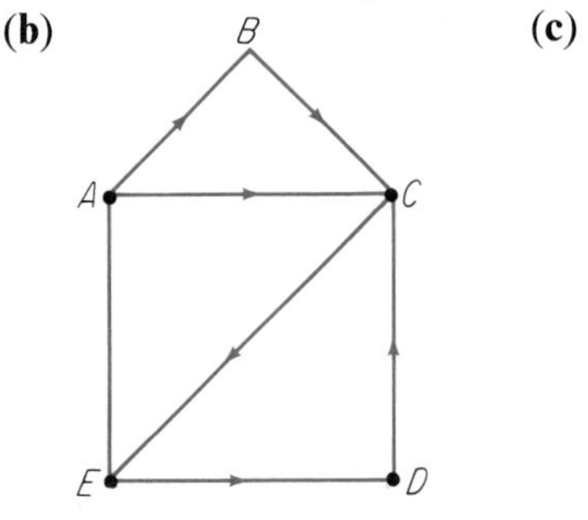

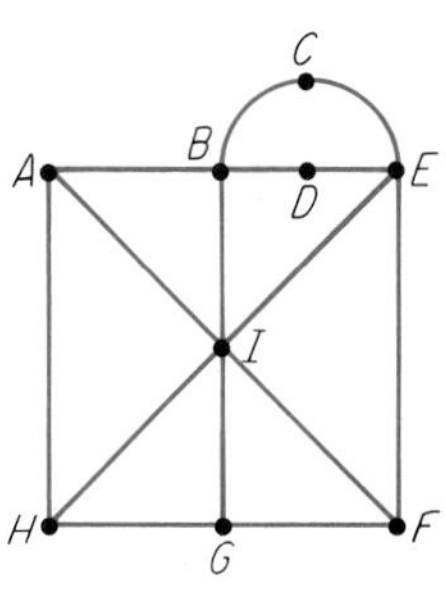

2. Numbers may be used on edges of graphs to indicate distances, travel times, costs, and so on. To find the most desirable (shortest, least expensive, and so on) path from a given initial vertex to a given terminal vertex, it is helpful to label each of the intermediate vertices with the least of its measures from the initial point. Find the measure of the most desirable path from A to B in the given graph.

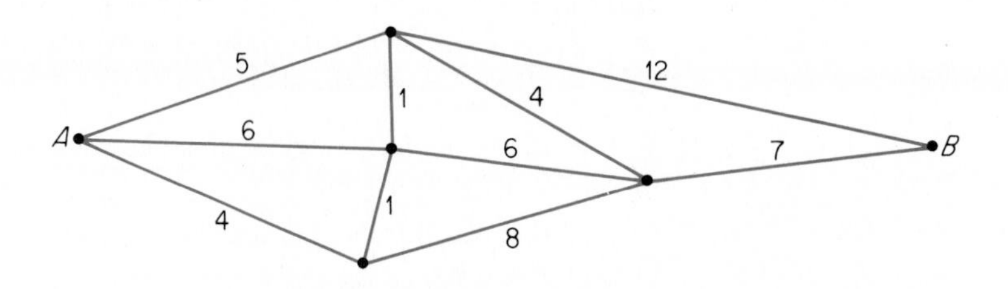

3. Repeat Exploration 2 for this graph.

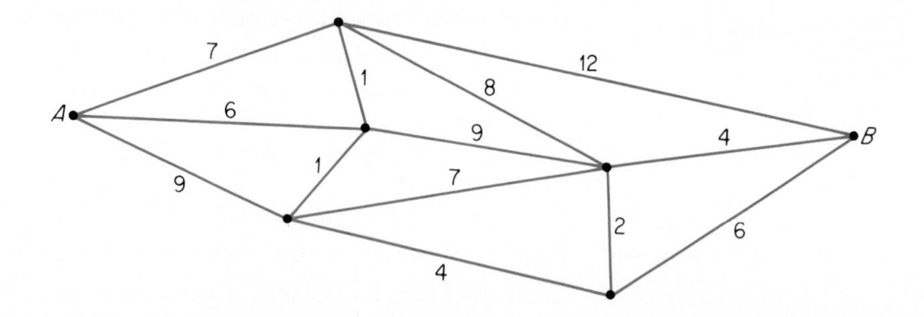

There are many figures that have very unusual properties. One of these figures is a surface that is one-sided. A fly can walk from any point on this surface to any other point without crossing an edge. Unlike a table top or a wall this surface does not have a top and a bottom or a front and a back. This surface is called a **Möbius strip**, and it may be constructed easily from a rectangular piece of paper such as a strip of gummed tape. Theoretically, size is unimportant, but a strip 3 to 5 centimeters wide and about 40 centimeters long is easy to handle. We may construct a Möbius strip by twisting the strip of gummed tape just enough (one half-twist) to stick the gummed side at one end to the gummed side at the other end.

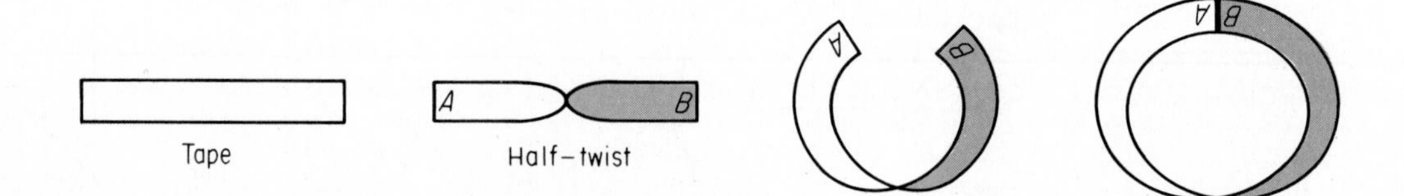

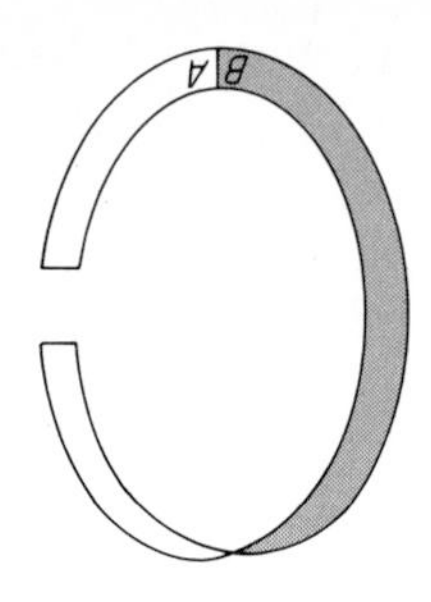

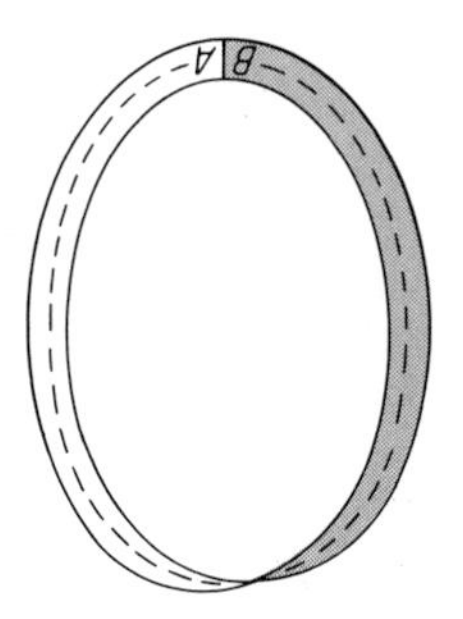

If we cut across a Möbius strip, we again get a single strip similar to the one that we started with. But if we cut around the center line of the Möbius strip (see the dashed line in the second figure in the margin), we do not get two strips. Instead, we get one strip with two half-twists in it.

On one occasion one-sided surfaces of this sort were used as place cards at a 7-year-old's birthday party. While waiting for dessert, the youngsters were encouraged to cut the strip down the middle while guessing what the result would be. They were suitably impressed when they found only one piece, and were anxious to cut it again. Once more they were impressed when they found two pieces linked together. Almost a year after the party, one of the children asked about "the piece of paper that was in only one piece after it was cut in two."

When confronted with unusual properties such as those of the Möbius strip, both children and adults may ask very difficult questions. This is good for all concerned, since it impresses upon them that there is more to mathematics than formal algebraic manipulations and classical geometric constructions.

4. Construct a Möbius strip and cut around the center line to obtain one strip with two half-twists in it.
5. Repeat Exploration 4 and then cut around the center line of the new surface.
6. Construct a Möbius strip and cut along a path that is about one-third of the distance from one edge to the other.
7. Construct a Möbius strip, mark a point A on it, and draw an arc from A around the strip until you return to the point A.
8. Explain why a Möbius strip is called a one-sided surface.
9. Does a Möbius strip have one or two simple closed curves as its edge? Explain the reason for your answer.
10. Be on the alert for applications of Möbius strips as you go to repair shops and other places where machines are operated by belts and pulleys. Describe to your class any applications that you observe and explain their usefulness.

Chapter Review

Sketch the indicated figure and two of its possible shadows, with their dimensions labeled.

1. A pyramid 6 centimeters high with a square base having an edge of 3 centimeters.
2. A right circular cylinder with height 10 centimeters and radius 2 centimeters.
3. A right circular cone with height 10 centimeters and radius 2 centimeters.

Identify as true (always true) or false (not always true).

4. **(a)** There is one and only one line that is perpendicular to a given plane and contains a given point.
   (b) There is one and only one plane that is perpendicular to a given plane and contains a given line.
5. **(a)** Any two lines intersect or are parallel.
   (b) Any two planes intersect or are parallel.
6. **(a)** Every shadow of a cube is a square region.
   (b) Every shadow of a sphere is a circular region.

Use paper folding to demonstrate the given statement.

7. Any given line segment has one and only one perpendicular bisector.
8. For any given triangle the three lines that contain a vertex and the midpoint of the opposite side are concurrent.

Use straightedge and compass for all constructions.

9. Draw an acute angle ABC and construct its angle bisector.
10. Draw a line AB and construct a line that is perpendicular to the line AB at the point B.

Sketch.

11. A convex polygon with five sides.
12. A trapezoid that is not a parallelogram.
13. A quadrilateral with diagonals PQ and CD.
14. A rectangular box (also label its vertices and list one set of parallel edges).
15. A figure that includes a line perpendicular to a plane (in the figure, identify such a line and plane).
16. A figure that includes two perpendicular planes (in the figure, identify two such planes).
17. A triangular pyramid with base ABC and vertex D. Then identify the lines that **(a)** intersect the line AB in a single point **(b)** form skew lines with the line AB.

In Exercises 18 through 21 identify **(a)** *the number of even vertices* **(b)** *the number of odd vertices* **(c)** *whether or not the network is traversable* **(d)** *if it is traversable, the vertices that are possible starting points.*

18. **19.**

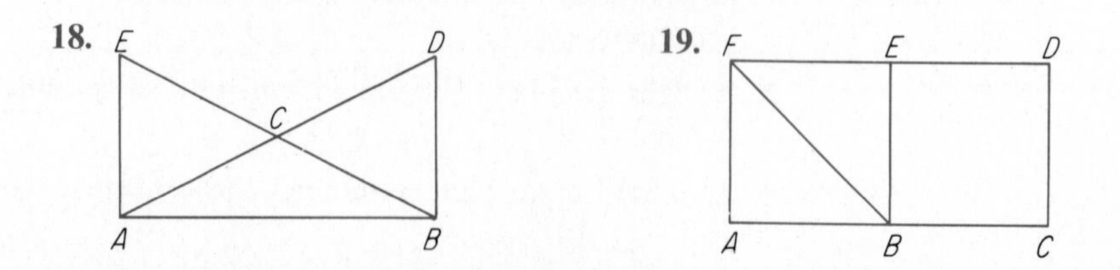

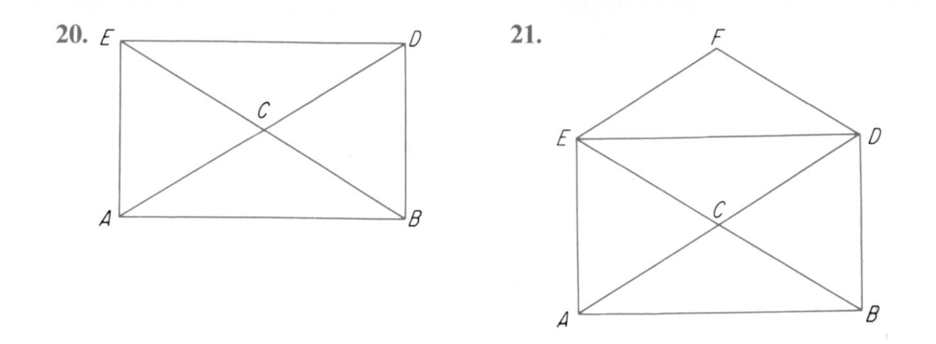

22. For each network that is not traversable in Exercises 18 through 21 copy the network and add one more arc to obtain a traversable network.

State whether the given graph has **(a)** *an Euler path* **(b)** *an Euler circuit* **(c)** *a Hamiltonian path* **(d)** *a Hamiltonian circuit.*

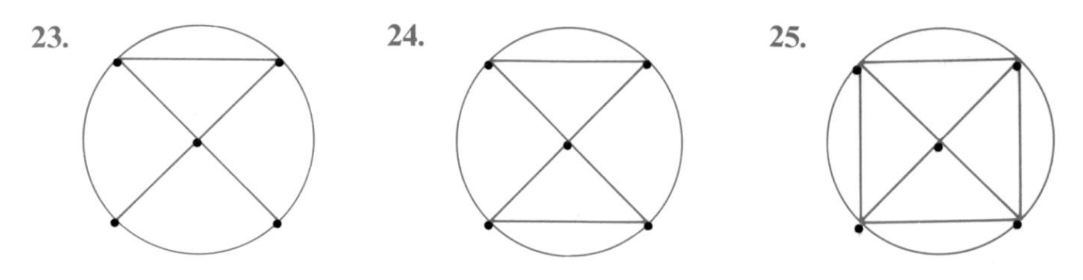

LOGO COMMANDS

The ability to make graphic displays with BASIC varies greatly from one microcomputer to another. In addition, the generation of the desired graphical display often calls for a strong command of the BASIC language and, usually, fairly intricate knowledge of the operating system for the individual computer involved.

These pitfalls can all be avoided through the use of the programming language Logo. Logo allows for graphics, and a great deal more, with a minimum of problems on the part of the computer user. This particular form of graphics is called **turtle graphics** because the device originally used in creating such graphics resembled a turtle. The first use of the Logo language for such graphics occurred in the MIT Artificial Intelligence Laboratories with children sending commands in Logo to turtlelike machines which drew pictures on the floor as they moved about in response to the children's directions to the computer.

The commands are amazingly simple. On the screen the turtle is represented by a shaded triangle. To move the turtle about on the screen, all one has to do is provide it with a direction and a distance. One can tell the turtle to move FORWARD (FD) or BACKWARD (BK), take a LEFTTURN (LT) or RIGHTTURN (RT), or do a number of other things.

Graphic displays are obtained by sequences of commands given the turtle. The first of the four figures shows the turtle in HOME position, that is, pointed upward. The remaining figures show the stages as the turtle traces a "pointed finger" in response to these commands:

FD 60 (forward 60 units)
RT 55 (right turn 55 degrees)
FD 20 (forward 20 units)

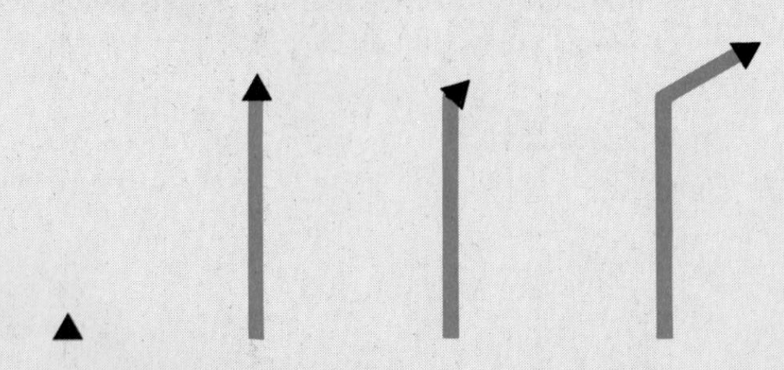

The turtle can be used to draw a variety of geometric figures, such as sketches of the standard polygons, angles, and three-dimensional figures. To draw a square with side 10, the turtle can be given the commands:

```
FD 10
RT 90
FD 10
RT 90
FD 10
RT 90
FD 10
RT 90
```

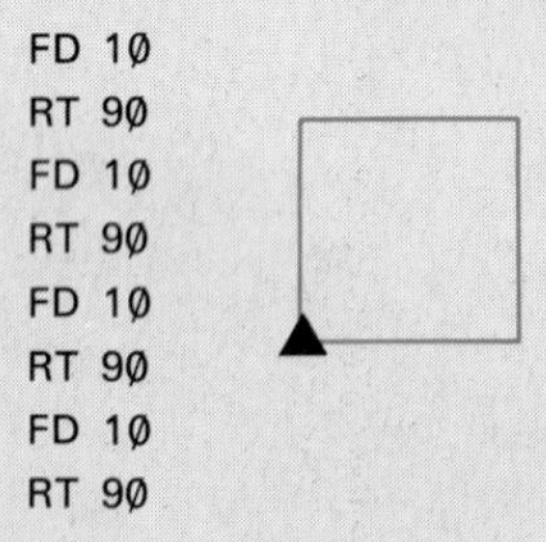

The last right turn brings the turtle back into the original position with the leading point directed to the top of the monitor screen.

List, and if possible try on a computer, the commands for the turtle to draw an equilateral triangle, a rectangle, a parallelogram, and some other geometric figures. Then use the turtle to draw some other figures representing objects in your environment. You might wish to use the additional commands PU (PEN UP) and PD (PEN DOWN) to move the turtle without leaving a trace on the screen and then to reinstate the use of the pen to draw once again.

9

Measurement and Geometry

Thomas Jefferson
(1743–1826)

In addition to serving as the third President of the United States, Thomas Jefferson was an active supporter of mathematics in the new nation. Jefferson not only wrote the Declaration of Independence, but also wrote and published articles on surveying, astronomy, measures, and the theory of the pedometer. Some of his work on measurement was a forerunner of the present metric system of measurement. Jefferson also worked hard to upgrade the teaching of mathematics in colonial universities and academies.

9-1 The Metric System

About 300 years ago Gabriel Mouton, Vicar of St. Paul in Lyons, France, recognized the need for a standardized system of measurement that would be accepted throughout the world. In 1670 he proposed a decimal system of measurement that was based on a standard unit equal to the length of one minute of arc of a great circle of the earth. In 1790, in the midst of the French Revolution, the National Assembly of France asked the French Academy of Sciences to create a uniform standard for all weights and measures.

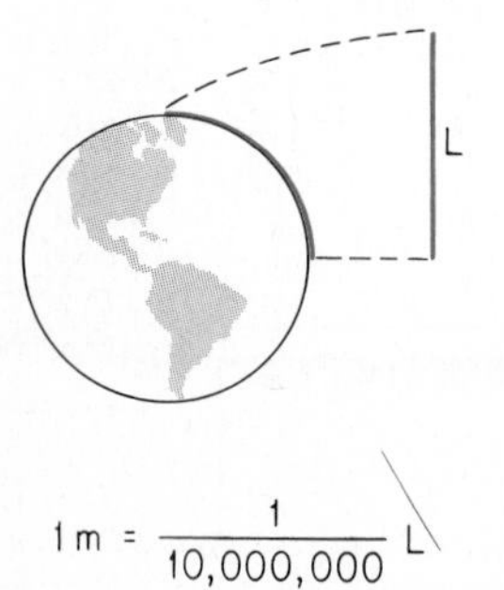

In establishing a uniform standard there are a number of fundamental principles that need to be kept in mind.

1. The standard unit used should be based on some invariant factor in the physical universe.
2. Basic units of length, capacity (volume), and weight (mass) should be interrelated.
3. The multiples and subdivisions of the basic unit should be in terms of the decimal system, that is, in terms of powers of ten.

The symbols m, cm, g, cc, and so forth, for metric units are used without periods.

With these principles in mind, the Academy proposed that the basic unit of length be one ten-millionth of an arc drawn from the North Pole to the equator. They called this basic unit the **metre (m)** as it is commonly known in most parts of the world and in scientific work. The word *metre* is derived from the Greek *metron*, "a measure," and is usually spelled as **meter** in the United States. One meter is a little longer than a yard. One-tenth of a meter is 1 **decimeter**; one-tenth of a decimeter is 1 **centimeter (cm)**. We can also think of these units of measure in terms of groups of ten in this way.

1 centimeter (cm)

1 meter = 10 decimeters

1 decimeter = 10 centimeters

We conclude that 1 meter = 100 centimeters, or that 1 centimeter is one-hundredth of a meter.

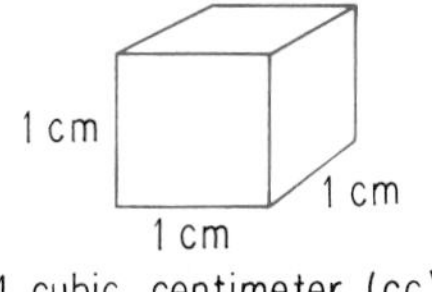

1 cubic centimeter (cc)

The name **kilogram (1000 grams)** was assigned to a metric unit of mass (weight). The weight of 1 **cubic centimeter** of water is about 1 **gram (g)**. A cubic centimeter is a cube with each edge of length 1 centimeter.

As a basic measure of capacity, the Academy chose the **liter** and defined this as a **cubic decimeter** that is, the volume of a cube with each edge of length 1 decimeter.

1 decimeter

On December 23, 1975, President Gerald Ford signed the Metric Conversion Act of 1975, which stated that there should be an increase in the use of the metric system in the United States through voluntary conversion. Such change has been extremely slow in coming into effect.

In 1840 France made it compulsory to use the metric system, and soon thereafter many other nations followed suit. In 1875 the Metric Convention met and 17 nations (including the United States) signed the "Treaty of the Meter," which established permanent metric standards for length and weight. By 1900 most of the nations of Europe and South America, a total of 35 in all, had officially adopted the metric system of measurement. There is now an international General Conference on Weights and Measures, to which the United States belongs.

About 1890 the International Bureau of Weights and Measures defined the meter as the distance between two marks on a platinum bar that was kept at a constant temperature of 32° Fahrenheit. This distance was approximately equal to one ten-millionth of the distance from the North Pole to the equator. However, in 1960 the Eleventh General Conference on Weights and Measures abandoned the meter bar as the international standard of length, and redefined the meter in terms of 1 650 763.73 wavelengths of the orange-red line in the spectrum of krypton 86. With this definition, lengths of one meter could be reproduced in scientific laboratories anywhere in the world. The 1960 Conference adopted a revision and simplification of the metric system, *Le Système International d' Unités* (International System of Units), now known as SI.

Commas are replaced by spaces in the metric system of notation for numbers.

The SI system of measure is really quite simple. First a basic unit, such as the meter, gram, or liter, is adopted. Thereafter, multiples and subdivisions of this unit are always given in terms of powers of ten. Greek prefixes are used to denote multiples, and Latin prefixes to denote subdivisions. Some of the basic prefixes used are shown with their symbols in the following table.

tera-	T	one trillion	1 000 000 000 000	10^{12}
giga-	G	one billion	1 000 000 000	10^{9}
mega-	M	one million	1 000 000	10^{6}
kilo-	k	one thousand	1 000	10^{3}
hecto-	h	one hundred	100	10^{2}
deka-	da	ten	10	10^{1}
BASE UNIT		one	1	10^{0}
deci-	d	one-tenth	0.1	10^{-1}
centi-	c	one-hundredth	0.01	10^{-2}
milli-	m	one-thousandth	0.001	10^{-3}
micro-	μ	one-millionth	0.000 001	10^{-6}
nano-	n	one-billionth	0.000 000 001	10^{-9}
pico-	p	one-trillionth	0.000 000 000 001	10^{-12}

There are several observations that should be made concerning the table. Note the increasing (or decreasing) powers of ten as one reads up (or down) the table. In the SI system, spaces are used instead of commas to separate groups of three digits starting from the decimal point in either or both directions. The reason for this is the use, in certain countries, of commas to represent decimal points.

We consider these prefixes first with the basic unit of 1 meter.

1 kilometer = 1000 meters	(1 km = 1000 m)
1 meter = 0.001 kilometer	(1 m = 0.001 km)
1 centimeter = 0.01 meter	(1 cm = 0.01 m)
1 meter = 100 centimeters	(1 m = 100 cm)

In the metric system any measurement in terms of one unit may be expressed in terms of a larger unit or a smaller unit very easily since the units correspond to the place values in our decimal notation

Tm	Gm	Mm	km	hm	dam	Meter	dm	cm	mm	μm	nm	pm
0	0	0	0	0	0	1.	0	0	0	0	0	0

Example 1 Change 358 centimeters to meters and change 9.26 meters to centimeters.

Solution 100 centimeters = 1 meter; 358 centimeters = 3.58 meters.
1 meter = 100 centimeters; 9.26 meters = 926 centimeters.
Alternatively, use the decimal place values and move the decimal point to the desired new unit.

	m	dm	cm	
	3	5	8	centimeters
is	3.	5	8	meters
	9.	2	6	meters
is	9	2	6.	centimeters ■

Example 2 Change 749 523 centimeters to kilometers.

Solution

	km	hm	dam	m	dm	cm	
	7	4	9	5	2	3	centimeters
is	7.	4	9	5	2	3	kilometers ■

The United States has initiated steps toward officially adopting the metric system during the decade of the 1980s. Almost all of the scientific laboratories in this country use the metric system and have done so for many years.

Today most of the civilized world uses the metric system of measurement. We make use of a number of metric units in our daily lives. For example, many people own or at least have heard of a 35-millimeter camera, and contents of vitamins are usually given in terms of multiples of a gram. Furthermore, some road signs give distances in both miles and kilometers. The general public needs to understand such measurements.

The metric system is based on powers of ten. Since our system of numeration is also based on powers of ten, computations using metric units are relatively simple. For practical everyday use, there are surprisingly few units of measure to learn. The basic unit for measuring lengths is the meter (m). A meter is approximately 39 inches in length. The following scale drawing shows the comparison between 1 meter and 1 yard.

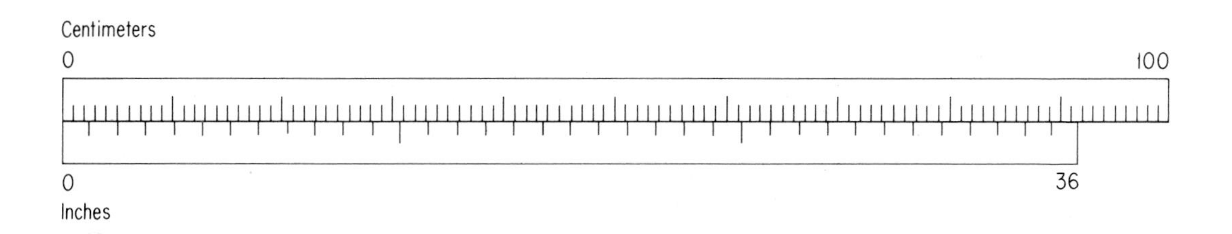

1 cm

Normally, the yard is divided into 36 parts, each 1 inch in length. The meter, as we have seen, is divided into 100 parts, each 1 centimeter in length. It is well to have an intuitive feeling for the size of 1 centimeter. The centimeter is about the width of the nail on your little finger. A nickel is about 2 centimeters in diameter (wide). An inch is now defined to be 2.54 centimeters.

The definition of one inch as 2.54 centimeters means that

1 inch = 2.54 cm

is an exact rather than an approximate statement of equality.

The figures that follow show the *actual size* of 1 centimeter, the *actual size* of 1 inch, and a comparison of the lengths of centimeters and inches on a part of a ruler that is marked in both centimeters and inches.

1 centimeter:

1 inch:

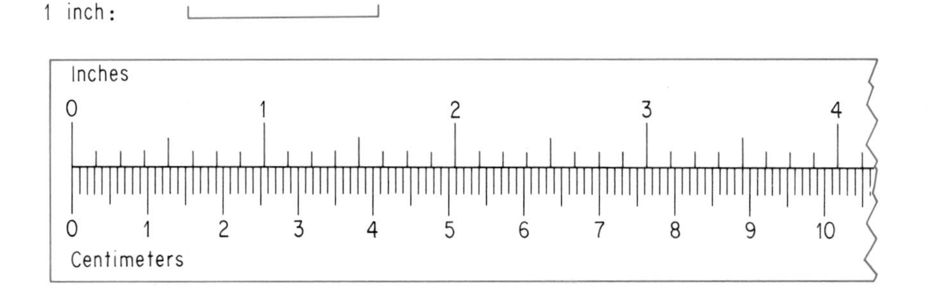

As in the previous figure each centimeter is divided into ten parts, each 1 millimeter (mm) in length.

1 centimeter = 10 millimeters (1 cm = 10 mm)

1 millimeter = 0.1 centimeter (1 mm = 0.1 cm)

To help you visualize the size of a millimeter note that a nickel is about 2 millimeters thick; 35 millimeter film is 35 millimeters in width.

For measuring greater distances we use the kilometer (km), which is equal to 1000 meters. A kilometer is approximately equal to 0.6 of a mile, so that 10 kilometers is equal to about 6 miles and a speed of 60 miles per hour is approximately 100 kilometers per hour.

For normal, everyday usage in a country that has adopted the metric system, the average citizen uses the centimeter for small measurements, the meter for somewhat larger ones, and the kilometer for greater distances. Thus, in countries that have adopted the metric system, distances between cities are given in terms of kilometers. However, the meter is the basic unit of length. Several common multiples of this unit are shown in the following table.

1 *kilo*meter	=	1000 meters
1 *hecto*meter	=	100 meters
1 *deka*meter	=	10 meters
1 meter	=	1 meter
1 *deci*meter	=	0.1 meter
1 *centi*meter	=	0.01 meter
1 *milli*meter	=	0.001 meter

Example 3 Normally, metric rulers are marked in centimeters and subdivided into millimeters. Read each indicated point on this metric ruler in millimeters and in centimeters.

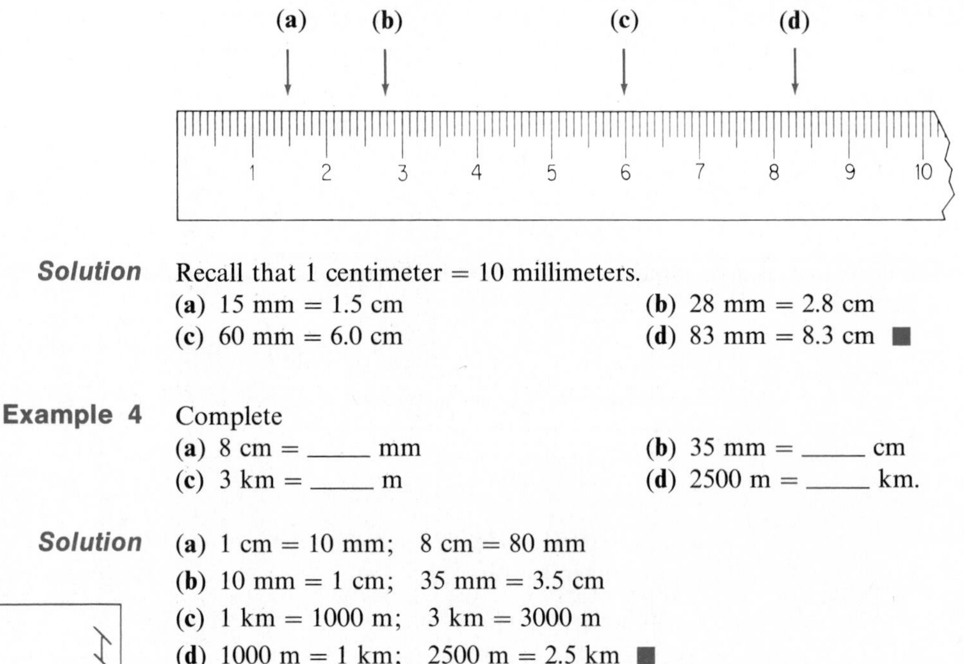

Solution Recall that 1 centimeter = 10 millimeters.

(a) 15 mm = 1.5 cm **(b)** 28 mm = 2.8 cm

(c) 60 mm = 6.0 cm **(d)** 83 mm = 8.3 cm ■

Example 4 Complete

(a) 8 cm = _____ mm **(b)** 35 mm = _____ cm

(c) 3 km = _____ m **(d)** 2500 m = _____ km.

Solution

(a) 1 cm = 10 mm; 8 cm = 80 mm

(b) 10 mm = 1 cm; 35 mm = 3.5 cm

(c) 1 km = 1000 m; 3 km = 3000 m

(d) 1000 m = 1 km; 2500 m = 2.5 km ■

Two football fields: approximately 1 hectare

The most common units of area measure are a square centimeter, a square meter, a square hectometer (called a **hectare**), and a square kilometer. Hectares are widely used. One hectare is a little less than 2.5 acres (one acre is 43,560 square feet, one square mile is 640 acres). Cubic units such as a cubic centimeter and a cubic meter are used for volumes. As noted earlier the basic unit for capacity is the *liter*, 1 cubic decimeter, which is a little more than a quart.

Various multiples of 1 liter are obtained by using the set of prefixes previously listed. Again, note the use of decimal notations, with each unit being ten times as great as the one listed beneath it in this table.

1 *kilo*liter	=	1000 liters
1 *hecto*liter	=	100 liters
1 *deka*liter	=	10 liters
1 liter	=	1 liter
1 *deci*liter	=	0.1 liter
1 *centi*liter	=	0.01 liter
1 *milli*liter	=	0.001 liter

One might encounter the **milliliter (ml)** in small measurements. For example, in cooking and baking, it would be important to know that 5 milliliters is approximately equivalent to 1 teaspoonful.

The *gram* is a unit of metric measure for *mass*. The *weight* of an object is determined by its mass and the force of gravity to which it is subjected. For example, an astronaut's body has the same mass on earth as when he or she is in a weightless condition orbiting the earth. However, most of us stay relatively close to the surface of the earth. Therefore, we are concerned with objects that are subjected to an approximately constant force of gravity. Thus we shall use the term *weight* instead of using the technical term and concept of *mass*. Then the *gram* is also a unit for weight; a nickel weighs about 5 grams. As before, other multiples of a gram are obtained as in the following table.

1 *kilo*gram	=	1000 grams
1 *hecto*gram	=	100 grams
1 *deka*gram	=	10 grams
1 gram	=	1 gram
1 *deci*gram	=	0.1 gram
1 *centi*gram	=	0.01 gram
1 *milli*gram	=	0.001 gram

In practice the kilogram is the basic unit of weight.

Of these various multiples of the gram, the one that is most likely to be encountered when the United States finally "goes metric" is the **kilogram (kg)**. A kilogram is a little over 2 pounds, about 2.2 pounds, and is generally referred to as a **kilo**. Thus a shopper ordering 1 kilo of apples can expect to receive a little more than 2 pounds of apples; by ordering one-half of a kilo, one would receive about 1 pound.

The **milligram (mg)** is a frequently used unit of measure for various medicines and vitamins. A typical multipurpose vitamin pill might contain 250 milligrams of vitamin C. Inasmuch as 1 milligram is equal to one-thousandth of a gram, 250 milligrams is equal to one-fourth of a gram in weight.

Example 5 Complete

(a) 3 kg = ____ g

(b) 2500 g = ____ kg

(c) 5 g = ____ mg

(d) 7000 mg = ____ g

Solution

(a) 1 kg = 1000 g; 3 kg = 3000 g

(b) 1000 g = 1 kg; 2500 g = 2.5 kg

(c) 1 g = 1000 mg; 5 g = 5000 mg

(d) 1000 mg = 1 g; 7000 mg = 7 g ■

EXERCISES

Classify each statement as true or false.

1. 100 meters = 1 kilometer
2. 1 meter = 100 centimeters
3. 1 centimeter = 100 millimeters
4. 10 millimeters = 1 centimeter
5. 1 decimeter = 10 centimeters
6. 1 meter = 1000 millimeters
7. 1 millimeter = 0.01 centimeter
8. 10 meters = 1 decimeter
9. 1 centimeter = 0.01 meter
10. 1 centimeter = 0.01 decimeter

Each of the following sentences uses a metric measure. State whether each seems likely, or unlikely.

11. A college football player weighs 200 kilograms.
12. A college basketball player is 2 meters tall.
13. An empty tank in a compact automobile will hold 40 liters of gasoline.
14. A quarter weighs 25 grams.
15. The average student drinks 5 liters of water daily.
16. The diameter of a nickel is about 5 centimeters.
17. A new pencil is about 20 centimeters long.
18. The distance between New York City and Boston is about 300 kilometers.
19. A typical man's wristwatch has a diameter of about 25 millimeters.
20. A student ate a steak that weighed 2 kilograms.

Select the most likely answer for each situation.

21. The weight of a quarter is approximately equal to
    (a) 1 gram (b) 5 grams
    (c) 15 grams (d) 30 grams
22. The weight of this book is approximately equal to
    (a) 10 grams (b) 50 grams
    (c) 1 kilogram (d) 10 kilograms
23. Ten gallons of gasoline is approximately equal to
    (a) 20 liters (b) 30 liters
    (c) 40 liters (d) 80 liters
24. The distance between New York City and Washington, D.C. is approximately equal to
    (a) 50 kilometers (b) 125 kilometers
    (c) 250 kilometers (d) 400 kilometers

25. Some doctors recommend that a person should spend an hour per day walking. In this time, one would probably walk
(**a**) 1 kilometer
(**b**) 5 kilometers
(**c**) 10 kilometers
(**d**) 15 kilometers

Change each of the following as indicated.

26. 358 centimeters to meters
27. 482 millimeters to centimeters
28. 3785 meters to kilometers
29. 7500 millimeters to meters
30. 85 kilometers to meters
31. 9.82 meters to centimeters
32. 5.8 centimeters to millimeters
33. 3.5 meters to millimeters
34. 3000 grams to kilograms
35. 7 kilograms to grams
36. 7500 milligrams to grams
37. 2500 milliliters to liters

For Exercises 38 *through* 42 *read each indicated point on this metric scale* (a) *in millimeters* (b) *in centimeters.*

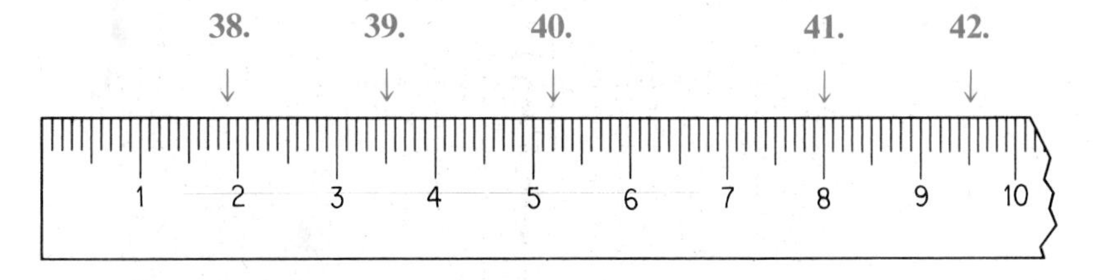

*43. The prefix *mega* represents 10^6, that is, 1 000 000. Thus 1 megameter = 1 000 000 meters. Complete each statement.
(**a**) 1 megaliter = _____ liters
(**b**) 1 megagram = _____ grams
(**c**) 1 megameter = _____ kilometers
(**d**) 5000 kilometers = _____ megameters

*44. The prefix *micro* represents 10^{-6}, that is, 0.000 001. Complete each statement.
(**a**) 1 micrometer = _____ meter
(**b**) 1 meter = _____ micrometers
(**c**) 5 grams = _____ micrograms
(**d**) 3 000 000 micrograms = _____ grams

EXPLORATIONS

1. Since it takes 100 pennies to equal one dollar, we may write such equivalents as 249¢ = \$2.49 and \$3.75 = 375¢. Discuss how this idea may be used to help illustrate translations from a number of centimeters to a number of meters and also from a number of meters to a number of centimeters.
2. Obtain a metric ruler and find the measure of the width of your palm, your hand span, and the distance from the tip of your elbow to the end of your outstretched middle finger, all to the nearest centimeter. Compare these measurements with those of other members of your class.

3. Prepare a collection of everyday objects that show measurements in metric units.
4. Prepare a multiple choice test of ten items that endeavor to evaluate familiarity with metric measures. Use Exercises 11 through 15 of this section as a guide.
5. Prepare a bulletin board display that serves to generate interest in the metric system.
6. Use a metric ruler and find the dimensions of a one-dollar bill to the nearest millimeter.
7. Use a metric ruler and find the diameter of a quarter to the nearest millimeter.
8. Mark off, as carefully as possible, a 10-centimeter scale on a strip of paper about 2 centimeters wide. Then make nine more such scales and fasten them together to form a strip one meter long and subdivided into centimeters.
9. Bring to class three objects that are approximately 10 centimeters each in length.
10. Construct an open box with each edge 10 centimeters long. Then the volume of the box will be one cubic decimeter, which is, by definition, one liter.
11. Explore a recently published set of elementary school mathematics textbooks, or a series of workbooks devoted to metric measures, and report on the manner in which metric measures are introduced (**a**) at the primary grade level (**b**) at the intermediate grade level. Discuss the use of, or absence of, decimal notation at each of these levels.

9-2 Concept of Measurement

Most people expect mathematics to serve two basic needs: to compute and to measure. The typical consumer often needs to measure to find the dimensions of a room in order to buy furniture, the size of a lawn in order to buy seed, the area of a wall in order to buy wallpaper, and so on. We also need to understand measurement when we note the weight or capacity of a container or read road signs concerning distances. In fact, throughout the ages people have had a need to measure. As the Nile River overflowed its banks each year, the ancient Egyptians needed to survey the flooded area to determine boundary marks, and they are believed to have used ropes stretched to form right angles in doing so. Because of the process used, these surveyors were actually referred to as "rope-stretchers." From the Pythagorean theorem we know that if a rope is knotted to form 3, 4, and 5 units, respectively, a right triangle and therefore a right angle can be formed, as shown in the figure.

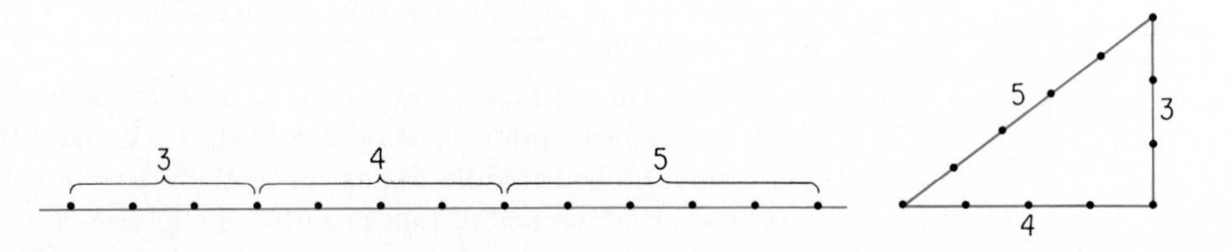

The *hand* used in measuring heights of horses is equal to 4 inches.

Various parts of the human body were used as early units of measure. Records of early Egyptian and Babylonian civilizations indicate that lengths were first measured by such units as a **palm** and a **span**, the distance covered by an outstretched hand.

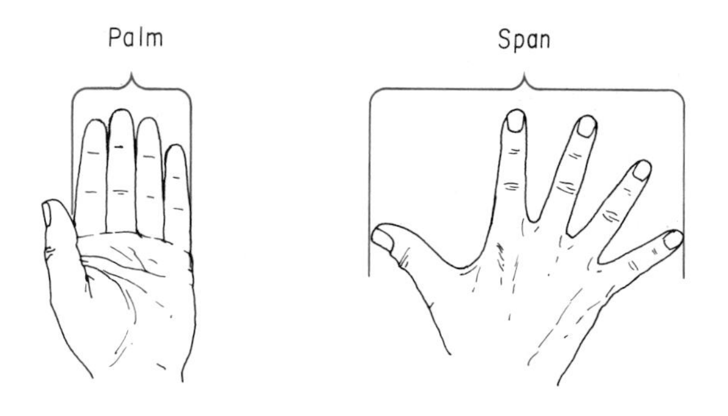

Mention of the cubit has been found in excavations that date back to 3000 B.C., and it is a unit of measure frequently mentioned in the Bible. Thus in the sixth chapter of Genesis we find a description of Noah's ark that states "the length of the ark shall be 300 cubits, the breadth of it 50 cubits, and the height of it 30 cubits."

Another widely used early unit of measurement was a **cubit**, the distance from one's elbow to the tip of the middle finger when the hand is held straight. For most people, this is a distance of approximately 18 inches.

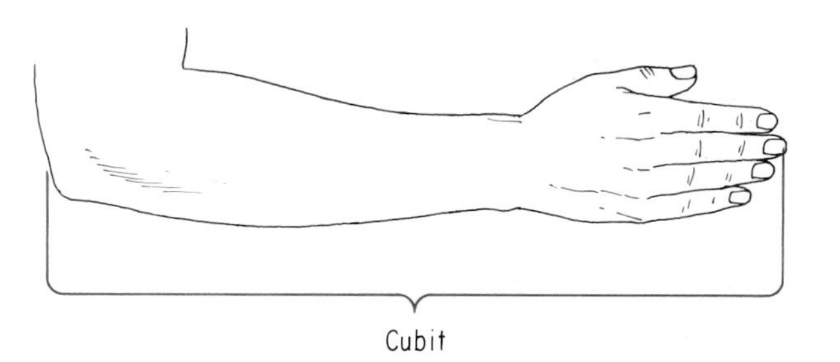

There are many interesting stories about the origins of units of measure. It is said that King Henry I of England decreed that the distance from the tip of his nose to the end of his thumb should be the official equivalent of one yard. Then in the thirteenth century, King Edward I declared that one-third of a yard should be called the foot. Later, in the sixteenth century, Queen Elizabeth I declared that the traditional *Roman mile* of 5000 feet would be replaced by one of 5280 feet. However it was the Romans who used 12 as a base for measurement, and our use of 12 inches to a foot is derived from them.

Although the use of a unit of measure such as the cubit was more meaningful than such units as "a bundle" or "a day's journey," it certainly was not a standardized unit and obviously varied from person to person. There was a great need throughout the world for the establishment of a uniform standard for all weights and measures. The metric system meets most of this need but a variety of units remain in a few countries, especially the United States.

The basic concepts of measurement are the same for all types of measurements. Consider the measurement of the line segment PQ in the next figure. A *unit of linear measure* is needed. As usual any convenient unit of measure can be used. The selected unit is repeated as indicated on the scale (*ruler*).

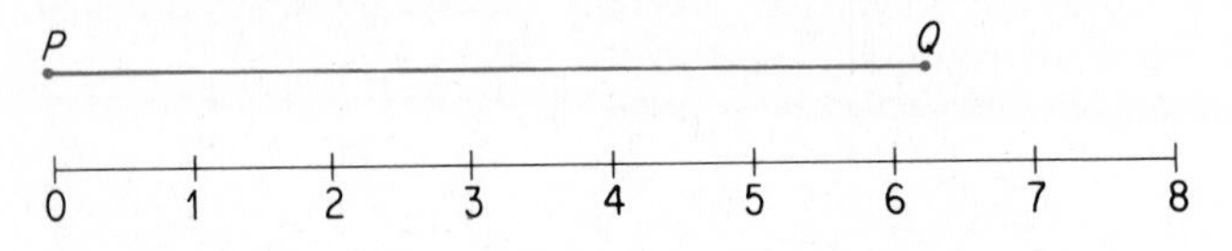

In geometry, particularly in applied situations, we often are concerned with the size of a figure. Thus we frequently measure distances, areas, volumes, and amounts of rotation.

To the nearest unit, $PQ = 6$ and we write $PQ \approx 6$. (Recall that as before, the symbol $\approx$ is read "is approximately equal to.") Some people mentally subdivide the unit of length on a scale and *estimate* measures to smaller units. For example, different people might give the following estimates.

$PQ \approx 6$, that is, 6 units.
$PQ \approx 6\frac{1}{4}$, that is, 25 quarters of the unit.
$PQ \approx 6.2$, that is, 62 tenths of the unit.

The length of the unit measure may be an irrational number of some standard unit of length. But the act of visually assigning a measure involves a rational number of the unit that is actually used.

Whatever subdivisions are used, the measure is based on someone's "reading the scale" and thus is approximate. All linear measurements are approximate since they are based on estimation from reading a scale. Indeed *all measurements from reading a scale are approximate*.

The *length* of a line segment, such as $\sqrt{2}$ for a diagonal of a unit square, may be an irrational number but any *measurement* of that line segment must be a rational number of the units used in making the measurement. Even though a measurement is made with a very precise instrument, there is always a final estimation and thus an approximation. The smallest unit that is actually used is the **unit of precision** and is used to indicate the **precision** of the measurement. Suppose that the selected unit that is marked on the scale is 1 centimeter. Then in a measurement to the nearest unit, the unit of precision would be one centimeter; in a measurement to the nearest tenth of a centimeter, 1 millimeter would be the unit of precision.

Example 1 Identify the precision of the given measurement.

(a) 5.2 kilometers **(b)** $3\frac{1}{8}$ units

(c) 5000 to the nearest hundred people

Solution **(a)** One-tenth kilometer **(b)** One-eighth unit

(c) 100 people ■

Note that the precision in the first two parts of Example 1 was indicated by the numerals but in the third part a verbal statement was needed. This need for clarification can be avoided by using scientific notation (Section 7-3). For example, the number of people in a crowd reported as 5000 people at a school game could be

5×10^3, that is, 5 thousand to the nearest thousand,
5.0×10^3, that is, 50 hundred to the nearest hundred,
5.00×10^3, that is, 500 tens to the nearest ten, or
5.000×10^3, that is, 5000 by actual count.

When using fractions or decimals to represent measurements, equivalent fractions have different interpretations and zeros are not added to decimals unless they are meaningful relative to the unit of precision used.

$5\frac{1}{4}$	has unit of precision	$\frac{1}{4}$ unit.
$5\frac{2}{8}$	has unit of precision	$\frac{1}{8}$ unit.
5.25	has unit of precision	0.01 unit.
5.250	has unit of precision	0.001 unit.

If the length of a segment is 6 centimeters to the nearest centimeter, then the *greatest possible error* is 0.5 centimeter (or 5 millimeters). Thus each of the segments shown below has a measure of 6 centimeters to the nearest centimeter. That is, each segment has a length that is closer to 6 than it is to 5 or 7 centimeters; each has a length between 55 and 65 millimeters.

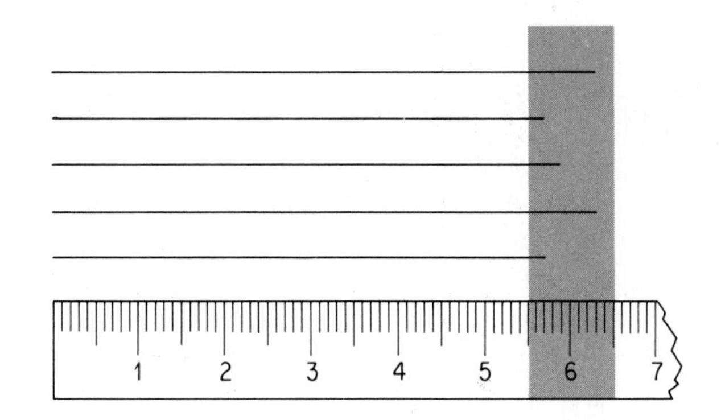

If a length is measured as 45 meters to the nearest meter, then the greatest possible error is 0.5 meter (or 5 decimeters). That is, the true length is somewhere between 44.5 and 45.5 meters. For any measurement, the **greatest possible error** is one-half of the smallest unit used, that is, one-half of the unit of precision. In general, for any measurement a unit of measure is needed, the measurement is approximate, and the positive difference between the actual measure and the stated measure is at most one-half of the unit of precision.

A *count* of individual items is a special type of measurement. Whenever a count is actually made, the measurement is *exact* and the greatest possible error is zero. For example, the number of registered students who are in your class and eventually obtain credit for the course is an exact number. The number of registered students who have studied from one of the editions of this book and obtained college credit for the course before January 1, 1987 is an exact number, although we cannot readily count the students to determine that number. It should be possible to identify any given measurement as either an **exact measurement** based upon counting or an **approximate measurement** with a positive greatest common error and therefore a positive unit of precision.

Example 2 Identify the given measurement as exact or approximate.

(a) June lives *2 kilometers* from school.

(b) Pedro's next class is *3 doors* down the corridor from this class.

(c) Susan bought *two new tires.*

(d) Bob bought *1 kilogram* of grapes.

Solution **(a)** Approximate. **(b)** Exact. **(c)** Exact. **(d)** Approximate. ■

In applications of mathematics the distinction between exact and approximate measurements can require careful consideration. For example, the measurement involved in buying *two one-quart packages* of ice cream is exact but the measurement involved in buying *two quarts* of ice cream that is packaged into cartons is an approximate measurement since the measurement of each quart is approximate.

Any **measurement** is a number of units and may be expressed as an *integral number* of its units of precision. The number is considered to be the **measure**, as distinct from the *measurement*, which is a number of units. A number associated with an approximate measurement is often called an **approximate number**. Computations with approximate numbers require special rules and are often called **approximate computations**.

Suppose that two friends see an item that they would like to purchase together. If the item costs $12.98, one has $7 to the nearest dollar and the other has $6.12, do they have enough together to make the purchase? One friend has $6.12 and the other has between $6.50 and $7.50. Together they have between $12.62 and $13.62. There is not sufficient information to answer the question. Without further information the sum is taken as $13 to the nearest dollar since one of the addends is to the nearest dollar. This is only an approximation for the sum, but we can expect only approximate answers when working with approximate numbers. In general we cannot expect the precision of a measure to be increased by addition or subtraction. In approximate computation we assume that

> A sum or difference of approximate numbers is no more *precise* than the *least precise* given number.

For multiplication and division involving approximate numbers another rule and the concept of *accuracy* are needed. Consider the measurements

5.061 m   0.007 m   3.0 cm

with the same precision (millimeters)

5061 mm   7 mm   30 mm

and the same greatest possible error (0.5 mm). The greatest possible errors are divided by the measures to obtain the respective greatest possible **relative errors**

$$\frac{0.5}{5061} = \frac{1}{10{,}122} \qquad \frac{0.5}{7} = \frac{1}{14} \qquad \frac{0.5}{30} = \frac{1}{60}$$

Of these, the greatest possible relative error for the measurement 5.061 m is the smallest, and that measurement is considered to be the most *accurate*. In general the number of decimal digits needed to express the measurement as an integral number of its unit of precision, that is, the number of **significant digits**, is taken as a measure of the **accuracy** of the measurement. Thus

5.061 m has four-digit accuracy.
0.007 m has one-digit accuracy.
3.0 cm has two-digit accuracy.

The digits used to express an approximate number in scientific notation are its *significant digits*. All nonzero digits are significant. Any zero between two nonzero digits, as in 5061, is significant. Any zero on the right of a significant digit and after the decimal point, as in 3.0, is significant. Zeros used only to locate the decimal point, as in 0.007, are not significant.

Exact numbers may be assumed to have infinite precision and infinitely many significant digits. Thus exact numbers do not affect computations with approximate numbers. For example, if the side s of a square has three-digit accuracy, then the perimeter of the square $4s$ also has three-digit accuracy. In general we assume that

A product or quotient is no more *accurate* than the least accurate of the given approximate numbers.

Example 3 Assume that each number is approximate, perform the indicated operations, and state the precision and accuracy of the answer.

(a) $\begin{array}{r} 12.35 \\ -\ \underline{7.25} \end{array}$ **(b)** $\begin{array}{r} 5.3 \\ +\ \underline{0.03678} \end{array}$

(c) $\begin{array}{r} 3.1 \\ \times\ \underline{0.2} \end{array}$ **(d)** $\begin{array}{r} 456 \\ \times\ \underline{0.20} \end{array}$

Solution **(a)** 5.10, nearest hundredth, three-digit accuracy
(b) 5.3, nearest tenth, two-digit accuracy
(c) 0.6, that is, 6×10^{-1}, nearest tenth, one-digit accuracy
(d) 91, that is, 9.1×10^{1}, nearest unit, two-digit accuracy ■

1
Linear unit

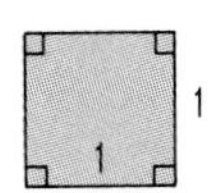

Unit square

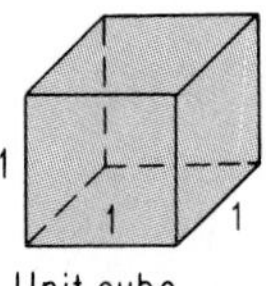

Unit cube

The unit of area measure is a *unit square*, that is, a square with side one linear unit. The unit of volume measure is a *unit cube*, that is, a cube with edge one linear unit. The common units of angular measure are defined so that

$$\begin{aligned} 1 \text{ revolution} &= 360 \text{ degrees} = 4 \text{ right angles} \\ &= 2 \text{ straight angles} = 2\pi \text{ radians} \end{aligned}$$

where a **radian** is the *SI* unit of angle measure.

An angle has been defined as the union of two rays with a common endpoint. To measure an angle it is customary to select one ray as the **initial side**, the other ray as the **terminal side**, and a direction of rotation (clockwise, −, or counterclockwise, +) about the vertex of the angle. The measure of the angle is the amount of rotation needed to rotate a ray from the position of the initial side to the position of the terminal side. When an angle is represented but an initial side is not indicated, the measurement is usually taken as a number n of degrees, $0 \le n \le 180$. Thus, in the first figure, $\angle ABC = 60°$ unless curved arrows are added to indicate the initial side and a direction of rotation that requires a different angle measurement.

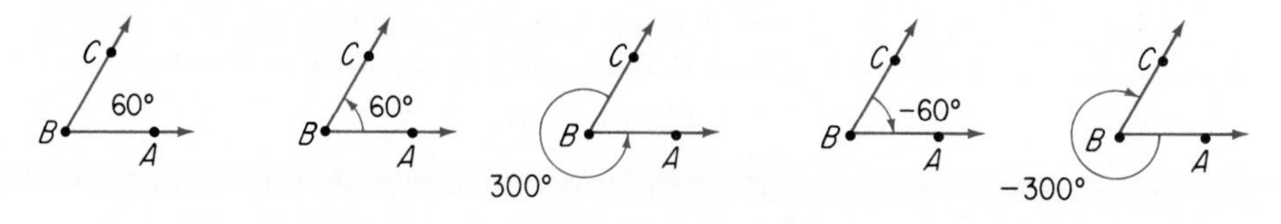

For all measurements (60°, 20 centimeters, and so on) the numbers (60, 20, and so on) are the *measures* and the numbers of units are the *measurements*, as noted earlier. However, the distinction—like that for number and numeral—is emphasized only when it is necessary for effective communication. Similarly, the symbols "$\angle ABC$" and "$\angle B$" are used for an angle (a set of points), a measure of the angle (a number), or a measurement of the angle (a number of units) according to the context.

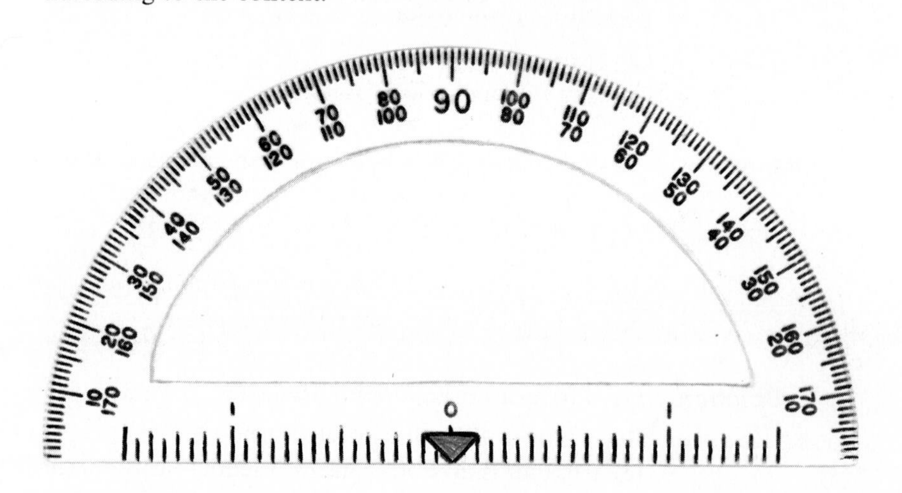

Protractors are used to measure angles just as rulers are used to measure line segments. One common type of **protractor** is shown in the figure. Several types of angles are defined in terms of their measures in degrees. An angle ABC is

an **acute angle** if $0° < \angle ABC < 90°$

a **right angle** if $\angle ABC = 90°$

an **obtuse angle** if $90° < \angle ABC < 180°$

a **straight angle** if $\angle ABC = 180°$

Some textbooks postpone the use of the term straight angle and refer to two rays such as BA and BC, that have the same endpoint, as *opposite rays* when their union is a straight line.

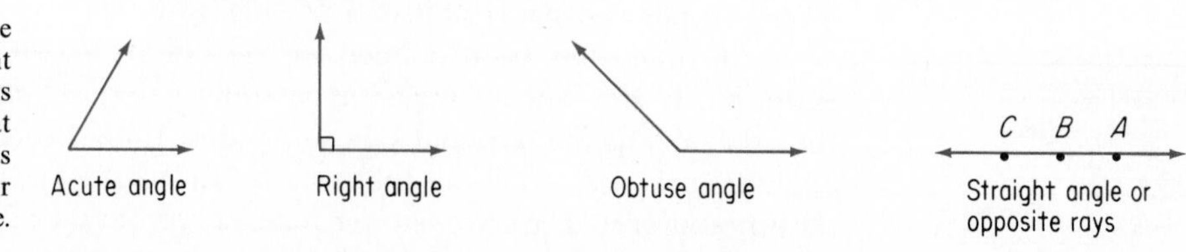

Measurements enable us to extend our paper-folding definitions of the congruence of line segments and the congruence of angles. Any two line segments with equal measurements are **congruent line segments**; any two angles with equal measurements are **congruent angles**.

When working with metric (SI) units, and in most advanced work in mathematics, the common unit of angular measure is a radian. As in the adjacent figure, an angle with its vertex at the center of a circle (a **central angle**) and with the length of the intercepted arc of the circle equal to the length of the radius of the circle has a measurement of 1 **radian**.

We can convert from degrees to radians and from radians to degrees by recalling a formula from elementary geometry for the *circumference* of a circle, that is, the distance around the circle. The formula states that the circumference C is equal to 2π times the radius r, where π is a constant irrational number. (We often approximate the value of π as 22/7 or as 3.1416.) In symbols, this formula may be written as

$$C = 2\pi r$$

It therefore follows that a complete rotation of 360° is equal to 2π radians. This provides us with the following conversion formulas.

$$360^\circ = 2\pi \text{ radians}$$

$$1^\circ = \frac{\pi}{180} \text{ radians}$$

$$1 \text{ radian} = \frac{180}{\pi} \text{ degrees}$$

If we use 3.14 as π and divide 180 by 3.14, we find that 1 radian is about 57 degrees. Similarly if we divide 3.14 by 180, we can show that 1 degree is approximately 0.0175 radian.

Example 4 (**a**) Change 30° to radians. (**b**) Change $\frac{\pi}{6}$ radians to degrees.

Solution (**a**) $1^\circ = \frac{\pi}{180}$ radians; $30^\circ = 30\left(\frac{\pi}{180}\right) = \frac{\pi}{6}$ radians

(**b**) $1 \text{ radian} = \left(\frac{180}{\pi}\right)^\circ$; $\frac{\pi}{6} \text{ radians} = \left(\frac{\pi}{6}\right)\left(\frac{180}{\pi}\right) = 30^\circ$ ■

EXERCISES *Without measuring, choose the length that is your best estimate for the length of the given line segment.*

1. ____

(**a**) 2 inches, 3 inches, $2\frac{1}{2}$ inches, $1\frac{1}{2}$ inches

(**b**) 4 cm, 5 cm, 6 cm, 7 cm, 8 cm

2. ______________________________

(a) $3\frac{1}{2}$ inches, $3\frac{3}{4}$ inches, 4 inches, $4\frac{1}{4}$ inches, $4\frac{1}{2}$ inches

(b) 9 cm, 9.5 cm, 10 cm, 10.5 cm, 11 cm

State the unit of precision for the given measurement.

3. **(a)** 3 kilometers **(b)** 1.02 centimeters
4. **(a)** $2\frac{1}{4}$ meters **(b)** 2.25 meters
5. **(a)** $3\frac{1}{2}$ liters **(b)** 3.5 liters
6. **(a)** $22\frac{1}{2}$ degrees **(b)** $22\frac{2}{4}$ degrees

Identify the indicated measurement as exact or approximate.

7. **(a)** Doris has *three brothers.* **(b)** The oldest is *1.5 meters* tall.
8. **(a)** Art drove *500 miles* today. **(b)** He visited *seven customers.*
9. **(a)** Carol drove for *eight hours.* **(b)** She made only *three stops.*
10. **(a)** Sue painted *two rooms* last week.
    (b) She used *four and one-half gallons* of paint.

Identify the accuracy of the given approximate number or measurement.

11. **(a)** 2.05 **(b)** 3.1416 **(c)** 0.025
12. **(a)** 250.1 **(b)** 3.7×10^6 **(c)** 2.5×10^{-3}
13. **(a)** 72 inches **(b)** 2 yards **(c)** 6 feet
14. **(a)** 144 inches **(b)** 4 yards **(c)** 12 feet

Assume that each number is approximate, perform the indicated operation, and state the precision and accuracy of the answer.

15. $65.7 + 10.74$

16. $72.1 + 2.345$

17. $2.35 - 1.35$

18. $346.2 - 36.2$

19. $12.4 \div 3.1$

20. $66.0 \div 11$

21. 2.50×0.200

22. 756.3×1.1

23. $3.5 \times 10^3 + 1.8 \times 10^2$

24. $21.5 \times 10^{-2} + 8.4 \times 10^{-4}$

25. $6.7 \times 10^3 \times 2.0 \times 10^2$

26. $5.86 \times 10^{-2} \times 3.1 \times 10^3$

If possible, sketch and label the indicated figure. If not possible, explain why it is not possible.

27. An acute angle *ABC*.
28. Two acute angles *CDE* and *FGH* that are not congruent.

29. A right angle IJK.
30. Two right angles LMN and OPQ that are not congruent.
31. An obtuse angle RST.
32. Two obtuse angles UVW and XYZ that are not congruent.

Change the given angle measurement to radians.

33. $45°$ 34. $90°$ 35. $180°$ 36. $270°$
37. $360°$ 38. $75°$ 39. $225°$ 40. $-120°$
41. $-315°$ 42. $-135°$ 43. $720°$ 44. $-450°$

The angle measurement is given in radians; change it to degrees.

45. $\frac{\pi}{2}$ 46. $\frac{\pi}{4}$ 47. $\frac{3\pi}{2}$ 48. π
49. $\frac{7\pi}{4}$ 50. $\frac{2\pi}{3}$ 51. -2π 52. $\frac{-5\pi}{4}$
53. $\frac{-3\pi}{2}$ 54. 4π 55. -3π 56. $\frac{-5\pi}{2}$

EXPLORATIONS

1. Make a collection of illustrations of popular uses of measurements that are not standardized. For example, consider "a pinch of salt."
2. Estimate the length of your car in terms of cubits. Then use your arm to find its length using this measure.
3. Repeat Exploration 2 for the distance across the front of your classroom.
4. Suppose that you have two pieces of paper and a pencil. Explain how you can use just these tools to make a ruler with multiples and subdivisions of a selected unit.
5. Suppose that you have one piece of paper, a pair of scissors, and a pencil. Explain how you could use just these tools and make a protractor with which you could measure angles to about the nearest 6°.
6. Prepare a plan for a 30-minute lesson that can be used to teach an upper elementary school class how to use a protractor. Include in your plan both the use of a visual aid and student worksheets.
7. With some other people, explore several estimation activities, such as the following, and report on the results.
   (a) Ask each person to draw a line segment 4 centimeters long and then to check the error to the nearest centimeter.
   (b) Ask each person to draw an angle of 120° and then to check the error to the nearest 5°.

Four thousand decimal places of π may be found in Philip J. Davis's *The Lore of Large Numbers*, now published by the Mathematical Association of America. Here are a few.

π = 3.14159 26535 89793
23846 26433 83279
50288 41971 69399
37510 58209 74944
59230 78164 06286
20899 86280 34825
34211 70679 82148
08651 32823 06647
09384 46095 50582
23172 53594 08128
48111 74502 84102
70193 85211 05559
64462 29489 54930
38196

8. Find and report on a reference that explains the historical basis for our division of a circle into 360 congruent parts to obtain measures of angles.
9. Locate a number of circular objects. Measure the circumference C and the diameter d of each object. Then compute the ratio C/d for each object and see how closely you approximate the value of π. (Note that since $C = 2\pi r$, then $C = \pi(2r) = \pi d$, and $\pi = C/d$.)
10. Mnemonic devices are often used by students as an aid to memorization. A very famous one that gives 13 digits for π is this: "See, I have a rhyme assisting my feeble brain, its tasks ofttimes resisting." Replace each word by the number of letters in that word. Thus "See" = 3, "I" = 1, "have" = 4, and so on. This gives $\pi \doteq 3.141592653589$ correct to 12 places to the right of the decimal point. See what other mnemonic devices you can find, or invent, that are helpful in memorizing important mathematical facts.
11. Write a report on the history of the number π. For example, see "A Chronology of π" on pages 85 through 90 of *An Introduction to the History of Mathematics*, fifth edition, by Howard Eves, Saunders College Publishing, 1983.

9-3 Measures of Plane Figures

Any line segment, arc of a curve, or union of a finite number of line segments and arcs has a positive real number as its length. On a number line the length of any line segment RS may be expressed in terms of the coordinates of its endpoints.

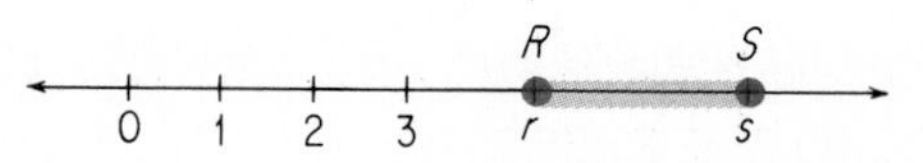

Let r be the coordinate of R and s the coordinate of S. The length of the line segment RS is

$$s - r \text{ if } s > r \quad \text{and} \quad r - s \text{ if } s < r$$

Any pair of opposites such as 3 and -3 or $s - r$ and $r - s$ have a common magnitude called their **absolute value**. We write

$$|3| = 3 \qquad |-3| = 3$$

$$|s - r| = s - r \text{ if } s > r \qquad |s - r| = r - s \text{ if } s < r$$

Note that the *distance* between two points is always a positive number and the *length* of a line segment is always positive regardless of the direction in which the line segment is measured or positioned.

For any real number x, we define

$$|x| = x \quad \text{if } x \text{ is a positive number or zero.}$$
$$|x| = -x \quad \text{if } x \text{ is a negative number.}$$

Then the length of the line segment RS is $|s - r|$ and may also be written as $|r - s|$. The absolute value of any real number x, $|x|$, may be defined as the distance from the origin to the point with coordinate x. The absolute value of any real number different from zero is a positive number; $|x| = |-x|$; $|0| = 0$, that is, a point is considered to have length 0.

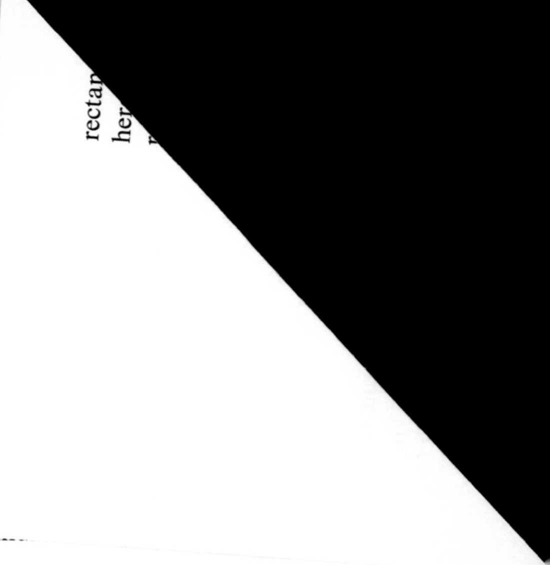

Example 1 Find the length of the line segment hav
as endpoints.
(a) 0 and -5 **(b)** -2 a

Solution **(a)** $|(-5) - 0| = |-5| = 5$
(b) $|7 - (-2)| = |7 + 2| = 9$
(c) $|(-7) - (-11)| = |(-7) + 11| =$

Any polygon or other simple cl
of a bounded region (its **interior**) an
area $\mathscr{A}$ of the polygon is the number
to the interior of the polygon. It i
polygon" rather than to specify
polygon." Here are some common fo

Square $\mathscr{A} = s^2$

Rectangle $\mathscr{A} = bh$

Parallelogram $\mathscr{A} = bh$

Right triangle $\mathscr{A} = \frac{1}{2}bh$

Any triangle $\mathscr{A} = \frac{1}{2}bh$

Trapezoid $\mathscr{A} = \frac{1}{2}h(b_1 + b_2)$

The area of any square with a side of integral length can be found by counting the number of unit squares in the square region. For example, a square with side 3 can be seen to "contain" 9 unit squares. Similarly, the area of any

gle with sides of integral lengths can be obtained by counting. We assume e without formal proof that the given formulas hold for all squares and ectangles.

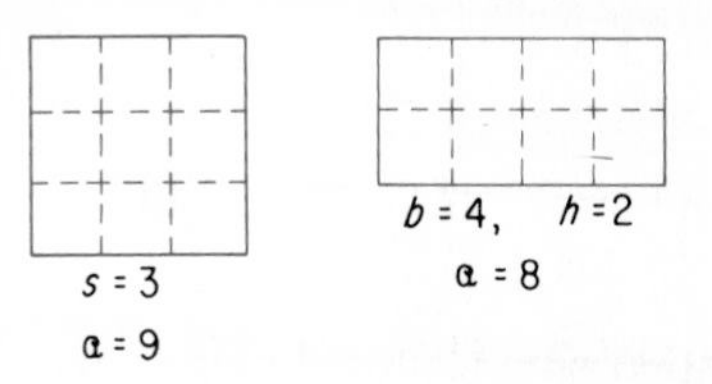

Any parallelogram is equivalent to a rectangle with the same base and height, as shown in the figures.

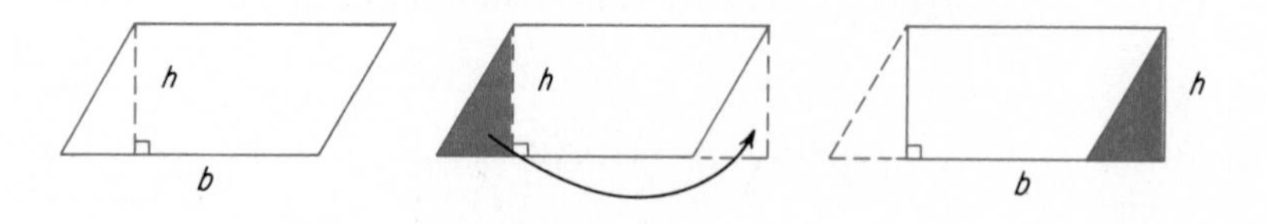

Any right triangle is half of a rectangle; any triangle is half of a parallelogram.

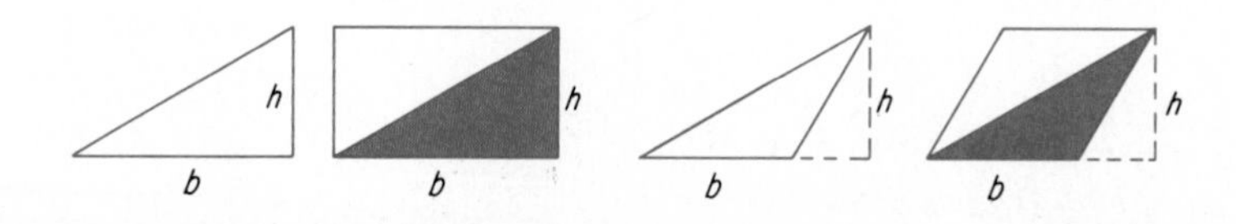

Any trapezoid is equivalent to a sum of two triangles.

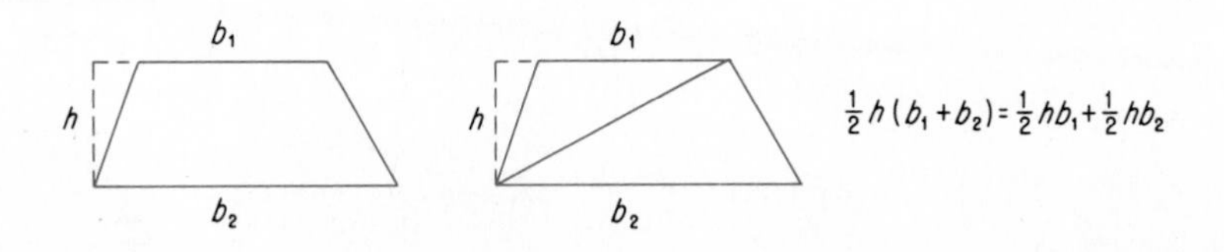

These visualizations of areas of parallelograms, triangles, and trapezoids may be demonstrated by constructions and also by cutting out and reassembling figures. Formal proofs are left for advanced courses. The word *area* is used for both the area measure (number) and the area measurement (number of units). However, the units should be specified in the answer to a problem whenever units have been designated.

For a given unit of length any line segment has a real number as its *length* and any polygon has a real number as its **perimeter**, the sum of the lengths of the sides of the polygon.

Example 1 Find the perimeter and the area of the given region.

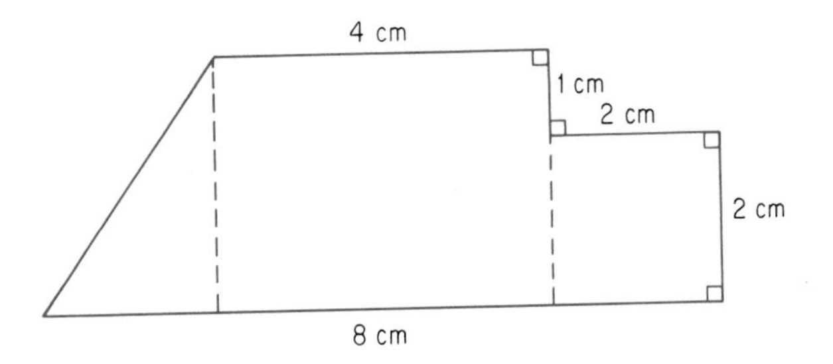

Solution We first subdivide the region into triangular and rectangular regions. Since the triangle is a right triangle, the Pythagorean theorem may be used to find the length of the hypotenuse of the triangle, $\sqrt{4 + 9} = \sqrt{13}$.

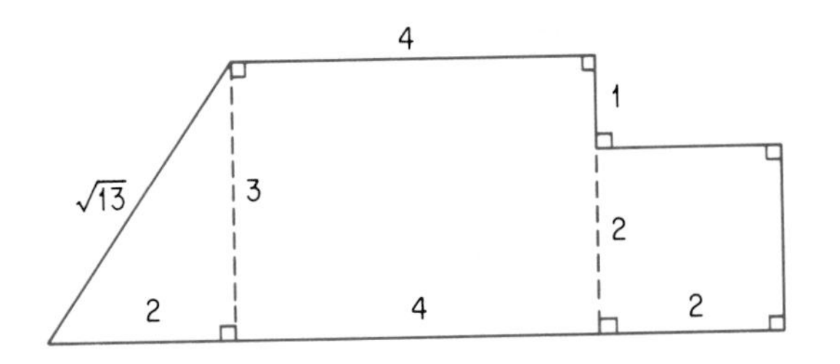

The perimeter is $8 + 2 + 2 + 1 + 4 + \sqrt{13}$, that is, $17 + \sqrt{13}$ cm. The triangular region has area (1/2)(2)(3), that is, 3; the rectangular regions have areas 3×4 and 2×2, that is, 12 and 4. Thus, in square centimeters, the area of the given region is 19 cm^2. ■

Any circle of radius r has **circumference** C and area $\mathscr{A}$ as given by these formulas

$$C = 2\pi r$$
$$\mathscr{A} = \pi r^2$$

We accept these formulas here without proof. The first formula may also be expressed as $C = \pi d$, where d is the diameter of the circle. As in Section 9-2, Exploration 9, this formula serves as a definition of π. The second formula may be explained as in Exploration 3 of this section. The length of a simple closed curve is called the *circumference* if the curve is a circle and the *perimeter* if the curve is a polygon. Often the words *perimeter* and *circumference* are used for other curves.

Example 2 Find the length and area of the given curve formed by line segments and semicircles. The line segments intersect at right angles.

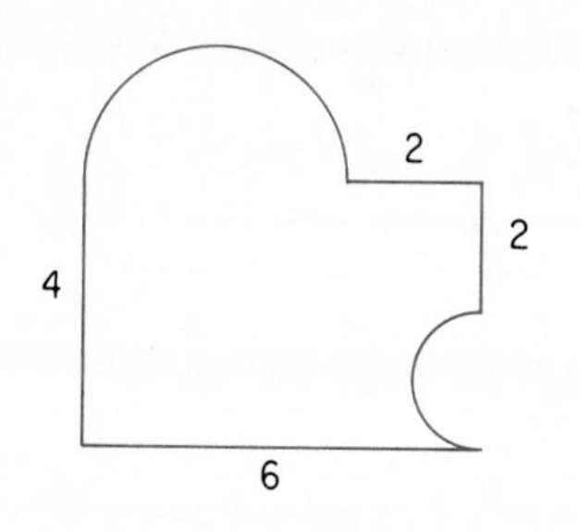

Solution The semicircles have radii 1 and 2 and thus have lengths π and 2π. The length of the curve is

$$4 + 6 + \pi + 2 + 2 + 2\pi, \quad \text{that is,} \quad 14 + 3\pi \text{ units}$$

To find the area we may think of a 4×6 rectangle with one semicircular region added and the other subtracted. Then the area is $24 + 2\pi - \pi/2$, that is, $24 + (3/2)\pi$ square units. ■

EXERCISES *Evaluate.*

1. **(a)** $|-5|$ **(b)** $|0|$ **(c)** $|6|$
2. **(a)** $|7|$ **(b)** $|-11|$ **(c)** $|2 - 3|$
3. **(a)** $|-4| + |4|$ **(b)** $|-4 + 4|$ **(c)** $|-4 - 4|$
4. **(a)** $|-8| + |8|$ **(b)** $|-8 + 8|$ **(c)** $|-8 - 8|$
5. **(a)** $|2 - 7|$ **(b)** $|7 - 2|$ **(c)** $|2| + |-7|$
6. **(a)** $|5 - 8|$ **(b)** $|8 - 5|$ **(c)** $|5| + |-8|$

For each given region assume that the given measures are exact and find **(a)** *the perimeter* **(b)** *the area.*

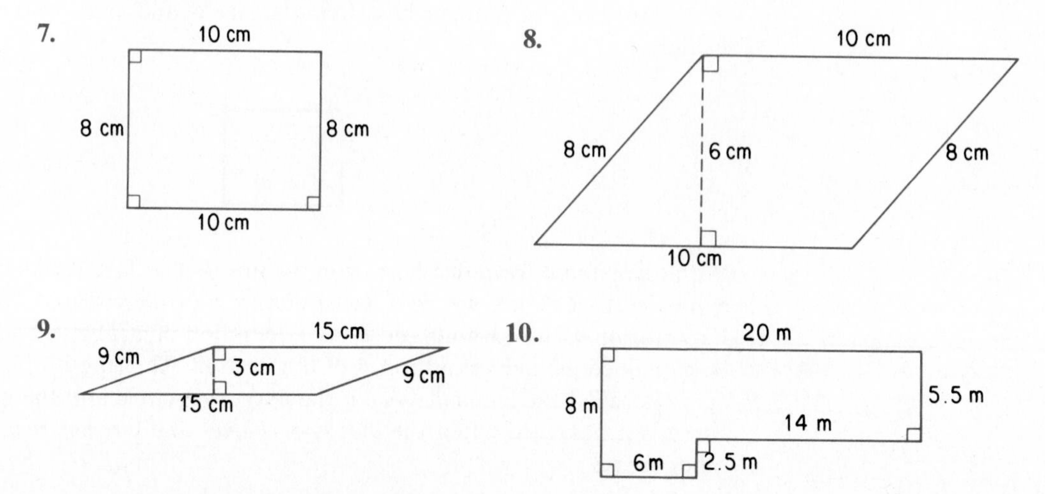

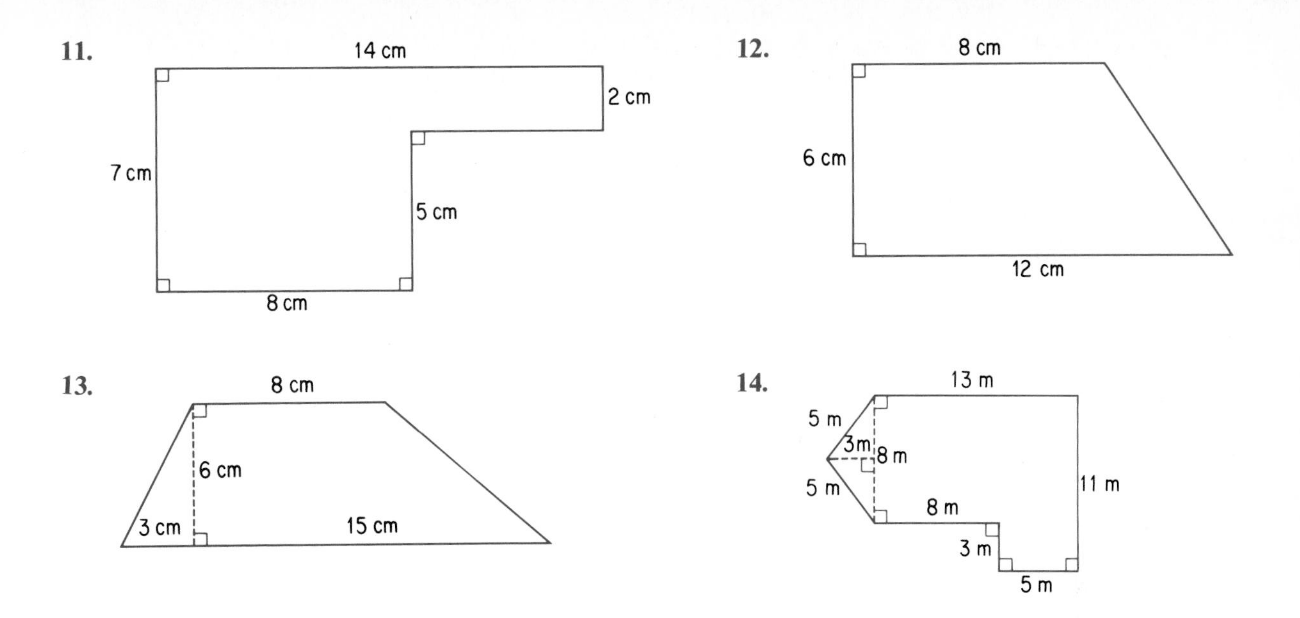

Recall that $\pi = 3.141\ 592\ 6\ldots$ and find the approximate area of any circle with the specified radius.

15. 3.0 cm **16.** 10.00 cm **17.** 9 mm

18. 5.00 mm **19.** 6.1 km **20.** 0.5 km

The curves in Exercises 21 and 22 are formed by line segments and circular arcs. For the given curve find **(a)** *the length* **(b)** *the area.*

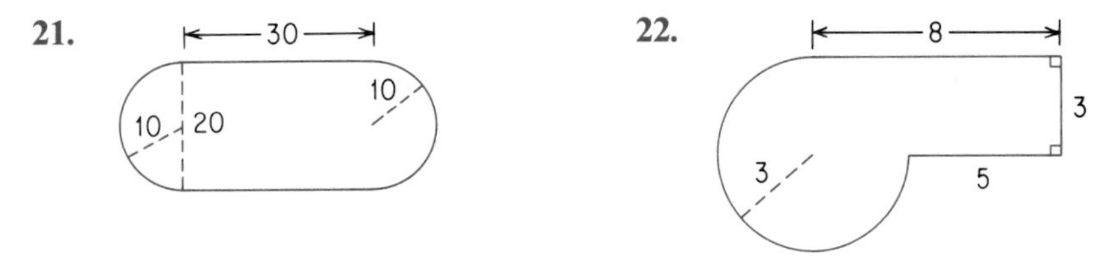

23. A border 2.0 centimeters wide is part of a rectangular sheet of paper 20 centimeters by 30 centimeters. Find the area of the border.

24. A walk 1.0 meter wide is to be built outside and as a boundary of a rectangular garden 15 meters by 20 meters. Find the area of the walk if the outside "corners" of the walk are
(a) square **(b)** arcs of a circle.

***25.** Suppose that the figure in Exercise 21 is the inside rail of a small track and the dimensions of the track are in meters. The surface of the track is 4 meters wide. Find the area of the surface of the track.

***26.** Suppose that the dimensions of the figure in Exercise 22 are in meters and a border one meter wide (rounded corners where appropriate) is constructed around the outside of the region. Find the approximate area of the border. Explain why your answer is approximate rather than exact.

*27. The circular rings of a certain dart board are formed by circles with the same center and with diameters 6″, 12″, 18″, and 24″. Express the area of each ring in terms of the area $\mathscr{A}$ of the circular region in the center of the board.

EXPLORATIONS

1. Prepare a classroom demonstration of the given statement. Use several different figures for each statement.
   (a) Any parallelogram is equivalent to a rectangle.
   (b) Any right triangle is half of a rectangle.
   (c) Any triangle is half of a parallelogram.
   (d) Any trapezoid is equivalent to two triangles.
   (e) Any trapezoid is half of a parallelogram.
2. We expect the area covered by polygonal regions to be independent of the arrangement of the regions.
   (a) Use graph paper and cut out a square with 16 units on each side. Note that the area of this square is 256 square units.
   (b) Mark and cut the square into four pieces as indicated in the figure.

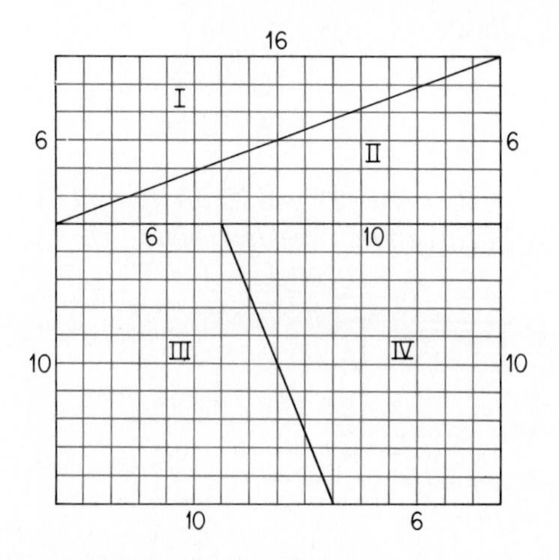

   (c) Rearrange the pieces as in this figure.

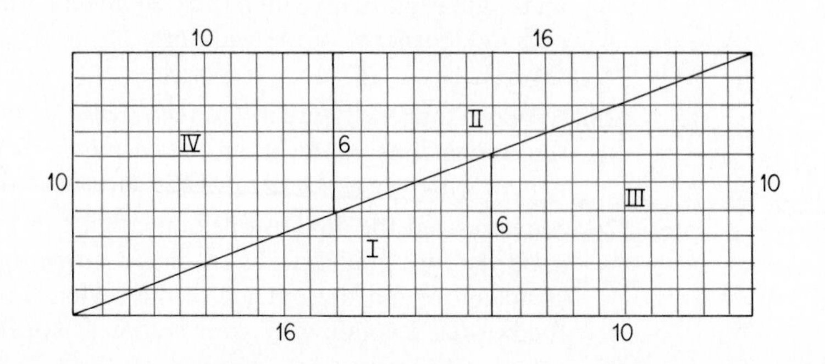

(d) Note that a rectangular region 10 by 26 units has area 260 square units. Explain the increase of 4 square units.

3. Think of a circular region as a union of a very large number of triangular regions with their bases on the circumference of the circle. Find the area of the circle as the sum of the areas of the triangles. Use the radius of the circle as the height of the triangles and the circumference of the circle as the sum of the bases of the triangles.

Many algebraic formulas were originally expressed in words as statements about areas of geometric figures. Explain a possible use of each figure as a basis for the given algebraic formula.

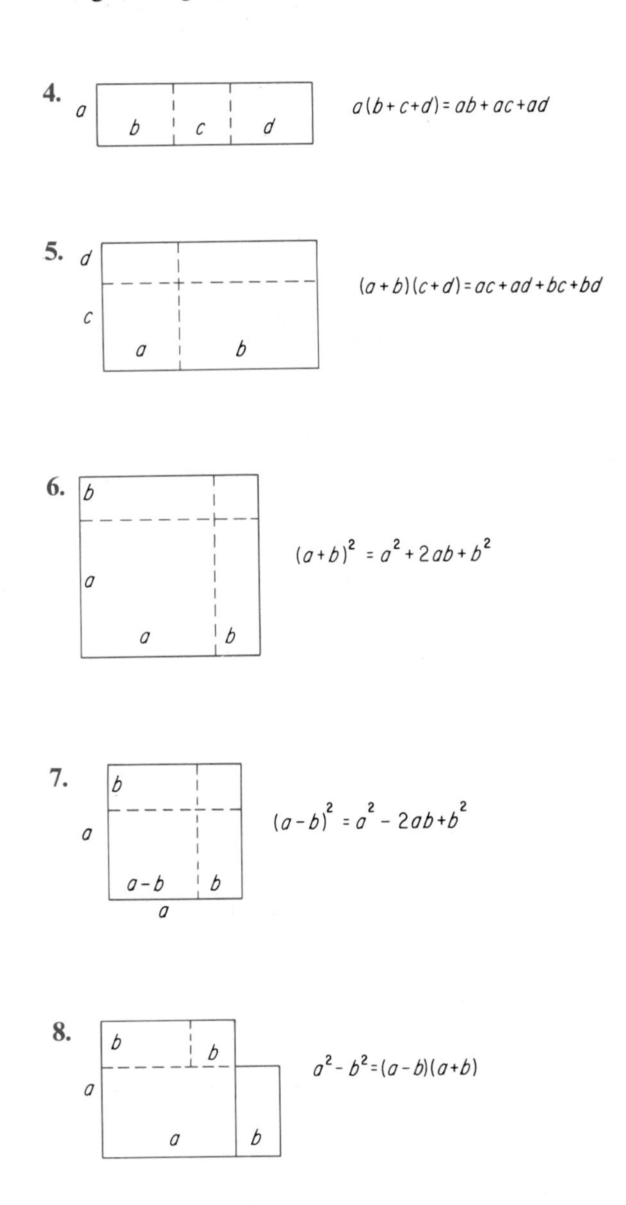

9-4 Geoboards

Geoboards enable us to obtain measurements of geometric figures by counting units. We can also use geoboards to introduce coordinate planes. Geoboards can be purchased or they can be homemade. Most **geoboards** are square arrays consisting of rows and columns of nails or pegs. There are also geoboards in other shapes. Circular geoboards are particularly useful for exploring theorems that involve circles. On any geoboard, geometric figures may be formed by stretching rubber bands around selected sets of pegs. **Dot paper** may be used to represent a geoboard, as illustrated in the figure.

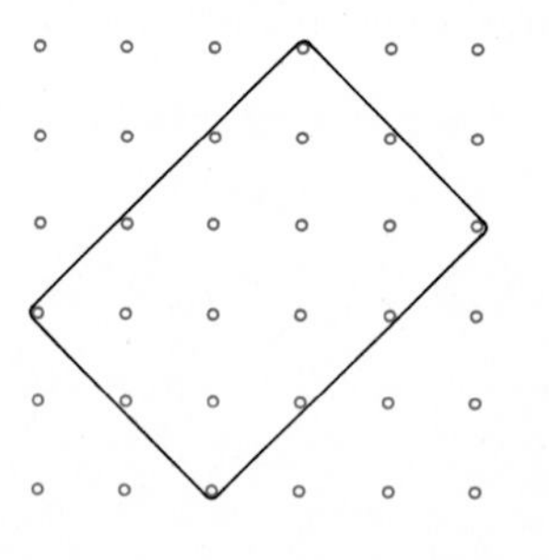

We restrict our basic discussion to a geoboard of six columns and six rows, that is, a **6-by-6 geoboard**. As an aid to our measurement of distances, it is convenient to start at 0. Thus the columns are numbered 0, 1, 2, 3, 4, 5 (from left to right) and the rows are numbered 0, 1, 2, 3, 4, 5 (from bottom to top)—so that the lower left corner becomes a common starting point for numbering both columns and rows. The column numbered 0 is the **initial column**; the row numbered 0 is the **initial row** of the geoboard.

Row numbers: 5, 4, 3, 2, 1, 0

Column numbers: 0, 1, 2, 3, 4, 5

A 6-by-6 geoboard has 36 pegs; "6-by-6" is often written as "6 × 6." Each peg of a geoboard represents a *point*. In any discussion of a geoboard, the term *point* refers to the position of a peg and not one of the numerous positions that might be imagined as existing between pegs. Each point may be identified by its column number and its row number.

Example 1 Make dot paper to represent a 6×6 geoboard and label the point in the specified position as indicated.

(a) Column 0 and Row 0 as O.
(b) Column 2 and Row 1 as A.
(c) Column 1 and Row 2 as B.
(d) Column 5 and Row 3 as C.
(e) Column 3 and Row 0 as D.
(f) Column 0 and Row 3 as E.

Solution

5	∘	∘	∘	∘	∘	∘
4	∘	∘	∘	∘	∘	∘
3	• E	∘	∘	∘	∘	• C
2	∘	• B	∘	∘	∘	∘
1	∘	∘	• A	∘	∘	∘
0	• O	∘	∘	• D	∘	∘
	0	1	2	3	4	5

■

The pattern that we have followed will be our standard pattern.

The initial column is at the left.

The other columns are numbered as they are counted from the initial column.

The initial row is at the bottom.

The other rows are numbered as they are counted from the initial row.

The distance between adjacent columns is taken as the *unit distance*. The same unit distance separates adjacent rows. Then the number of any column is its distance from the initial column and the number of any row is its distance from the initial row.

Each point of the geoboard is at the intersection of a column and a row. Thus the point may be identified by the column number and the row number. It is customary to state the column number first and the row number second, that is, to use an *ordered pair* of numbers. Then the points considered in Example 1 may be identified as

O: (0, 0) $\quad$ A: (2, 1) $\quad$ B: (1, 2) $\quad$ C: (5, 3) $\quad$ D: (3, 0) $\quad$ E: (0, 3)

Note that the ordered pairs (2, 1) and (1, 2) represent different points.

The numbering of the rows and columns of the geoboard enables us to use these numbers as coordinates of the points of the geoboard. This approach is extended to all points of the plane in Section 9-5.

The two numbers of an ordered pair that identify a point on the geoboard are called the **coordinates** of the point. In the terminology used for coordinate planes, the first number is the ***x*-coordinate** and the second number is the ***y*-coordinate**. Both coordinates are measures obtained by counting. The x-coordinate is the number of units (distance) of the column of the point from the initial column. The y-coordinate is the number of units of the row of the point from the initial row. The point (0, 0) is the **origin**.

Relations among the numbers of these ordered pairs (coordinates) appear as geometric patterns of the points represented by the pegs of the geoboard and as arithmetic patterns of the coordinates of the points. In the solution for Example 1, the points E, B, A, and D appear to be on a straight line. The coordinates of these points are, respectively,

$$(0, 3) \quad (1, 2) \quad (2, 1) \quad (3, 0)$$

Note the arithmetic patterns.

The first coordinates are 0, 1, 2, 3
The second coordinates are 3, 2, 1, 0
For each point the sum of the coordinates is 3.

The last of these three arithmetic patterns may be used to identify the "line" of points E, B, A, and D on the geoboard, that is, these are the points of the geoboard such that the sum of their coordinates is 3. For an arbitrary point (x, y), we write this condition as

$$x + y = 3$$

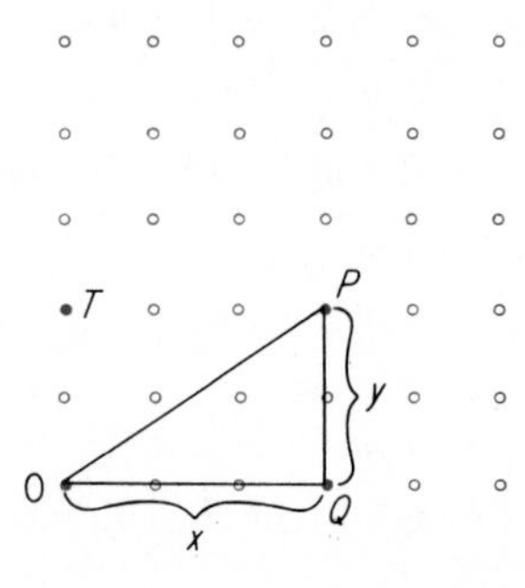

The determination of the distance of any point $P: (x, y)$ from the origin is a special case of the determination of the length of any line segment that can be represented on the geoboard. If $x = 0$, the point $T: (0, y)$ is in the initial column and at a distance y from the origin. If $y = 0$, the point $Q: (x, 0)$ is in the initial row and at a distance x from the origin. If $xy \neq 0$, then by the Pythagorean theorem $(OP)^2 = x^2 + y^2$ since the point $Q: (x, 0)$ is in the same column as $P: (x, y)$, triangle OQP is a right triangle, and the legs of the triangle have lengths x and y. The distance formula $(OP)^2 = x^2 + y^2$ holds also for $T: (0, y)$ since $y^2 = 0 + y^2$ and for $Q: (x, 0)$ since $x^2 = x^2 + 0$. That is, the square of the distance of any point from the origin is equal to the sum of the squares of the coordinates of the point.

Example 2 For a 6×6 geoboard, make an array showing the distance from the origin of each point in **(a)** Row 0 **(b)** Row 1.

Solution

(a)

Point	(0, 0)	(1, 0)	(2, 0)	(3, 0)	(4, 0)	(5, 0)
Distance	0	1	2	3	4	5

(b)

Point	(0, 1)	(1, 1)	(2, 1)	(3, 1)	(4, 1)	(5, 1)
Distance	1	$\sqrt{2}$	$\sqrt{5}$	$\sqrt{10}$	$\sqrt{17}$	$\sqrt{26}$ ■

The solution for Example 2 also could have been given as in the first figure below.

$1 \quad \sqrt{2} \quad \sqrt{5} \quad \sqrt{10} \quad \sqrt{17} \quad \sqrt{26}$

0 1 2 3 4 5

P T Q 0

This distance formula is extended in Section 9-5 to include all line segments on the plane.

Suppose that P: (a, b) and Q: (c, d) where $a \neq c$ and $b \neq d$ are two points on a 6×6 geoboard. These two points and the point T: (c, b) determine a right triangle PTQ with right angle at T and hypotenuse PQ. The length of side PT is $|c - a|$, the nonnegative difference of the x-coordinates of the points; the length of TQ is $|d - b|$. By the Pythagorean theorem,

$$(PQ)^2 = (c - a)^2 + (d - b)^2$$

The possible lengths of line segments with endpoints at points of the geoboard (Exercise 3) are

$$1, \quad \sqrt{2}, \quad 2, \quad \sqrt{5}, \quad 2\sqrt{2}, \quad 3, \quad \sqrt{10}, \quad \sqrt{13}, \quad 4, \quad \ldots$$

Consider the problem of representing a square of area 10 square units on a 6×6 geoboard. Each side of the square must have length $\sqrt{10}$. However, even though $OP = \sqrt{10}$ for O: $(0, 0)$ and P: $(3, 1)$ in the next figure, the line segment OP cannot be used as a side of a square with vertices at pegs of the geoboard.

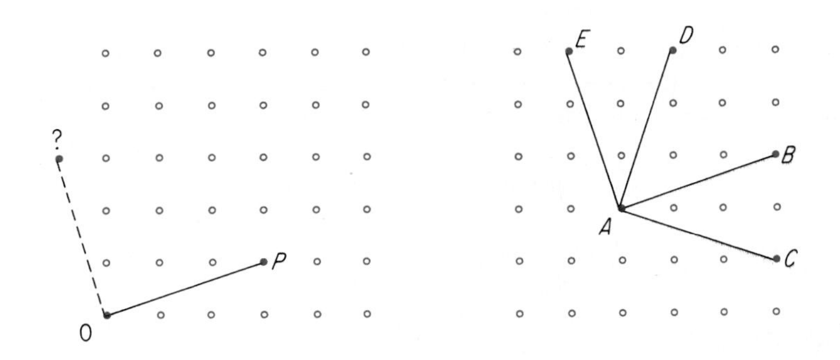

A square with the line segment OP as a side would need a vertex that is not on the 6×6 geoboard, as indicated by the dashed line in the first figure. However the position of the line segment OP does indicate a method for obtaining line segments of length $\sqrt{10}$. From any point A count three units to the right (or left) and one unit up (B) or down (C); also from A count three units

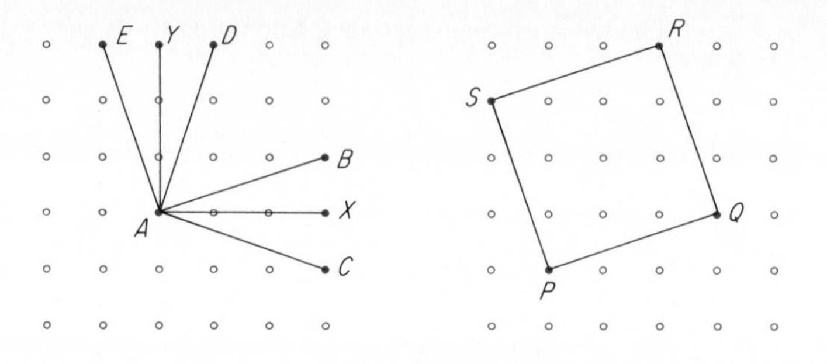

up (or down) and one unit to the right (D) or left (E). If the procedure terminates at a point of the geoboard, the distance is $\sqrt{10}$.

$$AB = AC = AD = AE = \sqrt{10}$$

As a guide to determining right angles for the desired square, label X: (5, 3) and Y: (3, 5). Then the points A and X are on a row AX; the points A and Y are on a column AY; and $\angle XAY$ is a right angle. Other right angles can be obtained by rotating rays AX and AY the same amount in the same direction. The equal amounts of rotation (the congruence of the angles) can be identified from the congruence of triangles

$$\triangle AXC \cong \triangle AXB \cong \triangle AYD \cong \triangle AYE$$

and the congruence of corresponding angles

$$\angle XAC \cong \angle XAB \cong \angle YAD \cong \angle YAE$$

Thus a clockwise turn of right angle XAY so that ray AX coincides with ray AC makes ray AY coincide with ray AD and $\angle CAD$ is a right angle. Similarly, $\angle BAE$ is a right angle.

Right angles CAD and BAE illustrate the procedure for obtaining sides that intersect at right angles. For $\angle CAD$, the counts from A were three units to the right and one down (a clockwise turn) to obtain AC, then from A three units up and one to the right (a second clockwise turn) to obtain AD. Similarly, counterclockwise turns may be used to obtain right angle BAE.

A square of area 10 square units may be positioned on the geoboard in any one of eight positions (Exercise 7). If we select P: (1, 1) as a vertex and copy right angle BAE, then Q: (4, 2), R: (3, 5), and S: (0, 4) are the other three vertices of square $PQRS$ with area 10 square units. Note that each side of the square is the hypotenuse of a right triangle (actually two right triangles) with perpendicular sides of lengths 3 and 1 along the rows and columns of the geoboard.

The area of any polygonal region with vertices at points of a geoboard can be determined by counting unit squares, using the fact that congruent figures have equal areas, and adding or subtracting areas. Any square with side s has area s^2 square units. The diagonal PR of any square $PQRS$ separates the square into triangles PQR and PSR, where $\triangle PQR \cong \triangle PSR$. Therefore, $\triangle PQR$ has area $s^2/2$.

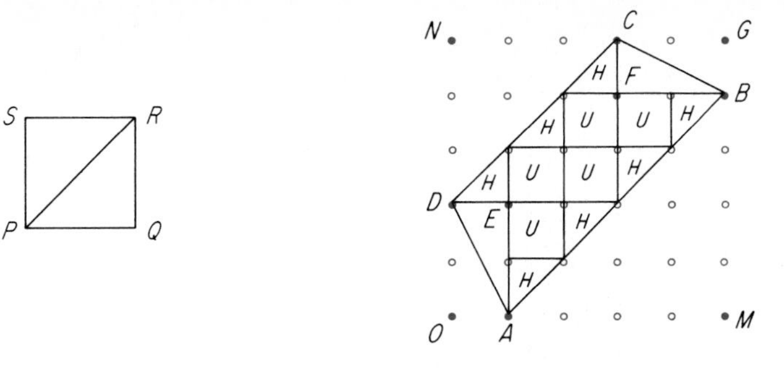

As a final example we find the perimeter and area of trapezoid $ABCD$ with vertices A: (1, 0), B: (5, 4), C: (3, 5), and D: (0, 2). The quadrilateral is a trapezoid since sides AB and CD are parallel; it is not a parallelogram since AD and BC are not parallel. The trapezoid is an isosceles trapezoid since $AD \cong BC$. The perimeter $DA + AB + BC + CD$ is

$$\sqrt{2^2 + 1^2} + \sqrt{4^2 + 4^2} + \sqrt{1^2 + 2^2} + \sqrt{3^2 + 3^2}$$
$$= \sqrt{5} + 4\sqrt{2} + \sqrt{5} + 3\sqrt{2}, \quad \text{that is,} \quad 7\sqrt{2} + 2\sqrt{5} \text{ units}$$

The trapezoidal region may be subdivided into these regions: five unit squares (indicated by U in the figure), seven triangles each with area 1/2 square unit (indicated by H), and two triangles that we designate as $\triangle AED$ for D: (1, 2) and $\triangle BFC$ for F: (3, 4). These two triangles are congruent; each is half of a rectangle such as $BGCF$ for G: (5, 5). Thus each triangle has area 1 square unit. Then the area of the trapezoid $ABCD$ is $5 + 7(1/2) + 2(1)$, that is, 10.5 square units.

To find the area of a region on a geoboard find the sum of the areas of the parts of the region or, alternatively, find the sum of the areas of the parts of the geoboard that are not parts of the desired region and subtract from the total area of the geoboard.

We found the area of trapezoid $ABCD$ by adding the areas of certain parts of the region. Other selections of parts could have been used; also the parts of the geoboard that are not in the trapezoidal region could have been used. For example, label O: (0, 0), M: (5, 0), and N: (0, 5). The entire geoboard is a square $OMGN$ with side 5 and area 25 square units. The triangular region AMB is half of a square of side 4 and thus has area 8 square units. Triangle CND is half of a square of side 3 and has area 4.5 square units. Triangles OAD and GBC are each half of a 1×2 rectangle and thus together have area 2 square units. Therefore, the area of the trapezoid $ABCD$ is $25 - 8 - 4.5 - 2$, that is, 10.5 square units.

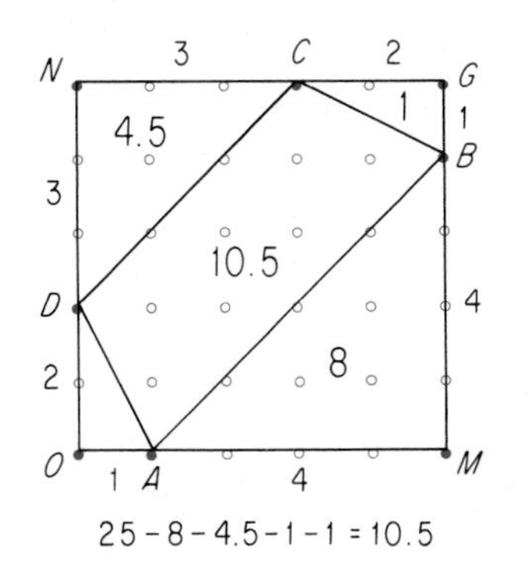

The points of a 6×6 geoboard have been represented by ordered pairs (x, y), where x and y are members of the set S with

$$S = \{0, 1, 2, 3, 4, 5\}$$

That is, the set of points of a 6×6 geoboard has been represented by the set of ordered pairs that are elements of the Cartesian product $S \times S$. Similarly, we may think of a 100×100 geoboard and in general an **extended geoboard** with points represented by ordered pairs (x, y), where x and y are members of the set W of whole numbers.

On an extended geoboard any rectangle with sides of integral lengths a and b may be represented with vertices A: (0, 0), B: $(a, 0)$, C: (a, b), and D: $(0, b)$. The rectangle has *area* ab and *perimeter*, $2a + 2b$, the sum of the lengths of the sides.

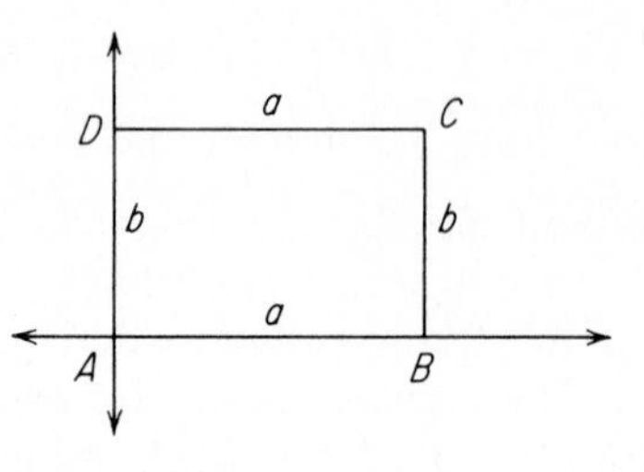

Example 3 Make a table of the coordinates of the vertices A: (0, 0), B: $(a, 0)$, C: (a, b), and D: $(0, b)$, the perimeter p, and the area $\mathscr{A}$, of the nine such rectangles with sides of integral lengths and area 36 square units.

Solution The nine expressions for 36 as a product $a \times b$ of counting numbers are

1×36 $\quad 2 \times 18$ $\quad 3 \times 12$ $\quad 4 \times 9$ $\quad 6 \times 6$

9×4 $\quad 12 \times 3$ $\quad 18 \times 2$ $\quad 36 \times 1$

In Example 3, note that although all of these rectangles have the same area, their perimeters vary from 24 to 74. If you were building a fence to enclose 36 square meters, you probably would prefer to buy as little fencing material as possible. As you may have observed in Example 3, the shape of the quadrilateral that makes the perimeter as small as possible for a given area can be predicted. Similar predictions can be made for the areas of figures with the same perimeters (Exercise 35).

A	B	C	D	p	$\mathscr{A}$
(0, 0)	(1, 0)	(1, 36)	(0, 36)	74	36
(0, 0)	(2, 0)	(2, 18)	(0, 18)	40	36
(0, 0)	(3, 0)	(3, 12)	(0, 12)	30	36
(0, 0)	(4, 0)	(4, 9)	(0, 9)	26	36
(0, 0)	(6, 0)	(6, 6)	(0, 6)	24	36
(0, 0)	(9, 0)	(9, 4)	(0, 4)	26	36
(0, 0)	(12, 0)	(12, 3)	(0, 3)	30	36
(0, 0)	(18, 0)	(18, 2)	(0, 2)	40	36
(0, 0)	(36, 0)	(36, 1)	(0, 1)	74	36 ■

EXERCISES

1. Make dot paper to represent a 6×6 geoboard and label the points $A: (0, 2)$, $B: (3, 4)$, $C: (1, 5)$, and $D: (4, 0)$.
2. Make dot paper to represent a 6×6 geoboard and label each point with its distance from the origin.
3. List in increasing order all possible lengths of line segments with endpoints at different points of a 6×6 geoboard.
4. The origin may be used as an endpoint for line segments of all lengths that are possible on a 6×6 geoboard. Identify by their coordinates all other points that may be used as an endpoint for line segments of all possible lengths.
5. Answer each question for line segments with endpoints at different points of a 6×6 geoboard.
   (a) What is the smallest nonintegral length?
   (b) What integral multiples of this smallest nonintegral length arise?
   (c) For line segments with lengths as in part (b), one endpoint at the origin, and the endpoint at $P: (x, y)$, give an equation to identify the points P.
6. Use the fact that $|r - s| = \sqrt{(r - s)^2}$ and explain why the formula $(PQ)^2 = (c - a)^2 + (d - b)^2$ holds if $b = d$, that is, if $P: (a, d)$ and $Q: (c, d)$ are in the same row and there is no right triangle.
7. Use the coordinates of the vertices to identify the eight possible positions of a square $PQRS$ of area 10 square units on a 6×6 geoboard.
8. For squares with vertices at points of a 6×6 geoboard what is the area of the specified figure?
   (a) The smallest possible square.
   (b) The next to the smallest possible square.
   (c) The largest possible square.
9. List in increasing order all possible areas of rectangles with vertices at points of a 6×6 geoboard and sides parallel to the edges of the board.
10. Repeat Exercise 9 for rectangles with sides that are integral multiples of $\sqrt{2}$.

In Exercises 11 and 12 the triangles have vertices at the points of a 6×6 geoboard.

11. **(a)** How many right triangles of different shapes have at least one side of length 1?
    (b) For each right triangle in part (a), identify the length of the other of the two perpendicular sides.
12. List the lengths of the congruent sides of the possible isosceles triangles such that the lengths of the congruent sides are
    (a) integers
    (b) multiples of $\sqrt{2}$
    (c) neither integers nor multiples of $\sqrt{2}$.

For triangles with vertices at points of a 6 × 6 geoboard, find each length.

13. The length of the base of an isosceles triangle with two sides of length
(**a**) 1 (**b**) $\sqrt{2}$ (**c**) 2 (**d**) $2\sqrt{2}$
(**e**) 3 (**f**) 4 (**g**) $\sqrt{41}$.

14. The lengths of the bases of two isosceles triangles of different shapes but with two sides of length
(**a**) $2\sqrt{5}$ (**b**) $\sqrt{26}$ (**c**) $\sqrt{29}$.

15. The lengths of the bases of four isosceles triangles of different shapes but with two sides of length
(**a**) $\sqrt{10}$ (**b**) $\sqrt{13}$ (**c**) $\sqrt{17}$ (**d**) 5.

16. The lengths of the bases of six isosceles triangles of different shapes but with two sides of length $\sqrt{5}$.

For the triangle with the specified vertices find (**a**) *the perimeter* (**b**) *the area.*

17. $A{:}\,(1, 0)$, $B{:}\,(5, 0)$, $C{:}\,(3, 1)$
18. $D{:}\,(0, 1)$, $E{:}\,(4, 1)$, $F{:}\,(5, 3)$
19. $G{:}\,(1, 2)$, $H{:}\,(5, 2)$, $I{:}\,(1, 5)$
20. $J{:}\,(1, 5)$, $K{:}\,(4, 5)$, $L{:}\,(2, 3)$
21. $M{:}\,(0, 0)$, $N{:}\,(4, 2)$, $O{:}\,(5, 4)$
22. $P{:}\,(0, 0)$, $Q{:}\,(4, 3)$, $R{:}\,(5, 4)$
23. $S{:}\,(0, 1)$, $T{:}\,(4, 4)$, $U{:}\,(5, 5)$
24. $W{:}\,(0, 5)$, $X{:}\,(2, 2)$, $Y{:}\,(4, 0)$

Find the area of the quadrilateral with the specified vertices.

25. $A{:}\,(0, 0)$, $B{:}\,(4, 0)$, $C{:}\,(5, 3)$, $D{:}\,(1, 3)$
26. $E{:}\,(1, 1)$, $F{:}\,(4, 1)$, $G{:}\,(4, 5)$, $H{:}\,(1, 5)$
27. $I{:}\,(0, 2)$, $J{:}\,(3, 2)$, $K{:}\,(3, 5)$, $L{:}\,(1, 5)$
28. $M{:}\,(0, 1)$, $N{:}\,(4, 1)$, $P{:}\,(3, 4)$, $Q{:}\,(1, 4)$
29. $R{:}\,(0, 5)$, $S{:}\,(5, 4)$, $T{:}\,(4, 2)$, $U{:}\,(2, 2)$
30. $W{:}\,(0, 1)$, $X{:}\,(4, 2)$, $Y{:}\,(3, 5)$, $Z{:}\,(2, 4)$

With one vertex at the origin and one side along the line $y = 0$, give coordinates for the vertices of at least one representation of each figure on an extended geoboard.

31. A triangle with sides of lengths
(**a**) 3, 4, 5 (**b**) 6, $3\sqrt{2}$, $3\sqrt{2}$ (**c**) 5, 5, 6.

32. A rectangle with sides of lengths
(**a**) 2, 5 (**b**) 1, 4 (**c**) 3, 8

33. A parallelogram that is not a rectangle but has sides of lengths
(**a**) 2, 5 (**b**) 5, 6 (**c**) $\sqrt{13}$, 4.

34. An isosceles trapezoid with sides of lengths
   (a) 2, 5, 5, 8 (b) 2, 5, 5, 10 (c) 1, 10, 10, 13.

Use an extended geoboard for Exercises 35 *through* 38.

35. A parallelogram $ABCD$ has base AB with A: (0, 0) and B: (6, 0). For the given point D, list the coordinates of the vertex C, the perimeter of the parallelogram, the length of the altitude to base AB, and the area of the parallelogram. Then for a parallelogram with a given base and perimeter, conjecture the shape that has the greatest possible area.
   (a) D: (0, 5) (b) D: (3, 4) (c) D: (4, 3)
36. Consider a triangle ABC with base AB where A: (0, 0) and B: (2, 0), area 1 square unit, and C: (x, y) where $0 < y$.
   (a) Give the coordinates of five of the possible points C.
   (b) If the set of all possible points C forms a geometric figure, identify that figure.
   (c) If there is a smallest possible perimeter for triangle ABC, identify that smallest perimeter.
   (d) If there is a largest possible perimeter, identify that largest perimeter.
37. Repeat Exercise 36 for an area of 3 square units.
38. Repeat Exercise 36 for the vertices C and D of the parallelograms $ABCD$ with base AB and area 4 square units.

EXPLORATIONS

1. Prepare a lesson plan on the representations of at least five different types of triangles on a geoboard.
2. Repeat Exploration 1 for quadrilaterals.

The formula that is developed in the next two explorations is **Pick's formula** for the area of any polygonal region represented on a geoboard.

3. For each of the following figures, count the number b of points on the boundary of the figure. Then compute the area A in terms of square units as in the table. Finally, conjecture a formula for A in terms of b.

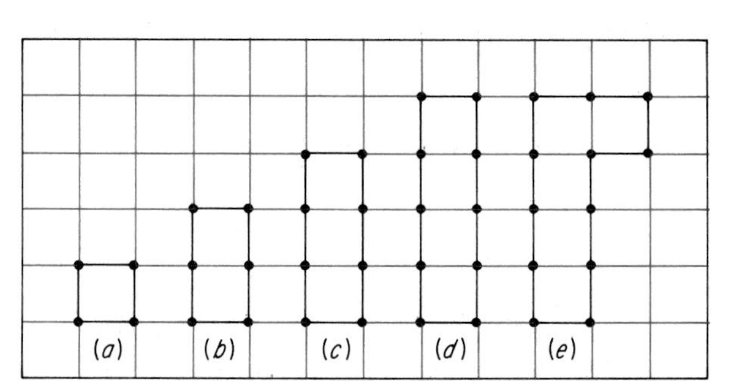

	(a)	(b)	(c)	(d)	(e)
Number (b) of boundary points	4	6	8	10	12
Area (A)	1	2			

4. For each of the following figures, count the number b of boundary points, and count the number i of interior points. Then compute the area A in square units as in the given table. Finally, conjecture a formula for A in terms of b and i.

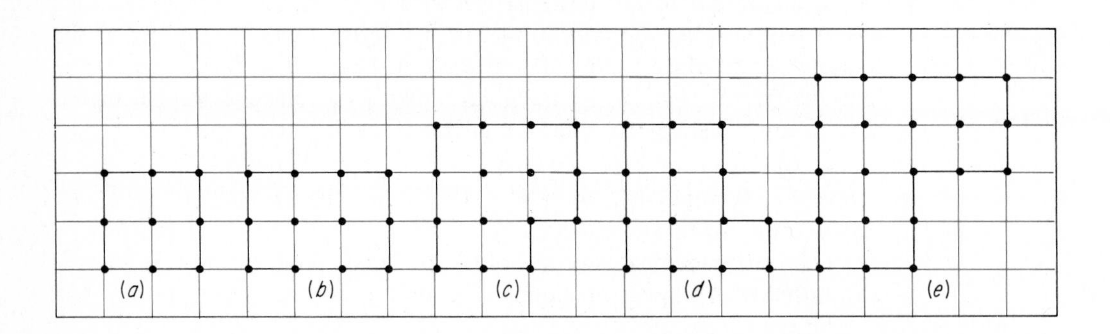

	(a)	(b)	(c)	(d)	(e)
Number (b) of boundary points	8	10	12		
Number (i) of interior points	1	2	3		
Area (A)	4	6			

9-5 The Coordinate Plane

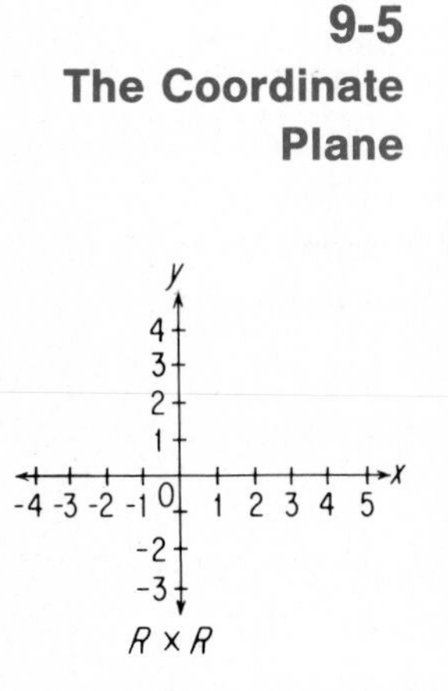

Any ordinary line may be considered as a *real number line*. Any ordinary plane may be considered as a **real coordinate plane**, that is, in Euclidean geometry the set of points of any plane may be represented by the set of ordered pairs (x, y) that are elements of the Cartesian product $R \times R$, where R is the set of real numbers. As on a geoboard the two numbers of an ordered pair that identify a point are the **coordinates** of the point. The first number x is the ***x*-coordinate** and the second number y is the ***y*-coordinate** of the point.

On a coordinate plane, two perpendicular number lines OX and OY are used as **coordinate axes**. Their intersection is at 0 on each number scale and this intersection is the **origin** of the real coordinate plane. The ordered pair approach that was used to identify any point of the geoboard may now be used to identify any point of the coordinate plane. The points of the bottom edge (initial row) of the geoboard all have y-coordinate 0 and are on the line

$$y = 0, \quad \text{that is,} \quad \text{the } x\text{-axis}$$

The lines $y = 0$, $y = 2$, $y = 3$, and $y = -2$ on a real coordinate plane are shown in the first figure. All lines with equations of the form $y = k$ are parallel to the x-axis, at a distance $|k|$ from the x-axis, above the x-axis if k is positive, and below if k is negative. The line $x = 0$ is the ***y*-axis**. The lines $x = 0$, $x = 2$, $x = 3$,

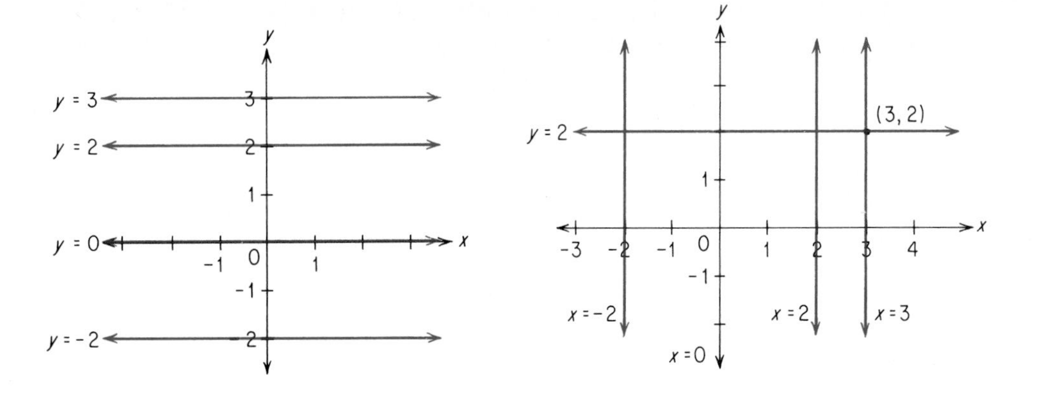

and $x = -2$ on a real coordinate plane are shown in the second figure. All lines with equations of the form $x = h$ are parallel to the y-axis, at a distance $|h|$ from the y-axis, on the right of the y-axis if h is positive, and on the left of the y-axis if h is negative. Each point (h, k) is at the intersection of the lines $x = h$ and $y = k$. For example, the point (3, 2) is at the intersection of the lines $x = 3$ and $y = 2$, as shown in the figure. The lines are the **graphs** of the equations; the equations are the **equations** of the lines. Similarly, the points are the **graphs** of ordered pairs of numbers; the ordered pairs of numbers are the **coordinates** of the points. To draw a figure, such as a line or a point, on a coordinate plane is **to graph** that figure.

Example 1 Graph $x = \frac{-3}{2}$, $y = 2\sqrt{2}$, and $\left(\frac{-3}{2}, 2\sqrt{2}\right)$.

Solution

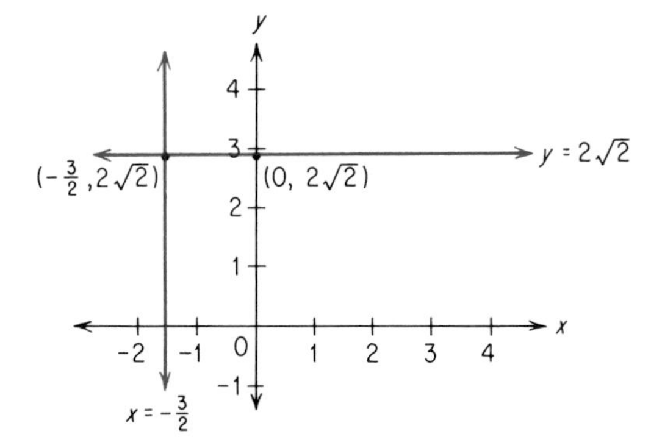

The line $x = -3/2$ is parallel to and $1\frac{1}{2}$ units to the left of the y-axis. The line $y = 2\sqrt{2}$ is parallel to and $2\sqrt{2}$ units above the x-axis. The point $(-3/2, 2\sqrt{2})$ is at the intersection of these two lines. ■

To graph the line $y = 2\sqrt{2}$ in the solution of Example 1, we approximated $2\sqrt{2}$ as 2.8, found the point (0, 2.8) on the y-axis, and drew the line that contains that point and is parallel to the x-axis. Even if we had found the point $(0, 2\sqrt{2})$

by straightedge and compass construction, the result would have been an approximation for the position of the point since our construction would have involved the visual fitting of the compass to a pair of points. In almost all cases, graphing involves approximations.

The coordinate axes separate the plane into four regions, **quadrants**, that are numbered counterclockwise as in the figure. A point $P:(x, y)$ is on a coordinate axis if and only if at least one of its coordinates is zero, that is, if and only if $xy = 0$. If $xy \neq 0$, the point P is in one of the quadrants and that quadrant may be identified by the signs of the coordinates of the point P as in the following array.

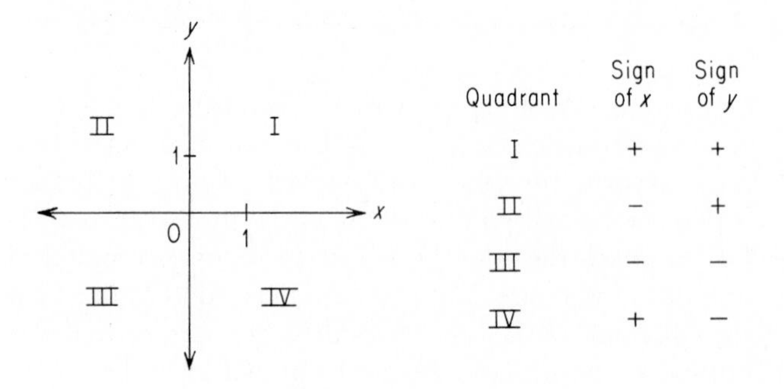

Quadrant	Sign of x	Sign of y
I	+	+
II	−	+
III	−	−
IV	+	−

Transformations of figures on a coordinate plane may be specified in terms of the coordinates of the points of the plane. Consider a *slide* (**translation**) 3 units in the positive x-direction. The coordinates $(x'\ y')$ of the **image** of any given point (x, y) under this transformation may be found by increasing the x-coordinate by 3 and leaving the y-coordinate unchanged. That is, each point $P:(x, y)$ is **mapped** onto a point $P':(x', y')$ where

$$x' = x + 3 \quad \text{and} \quad y' = y$$

As shown in the figure for the vertices of triangle ABC

$A:(0, 0)$ is mapped onto $A':(3, 0)$.
$B:(1, 0)$ is mapped onto $B':(4, 0)$.
$C:(2, 2)$ is mapped onto $C':(5, 2)$.

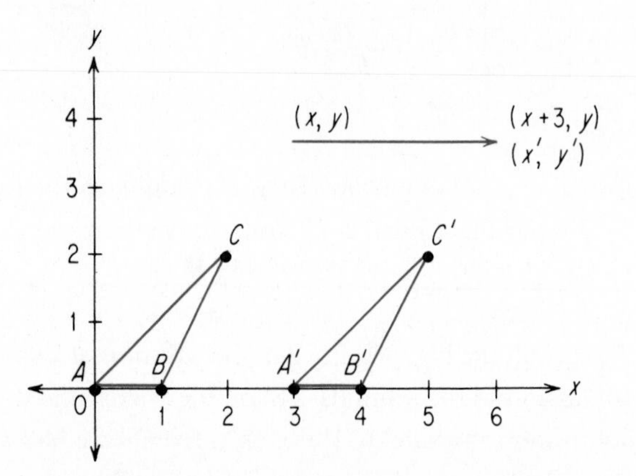

A translation of -2 units parallel to the y-axis would be given by the equations

$$x' = x \quad \text{and} \quad y' = y - 2$$

In general any translation (any slide) can be expressed as a translation of h units parallel to the x-axis and k units parallel to the y-axis and given by the equations

$$x' = x + h \quad \text{and} \quad y' = y + k$$

If one geometric figure is an enlargement or a reduction of another, then the two figures have the same shape and are called **similar figures**.

On a coordinate plane any figure may be *enlarged* by multiplying each coordinate by a real number $k > 1$ and *reduced* using a positive real number $k < 1$. The equations for such transformations are

$$x' = kx \quad \text{and} \quad y' = ky$$

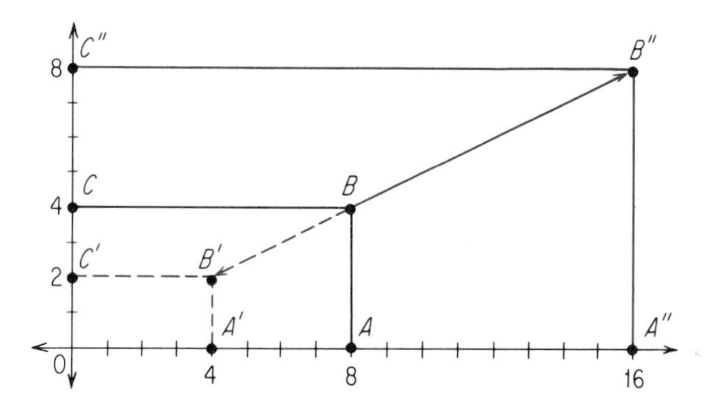

Consider the rectangle $OABC$ with O: (0, 0), A: (8, 0), B: (8, 4), and C: (0, 4). For $k = 1/2$ the rectangle $OABC$ is reduced to $OA'B'C'$ with vertices O: (0, 0), A': (4, 0), B': (4, 2), and C': (0, 2). For $k = 2$ the rectangle $OABC$ is enlarged to $OA''B''C''$ with vertices O: (0, 0), A'': (16, 0), B'': (16, 8), and C'': (0, 8). In each case the length of each side of the rectangle has been multiplied by k, and the area has been multiplied by k^2.

The study of figures under transformations such as translations, rotations, enlargements, and reductions is called **transformation geometry**. For many years it was thought that all geometry could be included in the study of properties of figures and their images under transformations. Even today the exceptions to this concept of geometry occur at only a very advanced and abstract level.

Our ordinary Euclidean geometry is often called the geometry of rigid motions, that is, the geometry of transformations under which size and shape are unchanged. Translations (Exploration 8), rotations (Exploration 9), and indeed all rigid motions can be performed by sequences of reflections in lines. Accordingly, we next combine techniques from paper folding and geoboards to consider reflections in lines and symmetry with respect to lines. The following statements would be true if a coordinate plane were folded along the y-axis.

The points on the y-axis would be folded onto themselves.

Each point $(h, 0)$ on the x-axis would be superimposed on the point $(-h, 0)$.

Each line $x = h$ would be superimposed on the line $x = -h$.

Each line $y = k$ would be folded onto itself.

Each point (h, k) would be superimposed on the point $(-h, k)$.

In the language of line reflections, each point $P: (x, y)$ would be *mapped* onto the point $Q: (-x, y)$. The mapping is a **reflection in the y-axis**, the y-axis is the **axis of symmetry**, the points P and Q are **symmetric with respect to the y-axis**, and the y-axis is the perpendicular bisector of the line segment PQ. In general, under a **line reflection** in a line m, each point of m is mapped onto itself and each point P that is not a point of m is mapped onto a point P' such that the line segment PP' has the line m as its perpendicular bisector.

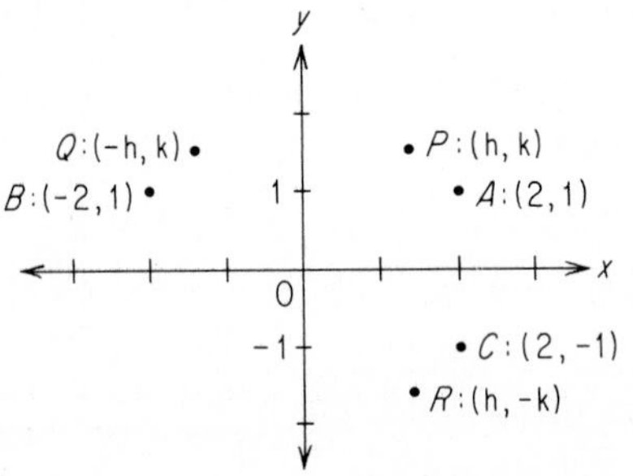

Similarly, a reflection in the x-axis maps $A: (2, 1)$ onto $C: (2, -1)$ and maps $P: (h, k)$ onto $R: (h, -k)$. The x-axis is the axis of symmetry and the perpendicular bisector of each line segment with corresponding points as endpoints. Reflections may be made in any line either by paper folding or using coordinates.

Example 2 Draw the line segment AB with $A: (0, 0)$ and $B: (2, 0)$. Reflect the line segment AB in the line $y = x$ that bisects angle XOY. Label the new line segment AD. Reflect the line segment AB in the line $y = 1$ and reflect the line segment AD in the line $x = 1$.

Solution Under a reflection in the line $y = x$, the point B is mapped onto the point $D: (0, 2)$, the point A is mapped onto itself, and the line segment AB is mapped onto AD.

Under a reflection in the line $y = 1$, the point A is mapped onto D, the point B is mapped onto C: (2, 2), and AB is mapped onto DC.

Under a reflection in the line $x = 1$, the point A is mapped onto B and B onto A, the point D is mapped onto C and C onto D, and the line segment AD is mapped onto BC. Note that also the square $ABCD$ is mapped onto $BADC$, that is, the square is mapped onto itself.

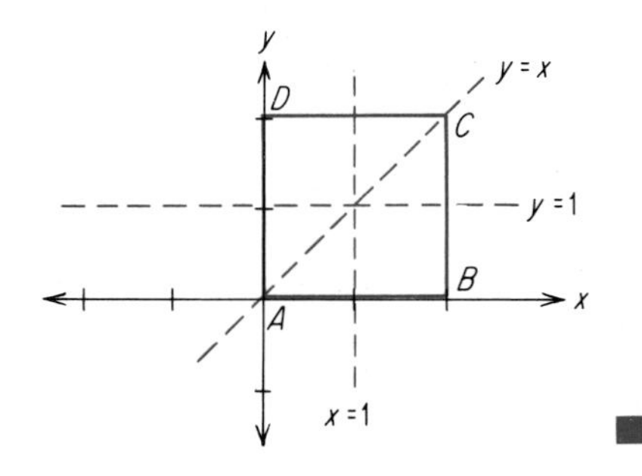

■

In the figure for Example 2 the square $ABCD$ is mapped

onto itself ($BADC$) when reflected in the line $x = 1$,
onto itself ($DCBA$) when reflected in the line $y = 1$,
onto itself ($ADCB$) when reflected in the line $y = x$,
onto itself ($CBAD$) when reflected in the line DB.

In other words, the square is symmetric with respect to each of these four lines and each line is an axis of symmetry of the square. Any square has the two perpendicular bisectors of its sides and the two diagonals as axes of symmetry.

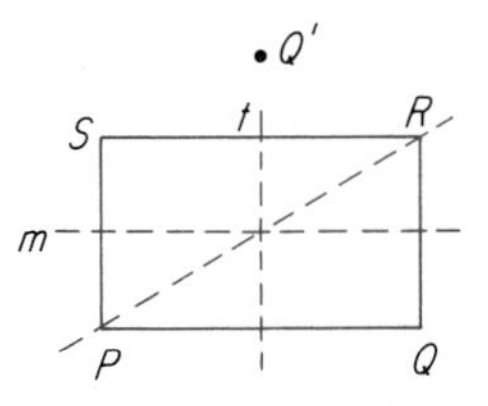

The quadrilateral $PQRS$ is a rectangle that is not a square, t is the perpendicular bisector of the side PQ, and m is the perpendicular bisector of PS. The rectangle is symmetric with respect to t. The rectangle is also symmetric with respect to m. The rectangle is not symmetric with respect to its diagonal PR since, for example, the image Q' of Q is not a point of the rectangle. Any rectangle has two axes of symmetry, the perpendicular bisectors of its sides. Axes of symmetry of other figures are considered in the exercises.

EXERCISES *On a real coordinate plane, graph and label the given figure.*

1. $x = 4, \quad y = -1, \quad P: (4, -1)$
2. $x = -3, \quad y = 2, \quad Q: (-3, 2)$
3. $x = \frac{-3}{2}, \quad y = \frac{5}{2}, \quad R: \left(\frac{-3}{2}, \frac{5}{2}\right)$
4. $x = 2, \quad y = \sqrt{2}, \quad S: (2, \sqrt{2})$
5. $A: (-2, 3), \quad B: (3, -3), \quad C: (-1, -2), \quad D: \left(\frac{1}{2}, \frac{3}{2}\right)$
6. $E: (-3, 2), \quad F: \left(\frac{2}{3}, \frac{4}{3}\right), \quad G: (\sqrt{2}, -2), \quad H: (-1, -\sqrt{2})$

Give coordinates for the vertices of at least one representation of each figure on a coordinate plane. Unless otherwise instructed place one vertex at the origin and one side along the positive x-axis.

7. A right triangle with sides of lengths a and b and with hypotenuse c
   (**a**) With the right angle at the origin.
   (**b**) With an acute angle at the origin.
8. A rectangle with sides of lengths a and b.
9. An isosceles triangle with sides of lengths a, b, and b.
   (**a**) With a vertex on the positive y-axis and its base on the x-axis.
   (**b**) With a base angle at the origin.
10. A square with sides s and with
    (**a**) Two sides on the coordinate axes.
    *(**b**) All four vertices on the coordinate axes.

State equations for the mapping of $P: (x, y)$ onto $P': (x', y')$ to make each specified translation (slide) of figures on a coordinate plane. Assume that the positive x-direction is to the right and the positive y-direction is upward.

11. 5 units to the right.
12. 3 units to the left.
13. 4 units down.
14. 5 units up.
15. 2 units to the right and 3 units up.
16. 3 units to the right and 2 units down.
17. 5 units to the left and 2 units down.
18. 4 units to the left and 5 units up.

Any figure, its enlargements, and its reductions have the same shape and are *similar figures.*

The enlargement or reduction of figures on a coordinate plane may be expressed using equations of the form $x' = kx$, $y' = ky$. For the square ABCD with $A: (0, 0)$, $B: (2, 0)$, $C: (2, 2)$, $D: (0, 2)$, state the coordinates of the vertices of the image square $A'B'C'D'$ under the specified transformation.

19. $k = 2$
20. $k = 4$
21. $k = 1/2$
22. $k = 1/4$

Draw each figure on a coordinate plane.

23. **(a)** Draw a square of side s where $s = 3$. Then draw a square of side $2s$ and find the ratio of the area of the larger square to that of the smaller.
(b) Conjecture the effect upon the area of a square when the length of each of its sides is multiplied by 2.

24. Draw figures as in Exercise 23 as needed and conjecture the effect upon the area of a square when the length of its sides is multiplied by
(a) 3
(b) 4
(c) k, for any natural number k.
(d) Then use the formula $\mathscr{A} = s^2$ and give an algebraic proof of your conjecture in part (c).

25. Conjecture the effect upon the area of any rectangle, right triangle, or other polygon when a similar figure is obtained by multiplying the length of each side of the given figure by a natural number k. Explain the basis for your conjecture.

26. **(a)** Draw the line segment with endpoints (0, 2) and (2, 0). Reflect this line segment in the y-axis. Reflect the two line segments in the x-axis.
(b) Identify the figure obtained in part (a) and two of its axes of symmetry.

For Exercises 27 *through* 30 *copy each figure and draw as many axes of symmetry as possible.*

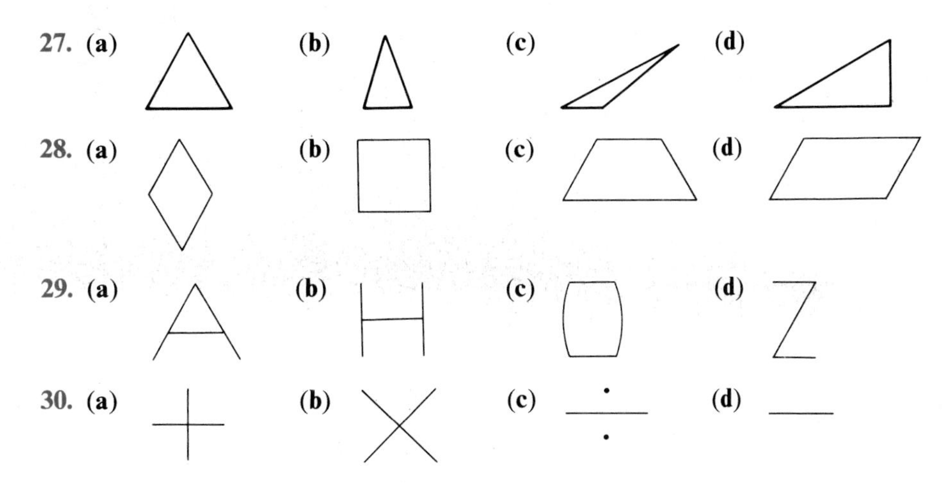

31. The axes of symmetry of a capital letter may depend upon the manner in which the letter is formed. For example, the symmetry of the letter E depends on whether the middle bar is centered vertically.
(a) On a coordinate plane print as many capital letters of the English alphabet as possible so that each letter is symmetric with respect to the y-axis.
(b) Repeat part (a) for symmetry with respect to the x-axis.

32. Print each capital letter of the English alphabet so that it will have as many axes of symmetry as possible, and then complete the table.

Letter	A	B	...
Number of axes of symmetry	1	1	...

EXPLORATIONS

1. 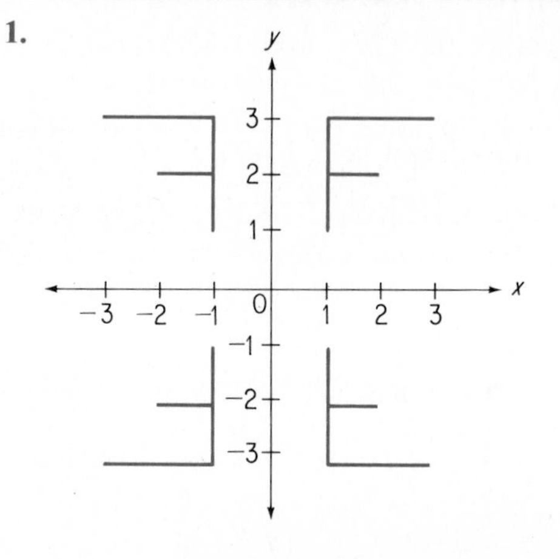

In the figure the letter F was drawn in the first quadrant and reflections in the coordinate axes were used to obtain its images in the other quadrants. Copy the figure, assign coordinates to each of the vertices in the first quadrant, and label each of the vertices in the other quadrants with their coordinates.

2. Capital letters of the English language are drawn in the first quadrant with images in the other quadrants from reflections in the coordinate axes. Which letters can be formed so that the figures appear the same in
   (a) the first and second quadrants
   (b) the first and fourth quadrants
   (c) the first and third quadrants?
3. Repeat Exploration 2 for decimal digits and write a five-digit numeral that would appear the same in all four quadrants.
4. Try to write the word SEVEN as it would appear if held up to a mirror. Then use a mirror to check your results. Try other positions of the mirror as needed until you can describe the mirror images in terms of a line reflection.
5. Make a list of places in which you encounter mirror images in daily life. This list may range from such activities as simply combing your hair while

looking in a mirror, to seeing the word "ambulance" printed on the front of an ambulance so that it reads correctly when you see it in your rear-view mirror.

6. Select an appropriate elementary school grade level and make a worksheet on line reflections. Use words and numbers such as MOM and 101 that are unchanged when reflected in the y-axis, COB and 108 that are unchanged when reflected in the x-axis, and OH and 108 with images that are different but familiar when reflected in the y-axis or when reflected in both axes.
7. When students are first introduced to the concept of rectangular coordinates, they enjoy an exercise that involves plotting points that can be used to form recognizable figures. For example, a set of ordered pairs of numbers can be given, graphed, and connected in order. The resulting figure should be familiar, for instance, an animal or a tree. Prepare a set of such ordered pairs of numbers.
8. Study the effect upon a triangle ABC of two successive line reflections in parallel lines m and n. Express the result in terms of a translation, the direction of the parallel lines, and the distance between the parallel lines.

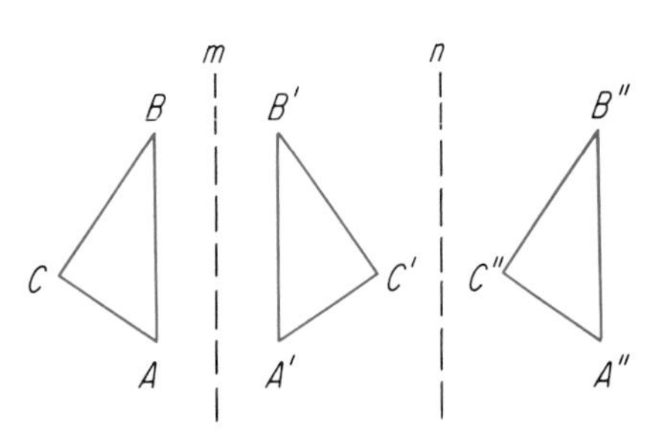

9. The result of two successive line reflections of a figure in two intersecting lines is a rotation about the point of intersection of the lines and of twice the angle formed by the lines. Draw several figures to illustrate this basic role of line reflections and explain why the specified rotation is obtained.

9-6 Measures of Space Figures

Any polyhedron or other simple closed surface is the common boundary of a bounded solid region (its *interior*) and an unbounded solid region (its *exterior*). The **volume** V of the polyhedron is the number of solid unit cubes that are "equivalent" to the interior of the polyhedron. It is customary to speak of the "volume of the polyhedron" rather than specifying the "volume of the solid region bounded by the polyhedron." The surface area S of any polyhedron is the sum of the areas of its faces. In drawings of space figures, hidden edges are indicated by dashed line segments, as in the adjacent drawing of a unit cube.

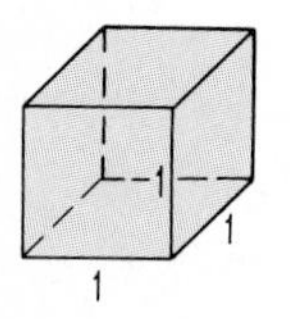

Volumes are considered in a manner very similar to areas. We define any rectangular box with edges of length a, b, and c on a common endpoint to have volume abc (see the next figure). A cube with edge of length e has volume e^3 and surface area $6e^2$. The rectangular box with edges a, b, and c has surface area $2ac + 2ab + 2bc$. As for lengths and areas the word *volume* is used to refer to both the *volume measure* (number) and the *volume measurement* (number of units). The units are specified when they are known.

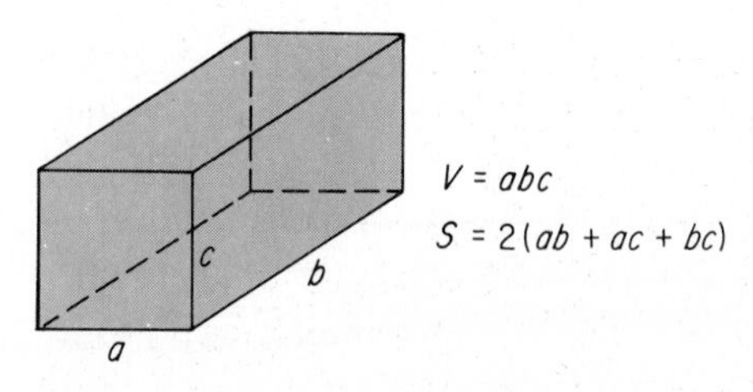

Example 1 For a rectangular box 20 centimeters by 10 centimeters by 4 centimeters, find **(a)** the volume **(b)** the surface area.

Solution **(a)** Sketch the box. The number of cubic centimeters that could be placed in the box can be counted. The height of 4 centimeters indicates that there could be four layers of unit cubes. Each layer must cover a 10 centimeter by 20 centimeter rectangular region and thus contain 200 unit cubes. Then the four layers contain 4×200, that is, 800 cubic centimeters (cm^3). As in the formula, $V = 20 \times 10 \times 4 = 800$ cm^3.

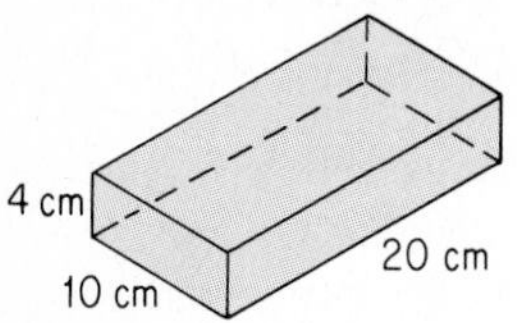

(b) The box has six rectangular regions as faces. We may think of the top and bottom as each 10 centimeters by 20 centimeters, that is, with area 200 cm^2 each; the two ends as 10 centimeters by 4 centimeters, that is, with area 40 cm^2 each; and the two sides as 20 centimeters by 4 centimeters, that is, with area 80 square centimeters (cm^2). Thus the surface area of the box is, as in the formula $S = 2ac + 2ab + 2bc$,

$$2(10 \times 20) + 2(10 \times 4) + 2(20 \times 4), \quad \text{that is,} \quad 640 \text{ cm}^2 \quad ■$$

1 dm = 10 cm
1 dm^2 = 100 cm^2
1 dm^3 = 1000 cm^3

To change units for areas (surfaces) "jumps" of 2 decimal places are needed for each change of 1 decimal place for distances (Section 9-1).

640 cm^2 = 6.4 dm^2

To change units for volumes "jumps" of 3 decimal places are needed for each change of 1 decimal place for distances.
800 cm^3 = 0.8 dm^3

Several solid figures were considered in Section 8-5. The bases of any prism or cylinder are congruent figures in parallel planes. The height h of a prism or cylinder is the perpendicular distance between the planes of its bases. The volume V of a prism or cylinder is

$$V = Bh$$

where B is the area of the base and h is the height. If the lateral edges of a prism are perpendicular to the plane of a base, then the prism is a right prism and the length of an edge is the height. The volumes V and the surface areas S of some common solids are shown in the figures.

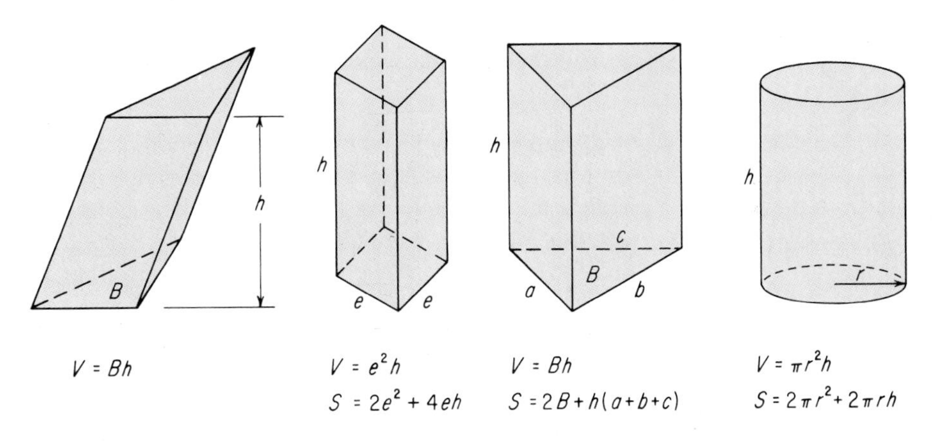

$V = Bh$

$V = e^2h$
$S = 2e^2 + 4eh$

$V = Bh$
$S = 2B + h(a+b+c)$

$V = \pi r^2 h$
$S = 2\pi r^2 + 2\pi rh$

Example 2 For a right circular cylinder with radius 3 centimeters and height 10 centimeters, find **(a)** the volume **(b)** the surface area.

Solution **(a)** $V = \pi(3^2)(10) = 90\pi \text{ cm}^3$

(b) $S = 2\pi(3^2) + 2\pi(3)(10) = 78\pi \text{ cm}^2$
Note that the lateral (curved) surface is like the paper label around a cylindrical can. If it were cut along a line perpendicular to the base, it would unroll into a rectangular region with the circumference of the base as one side and the height of the cylinder as the other side. ■

Think of ways in which you use measures of space figures. Consider, for example, distances in space (such as the distance between two classes), areas of space figures (such as the surface of a lawn or the walls of a room that is to be painted), and volumes or capacities (such as the amount of gasoline now in the car).

Pyramids and cones were also considered in Section 8-5. Each has a base and a vertex that is not on the plane of the base. The height is the distance of that vertex from the plane of the base as measured along a line that is perpendicular to the plane of the base. The volume of a pyramid or cone is

$$V = \frac{1}{3}Bh$$

where B is the area of the base and h is the height. Formulas for volumes of pyramids, cones, and spheres are shown with the following figures and accepted here without formal proof.

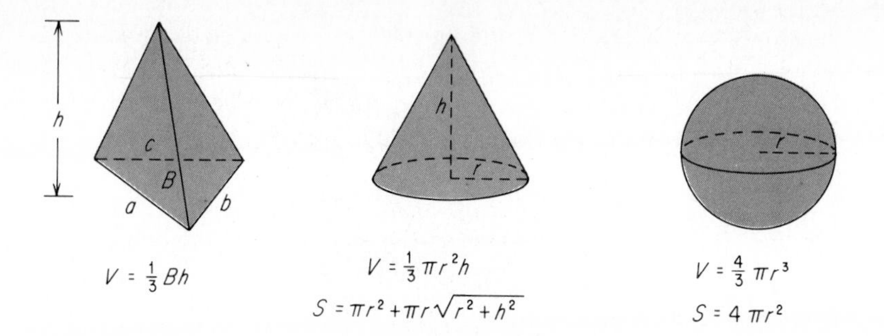

The **surface area** of a pyramid is the sum of the areas of its faces. The volumes and surface areas of many other space figures may be found by adding or subtracting those measures of common figures.

Example 3 A certain building has the shape of a rectangular box with a hemispherical dome on top. The box is 20 meters by 30 meters by 10 meters. The dome has radius 6 meters and the height of the building is 16 meters. Find

(a) the volume of the building

(b) the exposed surface area of the building, that is, the area above the level ground on which the building is located.

Solution **(a)** The rectangular box has volume $20 \times 30 \times 10$, that is, 6000 m^3. The dome has volume $(1/2)[(4/3)\pi \times 6^3]$, that is, 144π m^3. Thus the building has volume $(6000 + 144\pi)$ m^3.

(b) The sides of the building have area $200 + 300 + 200 + 300$, that is, 1000 m^2. The flat roof has area $(600 - 36\pi)$ m^2. The dome has area $(1/2)(4\pi \times 36)$, that is, 72π m^2. Thus the exposed surface area is $1000 + (600 - 36\pi) + 72\pi$, that is, $(1600 + 36\pi)$ m^2. ■

EXERCISES

For a cube with edges of the given length, find
(a) *the volume* **(b)** *the surface area.*

1. 5 cm **2.** 8 dm **3.** 7 cm
4. 2 m **5.** 15 mm **6.** 25 cm

For the given rectangular box, find
(a) *the volume,* **(b)** *the surface area.*

7.

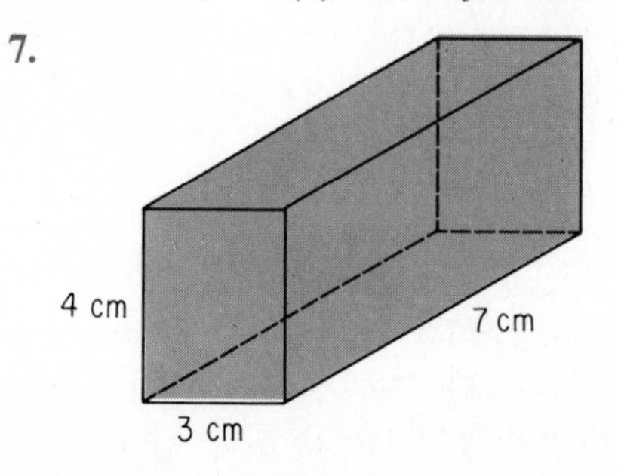

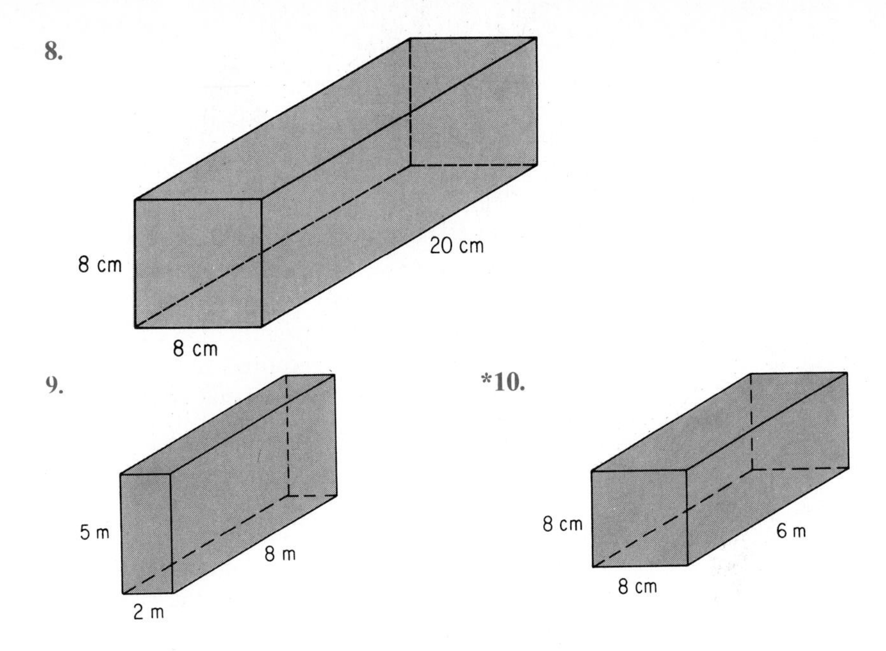

Find the volume of the indicated pyramid.

11.

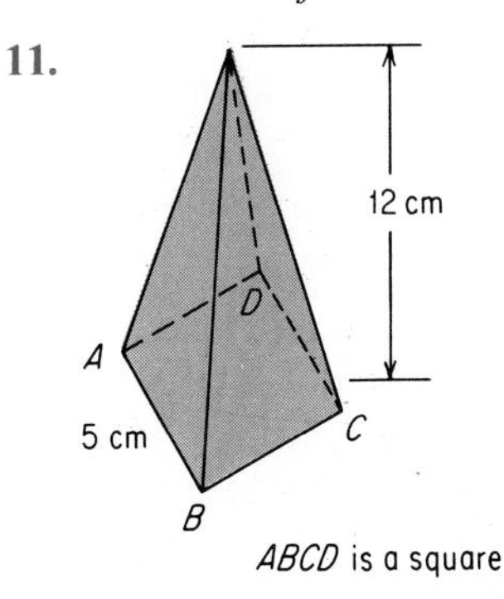

12.

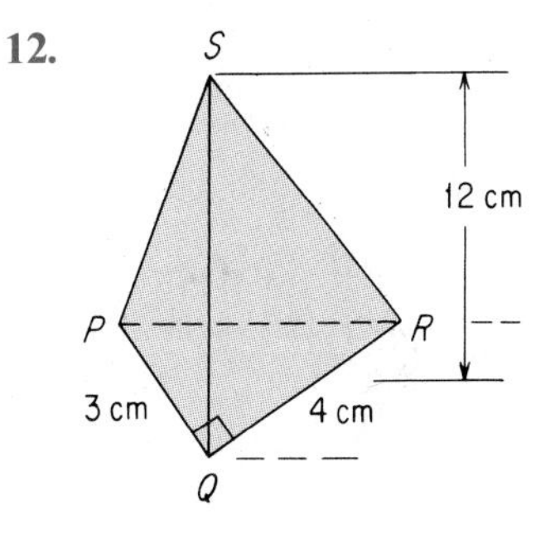

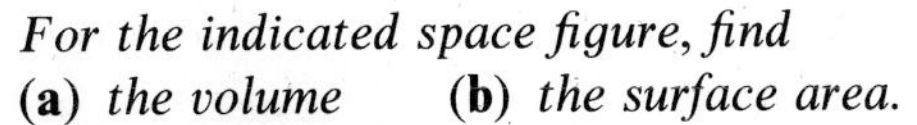

For the indicated space figure, find
(a) *the volume* **(b)** *the surface area.*

15. A spherical ball with radius 10 centimeters.
16. A right circular cylinder 20 centimeters long, with radius 3 centimeters, and with hemispherical caps on both ends.
17. A cone with radius 5 centimeters and with height 12 centimeters.
18. A cone with radius 4 centimeters, a hemispherical base of the same radius, and total height of cone and hemisphere 20 centimeters.

In Exercises 19 *through* 22, *describe the change in volume and the change in the surface area of the given figure.*

19. A cube when the length of each edge is multiplied by
(**a**) 2 (**b**) 3 (**c**) 1/2 (**d**) a positive number k.

20. Repeat Exercise 19 for a rectangular box.

21. Repeat Exercise 19 for the radius and height of a right circular cylinder.

22. Repeat Exercise 19 for the radius and height of a cone.

23. A given spherical ball required 36 cm^2 of gold leaf to cover it. How much gold leaf would be required to cover a spherical ball whose diameter is (**a**) half (**b**) twice that of the given ball?

24. If a given solid spherical ball weighs 64 pounds, how much does a spherical ball of the same material weigh if its diameter is (**a**) half (**b**) twice that of the given ball?

25. If a given solid cube of gold costs $8000, approximately how much should you expect to pay for a cube of the same quality gold with edge (**a**) half (**b**) twice that of the given cube?

26. If it costs $2.25 to polish all faces of a given cube, approximately how much should you expect it to cost to polish a cube of the same material with edge (**a**) half (**b**) twice that of the given cube?

***27.** All of the formulas for volumes that we have considered are special cases of the **prismoidal formula**.

$$V = \frac{h}{6}(B_1 + 4M + B_2)$$

where B_1 and B_2 are the areas of the bases and M is the area of the intersection of the figure with a plane halfway between the bases. In this case the vertex of a pyramid or a cone is considered a base of area zero. Show that the prismoidal formula gives the usual formula for the volume of a cube.

***28.** Repeat Exercise 27 for
(**a**) a prism with base of area B and height h
(**b**) a cylinder with radius r and height h.

***29.** Think of a sphere as in the figure and use the prismoidal formula to find an expression for the volume of a sphere of radius r.

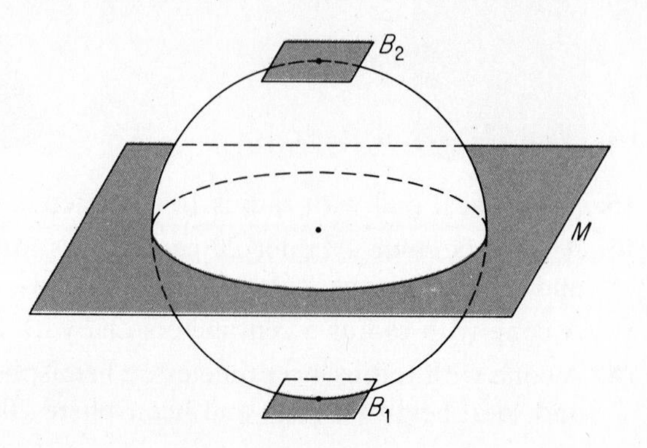

*30. For a square pyramid with height h and a base with an edge of linear measure e the midsection is a square with an edge $(1/2)e$. Use the prismoidal formula to find an expression for the volume of the square pyramid.

*31. Use the prismoidal formula to find an expression for the volume of a circular cone with height h if the base has radius r and the midsection has radius $(1/2)r$.

EXPLORATIONS

Each of the Explorations 1 through 5 involves the determination of a shortest path a spider has to crawl along the outside of a box to reach a fly F that is assumed to stay in one place. As in the figure, we consider a closed rectangular box 12 centimeters wide, 18 centimeters long, and 8 centimeters tall.

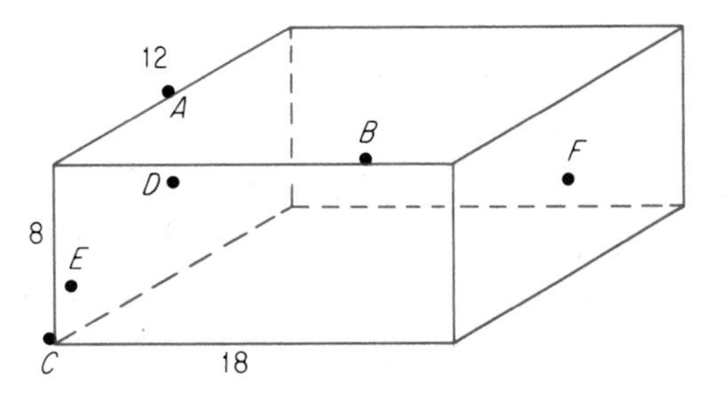

For each exploration assume that the fly is at the center F (intersection of the diagonals) of an end of the box. Copy the figure, draw the shortest path(s) for the spider, and find the length of the shortest path. If there are two or more paths of minimal length, draw them all.

1. The spider is at position A in the center of the top edge of the opposite end of the box.
2. The spider is at position B at the top of an adjoining side and 4 centimeters from the end containing the fly. (*Hint*: Think of the box as flattened out in some way.)
3. The spider is in the corner C at the bottom of the opposite end of the box.
4. The spider is at the middle D of the end of the box opposite the fly.
5. The spider is on the opposite end of the box at position E 1 centimeter from the vertical edge of the box over C and 2 centimeters from the bottom of the box.

Volumes of space figures may be used to develop the concept of large numbers. For Explorations 6 through 9 conjecture answers for the stated questions relative to your classroom or another room that is accessible to you.

Then describe procedures that could be used for verifying, or correcting, your conjectures.

6. Would 1 million pennies fit in the room? One thousand? One billion?
7. Repeat Exploration 6 for quarters.
8. Would 1 million Ping-Pong balls fit in the room? Tennis balls? Basketballs?
9. Would 1 million jelly beans fit in the room? Eggs? Quart cartons of milk?

10. Cut out a circular region (disk) and remove a sector by cutting along two radii. Use the remaining part of the disk as a model for the lateral surface of a cone. Explore ways of predicting the radius and the height of the resulting cone.
11. Examine at least one series of mathematics textbooks for grades 1 through 6. Summarize the treatment of space figures and related formulas in the series of textbooks.
12. Select a grade level and prepare a short lesson plan for introducing the volume of a right circular cylinder.

Chapter Review

1. Change each of the following as indicated.
   (a) 2 centimeters to millimeters
   (b) 5000 meters to kilometers
   (c) 12 grams to milligrams
   (d) 1500 milliliters to liters

2. State the unit of precision for the given measurement.
   (a) 2.5 cm
   (b) $2\frac{1}{2}$ cm

3. Identify the indicated measurement as exact or approximate.
   (a) Spot weighs 22 *kilograms*
   (b) Jack is carrying 3 *books*.

4. Find
   (a) The radian measure for an angle of 120°.
   (b) The degree measure for an angle of $\pi/6$ radians.

In Exercises 5 *through* 8 *construct each figure for a given line segment AB.*

5. An acute angle BAC and its angle bisector.
6. An equilateral triangle ABE.
7. A right isosceles triangle ABD with the right angle at B.
8. A square $ABFG$.

9. Make dot paper to represent a 6×6 geoboard and label the point in the specified position as indicated.
   (a) Column 0 and Row 3 as A.
   (b) Column 2 and Row 4 as B.
   (c) Column 3 and Row 1 as C.
   (d) Column 5 and Row 3 as D.
10. Make dot paper to represent a 6×6 geoboard and label each point in Row 3 with its distance from the origin.
11. For triangles with vertices at points of a 6×6 geoboard state the lengths of the bases of four isosceles triangles of different shapes but with two sides of length $\sqrt{10}$.
12. Find the area of the quadrilateral $ABCD$ with A: (2, 0), B: (5, 3), C: (4, 5), and D: (1, 4).
13. On a real coordinate plane, graph and label $x = -2$, $y = 3$, and P: $(-2, 3)$.
14. For triangles ABC with A: (2, 0) and B: (6, 0), positive integral coordinates for C, and area 6, find the coordinates for C such that the perimeter of the triangle is as small as possible.
15. Draw a square $PQRS$ and as many axes of symmetry as possible.
16. Draw **(a)** the line segment AB with A: $(-1, 0)$ and B: (0, 1) **(b)** the reflection in the x-axis of the line segment AB **(c)** the reflection in the y-axis of the line segment AB and the line segment obtained in part (b).
17. Approximate π as 3.14 and find the approximate area of any circle with radius 4 centimeters.

Find the area of the given figure.

18.

19.

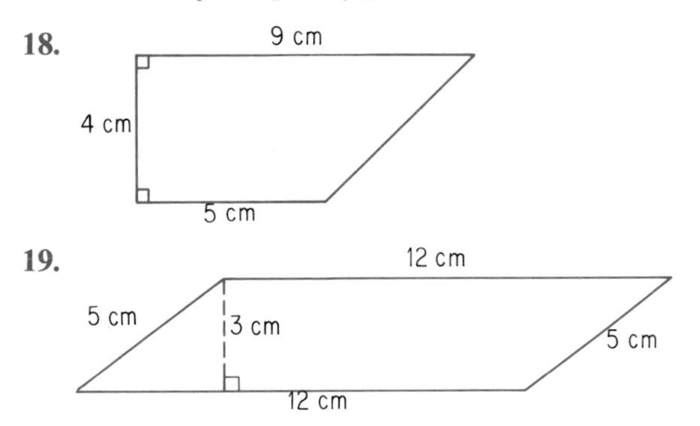

20. For the given curve, find **(a)** the length **(b)** the area. The curve is formed by line segments and arcs of circles.

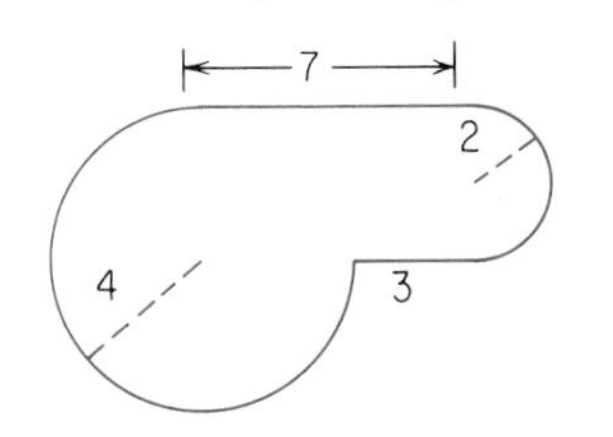

For each figure in Exercises 21 *through* 24, *find* **(a)** *the volume* **(b)** *the surface area.*

21. A cube with an edge of length 3 meters.
22. A rectangular box with edges of lengths 3 meters, 5 meters, and 7 meters.
23. A right circular cylinder of radius 2 centimeters and height 7 centimeters.
24. A spherical ball with radius 5 centimeters.
25. If each of the edges of a cube is multiplied by 5, describe the change in **(a)** the surface area **(b)** the volume.

LOGO PROCEDURES

In the computer applications section of Chapter 8, we saw the use of four repetitions of the command FD 1Ø RT 9Ø to complete the drawing of a square. There are two ways in which this repetitive process can be shortened. The first of these is through the use of a REPEAT command. This command allows one to define a certain sequence of steps and then specify the number of times they are to be repeated. For example, the command

```
REPEAT 4[FD 1Ø RT 9Ø]
```

directs the computer to carry out the sequence of commands FD 1Ø RT 9Ø four times and then stop. The result is the drawing of a square.

Consider the use of the command REPEAT 6[FD 1Ø RT 6Ø]. Describe the figure obtained from the execution of this command. Make up some additional commands of this type and try them out on your computer. The [and] symbols, if not on your keyboard, can often be accessed by typing the capital letters *N* and *M*, respectively, when the Logo language is loaded in the microcomputer.

The second method for executing a specified set of commands involves the designation of the set of commands as a PROCEDURE. For example, the entire set of commands to draw a square in Chapter 8 could have been saved in total as a procedure, identified by a name such as TO SQUARE, and recalled by name when needed.

To develop a procedure to draw a polygon of 12 sides, a dodecagon, we could enter the following Logo commands to the computer:

```
TO DODECAGON
REPEAT 12[FD 1Ø RT 3Ø]
END
```

The execution of the procedure would be as shown. To get the figure after the procedure has been defined all we have to do is enter the command DODECAGON, the name of the procedure. This will cause the turtle to start immediately to execute the design.

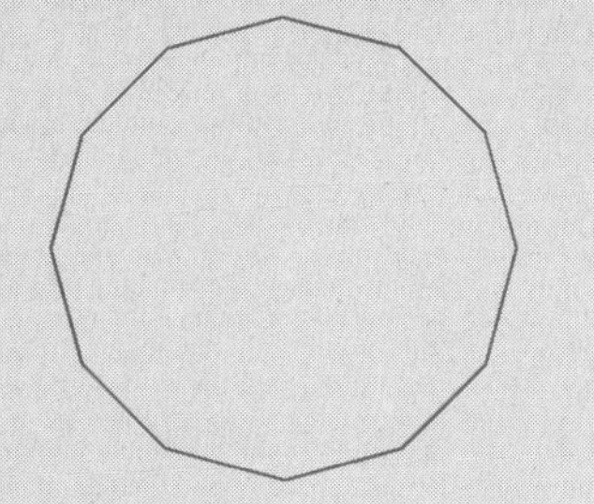

This form of giving Logo commands can be extended to handle even more difficult figures, such as the six-pointed star shown beside its Logo commands. This procedure is given by:

```
TO STAR
REPEAT 3[FD 6Ø RT 12Ø]
FD 2Ø
LT 6Ø
FD 2Ø
REPEAT 3[RT 12Ø FD 6Ø]
END
```

Make up a few procedures of your own, predict their outputs, and then, if possible, enter them and execute them under Logo. Were the results what you expected? If not, can you make the necessary changes?

10

An Introduction to Probability

Pierre de Fermat
(1601–1665)

The French mathematician Fermat made significant contributions to analytic geometry and number theory. Along with his contemporary Blaise Pascal, Fermat laid the foundations for the modern theory of probability. This work began with the question of the equitable distribution of the stakes in a game of chance if the game is interrupted before it has had an opportunity to terminate according to the rules. These contributions are even more amazing when one recognizes that Fermat, a lawyer, worked on mathematics only as a recreation.

10-1 Concept of Probability

We make frequent reference to probability problems in everyday language.

The probability of rain today is 20%.
The odds are in her favor.
His chances are 50–50.

One of the authors of this book has two children. One of these children is a boy. What is the probability that they are both boys? Believe it or not the answer is *not* 1/2. This problem appears in the Explorations for Section 10-2.

Predictions of probabilities are usually based upon some form of counting of past experiences and a comparison of the number of favorable outcomes with the total number of outcomes. For example, since 20% = 1/5, the statement that the probability of rain is 20% is a statement that in past situations in which conditions were like the present, the ratio of the number of such days in which it has rained to the total number of these days is 1 to 5. In general, the *probability* of an event is a number, usually expressed as a ratio:

$$\frac{\text{Number of favorable outcomes}}{\text{Total number of possible outcomes}}$$

The task of counting is an essential part of the study of probability. To illustrate various problems in this chapter, we shall invent a fictitious club consisting of a set M of members, $M = \{\text{Betty, Doris, Ellen, John, Tom}\}$.

Let us form a committee that is to consist of one boy and one girl, each selected from the set M of club members. How many such committees are possible? One way to find the answer to this question is by means of a *tree diagram.*

Tree diagrams are very useful when solving many problems that involve probabilities.

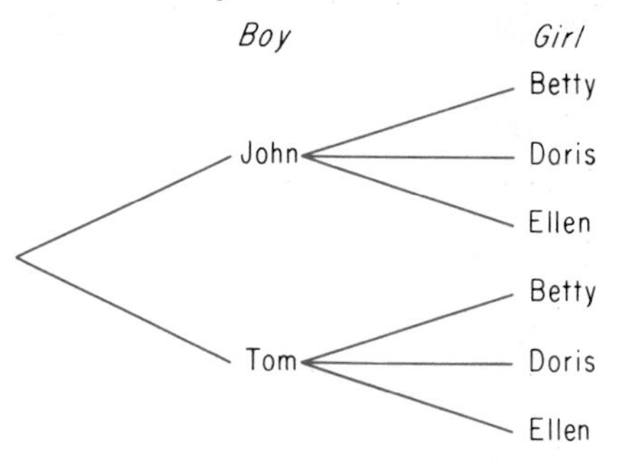

For each of the two possible choices of a boy, there are three possible choices of a girl. Thus the following six distinct possible committees can be formed and can be read from the tree diagram.

John-Betty	Tom-Betty
John-Doris	Tom-Doris
John-Ellen	Tom-Ellen

Suppose that we had selected a girl first. Then the tree diagram would be as shown in the next figure and there would still be six possibilities, the same six committees as before.

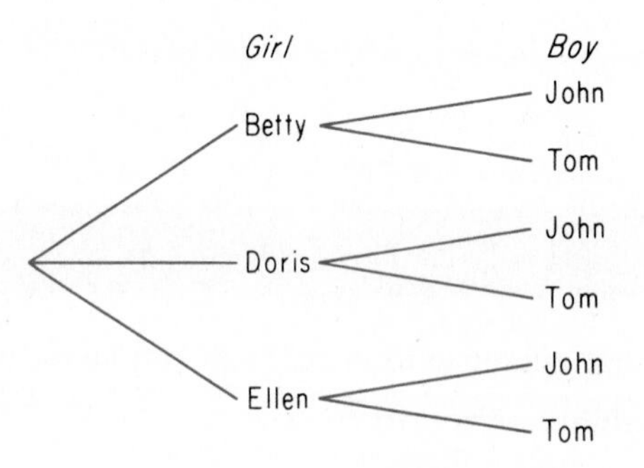

Example 1 How many different selections of two officers, a president and a vice-president, can be elected from the set M of club members?

Solution Let us select the officers in two stages. There are five possible choices for the office of president. Each of these five selections may be paired with any one of the remaining four members. Thus there are 20 possible choices. These can be read from the following diagram.

Here is a problem for you to explore. Fifty teams enter a basketball tournament. If a team is eliminated by a single loss, how many games must be played to determine a winner?

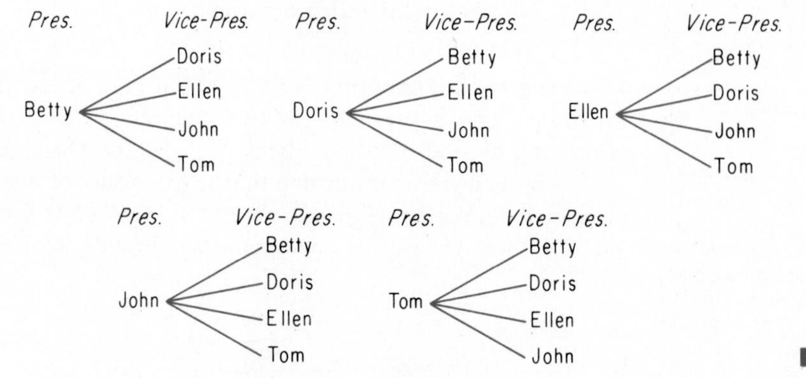

■

Principle of counting

In general, if one task can be performed in m different ways and a second task can be performed in n different ways, then the first and second tasks together can be performed in $m \times n$ different ways. This **general principle of counting** can be extended if there are additional tasks.

$$m \times n \times r \times \cdots \times t$$

Example 2 The club M must send a delegate to a meeting tomorrow and also a delegate to a different meeting next week. How many different selections of these delegates may be made if any member of the club may serve as a delegate to either, or both, of these meetings?

Solution There are five possible choices of a delegate to the first meeting. Since no restriction is made, we assume that the same member may attend each of the two meetings. Thus, there are five choices for the delegate to next week's meeting. In all, there are 5×5, that is, 25 choices. ■

Example 3 How many three-letter "words" may be formed from the set of vowels $V = \{a, e, i, o, u\}$ if no letter may be used more than once? (A word in this sense is any arrangement of three letters, such as *aeo*, *iou*, and so on.)

Solution There are five choices for the first letter, four for the second, and three for the third. In all, there are $5 \times 4 \times 3$, that is, 60 possible words. Notice that 125 words are possible if repetitions of letters are permitted. ■

When a normal coin is tossed, we know that there are two distinct and **equally likely** ways in which it may land, heads or tails. We say that the probability of getting a head is one out of two, or simply 1/2.

In throwing one of a pair of normal dice, there are six equally likely ways in which the die may land. We say that the probability of throwing a 5 on one throw of a die is one out of six, or 1/6.

In each of these two examples, the **events** that can occur are said to be **mutually exclusive**. That is, one and only one of the events can occur at any given time. When a coin is tossed, there are two possible events (heads and tails); one and only one of these can occur. When a single die is thrown, there are six events $\{1, 2, 3, 4, 5, 6\}$; one and only one of these can occur. In general, if an event can occur in any one of n mutually exclusive and equally likely ways and if m of these ways are considered favorable, then the **probability** $P(A)$ that a favorable event A will occur is given by the formula $P(A) = m/n$.

Informally we define the probability of success as the ratio of the number of successes of an event to the number of possible outcomes of that event;

$$P(A) = \frac{m}{n}$$

The notation $n(S)$ for the number of elements in any set S (Section 2-2) enables us to state this definition of probability as follows:

> For any set S of all possible outcomes of an event, if the outcomes are mutually exclusive and equally likely, then the probability of an outcome from a subset A is
>
> $$P(A) = \frac{n(A)}{n(S)}$$

The probability m/n satisfies the relation $0 \leq m/n \leq 1$, since m and n are integers and $m \leq n$.

> When success is inevitable, $m = n$ and the probability is 1.
>
> When an event cannot possibly succeed, $m = 0$ and the probability is 0.

For example, the probability of getting either a head or a tail on a single toss of a coin is 1, assuming that the coin does not land on an edge. The probability of throwing a sum of 13 with a single throw of a pair of normal dice is 0. (Always assume, unless otherwise instructed, that normal dice are used.)

The sum of the probability that an event occurs and the probability that the same event does not occur is 1.

$$\text{If } P(A) = \frac{m}{n}, \quad \text{then} \quad P(\text{not } A) = 1 - \frac{m}{n}.$$

Example 4 A single card is selected from a deck of 52 bridge cards. What is the probability that it is a spade? What is the probability that it is not a spade? What is the probability that it is an ace or a spade?

Solution Of the 52 cards, 13 are spades. Therefore, the probability of selecting a spade is 13/52, that is, 1/4. The probability that the card selected is not a spade is 1 − 1/4, that is, 3/4. There are four aces and 12 spades besides the ace of spades. Therefore, the probability that the card selected is an ace or a spade is (4 + 12)/52, that is, 16/52, which we express as 4/13. ■

It is very important that only equally likely events be considered when the probability formula is applied; otherwise, faulty reasoning can occur.

Consider again the first question asked in Example 4. One might reason that any single card drawn from a deck of cards is either a spade or is not a spade; thus there are two possible outcomes, and the probability of drawing a spade must therefore be 1/2. It is correct to say that there are these two possible outcomes, but of course they are *not* equally likely since there are 13 spades in a deck of cards and 39 cards that are not spades. If all possible events are equally likely, the events are said to occur **at random**.

Example 5 A committee of two is to be selected at random from the set

$$M = \{\text{Betty, Doris, Ellen, John, Tom}\}$$

by drawing names out of a hat. What is the probability that both members of the committee will be girls?

Solution We solve this problem by first listing all of the possible committees of two that can be formed from the set M.

[Betty – Doris]	Doris – John
[Betty – Ellen]	Doris – Tom
Betty – John	Ellen – John
Betty – Tom	Ellen – Tom
[Doris – Ellen]	John – Tom

Making a list continues to be a very important problem-solving strategy.

Of the ten possible committees, there are three (those boxed) that consist of two girls. Thus the probability that both members selected are girls is 3/10. What is the probability of selecting a committee to consist of two boys? ■

EXERCISES

1. If no letter may be used more than once in a given "word," how many three-letter "words" may be formed from the given set?
   (a) $\{m, a, t, h\}$ **(b)** $\{m, e, t, r, i, c\}$
2. Repeat Exercise 1 for four-letter "words."
3. Repeat Exercise 1 if repetitions of letters are allowed.
4. If repetitions of letters are allowed, how many four-letter "words" may be formed from the given set?
   (a) $\{h, o, p, e\}$ **(b)** $\{a, e, i, o, u\}$
5. How many different batteries consisting of a pitcher and a catcher may a baseball team form from the specified group?
   (a) Four pitchers and two catchers.
   (b) Five pitchers and three catchers.
   (c) Six pitchers and three catchers.
6. How many different outfits can Roberto assemble if he can wear any combination of the specified shirts and slacks?
   (a) Three sport shirts and four pairs of slacks.
   (b) Five sport shirts and four pairs of slacks.
   (c) Six sport shirts and four pairs of slacks.
7. How many different outfits can Maria assemble if she can wear any combination of the specified dresses, hats, and pairs of shoes?
   (a) Four dresses, three hats, and two pairs of shoes.
   (b) Five dresses, three hats, and two pairs of shoes.
   (c) Six dresses, four hats, and three pairs of shoes.
8. Assume that no person is allowed to hold more than one office at a time and the number of members of the Portland Swim Club is
   (a) 15 **(b)** 20 **(c)** 50 **(d)** 100.
   How many different sets of officers consisting of a president, a vice-president, and a secretary are possible?

What is the probability that the specified event will occur in a single throw of one die?

9. An even number.
10. An odd number.
11. A number greater than 2.
12. A number less than 4.
13. A number different from 4.
14. A number different from 0.
15. The number 0.
16. A number less than 7.

What is the probability that the specified event will occur in a single draw from an ordinary deck of 52 bridge cards?

17. An ace.
18. A king.
19. A spade.
20. A red card.

How many two-digit numbers may be formed from the given set of digits **(a)** *if repetitions are not allowed* **(b)** *if repetitions are allowed?*

21. $\{1, 2, 3, 4\}$
22. $\{1, 2, 3, 4, 5\}$
23. $\{1, 2, 3, 4, 5, 6\}$
24. $\{1, 2, 3, 4, 5, 6, 7\}$
25. $\{1, 2, 3, 4, 5, 6, 7, 8\}$
26. $\{1, 2, 3, 4, 5, 6, 7, 8, 9\}$

How many three-digit numbers may be formed from the given set of digits if zero is not to be used as the first digit and
(a) *repetitions are not allowed* **(b)** *repetitions are allowed?*

27. $\{0, 1, 2, 3, 4\}$
28. $\{0, 1, 2, 3, 4, 5\}$
29. $\{0, 1, 2, 3, 4, 5, 6\}$
30. $\{0, 1, 2, 3, 4, 5, 6, 7\}$
31. $\{0, 1, 2, 3, 4, 5, 6, 7, 8\}$
32. $\{0, 1, 2, 3, 4, 5, 6, 7, 8, 9\}$

In Exercises 33 *through* 36 *consider the appointment of a committee of two from the set* $N = \{Alice, Bob, Carolyn, Doug, Ellen, Frank\}$.

33. How many different committees of two can be formed?

34. What is the probability that a committee of two will consist of two boys?

35. What is the probability that a committee of two will consist of two girls?

36. What is the probability that a committee of two will consist of one boy and one girl?

Hint for Exercise 37(a): Think of the words in this way: $\underline{i}$ — — —.

37. Find the number of "words" of three different letters that may be formed from the set of vowels $V = \{a, e, i, o, u\}$ if
(a) the first letter must be i
(b) the first letter must be e and the last letter must be i.

38. How many different license plates can be made using a letter from our alphabet followed by four decimal digits
(a) if the first digit must not be zero
(b) if the first digit must not be zero and no digit may be used more than once?

39. What is the probability that the first-named author of this text was born in the month of December?

40. What is the probability that your instructor's telephone number has a 7 as its final digit?

41. The probability of obtaining all heads in a single toss of three coins is 1/8. What is the probability that not all three coins are heads in such a toss?

42. What is the probability that the next person you meet was not born on a Sunday?

In Exercises 43 *through* 48, *how many three-digit numbers may be formed from the set* $\{0, 1, 2, 3, \ldots, 9\}$ *if zero is not an acceptable first digit and the given conditions must be satisfied?*

***43.** Repetitions are allowed and the number must be divisible by
(a) 5 **(b)** 25.

*44. Repetitions are allowed and the number must be
(a) even (b) divisible by 10.

*45. Repetitions are not allowed and the number must be
(a) divisible by 10 (b) even.

*46. Repetitions are not allowed and the number must be divisible by
(a) 5 (b) 25.

*47. The number must be odd and less than 600 with repetitions
(a) allowed (b) not allowed.

*48. Repeat Exercise 47 for numbers less than 800.

*49. Find the number of different possible license plates if each one is to consist of two letters of our alphabet followed by four decimal digits; the first digit may not be zero, and no repetitions of letters or numbers are permitted.

*50. Repeat Exercise 49 for license plates consisting of four consonants followed by three decimal digits.

*51. In a certain combination lock, there are 60 different positions. To open the lock you move to a certain number in one direction, then to a different number in the opposite direction, and finally to a third number in the original direction.
(a) What is the total number of such "combinations" if the first turn must be clockwise?
(b) What is the total number of such "combinations" if the first turn may be either clockwise or counterclockwise?

*52. Two distinct integers from the set 1 through 7 are selected at random.
(a) What is the probability that the first integer selected is even?
(b) What is the probability that both integers are even?

*53. Repeat Exercise 52 for the set of integers 1 through 13.

EXPLORATIONS

1. Identify at least one item that appears regularly in the daily newspaper and is based upon probability.

There is evidence of dice having been used as early as 3000 B.C.

Find (or make) two boxes to use as two cubes. Cover each cube with paper so that the cubes are of different colors, such as red and green. To introduce the vocabulary of dice games think of one cube as a red die and the other cube as a green die. Number the faces of each die 1 through 6. Have two students hold the "dice" so that the rest of the class can see only one number on each die.

2. Suppose that the number on the green die is 5. If possible identify a number that can be shown on the red die so that the two numbers have the indicated sum.
(a) 5 (b) 6 (c) 7
(d) 11 (e) 3 (f) 12

3. If the number on the green die is 2, what is the set of numbers that can be obtained as sums by suitable selections of numbers on the red die?

4. What is the set of numbers that can be obtained as sums by suitable selections of a number on each of the two dice?
5. Which of the sums obtained in Exploration 4 can be obtained in only one way? List this way for each of the numbers.
6. Which of the sums obtained in Exploration 4 can be obtained in only two ways? List the ways for each of the numbers.
7. Which of the sums obtained in Exploration 4 can be obtained in the largest number of ways? List these ways.

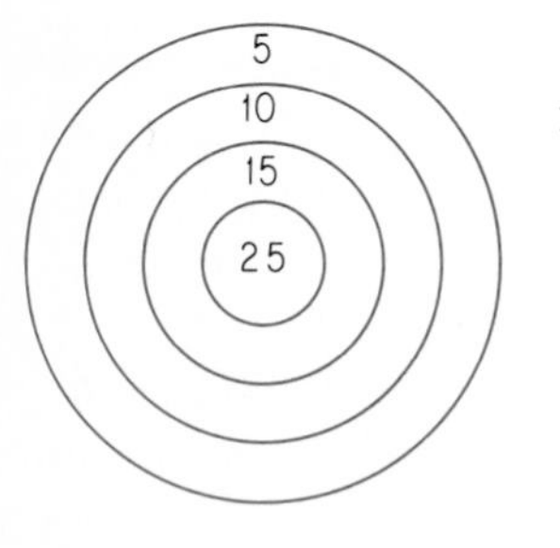

Counting problems arise in many mathematical puzzles. See the adjacent figure and consider, for example, the following dart puzzles.

8. You are allowed to throw four darts, and we shall assume that there are no misses. In how many different ways can you obtain a score of 60?
9. Draw six concentric circles similar to the four circles in the figure for Exploration 8. Starting at the center and working outward, write the numbers 20, 19, 13, 12, 8, and 7 in order, one in each region. Use four darts, assume no misses, and show two ways of scoring 40.
10. Find a floor with boards of the same width (or draw equally spaced parallel lines on the floor) and cut a piece of wire so that the distance between the lines separating the boards is twice the length of the wire. This exploration is a special case of the **Buffon needle problem** with the wire serving as the needle. Each trial consists of dropping the needle and observing whether the needle touches a line. Let T be the number of trials and N the number of these trials in which the needle touches a line. It is known from advanced mathematics that for very large values of T the probability of the needle touching a line is $1/\pi$ and is approximately N/T. Make at least 100 trials and compute T/N as an approximation for π.
11. Ask at least 100 people to select an integer between 1 and 10. Tabulate the number of times that each integer is selected. Discuss the results relative to a recent conjecture that 7 is the most frequently selected number under such circumstances.
12. Repeat Exploration 11 for a two-digit number that is less than 50 and has odd numbers for both of its digits. Compare your results with the conjecture that the number most frequently selected is 37.
13. Observe and list in order the last two digits of 20 license plates on automobiles in any large parking lot. Repeat this procedure for at least five sets of 20 two-digit numbers. For each set of 20 numbers note how often you find a repetition of pairs of digits (the same two-digit number appearing at least twice).
14. Repeat the preceding exploration by opening a telephone book and selecting 20 telephone numbers at random. Record the last two digits only, and note the frequency with which you find a repetition of pairs of digits within a set of 20 two-digit numbers.
15. Prepare a bulletin board display showing the use of different concepts of probability that appear in newspaper and magazine articles.

If you are enrolled in a class of 30 students, what would you guess is the probability that there are two members of the group that celebrate their birthdays on the same date? Normally you would expect the probability to be low, inasmuch as there are at least 365 different days of the year on which an individual's birth date could occur, and 30 is small relative to 365.

Although the mathematical explanation is beyond our scope at this time, it can be shown that the actual probability of at least two birth dates occurring on the same day of the year in a random group of 30 students is 0.71. That is, you can safely predict that 71% of the time such an occurrence will take place. Or, alternatively, if you bet that at least two members of the group have the same birth date, then you can expect to win such a bet approximately 71 times out of 100.

The graph shows the probability of two common birth dates for people in groups of different sizes. Note that the probability in a group of 23 is approximately 0.50. That is, for a group of 23 individuals, there is a 50% chance of having such an occurrence take place. For a group of 60 the probability is 0.99, or almost certainty!

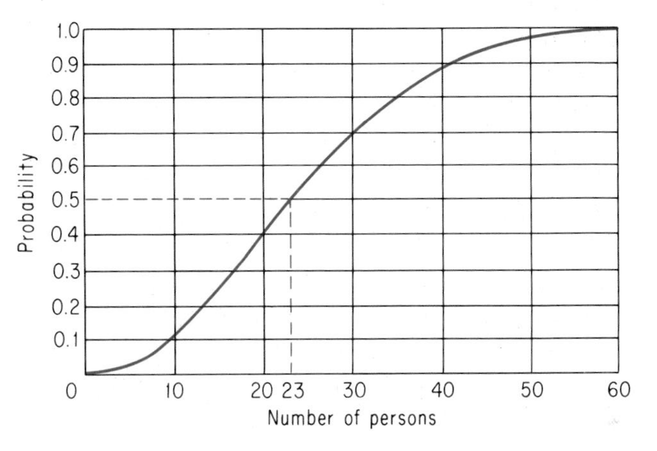

16. Use the given graph to determine the probability of two or more common birth dates for the members of a class of
    (a) 20 students (b) 40 students
17. Use the members of your class to test the results shown in the graph.
18. Test the results shown in the graph by going to an encyclopedia and using the dates of birth of 30 individuals selected at random.

10-2 Sample Spaces

A list of possible outcomes is often useful when solving problems that involve probabilities. Such a listing with each outcome included exactly once is called a **sample space**. Consider for example the problem of tossing two coins. The set of all possible outcomes for this problem may be designated as

$$S_1 = \{(H, H), (H, T), (T, H), (T, T)\}$$

and represented as in the following figure.

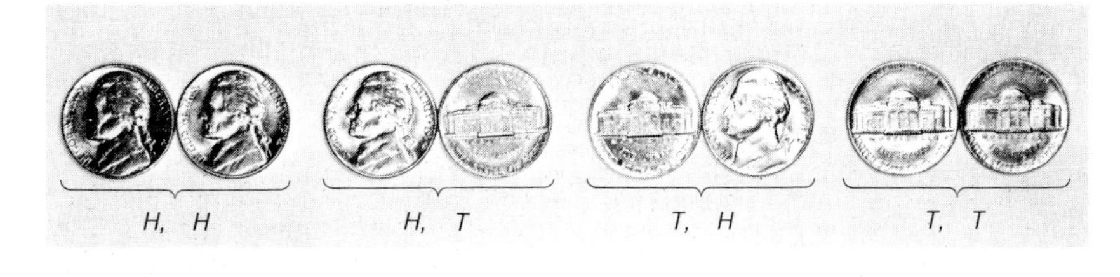

These four possible outcomes may be obtained by using a tree diagram to list all possible cases or by listing the possibilities as in the following array.

From the array we may observe the following.

Two heads occur in one event.

One head and one tail occur in two events.

No heads, that is, two tails, occur in one event.

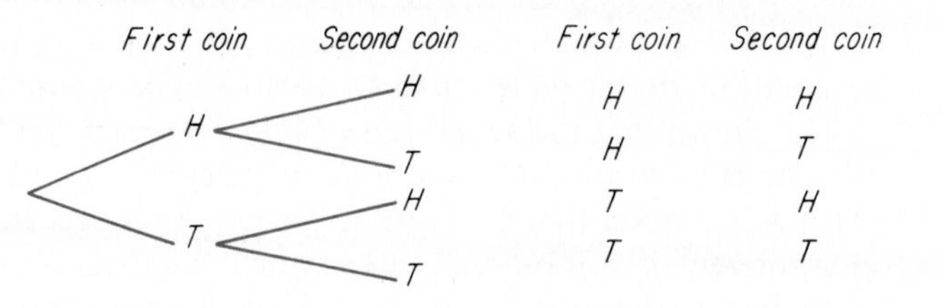

Then we may list the various probabilities regarding the tossing of two unbiased coins. Since all possibilities have been considered, the sum of the probabilities is 1. This provides a check on our computation. For two unbiased coins the four elements of the sample space (*events*) are equally likely and each event has probability 1/4. A listing of the elements of a sample space with their probabilities is sometimes called a **probability distribution**. A sample space in which the elements have equal probabilities is a **uniform sample space**.

The sample space S_1 provides complete information about the problem of tossing two coins and is a uniform sample space. However, for particular problems other sample spaces may provide all the information that is needed. For example, the two coins must either match, M, or be different, D. The sample space $S_2 = \{M, D\}$ is based upon this classification and is also a uniform sample space. Another classification may be established in terms of the numbers of heads. If we use

$$H_2 = \{(H, H)\}, \qquad H_1 = \{(H, T), (T, H)\}, \qquad H_0 = \{(T, T)\}$$

then a third sample space may be represented by

$$S_3 = \{H_2, H_1, H_0\}$$

Event	*Probability*
H_2	1/4
H_1	1/2
H_0	1/4

Note that the sample space S_3 is not a uniform sample space.

For the case of three coins, the following tree diagram and array may be made.

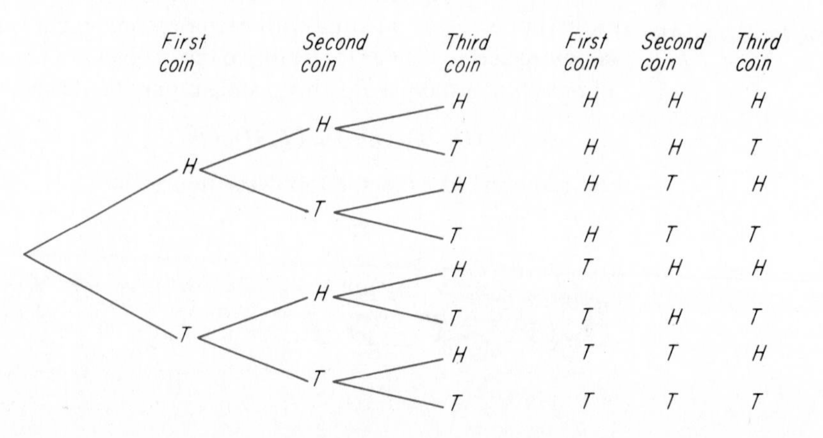

From the tree diagram we have the following sample space for the tossing of three coins.

$$\{HHH, HHT, HTH, HTT, THH, THT, TTH, TTT\}$$

Event	*Probability*
0 heads	1/8
1 head	3/8
2 heads	3/8
3 heads	1/8

We may also list the probabilities of specific numbers of heads as in the accompanying array. Both for two coins and for three coins the sum of the probabilities is 1. That is, all possible events have been listed and these events are mutually exclusive. Note also that for two coins there were four possible outcomes and for three coins there were eight possible outcomes. For n coins there would be 2^n possible outcomes.

Example 1 A box contains two red and three white balls. Two balls are drawn in succession without replacement. List a sample space for this experiment.

Solution To identify individual balls, we denote the red balls as R_1 and R_2, and the white balls as W_1, W_2, W_3. Then the sample space is

Note in the example that a sample space may be presented as the elements of an array as well as in set notation.

R_1R_2	R_2R_1	W_1R_1	W_2R_1	W_3R_1
R_1W_1	R_2W_1	W_1R_2	W_2R_2	W_3R_2
R_1W_2	R_2W_2	W_1W_2	W_2W_1	W_3W_1
R_1W_3	R_2W_3	W_1W_3	W_2W_3	W_3W_2 ■

An element of a sample space is said to be **selected at random** if all elements of the sample space are equally likely to be selected. For example, an unbiased coin may be tossed to select an element of the sample space $\{H, T\}$ at random. Similarly, an unbiased spinner may be used to select elements at random from a sample space of congruent segments (or arcs) of a circle.

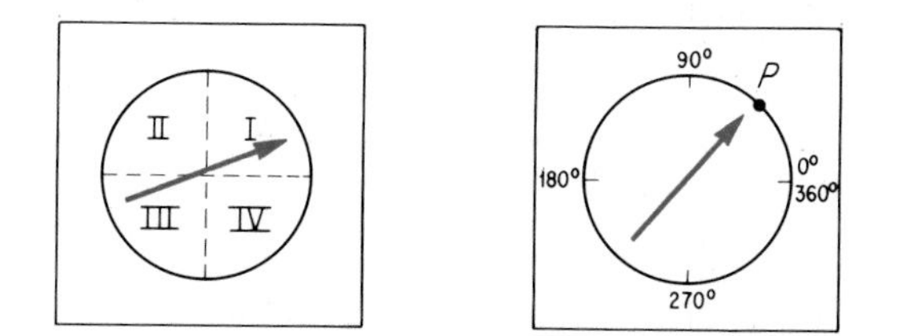

In the first figure the four quadrants of the circular region are shown and the probability of an unbiased spinner selecting any particular quadrant is 1/4. In the second figure the circular arc is marked in degrees. Each point P selected by the spinner may be identified by a real number p, where $0 \leq p < 360$. For example, P is in the first quadrant if $0 \leq p < 90$ and the probability of P being in the first quadrant is 90/360, that is, 1/4. Since there are infinitely many real numbers on the interval from 0 to 360, we calculate probabilities in terms of measures of arcs. Since a point may be considered to be an arc of length 0, the probability of the spinner selecting any particular point is 0.

Example 2 Suppose that Jane has a compass on the dashboard of her car and lives in a part of the country where a driver is as apt to be headed in one direction as another, that is, all compass directions are equally likely. Find the probability that at a specified time Jane's car is headed

(a) Between southwest and southeast.
(b) Between south-southwest and south-southeast.
(c) Within one degree of south.
(d) Exactly south.

Solution **(a)** Southwest is 45° west of south; southeast is 45° east of south. The target arc is 90° long and the probability is 90/360, that is, 1/4.
(b) South-southwest is $22\frac{1}{2}°$ west of south; south-southeast is $22\frac{1}{2}°$ east of south. The target arc is 45° long and the probability is 45/360, that is, 1/8.
(c) The target arc is 2° long and the probability is 2/360, that is, 1/180.
(d) The target arc is 0° long and the probability is 0/360, that is, 0. ■

EXERCISES

1. List a sample space and assign a probability for each outcome using an unbiased spinner. Assume that regions that appear congruent are congruent.

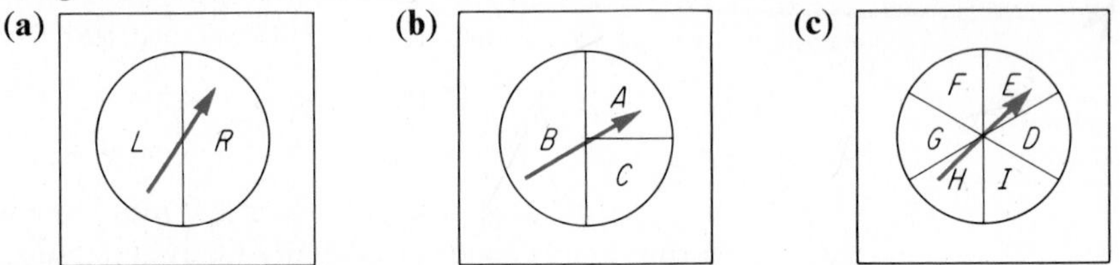

2. List a sample space for the experiment of spinning the two spinners in parts (a) and (b) of Exercise 1.

Use the sample space in illustrative Example 1 *in this section to find the probability of each event.*

3. Both balls are red.
4. Both balls are white.
5. The first ball is red.
6. The first ball is red and the second ball is white.
7. One ball is red and the other is white.

In Exercises 8 *through* 10 *use the sample space for the outcomes when three coins are tossed to find the probability of each event.*

8. All three coins are heads.
9. At least two coins are heads.
10. At most one coin is tails.

11. Make a tree diagram and give the sample space for the tossing of four coins.

In Exercises 12 *through* 14 *use the sample space for the outcomes when four coins are tossed* (*Exercise* 11) *to find the probability of each event.*

12. (**a**) No coins are heads.
(**b**) All four coins are tails.

13. (**a**) At least three coins are heads.
(**b**) At most one coin is tails.

14. (**a**) At most two coins are tails.
(**b**) At least two coins are heads.

15. Make a tree diagram and give a sample space for the throwing of a pair of dice. Represent each outcome by an ordered pair of numbers. For example, let (1, 3) represent a 1 on the first die and a 3 on the second die.

In Exercises 16 *through* 22 *use the sample space for the outcomes when a pair of dice are thrown* (*Exercise* 15) *to find the probability of each event.*

16. (**a**) The number on the first die is 2.
(**b**) The number on the first die is not 2.

17. (**a**) The number 1 is on both dice.
(**b**) It is not true that the number 1 is on both dice.

18. (**a**) The same number is on both dice.
(**b**) There are different numbers on the two dice.

19. (**a**) The sum of the numbers obtained is 11.
(**b**) The sum of the numbers obtained is not 11.

20. (**a**) The sum of the numbers obtained is 7.
(**b**) The sum of the numbers obtained is not 7.

21. (**a**) The number on the second die is twice the number on the first die.
(**b**) The number on one die is twice the number on the other die.

22. (**a**) The number on one die is three more than the number on the other die.
(**b**) The number on one die is two less than the number on the other die.

A number is selected at random from the set {0, 1, 2, 3, 4, 5, 6, 7, 8, 9} *of whole numbers that can be represented using one decimal digit. Find the probability that the selected number has the specified property.*

Hint: The numbers 0 and 1 are neither prime nor composite.

23. (**a**) An even number. (**b**) The square of an integer.

24. (**a**) A prime number. (**b**) A composite number.

25. (**a**) Greater than 9. (**b**) Less than 10.

A number is selected at random from the set {10, 11, 12, 13, . . . , 98, 99} *of whole numbers that are ordinarily written with two decimal digits. Find the probability that the selected number has the specified property.*

26. (**a**) A multiple of 10. (**b**) An even number.

27. (**a**) A multiple of 5. (**b**) The square of an integer.

28. (**a**) A multiple of 3. (**b**) A multiple of 7.

29. **(a)** A prime number. **(b)** A composite number.

30. **(a)** A multiple of 2 or 3. **(b)** A multiple of 2 and 3.

A number is selected at random from the nonnegative real numbers less than or equal to 12. In Exercises 31 through 36 find the probability that the selected number x has the specified property.

31. **(a)** $0 \leq x \leq 2$ **(b)** $4 \leq x \leq 8$

32. **(a)** $x \geq 0$ **(b)** $x > 12$

33. **(a)** $x < 6$ **(b)** $x \leq 6$

34. **(a)** $x = 8$ **(b)** $x \neq 8$

35. **(a)** $x = 6$ **(b)** $x \neq 6$

36. **(a)** $x < 12$ **(b)** $x \geq 12$

37. A certain dart board has an area of 400 cm^2. A target with area 50 cm^2 is placed on the dart board.
    (a) If a dart hits the dart board at random, what is the probability that the dart hits the target?
    (b) In part (a) does the shape of the target affect the probability? Explain.

38. A target of unknown area is placed on a dart board with area 300 cm^2. If a method has been found for throwing darts at the board at random, a large number of darts have been thrown, and it has been observed that one-sixth of the darts hit the target, what is the approximate area of the target?

39. A box contains two red balls R_1 and R_2 and two white balls W_1 and W_2. List a sample space for the outcomes when two balls are drawn in succession without replacement. Find the probability that both balls are red.

40. Repeat Exercise 39 for the case in which the first ball is replaced before the second ball is drawn.

41. Repeat Exercise 39 for a box that contains four red balls and two white balls.

42. Repeat Exercise 41 for the case in which the first ball is replaced before the second ball is drawn.

43. Make a tree diagram and give a sample space for a toss of a coin followed by the throw of a die.

In Exercises 44 through 46 use the sample space for the outcomes when a coin is tossed and a die is thrown (Exercise 43) to find the probability of each event.

44. **(a)** The coin is heads and the number on the die is even.
    (b) The coin is tails and the number on the die is odd.

45. **(a)** The coin is heads and the number on the die is 5.
    (b) The coin is tails and the number on the die is not 5.

46. **(a)** The coin is tails and the number on the die is greater than 4.
    (b) The coin is heads and the number on the die is prime.

EXPLORATIONS

Assume that a red die and a green die are thrown and the sum S of the numbers shown on the dice is noted.

1. Give the sample space of the possible values of S.
2. List the number r on the red die and the number g on the green die as (r, g) and make a table of all possible such ordered pairs of numbers for each possible value of S.
3. List each possible value of S with its probability.
4. Throw a pair of dice at least forty times and see if the probabilities of the values of S are *approximately illustrated* by the results obtained.

There are many interesting probability questions whose answers are not intuitively obvious. Here are two that are best solved by means of a sample space.

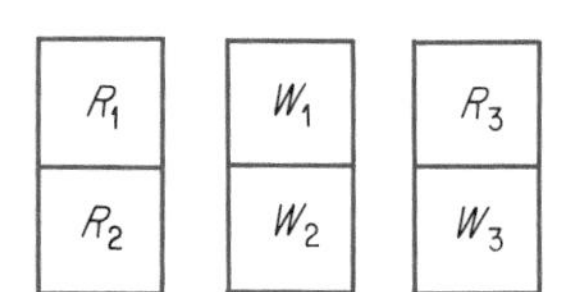

5. Three cards are in a box. One is red on both sides, one is white on both sides, and one is red on one side and white on the other. A card is drawn at random and placed on a table. The card has a red side showing. What is the probability that the side not showing is also red? (Contrary to popular belief, the answer is not 1/2).

   As an aid to the solution of this problem, let us identify the three cards as in the adjacent diagram. Now consider the set of possible outcomes when a card is drawn at random and placed on the table. In each of the following ordered pairs, the first side noted represents the one placed face up on the table whereas the second side indicates the face hidden from view.

$$\{R_1R_2, \quad R_2R_1, \quad R_3W_3, \quad W_1W_2, \quad W_2W_1, \quad W_3R_3\}$$

   Inasmuch as we are told that a red side is showing, we may narrow down the sample space to the first three pairs only. Of these three possibilities, if the first side is red can you now tell the probability that the second side is also red?

Hint for Exploration 6: List the sample space using B for boy and G for girl. Then note that GG is *not* a possibility.

6. One of the authors of this book has two children. One of these children is a boy. What is the probability that they are both boys? Believe it or not, the answer is *not* 1/2!
7. Find several newspaper articles that involve sample spaces and note the importance of the selection of the sample space. For example, suppose that the first paragraph of an article contains a report that a small company employs 10 men and 10 women with average salaries at \$18,000 and \$14,600, respectively (discrimination?). Suppose also that a detailed reading of the article reveals that when length of service is considered, the salaries may be described as in the array.

Years of service \ Sex	Male		Female	
	Number	Average salary	Number	Average salary
Less than 5	2	$10,000	8	$12,000
At least 5	8	20,000	2	25,000
Totals	10	18,000	10	14,600

Although the women's salaries appear better under this more detailed comparison, the purpose of this example is not to enter the arguments about discrimination. Rather, our goal is to emphasize the need for considering background details in selecting a sample space for use in a comparison. In this regard discuss the apparent reliability of the newspaper articles that you find.

10-3 Computation of Probabilities

Consider the set S of natural numbers less than 100

$$S = \{1, 2, 3, \ldots, 99\}$$

Suppose that an element of S is selected at random and these four events are considered.

E_1: the number is a multiple of 2.
E_2: the number is a multiple of 5.
E_3: the number is not a multiple of 5.
E_4: the number is a multiple of 57.

Then the sample spaces for these events are

$$S_1 = \{2, 4, 6, \ldots, 98\}$$

$$S_2 = \{5, 10, 15, 20, \ldots, 95\}$$

$$S_3 = \{1, 2, 3, 4, 6, 7, 8, 9, 11, \ldots, 99\}$$

$$S_4 = \{57\}$$

The definition of probability may be used to find the probability of each event.

$$P(E_1) = \frac{49}{99} \qquad P(E_2) = \frac{19}{99} \qquad P(E_3) = \frac{80}{99} \qquad P(E_4) = \frac{1}{99}$$

Note that the sets S_2 and S_3 are *complementary sets*

$$S_2 \cup S_3 = S \qquad S_2 \cap S_3 = \varnothing$$

$$P(E_2 \text{ or } E_3) = P(S) = 1 \qquad P(E_2 \text{ and } E_3) = P(\varnothing) = 0$$

The events E_2 and E_3 are called **complementary events.**

The probabilities of several other related events may be found. For example,

$$P(E_1 \text{ and } E_2) = P(\{10, 20, 30, \ldots, 90\}) = \frac{9}{99} = \frac{1}{11}$$

$$P(E_1 \text{ or } E_2) = P(E_1 \text{ or } \{5, 15, 25, \ldots, 95\}) = \frac{49}{99} + \frac{10}{99} = \frac{59}{99}$$

$$P(E_1 \text{ or } E_4) = P(E_1 \text{ or } \{57\}) = \frac{49}{99} + \frac{1}{99} = \frac{50}{99}$$

$$P(E_1 \text{ and } E_4) = P(\varnothing) = 0$$

We use the probabilities of these events to emphasize an important role of **mutually exclusive events** (events with sample spaces that do not have any common elements) in the computation of probabilities. For example, E_1 and E_4 are mutually exclusive events

$$S_1 \cap S_4 = \varnothing \qquad P(E_1 \text{ and } E_4) = 0 \qquad P(E_1 \text{ or } E_4) = P(E_1) + P(E_4)$$

> For any two mutually exclusive events A and B
>
> $$P(A \text{ and } B) = 0 \qquad P(A \text{ or } B) = P(A) + P(B)$$

The events E_1 and E_2 are not mutually exclusive, since $S_1 \cap S_2 \neq \varnothing$. If we think of the sample spaces as we determine probabilities, the elements of $S_1 \cap S_2$ are included once when $P(E_1)$ is found and a second time when $P(E_2)$ is found. For the events E_1 and E_2

$$S_1 \cap S_2 = \{10, 20, 30, 40, 50, 60, 70, 80, 90\}$$

$$P(E_1 \text{ and } E_2) = \frac{9}{99}$$

$$P(E_1 \text{ or } E_2) = \frac{59}{99} = \frac{49}{99} + \frac{19}{99} - \frac{9}{99} = P(E_1) + P(E_2) - P(E_1 \text{ and } E_2)$$

> For any two events A and B
>
> $$P(A \text{ or } B) = P(A) + P(B) - P(A \text{ and } B)$$

Note that the previous rule is a special case of this one and this rule holds whether the events are mutually exclusive or not mutually exclusive.

Example 1 A bag contains three red, two black, and five yellow balls. Find the probability that a ball drawn at random (each is equally likely) will be red or black.

Solution The probability of drawing a red ball, $P(R)$, is 3/10. The probability of drawing a black ball, $P(B)$, is 2/10. Then

$$P(R \text{ or } B) = P(R) + P(B) = \frac{5}{10} = \frac{1}{2} \quad \blacksquare$$

This process can be extended to find the probability of any finite number of mutually exclusive events.

$$P(A_1 \text{ or } A_2 \text{ or } A_3 \text{ or } \cdots A_n) = P(A_1) + P(A_2) + P(A_3) + \cdots + P(A_n)$$

Example 2 A single die is thrown. What is the probability that either an odd number or a number greater than 3 appears?

Solution There are three odd numbers possible, {1, 3, 5}, so the probability of throwing an odd number is 3/6. The probability of getting 4, 5, or 6, that is, a number greater than 3, is also 3/6. Adding these probabilities gives 1. Something is obviously wrong, since a probability of 1 implies certainty and we can see that an outcome of 2 is neither odd nor greater than 3. The difficulty lies in the fact that the events are *not* mutually exclusive; a number may be both odd and also greater than 3. In particular, 5 is both odd and greater than 3. Thus $P(5)$ has been included twice. Since $P(5) = 1/6$, the answer is

$$\frac{3}{6} + \frac{3}{6} - \frac{1}{6}, \quad \text{that is,} \quad \frac{5}{6}$$

By an actual listing we can see that five of the six possible outcomes in Example 2 are either odd or greater than 3, namely 1, 3, 4, 5, and 6. The only "losing" number is 2. Thus the probability $P(A \text{ or } B)$ must be 5/6.

as implied by our general rule and illustrated by the adjacent Venn diagram. ■

Next we turn our attention to the probability that several events will occur, one after the other. Consider the probability of tossing a coin twice and obtaining heads on the first toss and tails on the second toss. From a sample space we see that the probability is 1/4.

$\{HH, \textcircled{HT}, TH, TT\}$

Furthermore, we see that the probability $P(A)$ that the first coin is heads is 1/2. The probability $P(B)$ that the second coin is tails is 1/2. Then

$$P(A \text{ and } B) = \frac{1}{2} \times \frac{1}{2} = \frac{1}{4}$$

Note that the outcome of the first toss does not affect the second toss.

Example 3 Urn A contains three white and five red balls. Urn B contains four white and three red balls. One ball is drawn from each urn. What is the probability that they are both red?

Solution Let $P(A)$ be the probability of drawing a red ball from urn A and $P(B)$ be the probability of drawing a red ball from urn B. Then

$$P(A) = \frac{5}{8} \qquad P(B) = \frac{3}{7} \qquad P(A \text{ and } B) = \frac{5}{8} \times \frac{3}{7} = \frac{15}{56}$$ ■

Let the probability that a second event B occurs after an event A has occurred be $P(B \text{ given } A)$, the **conditional probability** of B given A. Then

For any two events A and B

$$P(A \text{ and } B) = P(A) \times P(B \text{ given } A)$$

Example 4 Two cards are selected in succession, without replacement, from an ordinary bridge deck of 52 cards. What is the probability that the second card is an ace given that the first card is an ace, that is, what is the probability that both cards are aces?

Solution The probability that the first card is an ace is 4/52. If it is an ace, then the probability that the second card is an ace is 3/51. The probability that both cards are aces is $(4/52) \times (3/51)$, that is, 1/221. ■

EXERCISES

What is the probability that the specified event will occur in a single throw of one die?

1. An even number or a number greater than 4.
2. An odd number or a number less than 4.
3. An odd number or a number greater than 4.
4. An even number or a number greater than 6.
5. An odd number or a number greater than 7.
6. An odd number or a number less than 6.
7. An even number or a number less than 7.

A single card is drawn from an ordinary deck of 52 bridge cards. Find the probability that the card selected is as described.

8. An ace or a queen.

9. A spade or a diamond.

10. A spade or a queen.

11. A spade and an ace.

12. A spade and a queen.

13. A heart or a king or a queen.

14. A club or an ace or a king.

15. A club and a spade.

Two cards are drawn in succession from an ordinary deck of 52 bridge cards without replacement of the first card. Find the probability of the specified event.

16. **(a)** Both cards are spades.
(b) Both cards are the ace of spaces.

17. **(a)** The first card is a spade and the second card is a heart.
(b) The first card is an ace and the second card is the king of hearts.

18. **(a)** The two cards are of the same suit.
(b) The two cards are of different suits.

Two cards are drawn in succession from an ordinary deck of 52 bridge cards with the first card replaced before the second card is drawn. In Exercises 19 *and* 20 *find the probability of the specified event.*

19. **(a)** Both cards are diamonds.
(b) Both cards are the ace of diamonds.

20. **(a)** The first card is a heart and the second card is a club.
(b) The first card is an ace and the second card is the king of hearts.

21. A coin is tossed six times. What is the probability that all six tosses are heads?

Hint: First find the probability of getting no heads.

22. A coin is tossed six times. What is the probability that at least one head is obtained?

23. A coin is tossed and then a die is thrown. Find the probability of obtaining
(a) a head and a 3
(b) a head and an even number
(c) a head or a 3
(d) a head or an even number.

24. A bag contains four red balls and seven white balls.
(a) If one ball is drawn at random, what is the probability that it is white?
(b) If two balls are drawn at random, without replacement, what is the probability that they are both white?

25. Repeat Exercise 24 for a bag with five red balls and seven white balls.

26. Repeat Exercise 24 for a bag with five red balls and eight white balls.

27. Five cards are drawn at random, without replacement, from an ordinary bridge deck of 52 cards. Find the probability that all five cards drawn are spades.

28. A box contains three red, four white, and six green balls. Three balls are drawn in succession, without replacement. Find the probability that
(a) all three are red

(b) the first is red, the second is white, and the third is green
(c) none are green
*(d) all three are of the same color.

29. Repeat Exercise 28 if each ball is replaced after it is drawn.
30. Repeat Exercise 28 for a box that contains four red, three white, and five green balls.
31. Repeat Exercise 28 for a box that contains four red, three white, and six green balls.
32. A die is thrown three times. Find the probability that
    (a) a 6 is obtained on the first throw
    (b) a 6 is obtained on each of the first two throws
    (c) a 6 is obtained on each of three throws
    (d) a 6 is obtained on the first throw and not obtained on the second or third throws.
33. A die is thrown three times. Find the probability that
    (a) an even number is obtained on all three throws
    (b) an even number is obtained on the first two throws and an odd number on the third throw
    (c) an even number is obtained on the first throw and an odd number on the second and third throws
    (d) exactly one even number is obtained.

*34. A die is thrown four times. Find the probability that an even number is obtained on
    (a) all throws
    (b) exactly one throw
    (c) exactly two throws.

*35. Repeat Exercise 34 for a die that is thrown five times.

*36. A die is thrown three times. Find the probability that
    (a) at least one 6 is obtained
    (b) exactly one 6 is obtained.

EXPLORATIONS

These explorations include several experiments that can be performed to illustrate basic concepts of probability.

1. Consider the network of streets shown in the adjacent figure. You are to start at the point S and move three "blocks." Each move is determined by tossing a coin. If the coin lands *tails*, move one block to the right; for *heads*, move one block up. Your terminal point will be at A, B, C, or D. Try to predict the number of times you will land at each point if the experiment is to be repeated 16 times. Then complete 16 trials, keeping a tally of the number of times you land at each point. Compare your actual results with your predictions as well as with the results obtained by your classmates. Finally, if necessary, revise your predictions on the basis of your experimentation.

A
B
C
S
D

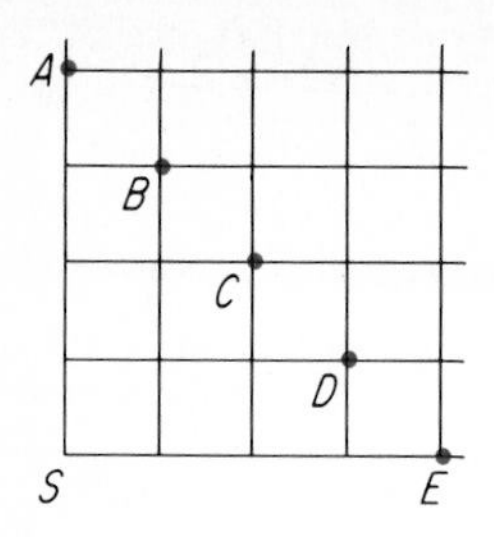

2. Repeat Exploration 1 for the adjacent network. This time you are to make four moves and will land at point A, B, C, D, or E.

3. If a pair of normal dice is thrown repeatedly, a sum of 7 can be expected theoretically to occur in one out of six throws. Thus if a pair of dice is thrown 36 times, on the average six of the throws will give a sum of 7. Throw a pair of dice 36 times and count the frequency with which a sum of 7 appears. Compare your actual results with the theoretical probability of 1/6.

4. As the number of throws becomes very large, that is, **in the long run**, the experimental results approach the theoretical results. Collect the data for Exploration 3 from each member of your class. Tabulate the results of your own experiment as for one student, your results and those of one of your classmates as for two students, your results and those of two classmates as for three students, and so forth as suggested in the array. Note the tendency of S/N to approach 1/6 as N increases.

Number of students	1	2	3	4	5	6	...
Number N of throws	36	72	108	144	180	216	...
Number S of sums of 7	____	____	____	____	____	____	...
S/N	____	____	____	____	____	____	...

5. Make 72 throws of a pair of dice. On graph paper record each throw by placing an X in the column that represents the sum. The final result should be a bar graph that is fairly "normal" in shape. In particular the height of each bar should be approximately equal to the theoretical expectation of the corresponding sum for 72 throws. The following figure shows the results obtained after one experiment of 18 throws.

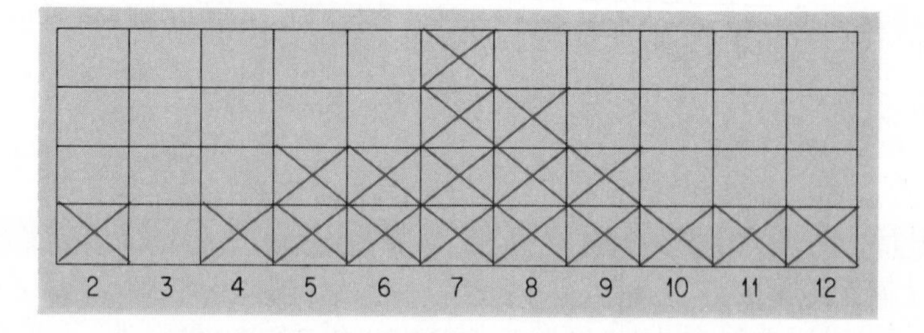

6. Make bar graphs as in Exploration 5 for the results obtained by
   (**a**) you and one classmate
   (**b**) you and two classmates
   (**c**) you and three classmates
   (**d**) you and four classmates.
   Compare your bar graph for Exploration 5 and these bar graphs. Describe the shape of the curve and the effect of increasing the number of throws.

7. Shuffle an ordinary deck of cards. Conjecture the number of cards that one must turn over, on the average, to reach the first ace. Then test your conjecture by performing the experiment at least 20 times, thoroughly shuffling the deck each time. Report on your results.

10-4 Odds, Mathematical Expectation, and Simulation

Consider the problem of finding the odds in favor of obtaining a 3 in one throw of a die. Since the probability of obtaining a 3 is 1/6, many people are tempted to state the odds as 1 to 6, which is not correct. In the long run we expect a 3 on one of every 6 throws. That is, for each throw of a 3 we expect five throws that do not produce a 3. The correct odds in favor of a 3 in one throw of a die are 1 to 5. The odds against obtaining a 3 are 5 to 1. Formally we define odds as follows.

> The **odds in favor** of an event are the ratio of the probability that an event will occur to the probability that the event will not occur.
>
> $$\frac{P(A)}{1 - P(A)}$$
>
> The **odds against** the occurrence of the event are the reciprocal of this ratio.

The sports section of a newspaper often includes statements regarding the "odds" in favor of, or against, a particular team or individual's winning or losing some encounter. For example, we may read that the odds in favor of the Cardinals' winning the pennant are "4 to 1." In this section we shall attempt to discover just what such statements really mean.

Since $P(A)$ and $1 - P(A)$ may be expressed as like fractions and their quotient simplified, the odds in favor of an event that may occur in several equally likely ways may be expressed as the ratio of the number of favorable ways to the number of unfavorable ways.

Notice that odds and probabilities are very closely related. Indeed, if either the odds for or the probability of an event is known, then the other can be found. For example, if the odds for an event are 1 to 2, then we have the ratio

$$\frac{\text{Number of favorable ways}}{\text{Number of unfavorable ways}} = \frac{1}{2}$$

and the probability is the ratio

$$\frac{\text{Number of favorable ways}}{\text{Total number of all ways (favorable and unfavorable)}} = \frac{1}{3}$$

Similarly, if the probability is 2/5, then the odds are 2 to 3.

Example 1 Find the *odds in favor* of drawing a spade from an ordinary deck of 52 bridge cards and the *odds against* drawing a spade from an ordinary deck of 52 bridge cards.

Solution Since there are 13 spades in a deck of cards, the probability of drawing a spade is 13/52, that is, 1/4. The probability of failing to draw a spade is 3/4. The odds in favor of obtaining a spade are $(1/4) \div (3/4)$, that is, 1/3.

The *odds in favor* of drawing a spade are stated as 1/3. They may also be stated as 1 to 3 or as 1:3. Similarly, the *odds against* drawing a spade are 3/1, which may be written as 3 to 1 or 3:1. ■

Mathematical expectation is closely related to odds and is defined as the product of the probability that an event will occur and the amount to be received upon such occurrence. Suppose that you are to receive \$2.00 each time you obtain two heads on a single toss of two coins. You do not receive anything for any other outcome. Then your mathematical expectation will be one-fourth of \$2.00 and three-fourths of \$0. In other words, your mathematical expectation from the game is

$$\frac{1}{4}(\$2.00) + \frac{3}{4}(\$0), \quad \text{that is,} \quad \$0.50$$

This means that you should be willing to pay \$0.50 each time you toss the coins if the game is to be a fair one. *In the long run* both you and the person who is running the game would break even. Note that for each game that you pay \$0.50 and win \$2.00, your net gain is \$1.50; for each game that you pay \$0.50 and lose, your net gain is a loss of \$0.50. The game has two possible outcomes.

Pay \$0.50 and win \$2.00 with probability 1/4.
Pay \$0.50 and receive nothing with probability 3/4.

If you pay \$0.50 to play, your mathematical expectation of net gain in playing the game is

$$\frac{1}{4}(\$2.00 - \$0.50) + \frac{3}{4}(\$0 - \$0.50), \quad \text{that is,} \quad \$0.$$

The zero expectation indicates that the game is a fair game to both you and the person running the game.

Whenever you use the formula for mathematical expectation, it is worthwhile to check that all possible outcomes have been considered. To do so, show that the sum of the probabilities is equal to 1.

If an event has several possible outcomes that occur with probabilities p_1, p_2, p_3, and so forth, and for each of these outcomes one may expect the amounts m_1, m_2, m_3, and so on, then the mathematical expectation E may be defined as

$$\boxed{E = m_1p_1 + m_2p_2 + m_3p_3 + \cdots}$$

Example 2 Suppose that you pay 5¢ to play a game in which a coin is tossed twice. You receive 10¢ if two heads are obtained, 5¢ if exactly one head is obtained, and nothing if there are no heads. What is the expected value to you of this game?

Solution Remember that you pay 5¢ to play the game. Thus for two heads you get your nickel back and gain another 5¢; for one head you break even; for no heads you lose the 5¢ that you have paid to play the game. In other words, the possible events, their probabilities, and their values to you are

Two heads	1/4	10 − 5,	that is,	5¢
One head	1/2	5 − 5,	that is,	0¢
No heads	1/4	0 − 5,	that is,	−5¢

Your expected value E, in cents, is

$$E = (5)\left(\frac{1}{4}\right) + (0)\left(\frac{1}{2}\right) + (-5)\left(\frac{1}{4}\right) = 0$$

The zero value implies that *if you played a large number of games*, your gains should equal your losses. If you paid less than 5¢ per game, you could expect to make a profit. If you paid more than 5¢ per game, you should expect to have a loss. ■

In Example 2, if you were not told the price for playing this game, then you could determine the fair price by computating your expectation in this way:

$$E = (10)\left(\frac{1}{4}\right) + (5)\left(\frac{1}{2}\right) + (0)\left(\frac{1}{4}\right) = 5$$

Thus, as previously observed, 5¢ is the fair price to pay to play this game.

Many games and experiments may be simulated using a table of random numbers instead of actually playing the game or performing the experiment.

Tables of *random numbers* may be generated by selecting numbers from a set with replacement and at random. Here is a table of 100 two-digit random numbers.

80	68	30	67	70	21	62	01	79	75
18	53	29	65	19	85	68	11	62	56
63	64	39	34	88	25	76	42	66	21
82	25	11	76	63	67	55	03	57	77
27	14	60	76	72	25	64	62	12	64
11	73	60	93	75	07	05	77	42	57
78	61	69	29	36	65	82	92	16	28
45	92	63	01	62	95	91	92	13	18
11	84	97	48	73	95	49	84	34	65
23	76	77	14	15	10	12	58	79	93

Three steps are involved in the use of a table of random numbers to **simulate** a game or an experiment.

1. Decide on a method for selecting numbers from the table. For example, select the first 40 numbers left to right by rows top to bottom.
2. Assign sets of numbers to each element of the sample space of the experiment to be performed. This is done so that the probability of a number from the set from which the table was formed being in the assigned set is the same as the probability of the element to which the set is assigned. For example, assign even numbers to heads and odd numbers to tails for the tossing of a coin.
3. Consider the numbers selected in (1). If the number is not one of those assigned in (2), proceed to the next number.

Example 3 Use the given table of random numbers to simulate the tossing of a coin 100 times. How many heads are obtained?

Solution Among others:
(a) Use the 100 numbers in the table for the 100 tosses of the coin.
(b) Assign even numbers to heads and odd numbers to tails.

Different assignments and different tables, like different instances of performing an experiment may produce different answers. For the above assignment, the 46 even numbers in the table imply 46 heads for this simulation of the experiment. ■

EXERCISES *What are the odds in favor of the event?*

1. Two heads in a single toss of two coins.
2. At least two heads in a single toss of three coins.
3. Two heads when a single coin is tossed twice.
4. At least two heads when a single coin is tossed three times.

In Exercises 5 through 8 what are the odds against the event?

5. Two heads in a single toss of two coins.
6. An ace in a single draw from a deck of 52 bridge cards.
7. An ace or a king in a single draw from a deck of 52 bridge cards.
8. A 7 or an 11 in a single throw of a pair of dice.

9. For the event of obtaining a 7 or an 11 in a single throw of a pair of dice, what are the odds in favor of the event?
10. One hundred tickets are sold for a lottery. The grand prize is $2000. What is your mathematical expectation if you are given a ticket?
11. Repeat Exercise 10 for the case where 250 tickets are sold.
12. What is your mathematical expectation when you are given one of 400 tickets for a single prize worth $1200?
13. In the "long run" what is your expected profit, or loss, per game if you pay $2 to play each game and receive $10 if you throw a "double" (the same number on both dice) on a single toss of a pair of dice?
14. A box contains three dimes and two quarters. You are to reach in and select one coin, which you may then keep. Assuming that you are not able to determine which coin is which by its size, what would be a fair price for the privilege of playing this game?
15. There are three identical boxes on a table. One contains a five-dollar bill, one contains a one-dollar bill, and the third is empty. A man is permitted to select one of these boxes and to keep its contents. What is his expectation?
16. Three coins are tossed. What is the expected number of heads?

*17. Two bills are to be drawn without replacement from a purse that contains three five-dollar bills and two ten-dollar bills. What is the mathematical expectation for this drawing?

18. If there are two pennies, a nickel, a dime, a quarter, and a half-dollar in a hat, what is the mathematical expectation of the value of a random selection of a coin from the hat?

*19. Repeat Exercise 18 for the selection of two coins from the hat without replacing the first coin selected.

20. Suppose that the probability of your obtaining an A in this course is 0.3. What are the odds
    (a) in favor of your obtaining an A
    (b) against your obtaining an A?

21. If the probability of an event is p, what are the odds
 (a) in favor of the event
 (b) against the event?

Hint: This value is the algebraic sum of the expected loss and the expected gain.

*22. Suppose that n tickets are sold at \$1 each for a single prize of \$1000. Express the theoretically expected value of each ticket in terms of n.

*23. Repeat Exercise 22 if there is a first prize of \$5000 and a second prize of \$1000.

Use the table of random numbers given in the text and simulate each experiment.

24. Toss a coin 100 times. Assign numbers 00 through 49 to heads and 50 through 99 to tails.

25. Toss two coins 100 times. Use the assignment in Example 3 for one coin and that in Exercise 24 for the other.

*26. Toss three coins 100 times. Use the assignments in Exercise 25 and the evenness of the sum of the digits for heads of the third coin.

EXPLORATIONS

Games, newspaper clippings, and classroom experiments may be used very effectively for teaching concepts of probability, odds, and mathematical expectation.

1. Name at least three games that involve probability, that is, games that are "games of chance" rather than "games of skill."
2. Describe at least one classroom experiment that can be used effectively for one or more of these purposes.
3. Start a collection of newspaper clippings that can be used with an elementary school class for one or more of these purposes.
4. Try to obtain information from one of the state or provincial lotteries and estimate the mathematical expectation of a ten-dollar purchase of tickets.

10-5 Permutations

Suppose that three people, Ruth, Joan, and Debbie, are waiting to play singles at a tennis court. Two of the three can be selected in six different ways if the order in which they are named is significant, for example, if the first person named is to serve first. We may identify these six ways from a tree diagram.

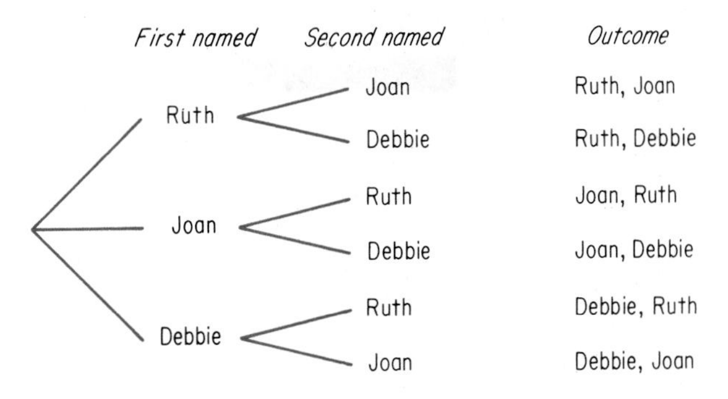

We say that there are 3×2, that is, 6, *permutations* of the set of three people selected two at a time. In each case the *order* in which the two people are named is significant. A **permutation** of a set of elements is an *arrangement* of certain of these elements *in a specified order*. In the problem just discussed, the number of permutations of three things taken two at a time is 6. In symbols we write

Note that *order* is important in this discussion.

$_3P_2 = 6$, read as "the number of permutations of three things taken two at a time is six."

To find a general formula for $_nP_r$, the number of permutations (arrangements) of n things taken r at a time, we use the *general principle of counting* (Section 10-1). Note that we can fill the first of the r positions in any one of n different ways. Then the second position can be filled in $n - 1$ different ways, and so on.

Position:	1	2	3	4	$\cdots$	r
	↓	↓	↓	↓		↓
Number of choices:	n	$n-1$	$n-2$	$n-3$	$\cdots$	$n-(r-1)$, that is, $n-r+1$

The product of these r factors gives the number of different ways of arranging r elements selected from a set of n elements, that is, the permutations of n things taken r at a time.

No repetitions are allowed when n things are taken r at a time.

$$_nP_r = (n)(n-1)(n-2)\cdots(n-r+1)$$

where n and r are integers and $n \geq r$.

Example 1 Find $_8P_4$.

Solution Here $n = 8$, $r = 4$, and $n - r + 1 = 5$. Thus

$$_8P_4 = 8 \times 7 \times 6 \times 5 = 1680$$

Note that there are r, in this case 4, factors in the product. ■

Example 2 How many different three-letter "words" can be formed from the 26 letters of the alphabet if each letter may be used at most once?

Solution We wish to find the number of permutations of 26 things taken three at a time.

The solution to Example 2 is based on the *general principle of counting.*

$$_{26}P_3 = 26 \times 25 \times 24 = 15{,}600$$ ■

A special case of the permutation formula occurs when we consider the permutations of n things taken n at a time. For example, let us see in how many different ways we may arrange in a row the five members of a given set. Here we have the permutations of five things taken five at a time.

$$_5P_5 = 5 \times 4 \times 3 \times 2 \times 1$$

In general, for n things n at a time, $n = n$, $r = n$, and $n - r + 1 = 1$.

$${}_nP_n = (n)(n-1)(n-2)\cdots(3)(2)(1)$$

We use a special symbol, $n!$, read "n factorial," for this product of integers 1 through n. The following examples should illustrate the use of this new symbol.

Guess how large 10! is. Then use a calculator to check your estimate.

$1! = 1$	$5! = 5 \times 4 \times 3 \times 2 \times 1$
$2! = 2 \times 1$	$6! = 6 \times 5 \times 4 \times 3 \times 2 \times 1$
$3! = 3 \times 2 \times 1$	$7! = 7 \times 6 \times 5 \times 4 \times 3 \times 2 \times 1$
$4! = 4 \times 3 \times 2 \times 1$	$8! = 8 \times 7 \times 6 \times 5 \times 4 \times 3 \times 2 \times 1$

Also, we *define* $0! = 1$ so that $(n-r)!$ may be used when $r = n$.

Using this **factorial notation**, we are now able to provide a different, but equivalent, formula for ${}_nP_r$.

${}_nP_n = n!$

${}_nP_r = \dfrac{n!}{(n-r)!}$

$${}_nP_r = n(n-1)(n-2)\cdots(n-r+1)$$

$$\times \frac{(n-r)(n-r-1)(n-r-2)\cdots(3)(2)(1)}{(n-r)(n-r-1)(n-r-2)\cdots(3)(2)(1)} = \frac{n!}{(n-r)!}$$

Then for $r = n$ we have ${}_nP_n = n!$

Example 3 Evaluate ${}_7P_3$ by two different methods.

Solution **(a)** ${}_7P_3 = 7 \times 6 \times 5 = 210$

(b) $${}_7P_3 = \frac{7!}{4!} = \frac{7 \times 6 \times 5 \times 4 \times 3 \times 2 \times 1}{4 \times 3 \times 2 \times 1}$$

$$= 7 \times 6 \times 5 = 210 \quad \blacksquare$$

Example 4 A certain class consists of 10 girls and 12 boys. They wish to elect officers so that the president and treasurer are girls and the vice-president and secretary are boys. How many such sets of officers may be obtained?

Solution The number of selections of the president and the treasurer is ${}_{10}P_2$. The number of selections of the vice-president and secretary is ${}_{12}P_2$. By the general principle of counting, the total number of possible selections of a set of officers is

$$({}_{10}P_2) \times ({}_{12}P_2) = (10 \times 9) \times (12 \times 11) = 11{,}880 \quad \blacksquare$$

Example 5 **(a)** Find the number of three-digit numbers that can be formed using the digits 7, 8, 9 if no digit may be used more than once in a number.

(b) How many of these numbers will be even?

(c) What is the probability that a number selected at random will be odd?

Solution **(a)** ${}_3P_3 = 3! = 6$

(b) For the number to be even, the units digit must be even. We have only one choice for this digit, that is, 8. Thus two of the numbers will be even, namely, 798 and 978.

(c) Since two of the numbers are even, the other four numbers will be odd. Thus the probability that a number selected at random will be odd is 4/6, that is, 2/3. ■

At times there may be items in a set that are not distinguishable from one another. Consider, for example, the number of different arrangements possible using the letters of the word ERROR. If the three R's were distinguishable letters, then the answer would be ${}_5P_5 = 5! = 120$. In that case, each of the following six different words listed below would appear in the listing.

$ER_1R_2OR_3$	In the absence of subscripts,
$ER_1R_3OR_2$	these six words would appear
$ER_2R_1OR_3$	as the single word ERROR. That is,
$ER_2R_3OR_1$	they could not be distinguished
$ER_3R_1OR_2$	from one another.
$ER_3R_2OR_1$	

Try to find a systematic manner for listing all 20 possible arrangements.

Now note that the three R's, if distinguishable by the subscripts, can be arranged in ${}_3P_3 = 3!$ ways. Therefore, if the subscripts are erased, the number of possible distinguishable arrangements of the letters of the word ERROR is $5!/3! = 5 \times 4 = 20$. In general

> For a set of n objects of which r are alike, the number of distinguishable permutations of the n objects is $\frac{n!}{r!}$

Example 6 Find the number of distinguishable permutations of the letters of the word OHIO. List all such possible arrangements.

Solution We have $n = 4$ and $r = 2$. Thus $\frac{n!}{r!} = \frac{4!}{2!} = 12$. These may be listed as follows:

OHIO	HOIO	OOHI	OOIH
IHOO	HIOO	OIHO	OIOH
OHOI	HOOI	IOHO	IOOH ■

The preceding discussion can be extended to form this general result.

> Given a set of n objects of which n_1 are alike, n_2 are another kind that are alike, ..., and n_k are yet another kind that are alike, then the number of distinguishable permutations of the n objects is
>
> $$\frac{n!}{n_1!n_2!\cdots n_k!}$$

EXERCISES

1. Consider the set $S = \{\text{bat, ball}\}$ and list the permutations of the elements of S taken
   (a) one at a time
   (b) two at a time.
2. Consider the set $R = \{\text{reading, writing, arithmetic}\}$ and list the permutations of the elements of R taken
   (a) one at a time
   (b) two at a time
   (c) three at a time.
3. Consider the set $T = \{A, B, C, D\}$ and list the permutations of the elements of T taken
   (a) one at a time
   (b) two at a time
   (c) three at a time
   (d) four at a time.

Evaluate.

4. $5!$
5. $6!$
6. $\frac{8!}{6!}$
7. $\frac{11!}{7!}$
8. ${}_7P_2$
9. ${}_7P_3$
10. ${}_{10}P_1$
11. ${}_{10}P_{10}$
12. ${}_{12}P_{12}$
13. ${}_{12}P_3$
14. ${}_{10}P_3$
15. ${}_{10}P_7$

Express in terms of n.

16. ${}_nP_1$
17. ${}_nP_{n-1}$
18. ${}_nP_2$
19. ${}_nP_{n-2}$
20. ${}_nP_n$
21. ${}_nP_0$

Solve for n.

22. ${}_nP_1 = 7$
23. ${}_nP_1 = 21$
24. ${}_nP_{n-1} = 6$
25. ${}_nP_{n-1} = 120$
26. ${}_nP_2 = 20$
27. ${}_nP_2 = 30$

28. Find the number of different arrangements of the set of five letters $V = \{a, e, i, o, u\}$ taken
    (a) two at a time
    (b) five at a time.
29. (a) Find the number of four-digit numbers that can be formed using the digits 1, 2, 3, 4, 5 if no digit may be used more than once in a number.
    (b) How many of these numbers will be even?
    (c) What is the probability that such a four-digit number will be even?
    (d) What is the probability that such a four-digit number will be odd?
30. Find the number of different signals that can be formed by running up three flags on a flagpole, one above the other, if seven different flags are available.
31. How many different signals can be formed by running up seven flags on a flagpole, one above another, if four of the flags are of the same color?
32. Repeat Exercise 31 if three of the flags are red, two are blue, and two are green.

33. Find and list the number of distinguishable permutations of the letters of the word PAPA.
34. Repeat Exercise 33 for the word EERIE.
35. How many eleven-letter "words" can be formed from the letters of the word MISSISSIPPI?
36. Repeat Exercise 35 for the word MATHEMATICS.
37. An interior designer wishes to use books of different colors side by side on a shelf as a decoration. Considering color only, how many different arrangements can be formed with 10 books if one is black, two are red, three are green, and four are blue?
38. Repeat Exercise 37 for 12 books if two are red, two are green, four are blue, and four are black.

EXPLORATIONS

1. Make a worksheet of at least 20 problems for elementary school students to use in exploring the number of permutations of five things taken three at a time.

Permutations are assumed to be linear (along a line) unless otherwise specified. Suppose that we consider the permutation (**circular permutation**) of seating people at a circular table. The various places at the table are assumed to be indistinguishable and only the relative positions of the people are considered. Then there is only one arrangement for seating one person at a circular table and only one arrangement for seating two people at a circular table with two equally spaced chairs.

Use a circular region to represent a circular table, identify people by letters A, B, $C, \ldots,$ and sketch each of the possible arrangements for seating the indicated number of people at a circular table with the indicated number of equally spaced chairs.

2. Three.
3. Four.

In Explorations 4 and 5, how many arrangements are possible for seating the indicated number of people at a circular table with the indicated number of equally spaced chairs?

4. Five.
5. Six

6. Explain the appropriateness of the expression ${}_nP_n/n$, that is, $(n - 1)!$, for the number of permutations of n people in n seats around a circular table.

Unlike the seating of people around a table, a key ring can be turned over. Consider the effect of this phenomenon on the number of distinguishable permutations of keys on a circular key ring.

7. How many arrangements are possible for five keys on a circular key ring?

8. How many arrangements are possible for seven keys on a circular key ring?
9. Give an expression for the number of possible arrangements of n keys on a circular key ring.

10-6 Combinations

A fictitious club with five members was considered in Section 9-1. The set M of members is {Betty, Doris, Ellen, John, Tom}. The number of ways in which a president and a vice-president can be selected is the number ${}_5P_2$ of permutations of five things taken two at a time. Here order is important, since Betty as president and Doris as vice-president is a different set of officers than Doris as president and Betty as vice-president.

Now, suppose we wish to select a committee of two members from the set M without attaching any meaning to the order in which the members are selected. Then the committee consisting of Betty and Doris is certainly the same as the one consisting of Doris and Betty. In this case, we see that *order is not important*, and we call such a set a **combination**. One way to determine the number of possible committees of two to be formed from the set M is by a listing of all possible *subsets* of two. There are ten possible subsets (committees) of two.

Betty-Doris	Doris-John
Betty-Ellen	Doris-Tom
Betty-John	Ellen-John
Betty-Tom	Ellen-Tom
Doris-Ellen	John-Tom

We summarize this discussion by saying that the number of combinations of five things taken two at a time is ten. In symbols we write

${}_5C_2 = 10$, read as "the number of combinations of five things taken two at a time is ten."

Note that ${}_5P_2 = 20$ and ${}_5C_2 = 10 = {}_5P_2 \div 2$ since each combination of two elements such as {Betty, Doris} could have come from either of two permutations of those elements, in this case Betty-Doris or Doris-Betty. Here is the basic distinction between permutations and combinations.

A **permutation** is an *ordered set* of elements and *order is important.*

A **combination** is a *set* of elements and *order is not important.*

Consider a set of three elements; then

$${}_3P_3 = 3 \times 2 \times 1 = 6 \quad \text{and} \quad {}_3C_3 = 1$$

In general ${}_nP_n = n!$ but ${}_nC_n = 1$ for any counting number n. Also, for n greater than or equal to 2

$${}_nC_2 = \frac{{}_nP_2}{{}_2P_2} = \frac{n(n-1)}{2 \times 1}$$

where ${}_2P_2$ is the number of permutations (arrangements) associated with each combination of two elements.

To find ${}_5C_3$ consider again the specific problem of selecting committees of three from the set M. There are ten such possibilities, and we list them for $M = \{B, D, E, J, T\}$ using only the first initial of each name.

B, D, E	B, D, J	B, D, T	B, E, J	B, E, T
B, J, T	D, E, J	D, E, T	D, J, T	E, J, T

Note that selecting committees (subsets) of three is equivalent to selecting subsets of two to be omitted. That is, omitting J and T is the same as selecting B, D, and E. Therefore, we find that ${}_5C_3 = {}_5C_2 = 10$.

Inasmuch as we wanted only committees, and assigned no particular jobs to the members of each committee, we see that order is not important. However, suppose that each committee is now to elect a chairperson, secretary, and historian. In how many ways can this be done within each committee? This is clearly a problem in which order is important; we must therefore use permutations. The number of such possible arrangements within each committee is ${}_3P_3$, that is, 3!. For example, the committee consisting of B, D, and E can rearrange themselves as chairperson, secretary, and historian, respectively, as follows.

B, D, E   B, E, D   D, E, B   D, B, E   E, B, D   E, D, B

All six of these permutations are associated with just one combination. Similarly, the elements of each of the 10 combinations can be arranged in 6 ways. Thus

$$ {}_5C_3 \times {}_3P_3 = {}_5P_3 \quad \text{and} \quad {}_5C_3 = \frac{{}_5P_3}{{}_3P_3} $$

In general, each combination of three elements is associated with ${}_3P_3$ permutations of these elements.

${}_3P_3 = 3! = 3 \times 2 \times 1$

$$ {}_nC_3 = \frac{{}_nP_3}{{}_3P_3} = \frac{n(n-1)(n-2)}{3 \times 2 \times 1} $$

As in the cases of ${}_nC_2$ and ${}_nC_3$, the general form of ${}_nC_r$ may be expressed as a quotient and is frequently written in symbols in any one of these forms.

$$ {}_nC_r = \binom{n}{r} = \frac{{}_nP_r}{{}_rP_r} = \frac{{}_nP_r}{r!} = \frac{n(n-1)(n-2)(n-3)\cdots(n-r+1)}{r(r-1)(r-2)(r-3)\cdots 1} = \frac{n!}{r!(n-r)!} $$

Note that when ${}_nC_r$ was first expressed as a quotient of numbers, the denominator was $r!$, the product of r successive integers starting with r and decreasing; also the numerator was the product of r successive integers but in this case starting with n and decreasing.

Example 1 Evaluate ${}_7C_2$ by two different methods.

Solution Use ${}_nC_r = \dfrac{{}_nP_r}{r!}$ and ${}_nC_r = \dfrac{n!}{r!(n-r)!}$

(a) $_7C_2 = \frac{_7P_2}{2!} = \frac{7 \times 6}{2 \times 1} = 21$

(b) $_7C_2 = \frac{7!}{2!5!} = \frac{7 \times 6 \times 5 \times 4 \times 3 \times 2 \times 1}{2 \times 1 \times 5 \times 4 \times 3 \times 2 \times 1} = 21$ ■

Example 2 A box contains six red beads and four green beads. How many sets of four beads can be selected from the box so that two of the beads are red and two are green?

Solution We are dealing with a problem that involves combinations since the order in which the beads are selected does not affect the set that is obtained. The number of subsets of two red beads is $_6C_2$; the number of subsets of two green beads is $_4C_2$. By the general principle of counting, the total number of subsets of four beads with two red beads and two green beads is the product $_6C_2 \times {_4C_2}$.

$$_6C_2 = \frac{6!}{2!4!} = \frac{6 \times 5 \times 4 \times 3 \times 2 \times 1}{2 \times 1 \times 4 \times 3 \times 2 \times 1} = 15$$

$$_4C_2 = \frac{4!}{2!2!} = \frac{4 \times 3 \times 2 \times 1}{2 \times 1 \times 2 \times 1} = 6$$

$$_6C_2 \times {_4C_2} = 15 \times 6 = 90$$

Ninety different sets of four beads with two red and two green can be selected. ■

Example 3 How many different hands of five cards each can be dealt from a deck of 52 cards? What is the probability that a particular hand contains four aces and the king of hearts?

Solution The order of the five cards is unimportant, so this is a problem involving combinations.

$$_{52}C_5 = \frac{52!}{5!47!} = \frac{52 \times 51 \times 50 \times 49 \times 48}{5!} \times \frac{(47!)}{(47!)} = 2{,}598{,}960$$

The probability of obtaining any one particular hand, such as that containing the four aces and the king of hearts, is 1/(2,598,960). ■

Many problems in probability are most conveniently solved through the use of the concepts of combinations presented here. For example, let us consider again Example 4 of Section 10-3. There we were asked to find the probability that both cards would be aces if two cards are drawn from a deck of 52 cards. This problem can be solved by noting that $_{52}C_2$ is the total number of ways of selecting two cards from a deck of 52 cards. Also, $_4C_2$ is the total number of ways of selecting two aces from the four aces in a deck. The required probability is then given as

$$\frac{_4C_2}{_{52}C_2} = \frac{\frac{4!}{2!2!}}{\frac{52!}{2!50!}} = \frac{4 \times 3}{52 \times 51} = \frac{1}{221}$$

EXERCISES *Evaluate.*

1. $\frac{7!}{4!3!}$
2. $\frac{8!}{4!4!}$
3. $\frac{10!}{3!7!}$
4. $\frac{12!}{8!4!}$
5. $\frac{20!}{18!2!}$
6. $\frac{52!}{50!2!}$
7. ${}_8C_2$
8. ${}_{10}C_4$
9. ${}_8C_5$
10. ${}_9C_5$
11. ${}_{11}C_3$
12. ${}_{15}C_4$

13. List the ${}_3P_2$ permutations of the set $\{r, s, t\}$. Then identify the permutations associated with each combination and find ${}_3C_2$.
14. List the ${}_4P_3$ permutations of the set $\{w, x, y, z\}$. Then identify the permutations associated with each combination and find ${}_4C_3$.
15. List the elements of each of the ${}_4C_3$ combinations of the set $\{a, b, c, d\}$. Then match each of these combinations with a ${}_4C_1$ combination to illustrate the fact that ${}_4C_3 = {}_4C_1$.
16. Find a formula for ${}_nC_n$ for any positive integer n.
17. Find the value and give an interpretation of ${}_nC_0$ for any positive integer n.
18. Evaluate ${}_3C_0$, ${}_3C_1$, ${}_3C_2$, ${}_3C_3$ and check that the sum of these combinations is 2^3, the number of possible subsets that can be formed from a set of three elements.
19. Evaluate ${}_5C_0$, ${}_5C_1$, ${}_5C_2$, ${}_5C_3$, ${}_5C_4$, ${}_5C_5$ and check the sum as in Exercise 18.
20. Use the results obtained in Exercises 18 and 19 and conjecture a formula for any positive integer n for

$${}_nC_0 + {}_nC_1 + {}_nC_2 + {}_nC_3 + \cdots + {}_nC_{n-1} + {}_nC_n$$

21. How many sums of money (include the case of no money) can be selected from a penny, a nickel, a dime, a quarter, a half-dollar, and a one-dollar bill?
22. A man has a penny, a nickel, a dime, a quarter, and a half-dollar in his pocket. How many different amounts can he leave as a tip if he wishes to use exactly two coins?
23. A class consists of 8 boys and 12 girls. How many different committees of four can be selected from the class if each committee is to consist of two boys and two girls?
24. How many different hands of 13 cards each can be selected from a bridge deck of 52 cards?
25. How many choices of three books to read can be made from a set of nine books?
26. Explain why a combination lock should really be called a permutation lock.
27. Urn A contains five balls and urn B contains eight balls. How many different selections of ten balls each can be made if three balls are to be selected from urn A and seven from urn B?

28. An urn contains three black balls and three white balls.
   (a) How many different selections of four balls each can be made from this urn?
   (b) How many of these selections will include exactly three black balls?
29. Repeat Exercise 28 for an urn that contains ten black balls and six white balls.
30. Three arbiters are to be chosen by lot from a panel of 12. What is the probability that a certain individual on the panel will be one of those chosen?
31. Six cards are drawn at random from an ordinary bridge deck of 52 cards. Find the probability that all four aces are among the cards drawn.

State whether each question involves a permutation, a combination, or a permutation and a combination. Then answer each question.

32. The 30 members of the Rochester Tennis Club are to play on a certain Saturday evening. How many different pairs can be selected for playing singles?
33. How many different selections of a set of four records can be made by a disc jockey who has 12 records?
34. In how many orders can eight people line up at a single theater ticket window?
35. How many lines are determined by 12 points if no three points are collinear?
36. How many selections of nine students to play baseball can be made from a class of
   (a) nine students   (b) 12 students   (c) 20 students?
37. How many different hands can be dealt from a deck of 52 bridge cards if each hand contains
   (a) four cards   (b) seven cards
38. A committee of five is to be selected from a group that includes six Democrats and five Republicans. In how many ways can this be done if the committee is to consist of three Democrats and two Republicans?
39. How many sets of five cards can be selected from a deck of 52 cards if two of the cards are to be spades and three of the cards are to be hearts?
40. Five cards are dealt from a deck of 52 cards. What is the probability that the hands contain
   (a) four aces
   (b) three aces and two kings
   (c) five hearts?
41. All 25 members of a class shake hands with one another on the opening day of class.
   (a) How many handshakes are there in all?
   (b) How many will there be if, in addition, each shakes hands with the instructor as well?

***42.** A class is to be divided into two committees of at least one student each. How many different pairs of committees are possible from a class of eight students?

***43.** How many different pairs of committees of four students each can be formed from a class of eight students?

EXPLORATIONS

Blaise Pascal (1623–1662) was a young French mathematical genius who invented the first "computing machine" before he was 20. Poor health convinced him to devote most of his life to religion. Although Pascal's work on the arithmetic triangle was printed in 1665, the triangle was actually referred to in 1303 by the Chinese algebraist Chu Shih-chieh.

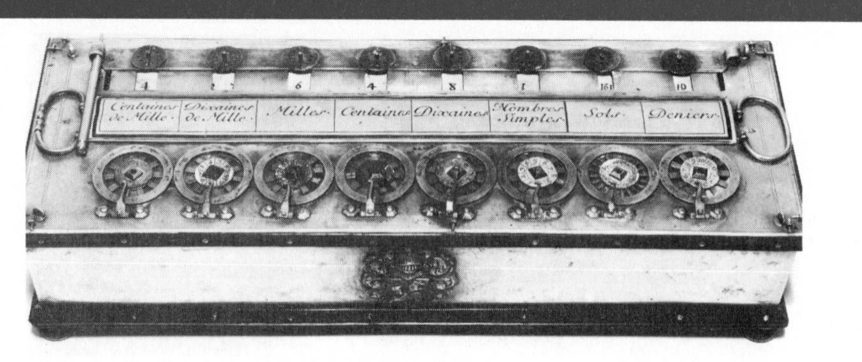

Pascal's adding machine

Recall that the notation ${}_nC_r$ may also be written in the form $\binom{n}{r}$. Consider the following array for sets of n elements where $n = 0, 1, 2, 3, \ldots$.

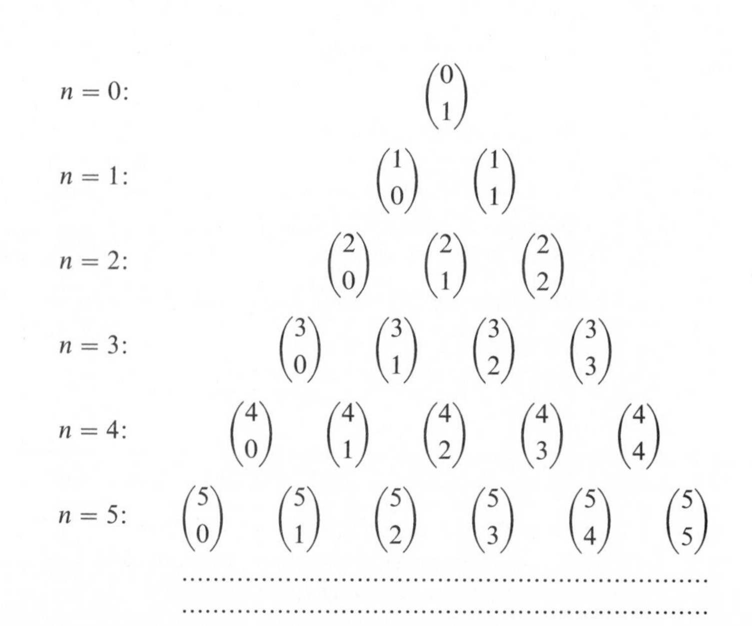

$n = 0$: $\binom{0}{1}$

$n = 1$: $\binom{1}{0}$ $\binom{1}{1}$

$n = 2$: $\binom{2}{0}$ $\binom{2}{1}$ $\binom{2}{2}$

$n = 3$: $\binom{3}{0}$ $\binom{3}{1}$ $\binom{3}{2}$ $\binom{3}{3}$

$n = 4$: $\binom{4}{0}$ $\binom{4}{1}$ $\binom{4}{2}$ $\binom{4}{3}$ $\binom{4}{4}$

$n = 5$: $\binom{5}{0}$ $\binom{5}{1}$ $\binom{5}{2}$ $\binom{5}{3}$ $\binom{5}{4}$ $\binom{5}{5}$

....

....

If we replace each symbol by its equivalent number, we may write the following array, known as *Pascal's triangle.*

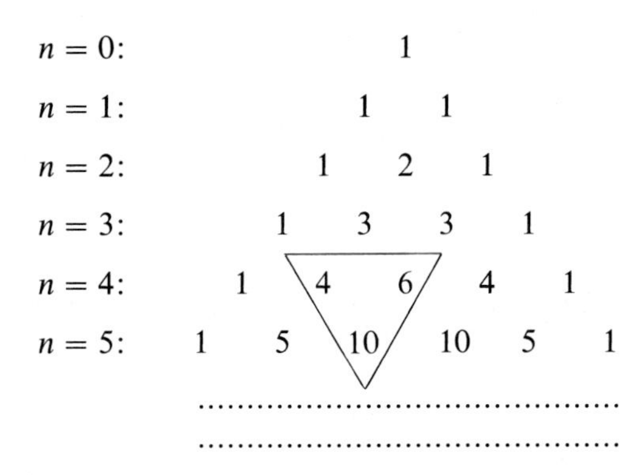

We read each row in this array by noting that the first entry in the nth row is $\binom{n}{0}$, the second is $\binom{n}{1}$, the third is $\binom{n}{2}$, and so on until the last entry, which is $\binom{n}{n}$. Since $\binom{n}{0} = \binom{n}{n} = 1$, each row begins and ends with 1.

There is a simple way to continue the array with very little computation. In each row the first number is 1 and the last number is 1. Each of the other numbers may be obtained as the sum of the two numbers appearing in the preceding row to the right and left of the position to be filled. Thus, to obtain the row for $n = 6$, begin with 1. Then fill the next position by adding 1 and 5 from the row for $n = 5$. Then add 5 and 10 to obtain 15, add 10 and 10 to obtain 20, and so forth as in this diagram.

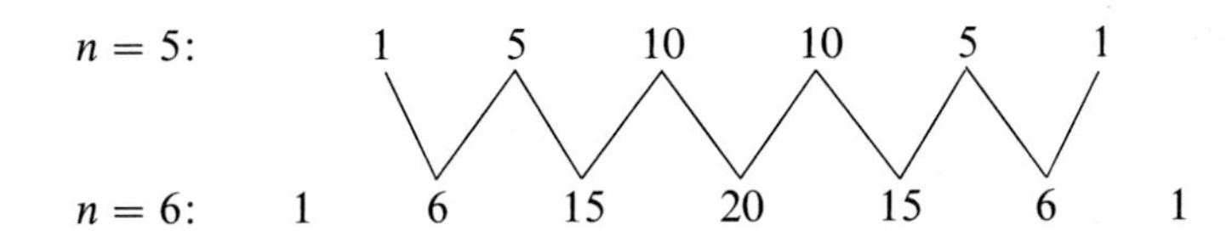

Pascal's triangle may be used in a routine manner to compute probabilities. The elements of the row for $n = 2$ are the numerators for the probabilities when two coins are tossed; the elements of the row for $n = 3$ are the numerators when three coins are tossed; and so on. The denominator in each case is found as the sum of the elements in the row used. For example, when three coins are tossed, we examine the row (1, 3, 3, 1) for $n = 3$. The sum is 8. The probabilities of 0, 1, 2, and 3 heads are then given respectively as

$$\frac{1}{8} \quad \frac{3}{8} \quad \frac{3}{8} \quad \frac{1}{8}$$

Note that the sum of the entries in the row for $n = 2$ is 2^2, that is, 4; the sum in the row for $n = 3$ is 2^3, that is, 8; the sum in the row for $n = 4$ is 2^4, that is, 16. In general the sum in the row for n is 2^n.

1. Use the row of Pascal's triangle for $n = 4$ to find the probabilities of 0, 1, 2, 3, 4 heads in a single toss of four coins.
2. Construct Pascal's triangle for $n = 0, 1, 2, 3, 4, 5, 6, 7, 8, 9, 10$.
3. List the entries in the row of Pascal's triangle for $n = 11$.
4. Repeat Exploration 3 for $n = 12$.
5. See how many different patterns of numbers you can find in Pascal's triangle. For example, find the sequence 1, 2, 3, 4, ...; find the triangular numbers (Section 8-2, Exploration 2); and others.
6. Compare the entries in the row for n of Pascal's triangle and the coefficients in the expression for $(a + b)^n$. Demonstrate this comparison for $n = 6$. Note that

$$(a + b)^2 = 1a^2 + 2ab + 1b^2$$
$$(a + b)^3 = 1a^3 + 3a^2b + 3ab^2 + 1b^3$$

7. Pascal's triangle is used as the basis for a very interesting and mystifying card trick. Learn how to perform this trick by reading Chapter 15, "Pascal's Triangle," in Martin Gardner's book *Mathematical Carnival*, published by Alfred A. Knopf, Inc., 1975. Then perform the trick in class and explain how it works.

Chapter Review

1. Evaluate **(a)** $\dfrac{17!}{13!}$ **(b)** $\dfrac{101!}{99!}$.
2. Evaluate **(a)** ${}_{20}P_2$ **(b)** ${}_{20}C_2$.
3. Consider the set $S = \{c, o, w\}$ and make a tree diagram to show the permutations of the letters of S taken 2 at a time without repetitions.
4. How many different four-letter "words" can be formed from the set $B = \{p, e, n, c, i, l\}$ if no letter may be used more than once?
5. Repeat Exercise 4 if repetitions of letters are permitted.
6. One card is selected from a deck of 52 bridge cards. Find the probability that the selected card is
   (a) not a spade **(b)** a spade or a heart.
7. Find the number of even three-digit numbers that may be formed from the set of digits $\{1, 2, 3, 4, 5, 6, 7\}$ if repetitions are not allowed.
8. A man has a nickel, a dime, a quarter, and a half-dollar. If he leaves a tip of exactly two coins, how many values are possible for such a tip?
9. Use a sample space to represent the possible outcomes described in Exercise 8 and find the probability that the tip is at least 30¢ when the selection of the two coins is at random.
10. What is the probability of obtaining on a single throw of two dice
    (a) a 7 **(b)** a number greater than 7?
11. Two integers between 0 and 10 are selected at random. If there is no replacement, what is the probability that at least one of the numbers is odd?
12. Make a sample space for a toss of a coin followed by the throw of a die.

A single card is drawn from an ordinary deck of 52 bridge cards. Find the probability that the card selected is as described.

13. **(a)** The ace of spades. **(b)** An ace or a spade.
14. **(a)** An ace or a red card. **(b)** An ace and a red card.

In Exercises 15 and 16 two cards are selected in succession without replacement from an ordinary deck of 52 bridge cards. Find the probability of the specified event.

15. **(a)** Both black cards **(b)** Both black cards or both red.
16. **(a)** Both kings. **(b)** Both kings or both queens.

17. What are the odds in favor of obtaining an ace when one card is drawn from an ordinary deck of 52 bridge cards?
18. What is your mathematical expectation when you are given 3 tickets out of a total of 6000 tickets for a single prize worth $1000?
19. Four coins are tossed.
    (a) What is the expected number of heads?
    (b) What are the odds in favor of at least one head?
20. A bag contains five red balls and three green balls. Two balls are drawn in succession without replacement. Find the probability that
    (a) both balls are red
    (b) the first ball is red and the second one is green.
21. Assume that n is greater than or equal to 5 and give an expression for
    (a) ${}_nP_5$ **(b)** ${}_nC_5$.
22. Find and simplify a formula for the given expression.
    (a) ${}_nP_0$ **(b)** $\frac{{}_nC_r}{{}_nP_r}$
23. Three judges are to be selected from a panel of ten. What is the probability that a particular member of the panel will not be chosen?
24. Nine people draw lots for their positions (1 through 9) in a line for concert tickets. How many outcomes are possible?
25. A class of ten students is divided into pairs of committees of at least one student each. How many such pairs of committees are possible?

COMPUTER SIMULATION OF ROLLING DICE

One of the very practical types of computer applications in the classroom is the simulation of experiments that would take a great deal of time if carried out by hand. For example, this is usually the case in sampling outcomes in probability experiments in which the exact probabilities of the outcomes are already known. Consider the rolling of a die. We know that there are six possible outcomes for each roll of the die. Further, we know that these six outcomes are equally likely on each roll of the die.

The following BASIC program to simulate the rolling of the die uses the greatest integer (INT) and the random number (RND) functions to simulate the value shown on the die. The RND function is used to generate a real number between 0 and 1. This number is multiplied by 7 to obtain a number between 0 and 7. If the product is between 0 and 1, new numbers are generated until the product of a number with 7 is greater than or equal to 1 and less than 7. Then the greatest integer function is used to obtain one of the positive integers 1 through 6—the outcomes for rolling a die. Note that these six numbers are obtained with equal probability through the use of the random number generating capability of the computer.

```
  5  HOME
 10  PRINT "THIS PROGRAM SIMULATES 72 ROLLS OF A DIE"
 12  PRINT
 15  A = 0
 16  B = 0
 17  C = 0
 18  D = 0
 19  E = 0
 20  F = 0
 30  FOR N = 1 TO 72
 40  X = INT(7 * (RND(1)))
 50  IF X < 1 THEN GOTO 40
 60  IF X = 1 THEN A = A + 1
 70  IF X = 2 THEN B = B + 1
 80  IF X = 3 THEN C = C + 1
 90  IF X = 4 THEN D = D + 1
100  IF X = 5 THEN E = E + 1
110  IF X = 6 THEN F = F + 1
120  NEXT N
130  PRINT  "OUT OF 72 ROLLS THERE WERE ";A;" ONES."
131  PRINT
132  PRINT "OUT OF 72 ROLLS THERE WERE ";B;" TWOS."
133  PRINT
134  PRINT "OUT OF 72 ROLLS THERE WERE ";C;" THREES."
135  PRINT
136  PRINT "OUT OF 72 ROLLS THERE WERE ";D;" FOURS."
137  PRINT
138  PRINT "OUT OF 72 ROLLS THERE WERE ";E;" FIVES."
139  PRINT
140  PRINT "OUT OF 72 ROLLS THERE WERE ";F;" SIXES."
150  END
```

When the program is run the computer performs the simulation of 72 rolls of the die to establish the number of times each of the six different values occurs in the 72 rolls of the die. The results obtained allow one to compare the observed outcomes with the expected outcomes of 12 of each of the 6 numbers. In the following sample run of the program, the number of each of the possible outcomes varies from a low of 7 to a high of 18.

```
RUN
THIS PROGRAM SIMULATES 72 ROLLS OF A DIE

OUT OF 72 ROLLS THERE WERE 9 ONES.

OUT OF 72 ROLLS THERE WERE 18 TWOS.

OUT OF 72 ROLLS THERE WERE 11 THREES.

OUT OF 72 ROLLS THERE WERE 13 FOURS.

OUT OF 72 ROLLS THERE WERE 7 FIVES.

OUT OF 72 ROLLS THERE WERE 14 SIXES.
```

If possible run the program several times yourself. Combine the results of several runs to obtain simulations for 144 rolls, 216 rolls, and so forth. As the number of rolls considered increases, do the sums for the respective face values of the die approach your expectations?

11

An Introduction to Statistics

Frederick Mosteller
(1916–   )

Frederick Mosteller is a contemporary American statistician known for his consulting work in public health, science, and other areas where his specialty is a major tool of communication and research. Known for his clear writing and teaching ability, Professor Mosteller has done much to develop new methods of explaining statistical data in easily understood ways, such as the stem-and-leaf plots and the box-and-whisker plots found in contemporary statistical texts. These methods are tools in the emerging statistical area known as exploratory data analysis.

11-1 Uses and Abuses of Statistics

Because of the widespread use of statistics for the consumer, we find that many adults are ready to quote "facts" that they read or hear, often without real understanding. Therefore, in this chapter we make an effort to acquaint you with a sufficient number of basic concepts so that you may better understand and interpret statistical data.

The average consumer is besieged daily with statistical data that are presented over radio and television, in newspapers, and in various other media. We are urged to buy a certain commodity because of statistical evidence presented to show its superiority over other brands. We are cautioned *not* to consume a particular item because of some other statistical study carried out to show its danger. We are told to watch certain programs, read certain magazines, see certain movies, and eat certain foods because of evidence produced to indicate the desirability of these acts as based on data gathered concerning the habits of others.

Almost every issue of the daily newspaper presents data in graphical form to help persuade the consumer to follow certain courses of action. Thus we are told what the "average" citizen eats, reads, earns, and even how the "average" person's leisure time is spent. Unfortunately, too many of us are impressed by statistical data regardless of their source.

Statistical data and concepts affect each of us and often have a profound influence upon the opportunities available to us. For example, among the items considered when you applied for admission to college were

Reminder:

LATIN		ENGLISH
datum	~	fact
data	~	facts

Your rank (first, second, third, ...) in your school class.
Your verbal Scholastic Aptitude Test score.
Your mathematics Scholastic Aptitude Test score.

Statistical computations determine the cost of your life insurance, health insurance, and automobile insurance—to mention only a few such items.

It has become fashionable, as well as informative and impressive, to support all sorts of predications and assertions with statistical computations. Our main purposes in this chapter are to help you become aware of the extensive use of statistics and to recognize some of its abuses. Many of the most blatant abuses of statistics occur in advertisements. For example, we frequently read or hear statements such as

Note vague comparisons.

Brush your teeth with GLUB and you will have fewer cavities.

The basic question that this statement fails to answer is: "Fewer cavities than what?" Regardless of the merits of GLUB, it is probably safe to say the following of every brand of toothpaste: "Brush your teeth with XXXX and you will have fewer cavities than *if you never brush your teeth at all.*" In other words, the example given is misleading in that it implies the superiority of GLUB without presenting sufficient data to warrant comparisons.

Many types of statements must be examined very carefully to separate their statistical implications from the impressions they are intended to make. Consider these statements.

Note false conclusions.

> More people are killed in automobile accidents than in wars. Therefore, it is safer to be on the battlefield.

Although the first statement may be true, the second one does not necessarily follow. Since there are generally many more people at home driving cars than there are in combat, the *percent* of deaths on the battlefield is actually much higher than in automobiles.

Frequently, graphs are presented in a style that misleads the reader. There are two common techniques for misleading casual readers of graphs. A scale may be started at a number different from zero. For example, if the scale on a graph to show trends in the circulation of a magazine does not start at zero, then the relative changes may appear much more significant than they really are. The first graph has been cut off so that the changes or trends appear to have greater significance than is actually the case. The same facts are shown in the second graph where the increase in circulation is recognizable as a very modest one of 400 readers, that is, 0.4%.

Note incomplete scales on graphs.

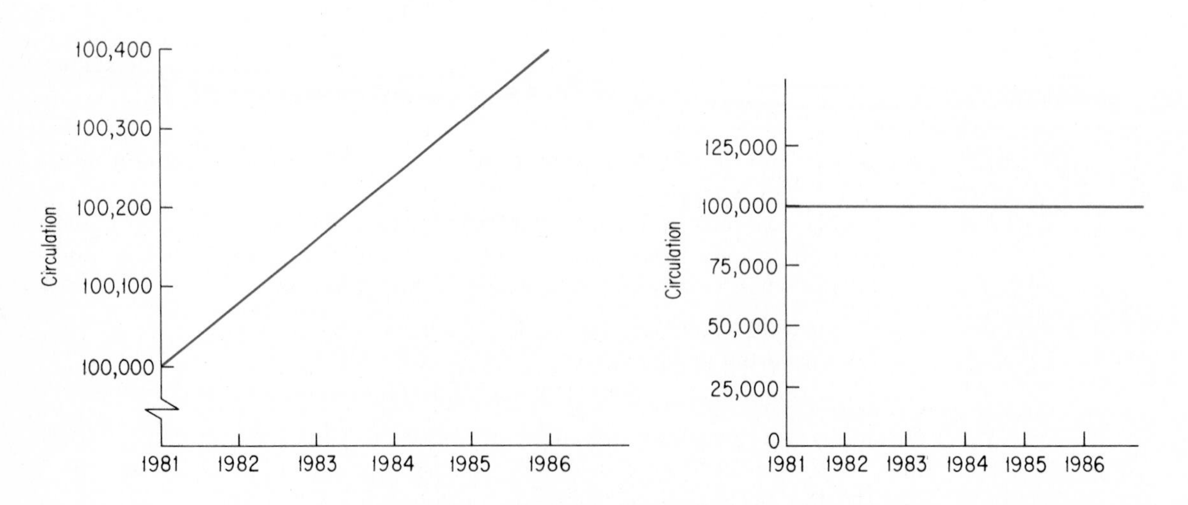

Another common abuse of graphical presentations is based upon the fact that for similar figures that are not congruent the constant ratio k of the lengths of corresponding line segments is different from the constant ratio k^2 of corresponding areas and also different from the constant ratio k^3 of corresponding volumes. For example, if the ratio of lengths of corresponding line segments is 3 to 2 (= 1.5), then the ratio of corresponding areas is 9 to 4 (= 2.25) and the

Compare these representations.

ratio of corresponding volumes is 27 to 8 (= 3.375). Area and volume measures are frequently misused to provide a visual misinterpretation of the actual data.

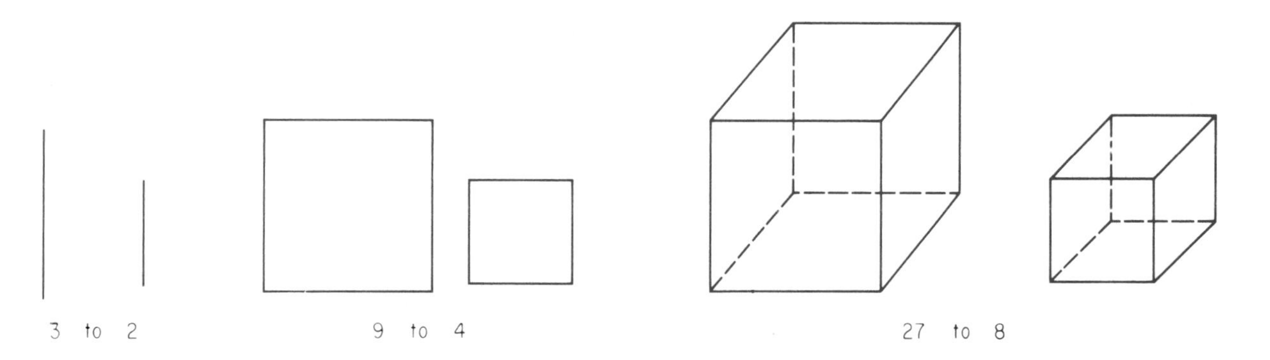

Collecting actual examples of uses of statistics to mislead readers and analyzing advertising claims can be enlightening. The intelligent consumer needs to ask basic questions about statistical statements that are presented to the public. Unfortunately, too many people tend to accept as true any statement that contains numerical data, probably because such statements sound impressive and authentic. Consider, for example, this statement:

Two out of three adults no longer smoke.

Before accepting and repeating such a statement, you should ask such questions as the following.

Does the statement sound reasonable?
What is the source of these data?
Can the facts be verified?
Do the facts appear to conform to your own observations?

You will undoubtedly think of other questions to raise. The important thing is that you raise these questions and do not accept all statistical facts as true in general. This is not to imply that all such facts are biased or untrue. Rather, as we have seen, and shall see later in this chapter, even true statements can be presented in many ways so as to provide various impressions and interpretations.

You should examine the logical structure of each statement; for example, is the statement or its converse suggested? You should also examine each statement to search for ambiguous words or meaningless comparisons. For example, consider the following assertion.

If you take vitamin C daily, your health will be better.

In addition to the questions you might raise about the scientific basis for this statement, note that the word *better* is ambiguous. We need to ask the question: "Better than what?" That is, we need to know the basis for comparison. As an extreme case, we certainly can say that your health will be better than the health of someone who is desperately ill! The exercises that follow will provide you with an opportunity to test your powers of reasoning and questioning.

EXERCISES *Discuss each of the following examples and tell what possible misuse or misinterpretation of statistics each one displays.*

Consult available newspapers and try to locate similar statements that present numerical data. Analyze these statements for possible misuses of statistics.

1. Most automobile accidents occur near home. Therefore, one is safer taking long trips than short ones.
2. Over 95% of the doctors interviewed endorse SMOOTHIES as a safe cigarette to smoke. Therefore, it is safe to smoke SMOOTHIES.
3. More college students are now studying mathematics than ever before. Therefore, mathematics must be a very popular subject.
4. Professor X gave out more A's last semester than Professor Y. Therefore, one should try to enroll in X's class next semester rather than in Y's class.
5. At a certain college 100% of the students bought *Contemporary Mathematics* last year. Therefore, this must be a very popular book at that school.
6. Most accidents occur in the home. Therefore, it is safer to be out of the house as much as possible.
7. Over 75% of the people surveyed favor the Republican candidate. Therefore, that candidate is almost certain to win the election.
8. A psychologist reported that the American girl kisses an average of 79 men before getting married.
9. A magazine reported the presence of 200,000 stray cats in the city of New York.
10. A recent report cited the statistic of the presence of 9,000,000 rats in the city of New York.
11. Arizona has the highest death rate from asthma in the nation. Therefore, if you have asthma, you should not go to Arizona.
12. During World War II more people died at home than on the battlefield. Therefore, it was safer to be on the battlefield than to be at home.
13. In a pre-election poll, 60% of the people interviewed were registered Democrats. Thus the Democratic candidate will surely win the election.
14. Over 90% of the passengers who fly to a certain city do so with airline X. It follows that most people prefer airline X to other airlines.
15. A newspaper exposé reported that 50% of the children in the local school system were below average in arithmetic skills.

For the given statement, list two or three questions that you would wish to raise before accepting the statement as true. Identify words, if any, that are ambiguous or misleading.

16. Most people can swim.
17. Short men are more aggressive than tall men.
18. If you walk 3 miles a day, you will live longer.
19. Brand A aspirin is twice as effective as brand B.
20. Teenagers with long hair are happier individuals.
21. People who swim have fewer heart attacks.

22. If you sleep 8 hours the night before an exam, you will do better.
23. Most students have success with this textbook.
24. Teachers tend to be less conscious about social ills.
25. About three out of four college students marry within one year of graduation.

Use the following graph for Exercises 26 *and* 27.

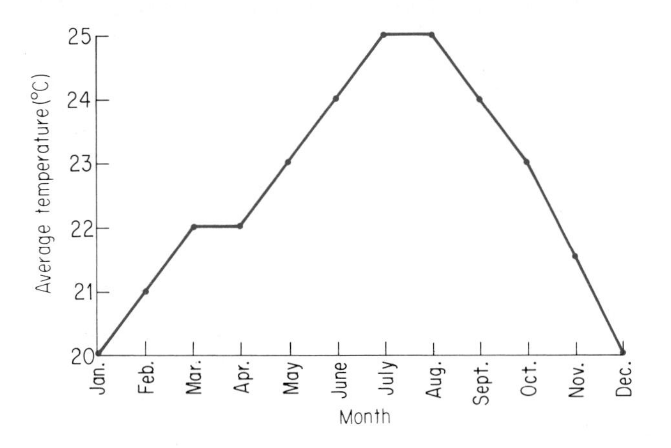

26. Disregarding the scale, what visual impression do you get from this graph about the fluctuation of temperatures during the year?
27. In what ways, if any, is the graph misleading?

Use the adjacent bar graph to answer Exercises 28 *through* 30 *and explain your answers.*

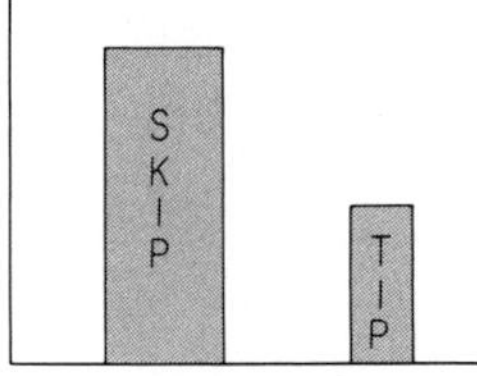

Circulation of two leading magazines

28. Can one deduce from the graph that the circulation of SKIP is greater than the circulation of TIP?
29. Can one deduce from the graph that the circulation of SKIP is approximately twice as great as the circulation of TIP?
30. What, if anything, could possibly be misleading or deceiving about the graph?

The following graphs for Exercises 31 *through* 33 *represent the average daily circulation in* 1975 *and the increase in circulation from* 1975 *to* 1985 *for each of the two newspapers. Explain your answers.*

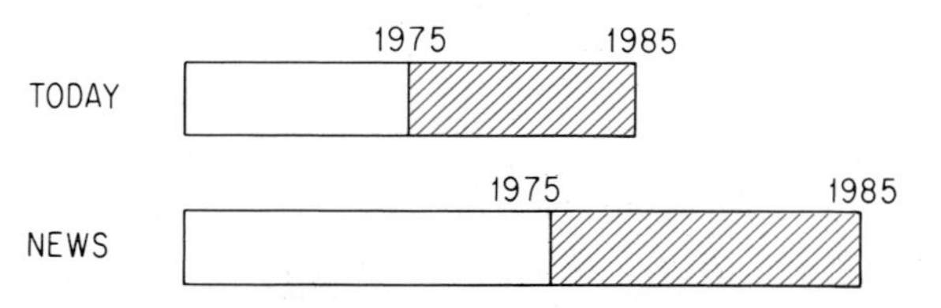

Circulation of two leading newspapers

31. Can one deduce from the graph that the increase in the circulation of TODAY from 1975 to 1985 was greater than the increase for NEWS?
32. Can one deduce from the graph that the percent of growth of the circulation of TODAY was greater after 1975 than the percent of growth for NEWS?
33. What, if anything, could possibly be misleading or deceiving about the graphs?

EXPLORATIONS

Begin a collection from newspapers, magazines, and other media of examples of misuses of statistics. In particular find examples that use each of the following.

1. Scales that have been cut off to mislead the reader.
2. Areas or volumes used to exaggerate comparisons.
3. Parts of the figure enlarged to provide undue emphasis.

For Explorations 4 and 5, discuss the given classic quotations.

4. Facts are facts.
5. Figures don't lie but liars figure.

6. Assume that ten percent of your regular class time for teaching is disrupted by extracurricular activities. Make a graph that presents the data correctly. Then make a graph that might be used to mislead the PTA into thinking that the situation is much worse than it actually is.
7. Read and report to your class on at least one chapter of a book such as *How To Use (and Misuse) Statistics* by Gregory A. Kimble and published by Prentice-Hall, Inc., 1978.

11-2 Descriptive Statistics

Descriptive statistics is the branch of statistics that is concerned with characterizing or summarizing a given set of data. Such condensations of data need to be carefully examined, as we have just observed. The consideration of methods of presentation of data helps us interpret the numerous statistical presentations that we encounter in newspapers and elsewhere. Accordingly, the present introduction to statistics is primarily concerned with descriptive statistics. The other main branch of statistics, **inferential statistics**, is concerned with the

Norway issued these stamps on the 100th anniversary of its Central Bureau of Statistics.

prediction of future events (election results, weather, stock market activity, and so on), and, in general, the testing of hypotheses.

As consumers our basic statistical needs are for the interpretation of statistical presentations made by others. Several of the methods used for statistical presentations are illustrated for the tossing of four coins. Assume that the coins have been tossed 32 times and that each time the number of heads obtained has been recorded. These **raw data** are listed left to right in the order of the occurrence of the events.

Number of Heads

1	2	3	2	4	2	2	1
0	2	4	2	2	1	3	3
2	3	1	1	2	2	3	0
4	1	3	2	4	2	1	2

The information is somewhat more meaningful when the number of heads is tallied and summarized in tabular form as a **frequency distribution**.

Use the given raw data to verify the entries in this table.

Number of Heads	*Tally*	*Frequency*
0	||	2
1	~~||||~~ ||	7
2	~~||||~~ ~~||||~~ |||	13
3	~~||||~~ |	6
4	||||	4

This information can now be treated graphically in a number of different ways. One common form of presenting data is a **histogram**, a bar graph without spaces between the bars.

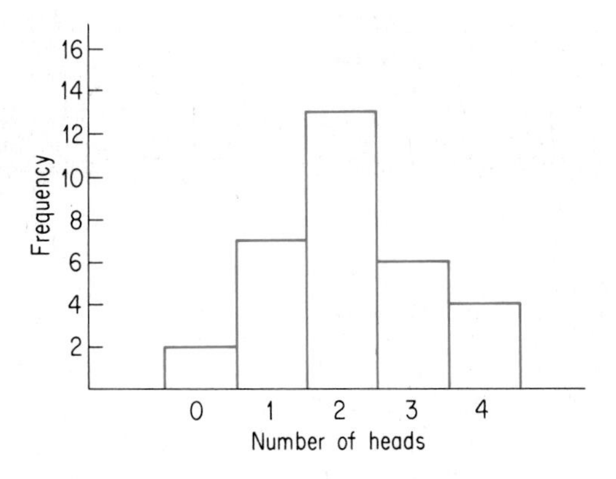

Frequently, a histogram is used to construct a **frequency polygon** or **line graph**. Thus the histogram can be approximated by a line graph obtained by connecting the midpoints of the tops of successive bars. Then the graph is extended to the base line as in the next figure.

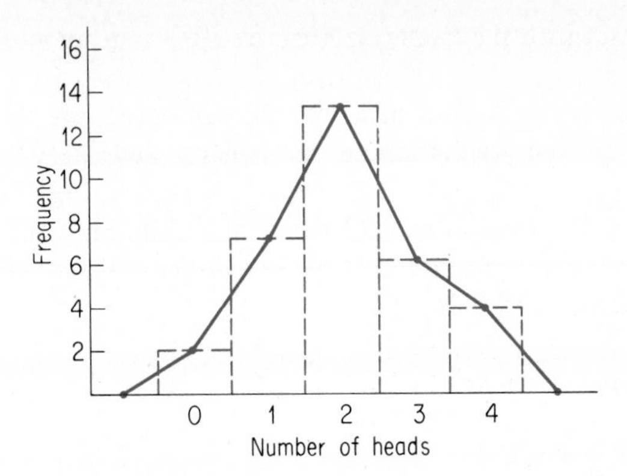

Line graphs appear to associate values with all points on a continuous interval rather than just with points that have integers as coordinates. Thus such line graphs are more appropriate for **continuous data** (all intermediate values have meaning) rather than with **discrete data** (only isolated values have meaning). For example, consider a graph of the temperatures in a particular city. Since it was 10 °C at 5 P.M. and 5 °C at 6 P.M., the temperature must have passed through *all* possible values between 10 and 5 in one hour. This, then, is an example of continuous data. Temperatures may be plotted and connected to produce a line graph as in the figure. Since gradual changes are expected, we "smooth out" the graph as a curve without "sharp turns."

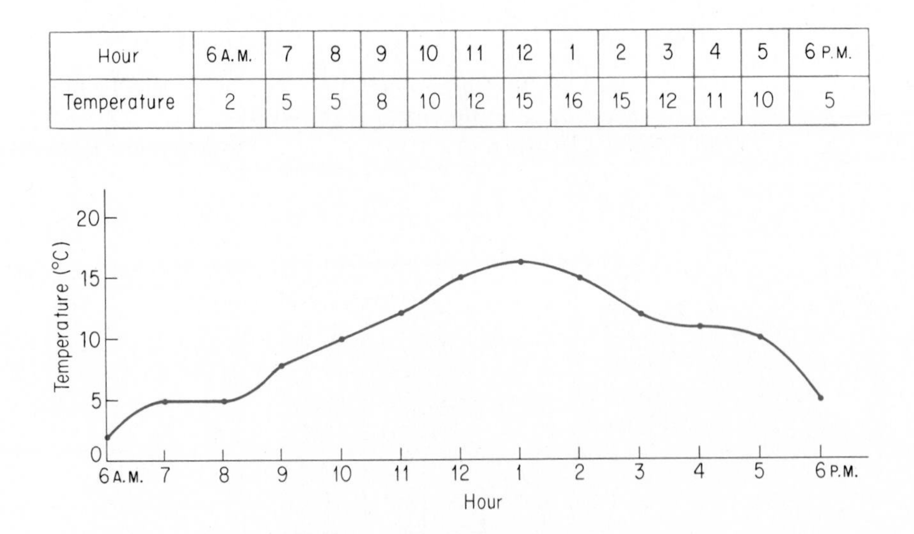

Hour	6 A.M.	7	8	9	10	11	12	1	2	3	4	5	6 P.M.
Temperature	2	5	5	8	10	12	15	16	15	12	11	10	5

Two other representations of frequency distributions deserve special mention. As our data consider these scores on an hour examination in a class of one of the authors.

79	82	57	82	83	66	61	79	70
92	84	90	90	67	79	86	80	93
86	91	83	83	88	72	67	97	87

Since there were no 100's and no scores below 50, the usual intervals

90–99   80–89   70–79   60–69   50–59

may be used to obtain a **grouped frequency distribution**.

Interval	*Tally*	*Frequency*
90–99	卌 I	6
80–89	卌 卌 I	11
70–79	卌	5
60–69	IIII	4
50–59	I	1

A much more informative representation may be obtained for essentially the same effort by using the following modification of this tallying procedure. We list the tens digits of the scores in a column on the left of a vertical bar and enter the units digit for each score in the appropriate row on the right of the vertical bar as in the first diagram below. Then we may rearrange the units digits in increasing order as in the second diagram.

Unordered leaves

```
9 | 2 0 0 3 1 7
8 | 2 2 3 4 6 0 6 3 3 8 7
7 | 9 9 0 9 2
6 | 6 1 7 7
5 | 7
```

Ordered leaves

```
9 | 0 0 1 2 3 7
8 | 0 2 2 3 3 3 4 6 6 7 8
7 | 0 2 9 9 9
6 | 1 6 7 7
5 | 7
```

These diagrams are called **stem-and-leaf diagrams**. Each row is a **stem** with the number on the left of the vertical bar as the **stem end** and the numbers (usually one-digit numbers) on the right as **leaves**.

There are many different types of graphs used to present statistical data. One very popular type is the **circle graph**, which is especially effective when one wishes to show how an entire quantity is divided into parts. Local and federal government documents frequently use this type of graph to show the distribution of tax money and various budget distributions. As another example, consider the following family budget.

Food	40%
Household	25%
Recreation	5%
Savings	10%
Miscellaneous	20%
Total	100%

To draw a circle graph for these data, we first recognize that there are 360° in a circle. Then we find each of the given percents of 360, and with the aid of a protractor construct central angles of appropriate sizes as in the following figure.

Verify each entry in the table. For example, show that 40% of 360° is 144°.

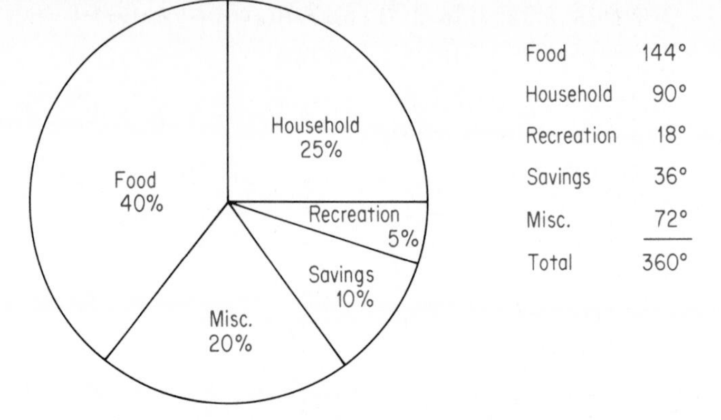

The same data can also be presented in the form of a **divided bar graph**. Here we select an arbitrary unit of length to represent 100%, and divide this in accordance with the given percents. In practice, even though the rectangular region (bar) has an arbitrary length, it is highly desirable to select that length so that it can be subdivided easily to obtain the desired parts. In the case of the data for our example all parts are integral multiples of 5%, that is, of one-twentieth of the 100% total. Thus it is very convenient to select the length of the bar so that twentieths of that length are easily identified. For example, if the total length were 10 centimeters, then the lengths of the parts would be 4 centimeters, 2.5 centimeters, 0.5 centimeter, 1 centimeter, and 2 centimeters.

There is an almost endless supply of graphs that could be used to illustrate this method of presenting data. Indeed, the reader is constantly subjected to graphical presentations in the daily newspapers and in other periodicals. Other examples of graphs are considered in the exercises.

Food 40%	Household 25%	Rec. 5%	Savings 10%	Misc. 20%

EXERCISES

1. Toss four coins simultaneously and record the number of heads obtained. Repeat this for a total of 32 tosses of four coins. Present your data in the form of a frequency distribution.
2. Present the data for Exercise 1 in the form of a bar graph.
3. Present the data for Exercise 1 in the form of a line graph.
4. Repeat Exercise 1 for 32 tosses of a set of five coins.
5. Present the data for Exercise 4 in the form of a bar graph.
6. Present the data for Exercise 4 in the form of a line graph.
7. Here is the theoretical distribution of heads when four coins are tossed for a total of 64 times. Present these data in the form of a bar graph.

Number of heads:	0	1	2	3	4
Frequency:	4	16	24	16	4

8. Here is the theoretical distribution of heads when five coins are tossed for a total of 64 times. Present these data in the form of a bar graph.

Number of heads:	0	1	2	3	4	5
Frequency:	2	10	20	20	10	2

9. Construct a circle graph to show this distribution of time spent by one student.

Sleep	8 hours
School	6 hours
Homework	4 hours
Eating	2 hours
Recreation	4 hours
Total	24 hours

10. Construct a divided bar graph for the data of Exercise 9.

11. Make a stem-and-leaf diagram with ordered leaves for the following set of test scores.

73	97	94	95	63	86	99	79	86	77
91	94	80	80	96	92	38	68	84	71
69	55	75	66	81	90	71	82	81	85

12. Repeat Exercise 11 for the following set of test scores.

76	88	82	95	85	96	69	88	87	91
70	27	83	58	90	83	92	88	87	72
79	91	73	78	84	90	92	88	60	64
84	87	80	78	76					

13. The U.S. Office of Management and Budget prepares a summary of sources of income and expenses for each fiscal year. The items are represented as cents of "The Government Dollar." For fiscal 1974 and 1984 with minor adjustments to accomodate the use of different headings we have the following data.

Where It Came From	***1974***	***1984***
Individual income taxes	44¢	35¢
Social insurance taxes and contributions	29¢	29¢
Corporate income taxes	14¢	7¢
Excise taxes	6¢	4¢
Other (includes 21¢ borrowing in 1984)	7¢	25¢
Where It Went	***1974***	***1984***
National defense	29¢	27¢
Income security, social security, medicare	31¢	39¢
Health	8¢	4¢
Education, training and social services	4¢	3¢
Veterans	5¢	3¢
Commerce and transportation	5¢	3¢
Net interest	8¢	13¢
Other	10¢	8¢

What percent of the government dollar came from individual income taxes
(**a**) In fiscal 1974? (**b**) In fiscal 1984?

Use the data given in Exercise 13 *to answer Exercises* 14 *through* 20.

14. What percent of the government dollar went to income security, social security, and medicare
    (**a**) In fiscal 1974? (**b**) In fiscal 1984?
15. Does the given data suffice for answering these questions? Explain your answers.
    (**a**) What percent of the government expenses was for health matters?
    (**b**) How many dollars were spent on health matters?
16. Income security, social security, medicare, health, education, training and social services, and veterans expenses constitute expenses for *human resources*. What percent of the government dollar was spent on human resources
    (**a**) In fiscal 1974? (**b**) In fiscal 1984?
17. Use 21¢ for borrowing and 4¢ for other in fiscal 1984 and make a circle graph for the sources of that government dollar.
18. Use 11¢ for income security and 28¢ for social security and medicare in fiscal 1984 and make a circle graph for the uses of that government dollar.
19. From fiscal 1974 to fiscal 1984 find the percent of decrease in the part of the government dollar that came from
    (**a**) individual income taxes
    (**b**) corporate income taxes.
    (**c**) Does this data mean that individuals paid fewer dollars in taxes in 1984 than in 1974? Explain.
20. From fiscal 1974 to fiscal 1984 find the percent of increase, or decrease, of the part of the government dollar that was spent on
    (**a**) national defense
    (**b**) income security, social security and medicare
    (**c**) veterans.

EXPLORATIONS

1. In one kindergarten room the names of the months of the year were listed across the bottom of a long sheet of paper. Each child put an "X" in a column over the name of the month in which the child's birthday occurred. What kind of graph was being constructed? What are some of the questions that you would raise with the children after the graph was completed?
2. Describe several other primary school activities that result in graphs.
3. Study one or more recently published series of elementary school mathematics textbooks, describe the treatments given to the construction and interpretation of graphs, and identify the different types of graphs considered in grades K through 8.

4. Select an elementary school grade level and prepare a lesson plan for organizing data from measurements made by the students.
5. Give at least two examples of types of data that could be collected from a primary class for the purpose of making a graph of the specified type.
   (**a**) Bar graph. (**b**) Divided bar graph.
   (**c**) Circle graph. (**d**) Line graph.
6. Start a collection of graphs from newspapers, magazines, and other sources. Be sure to include sufficient descriptive material to enable you to use the graphs in an elementary school class. Also see how many different types of graphs you can find.
7. Report on contemporary statistical studies of a controversial subject, such as smoking, heart disease, or the safety of nuclear generators of electricity.
8. Read and report to your class on at least one of the following selections from *Statistics: A Guide to the Unknown.* This book was edited by Judith M. Tanur et al. for a joint committee of the American Statistical Association and the National Council of Teachers of Mathematics. The book is published by Holden-Day, Inc., 1972.

   "Setting Dosage Levels" by W. J. Dixon on pages 34 through 39.

   "Statistics, Scientific Method, and Smoking" by B. W. Brown, Jr. on pages 40 through 51.

   "The Importance of Being Human" by W. W. Howells on pages 92 through 100.

   "Parking Tickets and Missing Women: Statistics and the Law" by Hans Zeisel and Harry Kalven, Jr. on pages 102 through 111.

   "Deciding Authorship" by Frederick Mosteller and David L. Wallace on pages 164 through 175.

11-3 Measures of Centrality

Most of us have neither the ability nor the desire to digest large quantities of statistical data. Rather, we prefer to see a graphical presentation of such data, or some number cited as representing the entire collection of data. Thus we are often faced with statements such as these.

The word *average* is often used loosely and may have a number of different meanings.

On the average, 9 out of 10 doctors recommend H_2O.

The average family earns $15,000 per year and has 1.8 children.

In an average college class 21% of the grades are A's.

In each case the word *average* is used in an effort to provide a capsule summary of a collection of data by means of a single number.

Assume that the following set of data represents the number of written reports required of ten students last year.

$\{6, 7, 8, 9, 10, 12, 15, 15, 20, 28\}$

Then each of the following statements is correct for this particular set of ten students.

1. *On the average each student was required to write* 13 *reports last year.*

In this case the average is computed by finding the sum of the given numbers and dividing the sum by 10, the number of students. This is the most commonly used type of average. For any given set of data it is referred to as the **arithmetic mean**, or simply the **mean**, of the data, and is generally denoted by $\overline{x}$ (read as "x bar").

$$\begin{array}{r} 6 \\ 7 \\ 8 \\ 9 \\ 10 \\ 12 \\ 15 \\ 15 \\ 20 \\ \underline{28} \\ 130 \end{array} \qquad \overline{x} = \frac{130}{10} = 13$$

Arithmetic mean: 13

2. *On the average each student was required to write* 11 *reports last year.*

In this case the word *average* is used to denote the number that is in the middle of the ordered set of data or, if there is an even number of elements in the set, the number that is halfway between the two middle numbers. The number that divides a set of scores in this way is called the **median** of the set of data. In this case it is determined as the number midway between 10 and 12. Note that the median is not necessarily one of the scores and that the data must be considered in order of size before the median can be found.

$$\left.\begin{array}{r} 6 \\ 7 \\ 8 \\ 9 \\ 10 \end{array}\right\} \text{Five scores below 11}$$

$$\text{—} \qquad \text{Median: 11}$$

$$\left.\begin{array}{r} 12 \\ 15 \\ 15 \\ 20 \\ 28 \end{array}\right\} \text{Five scores above 11}$$

3. *On the average each student was required to write* 15 *reports last year.*

In this case the average is the number that appears most frequently in the set of data. That is, it is correct to say that more of these students were required to write 15 reports than any other number. Such an average is referred to as the **mode** of a set of data. The mode is the only type of average that is always an observable element of the set.

4. *On the average each student was required to write 17 reports last year.*

In this case the average is computed by first adding the smallest and largest numbers and then dividing by 2, as in finding the midpoint of an interval on a number line.

$$\frac{6 + 28}{2} = \frac{34}{2} = 17$$

In the example regarding written reports, the midrange would be the best type of average to use to persuade someone that too many reports are required. Actually, however, the median is more representative of the number of reports required of a typical student.

This is the **midrange**, the midpoint of the interval (range) from the smallest number to the largest number. In cold climates, the midrange of the daily temperature is found each day. If the midrange (M) is less than 65 °F on a particular day, then the difference $65 - M$ is the number of **degree-days** for that particular date. Fuel dealers use the number of degree days to estimate the amount of fuel that their customers are using and thus the time at which additional fuel should be delivered.

Each of the four types of averages that we have discussed is known as a **measure of central tendency**. That is, each average is an attempt to describe a set of data by means of a single representative number. Note, however, that not every distribution has a mode, whereas some may have more than one mode.

The set of scores $\{8, 12, 15, 17, 20\}$ does not have a unique mode; indeed, each score could be considered a mode, although this would be pointless.

The set $\{8, 10, 10, 12, 15, 15, 17\}$ is **bimodal** and has both 10 and 15 as modes.

Example 1 For the given set of test scores, find

(a) the mean
(b) the median
(c) the mode
(d) the midrange.

$$\{72, 80, 80, 82, 88, 90, 96\}$$

Solution **(a)** The sum of the seven scores is 588.

$$\bar{x} = \frac{588}{7} = 84$$

(b) The median is 82. This is the middle score; there are three scores below 82 and three above it.
(c) The mode is 80, since this score appears more frequently than any other.
(d) The midrange is 84 since $(72 + 96)/2 = 168/2 = 84$. ■

Example 2 Repeat Example 1 for these scores.

$$\{30, 80, 80, 82, 88, 90, 96\}$$

Solution **(a)** $\bar{x} = 546/7 = 78$.
(b) The median is still 82.
(c) The mode is still 80.
(d) The midrange is 63 since $(30 + 96)/2 = 126/2 = 63$. ■

A comparison of Examples 1 and 2 indicates that the test scores are the same except for the first one. The median and the mode are the same for the two sets of scores. The arithmetic mean and the midrange are not the same.

The mean and the midrange are affected by extreme scores and one of these two measures of central tendency should be used as a representative of a set of data when extreme (high or low) scores should be reflected in the average. Otherwise the median or mode should be used to describe a set of data that contains extreme scores unless deception is a major objective.

Consider this example of the earnings of the employees of a small business run by a supervisor and three other employees. The supervisor earns \$50,000 per year. The others earn \$9000, \$10,000, and \$11,000, respectively. To impress the union, the owner claims that the employees are paid an average salary of \$20,000. The arithmetic mean is selected as the representative salary.

$$50{,}000 + 9000 + 10{,}000 + 11{,}000 = 80{,}000$$

$$80{,}000 \div 4 = 20{,}000$$

Actually, the median, \$10,500, would present a fairer picture of the average, or typical, salary.

As a more extreme example, consider a group of 49 people with a mean income of \$10,000. Let us see what happens to the average income of the group when an additional person with an income of \$300,000 joins the group.

$$49 \times 10{,}000 = 490{,}000 \qquad \text{(total income of 49 people)}$$

$$1 \times 300{,}000 = 300{,}000$$

$$\text{Sum} = 790{,}000 \qquad \text{(total income of 50 people)}$$

$$\text{Average} = \frac{790{,}000}{50} = 15{,}800$$

In reading or hearing advertisements, the consumer must always raise the question of the type of average being used. Furthermore, the source and the plausibility of the data should be questioned.

The average (arithmetic mean) salary is now \$15,800. The addition of one extreme salary raised the mean by \$5800, whereas the median and the mode of the incomes are unaffected.

EXERCISES

For the given set of data, find

(a) *the mean* **(b)** *the median*
(c) *the mode* **(d)** *the midrange.*

1. {60, 61, 65, 65, 70, 73, 73, 79, 84}
2. {73, 79, 80, 82, 84, 84, 92}
3. {10, 15, 18, 19, 21, 24, 26, 27}
4. {10, 11, 14, 14, 14, 17, 18}
5. {85, 61, 68, 73, 91, 68, 93}
6. {6, 7, 7, 8, 8, 8, 9, 9, 10, 10}
7. {9, 8, 9, 8, 6, 7, 8, 9, 10, 8}
8. {61, 69, 73, 78, 81, 86, 88, 88, 92, 97}

In Exercises 9 through 11 state which one, if any, of the four measures of central tendency seems most appropriate to represent the data described.

9. The average salary in a shop staffed by the owner and five employees.
10. The average salary of the workers in a factory that employs 100 people.

11. The average number of cups of coffee ordered by individual diners in a restaurant.

12. What relationship does the mode have to the use of this word in everyday language?

13. The mean score on a set of 15 tests is 75. What is the sum of the 15 test scores?

14. The mean score on a set of 30 tests is 84. What is the sum of the test scores?

15. The mean score on a set of 40 tests is 86. What is the sum of the test scores?

16. The mean score on nine of a set of ten tests is 70. The tenth score is 50. What is the sum of the test scores?

17. The mean score on 25 of a set of 27 tests is 80. The other two scores are 30 and 35. What is the sum of the scores?

18. Two sections of a course took the same test. In one section the mean score on the 25 tests was 80. In the other section the mean score on the 20 tests was 75.
    (a) What is the sum of the 45 test scores?
    (b) To the nearest integer what is the mean of the 45 test scores?

19. The mean score on five tests is 85. The mean on three other tests is 78. What is the mean score on all eight tests?

20. A student has a mean score of 85 on seven tests taken to date. What score must the student achieve on the eighth test in order to have a mean score of 90 on all eight tests? Comment on your answer. (Assume that 100 is a perfect score on each test.)

*21. An interesting property of the arithmetic mean is that the sum of the deviations (considered as signed numbers $x_i - \overline{x}$) of each score x_i from the mean $\overline{x}$ is 0. Show that this is true for the set of scores: {8, 10, 13, 17, 22}.

*22. Show that the sum of the deviations from the mean is 0 for the sets of data given in Exercises 2, 4, 6, and 8.

*23. Repeat Exercise 22 for the sets of data given in Exercises 1, 3, 5, and 7.

24. During a semester four students obtained the following grades on their hour tests.

M:	60,	70,	70,	75,	80
P:	50,	50,	80,	90,	100
S:	40,	75,	80,	85,	95
T:	40,	40,	85,	90,	100

(a) Copy and complete the following summary table showing the students' averages for the semester.

	Mean	Median	Mode	Midrange
M				
P				
S				
T				

(b) Which student had the highest grade for each of these averages?
*(c) Which of these students would you rank as the best one of the four?
*(d) Provide an argument for each of the students to claim that each should be rated as the best of the four.

EXPLORATIONS

Make any necessary inquiries and describe the way in which the given average is usually determined.

1. For your college class, the average grade on a particular test.
2. The batting average for a professional baseball player.
3. The average number of points per game during a particular season for a basketball player.
4. The average number of points per game during a particular season for a football team.
5. The average number of miles per gallon on the highway for a new car.
6. The average seasonal snowfall at a particular ski resort.
7. The average daily temperature at a weather station.
8. The average monthly rainfall at a weather station.

Consider newspapers, magazines, books that are not mathematics books, and radio and television programs to identify as many instances as you can in which the word *average* appears to be used in the specified sense.

9. For the arithmetic mean.
10. For the median.
11. For the mode.
12. For the midrange.
13. To encourage different interpretations by different readers or listeners.

11-4 Measures of Dispersion

In this section we consider ways of obtaining information about sets of data without considering all of the elements of the data individually.

A measure of centrality describes a set of data through the use of a single number. However, as in the examples considered in Section 11-3, the use of a single number without other information can be misleading. Some information can be obtained by comparing the arithmetic mean and the median, since the mean is affected by extreme scores and the median is relatively stable.

A **measure of dispersion** is a number that provides some information on the variability of a set of data. The simplest such measure to use for a set of numbers is the **range**, the difference between the largest number and the smallest number in the set.

Consider the test scores of these two students.

Betty: $\{68, 69, 70, 71, 72\}$ $\quad \bar{x} = 70$

Jane: $\{40, 42, 70, 98, 100\}$ $\quad \bar{x} = 70$

The range for Betty's test scores is $72 - 68$, or 4; the range for Jane's test scores is $100 - 40$, or 60. The average and ranges provide a more meaningful picture of the two students' test results than the averages alone.

Betty: $\overline{x} = 70$ range, 4

Jane: $\overline{x} = 70$ range, 60

Although the range is an easy measure of dispersion to use, it has the disadvantage of relying on only two extreme scores—the lowest and the highest scores in a distribution.

Consider Bob's test scores:

$\{40, 90, 95, 95, 100\}$

Since Bob is usually quite consistent in his work there may well be some good explanation for his one low grade. However, in summarizing his test scores one would report a range of 60, since $100 - 40 = 60$. The midrange 70 and the arithmetic mean 84 of Bob's test scores are also seriously affected by the extreme score. However, the mode 95 is not affected by the low score; also, the median 95 would be the same if the score of 40 were replaced by any other score.

To disperse is to spread out or to distribute. A measure of dispersion is a number that describes the amount of spreading out, or the density of the distribution, of the data.

There is another measure of dispersion that is widely used in describing statistical data. This measure is known as the **standard deviation**. The average consumer of statistics needs to be able to understand and interpret standard deviations. This need is particularly acute for teachers since they need to interpret the performance of their students on standardized tests.

For any given set of data the standard deviation, generally denoted by s, is relatively cumbersome to compute, especially for data grouped in a table. However, it may be instructive to note the computation of s for a small set of scores. For such data we generally use the formula

Think of the Greek letter Σ (capital sigma) as an abbreviation for *sum of.* Then to find Σd^2 you determine the deviation of each score from the mean, square each deviation, and then *add* these squares.

$$s = \sqrt{\frac{\Sigma d^2}{n}}$$

where Σd^2 represents the sum of the squares of the deviations of each score from the mean, and n is the number of scores.

Consider the set of scores

$\{8, 10, 12, 16, 19\}$

We first compute the mean as 13. In the column headed d we find the differences when the mean $\overline{x}$ is subtracted from each score. The sum of these deviations is 0, as noted in Section 11-3, Exercise 21. Thus to obtain a meaningful average of these deviations we may use either absolute values or the squares of the deviations. The standard deviation is based on the squares of the deviations, as in the column headed d^2.

Scores	d	d^2
8	-5	25
10	-3	9
12	-1	1
16	3	9
19	6	36
$5\overline{)65}$		$80 = \Sigma d^2$
$\overline{x} = 13$		

$s = \sqrt{80/5} = \sqrt{16}$

$s = 4$

Example 1 Compute s for the set of scores $\{10, 11, 13, 14, 17\}$.

Solution

Recall that s is a measure of dispersion. In Example 1 we have a set of scores with the same mean as the set above, but with a much smaller range. Note that the standard deviation is correspondingly smaller.

The computation of $\sqrt{6}$ can be completed by using a calculator or a table of square roots.

Scores	d	d^2
10	−3	9
11	−2	4
13	0	0
14	1	1
17	4	16
$5\overline{)65}$		$30 = \Sigma d^2$
$\overline{x} = 13$		

$s = \sqrt{30/5} = \sqrt{6}$

$s \approx 2.5$

■

To understand the significance of standard deviations as measures of dispersion, we first turn our attention to a discussion of **normal distributions**. The line graphs of these distributions are the familiar bell-shaped curves that are used to describe distributions for so many physical phenomena. For example, the distribution of intelligence quotient (IQ) scores in the entire population of the United States can be pictured by a **normal curve**.

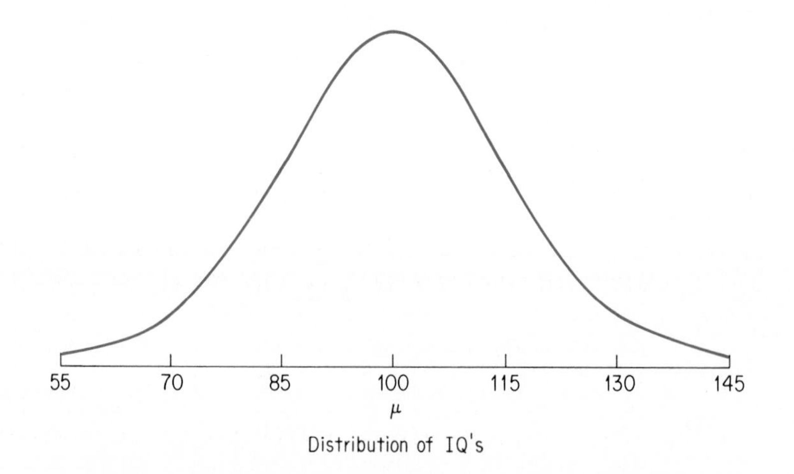

Distribution of IQ's

The area under a normal curve represents the entire population. According to psychologists, an IQ score over 130 is considered to be superior. At the other end of the scale, a score below 70 generally indicates some degree of academic retardation.

A normal curve should be expected only for data that involve a large number of elements and are based upon a general population. For example, the IQ scores of the honor society members in a school should not be expected to fit the distribution for the school as a whole. For any theoretical population, such as the set of all U.S. citizens, the mean is represented by the Greek letter μ (read as "mu") and the standard deviation by the Greek letter σ (read as "sigma").

$\overline{x}$ and s refer to results of calculations concerning a given set of data; μ and σ are their analogs for theoretical distributions.

If three standard deviations are added to and subtracted from the mean of a normal distribution, practically all (99.7%) of the data will fall on the interval from $\mu - 3\sigma$ to $\mu + 3\sigma$. If two standard deviations are added to and subtracted from the mean, approximately 95% of all data will fall on the interval from $\mu - 2\sigma$ to $\mu + 2\sigma$. An interval of one standard deviation about the mean includes approximately 68% of all data in a normal distribution. We may summarize these statements as follows.

In a normal distribution the mean, the median, and the mode all have the same value. This common value is associated with the axis of symmetry of the normal curve.

$\mu \pm 1\sigma \approx 68\%$

$\mu \pm 2\sigma \approx 95\%$

$\mu \pm 3\sigma \approx 100\%$

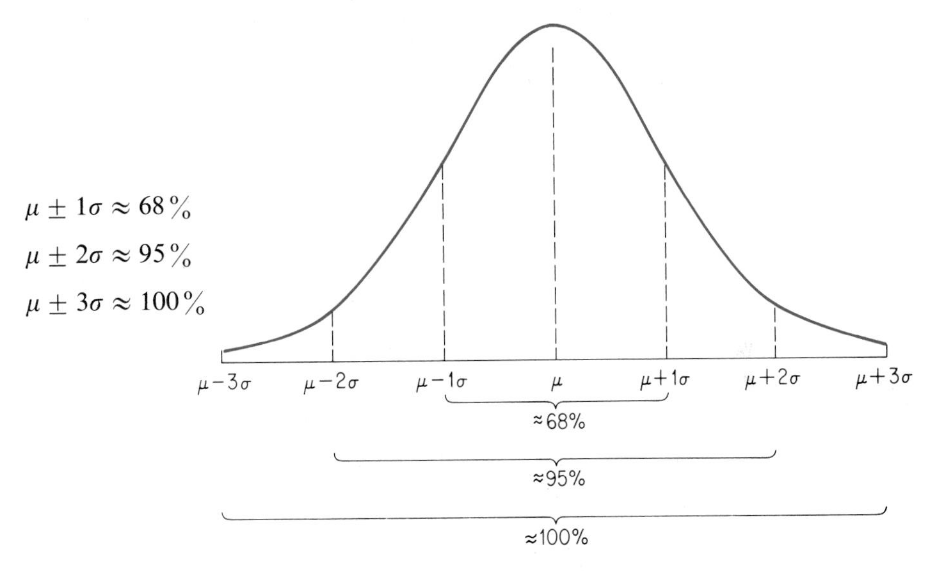

The set of five scores for which the standard deviation was computed in Example 1 is much too small to expect a close match with a normal distribution. In this case all scores are within $1\frac{1}{2}$ standard deviations of the mean.

Let us now return to the graph of IQ scores. Suppose we are told that for this distribution $\sigma = 15$. We may then show these standard deviations on the base line of the graph. According to our prior discussion, we may now say that approximately 68% of the population have IQ scores between 85 and 115, that is, on the interval $100 \pm 1\sigma$. Approximately 95% of the population have IQ scores between 70 and 130, that is, on the interval $100 \pm 2\sigma$. Finally, almost everyone has an IQ score between 55 and 145, that is, on the interval $100 \pm 3\sigma$.

Here "almost everyone" really means 99.7%; that is, we might expect 0.3% (3 in 1000) of the population to have scores below 55 or above 145.

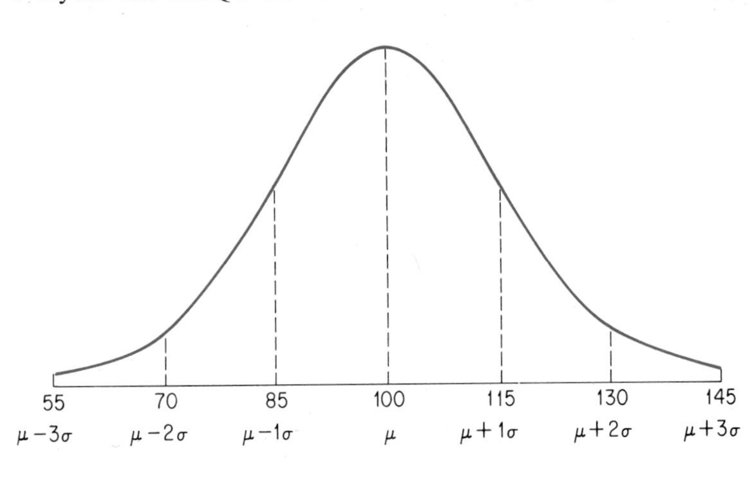

Example 2 What percent of the population have IQ scores below 115?

Solution We know that 50% of the population have IQ scores below 100, the mean. Also, 68% have scores on the interval $\mu \pm 1\sigma$. Then by the symmetry of the normal curve, 34% have scores between 100 and 115, and, as in the figure, 84% have scores below 115.

Note that when considering questions such as those in Example 2 we do not concern ourselves with the number of people who make any particular score, such as 100 or 115.

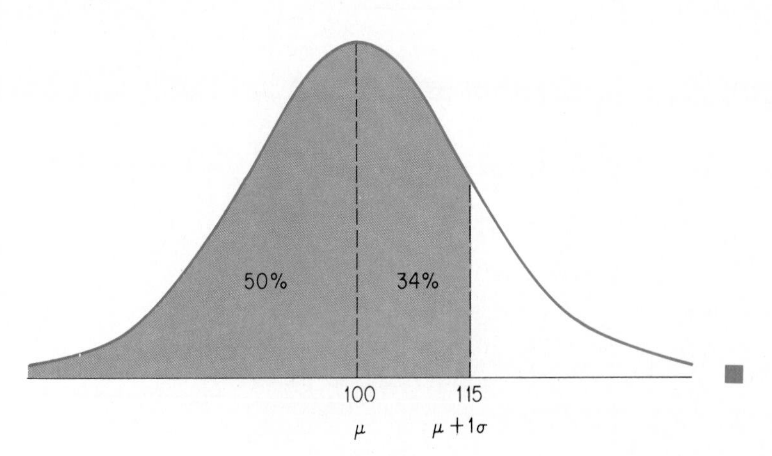

The z-score of any given raw score indicates the deviation of the raw score from the mean in terms of standard deviations. Tables of the percent of a normal population between the mean and any specified z-score are often used to interpret the relative positions of raw scores.

Scores such as an IQ score of 115 or a College Entrance Examination Board score of 600 are raw scores. Each of these two raw scores is one standard deviation above the mean and is said to have *z-score* 1 in the scale $-3, -2, -1, 0, 1, 2, 3$ of standard deviations from the mean. In general, a score x in a set of scores with population mean μ and standard deviation σ has **z-score** z where

$$z = \frac{x - \mu}{\sigma}$$

The scale $-3, -2, -1, 0, 1, 2, 3$ in standard deviations enables us to use the properties of a normal distribution such as the 68% within one standard deviation of the mean, for a wide variety of situations. For example, we found in Example 1 that 84% of the population have IQ scores below 115, which is one standard deviation above the mean. This same result can be used in many other situations. Scholastic Aptitude Test (SAT) scores have a mean of 500 and a standard deviation of 100. Thus 84% of the SAT scores on a particular test are below 600. College Entrance Examination Board (CEEB) scores also have a mean of 500, a standard deviation of 100, and 84% of the scores below 600.

For a set of scores with sample mean $\overline{x}$ and standard deviation s we use

$$z = \frac{x - \overline{x}}{s}$$

Example 3 What percent of the population have IQ scores above 130?

Solution Find the difference between the given score (130) and the mean (100). Then divide this difference by the standard deviation (15) to obtain the z-score.

$$z = \frac{130 - 100}{15} = 2$$

The scores above 130 are the scores above $\mu + 2\sigma$. The interval $\mu \pm 2\sigma$ includes the scores of 95% of the population. Then one-half of 95%, that is, 47.5% of the population have scores to the right of the mean on the interval from 100 to 130. Thus 97.5% of the population have scores below 130, and 2.5% of the population have IQ scores above 130.

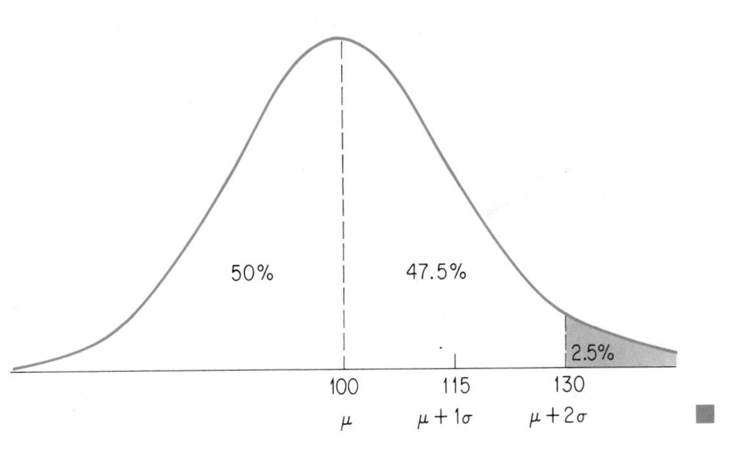

Note that we have discussed only normal distributions. However, the standard deviation can be computed for other distributions as well. In general, many sets of data tend to be approximately normal, and the standard deviation is a very useful measure of dispersion.

The result obtained in Example 3 as applied to SAT and CEEB scores shows that 97.5% of the scores are below 700 and 2.5% of the scores are over 700.

EXERCISES

1. Find the range for the given set of scores.

   {55, 67, 80, 85, 90, 92, 98}

2. Give a set of the same number of scores with the same mean as in Exercise 1 but with a smaller range.
3. Repeat Exercise 2 for a larger range.
4. Give a set of the same number of scores with the same range as in Exercise 1 but with a smaller mean.
5. Repeat Exercise 4 for a larger mean.
6. Give a set of ten scores with the same mean and the range as in Exercise 1.

Repeat the indicated exercise for this set of scores.

{45, 53, 69, 77, 85, 87, 90, 92, 95}

7. Exercise 1.
8. Exercise 2.
9. Exercise 3.
10. Exercise 4.
11. Exercise 5.
12. Exercise 6.

In Exercises 13 *through* 15, *state which of the distributions can be expected to be approximately normal.*

13. The scores of all graduating high school seniors on a particular college board examination.

14. The weights of all college freshmen males.
15. The number of heads obtained if 100 coins are tossed by each college graduate in the country.

16. For a normal distribution of 10,000 test scores the mean is found to be 500 and the standard deviation is 100.
 (a) What percent of the scores will be above 700?
 (b) What percent of the scores will be below 400?
 (c) About how many scores will be above 600?
 (d) About how many scores will be below 300?
 (e) About how many scores will be between 400 and 700?
17. If 100 coins are tossed repeatedly, the distribution of the number of heads is a normal one with a mean of 50 and a standard deviation of 5. What percent of the number of heads will be
 (a) greater than 60
 (b) less than 45
 (c) between 40 and 60?

Compute the standard deviation for the given set of scores.

18. {7, 9, 10, 11, 13}
19. {11, 12, 13, 15, 20, 20, 21}
20. {81, 83, 85, 88, 93}
21. {78, 82, 83, 85, 92}
***22.** {68, 74, 80, 82, 83, 85, 86, 88, 91, 93}
***23.** {58, 62, 65, 67, 70, 75, 77, 80, 85, 91}

EXPLORATIONS

In reading educational literature one frequently encounters the term **coefficient of correlation**, usually denoted by the letter r. The coefficient of correlation is given as a decimal on the interval -1.00 to $+1.00$ and provides an indication of how two variables are related.

A perfect positive correlation of 1.00 indicates that two sets of data are related so that the changes in one set are a positive multiple of the corresponding changes in the other set. For example, here is a set of ages and weights for a group of six individuals. Note that for each increase of 5 years of age there is a corresponding increase of 10 pounds of weight.

Age	*Weight*
20	130
25	140
30	150
35	160
40	170
45	180

The coefficient of correlation is 1.00.

Now consider the following table for six other individuals. For each increase of 5 years of age, there is a corresponding decrease of 10 pounds of weight.

Age	*Weight*
20	180
25	170
30	160
35	150
40	140
45	130

The coefficient of correlation is -1.00.

It is important to note that correlation does not imply causation. For example, there might be a positive correlation between the size of shoe that a student wears and the students' handwriting ability. This would not mean that big feet improve one's handwriting.

A coefficient of correlation of 0 indicates no uniform change of either variable with respect to the other. In general practice such extreme cases seldom occur. Furthermore, one has to read the literature accompanying any particular test or research study to determine whether any particular coefficient of correlation can be considered as significant.

1. There has been a high positive correlation between expenditures for alcohol and for higher education in recent years. Does this mean that drinking alcohol provides one with the thirst for knowledge? Does it mean that education leads one to drink? How can you explain this high correlation?

In each exploration explain why you would expect to find a high or a low correlation between the two sets.

2. IQ scores and scores on college entrance examinations of all graduating high school seniors.
3. Scores made by elementary school students in reading and in arithmetic.
4. Age and physical abilities of mentally retarded individuals.
5. Manual dexterity and age of normal elementary school children.
6. Grades in arithmetic and number of hours spent by elementary school children watching television.
7. Weights, relative to normal for their heights, of elementary school children and their mothers.
8. Academic grades and extent of participation in extracurricular activities of college students.
9. Effectiveness in teaching and years of college training of teachers.

11-5 Measures of Position

Rank in class, or class rank, is determined by counting down from the top of the class.

Anne, Bill and Carol were all classmates in a high school class of 400 students. Anne was 25th from the top of her class, that is, her **class rank** was 25. Bill's class rank was 100 but he preferred to think that three-quarters of his classmates were ranked below him. Thus Bill describe his position in the class as at the *third quartile* or as at the 75th *percentile*. Imagine that all 400 students were lined up according to their rank in class.

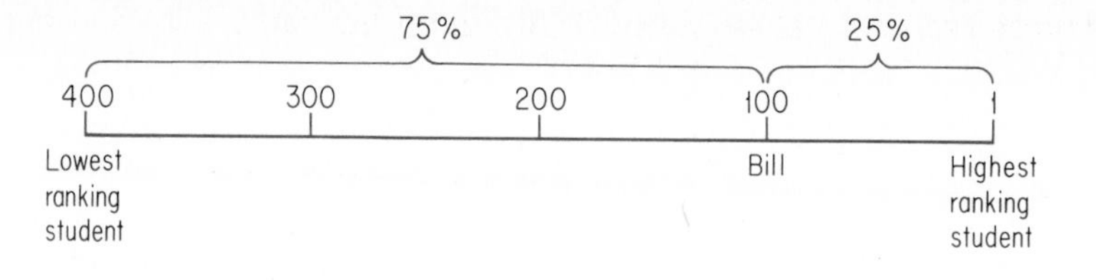

Then counting from the top of the class Anne would be 25th in line and Bill would be 100th. Bill is in the upper 25% of the class because 75% of the students are ranked below him.

The position of any element in an ordered set of data is often described in terms of class rank, quartiles, or percentiles. The class rank is the number of the element counting from the top of the list or class. The other measures of position are determined by the number of quarters or hundredths of the data that precede the specified element in the ordered set.

The ninety-nine **percentiles** are denoted by $P_1, P_2, P_3, \ldots, P_{98}, P_{99}$. As in the case of Bill at the 75th percentile, the kth percentile P_k has k percent of the data preceding it. A score or element at the kth percentile is said to have **percentile rank** k. To find the percentile rank of a given score or element in an ordered set of data first express the number of elements at or above the given score as a percent of the total number of elements in the data and then subtract that percent from 100%. For example, Carol's rank in class was 40. The 40 students at or above Carol's rank are 10% of the class of 400. Thus Carol may describe her position as at the 90th percentile.

The three **quartiles** are denoted by Q_1, Q_2, Q_3 and may be defined in terms of percentiles.

$$Q_1 = P_{25} \qquad Q_2 = P_{50} \qquad Q_3 = P_{75}$$

The second quartile, Q_2, is the median (see Section 11-3). If the data have an even number of scores the median is taken halfway between the two middle scores. Although we usually avoid such situations, other quartiles and percentiles are taken between scores when necessary.

Example 1 In a class of 500 students, how many students are ranked below the 80th percentile?

Solution The 80th percentile is preceded by 80% of the students. Since 80% of 500 is 400, there are 400 students who rank below the 80th percentile.

80% (400 students)
20% (100 students)
Lowest ranking student
P_{80}
Highest ranking student ■

Example 2 Don is ranked 21st in a class of 300. What is Don's percentile rank in his class?

Solution There are 279 members of the class ranked below Don. Since 279 is 93% of 300, there are 93% of the class ranked below Don. Thus Don's percentile rank is 93. ■

Verify that 279 is 93% of 300.

Example 3 For the given set of quiz scores, find
(a) the 50th percentile
(b) Q_1
(c) the 90th percentile.

$$\{8, 5, 9, 7, 10, 8, 6, 9, 8, 10, 7, 9\}$$

Solution The 12 scores must be considered in order before finding measures of position.

$$\{5, 6, 7, 7, 8, 8, 8, 9, 9, 9, 10, 10\}$$

(a) 50% of 12 is 6. The 50th percentile is 8.
(b) Q_1 is the 25th percentile; 25% of 12 is 3. The first quartile, Q_1, is 7.
(c) 90% of 12 is 10.8. The 90th percentile is 9.8. Note that 9.8 is the 10.8th score and is 0.8 of the way from the 10th score 9 to the 11th score 10. We have $9.8 = 9 + 0.8(10 - 9)$ using the difference of the two scores between which the desired percentile is located. ■

Percentiles for a normal distribution may be expressed in terms of z-scores, the numbers of standard deviations from the mean. The most commonly used values are given in the following table.

PERCENTILE	DEVIATIONS FROM THE MEAN
1	−2.33
5	−1.65
10	−1.28
20	−0.84
30	−0.52
40	−0.25
50	0
60	0.25
70	0.52
80	0.84
90	1.28
95	1.65
99	2.33

Measures of position are important since individual scores mean very little. For example, a test score of 49 is excellent if the total possible score is 50 but disappointing if the total possible score is 100.

Example 4 Find the 95th percentile for a normal distribution of scores with mean 800 and standard deviation 200.

Solution From the table the 95th percentile is 1.65 standard deviations above the mean.

Reading a table is an important problem-solving skill. Be sure that you see how the preceding table is used in the solution for Example 4.

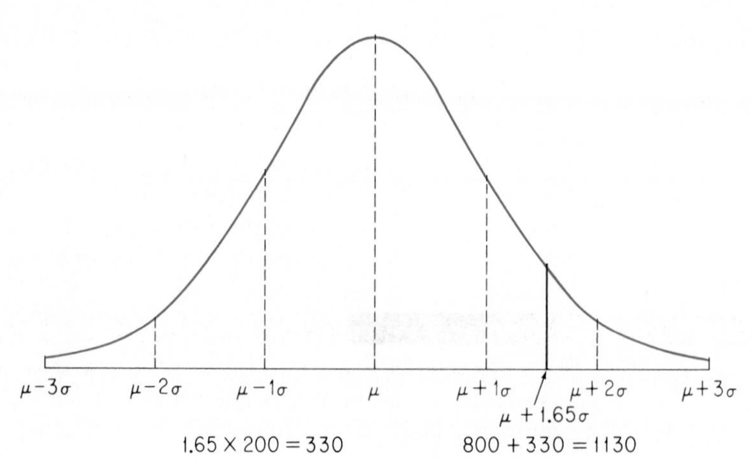

The 95th percentile is 1130. That is, for this distribution 95% of the scores are below 1130. ■

EXERCISES *In a high school class of 550 students how many students are ranked below the indicated position?*

1. P_{90}
2. Q_3
3. Q_2
4. P_{60}
5. Q_1
6. P_{80}

Find the percentile rank of the indicated student in a class of 400. Each student's rank in class is given.

7. Ann, 4th.
8. Barbara, 8th.
9. Charles, 12th.
10. David, 16th.
11. Eleanor, 24th.
12. Frank, 28th.

In each of the Exercises 13 *and* 14, *which student ranks highest in a class of* 300 *students? A measure of position is given for each student.*

13. Albert, P_{88}; Ben, Q_3; Charles, 25th.
14. George, Q_3; Harold, P_{70}; Irvine, 100th.

Repeat the indicated exercise for a class of 500 *students.*

15. Exercise 13.
16. Exercise 14.

Consider a normal distribution of SAT scores with arithmetic mean 500 *and standard deviation* 100. *Use the table for percentiles and deviations from the mean.*

17. Jack scored at the 95th percentile. What was his score?

18. Katherine scored 584. What was her percentile rank?

19. Find the percentile rank of each score.

(**a**) 500 (**b**) 665 (**c**) 475
(**d**) 416 (**e**) 335 (**f**) 733

20. Find the score at each percentile rank.

(**a**) 20 (**b**) 40 (**c**) 60
(**d**) 70 (**e**) 5 (**f**) 95

EXPLORATIONS

In Explorations 1 through 3 find as many examples as you can of each type of measure of position in newspapers, magazines, or books that are not mathematics books.

1. Quartiles.

2. Percentiles.

3. Rank in class.

4. Recent trends in SAT scores (see Section 11-4) and other test scores have caused numerous comments, suggestions, charges, and countercharges. Prepare a 10-minute report on some of the trends, causes, and implications during the last two decades.

11-6 Distributions and Applications

Many states require that all students take certain standardized tests. The test scores help the teachers identify the aptitudes and interests of the students. Then teachers have some objective bases for the informal day-to-day guidance and individual counseling that is an important part of teaching. For example, which students should be highly complimented and which should be challenged to seek a deeper understanding of a subject when a certain behavioral objective has been met? Which students are working close to their peak abilities and which are 'coasting' unchallenged and unmotivated while performing well above the class average? (See Explorations 7 and 8.)

Many statistical results are based upon simple yes-no or true-false responses. A manufacturer may sample preferences of potential customers for a product versus a competitor's product. A teacher may give a true-false quiz or test. The result, the distribution of the data from any such experiment in which there are only two possible outcomes for each event, is a **binomial distribution**.

The classical binomial experiment is tossing a coin. For a "fair" coin, the ratio of the number of heads to the total number of tosses is expected to be about

1/2 since the probability of heads is 1/2 on each toss of the coin (Section 10-2). If the experiment of tossing a coin n times is repeated over and over, the expected mean of the distribution of heads on these trials of the experiment is $(1/2)n$. The result of tossing one coin n times is the same as that of tossing n coins once. For example, if one coin is tossed 100 times or 100 coins are tossed once, we have probability $p = 1/2$ and $n = 100$. Therefore, the mean or average number of heads expected for each experiment is $(1/2)(100)$, that is, 50.

The **standard deviation of a binomial distribution** can be found by the following formula, which we shall not prove.

Using this formula for heads and 100 coins, we have $p = 1/2$, $1 - p = 1/2$, $n = 100$, and

$$\sigma = \sqrt{(1/2)(1/2)(100)} = \sqrt{25} = 5$$

$$\sigma = \sqrt{p \times (1 - p) \times n}$$

where p is the probability of success and n is the number of trials.

For large values of n, a binomial distribution is approximately normal. It has been shown for a normal distribution that if three standard deviations are added to and subtracted from the mean, we have the limits within which almost all of the data will fall. For the example of 100 coins, the mean is 50, and three times the standard deviation is 15, thus giving the limits 35 to 65. This is frequently stated in terms of **confidence limits**: we may say with "*almost* 100% *confidence*" that the number of heads will be between 35 and 65. Furthermore, we may say with approximately 95% confidence that the number of heads will be between 40 and 60, that is, within two standard deviations of the mean. We may say with approximately 68% confidence that the number of heads will be between 45 and 55, that is, within one standard deviation of the mean.

In most applications the assumed mean μ is the expected value of $\overline{x}$.

A **control chart** is a graph on which lines are drawn to represent the limits within which all data are expected to fall a given percent of the time. In industry, for example, a chart like the following may be drawn on which data from samples are plotted.

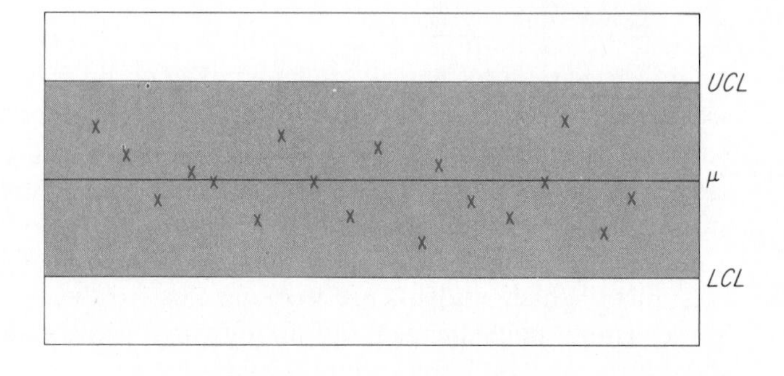

The middle horizontal line represents the **mean** (μ) or average around which these data are expected to lie. The other horizontal lines, called the **upper control limit** (UCL) and **lower control limit** (LCL), represent the limits within which all the data are expected to fall most of the time. Occasionally data may fall outside of these limits; as long as most of the samples tested produce data which lie between these lines, the process is said to be "in control."

For the case of 100 coins, the theoretical average is 50. The lower control limit would be $\mu - 3\sigma$, or 35. The upper control limit would be $\mu + 3\sigma$, or 65. We would suspect that something was wrong if a toss of 100 coins were to produce fewer than 35 or more than 65 heads. Such an event *can* take place just by chance, but it would be very unusual, occurring less than 3 times in 1000.

Example 1 An experiment consists of 180 throws of a single die.

(a) What is the mean number of fives to be expected?

(b) What is the theoretical standard deviation for the distribution?

Solution **(a)** These are the possible results when a die is thrown.

For a "fair" die these six possibilities are equally likely and we say that the probability of a 5 is 1 out of 6, that is, 1/6. Thus we have $p = 1/6$ and $n = 180$. Therefore, the average number of fives expected is $(1/6)(180) = 30$. Theoretically, we can expect 1 out of every 6 throws to produce a 5 and expect 30 fives in 180 throws.

(b) The probability of obtaining a five is 1/6. Thus $p = 1/6$, $(1 - p) = 5/6$, $n = 180$, and the standard deviation is $\sqrt{(1/6)(5/6)(180)} = \sqrt{25} = 5$. ■

Example 2 Suppose that a certain lecturer had a feeling that the 200 students in the lecture hall were not paying attention and asked two true-false questions about the lecture. Each student responded, possibly after tossing a coin, to each question with probability 1/2 of having a correct answer. Approximately how many students would give correct answers for

(a) the first question?

(b) Both questions?

(c) At least one question?

Solution **(a)** (1/2)(200), that is, 100.

(b) (1/2)(1/2)(200), that is, 50.

(c) About 50 would have both answers wrong; the remaining 150 would have at least one answer correct. ■

Note the impact of the increase of the probability from 1/2 to 3/4 on the answers for the parts of Example 2.

If the instructor in Example 2 had asked only one question with a show of hands for one of the possible responses, the resulting show of hands might have appeared impressive without the student having the least idea what the question was. Note that the standard deviation is

$$\sqrt{\left(\frac{1}{2}\right)\left(\frac{1}{2}\right)(200)} = \sqrt{50} \approx 7$$

Therefore, we can assert with about 95% confidence that at least 86 and at most 114 hands would have been raised for the correct response, just by chance.

Most students have some idea of the correctness of answers to questions even if they have not been paying attention. Suppose that the students in Example 2 could answer correctly with probability 3/4. Then the solution would be

(a) (3/4)(200), that is, 150.

(b) (3/4)(3/4)(200), that is, 112 or 113.

(c) 200 − (1/4)(1/4)(200), that is, 187 or 188.

The distribution for a single student responding to 200 questions is the same as the distribution for 200 students responding to one question.

Statistical results must be interpreted very carefully. In the case of the fourth toss of a coin that has shown heads three times in a row, either the probability of tails is still 1/2 on the fourth toss or the coin has a bias. Similarly, in the case of test scores neither a sequence of successes nor a sequence of failures provides more than a mild indication of the capabilities of the individual. Physical limitations (exhaustion, sickness, allergies), emotional limitations (excitement or distress over past or forthcoming events), and attitudes (expectations, desires, compatibility with other students or the teacher) can drastically affect student performance.

Several studies have shown that students tend to perform at the level that they feel their teacher expects of them!

Statistical data have many constructive pedagogical uses. However, one fact deserves special recognition: *Standardized test scores are not exact rankings* of the students. Each score indicates that the student's actual position is probably on an interval about that score. The size of the interval will vary from one test to another and for any *standardized test* should be given in the instructions to the person who is to interpret the test. For example, on College Entrance Examination Board tests a difference of test scores such as 598 and 601 in different subject areas is insufficient evidence for any major decisions as to the student's preferences or potential success. (Such a difference was once used by a student to decide upon a major when entering college. This was an outright abuse of the statistics.) A score of 598 might mean that the chances are 2 to 1 that the student's true score lies within the interval from 586 to 610; similarly, for 601 the chances might be 2 to 1 that the true score lies within the interval from 588 to 614. Note that these intervals, even for only 2 to 1 probability of correctness, have ranges of 24 and 26 points, respectively. Also the intervals overlap for about two-thirds of their extent. Even IQ scores may vary by 10, 15, or more points for the same person on tests taken as little as two weeks apart. Drastic differences are unusual but do occur.

The role of intervals may be illustrated by throws of a die. Consider the following graph of the theoretical results for the experiment in Example 1.

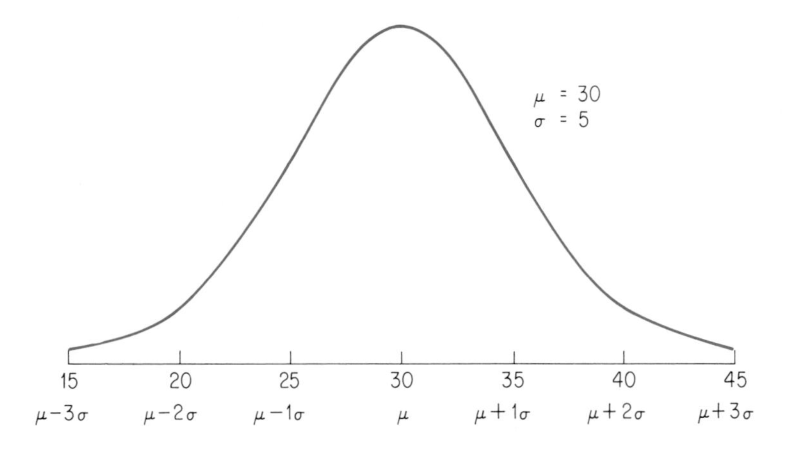

Thus, in a set of 180 throws of a die, we would expect with almost 100% confidence to have between 15 and 45 fives. With 95% confidence we would expect to have between 20 and 40 fives; and with 68% confidence we would expect between 25 and 35 fives.

These examples should serve to show that fluctuation is normal. That is, although the mean of the distribution is 30, we expect some variability. Not every set of 180 throws will produce exactly 30 fives—in fact, very few will. However, on the other hand, we expect this fluctuation to be within limits that can be described by our knowledge of the behavior of normal distributions. Someone could conceivably throw a die 180 times and produce 100 fives. However, an understanding of the principles of this chapter would dictate that you challenge the thrower of the die and examine the die quite carefully for evidence of foul play. In other words, although it is possible to obtain 100 fives, it is far from probable that this would happen.

EXERCISES *In Exercises* 1 *through* 5 *consider a single die that is thrown* 720 *times.*

1. What is the mean number of sixes to be expected?
2. What is the standard deviation for the distribution of sixes?
3. What is the maximum number of sixes we can expect to obtain with almost 100% confidence?
4. What is the minimum number of sixes we can expect to obtain with almost 100% confidence?
5. Within what limits can we expect the number of sixes to fall approximately 95% of the time?

6. Suppose that each member of your class tosses an unbiased coin 100 times and counts the number of heads. What is the expected mean of the numbers of heads obtained?

7. As in Exercise 6 on what interval would you expect about two-thirds of the numbers of heads to occur?
8. Repeat Exercise 7 for 95% of the heads.
9. Repeat Exercise 7 for practically all of the heads.
10. Suppose that you assigned each member of your class the experiment of tossing an unbiased coin 80 times. Then suppose that they reported the numbers of heads obtained as follows:

Alice 45	Bob 40	Charles 30
Doris 35	Eve 25	Fred 50
Gwen 42	Harry 37	Ike 69

Explain why you would suspect that one or more of the reports reflected either a failure to do the experiment or the use of a biased coin.

Consider a single die that is thrown 180 *times.*

11. What is the mean number of threes to be expected?
12. What is the maximum number of threes we can expect, with almost 100% confidence, to obtain?
13. What is the minimum number of threes we can expect, with almost 100% confidence, to obtain?
14. Within what limits can we expect the number of threes to fall approximately 95% of the time?

Explain your answer for the given question.

15. Jane has an IQ score of 130 and SAT scores of 610 and 590. Does she appear to be working up to her potential?
16. Jack has an IQ score of 95 and SAT scores of 450 and 500. Does he appear to be working up to his potential?
17. A student once remarked to the teasing of a classmate: "I am one of the people who make it possible for you to be in the upper half of the class." How would you explain the situation to an irate parent who considered the fact that half of your students were below average in arithmetic skills to be sufficient evidence for firing you and find a "good teacher?"
18. In a certain year the median SAT verbal score for students at Salem High School was 495. Was this cause for serious concern?

Find the approximate interval on which the indicated student can expect, with 95% *confidence, a grade on a test of* 25 *questions.*

19. José, who has an 80% chance of answering any given question correctly.
20. Dolores, who has a 90% chance of answering any given question correctly.

EXPLORATIONS

1. Use the concepts of this section to establish the limits within which the total number of heads can be expected to fall when a coin is tossed 64 times, using the 95% confidence limits. Then test these results by tossing a

coin 64 times, and asking as many other people as possible to repeat this experiment. (Note that instead of tossing a single coin for 64 tosses, you may also toss four coins for 16 tosses, 16 coins for 4 tosses, and so on.)

2. Repeat Exploration 1 for 180 throws of a single die, counting the number of sixes that appear.

Suppose that someone announces that he or she has just tossed a coin 19 times in a row and has obtained 19 heads. This person is about to toss the coin for the twentieth time. You are given an opportunity to place a bet on the outcome of this toss.

3. Would you bet on heads? On tails?
4. What is the probability of tossing 19 heads in a row with an unbiased coin?
5. What are the odds against tossing 19 heads in a row with an unbiased coin?
6. Discuss the probability that the coin and the method of tossing are both unbiased.

7. A student scores a grade of 127 on an aptitude test. What can you say about this student's aptitude as a result of this test score? Suppose that you are then given the additional information that the score of 127 is at the 95th percentile in a distribution of scores for all students who have taken this particular test. What can you conclude then about this student's aptitude?
8. Why does percentile rank appear to be more significant than rank (first, second, third, ...) in a class?

Select an elementary school grade level and prepare a worksheet, or set of worksheets, for helping the students understand the specified concept.

9. Measures of central tendency.
10. Measures of dispersion.

Explain why you would or would not expect a normal distribution of quiz (test) scores in each case.

11. For a class of ten students and on a topic that was covered while two students were absent.
12. For a class of ten students and on an arithmetic skill that has been the basis for considerable classroom practice.
13. For all 500 students entering a junior high school and on counting skills up to one thousand.
14. For all 500 students entering a junior high school and on a variety of arithmetic skills to identify areas which should be given special attention.
15. For all 500 students entering a junior high school and on reading skills.

Select an elementary school level and prepare a quiz of ten questions on which you would expect the scores to have the specified type of distribution.

16. A normal distribution.
17. A distribution with most of the scores clustered at the high end of the scale.
18. A *bimodal distribution*, that is, a distribution with two modes.

Chapter Review

1. Describe two common devices for making graphs misleading to the reader.
2. Make a frequency distribution for the grades of a class with grades: C, A, B, A, B, A, C, B, A, B, B, F, B, B, B, B, B, B, B, B, B, A, C, B, B, C, B, B, A, A, B, D, B, B, A, B, B.

Represent, in the form specified in Exercises 3, 4, *and* 5, *the following distribution of grades given last semester by Professor X.*

Grade	A	B	C	D	F
Percent	20	25	40	10	5

3. Histogram.
4. Circle graph.
5. Divided bar graph.

The given circle graph is based upon estimates of the population of China and the population of the world. Use these data for Exercises 6 *through* 8.

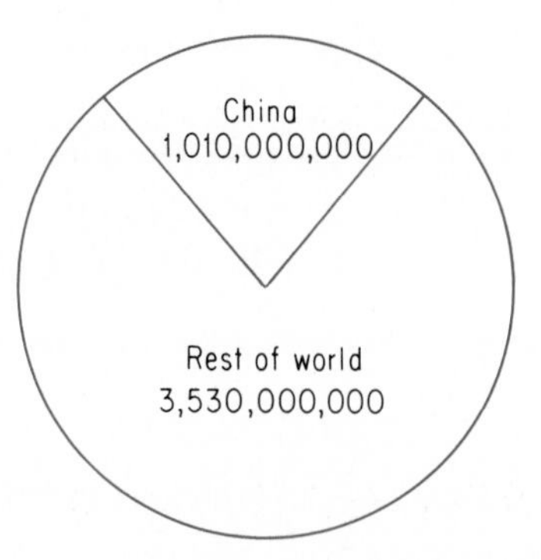

6. The population of China is approximately what percent (nearest whole number) of the population of the entire world? (Note that the population of the entire world includes the population of China as well.)
7. What is the size of the central angle (nearest degree) that represents the population of China?
8. Assume that you wish to draw a bar graph to show the same facts as does the circle graph. If the population of China is represented by a bar that is one centimeter high, approximately how long a bar (nearest 5 millimeters) would you need to show the population of the rest of the world?

Consider the set of scores {55, 62, 70, 74, 74, 79}. Then find the specified measure.

9. Arithmetic mean. **10.** Median. **11.** Mode.

For each of the following sets of scores find **(a)** *the range* **(b)** *the midrange.*

12. {75, 82, 64, 98, 79} **13.** {100, 45, 79, 82, 96, 68}

A collection of 1000 *scores forms a normal distribution with a mean of* 50 *and a standard deviation of* 10. *What percent of the scores are expected to be on the specified interval?*

14. Above 70.
15. Below 60.
16. Below 40.
17. Between 40 and 70.
18. Above 80.

19. Albert scored at the third quartile on a test in a class of 32 students. What was his class rank?

20. Beth ranked 30th in her high school class of 600. What was her percentile rank?

An unbiased coin is tossed 64 *times.*

21. What is the expected number of heads?

22. What is the standard deviation for results of this experiment when it is repeated many times?

23. Within what limits can we say, with almost 100% confidence, that we shall find the total number of heads?

24. If the experiment is repeated many times, what percent of the times should be expected to produce fewer than 36 heads?

25. As in Exercise 24 what percent of the times should be expected to produce more than 28 heads?

CALCULATING THE MEAN AND THE MEDIAN WITH A COMPUTER

The program listed in this application shows a method by which we can calculate the mean and the median of an arbitrary set of numbers. The steps in the program involve some advanced programming techniques in BASIC, but the program is easy to use. The person using the program is asked in line 14 how many numbers are in the set. Then, in lines 15 through 35, as the numbers are entered one by one, their sum is accumulated (line 3Ø). When all the numbers have been entered, the computer takes the number, N, of numbers in the list and divides their sum, S, to obtain their arithmetic mean in line 4Ø.

This part of the program uses INPUT and DIM(N) commands in BASIC. In line 14, the INPUT command is used to achieve two purposes. First, it commands the computer to print out the question contained in quotation marks, just as a

PRINT command would do. Then, it holds the cursor, or little square that moves on the screen, at the end of the line until the number of numbers in the list is entered. Then in line 15, the program gives the computer the dimensions of the list by using the command DIM(N). This command tells the computer to reserve space for N numbers in the list.

Line 16 instructs the computer to go through the numbers from 1 to N, inclusive, one at a time. In lines 25 and 26, the program asks for each of the numbers in the list to be entered. The command NEXT C in line 35 instructs the computer to change the value of C in the variable X(C) after each number in the list is entered. Also after each number is entered, the computer adds it to the previous sum S to get the new sum S in line 3∅. Line 35 then sends the computer back to find the next number in the list, unless the last number was N itself. The statement S = ∅ in line 5 just makes sure that the computer starts the process with a sum of 0 before any numbers are added. This is called *initializing* the variable S.

The remainder of the program involves the calculation of the median. The N numbers are arranged in order from large to small in lines 41 through 44. When this is completed, the program then instructs the computer to determine if the number, N, of numbers in the list is even or odd. If N is odd, there is a middle number in the list and, line 46, that number is automatically the median. If the number, N, of numbers in the list is even, the middle two numbers must be averaged, as on line 47, to determine the median. This part of the program uses several advanced commands in BASIC that can be used easily but whose explanations are beyond the scope of this book.

```
 5 S = ∅
1∅ PRINT "THIS PROGRAM FINDS THE MEAN AND THE MEDIAN"
11 PRINT
12 PRINT "OF THE DATA ENTERED IN THE PROGRAM."
13 PRINT
14 INPUT "HOW MANY NUMBERS ARE IN THE LIST?"; N
15 DIM X(N)
16 FOR C = 1 TO N
25 PRINT "ENTER X(";C;")"
26 INPUT "                 ";X(C)
3∅ S = S + X(C)
35 NEXT C
36 PRINT
4∅ PRINT "THE MEAN IS ";S/N;"."
41 FOR T = 1 TO N - 1: FOR J = N TO T + 1 STEP -1
42 IF X(J) > X(J - 1) THEN GOTO 44
43 S = X(J): X(J) = X(J - 1): X(J - 1) = S
44 NEXT J: NEXT T
45 M = (N + 1)/2
46 IF M = INT(M) THEN PRINT "THE MEDIAN IS ";X(M);"."
47 IF M <  > INT(M) THEN PRINT "THE MEDIAN IS ";
   (X(M - ∅.5) + X(M + ∅.5))/2;"."
48 END
```

Study the following output for the entered set of 12 numbers. Use your hand calculator to check the results. Then, if possible, try the program on a computer, enter some different sets of numbers, and check the results yourself.

```
RUN
THIS PROGRAM FINDS THE MEAN AND THE MEDIAN

OF THE DATA ENTERED IN THE PROGRAM.

HOW MANY NUMBERS ARE IN THE LIST? 12
ENTER X(1)
                    34
ENTER X(2)
                    752
ENTER X(3)
                    56
ENTER X(4)
                    34
ENTER X(5)
                    687
ENTER X(6)
                    34
ENTER X(7)
                    673
ENTER X(8)
                    523
ENTER X(9)
                    782
ENTER X(1Ø)
                    523
ENTER X(11)
                    741
ENTER X(12)
                    3211

THE MEAN IS 67Ø.833333.
THE MEDIAN IS 598.
```

12

An Introduction to Algebra

Emmy Noether
(1882–1935)

Emmy Noether was born and educated in Germany achieving a doctoral degree in mathematics in 1907. However, she was denied a professorial position due to being a woman. Undaunted, she continued to work in the area of algebra, bringing together seemingly unconnected areas of the subject and providing bases for further extensions. In 1933, she was forced to leave Germany and emigrated to the United States. At Bryn Mawr College, until her untimely death in 1935, she continued her research in algebra, developed strong ties with her students, and encouraged them in their work.

12-1 Algebra and Arithmetic

In arithmetic specific numbers are represented by such numerals as

$$2 \qquad 11 \qquad \frac{3}{4} \qquad 0 \qquad 1.4 \qquad -17 \qquad \sqrt{2}$$

Algebra is an extension of the arithmetic of integers, rational numbers, real numbers, and other sets.

In *algebra* an arbitrary element of a specified set of numbers is represented by a **numerical variable**, usually a letter from our alphabet; for example,

any natural number n
any prime number p
any real number x

The set of possible replacements for (*values of*) a variable is its **replacement set**. If no replacement set is specified for a variable, then the replacement set is assumed to be the set of real numbers.

Sentences such as

$2 + 3 = 5$ (true) $\qquad 5 - 3 = 7$ (false)

that can be identified as true or identified as false are called **statements**. In this sense a command such as

Stand up and be counted

is neither true nor false and is not considered to be a statement. A statement of equality, whether true or false, is an **equation.** A statement using, as in Section 6-2, one of the relations $\neq$, $<$, $\not<$, $\leq$, $\not>$, or $\geq$ is a **statement of inequality**.

In arithmetic the truth values of statements can be identified.

$7 + 3 = 3 + 7$ is a true statement of equality.
$7 - 3 = 3 - 7$ is a false statement of equality.
$7 + 2 > 7 - 2$ is a true statement of inequality.
$5 + 3 \neq 3 + 5$ is a false statement of inequality.
$8 - 2 < 2 - 8$ is a false statement of inequality.

In algebra a sentence in one variable may be true for all, none, or some but not all of the possible replacements of the variable. A sentence such as

$$x + 2 = 2 + x \quad \text{or} \quad x - 2 < x$$

that is true for all possible values of the variable is a *true statement* (**identity**). A sentence such as

$$x - 2 = x \quad \text{or} \quad x + 2 \neq 2 + x$$

that is false for all possible values of the variable is a *false statement*. A sentence such as

$$x + 5 = 3 \quad \text{or} \quad x^2 = 9$$

that is true for at least one value of the variable and false for at least one value of the variable is an **open sentence**.

For integers x the sentence $x + 5 = 3$ is an open sentence since it is true for $x = -2$ and false for $x \neq -2$. For whole numbers x the sentence $x + 5 = 3$ is a false statement of equality. That is, there is no whole number x for which $x + 5 = 3$. To *solve* an open sentence in one variable is to identify the set of possible replacements for which the sentence is a true statement, that is, to find the *solution set* of the given sentence. As in Section 7-6 the solution set may depend upon the replacement set of the variable. In practice the selection of a replacement set is usually determined by the conditions of the problem being solved.

Example 1 Solve.

(a) $x + 1 > 4$ for whole numbers x.
(b) $n + 2 = 2 + n$ for real numbers n.
(c) $x + 2 < x$ for integers x.
(d) $x^2 = 2$ for rational numbers x.

Solution **(a)** For $x = 3$ we have $3 + 1 = 4$. The solution set S consists of all whole numbers greater than 3 that is, $S = \{4, 5, 6, \dots\}$.

(b) The given sentence is a true statement of equality for all possible replacements of the variable, since the statement is an application of the commutative property of addition. The solution set is the set of all real numbers.

(c) Regardless of the integer selected as the replacement for x, the given sentence is *always* false. There are no solutions and the solution set is the empty set, that is,

$$\{x \mid x + 2 < x, x \text{ an integer}\} = \varnothing$$

(d) There is no rational number x for which the given statement is a true statement, since $\sqrt{2}$ and $-\sqrt{2}$ are not rational numbers (Section 7-1). The solution set is the empty set, that is,

$$\{x \mid x^2 = 2, x \text{ a rational number}\} = \varnothing \ \blacksquare$$

Any equation involving only numbers and numerical variables is an assertion that two *numerical expressions*, the two **members (sides)** of the equation, represent the same number. Since the numbers and numerical variables must all satisfy the commutative, associative, and other laws of arithmetic, there are procedures for expressing such equations in simplified form.

> For any numbers, numerical variables, or numerical expressions a, b, c, and d, where $d \neq 0$, if $a = b$, then
>
> $a + c = b + c \qquad a - c = b - c$
>
> $a \times d = b \times d \qquad a \div d = b \div d$

These procedures are often described by rules such as:

$=$'s $+$ $=$'s are $=$.

$=$'s $-$ $=$'s are $=$.

$=$'s $\times$ $=$'s ($\neq 0$) are $=$.

$=$'s $\div$ $=$'s ($\neq 0$) are $=$.

The same number may be added to, or subtracted from, both members of an equation.

Both members of an equation may be multiplied by, or divided by, any number different from zero.

In each case the operation does not affect the conditions (the set of solutions) under which the equation is true, that is, an **equivalent equation** is obtained.

The equation $x + 2 = 2 + x$ is equivalent to the equation $0 = 0$ and has the universal set as its solution set, that is, it is true for all values of x. Any equation with the universal set as its solution set may be called an *identity* and is equivalent to the equation $0 = 0$.

The equation $x + 2 = x$ is equivalent to the equation $1 = 0$ and has the empty set as its solution set. Any equation with the empty set as its solution set may be called an **impossible equation** and is equivalent to the equation $1 = 0$.

Equations that involve only the first power of the variable, for example,

$$2x + 1 = 7 \qquad x + 5 = 2x + 2 \qquad 3 = x$$

that are not identities and are not impossible equations are called **equations of the first degree**. Any given equation of the first degree in x can be *solved* by finding an equivalent equation of the form $x = m$. Usually the procedures just given are applied to obtain first an equation of the form $sx = t$ and then an equation of the form $x = m$.

Example 2 Solve (**a**) $2x + 5 = 19$ (**b**) $3x - 7 = 5$.

Solution (**a**)

$2x + 5 = 19$ (given)

$2x = 14$ (subtract 5)

$x = 7$ (divide by 2)

(**b**)

$3x - 7 = 5$ (given)

$3x = 12$ (add 7)

$x = 4$ (divide by 3) ■

The algebraic solutions in Example 2 may be replaced by *arithmetical solutions*. Without the algebraic notation the sentences are longer but the reasoning is the same. Consider these steps for Example 2(a).

> 5 more than twice a number is 19. Therefore,
> twice the number is 5 less than 19, that is, 14, and
> the number is half of 14, that is, 7.

Similarly, for Example 2(b).

> 7 less than three times a number is 5. Therefore,
> three times the number is 7 more than 5, that is 12, and
> the number is one-third of 12, that is, 4.

Such arithmetical solutions are useful when the variable occurs in only one member of the equation. When the variable occurs in both members, as, for example, in the equation.

$$2x + 6 = 3x - 2$$

the algebraic procedure must be applied to a term that involves the variable.

$6 = x - 2$ (subtract $2x$)

$8 = x$ (add 2)

$x = 8$ (symmetric property of $=$, Section 6-2)

Many problems can be solved using either arithmetic or algebra. Often sentences of equality provide the bases for the solutions. When necessary, the algebraic approach can be used to explain the method used in arithmetic. A systematic approach is helpful in the solution of any problem.

Example 3 The sum of two numbers is 20; their difference is 6. Find the numbers.

Solution Let the smaller number be n. Then $n + 6$ represents the larger number so that their difference is 6. The statement "the sum of the numbers is 20" can be expressed as an equation.

$$n + (n + 6) = 20$$

$n + n = 14$ (subtract 6)

$2n = 14$ (add; $n + n = 2n$)

$n = 7$ (divide by 2)

If the smaller number is n, then the larger number is $n + 6$. If $n = 7$, then $n + 6 = 13$. The numbers are 7 and 13.

Check: $7 + 13 = 20$
$13 - 7 = 6$ ■

Example 4 Find each number using algebra and then using arithmetic.

(a) If a number is doubled and then increased by 5, the result is 17.

(b) If a number is increased by 5 and that sum is doubled, the result is 18.

Solution (**a**) If the number is x, then

If only arithmetic is used, we have:

Twice the number is increased by 5 and the sum is 17.
Twice the number is 12.
The number is 6.

$2x$ is the double of (twice) the number

$2x + 5$ is the double increased by 5

$$2x + 5 = 17$$

$$2x = 12$$

$x = 6$ The number is 6.

(**b**) If the number is x, then

If only arithmetic is used, we have:

Twice the sum of the number and 5 is 18.
The sum of the number and 5 is 9.
The number is 4.

$x + 5$ is the number increased by 5

$2(x + 5)$ is twice that sum

$$2(x + 5) = 18$$

$$x + 5 = 9$$

$x = 4$ The number is 4. ■

Note that representations for numbers have been used in each of the previous examples.

n	$n + 6$	$2n + 6$
x	$2x$	$2x + 5$
x	$x + 5$	$2(x + 5)$

Such translations from verbal statements to expressions in the language of algebra are an important part of problem solving.

Each of the preceding problems also involved an equation in which two expressions (names) for the same number are equal to each other.

$$n + (n + 6) = 20$$

$$2x + 5 = 17$$

$$2(x + 5) = 18$$

Frequently such equalities are translations of verbal statements in which *is* means *is equal to.*

EXERCISES

For integers x identify the given sentence as an identity, a false statement, or an open sentence.

1. $x + 3 = 5$
2. $x + 5 = 3$
3. $x + 3 \neq 5$
4. $x + 5 \neq 3$
5. $x + 3 > x$
6. $x - 3 < x$
7. $x - 3 > x$
8. $x + 3 < x$
9. $x < x + 1$
10. $x + 2 < x + 4$
11. $x + 2 < x + 1$
12. $x > x - 1$

Solve the given sentence for whole numbers x.

13. $x + 3 = 8$
14. $x - 3 = 2$
15. $x + 2 < 6$
16. $x - 2 < 6$
17. $x - 2 < 5$
18. $x - 2 > 5$

Solve the given sentence for integers x.

19. $x + 1 < 5$
20. $x - 1 < 5$
21. $x - 2 < 1$
22. $x + 2 < 1$
23. $x + 5 > 2$
24. $x - 2 > 3$

Describe in words the solution set of the given sentence for real numbers x.

25. $x + 1 > 3$
26. $x + 1 < 3$
27. $x - 2 < 4$
28. $x - 2 > 4$
29. $x + 2 \neq 7$
30. $x + 3 \neq 3$

For a given number n, represent the given information as an algebraic expression.

31. Three more than twice the given number.
32. Twice the sum of the given number and three.
33. The given number decreased by two and that difference multiplied by five.
34. The given number multiplied by five and that product decreased by two.

Find each number using algebra and then using only arithmetic.

35. Three times a number is 99.
36. A number increased by 5 is 25.
37. Three times the sum of a number and 5 is 21.
38. Four times the difference of a number decreased by 2 is 12.
39. Four times the difference of 7 decreased by a number is 12.
40. Five times the difference of 11 decreased by a number is 35.
41. If the product of a number and 5 is increased by 2, the sum is 22.
42. If the quotient of a number divided by 7 is increased by 3, the sum is 24.
43. If the quotient of a number divided by 9 is decreased by 1, the difference is 17.
44. If the product of a number and 9 is decreased by 1, the difference is 17.

EXPLORATIONS

1. Elementary school students work with sentences and equations at various levels of abstraction. For example, each of the following is a sample of a type of question asked at an early grade level.

   The sum of 3 and some number is 5. What is the number?

   $3 + \square = 5$; what replacement should be used for $\square$ to obtain a true statement?

   $3 + ? = 5$; replace ? by a number to form a true sentence.

   $3 + \_\_ = 5$; fill in the blank to form a true sentence.

   $3 + n = 5$; find n.

   $3 + x = 5$; solve for x.

Review a recently published series of textbooks for grades 1, 2, and 3 and collect examples that illustrate this early introduction of the concept of a variable.

2. Do you believe that there is a specific age at which a child is first able to comprehend the concept of using a letter as a placeholder? That is, when does a child first understand the meaning of a sentence such as $3 + x = 5$? Explore this concept with several preschool children of various ages. Use blocks or other aids and try to build up such an abstraction. Repeat this procedure with children in the early elementary grades and discuss your results in class.

About 1650 B.C., an Egyptian scribe named Ahmes copied an earlier manuscript which he described as "the entrance into knowledge of all existing things and all obscure secrets." Ahmes' copy is often called the Rhind Papyrus. It contains 85 problems. These problems are from that ancient manuscript.

3. A number and its one-fourth added together become 15. What is the number?

Variables were not known 3000 years ago. Algebra did not exist. Problems were solved by arithmetic with a procedure that we now call the *method of false position.* This method can be used for some problems but is not useful for many others. For the preceding problem, we note that we need to take one-fourth of the number, so we try 4.

$$4 + \frac{1}{4}(4) = 4 + 1 = 5$$

If we try 4, we get 5. But we need 15, that is, 3×5. Therefore the answer is 3×4, that is, 12.

Try the method of false position for each of these problems from the Rhind Papyrus.

(a) A number and its one-fifth added together become 21. What is the number?

(b) A number, its one-third, and its one-quarter added together become 2. What is the number?

(c) If a number and its two-thirds are added together and from the sum one-third of the sum is subtracted, then 10 remains. What is the number?

12-2 Algebra and Geometry

Any real number may be graphed on a real number line (Section 7-1). The solution set of any sentence in one variable may be graphed on a real number line. We draw a graph to represent the set of points that correspond to the solution set of a sentence. We often refer to this graph simply as the *graph of the equation or inequality.*

Example 1 Graph the given sentence.

(a) $x + 3 \leq 5$ for whole numbers x.

(b) $x + 2 = 2 + x$ for real numbers x.

Solution **(a)** The solution set is $\{0, 1, 2\}$.

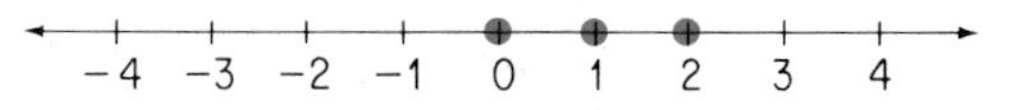

(**b**) The given sentence is true for all replacements of x; it is an identity. The solution set is the set of all real numbers. Thus the graph is the entire number line.

−4 −3 −2 −1 0 1 2 3 4 ■

On a real number line the graphs of inequalities are common geometric figures. Indeed, the inequalities may be used to define the geometric figures.

Example 2 Graph the given sentence for real numbers x.

(**a**) $x + 3 > 5$ (**b**) $x + 3 \geq 5$ (**c**) $x + 3 \not< 5$

Solution (**a**) The solution set consists of all real numbers greater than 2. The graph of the solution set is drawn by first placing a hollow dot at 2 on the number line to indicate that this point is not a member of the solution set. Then a heavily shaded arrow is drawn to show that all real numbers greater than 2 satisfy the given inequality.

−4 −3 −2 −1 0 1 2 3 4

(**b**) The given sentence is read "$x + 3$ is greater than or equal to 5." Therefore the given sentence is true if $x > 2$ and also if $x = 2$, that is, if x is greater than or equal to 2. In symbols the solution may be written as $x \geq 2$. The graph of the solution set is indicated by a solid dot at 2, to show that this point is a member of the solution set, and a shaded arrow as in the figure.

−4 −3 −2 −1 0 1 2 3 4

(**c**) The given sentence is read "$x + 3$ is not less than 5." This is equivalent to saying that $x + 3$ is greater than or equal to 5; $x + 3 \geq 5$. Thus the solution set and its graph are the same as for part (b) of this example. ■

For any real number b the graph of an inequality of the form $x > b$ (see Example 3(d)) is a half-line (Section 7-1). The graph of an inequality of the form $x < b$ is also a half-line, the *opposite half-line* of the graph of $x > b$. Similarly, the graph of an inequality of the form $x \geq b$ (see Example 2(b)) is a *ray*; the graph of $x \leq b$ is also a ray, the **opposite ray** of the graph of $x \geq b$. The point with coordinate b is the *endpoint* of each of these rays.

Example 3 Graph the given sentence for real numbers x.

(**a**) $-1 \leq x \leq 3$ (**b**) $-1 < x < 3$

Solution (a) The given sentence is read "-1 is less than or equal to x which is less than or equal to 3." In other words, "-1 is less than or equal to x and x is less than or equal to 3." That is, "x is greater than or equal to -1 and x is less than or equal to 3." Thus the solution set is the set of real numbers -1 *through* 3.

−4 −3 −2 −1 0 1 2 3 4

(b) The solution set consists of all the real numbers *between* -1 and 3.

−4 −3 −2 −1 0 1 2 3 4 ■

For any real numbers a and b where $a < b$, the graph of the sentence $a \leq x \leq b$ is a *line segment*. In Example 3(a), the points with coordinates -1 and 3 are *endpoints* of the line segment. The points of a line segment that are not endpoints are **interior points** of the line segment. The graph in Example 3(b) can be obtained from the graph in Example 3(a) by removing the endpoints of the line segment. When we wish to name each of these graphs, we call a line segment with its endpoints a **closed line segment** and a line segment without either endpoint, an **open line segment**; a line segment with one endpoint but not both is neither closed nor open. For example, the following graph of $-1 \leq x < 3$ is neither closed nor open; it is sometimes called a **half-open line segment.**

−4 −3 −2 −1 0 1 2 3 4

Consider the sentence

$$x + 1 > 2 \quad \text{and} \quad x - 2 < 1$$

Since no replacement set for x is specified, we assume that the replacement set is the set of real numbers. The sentence $x + 1 > 2$ is true for all x greater than 1; the sentence $x - 2 < 1$ is true for all x less than 3. The given sentence, is of the form p and q, is called a *compound sentence* (Section 1-3), and is true only when both parts of the sentence are true. Thus the given compound sentence is true for the set of elements in the *intersection* of the two sets. Graphically, we can show this as follows.

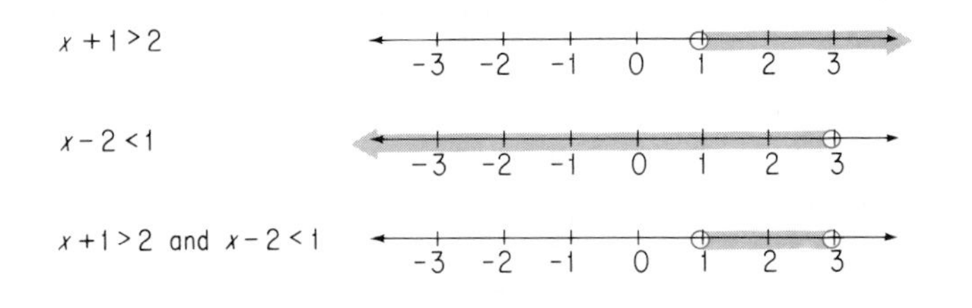

The graph of the compound sentence consists of an *open interval* (open line segment) and can be described as

$$1 < x < 3$$

The solution set of this compound sentence can be written in set-builder notation as $\{x \mid 1 < x < 3\}$ or as

$$\{x \mid x > 1\} \cap \{x \mid x < 3\}$$

This is read as "the intersection of the set of all x such that x is greater than 1 and the set of all x such that x is less than 3."

Example 4 Solve for integers x.

$$x \geq -2 \quad \text{and} \quad x + 1 \leq 4$$

Solution Here we want the set of integers that are greater than or equal to -2 but also are less than or equal to 3. (If $x + 1 \leq 4$, then $x \leq 3$.) The solution set is $\{-2, -1, 0, 1, 2, 3\}$. ■

Example 5 Solve for real numbers x.

$$x + 3 < 5 \quad \text{and} \quad x^2 = -5$$

Solution Note that the second part of this sentence, $(x^2 = -5)$ is false for all real numbers x. If part of a sentence of the form p and q is false, then the entire sentence is false. Thus the solution set is the empty set. ■

Next we consider a compound sentence involving the connective *or*.

$$x + 1 < 2 \quad \text{or} \quad x - 2 > 1$$

Note the distinction between the connectives *and* and *or* in a compound sentence. A sentence of the form *p and q* implies the *intersection* of two sets; a sentence of the form *p or q* implies the *union* of two sets.

The sentence $x + 1 < 2$ is true for all x less than 1; the sentence $x - 2 > 1$ is true for all x greater than 3. A sentence of the form *p or q* is true unless both parts are false. Thus the given compound sentence is true for the set of elements in the *union* of the two sets. Graphically, we have the following.

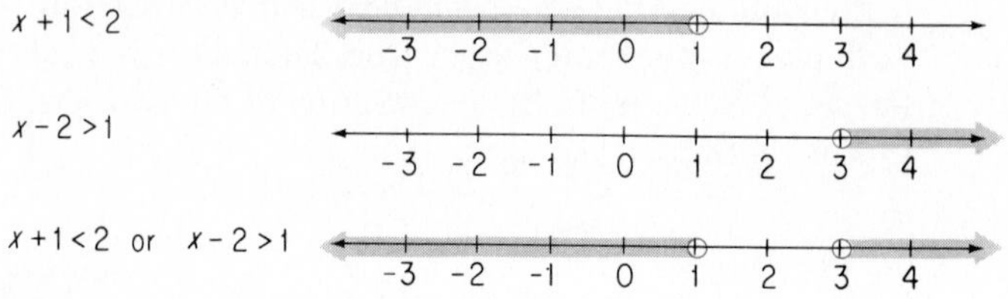

$x + 1 < 2$ — -3 -2 -1 0 1 2 3 4

$x - 2 > 1$ — -3 -2 -1 0 1 2 3 4

$x + 1 < 2$ or $x - 2 > 1$ — -3 -2 -1 0 1 2 3 4

The graph of the compound sentence consists of the union of two half-lines. The solution set can be written in set-builder notation as

$$\{x \mid x < 1\} \cup \{x \mid x > 3\}$$

This is read "the union of the set of all x such that x is less than 1 with the set of all x such that x is greater than 3."

Example 6 Solve for real numbers x.

(a) $x + 1 > x \quad \text{or} \quad x + 2 < 5$

(b) $x + 2 \neq 2 + x \quad \text{or} \quad x + 2 < x$

Solution **(a)** The compound sentence p or q is true if at least one of the parts is true. Since the first part of the given sentence is *always* true, the whole sentence is always true. That is, the compound sentence is true for *all* real numbers x; the solution set is the entire set of real numbers.

(b) Both parts of the sentence are always false; the solution set is the empty set. ■

Example 7 Graph **(a)** $x \leq -1 \quad \text{or} \quad x \geq 2$ **(b)** $x = -2 \quad \text{or} \quad x \geq 1$.

Solution **(a)** The graph is the union of two rays.

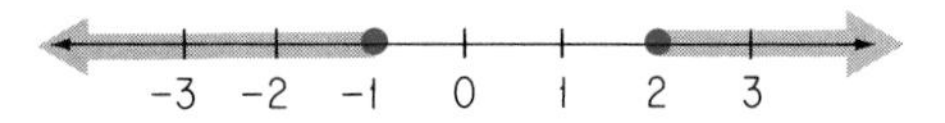

(b) The graph is the union of a point and a ray.

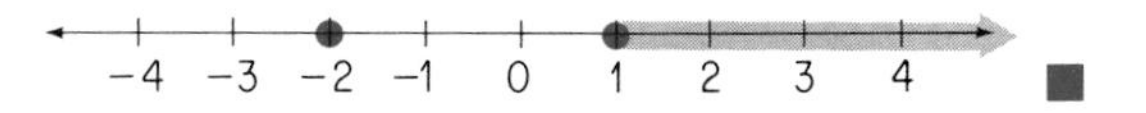

■

EXERCISES *Identify the graph of the solution set of the given sentence as a point, a half-line, a ray, a line segment, or a line.*

1. $x + 3 = 7$
2. $x - 2 > 7$
3. $x + 2 \geq 5$
4. $x < x + 3$
5. $-2 \leq x \leq 5$
6. $x + 3 \leq 5$
7. $x + 1 > x$
8. $x - 2 = 7$
9. $x - 1 < 5$
10. $-3 \leq x \leq 0$
11. $x + 2 \not> 5$
12. $x + 2 \not\leq x + 1$

Graph the given sentence for real numbers x.

13. $x + 1 > 3$
14. $x - 2 \leq 5$
15. $-3 \leq x \leq 4$
16. $-2 < x < 2$
17. $x + 3 \geq 5$
18. $x - 2 < 4$
19. $x + 2 > 4$
20. $x + 2 \not< 4$
21. $x + 3 > x$
22. $x + 3 = x$
23. $x + 2 \not> 5$
24. $x + 3 \neq 5$
***25.** $3x - 2 \geq -8$
***26.** $2x + 5 \leq 1$
***27.** $x^2 = 36$
***28.** $x^2 + 3 = 28$
***29.** $x^2 \leq 9$
***30.** $x^2 > 16$

Solve for integers x.

31. $x \geq 1$ and $x + 1 \leq 6$
32. $x \leq -2$ and $x > -5$
33. $x > 1$ and $x < 5$
34. $x \geq 2$ or $x \leq 1$
35. $x < 0$ or $x > 0$
36. $x < 1$ or $x > -1$

Graph for real numbers x.

37. $x \geq 0$ or $x < 0$
38. $x < 5$ or $x + 1 > 5$
39. $x = -1$ or $x \geq 0$
40. $x = 0$ or $x \geq 1$
41. $x + 1 < 5$ and $x > 5$
42. $x > 2$ and $x + 2 < 0$
43. $x + 2 = 5$ and $x < 3$
44. $x > 3$ and $x + 2 = 5$
45. $x \geq 2$ and $x \leq 5$
46. $x \leq 0$ or $x \geq 3$
47. $x + 2 < 5$ and $x \geq 0$
48. $x - 2 > 5$ or $x < 0$
49. $x + 3 \geq 5$ or $x - 1 < 0$
50. $x + 1 \geq 3$ or $x = -2$
51. $x + 3 \geq 5$ and $x - 1 < 0$
52. $x + 2 \geq 2$ and $x - 1 < 3$
53. $x - 1 > x$ and $x + 2 = 7$
54. $x < x + 1$ and $x - 3 > 5$
55. $x + 2 \neq x$ or $x + 2 \leq 5$
56. $x + 1 = x$ or $x^2 < 0$
57. $x^2 < 0$ or $x^2 + 1 > 0$
58. $x^2 \geq 4$

*59. $1 \leq x^2 \leq 9$

*60. $4 \leq x^2 \leq 25$

EXPLORATIONS

1. Prepare a set of transparencies suitable for demonstrating graphs of compound sentences to an upper elementary or junior high school mathematics class. Use a number line as a base, and then two other transparencies in different colors for the graphs of two other sentences. For example, one transparency could have the graph of $x \geq 3$ and the other one the graph of $x \leq 5$. Each of these can be shown separately, and then superimposed on the base to show their union and their intersection.
2. Compound sentences in various forms are common at many grade levels, as in the following examples. Find the solution set of each sentence where $n \in \{1, 2, 3, 4, 5, \ldots, 25\}$.
   (a) n is a multiple of 5 and an odd number.
   (b) n is less than 17 but greater than 11.
   (c) n is a multiple of 3 and $2n$ is a multiple of 12.
   (d) n is divisible by 3 and by 2.
   (e) n is divisible by 3 or by 11.

12-3 Equations and Inequalities

A sentence that involves one and only one variable and does not involve products or quotients of the variable is a sentence of the first degree in one variable.

$$2x - 3 = 7 \qquad \frac{1}{2}x - 2 > 5 \qquad 3 - 2x < 7$$

are sentences of the first degree. Any equation of the first degree may be solved as in Section 12-1.

$$2x - 3 = 7 \quad \text{(given)}$$

$$2x = 10 \quad \text{(add 3)}$$

$$x = 5 \quad \text{(divide by 2)}$$

A check of the solution in the given equation is desirable.

$$2(5) - 3 = 10 - 3 = 7$$

Inequalities may be defined in terms of equalities.

For all real numbers a and b, $a < b$ and $b > a$ if and only if $a + c = b$ for some positive number c.

This interrelation between inequalities and equalities may be taken as a definition of the inequalities. Then we may establish order relations by considering specific cases as follows. In each case verify that the first inequality listed is true. Then note that the accompanying inequality is also true.

$$2 < 5 \qquad 2 + 3 < 5 + 3$$

$$8 > 3 \qquad 8 + 5 > 3 + 5$$

$$-3 < 5 \qquad -3 + 1 < 5 + 1$$

These properties indicate that if the same number is added to both members of a statement of inequality, the "sense" of the inequality is preserved.

In general, we may list these **addition properties of order relations** for all real numbers a, b, and c:

ADDITION, $<$: If $a < b$, then $a + c < b + c$.

ADDITION, $>$: If $a > b$, then $a + c > b + c$.

If both members of a true statement of inequality are multiplied by the same positive number, the resulting sentence is also true. If both members of an inequality are multiplied by the same negative number, it is necessary to reverse the sense of the inequality to obtain an equivalent sentence. Thus in the following examples each member of the first inequality is multiplied by the same positive number, and the order of the inequality is maintained.

$3 < 7$ $\qquad$ $2 \times 3 < 2 \times 7$, that is, 6 is less than 14.

$5 > -1$ $\qquad$ $3 \times 5 > 3 \times (-1)$, that is, 15 is greater than -3.

In the next two examples each member of the first inequality is multiplied by the same negative number, and the order of the inequality is reversed.

$2 < 8$ $\qquad$ $-3 \times 2 > -3 \times 8$, that is, $-6 > -24$.

$3 > -1$ $\qquad$ $-2 \times 3 < -2 \times (-1)$, that is, $-6 < 2$.

We summarize our discussion by listing the **multiplication properties of order relations** for all real numbers a, b, and c:

MULTIPLICATION, $<$: If $a < b$ and $c > 0$, then $ac < bc$.
If $a < b$ and $c < 0$, then $ac > bc$.

MULTIPLICATION, $>$: If $a > b$ and $c > 0$, then $ac > bc$.
If $a > b$ and $c < 0$, then $ac < bc$.

Several uses of properties of order relations are illustrated in the examples that follow.

Example 1 Solve $\frac{1}{2}x - 2 > 5$ and explain each step.

Solution

$$\frac{1}{2}x - 2 > 5 \qquad \text{(given)}$$

$$\frac{1}{2}x > 7 \qquad \text{(add 2)}$$

$$x > 14 \qquad \text{(multiply by 2)} \blacksquare$$

Example 2 Solve $-2x + 3 < 7$ and explain each step.

Solution

$$-2x + 3 < 7 \qquad \text{(given)}$$

$$-2x < 4 \qquad \text{(subtract 3)}$$

$$x > -2 \qquad \text{(divide by } -2) \blacksquare$$

Note that dividing by -2 is the same as multiplying by $-1/2$.

EXERCISES *Explain each step in the following solutions.*

1. (a) $2x + 3 = 7$
   (b) $2x = 4$
   (c) $x = 2$

2. (a) $3x - 4 = 11$
   (b) $3x = 15$
   (c) $x = 5$

3. (a) $2x - 4 < 10$
   (b) $2x < 14$
   (c) $x < 7$

4. (a) $5x + 2 > 22$
   (b) $5x > 20$
   (c) $x > 4$

5. (a) $-3x + 5 > 17$
   (b) $-3x > 12$
   (c) $x < -4$

6. (a) $17 - 2x < 11$
   (b) $-2x < -6$
   (c) $x > 3$

Solve for x.

7. $3x + 2 = 14$
8. $2x - 5 = 9$
9. $4x - 3 = 17$
10. $5x + 7 = 17$
11. $-3x + 1 = 10$
12. $-2x + 3 = 7$
13. $5 - 2x = 13$
14. $1 - 3x = 16$
15. $7 - 3x = 1$
16. $9 - 4x = 5$
17. $3x - 2 < 10$
18. $3x - 1 > 8$
19. $2x + 1 > 7$
20. $5x + 3 < 18$
21. $-2x + 1 < 9$
22. $-3x + 2 < 8$
23. $6 - 2x > 0$
24. $20 - 3x > 8$
25. $\frac{1}{2}x - 3 = 8$
26. $\frac{2}{3}x - 7 = 5$
27. $\frac{3}{4}x + 1 = 10$
28. $\frac{5}{6}x + 18 = 3$
29. $-\frac{1}{2}x + 3 < 7$
30. $-\frac{2}{3}x - 4 < 8$

Determine whether each statement is always true for integers a, b, c, and d, where $a < b$ and $c < d$. If the statement is not always true, give an example for which it is false.

*31. $a + c < a + d$

*32. $a + c < b + d$

*33. $a - c < b - d$

*34. $a + d > b + c$

*35. $ac < bc$

*36. $ad < bc$

EXPLORATIONS

A sequence of operations may be used to describe an equation and this sequence may be modified to obtain a sequence of operations for solving the equation. Here are two examples of sequences of operations to describe equations.

Start with x.	x	Start with x.	x
	↓		↓
Multiply by 3.	$3x$	Divide by 2.	$x \div 2$
	↓		↓
Add 2.	$3x + 2$	Subtract 2.	$\frac{1}{2}x - 2$
	↓		↓
Equate to 17.	$3x + 2 = 17$	Equate to 3.	$\frac{1}{2}x - 2 = 3$

List a sequence of operations to describe each equation.

1. $2x + 3 = 7$

2. $3x - 5 = 10$

3. $\frac{1}{2}x - 1 = 7$

4. $\frac{1}{4}x + 2 = 6$

To obtain a sequence of operations for solving an equation consider in reverse order the sequence of steps used to describe the equation, replace "Equate to" with "Start with the equation," replace "Start with" with "The solution set is," and replace each operation with the opposite (inverse) operation. For the two previous examples these *reverse sequences* become

The solution set is	$\{5\}$	The solution set is	$\{10\}$
	↑		↑
Divide by 3.	$x = 5$	Multiply by 2.	$x = 10$
	↑		↑
Subtract 2.	$3x = 15$	Add 2.	$\frac{1}{2}x = 5$
	↑		↑
Start with the equation	$3x + 2 = 17$	Start with the equation	$\frac{1}{2}x - 2 = 3$

5. Use a reverse sequence to solve the equation in each of Explorations 1 through 4.

12-4 Algebra and Problem Solving

Algebraic notation and procedures enable us to use formulas for solving many problems. Any letter may be used as the variable when writing an equation. Often the letter is selected as the first letter of the word representing the quantity for which the variable is used. For example, in the metric system temperatures are measured on a scale of degrees Celsius where water freezes at 0 degrees and boils at 100 degrees. As of January 1, 1973, the American Society for Testing and Materials adopted the notation °C for degrees **Celsius** and °F for degrees **Fahrenheit**. A normal body temperature of 98.6 °F is about 37 °C.

The Celsius thermometer is named after the Swedish astronomer and scientist, Anders Celsius (1701–1744). He was the first to consider the separation of the distance between the freezing point and the boiling point of water into 100 equal parts.

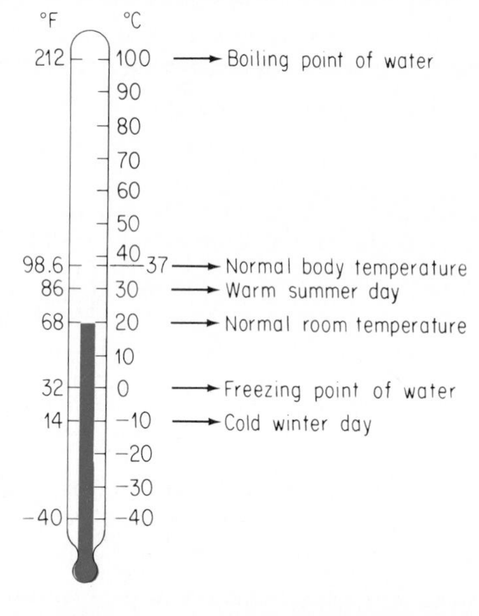

It is often helpful to consider degrees Celsius in terms of commonly recognized temperatures, as in the adjacent figure. Corresponding temperatures in degrees Fahrenheit are also shown in the figure.

There is a formula that relates temperatures in degrees Fahrenheit (F) to those in degrees Celsius (C). For temperatures above freezing, the number of degrees Fahrenheit above freezing is $F - 32$; the number of degrees Celsius is C. Furthermore, the interval from freezing to boiling is 180 °F but only 100 °C. Accordingly, a degree Fahrenheit is a smaller unit than a degree Celsius and the ratio of the number of degrees Fahrenheit above freezing to the number of degrees Celsius above freezing is given by this proportion:

This formula holds for all temperatures and involves the two variables F and C. The formula may be used to find the value of either of the variables when the value of the other is known.

$$\frac{F - 32}{C} = \frac{180}{100}$$

$$F - 32 = \frac{9}{5}C$$

$$F = \frac{9}{5}C + 32$$

Example 1 Convert

(a) 30 °C to the Fahrenheit scale **(b)** 77 °F to the Celsius scale.

Solution (a) $F = \frac{9}{5}(30) + 32$

$= 9(6) + 32$

$= 54 + 32$

$= 86$

$30\ °C = 86\ °F$

(b) $77 = \frac{9}{5}C + 32$

$45 = \frac{9}{5}C$

$5 = \frac{1}{5}C$

$25 = C$

$77\ °F = 25\ °C$ ■

There are many other useful algebraic formulas.

$d = 55t$ where d is the distance in miles traveled at 55 miles per hour in t hours.

$p = 4s$ where p is the perimeter of a square with side of length s.

$\mathscr{A} = s^2$ where $\mathscr{A}$ is the area of a square with side of length s.

$s = 16t^2$ where s is the distance in feet that a freely falling body falls in t seconds.

Problem solving requires a very careful reading and analysis of the statement of the problem. For instance, the problem in Example 4(b) of Section 12-1 could have been stated as

Twice the sum of a number and 5 is 18. Find the number.

The reader must recognize that for a number x the equation for this problem is $2(x + 5) = 18$ and is not $2x + 5 = 18$. In general, the following four steps are helpful in using an equation to solve a problem. We illustrate the steps for this problem involving a rectangle. The perimeter of a rectangle is 32 centimeters. The length of the rectangle is 1 centimeter greater than twice the width. Find the length and the width of the rectangle.

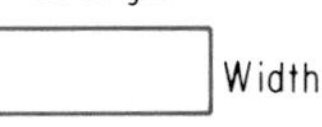

1. *Read the problem several times.* Be sure that you understand the terms that are used. Try to visualize what is given and what must be found. Where appropriate, draw a sketch as in the adjacent figure.

2. *Plan the solution.* Use a variable to represent one of the unknown quantities—usually one of the quantities that you are trying to find. Represent other unknown quantities in terms of the variable.

   Let x represent the width.
   Then $2x + 1$ represents the length.

3. *State and solve an equation or inequality.* This usually requires that a relation be found among the known and unknown quantities in the problem.

The perimeter is the distance around the rectangle:

$$(\text{width}) + (\text{length}) + (\text{width}) + (\text{length}) = \text{perimeter}$$

$$x + (2x + 1) + x + (2x + 1) = 32$$

$$6x + 2 = 32$$

$$6x = 30$$

$x = 5$ The width is 5 cm.

$2x + 1 = 11$ The length is 11 cm.

Steps in problem solving:
READ
PLAN
SOLVE
CHECK

Note that it is often necessary to find the value of more than one unknown quantity.

4. *Return to the original problem to check your answers.* Answers that have been correctly obtained may not be meaningful in the original problem.

The length: $11 = 2(5) + 1$

The perimeter: $5 + 11 + 5 + 11 = 32$

Example 2 Doris' class is renting a bus for a short field trip. If the members of the class pay \$1 each, there is a surplus of \$2. If the members pay 90¢ each, there is a shortage of \$1. How many members are in the class and what is the charge for renting the bus?

Solution Let x be the number of members in the class. Then the charge for renting the bus is $(1)x - 2$ and also $(0.90)x + 1.00$.

$$x - 2 = 0.9x + 1$$

$$0.1x = 3$$

$x = 30$ There are 30 members in the class.

$x - 2 = 28$ The charge for renting the bus is \$28.

Check: $30(1) = 28 + 2$ $30(0.90) + 1 = 27 + 1 = 28$ ■

Example 3 Ginny and Bill each drive small cars. Ginny averages 40 miles per gallon with her car; Bill averages 30 miles per gallon with his. They attend the same college and live the same distance from campus. If Bill needs 1 gallon more fuel than Ginny for five round trips to campus, how far does each live from campus?

Solution Let d be the distance in miles that each lives from campus. Then

$2d$ is the length of one round trip.
$10d$ is the length of five round trips.

At 40 miles per gallon, Ginny's car requires $10d/40$ gallons of fuel to travel $10d$ miles; at 30 miles per gallon, Bill's car requires $10d/30$ gallons of fuel.

$$\frac{10d}{40} + 1 = \frac{10d}{30}$$

$$\frac{d}{4} + 1 = \frac{d}{3}$$

$$3d + 12 = 4d \qquad \text{(multiply by 12)}$$

$$12 = d \qquad \text{Each lives 12 miles from campus.}$$

Check: 12 miles one way means 120 miles for five round trips. Ginny requires 3 gallons of fuel and Bill requires 4 gallons for 120 miles; $3 + 1 = 4$. ■

EXERCISES *Convert to the Fahrenheit scale.*

1. 0 °C
2. 20 °C
3. 25 °C
4. 40 °C
5. 60 °C
6. 100 °C

Convert to the Celsius scale.

7. 32 °F
8. 50 °F
9. 86 °F
10. 95 °F
11. 122 °F
12. 212 °F

Find each number.

13. Twice the sum of 6 and a number is 36.
14. The sum of 6 and twice a number is 36.
15. The product of 3 and a number is increased by 5. The sum is 26.
16. The product of 3 and a number increased by 5 is 27.

Solve each problem.

17. The perimeter of a rectangle is 60 centimeters. The length is twice the width. Find the length and the width.
18. The perimeter of a rectangle is 10 meters. The width is 1 meter less than the length. Find the length and the width.
19. The perimeter of a rectangle is three times the length. The width is 5 centimeters. Find the length and the perimeter.
20. The perimeter of a rectangle is five times the width. The length is 6 meters. Find the width and the perimeter.
21. Find a number such that the sum of twice the number and 5 is the same as the sum of the number and 17.
22. Find a number such that 50 more than three times the number is the same as the sum of the number and 80.
23. John's team needed cash for a project. At $5 each there was a surplus of $1; at $4.75 each there was a shortage of $1. How many people were on the team and how much money was needed?

24. The children of José and Dolores shared equally in the cost of a present for their mother. If each gave \$2, there was a surplus of \$1. If each gave \$1.50, there was a shortage of \$1. How many children were there and what was the cost of the present?
25. Joe and Ruth drive the same distance each week. Joe averages 25 miles per gallon and needs 6 gallons of fuel more than Ruth, who averages 35 miles per gallon. How far does Ruth drive each week?
26. Susan drives twice as far as Tom each week and uses 5 gallons of fuel more than Tom. If Susan averages 30 miles per gallon and Tom averages 35 miles per gallon, how far does Susan drive each week?

EXPLORATIONS

1. When the Celsius temperature scale became the official scale in the United Kingdom, school children were given the following algorithm for changing from degrees Celsius to degrees Fahrenheit.

   Double the number of degrees Celsius.

   Decrease the absolute value (the value without regard for sign) of this number by 10%.

   Add 32.

   Check that this algorithm holds for at least three cases. Explain why the algorithm works. Find a related algorithm for changing from degrees Fahrenheit to degrees Celsius.
2. Many mathematical games can be explained and developed through the use of algebraic techniques. Consider, for example, this set of directions.

   Think of a number.
   Multiply that number by 2.
   Add 9.
   Subtract 3.
   Divide by 2.
   Subtract the original number.
   The final answer will always be 3.

   Use n to represent the original number, and write a mathematical phrase to represent each step in the set of directions just given. For example, the first three steps are n, $2n$, and $2n + 9$. From this, show why the result is always 3, regardless of the number originally selected.
3. As in Exploration 2 write a set of directions for obtaining in order the numbers

   $$n \qquad 5n \qquad 5n + 7 \qquad 5n + 5 \qquad n + 1 \qquad 1$$
4. Prepare at least three sets of directions, as in Explorations 2 and 3, that can be used to provide students with practice in arithmetic and that can be quickly checked by the teacher.

5. Survey a recently published elementary mathematics textbook series and report on the attention given to problem solving. In particular, look for the treatment given to the four steps considered in this section.
6. Select an elementary school grade level and prepare a 10-minute introduction of problem solving for students at that level.

12-5 Sentences Involving Absolute Value

The **absolute value** of a number x may be defined as the distance from the origin to the point with coordinate x. Thus $|3| = 3$, $|-3| = 3$, and in general for any real number x

$\|x\| = x$	**if x is positive or zero**
$\|x\| = -x$	**if x is negative**

Note that if x is negative, then $-x$ is the opposite of x and therefore is a positive number. Thus if $x = -3$, $|-3| = -(-3) = 3$. Accordingly, the absolute value of any real number different from zero is a positive number; $|0| = 0$.

Example 1 Graph $|x| \geq 3$.

Solution The points whose distances from the origin are exactly 3 units ($|x| = 3$) are the points with coordinates 3 and -3. The points whose distances from the origin are greater than or equal to 3 units ($|x| \geq 3$) are the points whose distances from the origin are at least 3 units, that is, not less than 3 units.

"Is not less than," $\not<$, means the same as "is greater than or equal to," $\geq$.

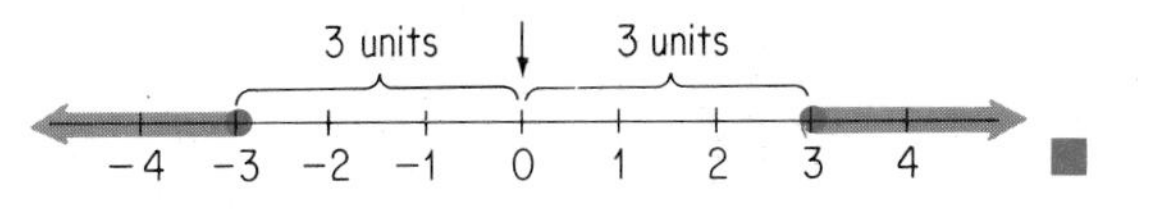

Note that the solution set of Example 1 is described by each of these compound sentences.

x is at most -3 or x is at least 3

$$x \leq -3 \quad \text{or} \quad x \geq 3$$

$$\{x \mid x \leq -3\} \cup \{x \mid x \geq 3\}$$

We may generalize the results of Example 1 to provide this definition of $|x| \geq k$ for any real number k.

$$\{x \mid |x| \geq k\} = \{x \mid x \leq -k\} \cup \{x \mid x \geq k\}$$

Example 2 Graph $|x| \leq 3$.

Solution The graph consists of the points that are at most three units from the origin.

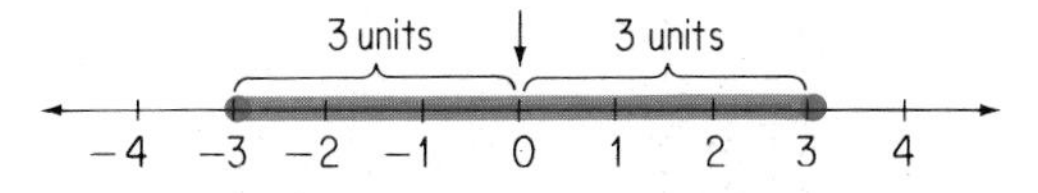

The solution set of Example 2 can be described in each of these ways.

$$x \geq -3 \quad \text{and} \quad x \leq 3$$

$$\{x \mid x \geq -3\} \cap \{x \mid x \leq 3\}$$

$$\{x \mid -3 \leq x \leq 3\}$$

We may generalize the results of Example 2 to provide this definition for $|x| \leq k$ for any nonnegative number k.

$$\{x \mid |x| \leq k\} = \{x \mid x \geq -k\} \cap \{x \mid x \leq k\}$$
$$= \{x \mid -k \leq x \leq k\}$$

Example 3 Solve $|x - 1| = 3$.

Solution If the absolute value of a number is 3, then that number is 3 or -3. If $x - 1 = 3$, then $x = 4$; if $x - 1 = -3$, then $x = -2$. Thus the solution set is $\{-2, 4\}$. ■

The points with coordinates -2 and 4 of the solution set for Example 3 are each at a distance of three units from the point with coordinate 1. Thus the given sentence

$$|x - 1| = 3$$

may be interpreted in terms of distance on a number line. The distance between the point with coordinate x and the point with coordinate 1 is 3. This interpretation is illustrated in the following figure.

3 units 3 units

-3 -2 -1 0 1 2 3 4 5

The absolute value notation may be used to indicate the distance between any two points on the number line. Thus $|x - h|$ represents the distance between the point with coordinate x and the point with coordinate h. Notice that the distance between the points with coordinates 2 and 5 may be denoted either as $|5 - 2|$ or as $|2 - 5|$, since $|5 - 2| = |3| = 3$ and $|2 - 5| = |-3| = 3$. In general, $|x - h| = |h - x|$, and the distance between any two distinct points is a positive number.

If we are told that $|x + 1| = 3$, we may consider this sentence in the form $|x - (-1)| = 3$; that is, the distance between the point with coordinate x and the point with coordinate -1 is 3. Thus, as shown in the figure, $x = -4$ or $x = 2$.

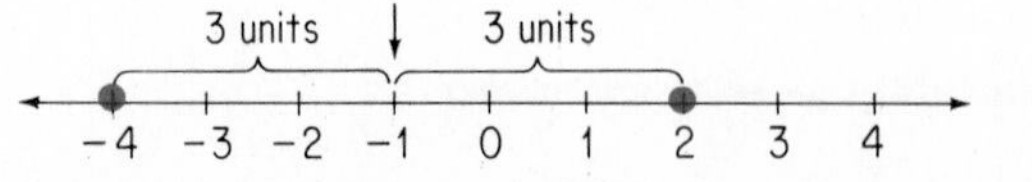

The problem $|x + 1| = 3$ can also be solved by using the fact that if the absolute value of a number is 3, then that number must be 3 or -3. If $x + 1 = 3$, then $x = 2$; if $x + 1 = -3$, then $x = -4$. Thus if $|x + 1| = 3$, then $x = -4$ or $x = 2$.

Example 4 Graph $\{x \mid |x - 1| \geq 3\}$.

Solution Consider the definition for $|x| \geq k$, and replace x by $x - 1$.

$$\{x \mid |x - 1| \geq 3\} = \{x \mid x - 1 \leq -3\} \cup \{x \mid x - 1 \geq 3\}$$

If $x - 1 \leq -3$, then $x \leq -2$; if $x - 1 \geq 3$, then $x \geq 4$. Thus the solution set is $\{x \mid x \leq -2 \text{ or } x \geq 4\}$; that is,

$$\{x \mid x \leq -2\} \cup \{x \mid x \geq 4\}$$

The graph of the solution set is

The graph of $|x - 1| \geq 3$ is the union of two rays and consists of the points on the number line that are at a distance of three units or more than three units from the point with coordinate 1. ■

Example 5 Graph $|x - 1| \leq 3$.

Solution The graph of $|x - 1| \leq 3$ consists of the points of the number line that are at a distance of at most three units from the point with coordinate 1. Thus the graph is a line segment with endpoints at -2 and 4.

■

Note that the sentence in Example 5 can be solved algebraically by applying the definition for $|x| \leq k$, and replacing x by $x - 1$.

$$\{x \mid |x - 1| \leq 3\} = \{x \mid x - 1 \geq -3\} \cap \{x \mid x - 1 \leq 3\}$$

If $x - 1 \geq -3$, then $x \geq -2$; if $x - 1 \leq 3$, then $x \leq 4$. Thus the solution set is

$$\{x \mid x \geq -2 \quad \text{and} \quad x \leq 4\} = \{x \mid x \geq -2\} \cap \{x \mid x \leq 4\}$$
$$= \{x \mid -2 \leq x \leq 4\}$$

EXERCISES *Find the set of integers for which the given sentence is true.*

1. $|x| = 5$ **2.** $|x| \leq 2$ **3.** $|x| < 3$

4. $|x| = -3$ **5.** $|x| < 0$ **6.** $|x| = 0$

7. $|x - 1| = 4$ **8.** $|x + 1| = 6$ **9.** $|x + 2| = 2$

10. $|x - 2| = 5$ **11.** $|x + 3| = 1$ **12.** $|x + 2| = 3$

Evaluate.

13. $|-9|$
14. $|-3| + |-5|$
15. $|(-3) + (-5)|$
16. $(|-6|)^2$
17. $|(-6)^2|$
18. $|-5| \times |-9|$

Identify the graph of the given sentence on a real number line as two points, a line segment, or the union of two rays.

19. $|x| \leq 7$
20. $|x - 2| = 5$
21. $|x| \geq 1$
22. $|x + 3| \leq 1$
23. $|x + 2| \leq 1$
24. $|x - 5| \geq 5$

Graph on a real number line.

25. $|x| \leq 3$
26. $|x| \geq 2$
27. $|x| > 2$
28. $|x| < 3$
29. $|x - 1| \geq 3$
30. $|x + 1| \leq 2$
31. $|x| = -2$
32. $|x| \geq 0$
33. $|x + 2| \geq 3$
34. $|x - 3| \leq 1$
35. $|x - 3| < 1$
36. $|x + 2| > 3$

*37. $2 \leq |x| \leq 4$

*38. $2 \leq |x - 1| \leq 5$

*39. $|x| = -x$

*40. $\dfrac{1}{|x - 3|} > 0$

*41. $\dfrac{x}{|x|} = 1$

*42. $\dfrac{|x - 2|}{x - 2} = -1$

Find each number.

*43. A number is equal to the absolute value of its difference from 5.

*44. A number is equal to twice the absolute value of its difference from 6.

*45. A number is equal to three times the absolute value of its difference from 12.

EXPLORATIONS

1. Assume that integers have been studied and graphed on a number line. Prepare a 10-minute introduction of absolute value for use with upper elementary school students with this background. Include the use of visual aids.
2. In a first course in algebra the following definition is usually given.

   $$\sqrt{x^2} = |x|$$

   For example, as a result of this definition we may write

   $$\sqrt{(3)^2} = |3| = 3 \quad \text{and} \quad \sqrt{(-3)^2} = |-3| = 3.$$

   Many students carelessly and incorrectly write $\sqrt{x^2} = x$. Show how this incorrect statement can contradict the use of the radical symbol to denote the nonnegative square root of a number.

12-6 Quadratic Equations

Any equation of the first degree in a variable x is a *linear equation* and can be solved as in Sections 12-1 and 12-3.

$$\text{If } ax + b = c \text{ and } a \neq 0, \text{ then } x = \frac{c - b}{a}.$$

Any equation of the form

$$ax^2 + bx + c = 0, \qquad a \neq 0$$

is of the *second degree*, a **quadratic equation**, in x and has as its solution (see Exercise 25).

$$\left\{\frac{-b + \sqrt{b^2 - 4ac}}{2a}, \frac{-b - \sqrt{b^2 - 4ac}}{2a}\right\}$$

The Norwegian mathematician Niels Henrik Abel (1802–1829) proved the impossibility of finding solutions for fifth-degree or higher-degree equations in terms of radicals, the operations of arithmetic, and the coefficients of the original equation. Read about his life in E. T. Bell's *Men of Mathematics*, Dover Publications, 1937.

Since the **discriminant** $b^2 - 4ac$ may be a negative number, the **roots** (members of the solution set) of a quadratic equation with real coefficients a, b, c may be imaginary numbers (Section 7-6). The formula

$$x = \frac{-b \pm \sqrt{b^2 - 4ac}}{2a}$$

may be used to find the roots of any quadratic equation $ax^2 + bx + c = 0$, $a \neq 0$, and is called the **quadratic formula**.

There exist formulas for solving equations of the third degree (cubic equations) and equations of the fourth degree (quartic equations):

$$ax^3 + bx^2 + cx + d = 0, \qquad a \neq 0$$

$$ax^4 + bx^3 + cx^2 + dx + e = 0, \qquad a \neq 0$$

There do not exist corresponding general formulas for solving equations of the fifth degree or for higher-degree equations.

Example 1 Use the quadratic formula to solve $2x^2 - 5x + 3 = 0$.

Solution We compare $ax^2 + bx + c$ and $2x^2 - 5x + 3$ to obtain $a = 2$, $b = -5$, and $c = 3$. Then

$$x = \frac{-(-5) \pm \sqrt{(-5)^2 - 4(2)(3)}}{2(2)} = \frac{5 \pm \sqrt{25 - 24}}{4} = \frac{5 \pm \sqrt{1}}{4}$$

$$x = \frac{5 + 1}{4} = \frac{6}{4} = \frac{3}{2} \quad \text{or} \quad x = \frac{5 - 1}{4} = \frac{4}{4} = 1$$

The solution set is $\{1, 3/2\}$. ■

The development of the quadratic formula (Exercise 25) is based upon the ancient method of completing the square (see the Explorations) to obtain an equation of the form $x^2 = b^2$. The recognition that if $x^2 = 9$, then $x = 3$ or $x = -3$ and, in general,

If $x^2 = b^2$, then $x = b$ or $x = -b$

may be based upon a property of real numbers.

The equivalence of $x^2 - b^2$ and $(x + b)(x - b)$ may be verified by multiplication.

$$\begin{array}{r} x + b \\ x - b \\ \hline - bx - b^2 \\ x^2 + bx \quad\quad \\ \hline x^2 \quad\quad - b^2 \end{array}$$

If the product of two or more numbers is zero, then at least one of the numbers must be zero.

Any equality $x^2 = b^2$ may be expressed as $x^2 - b^2 = 0$ and thus as $(x + b)(x - b) = 0$. Therefore if $x^2 = b^2$, then

$$x - b = 0 \quad \text{and} \quad x = b, \qquad \text{or}$$

$$x + b = 0 \quad \text{and} \quad x = -b$$

Example 2 Use the formula $s = 16t^2$ to find the time required for a freely falling body to fall 2500 feet.

Solution $s = 2500$ and $2500 = 16t^2$.

$$\frac{2500}{16} = t^2 \qquad \text{(divide by 16)}$$

$$t^2 = \frac{2500}{16} \qquad \text{(symmetric, =)}$$

$$t = \frac{50}{4} \quad \text{or} \quad -\frac{50}{4}$$

The nature of a problem often determines the replacement set of the variable.

The nature of the problem requires a positive answer. The time of fall is 12.5 seconds. ■

The methods used for linear equations may also be used for equations of forms such as $ax^2 + c = d$.

Example 3 Solve $17 - 2x^2 = 9$

Solution

$$\begin{array}{rl} 17 - 2x^2 = 9 & \text{(given)} \\ 8 - 2x^2 = 0 & \text{(subtract 9)} \\ 8 = 2x^2 & \text{(add } 2x^2\text{)} \\ 2x^2 = 8 & \text{(symmetric, =)} \\ x^2 = 4 & \text{(divide by 2)} \\ x = 2 \quad \text{or} \quad x = -2 & \text{(difference of squares)} \end{array}$$ ■

EXERCISES *Solve the given equation for x.*

1. $x^2 = 8^2$
2. $x^2 = (-11)^2$
3. $x^2 = (-25)^2$
4. $x^2 = 81$
5. $x^2 - 25 = 0$
6. $x^2 - 49 = 0$
7. $2x^2 - 72 = 0$
8. $3x^2 = 12$

9. $2x^2 + 5 = 23$

10. $3x^2 - 1 = 47$

11. $50 - 3x^2 = 23$

12. $75 - 2x^2 = 3$

Use the formula $s = 16t^2$ *to find the time for a freely falling body to fall the specified distance.*

13. (a) 4 feet. (b) 400 feet.

14. (a) 100 feet (b) 10,000 feet.

Find the height of a freely falling body that was dropped from 30,000 *feet and has fallen for the specified time.*

15. (a) 10 seconds. (b) 20 seconds.

16. (a) 30 seconds. (b) 40 seconds.

In Exercises 17 *through* 22 *use the quadratic formula and solve the given equation.*

17. $x^2 - x - 12 = 0$

18. $x^2 - 2x - 15 = 0$

19. $2x^2 + 3x - 5 = 0$

20. $3x^2 - 10x - 13 = 0$

21. $5x^2 + 7x - 6 = 0$

22. $2x^2 + 9x - 5 = 0$

23. Show how equations of the form $x^2 = d^2$ may be solved using the quadratic formula.

24. Show how equations of the form $x(x - n) = 0$ may be solved using the quadratic formula.

*25. The equation $ax^2 + bx + c = 0$, $a \neq 0$ is given. State a reason for each indicated step in this derivation of the quadratic formula.

(a) $x^2 + \frac{b}{a}x + \frac{c}{a} = 0 \times \frac{1}{a}$

(b) $x^2 + \frac{b}{a}x + \frac{c}{a} = 0$

(c) $\left(x^2 + \frac{b}{a}x + \frac{c}{a}\right) + \left(-\frac{c}{a}\right) = 0 + \left(-\frac{c}{a}\right)$

(d) $\left(x^2 + \frac{b}{a}x\right) + \left[\frac{c}{a} + \left(-\frac{c}{a}\right)\right] = 0 + \left(-\frac{c}{a}\right)$

(e) $\left(x^2 + \frac{b}{a}x\right) + 0 = 0 + \left(-\frac{c}{a}\right)$

(f) $x^2 + \frac{b}{a}x = -\frac{c}{a}$

(g) $x^2 + \frac{b}{a}x + \frac{b^2}{4a^2} = -\frac{c}{a} + \frac{b^2}{4a^2}$

(h) $\qquad = \frac{b^2}{4a^2} + \left(-\frac{c}{a}\right)$

$\qquad = \frac{b^2}{4a^2} + \left(\frac{-c}{a}\right) \quad \left(\text{since } -\frac{c}{a} = \frac{-c}{a}\right)$

(i) $$= \frac{b^2}{4a^2} + \frac{-4ac}{4a^2}$$

(j) $$= \frac{b^2 - 4ac}{4a^2}$$

(k) $$\left(x + \frac{b}{2a}\right)^2 + \left(-\frac{b^2 - 4ac}{4a^2}\right) = \frac{b^2 - 4ac}{4a^2} + \left(-\frac{b^2 - 4ac}{4a^2}\right)$$

(l) $$= 0$$

(m) $$\left(x + \frac{b}{2a}\right)^2 - \frac{b^2 - 4ac}{4a^2} = 0$$

$$\left(x + \frac{b}{2a}\right)^2 - \left(\frac{\sqrt{b^2 - 4ac}}{2a}\right)^2 = 0 \qquad \text{(properties of square root)}$$

$$x + \frac{b}{2a} = \frac{\sqrt{b^2 - 4ac}}{2a} \text{ or } \frac{-\sqrt{b^2 - 4ac}}{2a} \qquad \text{(difference of squares)}$$

$$x = \frac{-b}{2a} + \frac{\sqrt{b^2 - 4ac}}{2a} \text{ or } \frac{-b}{2a} - \frac{\sqrt{b^2 - 4ac}}{2a}$$

That is, $x = \dfrac{-b \pm \sqrt{b^2 - 4ac}}{2a}$.

EXPLORATIONS In ancient Greece the equivalent of a quadratic equation was a statement about areas. Zero and negative numbers had not yet been introduced. Thus a problem that we would represent by

$$x^2 + 2x - 24 = 0$$

would have been stated in words in a form equivalent to

$$x^2 + 2x = 24$$

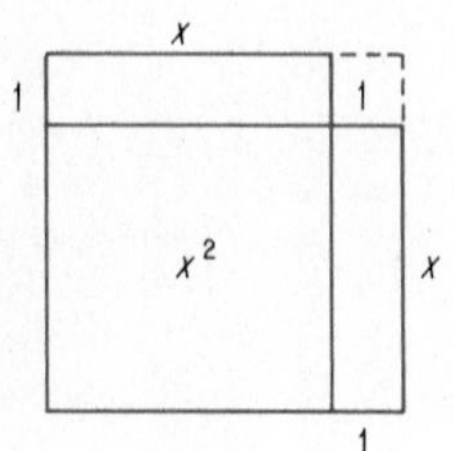

To solve this problem a square would be sketched and its area considered to be x^2. A unit of length would be selected initially unrelated to the side of the square, except that for convenience the unit would be less than x. Then two rectangular regions each 1 by x would be added to the square as in the adjacent figure. Finally a square of side $x + 1$ would be *completed* by adding a square of area 1. The square of side $x + 1$ has area $24 + 1$, that is, 25. The early Greeks knew the squares of the natural numbers and thus knew that if

$$(x + 1)^2 = 25 = 5^2$$

then

$$x + 1 = 5$$

$$x = 4$$

The use of the quadratic formula has been made possible during the last 400 years by the introduction of letters to represent variables. About 2300 years ago the classical Greeks represented both numbers (natural numbers) and magnitudes (other positive real numbers) by lengths of line segments. Products were represented by the areas of rectangular regions. Statements of equality were expressed in words.

The reasoning used to solve for x corresponded to the modern procedure of **completing the square**.

$$x^2 + 2x = 24$$

$$x^2 + 2x + 1 = 24 + 1$$

$$(x + 1)^2 = 25$$

$$x + 1 = 5 \text{ or } -5$$

$$x = 4 \text{ or } -6$$

However, only the positive value would have been obtained by the classical Greeks. This approach can be used for any equation of the form $x^2 + 2bx = c$, where b and c are positive numbers and a square of area b^2 is added.

Sketch and label the related figure. Find the positive solution for each equation by completing the square in the classical Greek manner.

1. $x^2 + 2x = 8$ **2.** $x^2 + 2x = 15$ **3.** $x^2 + 4x = 12$

4. $x^2 + 6x = 16$ **5.** $x^2 + 3x = 7/4$ **6.** $x^2 + 5x = 6$

Extend the reasoning just used for solving equations of the form $x^2 + 2bx = c$ to obtain a corresponding procedure for solving the following equations of the form $x^2 - 2bx = c$. For each equation in Explorations 7 through 12, solve by completing the square in the classical Greek manner.

7. $x^2 - 2x = 8$ **8.** $x^2 - 2x = 15$ **9.** $x^2 - 8x = 20$

10. $x^2 - 12x = 13$ **11.** $x^2 - 3x = 4$ **12.** $x^2 - 7x = 8$

13. Criticize the following "solution" of the given equation.

$$x^2 - 2x = 2$$

$$x(x - 2) = 2$$

$$x = 2 \quad \text{or} \quad x - 2 = 2$$

$$x = 2 \quad \text{or} \quad x = 0$$

Chapter Review

1. Solve for whole numbers x.
(**a**) $x + 7 < 11$ (**b**) $11 - x > 7$

2. Describe the solution set of the given sentence for real numbers x.
(**a**) $x + 2 < 7$ (**b**) $x + 7 \geq 5$

Graph for real numbers x.

3. $x + 3 \leq -1$ **4.** $x - 2 \nless 3$

5. $-1 \leq x \leq 2$ **6.** $x = 2$ or $x \geq 3$

7. $x + 3 < 4$ and $x \geq 0$ **8.** $x \leq 1$ or $x \geq 2$

9. $x + 3 < 4$ or $x \geq 0$ **10.** $x < x + 2$ and $x + 2 < 5$

11. $|x| > 2$ **12.** $|x - 2| \leq 3$

13. $|x + 1| = 1$ **14.** $|x + 2| < 1$

Solve for x.

15. $2x + 5 = 13$
16. $7 - 4x = 11$
17. $\frac{1}{2}x - 3 > 7$
18. $3 - \frac{2}{3}x < 1$
19. $5x + 6 = 4x + 5$
20. $|x - 3| = 5$
21. $2x^2 - 8 = 0$
22. $2x^2 - 3x - 5 = 0$

23. The sum of two numbers is 25; their difference is 7. Find the numbers.
24. The mother of one of Jane's students sent a box of cookies for the members of the class. If 2 cookies were given to each student, there was a surplus of 10 cookies. If 3 cookies were distributed to each student, there was a shortage of 5 cookies. How many students were there in the class?
25. Convert to the Fahrenheit scale **(a)** 15 °C **(b)** 35 °C.

CALCULATING THE ROOTS OF A QUADRATIC EQUATION

The quadratic formula was presented in the last section. The formula provides mathematicians with a procedure for finding the solutions, roots, of a quadratic equation. While the formula is a real boon to mathematicians, it still requires some calculations and manipulations. If the roots do not turn out to be rational numbers, some approximations are needed in order to establish the location of the roots on the real number line that forms the x-axis.

The coefficients A, B, and C are identified as in the quadratic equation $Ax^2 + Bx + C = 0$ and substituted in the quadratic formula. The roots are given exactly in radical form and also approximately in decimal form.

The program involves three techniques of programming in BASIC that we have not used previously. In line 14 the notation X ∧ 2, with all symbols on the same line, is used to represent x^2. In lines 22 and 23 a colon is used to introduce another command subject to the premise of the statement. In lines 26, 3Ø, and 34, SQR designates the square root function.

```
10  PRINT "THIS PROGRAM CALCULATES THE ROOTS OF"
11  PRINT
12  PRINT "A QUADRATIC EQUATION OF THE FORM:"
13  PRINT
14  PRINT "        A * X ∧ 2 + B * X + C = Ø."
15  PRINT
17  PRINT "ENTER THE VALUES FOR A, B, AND C."
18  PRINT
19  INPUT A, B, C
2Ø  PRINT
```

```
21  PRINT
22  IF B ∧ 2 − 4 * A * C < Ø THEN PRINT "NO REAL ROOTS.":GOTO 5Ø
23  IF A = Ø THEN PRINT "THE EQUATION IS NOT QUADRATIC.":GOTO 5Ø
24  PRINT "THE ROOTS ARE:"
25  PRINT
26  PRINT "(1/";2 * A;")(";−B;" + ";SQR(B ∧ 2 − 4 * A * C);")"
27  PRINT
28  PRINT "OR"
29  PRINT
3Ø  PRINT "(1/";2 * A;")(";−B;" − ";SQR(B ∧ 2 − 4 * A * C);")"
31  PRINT
32  PRINT "THAT IS"
33  PRINT
34  PRINT "(1/(2 * A)) * (−B + SQR(B ∧ 2 − 4 * A * C));" OR ";
              (1/(2 * A)) * (−B − SQR(B ∧ 2 − 4 * A * C))
5Ø  END
```

The following sample output of this program shows its application to the equation $12x^2 - 13x - 14 = 0$ where the coefficients are 12, -13, and -14, respectively. As the output shows, the roots are rational numbers and have decimal values, to nine places, of 1.75 and −Ø.666666667, that is, the roots are 7/4 and $-2/3$.

```
RUN
THIS PROGRAM CALCULATES THE ROOTS OF

A QUADRATIC EQUATION OF THE FORM:

     A * X ∧ 2 + B * X + C = Ø.

ENTER THE VALUES FOR A, B, AND C.

?12, −13, −14

THE ROOTS ARE:

(1/24)(13 + 29)

OR

(1/24)(13 − 29)

THAT IS

1.75 OR −Ø.666666667
```

If you have access to a computer, use this program to check your answers to the questions involving the roots of quadratic equations in the exercise set for Section 12-6.

13

Algebra and Geometry

Grace Chisholm Young
(1868–1944)

Grace Chisholm Young was born in England and was educated at home by her mother until she was 10 years of age. At that point she received private tutoring until she entered Cambridge University to study mathematics. Like other women in her time, she had to fight discrimination in mathematical circles. Following graduation at Cambridge, she matriculated at the university in Göttingen, Germany. Here she received the first doctorate in mathematics granted to a woman. Her work dealt with both algebra and geometry and the way in which they can be combined to describe motion and three-dimensional relationships.

13-1 Sentences in Two Variables

The algebra associated with sentences in one variable was considered in Chapter 12. The graphs of such sentences are subsets of a number line. We now consider sentences in two variables and in Section 13-5 a few sentences in three variables. The graph of a sentence in two variables is a subset of a coordinate plane; the graph of a sentence in three variables is a subset of coordinate space. In each case the geometric representations are very useful as aids to understanding and solving algebraic problems and the algebraic representations are very helpful as aids to understanding and solving geometric problems.

Consider a sentence in two real variables, for example,

$$x + y = 5$$

If a replacement, such as 3 for x, is made for one variable, then the resulting sentence

$$3 + y = 5$$

is an open sentence in the other variable. Thus a pair of replacements is needed before an open sentence in two variables can be identified as true or false for these replacements. As in Section 9-5, the two variables may be considered as an *ordered pair* such as (x, y). Then an ordered pair of numbers may be used as replacements for the variables. For example, if $(x, y) = (3, 4)$ in the sentence $x + y = 5$, then the resulting sentence

$$3 + 4 = 5$$

is a false statement. If $(x, y) = (7, -2)$ in the given sentence, then the resulting sentence

$$7 + (-2) = 5$$

is a true statement.

Any sentence in two variables may be considered as a sentence involving an ordered pair of variables such as (x, y). By convention the first variable of the ordered pair, in this case x, is called the **independent variable** and the second variable, in this case y, is the **dependent variable**. The *solution set* for any sentence in two variables is the set of ordered pairs of numbers for which the sentence is a true statement. As with sentences in one variable, the solution set depends upon the replacement sets used for the variables. If no replacement set is specified, the set of real numbers is used. Since any ordered pair of real numbers may be represented by a point on a coordinate plane, any sentence in two real variables may be represented by its *graph* (the set of points that represent the ordered pairs for which the sentence is a true statement) on a coordinate plane.

Example 1 Use the set of whole numbers as the replacement set for x and y and find the solution set for $x + y < 3$.

Solution When $x = 0$, y must be less than 3. Thus the ordered pairs of numbers (0, 0), (0, 1), and (0, 2) make the sentence true. Similarly, when $x = 1$, y must be less than 2; when $x = 2$, y must be less than 1. Is there a whole number y that satisfies the inequality when $x = 3$? When $x > 3$? The solution set is

$$\{(0, 0)\ (0, 1), (0, 2), (1, 0)\ (1, 1), (2, 0)\} \blacksquare$$

Example 2 Use the set of natural numbers as the replacement set for x and y and find the solution set for $x + y \leq 3$.

Solution $\{(1, 1), (1, 2), (2, 1)\}$ ■

If $\mathcal{U}$ is the replacement set for the variables, then any sentence in two variables has a subset of the *Cartesian product* $\mathcal{U} \times \mathcal{U}$ as its solution set. Graphical representations of solution sets on a coordinate plane are particularly convenient when the solution set has many elements. Either an integral coordinate plane or a real coordinate plane may be used.

Example 3 Use $\mathcal{U} = \{-3, -2, -1, 0, 1, 2, 3\}$ and graph $y \geq x + 2$, that is, graph $\{(x, y) \mid y \geq x + 2\}$.

Solution For the corresponding statement of equality, we have

$$\{(x, y) \mid y = x + 2\} = \{(-3, -1), (-2, 0), (-1, 1), (0, 2), (1, 3)\}$$

The graph of $y \geq x + 2$ consists of the points that are solutions of $y = x + 2$, and also all the points in $\mathcal{U} \times \mathcal{U}$ that are above the points of the graph of $y = x + 2$.

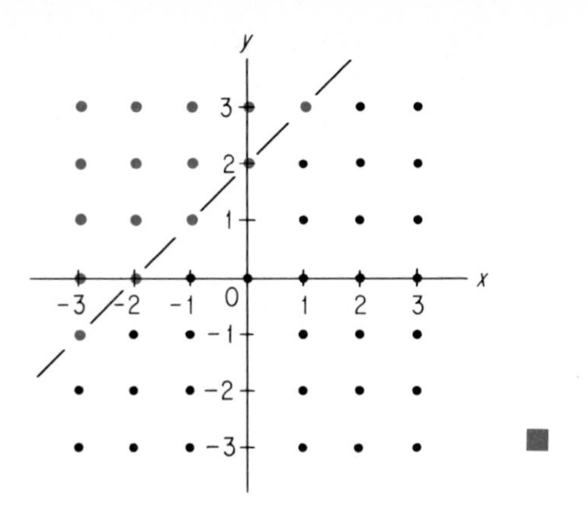

Any **relation** that is a set of ordered pairs of numbers may be graphed on a coordinate plane. The set of all first elements of the ordered pairs is called the **domain** of the relation; the set of all second elements is called the **range** of the relation. Consider, for example, $\{(x, y) \mid y > x - 1\}$, where $\mathscr{U} = \{1, 2, 3\}$. This set, which *is* the relation, is $\{(1, 1), (1, 2), (1, 3), (2, 2), (2, 3), (3, 3)\}$ with domain $\{1, 2, 3\}$ and range $\{1, 2, 3\}$. The relation may be graphed as in the adjacent figure.

A **function** is a set of ordered pairs (x, y) such that for each value of x there is exactly one value of y; that is, a function is a relation in which no first element appears with more than one second element. Any function is also called a **mapping** from the domain onto the range.

Consider, for example, $\{(x, y) \mid y = x\}$, if $\mathscr{U} = \{-3, -2, -1, 0, 1, 2, 3\}$. Then $\mathscr{U} \times \mathscr{U}$ consists of 49 ordered pairs of numbers. The following figure shows the graph of the function $\{(x, y) \mid y = x\}$.

$$\{(-3, -3), (-2, -2), (-1, -1), (0, 0), (1, 1), (2, 2), (3, 3)\}$$

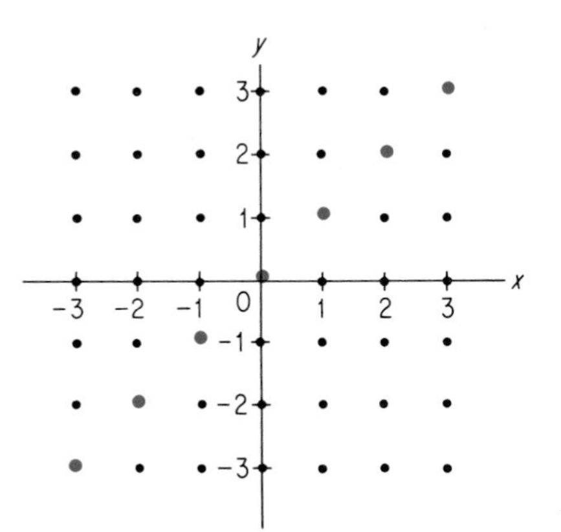

Example 4 Solve $y = x + 1$ if $\mathscr{U} = \{1, 2, 3, 4\}$. State the domain and the range. Is the solution set a function?

Solution For $x = 1$, $y = 2$; for $x = 2$, $y = 3$; for $x = 3$, $y = 4$. The solution set is $\{(1, 2), (2, 3), (3, 4)\}$. The domain is $\{1, 2, 3\}$ and the range is $\{2, 3, 4\}$. This set of ordered pairs is a function since for each value of x there is exactly one corresponding value of y. ■

If a function is graphed on an xy-plane, then the *domain* of the function is the set of values k for which the lines $x = k$ intersect its graph. Similarly, the *range* of the function is the set of values t for which the lines $y = t$ intersect its graph.

A formula such as $y = x^2$ defines a function but the formula is not itself a function. We have defined a function to be a set of ordered pairs (x, y), such as those obtained from the formula $y = x^2$ for the real variable x. Thus a formula may provide a rule by which the function may be determined. In other words, a formula may provide a means for associating a unique element in the range with each element in the domain. As soon as such distinctions are clearly understood it is customary to speak of the function $y = x^2$ and mean the function represented, or determined, by the formula $y = x^2$.

Example 5 Identify the domain and range of the given function.

(a) $y = x^2 - 1$ **(b)** $y = |x|$

Solution

(a) The variable x may assume any real number as a value, that is, the domain is the set of real numbers. Since $x^2 \geq 0$, then $x^2 - 1 \geq -1$ and the range is the set of real numbers greater than or equal to -1.

(b) *Domain*: Set of real numbers.
Range: Set of nonnegative real numbers, that is, $y \geq 0$. ■

Often other letters are used to indicate the quantities represented by the variables. For example, the formula

$$p = 4s$$

for the perimeter p of a square of side s indicates that the perimeter is determined by the length of the side. Similarly, the formula

$$d = 80h$$

indicates the distance that a person would drive in h hours at 80 kilometers per hour. The formula

$$A = P + 0.09P$$

indicates the amount A of money in a savings account in which P dollars have been deposited at 9% simple interest for one year.

Every function is a relation. However, not every relation is a function. When an equation such as $y = x^2 - 2x + 3$ is used to define a function, it is customary to think of y as *a function of* x and to write the formula in **functional notation**.

$$y = f(x) \quad \text{where} \quad f(x) = x^2 - 2x + 3$$

The symbol $f(x)$ may be read as "f evaluated at x," which is often abbreviated as "f at x." Some books use "f of x" for the "value of the function f

for any assumed value of x." If $y = f(x)$, then the value of y for any value b of x may be expressed as $f(b)$. For example,

$$f(b) = b^2 - 2b + 3$$

$$f(2) = 2^2 - 2(2) + 3 = 3$$

$$f(-1) = (-1)^2 - 2(-1) + 3 = 6$$

$$f(0) = 0^2 - 2(0) + 3 = 3$$

$$f(1) = 1^2 - 2(1) + 3 = 2$$

Other letters, such as $g(x)$ and $h(x)$, may be used in designating functions.

Example 6 If $g(x) = x^2 + 3x - 2$, find **(a)** $g(0)$ **(b)** $g(-2)$.

Solution **(a)** $g(0) = (0)^2 + 3(0) - 2 = 0 + 0 - 2 = -2$
(b) $g(-2) = (-2)^2 + 3(-2) - 2 = 4 - 6 - 2 = -4$ ■

For any function $y = f(x)$ the set of ordered pairs $(x, f(x))$ may be graphed. On a coordinate plane any vertical line intersects the graph of a function in at most one point. Here are the graphs of two relations that are also functions.

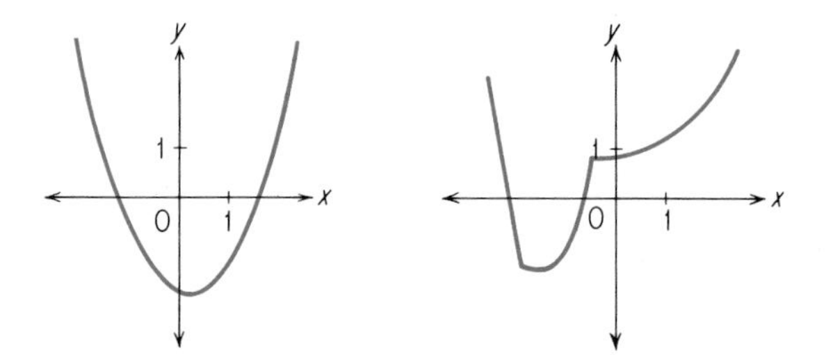

Notice that no vertical line can intersect the graph in more than one point.

Here are the graphs of two relations that are not functions.

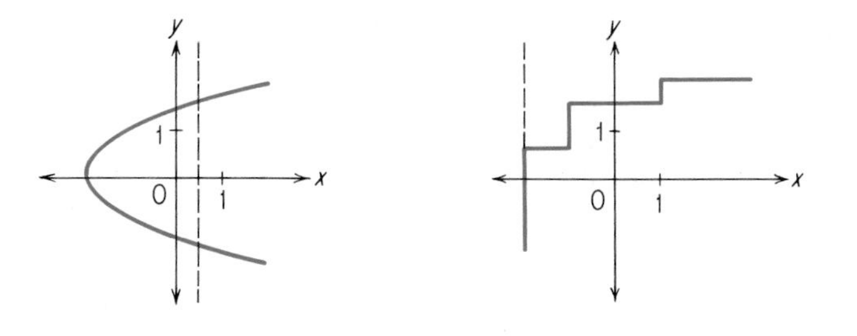

This use of graphs and vertical lines to determine whether or not a relation is a function is often called the **vertical line test**.

Notice that in each case there exists at least one vertical line that intersects the graph in two or more points.

EXERCISES *Use the replacement set* $\{1, 2, 3, 4\}$ *and find the solution set for each sentence.*

1. $x + y = 4$
2. $y = x + 1$
3. $x + y \leq 3$
4. $2x + y \leq 4$
5. $x + y < 2$
6. $y < x$

Use the set of natural numbers as the replacement set for x and y and find the solution set for the given sentence.

7. $x + y = 6$
8. $x + y < 5$
9. $y \leq 4 - x$
10. $y = 5 - x$
11. $x + 2y < 6$
12. $y \leq 6 - 2x$

Use the replacement set $\{-3, -2, -1, 0, 1, 2, 3\}$ *and find the solution set for the given sentence.*

13. $y = |x|$
14. $y = |-x|$
15. $y = -|x|$
16. $y = |x| - 2$
17. $y = x^2 + 1$
18. $y = x^2 - 3$

Use $\mathcal{U} = \{-3, -2, -1, 0, 1, 2, 3\}$ *and graph the set of points* (x, y) *that satisfy the given sentence.*

19. $y = x + 2$
20. $y = x - 2$
21. $y \geq x$
22. $y < x$
23. $y \leq x - 1$
24. $x + y \geq 3$
25. $y = x - 3$
26. $y = x + 3$
27. $y \geq x - 3$
28. $y \leq x + 3$
29. $y < 2 - x$
30. $y \leq -x$

State whether or not the relation shown in the given graph is a function.

31.

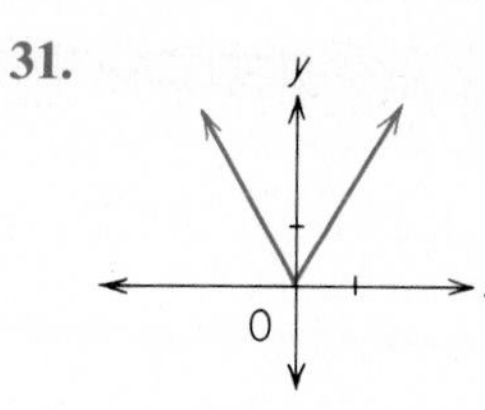

32.

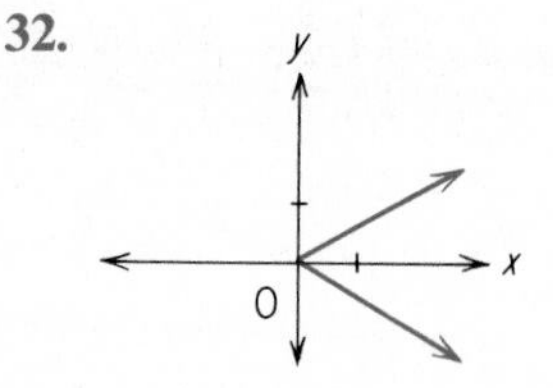

33.

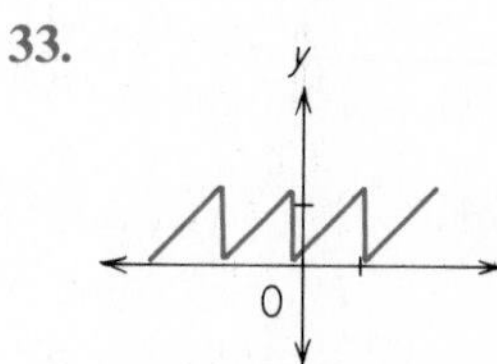

34.

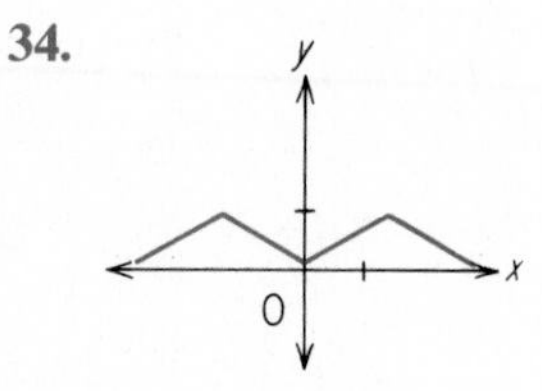

35.

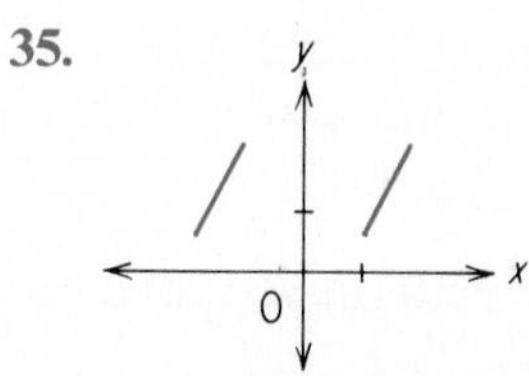

36.

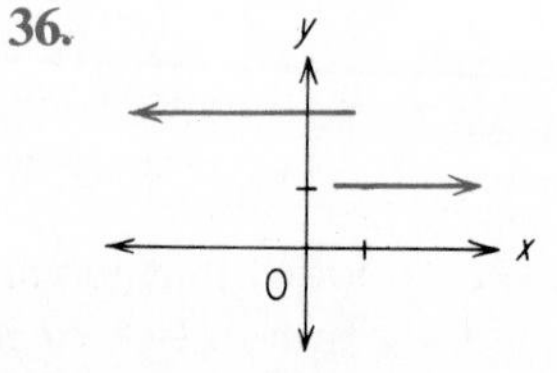

Use $\mathscr{U} = \{1, 2, 3, 4\}$. *State the domain and the range. Is the relation a function?*

37. $y = x$
38. $x + y = 4$
39. $y \leq x$
40. $x + y \neq 4$
41. $x + y < 3$
42. $x + y \leq 3$

In Exercises 43 *through* 58, *use the set of real numbers as the universal set* $\mathscr{U}$ *and proceed as in Exercises* 37 *through* 42.

43. $y = x + 2$
44. $y = x - 2$
45. $y = 2x + 1$
46. $y = 3 - x$
47. $y = 6 - 2x$
48. $y = 3x - 5$
49. $y = |x - 2|$
50. $y = |x + 2|$
51. $y = |x| + 2$
52. $y = |x| - 2$
53. $y < x$
54. $y \geq x$
55. $y \geq -x$
56. $y \leq x - 2$
57. $y = -x^2$
58. $y = (x + 2)^2$

59. If $f(x) = x - 3$, find **(a)** $f(1)$ **(b)** $f(3)$ **(c)** $f(10)$.
60. If $f(x) = x^2 - 4x + 9$, find **(a)** $f(0)$ **(b)** $f(2)$ **(c)** $f(-2)$.
61. If $g(x) = x^3 - 5$, find **(a)** $g(2)$ **(b)** $g(-2)$ **(c)** $g(-3)$.
62. If $h(x) = 3x^2 - 2x + 1$, find **(a)** $h(0)$ **(b)** $h(3)$ **(c)** $h(-3)$.
63. If $s(x) = -2x^3 + 3x - 1$, find **(a)** $s(-1)$ **(b)** $s(-2)$ **(c)** $s(3)$.

*64. An *ordered triple* of numbers is a set of three numbers in order, such as (1, 1, 2). List all the possible ordered triples that can be formed from the set $\mathscr{U} = \{1, 2\}$.

EXPLORATIONS

Ordered pairs may be used to show patterns. Some of these patterns may be expressed by algebraic sentences in two variables. Elementary school students usually enjoy the game of "trees" (originally developed by the University of Illinois Committee on School Mathematics). The teacher (or a selected student) thinks of a sentence in two variables. Initially two or three ordered pairs of integers that satisfy the sentence are stated and graphed. If a pair (x, y) satisfies the sentence, then "x trees y." If a pair does not satisfy the sentence, then "x does not tree y." Other pairs are suggested by members of the class as potential pairs for satisfying the sentence that is not known to the class. If for a suggested pair "x trees y," then the pair is graphed. Each student is encouraged to guess privately the sentence (or an equivalent) that is under discussion. To test a student who may know the sentence, a question such as the following is asked.

Does -5 tree 3?

Or, an instruction such as the following may be given.

Let x be 7 and give a value for y.

After many students think that they know the sentence, the sentence is discussed openly.

Try this game of "trees" with some of your classmates using sentences such as these.

1. $x + y = 6$
2. $y = x + 1$
3. $y = -x$
4. $y = 2 - x$
5. $y > x$
6. $y \leq x + 2$
7. $y = |x|$
8. $y > |x|$
9. $y \leq 1 + |x|$
10. $3 < 2 + |x|$

11. Some texts introduce the concept of function and relation on an intuitive basis through the use of arrow diagrams. For example, each of the following shows a correspondence (relation) between two sets. The arrow indicates which members of the second set (range) correspond to the given elements of the first set (domain). Which of the relations are functions?

(a)

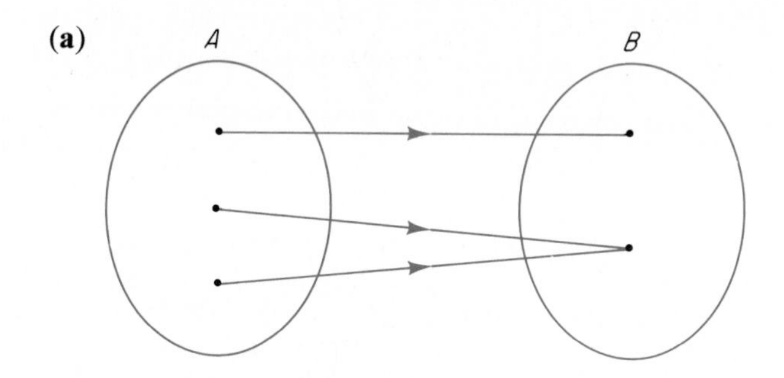

(b)

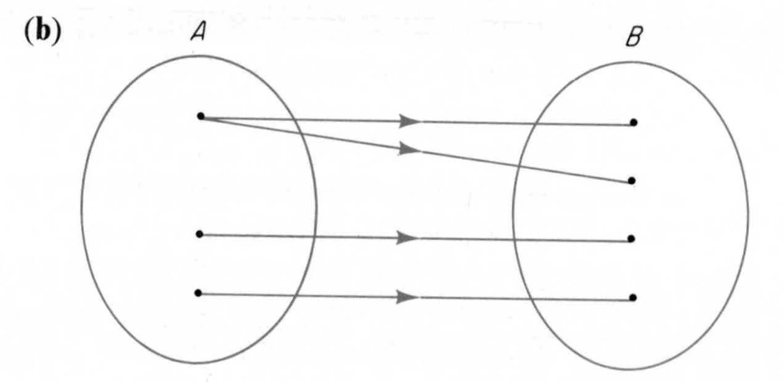

(c)

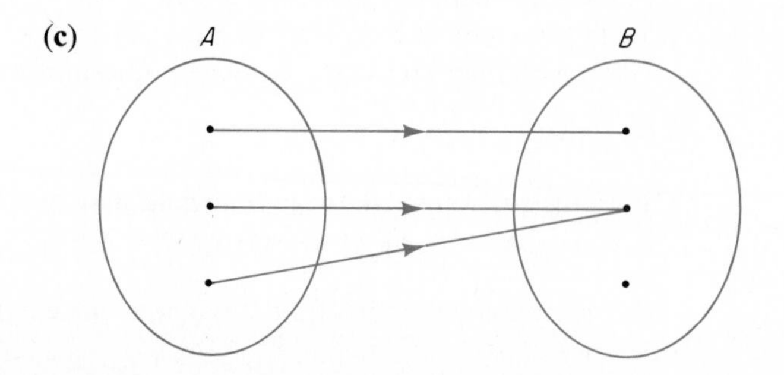

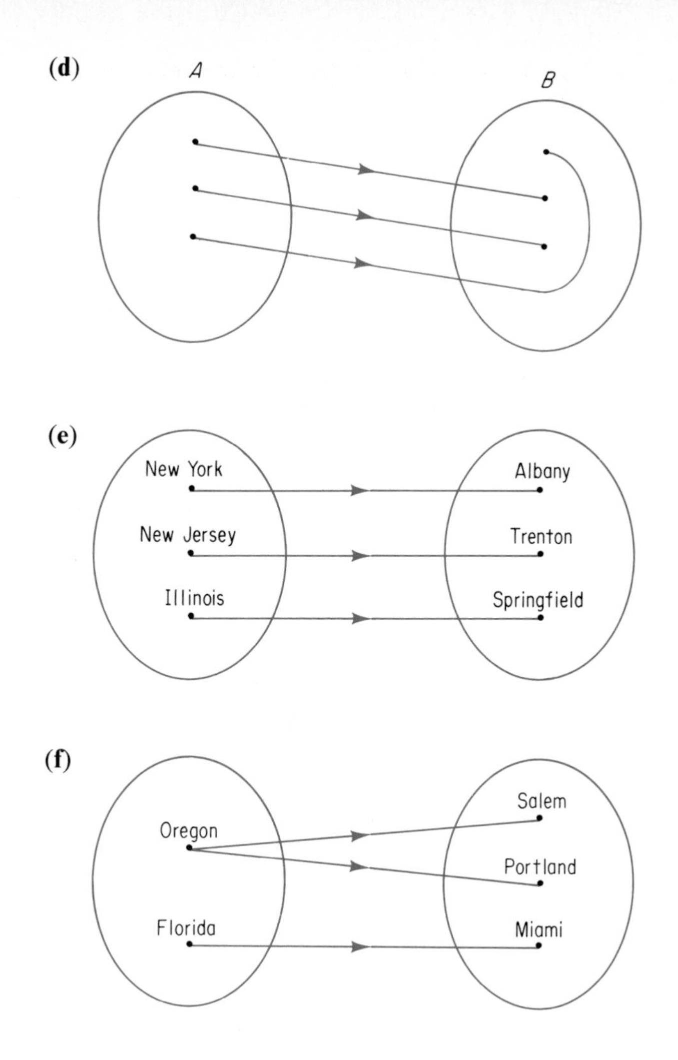

12. Any function may be viewed as a mapping from a set A (the domain) *onto* a set B (the range) such that each element of A is mapped onto a unique element of B and each element of B is the image of at least one element of A. Diagrams (a), (d), and (e) of Exploration 11 illustrate such mappings. Draw arrow diagrams for mappings of five elements *onto*
    (a) five elements **(b)** four elements **(c)** two elements.
13. In a **one-to-one mapping** of a set A onto a set B each element of set A corresponds to a unique element of set B, and each element of set B is the image of a unique element in set A. Draw an arrow diagram that illustrates such a mapping.

13-2 Linear Equations and Inequalities

Any equation of the first degree in one variable can be expressed in the form

$$ax + b = 0, \qquad a \neq 0$$

and is known as a *linear equation in one variable.* Such equations can be graphed

Note that any linear equation in one variable may be considered as a linear equation in two variables. For example, $ax + c = 0$ may be considered as $ax + 0y + c = 0$ and $by + c = 0$ may be considered as $0x + by + c = 0$.

on a number line as was done in Sections 12-1 and 12-3. Any equation of the form

$$ax + by + c = 0$$

where a and b are not both zero, may be considered as a *linear equation in two variables* and has a line as its graph on a coordinate plane (Section 9-5). Conversely, any line on a coordinate plane is the graph of a linear equation.

Consider the linear equation $2x + 3y - 6 = 0$. For any given value of y, the equation can be solved for x, that is, $x = (6 - 3y)/2$. In particular, if $y = 0$, then $x = 3$. The point (3, 0) is the intersection of the graph of the given equation and the line $y = 0$, that is, the x-axis. The number 3 is the **x-intercept** of the graph. Similarly, for any given value of x, the equation can be solved for y, that is, $y = (6 - 2x)/3$. In particular, if $x = 0$, then $y = 2$. The point (0, 2) is the intersection of the graph of the given equation and the line $x = 0$, that is, the y-axis. The number 2 is the **y-intercept** of the graph. To graph the line on a coordinate plane, graph the points (3, 0) and (0, 2), draw the line determined by the two points, and check that a third point of the solution set, such as (6, −2), is on the line. The line is the set (**locus**) of all points with coordinates that satisfy the equation.

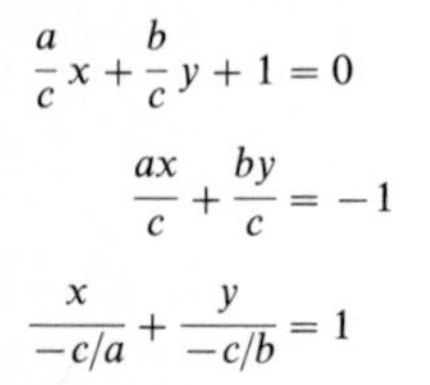

Any linear equation $ax + by + c = 0$ in which the *coefficients a, b,* and c are all different from zero may be written in **intercept form**

$$\frac{x}{-c/a} + \frac{y}{-c/b} = 1$$

$$\frac{a}{c}x + \frac{b}{c}y + 1 = 0$$

$$\frac{ax}{c} + \frac{by}{c} = -1$$

$$\frac{x}{-c/a} + \frac{y}{-c/b} = 1$$

and may be graphed using the intercepts $-c/a$ and $-c/b$ to identify the points $(-c/a, 0)$ and $(0, -c/b)$. One additional point as a check is desirable.

Example 1 Write in intercept form and graph $3x + 5y = 30$.

Solution

$$3x + 5y = 30$$

$$\frac{3x}{30} + \frac{5y}{30} = \frac{30}{30}$$

$$\frac{x}{10} + \frac{y}{6} = 1$$

The x-intercept is 10; the y-intercept is 6. As a check of the graph we note that the graph of another solution, such as (5, 3), is on the line.

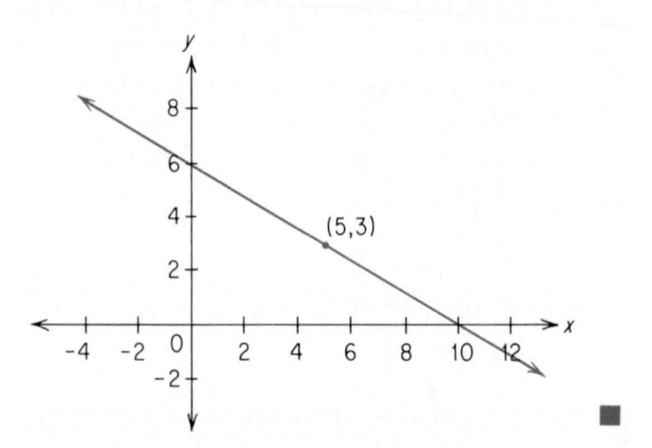

■

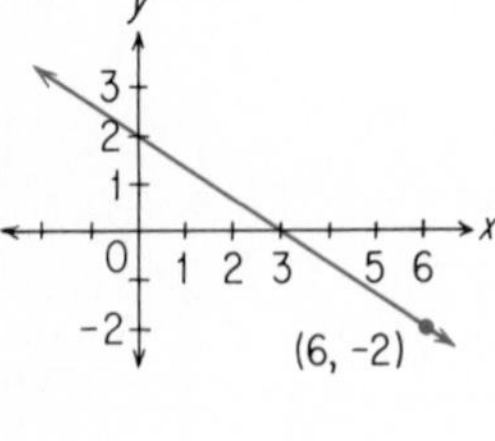

Example 2 Write in intercept form and graph $x - 2y = 4$.

Solution

$$\frac{x}{4} - \frac{y}{2} = 1, \qquad \text{that is,} \qquad \frac{x}{4} + \frac{y}{-2} = 1$$

Any equation of the form $ax + c = 0$ has as its root the x-intercept of the equation $y = ax + c$.

The x-intercept is 4; the y-intercept is -2. As a check of the graph we note that the graph of another solution, such as $(2, -1)$, is on the line.

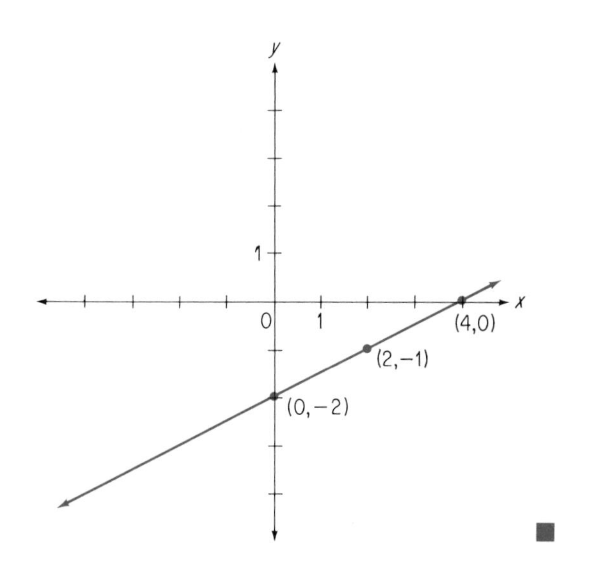

■

Consider the graph of any linear equation in two variables. For a given change in the value of x the change in the value of y is often called the *rise*. Then the change in the value of x is the *run* and the ratio of the rise to the run is the **slope** m of the line. For the change in x from 0 to 4 in Example 2 we note the points $(0, -2)$ and $(4, 0)$ and find

$$\frac{\text{Rise}}{\text{Run}} = \frac{0 - (-2)}{4 - 0} = \frac{2}{4} = \frac{1}{2}$$

As x increases from 0 to 2 we have $(0, -2)$ and $(2, -1)$

$$\frac{\text{Rise}}{\text{Run}} = \frac{-1 - (-2)}{2 - 0} = \frac{1}{2}$$

For any two points (x_1, y_1) and (x_2, y_2) of a line

$$m = \frac{\text{Rise}}{\text{Run}} = \frac{y_2 - y_1}{x_2 - x_1}$$

The slope of the line does not depend upon the choice of the points (x_1, y_1) and (x_2, y_2). If the equation of a line is solved for y and placed in **y-form,**

$$\boxed{y = mx + b}$$

then the coefficient of x is the slope and the *constant term b* is the y-intercept.

The equation given in Example 1 can be written in y-form as

$$y = -\frac{3}{5}x + 6$$

The graph has slope $-3/5$ and y-intercept 6. The slope can be observed as the decrease of y from 6 to 0 as x increases from 0 to 10, that is, $-6/10 = -3/5$. The equation given in Example 2 can be written in y-form as

$$y = \frac{1}{2}x - 2$$

The graph has slope $1/2$ and y-intercept -2.

Example 3 Find the slope and y-intercept of the graph of $x - 3y = 9$.

Solution

$$x - 3y = 9$$

$$3y = x - 9$$

$$y = \frac{1}{3}x - 3$$

The slope is $1/3$; the y-intercept is -3. The slope may also be obtained using points such as $(0, -3)$ and $(6, -1)$ of the line.

$$m = \frac{-1-(-3)}{6-0} = \frac{2}{6} = \frac{1}{3} \quad \blacksquare$$

Example 4 Find an equation for the line with
(a) slope 2 and y-intercept -1
(b) slope $1/2$ and y-intercept 3.

Solution **(a)** $y = 2x + (-1)$, that is, $y = 2x - 1$
(b) $y = (1/2)x + 3$ ■

The x-axis contains the points $(1, 0)$ and $(2, 0)$ and has slope 0.

$$\frac{0-0}{2-1} = 0$$

Any line that is parallel to the x-axis has an equation of the form $y = 0x + k$, that is, $y = k$, and has slope 0.

The y-axis contains the points (0, 1) and (0, 2) and does not have a real number as its slope.

$$\frac{2-1}{0-0} \quad \text{is undefined}$$

Any line such as $x = h$ that is parallel to the y-axis contains point $(h, 1)$ and $(h, 2)$ and also does not have a real number as its slope.

$$\frac{2-1}{h-h} \quad \text{is undefined}$$

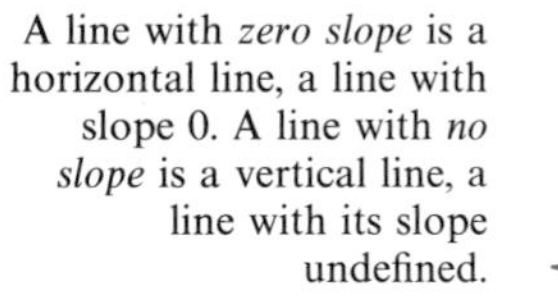

A line with *zero slope* is a horizontal line, a line with slope 0. A line with *no slope* is a vertical line, a line with its slope undefined.

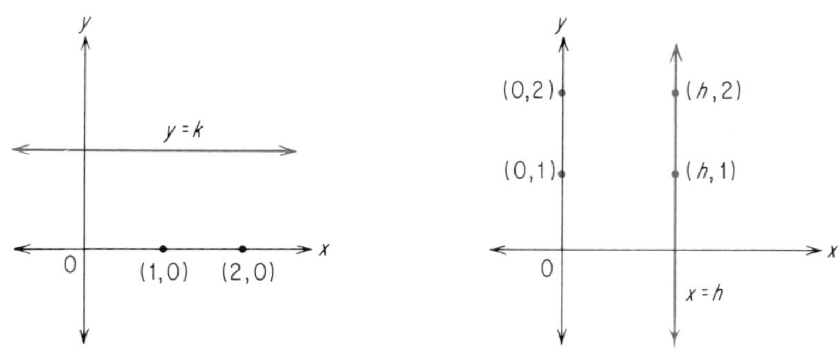

Two lines that are not parallel to the y-axis are **parallel** to each other if and only if they have the same slope. In advanced courses any line is considered to be parallel to itself.

Example 5 Find an equation for the line that contains the point (1, 4) and is parallel to the line $y = 2x - 1$.

Solution The line $y = 2x - 1$ has slope 2. The desired line has slope 2 and contains the point (1, 4).

$$\frac{y-4}{x-1} = 2$$

$$y - 4 = 2(x - 1)$$

$$y = 2x + 2 \quad ■$$

Alternative Solution

$y = 2x + c$

$4 = (2)(1) + c$

$c = 2$

$y = 2x + 2$

Inequalities of the first degree can be graphed, have a line as a boundary, and are called *linear inequalities*. Consider the sentence $y \leq x + 2$. Here the graph consists of all the points on the line $y = x + 2$, as well as the points in the *half-plane* below the line as indicated by the shaded portion of the graph. Note that for any value b of x the point $(b, b + 2)$ is on the line $y = x + 2$, and for any value of y less than $b + 2$ the point (b, y) is below the line. Thus the graph of $y \leq x + 2$ is the union of a line and a half-plane. A solid line is used in this graph since the points of the line are points of the graph. When the points of the line are not points of the graph a dashed line is used, as in Example 6.

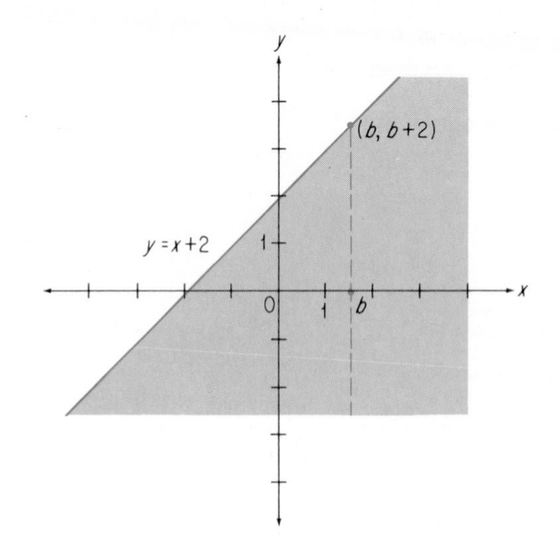

In the preceding figure, the line $y = x + 2$ separates the plane into two *half-planes*. One of these half-planes is part of the desired graph and satisfies the given inequality. One way to determine which half-plane is needed is to select a point and determine whether or not the point is part of the graph. For example, (0, 0) satisfies the inequality $y < x + 2$ and therefore the point with coordinates (0, 0) is in the desired half-plane. If the coordinates of any point satisfy the inequality, the point is in the desired half-plane. If the coordinates of any point, such as (0, 3), do not satisfy the inequality, the point is on the opposite side of the line from the desired half-plane.

Example 6 Graph $y > x - 1$.

Solution It is helpful to graph first the corresponding statement of equality, $y = x - 1$. The line is dashed since it is not part of the graph. This line separates the plane into two half-planes. The desired graph is the half-plane above the line. Note that the coordinates of the point (0, 0) satisfy the given inequality and the point is part of the desired half-plane; the coordinates of (2, 0) do not satisfy the given inequality and the point is on the opposite side of the line from the desired half-plane.

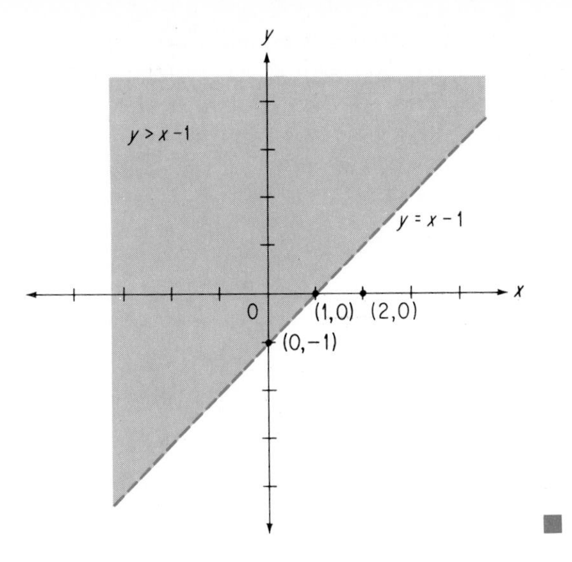

Example 7 Graph $x - 2y - 4 \geq 0$.

Solution The given inequality can be expressed in y-form.

An equation or inequality that is solved for y is said to be in y-form.

$$x - 2y \geq 4 \qquad \text{(add 4)}$$

$$-2y \geq -x + 4 \qquad \text{(subtract } x\text{)}$$

$$y \leq \frac{1}{2}x - 2 \qquad \text{(divide by } -2\text{)}$$

As illustrated by $-3 < -2$ but $3 > 2$, the "sense" of an inequality is changed whenever both sides of the inequality are multiplied, or divided, by a negative number.

The corresponding equality $y = (1/2)x - 2$ is graphed. Then, as indicated by the last inequality, the points on and below this line constitute the desired graph. Note that $(0, 0)$ does not satisfy the inequality; $(0, -3)$ does satisfy the inequality.

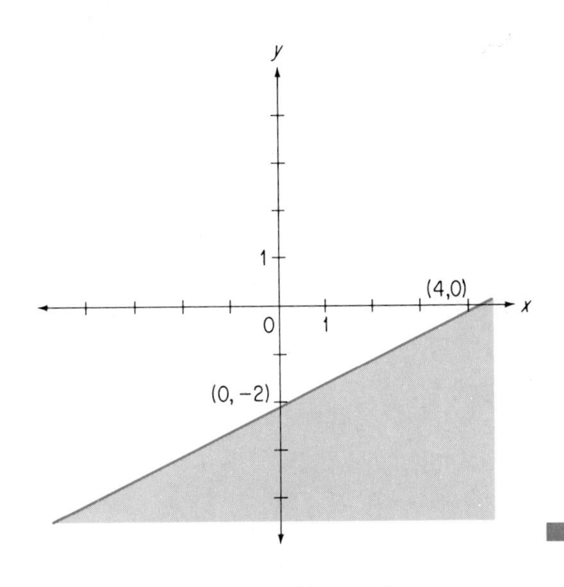

The half-plane that is to be included in the graph of a linear inequality can be determined either by testing the coordinates of points or from the inequality. Note that in each case the inequality was first solved for y. This corresponds to the y-form of an equation and is the *y-form of the inequality*. The graph of the inequality is *above* the line if the y-form is $y > \ldots$; the graph is *below* the line if the y-form is $y < \ldots$.

Example 8 Solve $3x - 4y \le 12$ for y.

Solution

$$3x - 4y \le 12 \qquad \text{(given)}$$

$$-4y \le 12 - 3x \qquad \text{(subtract } 3x\text{)}$$

$$y \ge \frac{3}{4}x - 3 \qquad \text{(divide by } -4\text{)} \ \blacksquare$$

EXERCISES *For the graph of each sentence, find*

(a) *the x-intercept* **(b)** *the y-intercept* **(c)** *the slope.*

1. $x - y = 3$
2. $2x - 3y = 12$
3. $3x - 2y = 12$
4. $y = 2x + 5$
5. $2x - y + 8 = 0$
6. $x + 3y - 6 = 0$

Find an equation for the line that has the given slope and intersects the y-axis at the given point.

7. $m = 4$, $(0, 5)$
8. $m = 3$, $(0, -2)$
9. $m = -1$, $(0, -3)$
10. $m = -2$, $(0, 3)$
11. $m = 0$, $(0, -2)$
12. $m = 0$, $(0, 0)$

Find an equation for the line that contains the given point and is parallel to the given line.

13. $(3, 5)$, $y = x$
14. $(-2, 1)$, $y = 3x + 5$
15. $(-2, 1)$, x-axis
16. $(3, -2)$, y-axis
17. $(4, 2)$, $2x + y = 3$
18. $(5, 7)$, $3x + 2y = 6$

In Exercises 19 through 29, the points are vertices of a quadrilateral. Determine the slopes, if any, of the sides of the quadrilateral and state whether or not the quadrilateral is a parallelogram.

19. $(1, 1)$, $(5, 1)$, $(5, 4)$, $(1, 4)$
20. $(-1, -2)$, $(3, -2)$, $(3, 5)$, $(-1, 5)$
21. $(0, 0)$, $(7, 0)$, $(9, 3)$, $(2, 3)$
22. $(0, -2)$, $(5, -2)$, $(3, 3)$, $(-2, 3)$
23. $(0, 0)$, $(3, -2)$, $(5, 1)$, $(2, 3)$

24. $(-1, 4)$, $(1, -2)$, $(4, -1)$, $(2, 5)$

25. $(-2, 4)$, $(0, -1)$, $(5, 1)$, $(3, 6)$

26. $(1, -5)$, $(7, 11)$, $(9, 13)$, $(3, -3)$

*27. (a, b), $(a + h, b + k)$, $(c + h, d + k)$, (c, d)

*28. (a, b), (c, d), (e, f), (g, h)

*29. $\left(\frac{a+c}{2}, \frac{b+d}{2}\right)$, $\left(\frac{c+e}{2}, \frac{d+f}{2}\right)$, $\left(\frac{e+g}{2}, \frac{f+h}{2}\right)$, $\left(\frac{a+g}{2}, \frac{b+h}{2}\right)$

(*Note*: These vertices are the midpoints of the sides of the general quadrilateral in Exercise 28.)

*30. Two lines are perpendicular if they are parallel to the coordinate axes or if the product of their slopes is -1. Which of the quadrilaterals in Exercises 19 through 26 are rectangles?

Write each sentence in y-form, that is, solve for y.

31. $2x + y = 9$
32. $4x - 2y = 12$
33. $2x + y \geq 8$
34. $x + 3y \leq 6$
35. $2x - y \leq 8$
36. $3x - 2y \geq 12$

For Exercises 37 through 42 select from the inequalities in **(a)** *through* **(i)** *the inequality that best describes the given graph.*

(a) $x + y - 1 > 0$ **(b)** $x + y - 1 \geq 0$ **(c)** $x + y - 1 \leq 0$
(d) $x + y - 1 < 0$ **(e)** $x + y + 1 \leq 0$ **(f)** $x - y + 1 \leq 0$
(g) $x - y + 1 \geq 0$ **(h)** $x - y - 1 \leq 0$ **(i)** $x - y - 1 \geq 0$

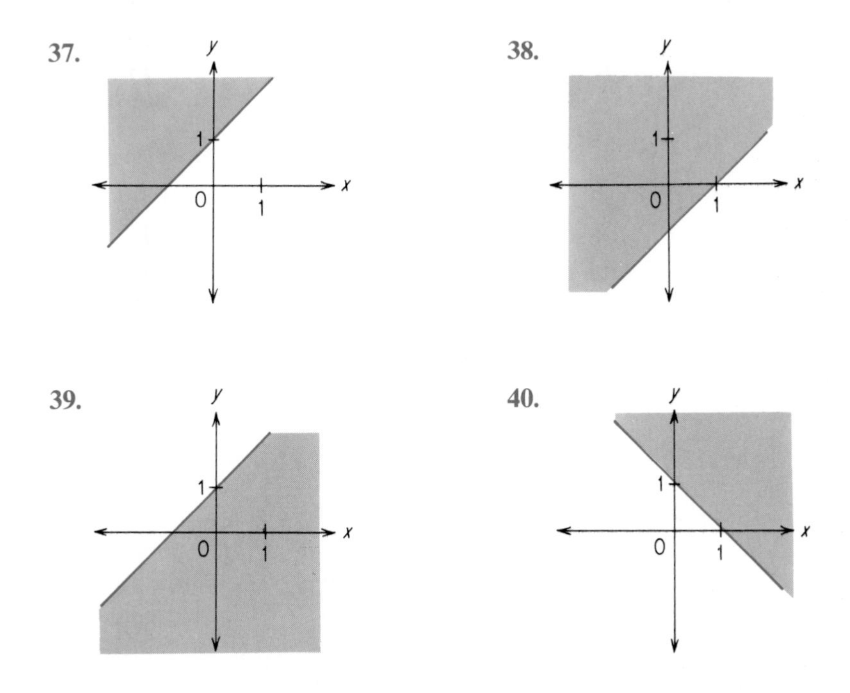

41.

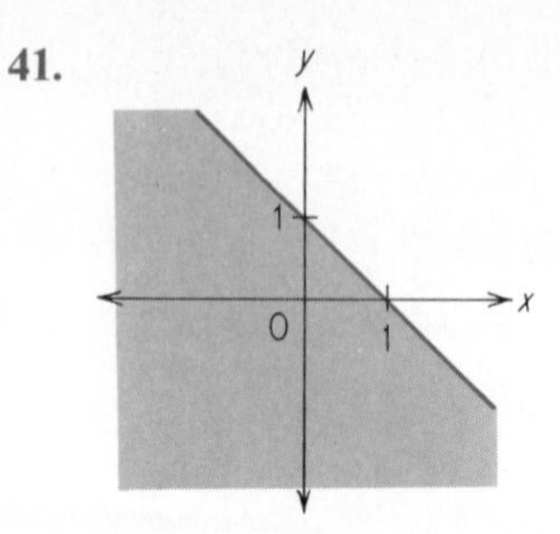

42.

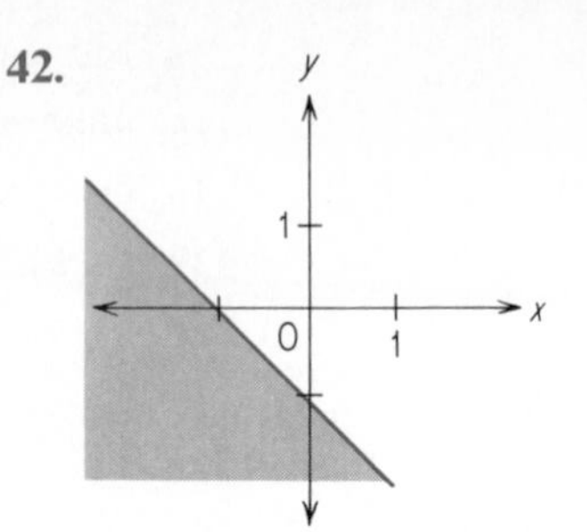

Graph each linear inequality.

43. $y \leq x + 2$

44. $y \geq x - 2$

45. $y > 2x - 1$

46. $y < 3x + 1$

47. $2x + y > 6$

48. $3x + 2y > 6$

49. $x + 2y - 4 \leq 0$

50. $2x + y \leq 4$

51. $x - y \leq 2$

52. $2x - y \geq 3$

53. $3x - 2y + 6 \geq 0$

54. $3x - 4y + 12 \geq 0$

EXPLORATIONS

1. On a coordinate plane the slope can be easily identified for any line or line segment that contains two points with integral coordinates. In particular, for any given union of line segments on a coordinate plane, a line along one side may be selected and the students may be asked to identify all parallel sides. Here is an example.

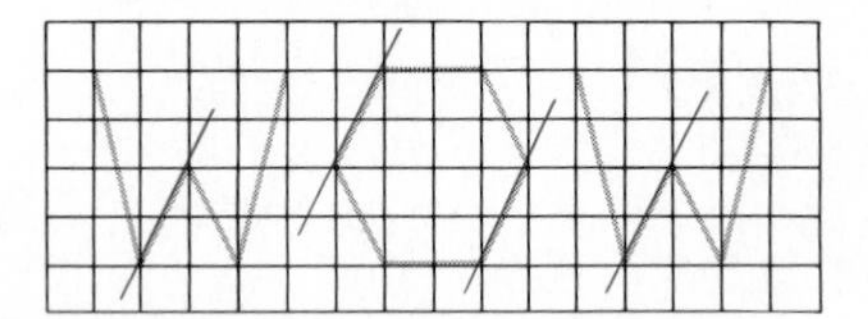

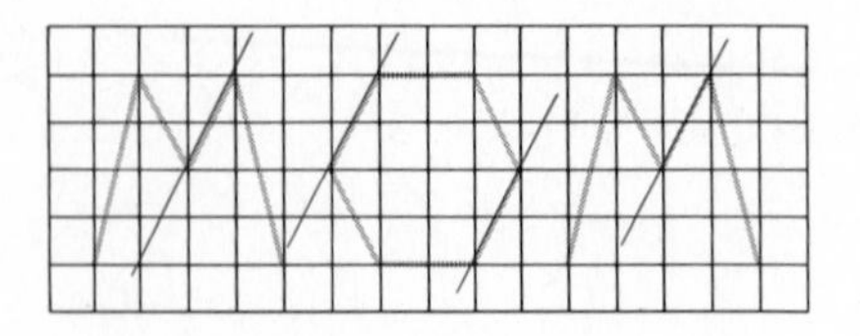

Make a set of worksheets for an elementary school class to use in the identification of parallel lines.

2. In the example in Exploration 1, every line segment is parallel to one of five line segments. Copy the example and use five different colors to show that only five sets of parallel line segments have been used. Explore the use of 1, 2, 3, 4, 5, and 6 sets of parallel line segments to form designs.

3. An algebra class ran into trouble when they tried to graph the solution set for the inequality $2x - 3y + 6 > 0$. First they constructed the graph of the corresponding equality as in the following figure.

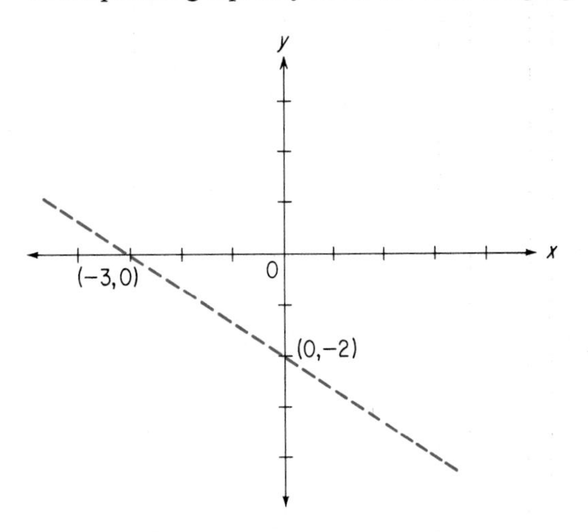

They then tested the point (0, 0) in the upper half-plane, which gives this true statement of inequality.

$$0 - 0 + 6 > 0$$

Thus they concluded that the required half-plane includes the origin, and must be *above* the line. As a check, they then went on to write the given inequality in y-form.

$$2x - 3y + 6 > 0 \qquad -3y > -2x - 6 \qquad y < \frac{2}{3}x + 2$$

This last statement indicates that the required half-plane must be *below* the line! Where did they make an error?

4. Prepare a set of exercises, with separate solutions, that could be used to help a student who made the error mentioned in Exploration 3 avoid such difficulties in the future.

The overhead projector is an effective device for showing graphs on a plane, especially through the use of translations and rotations. For example, prepare a set of coordinate axes and grid lines. On a separate sheet of acetate draw the graph of $y = |x|$.

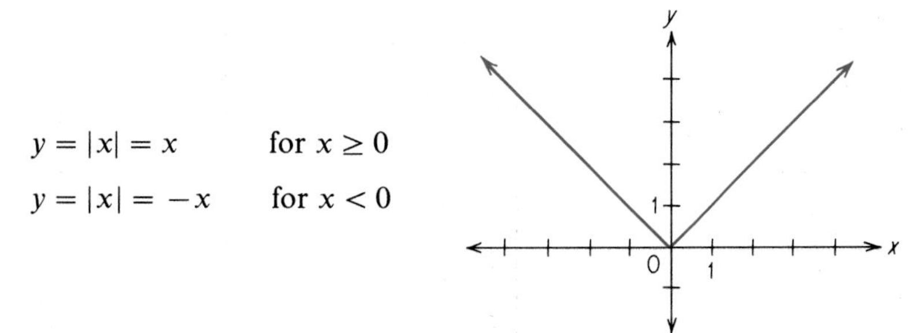

5. Demonstrate each of the following graphs by appropriate translations of the graph of $y = |x|$.

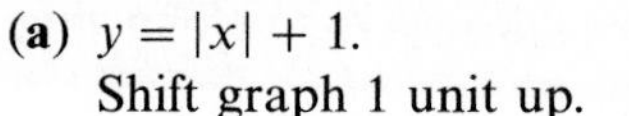

(a) $y = |x| + 1$.
Shift graph 1 unit up.

(b) $y = |x| - 1$.
Shift graph 1 unit down.

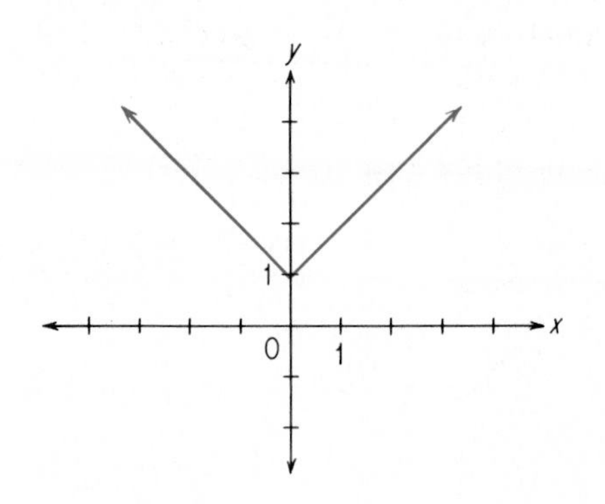

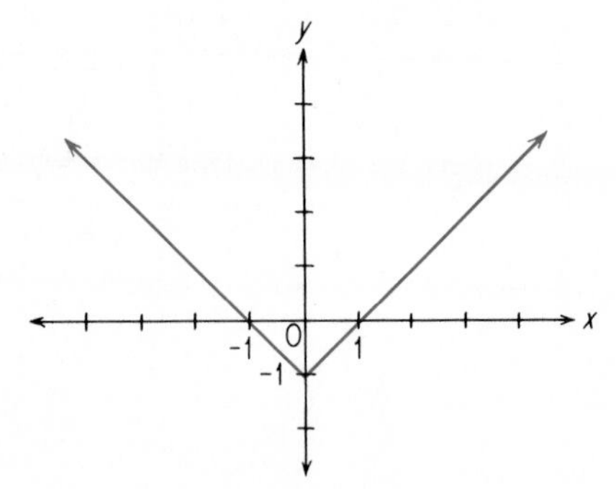

(c) $y = |x + 1|$.
Shift graph 1 unit to the left.

(d) $y = |x - 1|$.
Shift graph 1 unit to the right.

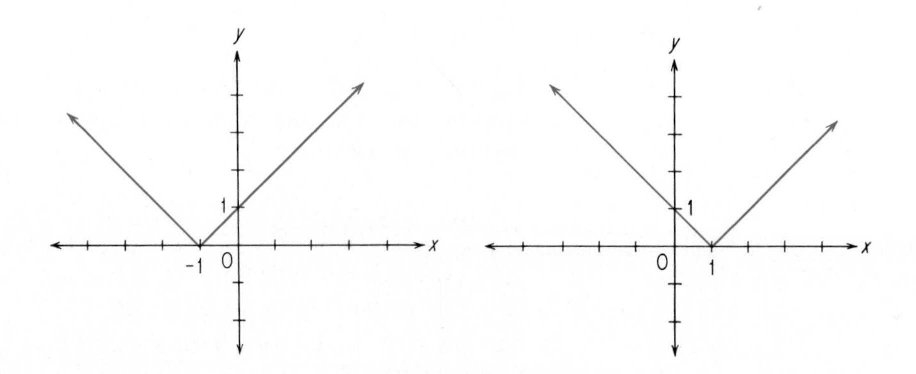

6. The graph of $y = -|x|$ may be obtained by a rotation of the graph of $y = |x|$ through 180° about the x-axis, that is, by a reflection in the x-axis. Demonstrate the use of rotations and translations of the graph of $y = |x|$ to obtain the graphs of each of these equations.

(a) $y = -\|x\|$	**(b)** $y = -\|x\| + 1$	**(c)** $y = -\|x\| - 1$
(d) $y = \|x\| + 2$	**(e)** $y = \|x\| - 2$	**(f)** $y = 2 - \|x\|$
(g) $y = \|x - 1\|$	**(h)** $y = -\|x - 1\|$	**(i)** $y = -\|x - 2\|$
(j) $y = -\|x + 1\|$	**(k)** $y = 2 - \|x - 1\|$	**(l)** $y = 1 - \|x + 1\|$

13-3 Linear Systems

Many applications make use of compound sentences involving two variables. Consider, for example, the compound sentence

$$x + y - 3 = 0 \quad \text{and} \quad x - y - 1 = 0$$

Frequently such sentences are written in the form

$$\begin{cases} x + y - 3 = 0 \\ x - y - 1 = 0 \end{cases}$$

(with or without the brace) and are referred to as a **system of linear equations** or as a set of **simultaneous linear equations**. To solve a set of two simultaneous equations in two variables, we find the set of ordered pairs that are solutions of both equations. Often a graphical approach is useful. For the given system of equations, we have the following graph.

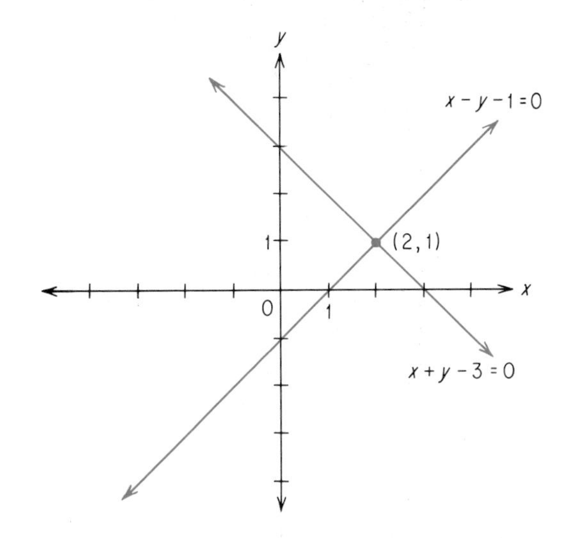

The point located at (2, 1) is on both lines and (2, 1) is the solution of the given system of equations. We may express the solution set of the given system in set-builder notation.

$$\{(x, y) \mid x + y - 3 = 0\} \cap \{(x, y) \mid x - y - 1 = 0\}$$

Each set is a set of ordered pairs. The solution set is the intersection of these two sets of ordered pairs, that is, $\{(2, 1)\}$.

Consider next the compound sentence

$$x + y - 3 = 0 \quad \text{or} \quad x - y - 1 = 0$$

The word *or* indicates that we are to find the set of ordered pairs of numbers that are solutions of either one or of both of the given equations.

$$\{(x, y) \mid x + y - 3 = 0\} \cup \{(x, y) \mid x - y - 1 = 0\}$$

The graph of the solution set consists of all the points that are on at least one of the two lines; that is, the graph is the union of the points of the two lines shown in the preceding figure.

Example 1 Graph

(a) $x - y - 1 = 0 \quad \text{or} \quad x - y + 2 = 0$

(b) $(x - y - 1)(x - y + 2) = 0$.

Solution (**a**) The graph consists of the union of the points of the two lines.

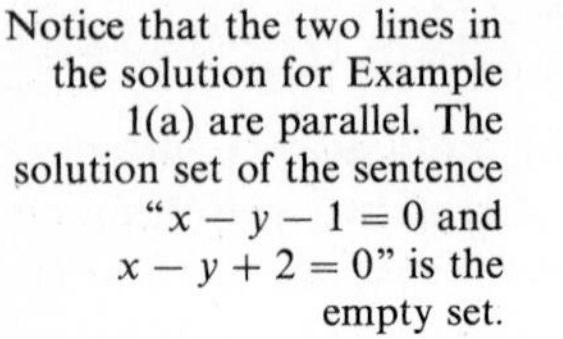

Notice that the two lines in the solution for Example 1(a) are parallel. The solution set of the sentence "$x - y - 1 = 0$ and $x - y + 2 = 0$" is the empty set.

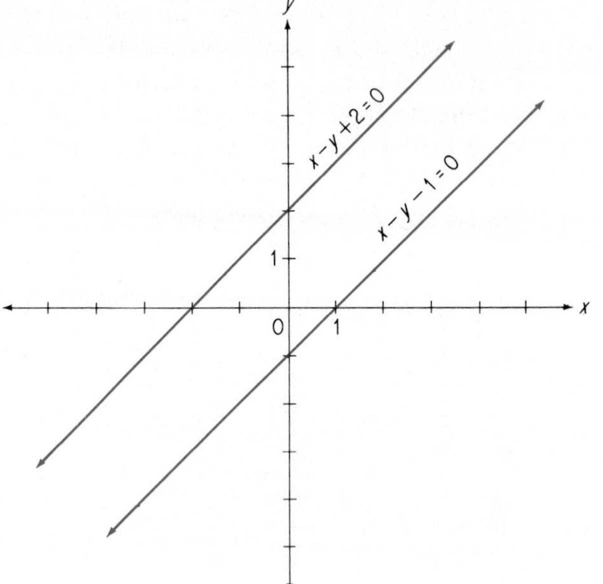

(**b**) Since any product $a \times b = 0$ implies that $a = 0$ or $b = 0$, the given sentence can be written as in part (a) and has the same graph. ■

Systems of inequalities can also be solved graphically as the union or intersection of half-planes. Consider the system

$$\begin{cases} x + y - 3 > 0 \\ x - y - 1 > 0 \end{cases}$$

The corresponding statements of equality have been graphed earlier in this section. The graphs of the inequalities are shown in the next two figures.

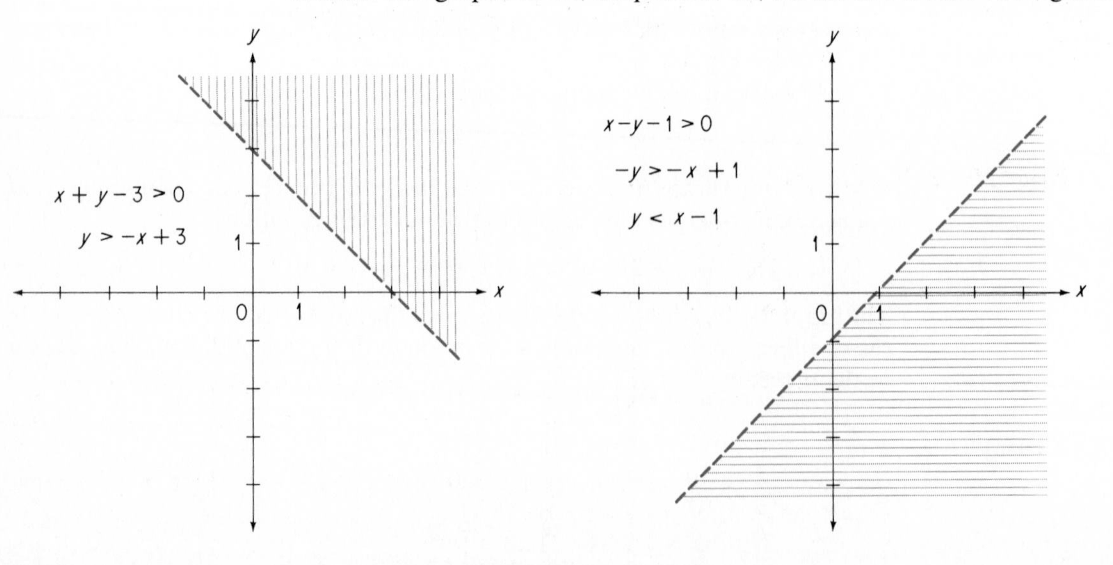

Note that the corresponding lines are dashed since the points of the lines are *not* included in the graphs. Also one graph has vertical shading and the other graph has horizontal shading. The graph of the system consists of the intersection of these two graphs, that is, the points of the region that is shaded both vertically *and* horizontally when the graphs of the two inequalities are drawn on the same coordinate plane.

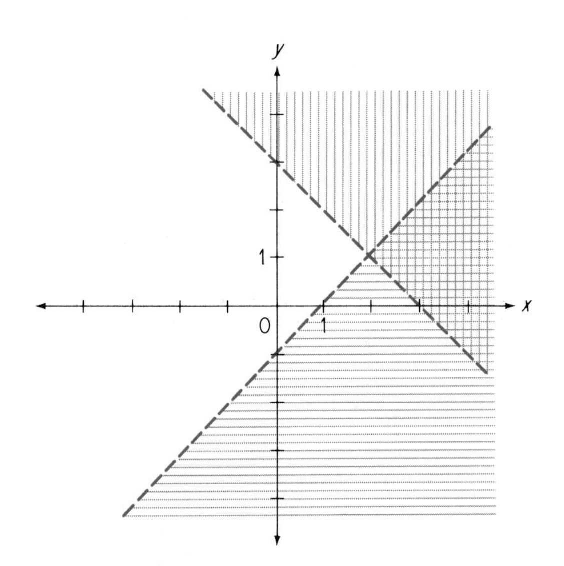

We can use the previous figure to determine the graph of the compound sentence $x + y - 3 > 0$ or $x - y - 1 > 0$, namely, the union of the points of the two shaded regions. The unshaded region (including adjacent parts of the dotted lines) represents the graph of the solution set of the system

$$\begin{cases} x + y - 3 \leq 0 \\ x - y - 1 \leq 0 \end{cases}$$

Example 2 Graph each set of points.

(a) $\begin{cases} x - y + 2 \leq 0 \\ 2x + y - 4 \geq 0 \end{cases}$

(b) $\{(x, y) \mid x - y + 2 \leq 0\} \cup \{(x, y) \mid 2x + y - 4 \geq 0\}$

Solution **(a)** In the figure at the top of the next page the graph of $x - y + 2 \leq 0$ has horizontal shading; the graph of $2x + y - 4 \geq 0$ has vertical shading. The graph of the solution set consists of the points in the region with both horizontal and vertical shading and includes the points of the two rays that serve as the boundary of this region.

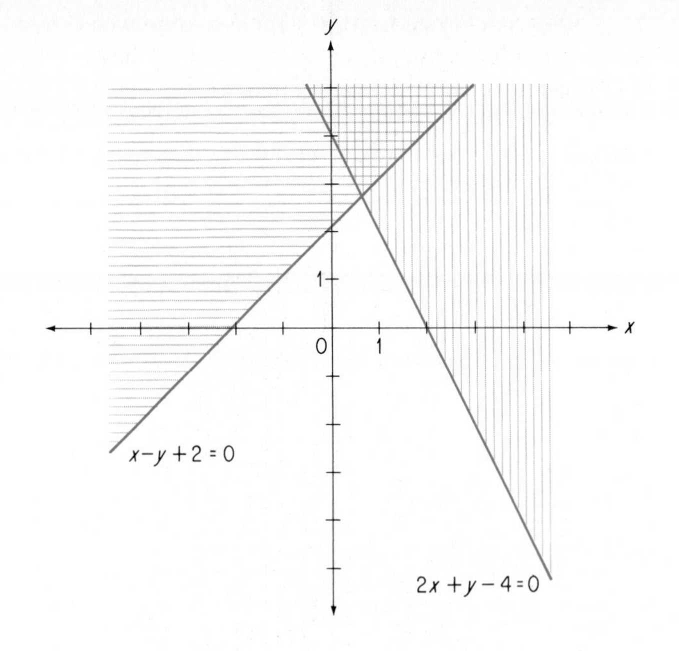

(b) Consider the graph for part (a). The solution set in part (b) consists of all the points in the regions that are shaded in any way, that is, all points except those in the unshaded region. ■

Many word problems that involve two unknown numbers can be solved graphically on a coordinate plane.

Example 3 Use the graph of a system of equations to find two numbers such that the sum of the numbers is 12 and the first number is twice the second number.

Solution Let the numbers be x and y.

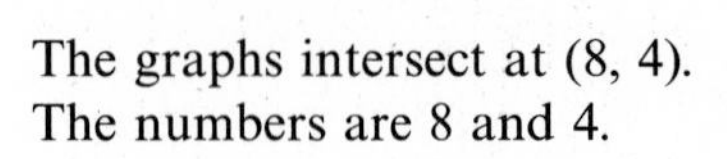

$$\begin{cases} x + y = 12 \\ x = 2y \end{cases}$$

The graphs intersect at (8, 4).
The numbers are 8 and 4.

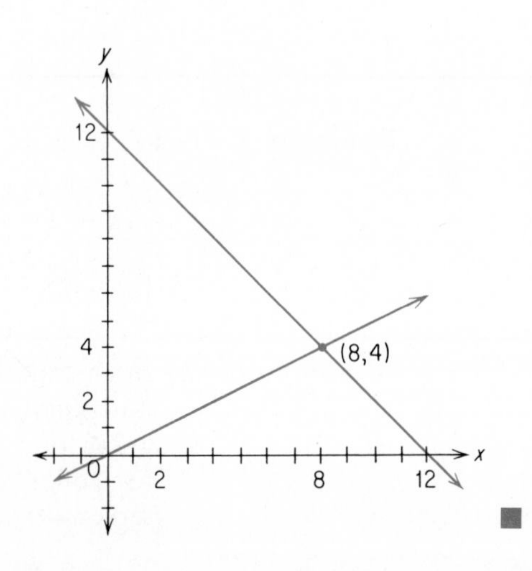

■

Graphs of linear sentences in two variables provide a very important tool called **linear programming**. This technique is used to solve a wide variety of problems in pure and applied mathematics. Although mathematicians often deal with sentences that are not necessarily linear, we shall consider only linear sentences. Thus we assume that the conditions of a problem have been represented by or approximated by linear sentences. Then the solution of the problem depends upon the solution of a system (that is, a set) of linear sentences. The usual method of solution is by graphing; that is, the method is geometric. Accordingly, we first consider two examples of graphs of systems of linear sentences in two real variables.

Example 4 Graph the system

$$\begin{cases} x \geq 0 \\ y \geq 0 \\ y \geq x - 1 \end{cases}$$

Solution The solution set of the given system consists of the ordered pairs of real numbers (x, y) that satisfy all three of the sentences. To graph the system, we graph each one of the three sentences and then take the intersection of their graphs. As in the previous two sections, we graph an inequality by first graphing the equality (using a solid line if its graph is part of the solution set, a dashed line otherwise).

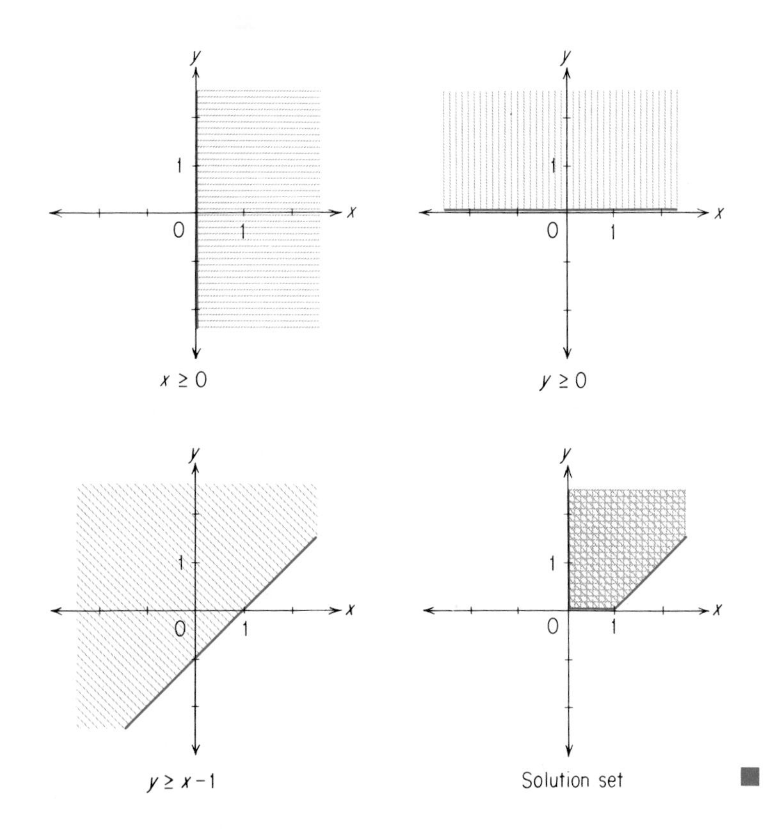

Example 5 Graph the system

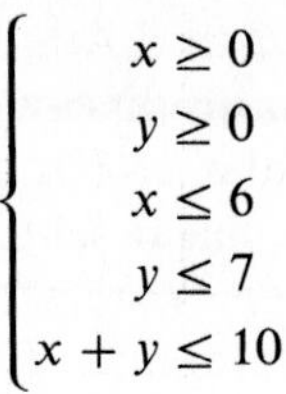

$$\begin{cases} x \geq 0 \\ y \geq 0 \\ x \leq 6 \\ y \leq 7 \\ x + y \leq 10 \end{cases}$$

Solution

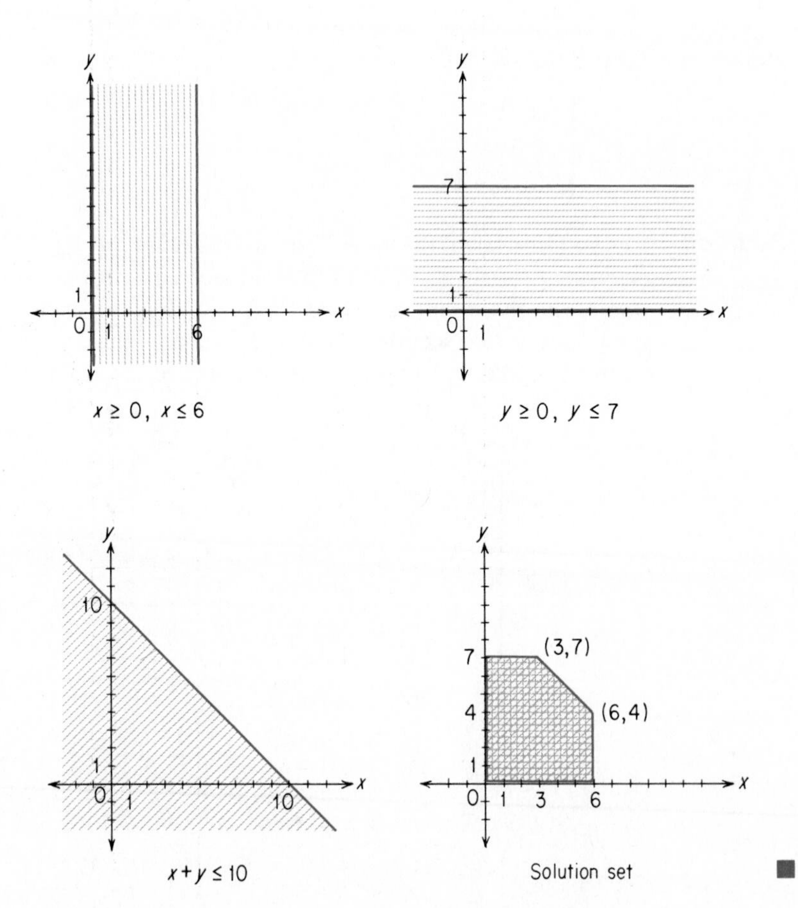

In Example 5 the solution set of the system of sentences is a **polygonal convex set**.

Suppose that we wished to maximize (make as large as possible) the value of the expression $x - y$ subject to the conditions

$$x \geq 0 \qquad y \geq 0 \qquad x \leq 6 \qquad y \leq 7 \qquad x + y \leq 10$$

graphed in Example 5. We could consider the coordinates of the points in the solution set for Example 5 and probably convince ourselves that the possible values of $x - y$ are at most 6 and that the maximum value of 6 for $x - y$ subject to the given conditions occurs for $x = 6$ and $y = 0$. In general, any linear programming problem involves maximizing or minimizing (making as small as possible) the value of a given linear expression, the **objective function**, subject to a set of linear conditions (**constraints**). Thus in a linear programming problem we not only set up our conditions, but we also maximize or minimize an expression for profit, cost, or other quantity. Suppose the conditions for Example 5 represent the manufacture of x metal boxes and y glass jars in a given time, and $x + 2y$ represents the manufacturer's profit. If the manufacturer wishes to have a profit of \$14, we then graph the equation $x + 2y = 14$ on the same coordinate axes as the solution set of Example 5.

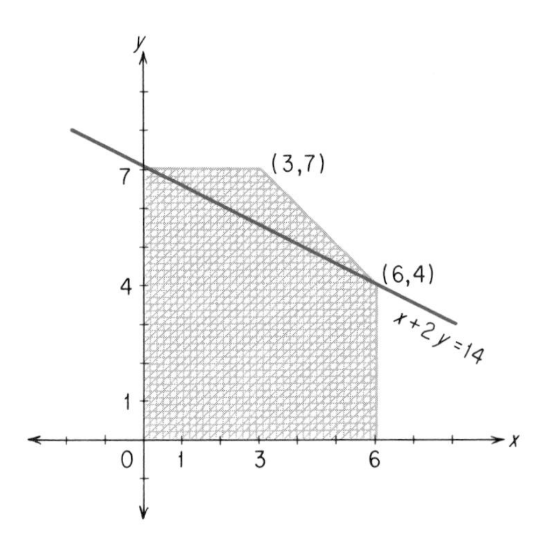

Note that there are now many ways in which the manufacturer can earn a profit of \$14. In particular, consider the points (0, 7) and (6, 4). A profit of \$14 can be earned by manufacturing 0 metal boxes and 7 glass jars, or 6 metal boxes and 4 glass jars. Indeed, any point that is within the polygonal region and also is on the line $x + 2y = 14$ represents an ordered pair (x, y) under the stated restrictions, and such that the profit is \$14.

Next consider the same example, but this time we wish the profit to be k. For each value k the graph of $x + 2y = k$ is a straight line. As k takes on different values, we have a set of parallel lines. When several of these lines are graphed with the solution of the conditions in Example 5, we see that under these conditions k may have any value from 0 to 17, inclusive. The maximum (largest) value that is possible for k under these conditions is 17 and occurs at (3, 7). The minimum (smallest) possible value for k is 0 and occurs at (0, 0). Recall that the conditions are for the manufacture of x metal boxes and y glass jars in a given time and that $x + 2y$ represents the manufacturer's profit. Then if we assume that the manufacturer can sell all the items produced, the most profit would be made by manufacturing 3 boxes and 7 jars per unit of time.

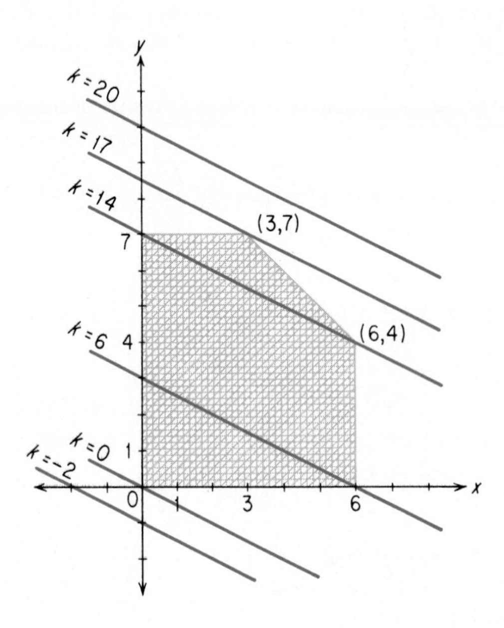

In any linear programming problem the maximum always occurs at a vertex (possibly at two vertices, that is, along a side) of the polygonal region. Also the minimum always occurs at a vertex or along a side of the region. Intuitively, the reason is that the region is convex and thus the lines of a set of parallel lines first intersect the region either by passing through a vertex, as in our example, or by passing along a side of the region. Accordingly, in our example, we could have found the maximum value of $x + 2y$ for points of the region by testing the values corresponding to the vertices (0, 0), (6, 0), (6, 4), (3, 7), and (0, 7) of the region. The corresponding values of $x + 2y$ are 0, 6, 14, 17, and 14, respectively. Thus, as we observed before, the minimum value of $x + 2y$ is 0 and occurs at (0, 0); the maximum value of $x + 2y$ is 17 and occurs at (3, 7).

Example 6 Graph the system

$$\begin{cases} x \geq 0 \\ x \leq 3 \\ y \leq 0 \\ x - y \leq 5 \end{cases}$$

Solution

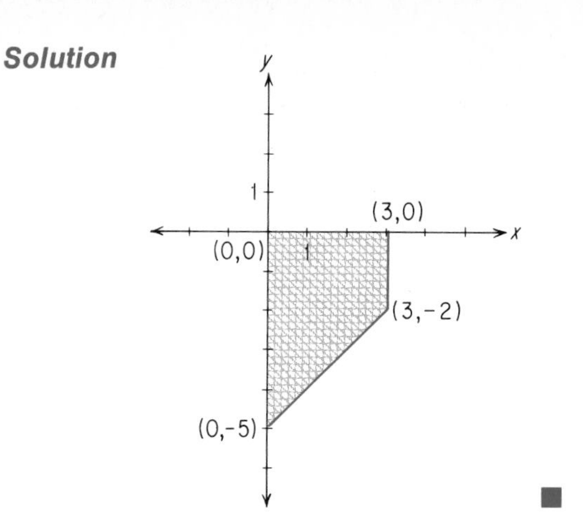

Example 7 Find the minimum and maximum values of the expression $2x + y$ defined over the solution set of the system of Example 6.

Solution Let f represent the expression $2x + y$. Then at $(0, 0)$, $f = 0$; at $(0, -5)$, $f = -5$; at $(3, 0)$, $f = 6$; at $(3, -2)$, $f = 4$. The minimum value, -5, occurs at $(0, -5)$. The maximum value, 6, occurs at $(3, 0)$. ■

Example 8 A manufacturer produces gidgets and gadgets, and has machines in operation 24 hours a day. To produce a gidget requires 2 hours of work on machine A and 6 hours of work on machine B. It takes 6 hours of work on machine A and 2 hours on machine B to produce a gadget. The manufacturer earns a profit of \$5 on each gidget and \$2 on each gadget. How many of each should be produced each day in order to earn the maximum profit possible?

Solution If we let x represent the number of gidgets to be produced, and y the number of gadgets, then the conditions of the problem may be stated and graphed as follows.

$$\begin{cases} x \geq 0 \\ y \geq 0 \\ 2x + 6y \leq 24 \\ 6x + 2y \leq 24 \\ \text{Profit} = P = 5x + 2y \end{cases}$$

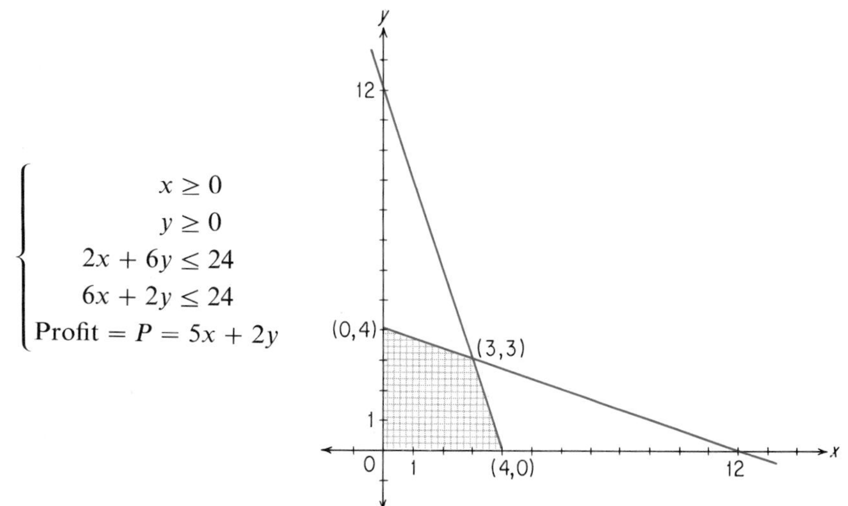

We test the profit expression, $5x + 2y$, at the vertices of the polygonal region. At $(0, 0)$, $P = 0$; at $(0, 4)$, $P = 8$; at $(3, 3)$, $P = 21$; and at $(4, 0)$, $P = 20$. Thus to insure a maximum profit, the manufacturer should produce 3 gidgets and 3 gadgets daily. ■

EXERCISES *Graph.*

1. $\begin{cases} x + y - 6 = 0 \\ x - y + 4 = 0 \end{cases}$
2. $\begin{cases} x + y - 5 = 0 \\ x - y - 3 = 0 \end{cases}$
3. $\begin{cases} x + y = 2 \\ x + 2y = 4 \end{cases}$
4. $\begin{cases} 2x - y = 8 \\ x + 2y = -4 \end{cases}$
5. $\begin{cases} 2x - y - 2 = 0 \\ 4x - 2y + 4 = 0 \end{cases}$
6. $\begin{cases} x + 2y - 4 = 0 \\ 2x + 4y - 8 = 0 \end{cases}$
7. $(x + y - 3)(x + y + 4) = 0$
8. $(x - y + 2)(2x + 3y + 6) = 0$
9. $(x - 2y + 6)(x + 2y - 6) = 0$
10. $(3x + 2y - 6)(x - 2y - 4) = 0$
11. $(x - 3y + 3)(2x + y - 2) = 0$
12. $(2x - 4y + 4)(x - 2y + 2) = 0$

Use the graph of a system of equations to find two numbers that satisfy the given conditions.

13. The sum of the numbers is 19. A difference of the numbers is 7.
14. The sum of the numbers is 21. A difference of the numbers is 13.
15. The sum of the numbers is 25. Twice the larger number is equal to three times the smaller.
16. The sum of the numbers is 23. The larger number is 5 more than twice the smaller.
17. The sum of the larger number and twice the smaller is 12. A difference of the numbers is 3.
18. The sum of the larger number and three times the smaller is 12. The larger is 2 less than four times the smaller.

Graph.

19. $\begin{cases} x - y + 3 > 0 \\ x + y - 3 > 0 \end{cases}$
20. $\begin{cases} 2x + y - 6 < 0 \\ x - 2y + 6 < 0 \end{cases}$
21. $\begin{cases} 3x - 2y - 6 \le 0 \\ 2x + 3y + 6 \ge 0 \end{cases}$
22. $\begin{cases} x + 2y - 6 \le 0 \\ 2x - y + 4 \ge 0 \end{cases}$
23. $\{(x, y) | 3x - y + 6 \le 0\} \cup \{(x, y) | 3x + 4y + 12 \ge 0\}$
24. $\{(x, y) | x + 4y - 8 \ge 0\} \cup \{(x, y) | 4x - 2y + 8 \ge 0\}$
25. $xy \ge 0$
26. $(x - 1)(y + 1) \ge 0$
27. $(x + 2)(y - 3) \le 0$
28. $x(x - y) \le 0$
29. $(2x - 3y + 6)(x + 2y - 4) \le 0$
30. $(x + 2y - 4)(x - y + 2) \le 0$

31. $\begin{cases} x \geq 0 \\ y \geq 0 \\ y \leq 3 \\ x + y \leq 5 \end{cases}$

32. $\begin{cases} x \geq 0 \\ y \geq 0 \\ y \leq x + 2 \\ x + y \leq 6 \end{cases}$

33. $\begin{cases} x \geq 2 \\ x \leq 4 \\ y \geq 0 \\ x + y \leq 5 \end{cases}$

34. $\begin{cases} x \geq 0 \\ y \geq 0 \\ x + 3y \leq 9 \\ 2x + y \leq 8 \end{cases}$

35. $\begin{cases} x \geq 0 \\ y \geq 0 \\ x \leq 4 \\ y \leq 2 \\ x + 2y \leq 6 \\ x + 2y \geq 2 \end{cases}$

36. $\begin{cases} x \geq 0 \\ y \geq 0 \\ x \leq 2 \\ x + y - 5 \leq 0 \\ 2x + y \geq 2 \\ 2x + y \leq 6 \end{cases}$

Find the values of x and y such that under the set of conditions in the specified exercise the given expression has **(a)** *a maximum value* **(b)** *a minimum value.*

37. $x + 2y$, Exercise 33

38. $x + y$, Exercise 34

39. $3x + y$, Exercise 33

40. $x + 5y$, Exercise 34

41. $2x + 3y$, Exercise 35

42. $2x + 3y$, Exercise 36

Use linear programming to solve these hypothetical problems.

43. A college is experimenting with a combination of teaching methods, using both teachers in the classroom and closed-circuit television. The college has facilities using closed circuit television for handling five sections of a class at once. For the five sections of a class that meets 3 clock hours per week, the conditions appear to be as follows: The cost per minute of regular teaching is \$5; the cost per minute of closed-circuit television is \$3. For a certain week at most \$750 can be spent on the instruction of these classes. Assume that the class hour can be spent in part with a teacher and in part with television. When neither the teacher nor the television is on, the students are free to discuss anything they wish. If the value to the students of x minutes of regular teaching and y minutes of closed-circuit television may be expressed as $3x + 2y$, how many minutes of regular teaching and how many minutes of closed-circuit television would be best for the students during the week?

44. Repeat Exercise 43 with the additional condition that the instructor must be present at least 30 minutes each week.

45. Use the conditions as for Exercise 44 and find the number of minutes of regular teaching and of closed-circuit television when the value to the students may be expressed as $x + 2y$.

46. Repeat Exercise 45 when the value to the students may be expressed as $2x + y$.

Find a linear system that has the indicated region as its graph.

*47. The triangular region with vertices (0, 0), (1, 0), and (0, 1).

*48. The rectangular region with vertices (0, 0), (2, 0), (2, 1), and (0, 1).

*49. The region with vertices (1, −2), (5, −2), (5, 1), and (1, 1).

*50. The region with vertices (1, 0), (3, 2), and (−1, 2).

*51. The region with vertices (0, 2), (−2, 0), (0, −2), and (2, 0).

*52. The region with vertices (0, 0), (2, −2), (4, 0), and (2, 2).

EXPLORATIONS

Mark a sheet of paper (or a transparency for use on an overhead projector) with points and axes for an integral coordinate plane.

1. Place three sheets of paper such that each sheet has an edge on at least two points of the integral coordinate plane and only a triangular region is left uncovered. Find a linear system for this triangular region. Repeat this exploration for several triangular regions.

2. Repeat Exploration 1 with four sheets of paper and a region with four sides.

3. Repeat Explorations 1 and 2 for a real coordinate plane such that the coordinates are known for at least two points on the edge of each sheet of paper.

4. In 1980 Mary was 5 years older than her brother John. Mary will be 16 in 1990. Use a coordinate plane with x the number of years after 1980 and y the person's age. Remember that each person gets one year older each year and graph the line that represents Mary's age. Graph the line that represents John's age.
   (a) How old was Mary in 1980?
   (b) How old was John in 1980?
   (c) Find and explain a method for determining when Mary will be twice as old as John.

5. Algebraic solutions of linear systems of two equations in two variables are of three types. For example, the system

$$\begin{cases} x + y = 5 \\ x - y = 1 \end{cases}$$

has a unique solution (3, 2). The system

$$\begin{cases} x + y = 5 \\ 2x + 2y = 6 \end{cases}$$

has no solution, that is, the system is impossible. The system

$$\begin{cases} x + \ \ y = 5 \\ 2x + 2y = 10 \end{cases}$$

has infinitely many solutions, since both equations are satisfied for any value b of x and the value $5 - b$ for y. Graph each of the equations in these three systems and identify the geometric situations associated with the three types of solutions.

Describe an algebraic procedure for identifying the type of solution of any given system.

Hint: Consider the slopes and intercepts of the lines.

$$\begin{cases} ax + by + c = 0 \\ ex + fy + g = 0 \end{cases}$$

Linear programming procedures may be used to explain many arithmetic games and puzzles as well as to solve problems in a wide variety of applications. In the arithmetic games some people consider the answer "obvious;" others find an explanation helpful; a few must "try" all the cases that they can imagine.

6. Think of natural numbers x and y, not necessarily distinct, but less than or equal to 12.
   (a) What is the largest possible value for $x + y$?
   (b) What is the largest possible value for $x - y$?
   (c) What is the largest possible value of y such that $x = 2y$?
   (d) What is the smallest possible value of y such that $x = 2y$?
   (e) What is the largest possible value of x such that $y = 3x$?
   (f) Use linear programming procedures to explain your answers for parts (a) and (b).

7. Repeat parts (a) through (e) of Exploration 6 for nonnegative real numbers less than or equal to 60.

The underlying principles of linear programming are used with 3, 4, 5, or even hundreds of variables.

8. Repeat parts (a) through (e) of Exploration 6 for nonnegative real numbers less than or equal to N.

*9. Dot enjoyed reading mystery stories, especially those of Agatha Christie, Patricia Wentworth, and Mary Stewart. At a certain used book sale of paperbacks, Agatha Christie books sold for 50¢, Wentworth books for 75¢, and Stewart books for $1. Dot had $5 to spend on books, wanted at least one book by each author, and wanted as many books as possible.
   (a) Let C be the number of Christie books bought, W the number of Wentworth books bought, and S the number of Stewart books bought. Find a linear system to express the conditions on Dot's book purchases and identify the expression to be maximized.
   (b) Explain why $S = 1$ for the solution of the problem.
   (c) Find the number of books by each author that Dot should buy to satisfy her expressed conditions.

13-4 Figures on a Coordinate Plane

The application of both algebra and geometry to the study of a single mathematical topic is illustrated by the many uses of coordinates.

Any two points R and S determine a line segment RS and a line RS. The line RS may be taken as a coordinate line.

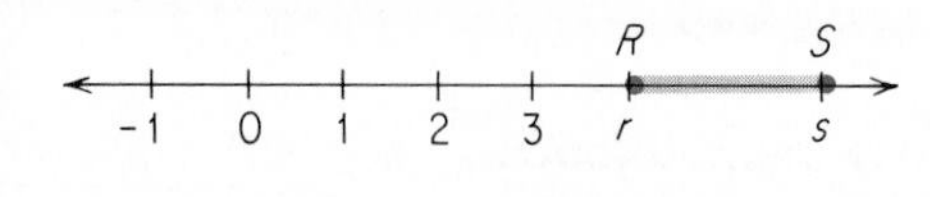

Let r be the coordinate of point R and let s be the coordinate of point S. If $s \geq r$, the length of the line segment RS is $s - r$; if $s < r$, the length is $r - s$. Thus in all cases the length of the line segment RS is $|s - r|$, the *absolute value* of $s - r$. If $R \neq S$, the line segment has positive length. The **directed distance** from R to S is $s - r$; the directed distance from S to R is $r - s$.

The *midpoint* M of any line segment RS on a number line may be found by starting at R with coordinate r and going halfway to S with coordinate s, that is, going $(1/2)(s - r)$ units from R toward S. Thus M has coordinate

$$r + \frac{1}{2}(s - r)$$

Since

$$r + \frac{1}{2}(s - r) = r + \frac{1}{2}s - \frac{1}{2}r = \frac{1}{2}r + \frac{1}{2}s = \frac{1}{2}(r + s)$$

the coordinate of M is $(1/2)(r + s)$. If R has coordinate -1 and S has coordinate 5, then the directed distance from R to S is 6,

$$s - r = 5 - (-1) = 6$$

and the coordinate of the midpoint M is $(1/2)[(-1) + 5]$, that is, 2.

If R has coordinate 7 and S has coordinate 3, then the directed distance from R to S is -4 and the coordinate of the midpoint M is $(1/2)(7 + 3)$, that is, 5.

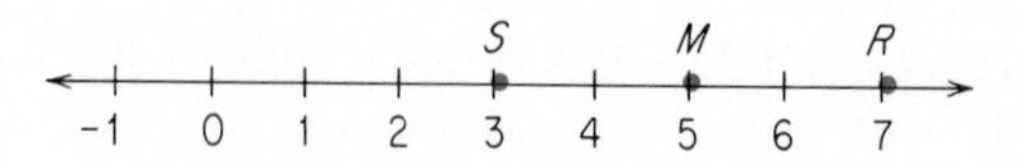

These procedures for finding the length and midpoint of any line segment on a coordinate line may be used on a coordinate plane for any line segments on lines that are parallel to a coordinate axis.

Example 1 Find the length and the midpoint of the line segment AB for $A\colon (1, 5)$ and $B\colon (7, 5)$.

Solution

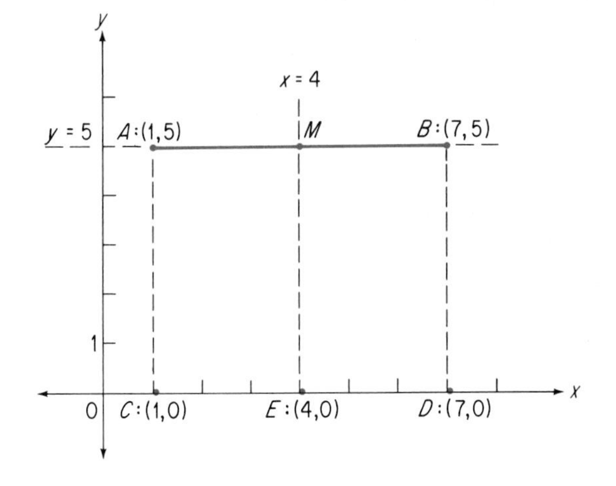

Graph the line segment AB. On the line $y = 5$, the line segment AB has length $7 - 1$, that is, 6. The points A, B, C: (1, 0), and D:(7, 0) are the vertices of a rectangle $BACD$ with $AB = CD = 6$. The y-coordinate of the midpoint M of the line segment AB must be 5, since every point of the line AB has y-coordinate 5. The midpoint E of the line segment CD has x-coordinate $(1 + 7)/2$, that is, 4. Since the three parallel lines AC, $x = 4$, and BD intercept (cut off) congruent line segments (CE and ED) on the x-axis, these parallel lines intercept congruent line segments (AM and MB) on the line $y = 5$. Therefore, M has coordinates (4, 5). ■

Any point on a coordinate plane may be identified by its coordinates. Thus in Example 1 the point M may also be identified as the point (4, 5). In general, any two points A and C that are on a line $y = y_1$ parallel to the x-axis are endpoints of a line segment with midpoint Q.

$$A:\ (x_1, y_1) \qquad C:\ (x_2, y_1) \qquad Q:\ \left(\frac{x_1 + x_2}{2}, y_1\right)$$

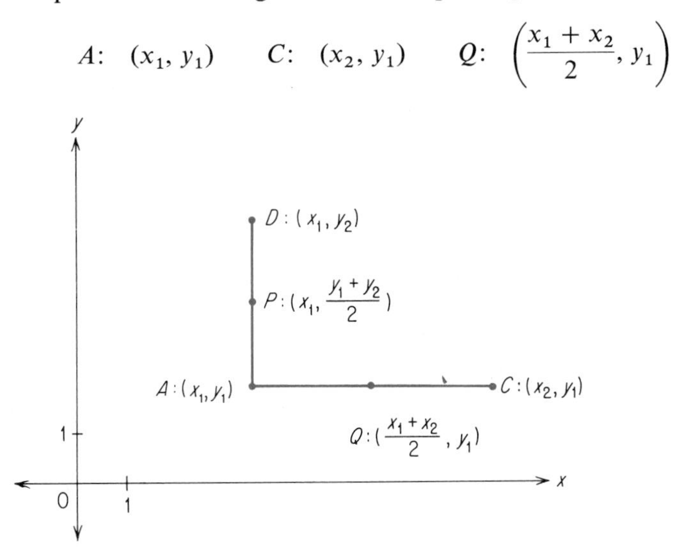

Similarly, any two points A and D that are on a line $x = x_1$ parallel to the y-axis are endpoints of a line segment with midpoint P.

$$A: \ (x_1, y_1) \qquad D: \ (x_1, y_2) \qquad P: \ \left(x_1, \frac{y_1 + y_2}{2}\right)$$

Example 2 Find the coordinates of the midpoints of the line segments with end-points (**a**) (1, 0) and (−3, 0) (**b**) (2, 1) and (2, 7).

Solution (**a**) $\left(\frac{1 + (-3)}{2}, 0\right) = (-1, 0)$ (**b**) $\left(2, \frac{1 + 7}{2}\right) = (2, 4)$ ■

An extension of the procedure in Example 1 may be used to obtain a midpoint formula for any line segment with end-points $A: (x_1, y_1)$ and $B: (x_2, y_2)$ that are not on a line parallel to a coordinate axis. Graph the points A and B. Draw the line AB, the lines $x = x_1$ and $y = y_1$ through A, and the lines $x = x_2$ and $y = y_2$ through B. The lines $x = x_2$ and $y = y_1$ intersect at a point $C: (x_2, y_1)$. On the line $y = y_1$ the line segment AC has midpoint $Q: ((x_1 + x_2)/2, y_1)$. The parallel lines

$$x = x_1 \qquad x = \frac{x_1 + x_2}{2} \qquad x = x_2$$

intercept congruent segments on the line AC and therefore on the line AB, that is, the line $x = (x_1 + x_2)/2$ contains the midpoint M of the line segment AB.

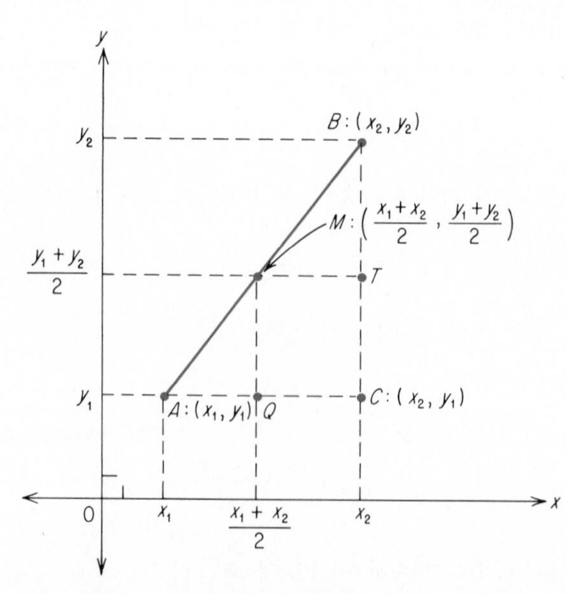

On the line $x = x_2$, the line segment CB has midpoint $T: (x_2, (y_1 + y_2)/2)$. The parallel lines

$$y = y_1 \qquad y = \frac{y_1 + y_2}{2} \qquad y = y_2$$

intercept congruent segments on the line CB and therefore on the line AB, that is, the line $y = (y_1 + y_2)/2$ contains the midpoint M of the line segment AB.

> Any line segment AB with endpoints $A: (x_1, y_1)$ and $B: (x_2, y_2)$ has midpoint $M: (x, y)$ where
>
> $$(x, y) = \left(\frac{x_1 + x_2}{2}, \frac{y_1 + y_2}{2}\right)$$

This is the **midpoint formula** for points on a coordinate plane. The formula holds whether or not the line segment is parallel to a coordinate axis.

Example 3 Find the midpoint M of the line segment AB with the given endpoints.

(a) $A: (6, 1)$ and $B: (-2, 5)$
(b) $A: (-3, 4)$ and $B: (1, -2)$
(c) $A: (5, 3)$ and $B: (-1, 3)$

Solution

(a) $\left(\frac{6 + (-2)}{2}, \frac{1 + 5}{2}\right) = (2, 3)$

(b) $\left(\frac{-3 + 1}{2}, \frac{4 + (-2)}{2}\right) = (-1, 1)$

(c) $\left(\frac{5 + (-1)}{2}, \frac{3 + 3}{2}\right) = (2, 3)$ ■

Example 4 For $A: (2, 4)$ and $M: (3, 2)$, find the coordinates of $B: (x, y)$ such that M is the midpoint of the line segment AB.

Solution

$$\left(\frac{2 + x}{2}, \frac{4 + y}{2}\right) = (3, 2)$$

$$\frac{2 + x}{2} = 3, \qquad x = 4$$

$$\frac{4 + y}{2} = 2, \qquad y = 0$$

The coordinates of B are $(4, 0)$. ■

For any two points $A: (x_1, y_1)$ and $B: (x_2, y_2)$ on a coordinate plane, there are the following three possibilities for the line segment AB.

The absolute value notation, $|y_2 - y_1|$, and $|x_2 - x_1|$, is used to give the lengths as positive numbers.

If $x_1 = x_2$, the line segment is parallel to the y-axis and has length $|y_2 - y_1|$.

If $y_1 = y_2$, the line segment is parallel to the x-axis and has length $|x_2 - x_1|$.

If $x_1 \neq x_2$ and $y_1 \neq y_2$, the line segment is not parallel to either axis, there is a point $C: (x_2, y_1)$ such that triangle ABC is a right triangle, and by the Pythagorean theorem

Graphical treatments of data are very common, used for many purposes, and generated by many types of machines. For example, electrocardiographs are used to record the action of a human heart. The following graph shows a typical electrocardiogram taken from a healthy person.

$$(AB)^2 = (AC)^2 + (BC)^2$$

$$(AB)^2 = (x_2 - x_1)^2 + (y_2 - y_1)^2$$

$$\boxed{AB = \sqrt{(x_2 - x_1)^2 + (y_2 - y_1)^2}}$$

This is the **distance formula** on a plane. The formula holds whether or not the line segment is parallel to a coordinate axis.

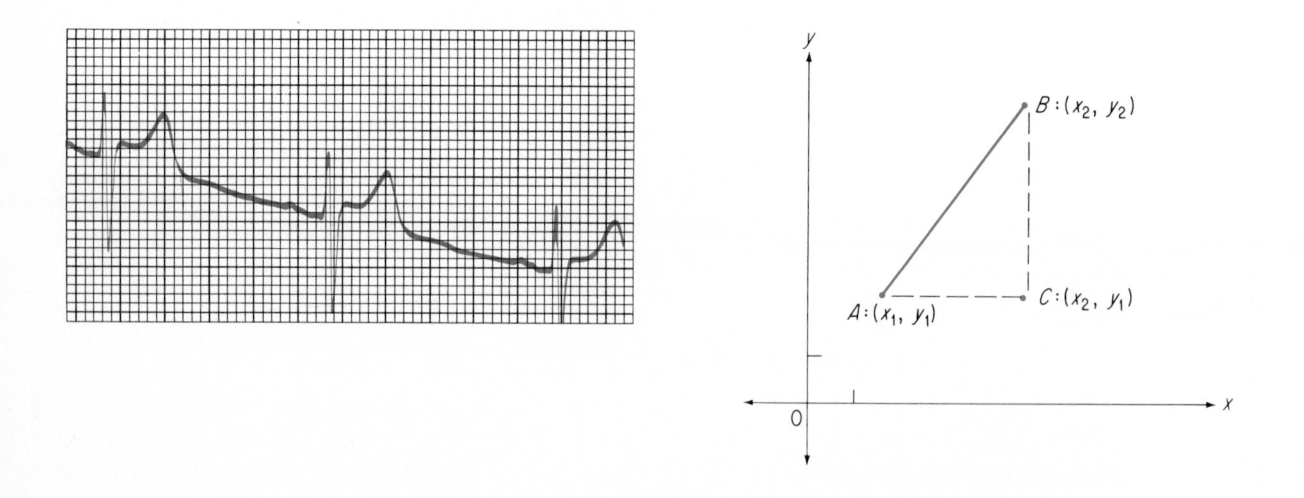

Example 5 Find the length of the line segment with the given endpoints.

(a) $A: (2, -3)$ and $B: (5, 1)$

(b) $D: (5, 3)$ and $E: (-1, 3)$

(c) $F: (4, 7)$ and $G: (4, -2)$

Solution

(a) $AB = \sqrt{(5 - 2)^2 + [1 - (-3)]^2} = \sqrt{3^2 + 4^2} = \sqrt{25} = 5$

(b) $DE = \sqrt{[(-1) - 5]^2 + (3 - 3)^2} = \sqrt{(-6)^2 + 0^2} = \sqrt{36} = 6$

(c) $FG = \sqrt{(4 - 4)^2 + [(-2) - 7]^2} = \sqrt{0^2 - (-9)^2} = \sqrt{81} = 9$ ■

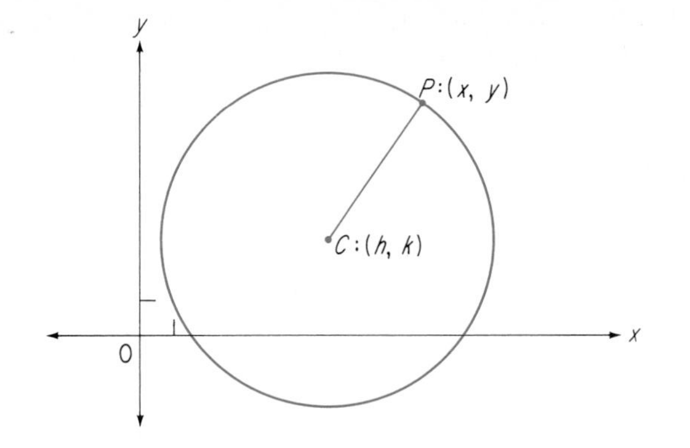

The distance formula may be used to obtain an equation for any circle on a coordinate plane. The circle with center C: (h, k) and radius r is the set of points P: (x, y) at a distance r from the center C.

$$\sqrt{(x-h)^2 + (y-k)^2} = r$$
$$(x-h)^2 + (y-k)^2 = r^2$$

Example 6 Describe the set of points (x, y) such that
(a) $x^2 + y^2 = 4$
(b) $(x-1)^2 + (y-3)^2 = 1$
(c) $(x+2)^2 + (y+1)^2 = 9$.

Solution **(a)** Since $x = x - 0$ and $y = y - 0$, the given equation can be expressed as

$$(x-0)^2 + (y-0)^2 = 2^2$$

Each point of the set is 2 units from the origin. Thus the set of points is a circle with center (0, 0) and radius 2.

(b) The set of points is a circle with center (1, 3) and radius 1.

(c) Since $x + 2 = x - (-2)$ and $y + 1 = y - (-1)$, the given equation can be expressed as

$$[x-(-2)]^2 + [y-(-1)]^2 = 3^2$$

and the set of points is a circle with center $(-2, -1)$ and with radius 3. ■

Example 7 Give an algebraic representation for the points that are on a coordinate plane and are at most 2 units from the point $(3, -5)$.

Solution The points form a circular region (the points of a circle and the interior points of the circle) with center $(3, -5)$ and radius 2. Thus we have

$$(x-3)^2 + (y+5)^2 \leq 4$$ ■

EXERCISES *Find the length and the midpoint of the line segment with the given endpoints.*

1. $(0, 3)$ and $(4, 3)$
2. $(2, 1)$ and $(5, 1)$
3. $(2, 1)$ and $(2, 5)$
4. $(5, -2)$ and $(5, 7)$
5. $(-3, 2)$ and $(-3, 7)$
6. $(-3, 2)$ and $(-7, 2)$

Find the midpoint of the line segment with the given endpoints.

7. $(0, 0)$ and $(6, -4)$
8. $(-4, -2)$ and $(0, 0)$
9. $(-4, -2)$ and $(-6, 8)$
10. $(-6, 4)$ and $(4, 2)$
11. $(5, 3)$ and $(-1, 7)$
12. $(-3, -7)$ and $(5, 7)$

Find the coordinates of the indicated point.

13. Endpoint B of line segment AB with $A: (1, 3)$ and midpoint $(2, 5)$.
14. Endpoint B of line segment AB with $A: (3, 0)$ and midpoint $(-1, 4)$.
15. Endpoint A of line segment AB with $B: (-5, 6)$ and midpoint $(-3, -1)$.
16. Vertex C of the square with $A: (0, 0)$, $B: (a, 0)$, and $D: (0, a)$.
17. Vertex S of rectangle $QRST$ with $Q: (0, 0)$, $R: (a, 0)$, and $T: (0, b)$.
18. Vertex C of quadrilateral $ABCD$ with $A: (0, 0)$, $B: (7p, 0)$, and $D: (2p, q)$ with positive numbers p and q, for $ABCD$
    (a) a parallelogram
    (b) an isosceles trapezoid that is not a parallelogram.
19. The vertex G of a quadrilateral $EFGH$ with $E: (0, 0)$, $F: (a, b)$, and $H: (0, 5b)$ with positive numbers a and b, for $EFGH$
    (a) a parallelogram
    (b) an isosceles trapezoid that is not a parallelogram
20. The vertex S of an isosceles trapezoid $QRST$ that is not a parallelogram, with $Q: (0, 0)$, $R: (a, 0)$, and $T: (b, c)$, where $0 < 2b < a$.

Find the length of the line segment with the given endpoints.

21. $A: (1, 2)$ and $B: (4, 6)$
22. $C: (-1, 6)$ and $D: (4, -6)$
23. $E: (2, 16)$ and $F: (9, -8)$
24. $G: (-2, 5)$ and $H: (7, 11)$
25. $I: (-1, -3)$ and $J: (2, 0)$
26. $L: (4, -5)$ and $M: (-7, -11)$

Describe the specified set of points (x, y).

27. $x^2 + y^2 = 16$
28. $(x + 3)^2 + (y - 2)^2 = 25$
29. $x^2 + y^2 < 36$
30. $(x - 5)^2 + (y + 1)^2 < 49$
31. $(x - 1)^2 + (y - 2)^2 \leq 4$
32. $(x + 1)^2 + (y + 2)^2 \leq 25$

Give an algebraic representation for the specified set of points on a coordinate plane.

33. The points of the circle with center (2, 5) and radius 3.
34. The points of the circle with center $(-3, 4)$ and radius 2.
35. The points of the circular region with center $(-3, -1)$ and radius 4.
36. The interior points of the circle with center $(6, -4)$ and radius 5.
37. The exterior points of the circle with center $(-4, 3)$ and radius 4.
38. The exterior points of the circle with center $(-3, -2)$ and radius 6.
39. The points at most 5 units from the point $(-1, 2)$.
40. The points at least 2 units and at most 5 units from the origin.
41. The points at least 3 units and at most 4 units from the point $(1, -2)$.
42. The points
    (**a**) 5 units from the y-axis
    (**b**) 2 units from the x-axis.
43. The points
    (**a**) 3 units from the line $x = 2$
    (**b**) 4 units from the line $y = -1$.
44. The points at most 2 units from the y-axis and at most 3 units from the origin.

Use properties of a coordinate plane and prove the given statement.

45. **If** A: (0, 0), B: $(a, 0)$, and C: (b, c) are the vertices of a triangle, then the midpoints of the sides AC and BC are endpoints of a line segment that is parallel to, and half as long as, the side AB.
46. **If** P: (0, 0), Q: $(a, 0)$, R: (a, b), and S: (c, b) are the vertices of a trapezoid $PQRS$ where $0 < c < a$ and $0 < b$, then the midpoints of the sides PS and QR are the endpoints of a line segment that is parallel to the bases and has length equal to one-half the sum of the lengths of the bases.

*47. The diagonals of a parallelogram bisect each other.

*48. The midpoints of the sides of any quadrilateral $ABCD$ are vertices of a quadrilateral $PQRS$ such that the diagonals of $PQRS$ bisect each other.

EXPLORATIONS

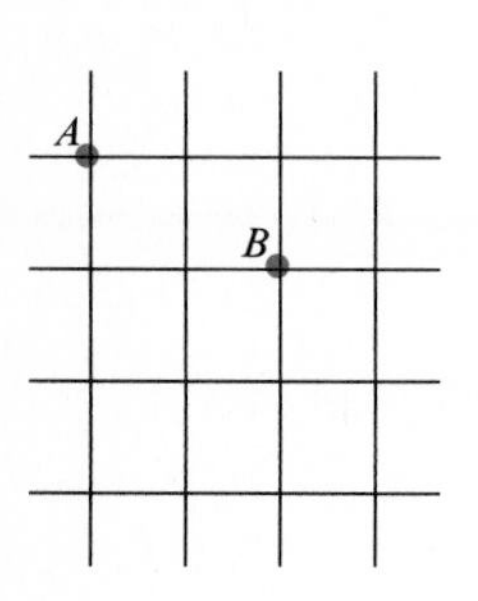

The taxicab distance from a street corner A to a street corner B is the same as the "straight line" distance only if A and B are on the same street. For a hypothetical modern city with streets one unit apart and intersecting at right angles, such *taxicab distances* may be represented on an *integral coordinate plane*, that is, on the *lattice* of lines with integral coordinates on any ordinary coordinate plane. Then the distance between any two locations $A: (x_1, y_1)$ and $B: (x_2, y_2)$ is given by the expression

$$|x_2 - x_1| + |y_2 - y_1|$$

1. Note that in this *taxicab geometry* you must stay on the streets; cutting "cross lots" is not allowed. Prepare for a middle school class an introduction of the taxi-distance between
   (a) any two given street corners
   (b) any two given points on streets.
2. Make a drawing of the city considered in Exploration 1. Select, and enclose in a small circle, a particular street corner as the location of a school. Label with A each of the street corners that are at taxi-distance 1 from the school. Label with B, C, and D each of the street corners that are at taxi-distances 2, 3, and 4, respectively, from the school. Draw the quadrilateral determined by the points labeled B. Draw the quadrilaterals for the points labeled A, C, and D. What is the shape of these "taxi-circles" of points at a taxi-distance n from the school? Express the number of street corners on each taxi-circle as a function of n. Express the area $\mathscr{A}$ of each taxi-circle as a function of n.
3. Select a point on a street but in the middle of a block as the location of the school. For $n = 1, 2, 3$, and 4, label points on streets as in Exploration 2, draw the taxi-circles; and for n greater than 1, express the number of labeled points on each taxi-circle as a function of n.
4. Suppose that Mary and Doris live at street corners with coordinates $(-3, 3)$ and $(3, -3)$ in the city considered in Exploration 1. Make a drawing of the city and mark at least ten street corners that are at equal taxi-distances from the homes of the two girls.
5. Prepare a report on taxicab geometry. For example, see "Taxicab geometry offers a free ride to a non-Euclidean locale" in Martin Gardner's "Mathematical Games" starting on page 18 of the November 1980 issue of *Scientific American.*

Describe and, if necessary, graph and label on an ordinary coordinate plane, the set of all possible endpoints of line segments that have length two units and have one endpoint on the given figure.

6. The origin.
7. The x-axis.
8. The graph of $xy = 0$.

9. The graph of $(x - y)(x + y) = 0$.
10. The lines $|x|$ equal to
    (a) 1 (b) 2 (c) 3
11. The circle with center at the origin and radius
    (a) 1 (b) 2 (c) 3
12. The square with vertices at $A: (0, 0)$, $B: (4, 0)$, $C: (4, 4)$, and $D: (0, 4)$.
13. The rectangle with vertices at $P: (0, 0)$, $Q: (10, 0)$, $R: (10, 20)$, and $S: (0, 20)$.

13-5 Figures in a Coordinate Space

We may think of the xy-plane in space with a line OZ perpendicular to the xy-plane at the origin. It is customary to select the coordinate axes OX, OY, and OZ as in the figure, where the tick marks indicate the unit points on each axis.

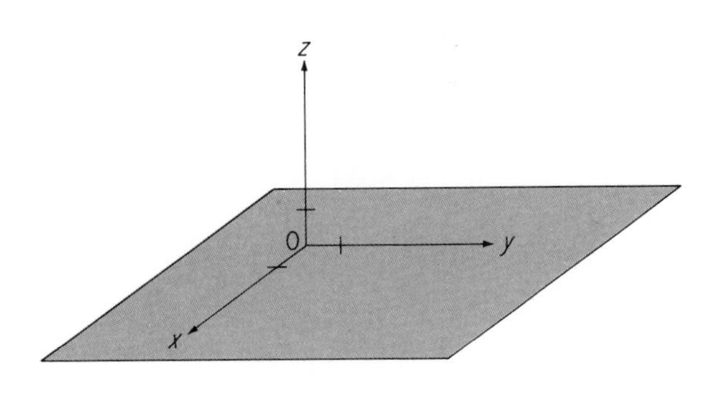

In space the set of points 1 unit from the xy-plane consists of two planes; that is, the plane $z = 1$ one unit above the xy-plane and the plane $z = -1$ one unit below the xy-plane. Any plane parallel to the xy-plane has an equation of the form

$$z = t$$

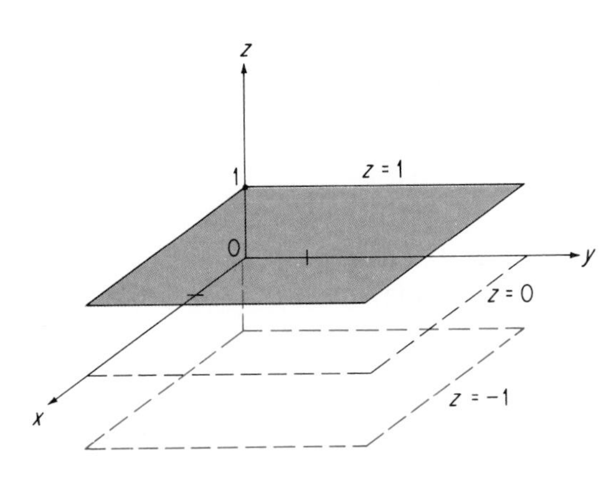

Similarly, any plane parallel to the xz-plane has an equation of the form

$$y = n$$

Any plane parallel to the yz-plane has an equation of the form

$$x = k$$

The location of any point (k, n, t) may be determined as the intersection of the three planes $x = k$, $y = n$, and $z = t$. By convention the x-coordinate is always listed first, the y-coordinate second, and the z-coordinate third in specifying the coordinates of a point in space as an **ordered triple** of real numbers (k, n, t).

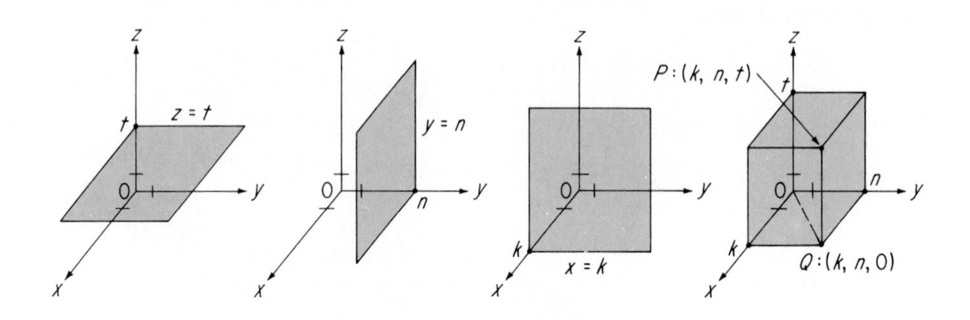

Each point in space has a unique ordered triple of real numbers as its *coordinates* with respect to a given set of coordinate axes; each ordered triple of real numbers has a unique point in space as its *graph*. When k, n, and t are each different from zero, we frequently can visualize the position of the point $P: (k, n, t)$ after sketching a rectangular box with a vertex at the origin, three edges on the coordinate axes, and a vertex at P.

Example 1 Graph and label each figure.

(a) The point $A: (1, 0, 0)$ and the plane $x = 1$.

(b) The point $B: (0, 2, 0)$, the plane $y = 2$, the point $C: (1, 2, 0)$, and the line t with equations

$$\begin{cases} x = 1 \\ y = 2 \end{cases}$$

(c) The point $D: (0, 0, 3)$, the plane $z = 3$, and the point $E: (1, 2, 3)$ as the graph of the system of equations

$$\begin{cases} x = 1 \\ y = 2 \\ z = 3 \end{cases}$$

Solution

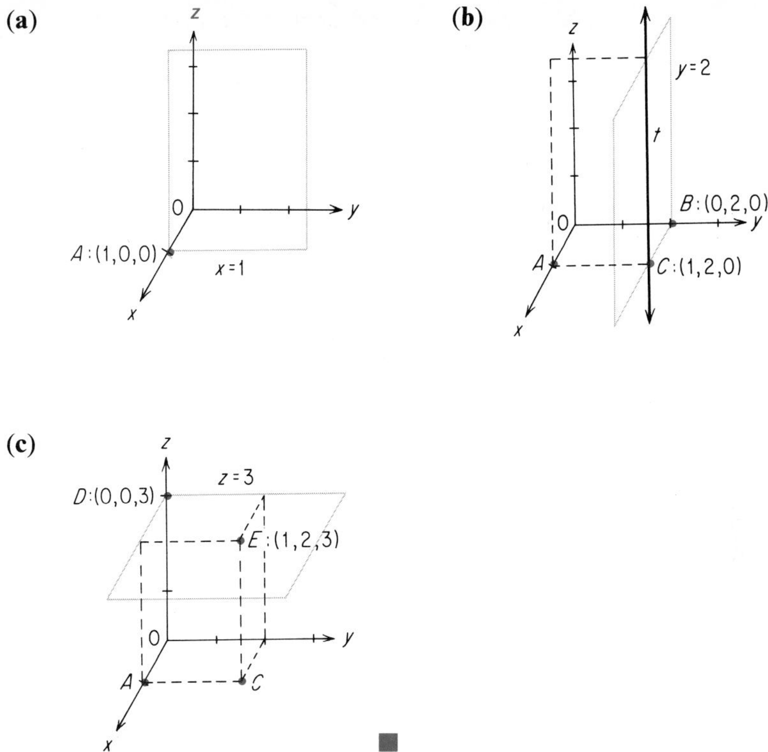

■

Any equation of the form

$$ax + by + cz + d = 0$$

where the coefficients a, b, and c are not all zero, may be considered as a *linear equation in three variables* and has a plane as its graph in a three-dimensional coordinate space. Conversely, any plane in three-dimensional coordinate space is the graph of a linear equation in three variables. Consider the linear equation

$$3x + 5y + 6z - 30 = 0$$

If $z = 0$, the graph intersects the xy-plane in the line (indicated in color in the next figure)

$$\begin{cases} 3x + 5y = 0 \\ z = 0 \end{cases}$$

As in Section 13-2 we may divide both members of the equation by 30 to find the *intercept equation.*

$$\frac{x}{10} + \frac{y}{6} + \frac{z}{5} = 1$$

Then the x-intercept is 10, the y-intercept is 6, and the z-intercept is 5. The plane may be graphed using the points (10, 0, 0), (0, 6, 0), and (0, 0, 5). Any fourth point such as (5, 3, 0) may be used as a check.

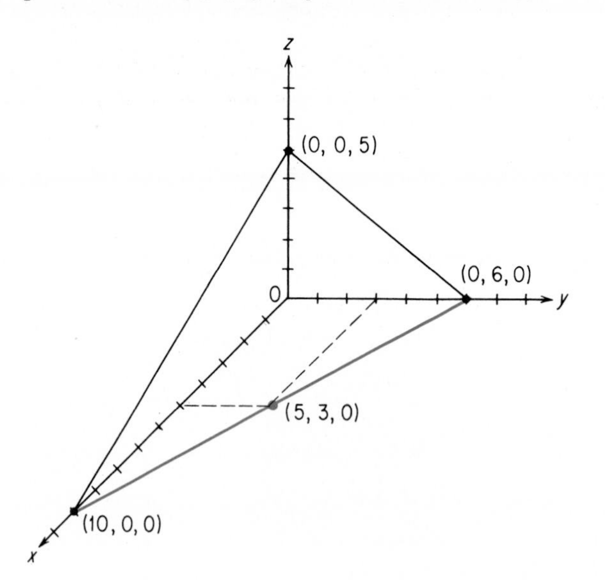

A point P: (k, n, t) in coordinate space is on a coordinate plane if one of its coordinates is zero. If all of its coordinates are different from zero, we consider also the points Q: $(k, n, 0)$ and R: $(k, 0, 0)$. Then on the xy-plane, there is a right triangle ORQ; by the Pythagorean theorem

$$OQ = \sqrt{k^2 + n^2}$$

Then since the points O, Q, and P are also the vertices of a right triangle,

$$OP = \sqrt{(OQ)^2 + t^2} = \sqrt{k^2 + n^2 + t^2}$$

This formula holds whether or not P is on a coordinate plane.

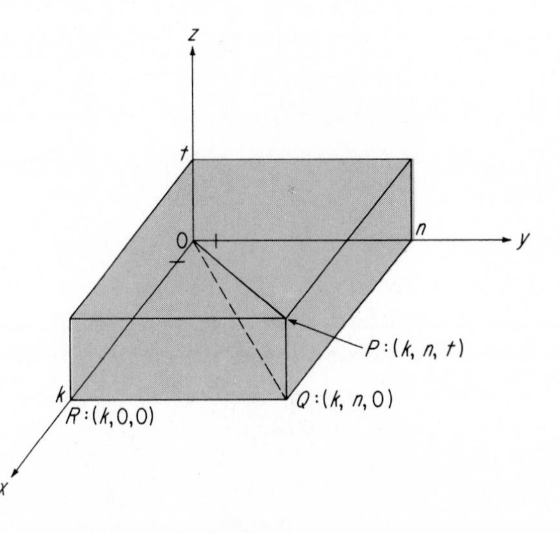

Example 2 Find OP for $P: (1, 5, 7)$.

Solution $OP = \sqrt{1^2 + 5^2 + 7^2} = \sqrt{75} = 5\sqrt{3}$ ■

Let $P_1: (x_1, y_1, z_1)$ and $P_2: (x_2, y_2, z_2)$ be any two points that are not on a plane that is parallel to a coordinate plane. Then, as in the figure, we can identify

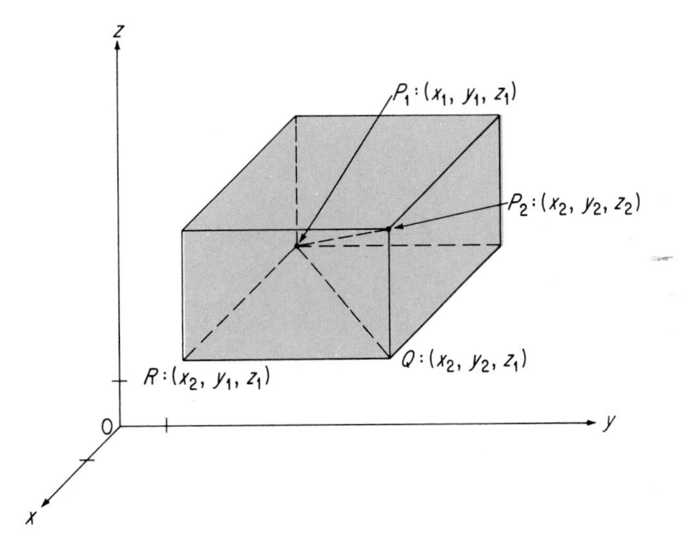

the vertices of a rectangular box with P_1 and P_2 as vertices and with its edges parallel to the coordinate axes. For $Q: (x_2, y_2, z_1)$ and $R: (x_2, y_1, z_1)$ we have a right triangle P_1RQ on the plane $z = z_1$. Then

$$P_1Q = \sqrt{(x_2 - x_1)^2 + (y_2 - y_1)^2}$$

and from the right triangle P_1QP_2 with $P_1P_2 = \sqrt{(P_1Q)^2 + (P_2Q)^2}$

$$\boxed{P_1P_2 = \sqrt{(x_2 - x_1)^2 + (y_2 - y_1)^2 + (z_2 - z_1)^2}}$$

This **distance formula** holds for any points P_1 and P_2 in space.

Example 3 Find AB for
(a) $A: (2, -3, 5)$ and $B: (7, 1, 11)$
(b) $A: (5, 4, 3)$ and $B: (5, 4, -2)$.

Solution **(a)** $AB = \sqrt{(7 - 2)^2 + [1 - (-3)]^2 + (11 - 5)^2}$
$= \sqrt{5^2 + 4^2 + 6^2} = \sqrt{25 + 16 + 36} = \sqrt{77}$

(b) $AB = \sqrt{(5 - 5)^2 + (4 - 4)^2 + [3 - (-2)]^2}$
$= \sqrt{0^2 + 0^2 + 5^2} = \sqrt{25} = 5$ ■

The distance PC of any point $P: (x, y, z)$ from a given point $C: (a, b, c)$ is given by the formula

$$PC = \sqrt{(x-a)^2 + (y-b)^2 + (z-c)^2}$$

The point P is

a *point of the sphere* with center C and radius r if $PC = r$,
an *interior point of the sphere* if $PC < r$,
an *exterior point of the sphere* if $PC > r$.

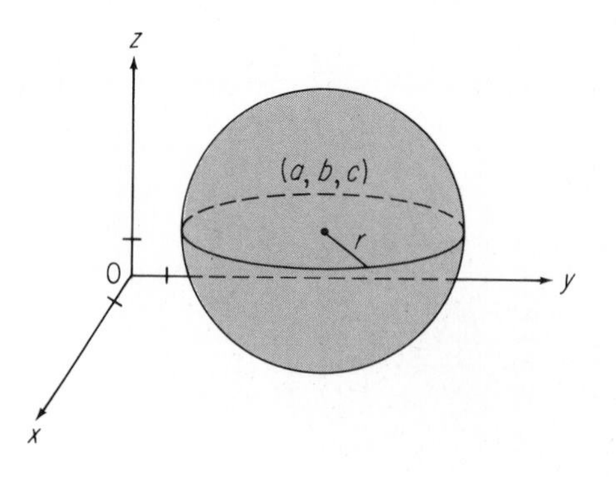

The sphere with center (a, b, c) and radius r has equation $$(x-a)^2 + (y-b)^2 + (z-c)^2 = r^2$$

Example 4 Find an equation for the sphere with
(a) center (0, 0, 0) and radius 3
(b) center (5, −7, 11) and radius 4.

Solution **(a)** From the formula, $(x-0)^2 + (y-0)^2 + (z-0)^2 = 9$, that is, $x^2 + y^2 + z^2 = 9$.
(b) $(x-5)^2 + (y+7)^2 + (z-11)^2 = 16$ ■

Example 5 Describe the set of points at most 2 units from the point (1, −2, 3) by an algebraic sentence.

Solution $(x-1)^2 + (y+2)^2 + (z-3)^2 \leq 4$ ■

EXERCISES *Sketch the given plane.*

1. $z = 3$ **2.** $x = 5$ **3.** $y = 1$
4. $y = -1$ **5.** $z = -1$ **6.** $z = -2$

Graph the given point as a vertex of a rectangular box with one vertex at the origin and three edges on the coordinate axes.

7. (3, 2, 1) **8.** (2, 1, 3) **9.** (1, 2, 2)
10. (3, −1, 2) **11.** (1, −1, 3) **12.** (1, 2, −1)

Describe the given set of points in space.

13. $x = 0$
14. $y = 0$
15. $z = 0$
16. $x = a$
17. $y = b$
18. $z = c$
19. $\begin{cases} x = 0 \\ y = 0 \end{cases}$
20. $\begin{cases} y = 0 \\ z = 0 \end{cases}$
21. $\begin{cases} x = 0 \\ z = 0 \end{cases}$
22. $\begin{cases} x = 1 \\ z = 0 \end{cases}$
23. $\begin{cases} x = 2 \\ y = 3 \end{cases}$
24. $\begin{cases} y = 2 \\ z = 1 \end{cases}$

Consider the rectangular box shown in the figure with $AB = 2$, $AE = 4$, and $AD = 3$. Then give the coordinates of each vertex under the specified conditions.

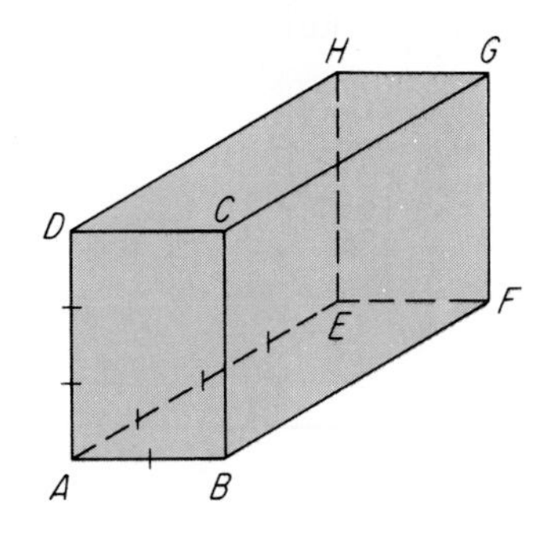

25. The origin is at E with ray EA as OX, ray EF as OY, and ray EH as OZ.
26. The origin is at A with the ray opposite AE as OX, ray AB as OY, and ray AD as OZ.
27. The origin is at F with ray FB as OX, the ray opposite FE as OY, and ray FG as OZ.
28. The origin is at H with ray HD as OX, ray HG as OY, and the ray opposite HE as OZ.

29–32. Repeat Exercises 25 through 28 for the next box with $AB = 5$, $AE = 7$, and $AD = 2$.

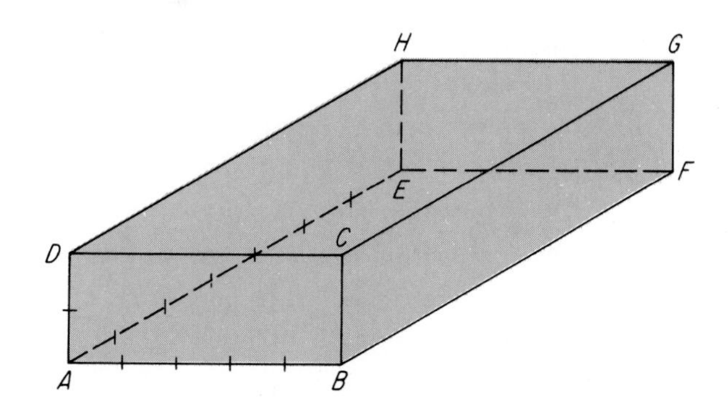

Find the length of the line segment AB with the given endpoints.

33. $A: (1, 0, 2)$ and $B: (1, 4, 5)$

34. $A: (2, 7, -3)$ and $B: (1, 5, 0)$

35. $A: (2, -1, 0)$ and $B: (3, 1, 2)$

36. $A: (1, 5, -6)$ and $B: (-2, 3, -1)$

37. $A: (-1, 5, 2)$ and $B: (1, -5, 7)$

38. $A: (4, -3, 5)$ and $B: (6, 2, 1)$

Find an equation for the sphere with the given center and radius.

39. Center $A: (1, -2, 5)$ and radius 2.

40. Center $B: (2, 3, -4)$ and radius 5.

41. Center $C: (-1, -2, 3)$ and radius 4.

42. Center $D: (5, -7, 11)$ and radius 6.

43. Center $E: (0, -2, 4)$ and radius 7.

44. Center $F: (-1, -2, 3)$ and radius 10.

Describe the given set of points.

45. $x^2 + y^2 + z^2 = 5$

46. $x^2 + y^2 + z^2 < 4$

47. $x^2 + y^2 + z^2 \leq 9$

48. $x^2 + y^2 + z^2 > 16$

49. $(x - 1)^2 + (y - 2)^2 + (z - 3)^2 = 4$

50. $(x - 1)^2 + (y + 1)^2 + (z + 2)^2 = 9$

51. $(x + 3)^2 + (y - 1)^2 + (z + 1)^2 < 1$

52. $(x + 2)^2 + (y - 3)^2 + (z + 5)^2 < 25$

53. $(x - 5)^2 + (y + 7)^2 + (z + 2)^2 \leq 16$

54. $(x + 4)^2 + (y - 5)^2 + (z + 2)^2 \leq 8$

55. $(x + 1)^2 + (y - 3)^2 + (z + 4)^2 > 10$

56. $(x + 3)^2 + (y + 1)^2 + (z - 5)^2 > 12$

Represent the indicated set of points in space by an equation, an inequality, or a system of such sentences.

57. The points 3 units above the xy-plane.

58. The points above the xy-plane.

59. The points at least 2 units above the xy-plane.

60. The points at least 1 and at most 3 units above the xy-plane.

61. The points 3 units from the origin.

62. The points at most 3 units from the origin.

63. The points at least 5 units from the origin.

64. The points at least 1 unit from $(1, 2, -3)$.

65. The points at most 2 units from $(2, 3, 4)$.

66. The points at least 1 unit and at most 2 units from the origin.

67. The points at least 2 units and at most 3 units from the point $(2, -1, 3)$.

68. The points at least 2 units and at most 5 units from the point $(1, 6, 4)$.

69. The points at least 5 units and at most 7 units from the point $(2, -3, 5)$.

***70.** The points at most 2 units from the yz-plane.

***71.** The points at most 3 units from the xz-plane.

EXPLORATIONS

1. Prepare a 10-minute lesson for introducing coordinates in space in a rectangular classroom and making use of the edges and a corner of the room.
2. Make a model of a $6 \times 6 \times 6$ integral coordinate space. Prepare a 10-minute lesson for introducing coordinates in space using this model.
3. Use coordinates in space as a reference and develop a problem set for helping students sketch three-dimensional figures.

Chapter Review

1. Use the replacement set $\mathscr{U} = \{-3, -2, -1, 0, 1, 2, 3\}$ and find the solution set for $y < x - 2$.

In Exercises 2 and 3 graph the given relation, state the domain and the range, and identify the relations that are functions.

2. $x + 2y \geq 6$
3. $y = |x| + 1$

4. Find an equation for the line that has slope 3 and contains $(1, -2)$.
5. Solve the given sentence for y.
   (a) $4x + 2y < 6$ **(b)** $2x - y \geq 4$

In Exercises 6 through 8 graph the given sentence or system.

6. $2x - 3y \leq 6$
7. $\begin{cases} x + y - 5 = 0 \\ x - y + 1 = 0 \end{cases}$
8. $\begin{cases} x + y - 3 \geq 0 \\ x - y + 2 \leq 0 \end{cases}$

9. Use the graph of a system of equations to find two numbers such that the first number is three times the second and the sum of the first number and twice the second number is 10.
10. Graph the given system and, under that set of conditions, find the values of x and y such that the expression $y - x$ has a maximum.

$$\begin{cases} x \geq 0 \\ y \geq 0 \\ x \leq 3 \\ y \geq 1 \\ y \leq 10 - 2x \end{cases}$$

11. Find **(a)** the length **(b)** the midpoint of the line segment with endpoints $(2, -5)$ and $(8, 7)$.
12. Find the vertex C of the rectangle $ABCD$ with $A\colon(-1, 0)$, $B\colon(3, 0)$, and $D\colon(-1, 2)$.
13. Find an algebraic representation for the circle with center $(5, -2)$ and radius 3.

14. Graph on a coordinate plane, or describe in words, the specified set of points.
(a) $4 < x^2 + y^2 < 16$ (b) $0 \leq (x-1)^2 + (y+2)^2 \leq 9$

In Exercises 15 *and* 16 *give an algebraic representation for the specified set of points on a coordinate plane.*

15. The points
(a) 3 units from the y-axis (b) 2 units from the line $x = 2$.
16. The points at least 1 unit and at most 2 units from the point (2, 3).

17. Graph the point (3, 1, 2).

Describe the given set of points in space.

18. $y = 2$

19. $\begin{cases} x = 2 \\ y = 3 \end{cases}$

20. $x^2 + y^2 + z^2 \leq 1$

21. $(x-1)^2 + (y+2)^2 + z^2 = 9$

Describe the indicated set of points in space by an algebraic sentence.

22. The y-axis
23. The points below the xy-plane.
24. The points of the sphere with center $(2, -1, 5)$ and radius 3.
25. The points at least 1 unit and at most 3 units from the point $(1, 0, -2)$.

SOLVING SYSTEMS OF LINEAR EQUATIONS ON A COMPUTER

In Section 13-3 the solution of a pair of linear equations was shown using a graphical means to find the solution. The solution can also be found using the addition and subtraction of the equations involved. Consider the solution of the system of equations:

$$\begin{cases} ax + by = c \\ dx + ey = f \end{cases}$$

If $a \neq 0$ and $d \neq 0$ we may select for each equation a multiple of the terms (members) on both sides of the equality symbol so that in the derived equations the coefficients of the variable x are opposites. Then these equations may be added to obtain an equation in which the variable x has coefficient zero. We may begin by multiplying the terms on both sides of the first equation by $-d$. This results in $-adx - bdy = -cd$. We then multiply the terms on both sides of the second equation by a. This results in $adx + aey = af$. Adding the two derived equations and using the distributive property, we have

$$(ae - bd)y = af - cd$$

If $ae - bd = 0$, the graphs of the original equations are either parallel lines or the same line and there is not a unique solution. If $ae - bd \neq 0$, we may divide both members of the equation by $ae - bd$ and obtain the value $(af - cd)/(ae - bd)$ for y. The value $(ce - bf)/(ae - bd)$ for x may be found in a similar fashion. These forms for calculating the values of x and y are used in lines 83 and 87 of the programs.

```
  5  HOME
 1Ø  PRINT "THIS PROGRAM FINDS THE SOLUTION OF A"
 11  PRINT
 12  PRINT "SYSTEM OF TWO EQUATIONS IN TWO UNKNOWNS."
 2Ø  PRINT
 3Ø  PRINT
 4Ø  PRINT "PLACE YOUR EQUATIONS IN THE FORMS:"
 44  PRINT
 45  PRINT "          A * X + B * Y = C"
 5Ø  PRINT "AND"
 55  PRINT "          D * X + E * Y = F."
 56  PRINT
 6Ø  INPUT "ENTER THE VALUES FOR A, B, AND C.        ";A,B,C
 62  PRINT
 63  INPUT "ENTER THE VALUES FOR D, E, AND F.        ";D,E,F
 64  PRINT
 65  IF A * E - B * D = Ø THEN PRINT "NO SOLUTION!":GOTO 1ØØ
 7Ø  PRINT "THE SOLUTION OF THE SYSTEM IS:"
 75  PRINT
 8Ø  LET  T = A * E - B * D
 81  XN = C * E - B * F
 82  YN = A * F - D * C
 83  PRINT "X = ";XN;"/";T;",  THAT IS, ";XN/T
 84  PRINT
 85  PRINT "AND"
 86  PRINT
 87  PRINT  "Y = ";YN;"/";T;",  THAT IS,  ";YN/T;"    ."
1ØØ  END
```

The sample run of the program shows its use in solving the system of equations

$$\begin{cases} 1x + 2y = 3 \\ 4x + 5y = 6 \end{cases}$$

The resulting solution is $x = -1$ and $y = 2$.

```
RUN
THIS PROGRAM FINDS THE SOLUTION OF A

SYSTEM OF TWO EQUATIONS IN TWO UNKNOWNS.


PLACE YOUR EQUATIONS IN THE FORMS:

      A * X + B * Y = C
AND
      D * X + E * Y = F.

ENTER THE VALUES FOR A, B, AND C.     1,2,3

ENTER THE VALUES FOR D, E, AND F.       4,5,6

THE SOLUTION OF THE SYSTEM IS:

X = 3/ - 3,  THAT IS, -1

AND

Y = -6/-3,  THAT IS, 2.
```

If you have access to a computer use the program to rework the problems on systems of linear equations in the exercise set for Section 13-3.

14

Computers in Elementary Mathematics

Ada Byron Lovelace
(1815–1852)

Ada Byron Lovelace was the daughter of the famous poet Lord Byron. Educated first by tutors and later by the English mathematician Charles Babbage, she became a strong mathematician. She and Babbage worked together on the construction of Babbage's calculating machines—the Analytical Engine and the Difference Engine. These devices were able to tabulate and, in the latter case, develop mathematical information on the basis of data entered into the machine. As such, they were forerunners of the modern computer. Ada Lovelace's major contribution was the establishment of the format for providing instructions to the machine. This contribution can be viewed as a first step in computer programming.

14-1 Introduction

Computers are playing an ever-increasing role in elementary school mathematics classrooms. Uses of microcomputers have been introduced and sample programs provided in the sections at the ends of each of the previous chapters. In this final chapter an independent introduction to the programming languages BASIC and Logo is provided, the writing of programs is considered, and the role of computers in solving problems and exploring various aspects of mathematics is examined.

The history of the modern computer spans the past 50 years and vividly captures the gigantic advances made in electronics and microtechnology. These advances, in turn, have made it possible to mass produce computers suitable for classroom use from the preschool level upward. Classroom uses include drill and practice with basic arithmetic facts, tutorial lessons on mathematical concepts, application-oriented simulation programs, and problem-solving microworlds. A **microworld** is a learning environment in which students can explore, discover, ask questions, and direct much of their own learning.

Computers provide both the teacher and student with a powerful tool for problem solving. Although hand calculators can carry out a large number of calculations, they must constantly be supplied with information as the computations are performed. The logical decisions are formed and then the single operational commands are entered one by one. The computer, on the other hand, makes it possible to think through the entire problem-solving sequence for a given problem and then enter the entire set of instructions at one time. This set of instructions, called a **program**, provides the computer with the ability both to make the necessary logical decisions and to carry out the appropriate arithmetic operations on the data provided the computer.

A computer usually consists of a number of different devices that are interconnected to form a **computer system**. The power of a computer system is based on the speed and accuracy with which it can perform a few elementary operations. These operations include the following.

Input operations such as accepting data and instructions for processing.

Arithmetic operations such as addition, subtraction, multiplication, and division.

Logical operations such as determining whether one number is greater than another.

Output operations such as producing displayed or printed information.

The ENIAC occupied a space 30 feet by 50 feet and weighed 30 tons. It was programmed by connecting various wires and setting up to 6000 switches. These wires and switches had to be changed each time a program was changed.

The most striking recent advances have been with **electronic digital computers**, that is, computers using sets of electronic impulses to represent digits in a system of numeration for numbers. The first large-scale electronic digital computer was called the Electronic Numerical Integrator And Computer (ENIAC). It was completed in 1946, contained 18,000 vacuum tubes, and could perform about 350 multiplications of two numbers in 1 second. Computers using vacuum tubes are called *first-generation computers.*

Growth of Approximate Number of Computers in the United States

1955	244
1958	2,550
1964	18,200
1965	26,000
1970	100,000
1980	500,000

The *second-generation computers* used transistors instead of tubes. These computers were faster, smaller, and less expensive than their predecessors. Transistors were invented at Bell Laboratories and the first transistorized computer (TRADIC) was built there in 1954. It has been estimated that there were 244 computers in use in the United States in 1955 and that altogether these 244 computers could do about 250,000 additions per second, that is, produce about the same output as one small modern computer. By 1959 the move to the faster and less-expensive transistorized computers was well established.

In 1964 IBM introduced its System/360 computers with their controlling circuitry stored on small chips instead of using transistors. These new components could be mass produced at low cost and rarely failed. The System/360 computers could perform 375,000 computations per second at a cost of $3\frac{1}{2}$ cents

Three generations of computer components

One of the most powerful computer systems today is the Cray-1 at the National Center for Atmospheric Research in Boulder, Colorado. It is used for analyzing atmospheric and weather information. It can process 80 million instructions per second and can store more than 1 million characters of data in its main storage unit.

for each 100,000 computations. A *third generation of computers* arose using this solid logic technology. This steady progress continues today with the application of microchip technology.

The early concept of computers as number processors does not do justice to the wide variety of applications of computers today. Modern computers can be programmed to accept words, even spoken words, and other nonnumerical data. The most rapid growth in the use of computers is in the area of information processing. Wherever large masses of data need to be routinely processed, computers are being increasingly relied upon for doing that processing. Here are a few examples:

Airline (railroad, hotel, concert, car rental) reservations.
Processing over 25 billion checks a year at banks.
Self-service automated teller terminals at banks.
Electronic funds transfer systems.
Computerized grocery checkout stands.
Credit card identification and records.
"Electronic mail" between business offices.
Computerized control of manufacturing processes.
Computer-assisted instruction in schools.

Computers are controlled by human beings and are being used to meet human needs. For example, many people are now typing reports at a computer terminal. The typed material can be read on a video screen. Insertions, deletions, and other corrections can be made. The computer adjusts the spacing and prints all lines the same length unless otherwise instructed. Some computers have a program that will check the spelling of each word, that is, the computer will match words against a file of acceptable words and display words that do not match. This method of checking spelling will not recognize the substitution of one recognizable word for another, such as "too" for "two" or "advice" for "advise."

Instructions and data were communicated to early computers in machine languages in which the activities of the computer were itemized step by step. The use of these languages was very difficult and tedious. Substantial improvements have been made, which allow people to communicate their desires to the computer without having to deal directly with a machine language. In essence the computer is provided in advance with instructions for converting the new programming language into a machine language. FORTRAN was developed at IBM and released in 1957. Among the other high-level programming languages COBOL was released in 1960, BASIC in 1965, PL/1 in 1966, and PASCAL in 1968. More than 200 languages have been developed during the past 30 years. The problem-solving orientation of the developers of these languages is indicated by some of the names selected for the languages,

ALGOL	**ALG**orithmic-**O**riented **L**anguage
BASIC	**B**eginner's **A**ll-purpose **S**ymbolic **I**nstruction **C**ode
COBOL	**CO**mmon **B**usiness-**O**riented **L**anguage
FORTRAN	**FOR**mula **TRAN**slation

BASIC was developed in 1965 by John Kemeny and Thomas Kurtz at Dartmouth College for the use of college students.

The first of these programming languages to have a major influence on education in schools was BASIC. Many *statements* in BASIC are very similar to those in ordinary arithmetic. Letters are used as in algebra.

Ordinary arithmetic:	$a + b$	$a - b$	$a \times b$	$a \div b$	$a < b$
BASIC:	A + B	A − B	A * B	A/B	A $<$ B

Capital letters are used for ease of reading. The symbols for multiplication and division are replaced to avoid confusion with the letter x and the symbol + for addition. The symbol $<$ is read "is less than" in both representations, for example, $3 < 5$. Also, as in the next display, parentheses may be used. The symbol ∅ is used for the numeral 0 to avoid confusion with the capital letter O. Notations for powers and roots are replaced so that all symbols may be written on the same horizontal line.

Ordinary arithmetic:	$2(3 + 5)$	50	a^2	b^3	$\sqrt{a}$
BASIC:	2 * (3 + 5)	5∅	A ∧ 2	B ∧ 3	SQR(A)

Example 1 Write in BASIC.

(**a**) $5(3 - 17)$

(**b**) $3^2 + 10$

(**c**) $54 \div (2^3 - 2)$

(**d**) $\sqrt{29} \div 3$

Solution (**a**) 5 * (3 − 17)

(**b**) 3 ∧ 2 + 10

(**c**) 54/(2 ∧ 3 − 2)

(**d**) SQR(29)/3 ■

All computers and some calculators have a memory. The **memory** (storage unit) of a computer can be used for both data (information) and instructions. The units for doing arithmetic operations (*arithmetic unit*), making decisions (*logic unit*), and controlling the flow of information and instructions (*control unit*) form a central processing unit which is the main part of the computer system.

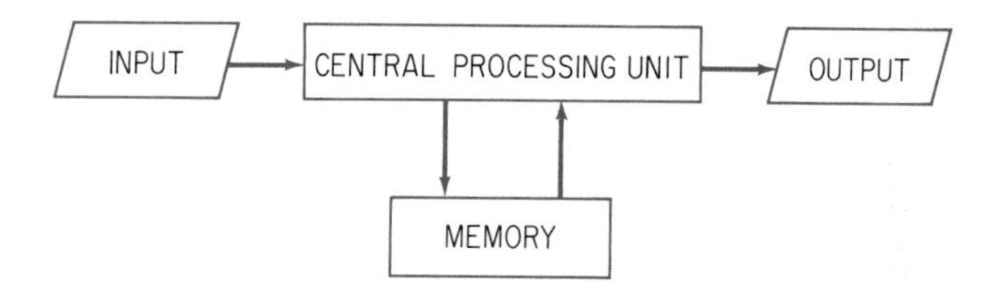

The memory makes it possible for the entire input to be scanned and the operations performed according to established rules. The following sequence of operations is followed in evaluating arithmetic expressions.

Expressions, if any, in parentheses are evaluated first working outward from the inner parentheses.

Then powers and roots, such as squares and square roots, are found.

Then multiplications and divisions are performed in order from left to right.

Finally, additions and subtractions are performed in order from left to right.

This is the same sequence of operations that you have always used in arithmetic. Practice in using these ordinary rules is provided in the exercises.

Example 2 List, in order, the operations that you would perform and evaluate each expression.

(a) $2 + 6 \times 4$ **(b)** $3 + 5^2$ **(c)** $54 \div (2^3 - 17)$

Solution **(a)** $6 \times 4 = 24, \quad 2 + 24 = 26$
(b) $5^2 = 25, \quad 3 + 25 = 28$
(c) $2^3 = 8, \quad 8 - 17 = -9, \quad 54 \div (-9) = -6$ ■

Example 3 Evaluate

(a) `(2 * 3) ∧ 2` **(b)** `5 + (12/4) ∧ 3`

Solution **(a)** $(2 \times 3)^2 = 6^2 = 36$
(b) $5 + (12/4)^3 = 5 + 3^3 = 5 + 27 = 32$ ■

EXERCISES *Use the rules for sequences of operations and evaluate.*

1. $(12 + 6) \div 2$
2. $12 + 6 \div 2$
3. $12 \div 6 \div 2$
4. $50 \div 8 \div 4$
5. $12 \div (6 \div 2)$
6. $50 \div (8 \div 4)$
7. $18 \div (3 \times 2)$
8. $18 \div 3 \times 2$
9. $54 \div 3^2 - 7$
10. $2 \times 3^2 - 4^2$
11. $54 \div (3^2 - 7)$
12. $2 \times (3^2 - 4)^2$
13. $2 - 2 + 2 - 2 + 2$
14. $2 \div 2 \times 2 \div 2 \times 2$
15. $30 \div 3 \div 3 \times 3 \div 3$
16. $30 - 3 \times 3 - 3 \div 3$
17. $60 \times 4 \div 2 \times 5 \div 3$
18. $60 \div 4 \times 2 \div 5 \times 3$

Write in BASIC and evaluate.

19. $5 \times 6 \div 3$
20. $2 + 3^2$
21. $6\sqrt{25 + 24}$
22. $6 - (8 \div 4)^2$
23. $8000 \div 2^5$
24. $75 - 2^6$

List in order the operations that you would perform and evaluate each expression.

25. `6 + 7 * 2`
26. `25 - 3 ∧ 2`
27. `6 * 7/3`
28. `4 * 5 ∧ 2`
29. `12 ∧ 2 - 1ØØ`
30. `25 + 3 * 2 ∧ 3`
31. `5 + 7 * 3 ∧ 2`
32. `11 + 3 * 2 ∧ 3`
33. `11 + SQR(169)`
34. `SQR(3 ∧ 2 + 4 ∧ 2)`
35. `17 + SQR(12 ∧ 2 + 5 ∧ 2)`
36. `SQR(36ØØ) - 1 ∧ 17`

EXPLORATIONS

1. Computers can sort, arrange in a specified manner, and provide printed reports based on data that have been received from a wide variety of sources. Describe briefly one such use of a computer. For example, consider the preparation of student grade reports, salary checks, or periodic inventory reports for a large store, the recording of airline reservations or reservations for a large chain of hotels, or the assignment of college classes to available classrooms.
2. Computers have the ability to read coded instructions and to perform operations very rapidly. Describe briefly one such use of a computer. For example, consider sorting checks for distribution to the banks on which they are drawn, sorting letters according to their zip codes, providing automatic pilots for the control of airplane landings, or providing weather predictions based on reports from a worldwide network of observation stations.
3. Collect pictures and information on individuals involved with the development of computing devices and make a time-line wall chart for use with students at the middle school level.
4. Consult a recent reference work on computers and find the most recent information on the rapidity with which the largest and fastest computers can process information in terms of operations per second.

14-2 BASIC Programs

The selection of line numbers is arbitrary. Numbers 1, 2, 3, ... could be used but would not allow later insertions. Numbers 10, 20, 30, ... are frequently used.

The language BASIC is widely used to state problems so that computers may be used in solving the problems. A set of instructions in BASIC consists of a set of lines called **statements**. Each statement starts with a *line number* and a word that indicates the type of statement. If a statement is too long to fit on a single line, it is split into two statements and a second line number is assigned to the part on the second line. A similar procedure is followed for additional lines. The line number serves both as a serial number for ordering the operations of the computer and as a label for the statement. Usually, consecutive numbers are avoided so that additional data (or instructions) may be inserted. New lines, with appropriate numbers, may be added at the end of the program. Lines that contain errors may be replaced simply by adding new lines with the same number as the lines to be replaced. The computer orders and uses the statements according to their serial numbers. The instruction LIST may be used by the operator to obtain a list of the lines of the program in their proper order. Then, if the program is satisfactory, the instruction RUN may be used to implement the program. The only additional procedure needed to operate the computer would be to satisfy the machine that you are a recognized user.

Many problems involve numerous repetitions of a simple task. A computer can be programmed to perform repetitive tasks any specified number of times or until a specified objective has been accomplished. The sequence of steps in such a step-by-step procedure can be represented in a flow chart. Arrows are used to show the flow of the steps. Different geometric figures are used to identify different types of steps. The symbols used include the following from the standards published by the American National Standards Institute.

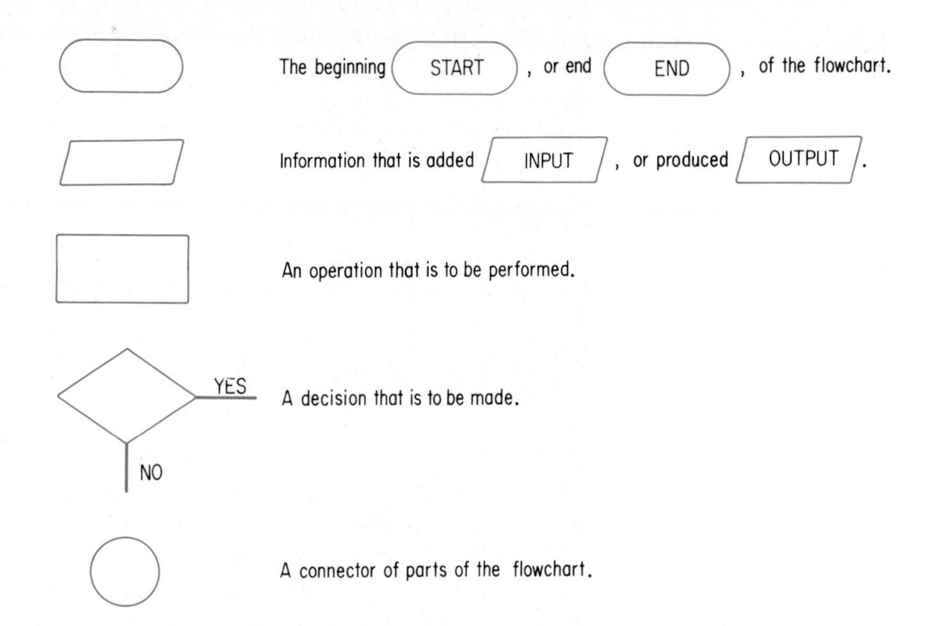

The steps needed to obtain a table of the cubes of the natural numbers 1 through 100 are indicated in the following flow chart.

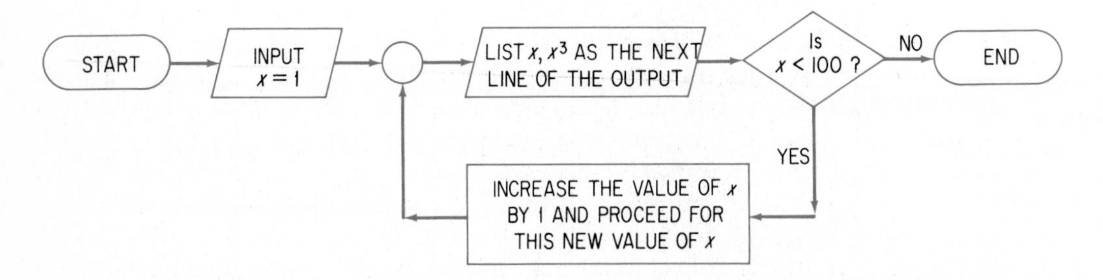

Instructions for a computer are often referred to as *commands.*

These steps may be modified slightly to obtain the following program which can be used to instruct a computer to produce a table of the cubes of the numbers 1 through 100.

```
10  FOR X = 1 TO 100
20  PRINT X, X * X * X
30  NEXT X
40  END
```

The FOR TO statement instructs the computer to start with $x = 1$ and to consider in turn the values 1, 2, 3, ..., 100 for the variable x. The PRINT statement instructs the computer to start a new line of output, to print the value of x, to leave a number of spaces before the next column starts, and then to print the value of x^3. Either X * X * X or X ∧ 3 may be used for x^3. The printed output will be in two columns. The NEXT X statement instructs the computer to return for more data, that is, in this program to return to line 1∅ and replace x by $x + 1$. Under this program the computer considers in order the instructions on lines

1∅, 2∅, 3∅, 1∅, 2∅, 3∅, ..., 1∅, 2∅, 3∅, ...

until all the data on line 1∅ have been used, that is, until lines 2∅ and 3∅ have been processed for $x = 100$. Then the computer proceeds to the next instruction, in this case line 4∅.

Example 1 State in ordinary notation the printed output for this program.

```
1∅  FOR X = 1 TO 5
2∅  PRINT X, 2 + X
3∅  NEXT X
4∅  END
```

Solution

```
1  3
2  4
3  5
4  6
5  7
```
■

The statements used in a BASIC program must be of specific types. These types are identified by the key words used. For example,

```
FOR X = 1 TO 5
```

is a FOR TO statement and

```
PRINT X, 2 + X
```

is a PRINT statement. The types of statements that may be used in BASIC programs also include the following.

- DATA — The statement includes a list of numbers or other representations of facts, concepts, or instructions ready for processing.
- READ — The computer is to read the first element, or specified number of elements, from the data.
- LET — A variable is assigned, or reassigned, as a name for a number or expression.
- GOTO — The computer is to go to the indicated line of the program.
- IF THEN — A condition is stated and the computer is to go to a specified line of the program if the condition is satisfied. If the condition is not satisfied, the computer goes to the next line of the program.
- LIST — The computer is to list the statements of the program in the order of their line numbers.
- RUN — The computer, for an acceptable program, is to provide the output of the program.

A DATA statement may be useful when the number of elements is not large or is not easily described to the computer. Here is a program for a table of the cubes of the first 20 natural numbers.

```
10  READ X
20  DATA 1, 2, 3, 4, 5, 6, 7, 8, 9, 10
30  DATA 11, 12, 13, 14, 15, 16, 17, 18, 19, 20
40  LET Y = X ∧ 3
50  PRINT X, Y
60  GOTO 10
70  END
```

The printout for this table of values will end with the statement

```
10  OUT OF DATA
```

If the DATA statements had consisted of the numbers 1 through 100, a table of the cubes of the numbers 1 through 100 would have been obtained. Thus a table of the cubes of the numbers 1 through 100 may be produced either using FOR TO and NEXT statements as in the program at the beginning of this section or using READ, DATA, and GOTO statements as in the preceding program.

A table of cubes may also be produced using LET and IF THEN statements. In the next program the first LET statement assigns the initial value 0 to the variable X; the second LET statement reassigns the variable X to the next natural number. In other words, the computer increases the value of X by 1 and renames the sum as X, that is, reassigns the variable X to the new value. We instruct the computer to do this by writing

```
LET X = X + 1
```

Such a statement cannot be true in ordinary algebra and should not be considered as an algebraic equation. Rather it is an instruction to the computer to rename the quantity X + 1 as X, that is, assign a new value to X. The following program instructs the computer to follow the steps that are described at the beginning of this section in the flowchart for obtaining a table of cubes of the numbers 1 through 100.

```
10  LET X = 0
20  LET X = X + 1
30  PRINT X, X ∧ 3
40  IF X < 100 THEN 20
50  END
```

We have considered three programs for tables of cubes in order to emphasize that often several different programs may be written to solve the same problem.

The pattern formed by the differences of the squares of the first eight successive whole numbers from the squares of their predecessors is shown in the output of the next program.

Printed Output

```
10  READ D
20  DATA 1, 2, 3, 4, 5, 6, 7, 8          1
30  LET P = D ∧ 2                         3
40  LET Q = (D − 1) ∧ 2                   5
50  LET R = P − Q                         7
60  PRINT R                               9
70  GOTO 10                              11
80  END                                  13
RUN                                      15
```

To test a conjecture regarding the pattern formed by the output of the previous program for some additional numbers, such as the first ten integers greater than one thousand, only the DATA statement needs to be changed and the new set of numbers inserted. Computers are frequently used to test conjectures of arithmetic patterns. A practical example of the use of computers by classroom teachers is considered in the following example.

Example 2 Suppose that you have three grades for each of your students and wish to average these grades. Prepare a BASIC program that may be used to do this.

Solution

```
10  READ G1, G2, G3
20  DATA   (List the grades for each student, being sure to list together
           three grades for each student; use extra lines 21, 22, 23, ...,
           as needed.)
60  LET A = (G1 + G2 + G3)/3
70  PRINT G1, G2, G3
80  PRINT A
81  PRINT
90  GOTO 10
95  END ■
```

The computer may be set so that the three grades and the average of these grades are printed with a separate line for each student.

Languages such as BASIC may be used on many different types of computers. A computer of any given type usually has special features that are intended to enhance its usefulness.

Programs for more difficult problems can be introduced after the problems are understood. Numerous problems in algebra, geometry, and especially advanced mathematics are routinely solved on computers throughout the industrialized countries of the world. The breadth of the problems for which computers are used is expanding daily. Most banks and big stores of all sorts use computers extensively. In some cities ordinary citizens use the touch-tone system of their telephones to instruct computers at their banks to pay their bills and complete other financial transactions.

EXERCISES

State in ordinary notation the printed output for each program.

1.
```
10  READ X
20  DATA 2, 4, 6, 8
30  PRINT X, X/2
40  GOTO 10
50  END
```

2.
```
10  READ X
20  DATA 3, 5, 7, 9
30  PRINT X, (X − 1)/2
40  GOTO 10
50  END
```

3.
```
10  READ X
20  DATA 1, 2, 3, 4
30  PRINT X, X ∧ 3
40  GOTO 10
50  END
```

4.
```
10  READ X
20  DATA 10, 11, 12, 13
30  PRINT X, X ∧ 2
40  GOTO 10
50  END
```

5.
```
10  READ X, Y
20  DATA 1, 2, 3, 4, 5, 6
30  PRINT X, Y, X + Y
40  GOTO 10
50  END
```

6.
```
10  READ X, Y
20  DATA 1, 2, 3, 4, 5, 6
30  PRINT X, Y, X * Y
40  GOTO 10
50  END
```

7.
```
10  READ X, Y, Z
20  DATA 1, 2, 3, 4, 5, 6
21  DATA 7, 8, 9, 10, 11, 12
30  PRINT X, Y, Z
40  PRINT X + Y + Z
41  PRINT
45  GOTO 10
60  END
```

8.
```
10  READ X, Y, Z
20  DATA 11, 12, 13, 14, 15, 16
30  PRINT X + Y, X + Z, Y + Z
40  GOTO 10
50  END
```

9.
```
10  FOR X = 10 TO 15
20  PRINT X, X ∧ 2
30  NEXT X
40  END
```

10.
```
10  FOR X = 1 TO 5
20  PRINT X, X * 3
30  NEXT X
40  END
```

11.
```
10  FOR X = 1 TO 6
20  LET Y = X + 1
30  PRINT X, Y, X * Y
40  NEXT X
50  END
```

12.
```
10  FOR X = 1 TO 6
20  LET Y = X + 1
30  LET Z = Y ∧ 2 − X ∧ 2
40  PRINT X, Y, Z
50  NEXT X
60  END
```

Assume that the computer has been preset to print answers with five places after the decimal point and write a BASIC program to print the specified data.

If a computer is available, modify your program as necessary and run it on the computer.

13. A table of the natural numbers 1 through 20 with their fifth powers.
14. A table of the natural numbers 1 through 50 with their square roots.
15. A table of the natural numbers 1 through 1000 with their squares and their cubes.
16. A table of the natural numbers 1 through 25 with their squares and their square roots.
17. A table of the natural numbers 1 through 10 with their squares, cubes, and square roots.

18. At 300 miles per hour a plane travels 5 miles each minute, that is, $5n$ miles in n minutes. Write a BASIC program that can be used to print a table of distances traveled at 300 miles per hour in n minutes for natural numbers 1 through 120.

*19. An investment of \$1000 at simple interest of 12% for n years amounts to $1000(1 + n \times 0.12)$ dollars. For any natural number n of years 1 through 20 write a BASIC program that could be used to obtain the amounts to which \$1000 would accumulate at 12% simple interest.

*20. An investment of \$1000 at 12% compounded monthly for n years amounts to $1000(1 + 0.01)^{12n}$ dollars. Assume that n may be any natural number from 1 to 20 and write a BASIC program that could be used to obtain a table of the amounts to which \$1000 would accumulate in n years at 12% compounded monthly.

Write a BASIC program to print the specified data.

*21. The semester average for students who take three 1-hour examinations and a 3-hour final examination. The hour examinations are equally important. The final examination and the average for the three 1-hour examinations each provide the basis for one-half of the student's final grade.

*22. Repeat Exercise 21 with the hour examinations providing the basis for two-thirds of the student's final grade and the final examination providing the basis for one-third.

EXPLORATIONS

1. The numbers of dots in square arrays are called **square numbers**.

Write a BASIC program to list the square numbers less than or equal to one million.

2. The numbers of dots in triangular arrays of the form shown below are called **triangular numbers**. Write a BASIC program to list the first 100 triangular numbers.

3. Write a BASIC program to produce a table of the natural numbers 1 through 100 and their factorials. For any natural number n the **factorial** of n, written $n!$, is equal to the product of the natural numbers that are less than or equal to n.

14-3 Improving BASIC Programs

The development and polishing of microcomputer programs requires constant attention and experience. The programs in the preceding section were rather short and did little to communicate their intent and purposes to the reader. In this section we focus on writing and improving programs in BASIC to communicate effectively with the user and to provide output that is easy to use.

The first improvement is the REMARK statement. This statement, represented by the term REM in the program, allows the programmer to make explanatory remarks in the program that do not appear in the printout (output). These statements are seen only when one asks for a LIST of the program. Type in the following program. The output for this program is shown in the margin.

```
10  REM THIS IS A PROGRAM TO SHOW THE USE OF REM
20  FOR C = 1 TO 12
30  PRINT C
40  NEXT C
50  END
```

```
1
2
3
4
5
6
7
8
9
10
11
12
```

The adjacent output for the program shows no evidence of the REM statement in line 10, but when one lists the program, the REM statement is included with the others in the statement of the program. Often the REM statement is used to describe what a particular variable stands for in the program, what a given portion of the program does, or what a user of the program might need to know about using the program.

```
10  FOR X = 1 TO 20
20  PRINT X, X ∧ 3
30  NEXT X
40  END
```

In the calculation of x^3 in the previous section, you may have gotten some unusual answers for the computer-derived values for some of the natural numbers involved. The adjacent values were obtained on an author's computer. These values result from the particular method that the computer uses to find the desired values. These special methods often are quite different from ordinary arithmetic. They are used to enable the computer to handle some very difficult problems and give quite accurate estimates of their answers. Sometimes they cause difficulty in situations in which various powers of the natural numbers are desired.

```
1     1
2     8
3     27
4     64
5     125
6     216
7     343
8     512
9     729.000001
10    1000
11    1331
12    1728
13    2197
14    2744
15    3375
16    4096
17    4913.00001
18    5832.00001
19    6859
20    8000
```

The unwanted decimal portions of an answer can sometimes be taken care of by changing the operations, for example, by replacing X ∧ 3 by X * X * X. Usually the problem can be corrected by rounding the answer to the nearest integer or appropriate decimal place. To do this, we use the greatest integer function $[x]$, which was discussed in the Computer Applications section in Chapter 3. The function $[x]$ is denoted by the command INT and produces the *greatest integer that is less than or equal to the given number*. This is equivalent to identifying the integer if the number is an integer or to giving the greatest integer that is less than the number if the number itself is not an integer. On the real number line $[x]$ is the integer with its graph at, or just to the left of, the point with the coordinate x.

For example,

$$[5.5] = 5$$

$$[17] = 17$$

$$[-4.5] = -5$$

Any number in decimal notation can be rounded to any specified unit of precision. Consider rounding the number 0.2417 to the nearest hundredth. To do this, we locate the hundredths digit, 4, look at the digit to its right, notice that it is less than 5, so we drop the digits to the right of the hundredths digit and record the rounded number as 0.24. If we were rounding to the nearest thousandth, we would have located the digit in the thousandths place, 1, looked at the digit to its right, noticed that 7 is greater than or equal to 5, added 1 to the digit in the thousandths place, and recorded the rounded number as 0.242.

Example 1 Round 34.563 to
(a) the nearest integer
(b) the nearest tenth
(c) the nearest hundredth.

Solution **(a)** 35 **(b)** 34.6 **(c)** 34.56 ■

Our next problem is to develop a way of carrying out this process on the computer. We use the following sequence of steps. Consider an arbitrary number x. Suppose we wish to round this number to the nearest integer. If x is not an integer, we need to determine whether the difference between x and the next larger integer is greater or less than the difference between x and the next smaller integer. This is akin to asking whether the distance of the point with coordinate x on a number line from the nearest point with an integer as coordinate on the left is less than 0.5 or the distance to the nearest point with an integer as coordinate on the right is less than 0.5. If both distances are 0.5 computer users habitually select the larger integer, that is, we round up.

a x b

The convention of rounding up when x is halfway between two integers enables us to use $[x + 0.5]$ to round to the nearest integer. Note that if x is near the next larger integer, then $x + 0.5$ is greater than that integer; if x is near the next smaller integer, then $x + 0.5$ is less than the next larger integer. Thus $[x + 0.5]$ provides the desired result in all cases.

Example 2 **(a)** Write a program to round 34.563 to the nearest integer.
(b) Describe the procedure used in the program.

Solution **(a)**

```
5  REM ROUNDING X TO NEAREST INTEGER
1Ø X = 34.563
2Ø Y = INT(34.563 + Ø.5)
3Ø PRINT Y
4Ø END
```

(b) The number 34.563 is entered (line 1Ø). Then $[x + 0.5] = [35.063] = 35$ is obtained (line 2Ø) and printed (line 3Ø). Thus, to the nearest integer, 34.563 is rounded to 35. ■

A number such as 541.37 may be rounded to the nearest tenth by

Multiplying by 10	5413.7
rounding this number to the nearest integer	5414
and dividing by 10	541.4

Similarly, 541.37 may be rounded to the nearest hundred by

multiplying by 0.01, that is, 10^{-2}	5.4137
rounding to the nearest integer	5
and dividing by 0.01	500

In general, to round a number to any specified place value we may

> multiply by a selected power of ten so that the digit in the specified place becomes the units digit,
>
> round to the nearest integer, and
>
> divide by the specified power of ten.

These steps are combined on line 2Ø in each of the programs in the solution of Example 3.

Example 3 Write a program to round 849.732
(a) to the nearest tenth
(b) to the nearest hundred.

Solution **(a)**

```
5  REM ROUNDING TO THE NEAREST TENTH
1Ø X = 849.732
2Ø Y = (INT(X * 1Ø + Ø.5))/1Ø
3Ø PRINT Y
4Ø END
```

(b)

```
5  REM ROUNDING TO THE NEAREST HUNDRED
1Ø X = 849.732
2Ø Y = (INT(X * Ø.Ø1 + Ø.5))/Ø.Ø1
3Ø PRINT Y
4Ø END
```
■

We now have the programming skills to write a program which will print out exactly what we wanted in the program to develop the listing of the first 20 natural numbers and their cubes. Here is one such program.

```
 5 REM PRINTING A LIST OF THE FIRST TWENTY
 6 REM NATURAL NUMBERS AND THEIR CUBES
1Ø FOR X = 1 TO 2Ø
2Ø PRINT X, INT(X ∧ 3 + Ø.Ø5)
3Ø NEXT X
4Ø END
```

The statements used in computer programs need to be understood by the computer and are sometimes condensed so drastically that they seem unusual to the general reader. One of the most striking examples of such terseness arises when one uses a variable C to count the number of times a particular part of a program is used. We start with C = 0 and increase the value of C by 1, that is, assign the new value C + 1 to the variable C, each time that part of the program is used. The convenient statement for making this change is simply "C = C + 1," which is to be interpreted as increasing the value of C by 1 rather than as an impossible algebraic statement.

Example 4 State the output of the program

```
 5  REM EXAMPLE OF A COUNTER
10  LET C = 0
20  FOR X = 1 TO 5.
30  C = C + 1
40  PRINT X
50  NEXT X
60  PRINT "C = ";C;"."
70  END
```

Solution

```
1
2
3
4
5
C = 5.
```

■

From the solution of Example 4 we observe that the program passed through line 30 five times during the execution of the list of commands. Counters are especially useful in setting up and running programs where a decision must be made after a certain event has been observed a given number of times. For example, a command such as

```
IF C = 5 THEN GOTO 65
```

may be used to shift to a different portion of the program.

Two special features of PRINT statements are used in line 60 of the program given in Example 4. The quotation marks around the symbols and spaces "C = " indicate that those symbols and spaces are themselves to be shown in the output. A semicolon indicates to the computer that there is more to be done before proceeding to the next line. On line 60 the first semicolon followed by C indicates PRINT C on this line, that is, the computer prints the value of C; the second semicolon followed by "." indicates that on the same line the value of C is to be followed by a period, as in the last line of the solution to Example 4. The use of PRINT statements composed of both symbols and numerical output do a great deal to make the results of a program's computations much more readily useful and understandable.

EXERCISES *Write the output associated with each of the BASIC statements.*

1. `INT(34.6 + Ø.5)`
2. `(INT(34.6 * 1Ø + Ø.5))/1Ø`
3. `(INT(321.2 * 1Ø + Ø.5))/1Ø`
4. `(INT(3.2856 * 1ØØ + Ø.5))/1ØØ`
5. `(INT(3.2851 * 1ØØ + Ø.5))/1ØØ`
6. `(INT(3.2856 * 1ØØØ + Ø.5))/1ØØØ`
7. `(INT(3.2851 * 1ØØØ + Ø.5))/1ØØØ`
8. `(INT(328516/1ØØ + Ø.5)) * 1ØØ`

In Exercises 9 through 15 give the output associated with each program.

9.
```
1Ø  FOR K = 1 TO 5
2Ø  PRINT "ANNE"
3Ø  NEXT K
4Ø  END
```
10.
```
1Ø  FOR K = 1 TO 5
2Ø  NEXT K
3Ø  PRINT "ANNE"
4Ø  END
```
11.
```
1Ø  FOR N = 1 TO 5
2Ø  PRINT INT((N ∧ 2)/2)
3Ø  NEXT N
4Ø  END
```
12.
```
1Ø  FOR N = 1 TO 6
2Ø  PRINT INT((4 * N)/3)
3Ø  NEXT N
4Ø  END
```
13.
```
1Ø  FOR D = 1 TO 1Ø
2Ø  IF 5Ø/D = INT(5Ø/D) THEN PRINT D
3Ø  NEXT D
4Ø  END
```
14.
```
1Ø  FOR N = 1 TO 5
2Ø  PRINT INT(((N ∧ 2) + N)/5)
3Ø  NEXT N
4Ø  END
```
15.
```
1Ø  PRINT "N", "Y"
2Ø  FOR N = 1 TO 5
3Ø  LET Y = (2 * N) ∧ 3 + 4 * N - 6
4Ø  PRINT N, Y
5Ø  NEXT N
6Ø  END
```
16. The BASIC statement SQR(X) represents the square root of X. Use this command and write a program having as its output the square roots of the first 20 natural numbers rounded to the nearest hundredth.

A multiple of 3 is a number of the form $3N$ where N is a natural number.

17. Write a program for finding the number of even numbers in the set of the first 50 positive multiples of 3.
18. For values of N from 1 through 100 find the number of times

```
INT(SQR(N)) = SQR(N)
```

Explain your answer in terms of numbers that are perfect squares.

The conversion formula is

$$F = \left(\frac{9}{5}\right)C + 32$$

19. Write a program to produce a conversion table that can be used to convert temperatures measured in degrees Celsius to temperatures measured in degrees Fahrenheit for temperatures in degrees Celsius ranging from -50 to $+50$.
20. Write a program for dropping all the digits in a decimal beyond the hundredths digit. This is called *truncating* the decimal at the hundredths place. Test your program with $X = 34.98234$.
21. Suppose that you are given \$1 to work on one day, \$2 to work on the next day, \$4 to work on the third day, ... If the pattern of twice as much each day as on the previous day continues, how much would you make on the 20th day? How much would you have made in all by the end of the 20th day? Write a program for solving this problem.
22. Develop a program to produce the following output.

```
XXXXX      XXXX    X    X
  X       X        X    X
  X       X        X    X
  X        X       X    X
  X         X      X    X
  X          X     X    X
  X           X    X    X
  X           X    X    X
XXXXX     XXXX     XXXXXX
```

23. Write a program to produce as its output a list of the integers between 1 and 100 that have a units digit of 4.
24. Write a program for listing the natural numbers from 1 through 20 that satisfy the equation 3 * X + 7 = 21.
25. Write a program for obtaining the average, rounded to the nearest tenth, of any ten natural numbers, where the numbers are entered in a DATA statement.

EXPLORATIONS

1. Write a program that provides the sum of any two given rational numbers. Use the PRINT statement with semicolons to present rational numbers in the form a/b, using the algorithm

$$\frac{a}{b} + \frac{c}{d} = \frac{ad + bc}{bd}$$

2. Extend the program developed in Exploration 1 to produce the sum as a rational number in lowest terms.
3. Write a program that provides the product of two rational numbers in lowest terms.
4. Develop a program that provides for any two natural numbers N and D their GCD, their LCM, and N/D in lowest terms.

14-4 Logo Programs

The Logo language was developed by Seymour Pappert at the Massachusetts Institute of Technology in the 1970's. This particularly rich language is especially noted for its graphics, called Turtle Graphics, as was noted in the Computer Applications section of Chapter 8. We now examine the main features of the language and use them in writing programs in Logo.

The Logo language is available for classroom use in a variety of forms. We focus on the APPLE Logo © form. To prepare to use Turtle Graphics, load the Logo language disk into your computer, and then type SHOWTURTLE. The screen is then cleared and the turtle (a small triangle) appears in the center of the screen pointed upward.

Logo commands, called **primitives**, can now be used to move the turtle about the screen, leaving a lighted trail of its path. The commands deal with a few elementary ideas such as direction and distance (magnitude). We consider first the commands FORWARD and BACKWARD. These commands may be entered along with a number to specify the distance the turtle should move.

The command FORWARD 1ØØ directs the turtle to move 100 units up the screen. The command BACKWARD 75 directs the turtle to move 75 units back down the screen. The commands FORWARD 1ØØ BACKWARD 75 direct the turtle to move forward 100 units and then back 75 units down its trail. Thus the turtle arrives at a point 25 units up from its initial position.

The command FORWARD can be abbreviated as FD and the command BACKWARD can be abbreviated by BK when typing commands. These commands can be typed on successive lines or all entered on a single line. In each case, the movements of the turtle are made in the order listed.

A second set of commands deals with turning the turtle from its initial position of pointing toward the top of the screen. This is accomplished through the use of RIGHTTURN and LEFTTURN commands. If we think of a giant circular protractor with 360° sweep in each direction, we see the possibilities for directing the turning of the turtle. A command of RIGHTTURN 9Ø or LEFTTURN 27Ø will direct the turtle to head for the right side of the screen. A command of either RIGHTTURN 18Ø or LEFTTURN 18Ø will direct the turtle to head for the bottom of the screen. These commands can be abbreviated through the use of RT and LT with the magnitude of the turn in degrees. These RT and LT commands can then be entered in succession with the FD and BK commands to give rise to various patterns on the screen.

Example 1 Develop a program to draw a rectangle which is 40 units by 70 units.

Solution

```
FD 40
RT 90
FD 70
RT 90
FD 40
RT 90
FD 70
```

Alternative Solution

```
FD 40 RT 90 FD 70 RT 90 FD 40 RT 90 FD 70
```
■

To clear the screen after drawing a given figure, type CLEARSCREEN (CS). The figure will vanish and the turtle returns to its HOME position. HOME brings the turtle to its initial position but leaves the figure in place on the screen and leaves the direction of the turtle unchanged.

The turtle can now be used to draw basic geometric figures and other designs. If we wish to move the turtle about the screen without leaving a track. We use the commands PENUP (PU) and PENDOWN (PD), which cause the turtle to cease and commence drawing, respectively. We can also make the turtle disappear and appear through the respective use of the commands HIDETURTLE (HT) and SHOWTURTLE (ST).

Example 2 Using pencil and paper, predict the figure that results from the following commands:

(a) FD 100 RT 120 FD 100 RT 120 FD 100 RT 120

(b) BK 75 LT 90 FD 75 RT 90 FD 75 LT 90 BK 75

Solution

(a) An equilateral triangle.

(b) A square ■

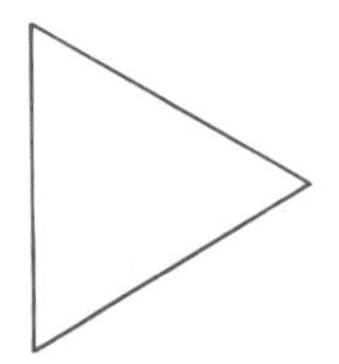

Note that other commands such as

```
FD 100 LT 60 BK 100 LT 60 FD 100 LT 60
```

for part (a) of Example 2 can be used to draw the same figure. In the original commands for part (a) the angles in an equilateral triangle required turns of 120° to draw, not the 60° angles one might expect. This happens due to the fact that we are turning through the **supplement**, $180° - \theta$, of each interior angle, θ, of the triangle. Note that in traversing completely the exterior of a triangle, we turn the turtle through 360° since the turtle must turn through the supplement of each of the three interior angles of the triangle, i.e., through the supplements of three angles whose sum is 180°. Since the three angles and their supplements have a sum of 540°, we have the turtle turning through 540° − 180°, that is, 360°. In general, the sum of the interior angles of a convex n-gon is $(n - 2)$ times 180°, and the sum of the exterior angles is 360°. Thus for the *exterior turns* of the turtle we can use this sum of 360° in checking the turns of the turtle in traversing the perimeter of the convex polygon and returning to its original direction.

Example 3 Write a Logo command sequence for drawing the following figure.

Solution Start at the lowest point and recall that the turtle is headed upward in its starting position. Let the length of the vertical segment be 75 units. Then the first command is FD 75.

The entire figure is not a convex polygon, but the figure may be viewed as a regular polygon (pentagon), also convex, with isosceles triangles on each of its sides. The exterior angles of the pentagon are each (360/5)°, that is, 72° in size.

These angles are also base angles of the triangles. Therefore the vertex angle of each triangle is 180° − 2(72)°, that is, 36°, and the turtle must turn through a supplement of 36°, that is, 144°. The Logo program may now be completed.

```
FD 75 RT 144 FD 75 RT 144 FD 75 RT 144 FD 75 RT 144 FD 75 RT 144
```

The final RT 144 leaves the turtle in its original position pointing to the top of the screen. Adding a command of HT, we can remove the turtle from the display, leaving the figure alone. ■

In both this example and the drawing of the equilateral triangle earlier, we have had to repeat a large number of moves of the turtle in drawing the figures. This often happens in dealing with regular polygons or with figures which have a great deal of symmetry in their formation. There is a special Logo statement, the REPEAT command, which helps in drawing such figures. Then the command for drawing an equilateral triangle may be expressed as

```
REPEAT 3[FD 6Ø RT 12Ø].
```

The command REPEAT tells the computer to repeat the set of commands enclosed in the brackets the number of times appearing in front of the brackets.

In this case, the number of times is three. If possible, type in this command and see the results yourself. If your computer keyboard does not have keys with the bracket symbols on them, you may be able to type a capital *N* for the left-hand bracket and a capital *M* for the right-hand bracket.

Example 4 Use the REPEAT command to write a program to draw the rectangle in Example 1.

Solution

```
REPEAT 2[FD 40 RT 90 FD 70 RT 90] ■
```

Example 5 If the segments in the drawing are each 50 units long, write a program that will generate the given figure.

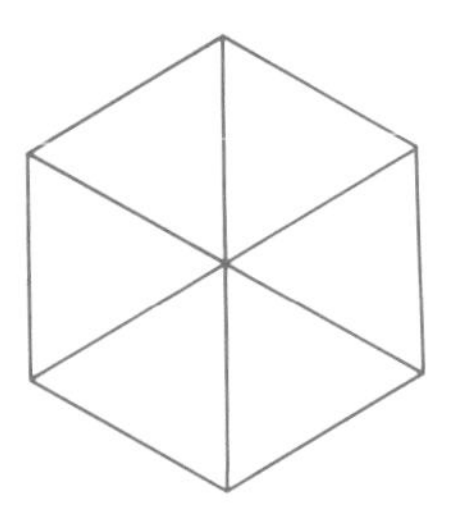

Solution Start at the center of the figure.

```
REPEAT 6[FD 50 RT 120 FD 50 RT 120 FD 50 RT 60] ■
```

Note in the solution of Example 5 that it is permissible to trace line segments more than once.

EXERCISES *Use pencil and paper to sketch the figure that results from each set of Logo instructions. Use dotted lines to show the path of the turtle with the pen up.*

1. FD 60 RT 30 BK 60 RT 90 FD 60
2. RT 90 FD 50 BK 50 LT 90 FD 100
3. BK 50 RT 45 FD 100 LT 45 FD 50 RT 45 BK 100
4. FD 90 RT 90 FD 90 RT 90 FD 90 RT 90 FD 90 RT 90 FD 90 RT 45 FD 100
5. FD 90 RT 90 FD 120 RT 90 FD 90 RT 90 FD 120
6. PU LT 90 FD 100 RT 90 PD BK 50 PU FD 50 RT 90 FD 50 LT 90 FD 50 PD FD 5
7. BK 50 LT 30 FD 100 LT 240 FD 100 LT 240 FD 100
8. REPEAT 3[FD 100 RT 60]
9. REPEAT 4[FD 60 RT 45]
10. REPEAT 6[FD 20 LT 30 PU FD 5 PD]

Select appropriate distances and angles and write a Logo program for drawing a figure of the given shape.

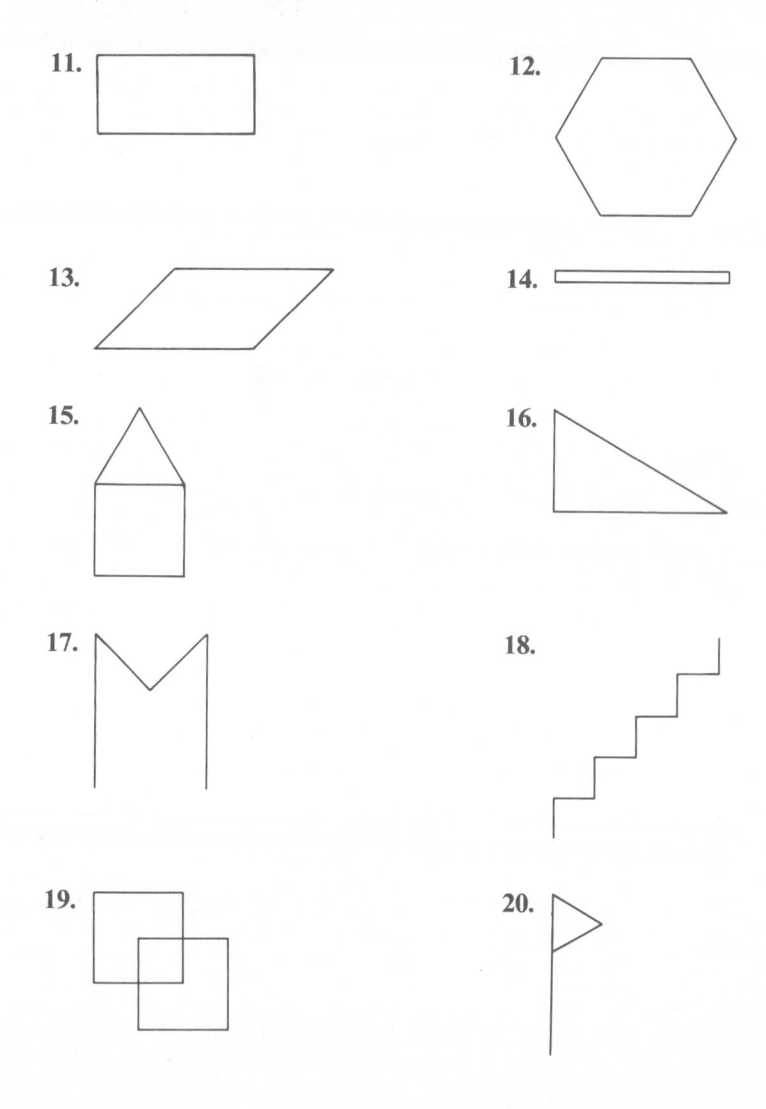

EXPLORATIONS

1. The REPEAT command allows the development of Logo programs which describe circles. Experiment with values of N in

```
REPEAT N[FD 1 RT 360/N]
```

2. Explore ways of modifying the results obtained in Exploration 1 to draw different circles by changing the number of units moved forward.

3. Write sets of Logo instructions for making the given designs.

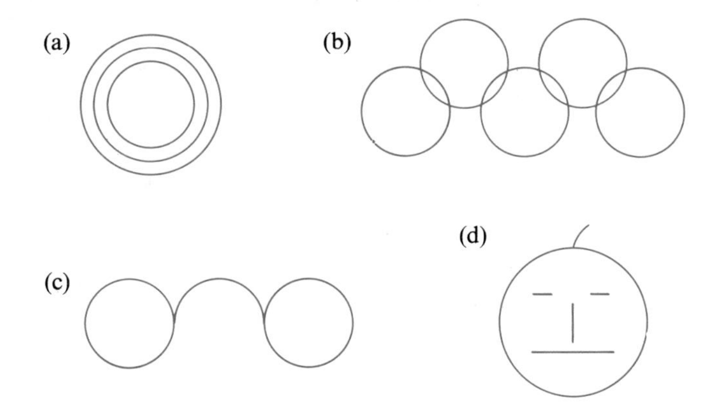

14-5 Logo Procedures

A procedure is a set of commands (subprogram) that we create, name, and enter into the computer's memory for use at a later time. In APPLE Logo © we initiate a procedure by typing the word TO followed by a name which contains no spaces. If you wish to separate words or other symbols in the name of your procedure, use periods for spacing. To name a procedure for drawing an equilateral triangle with side of length 100 we could use the line

```
TO TRIANGLE1ØØ
```

before listing the commands for the procedure.

When the computer receives the inserted line, headed by a TO, it stops executing commands for the turtle on the screen, enters its EDIT mode, and stores the commands making up the procedure in its memory for later use. After the return following the command END on the final line, the computer prints

```
TRIANGLE1ØØ DEFINED
```

and returns to the mode of executing commands on the screen as they are inserted.

```
TO TRIANGLE1ØØ
REPEAT 3[FD 1ØØ RT 12Ø]
END

TRIANGLE1ØØ DEFINED
```

```
TRIANGLE1ØØ
LT 6Ø
TRIANGLE1ØØ
END
```

We can now type in the command TRIANGLE1ØØ, touch return, and the turtle will automatically draw an equilateral triangle of side length 100. The program in the margin produces the figure shown.

Example 1 Write a procedure for drawing a square of side 75.

Solution

```
TO SQUARE75
REPEAT 4[FD 75 RT 90]
END ■
```

A large collection of such programs for various regular polygons with various side lengths can be developed and used with other programs. However a general procedure that allows for the entry of a variable denoting the length of the side or the number of degrees in a turn is much more convenient to use. The needed value(s) of the variable(s) can be entered when the procedure is called for execution. This is done by entering a variable name preceded by a space and a colon in naming the procedure.

For example, the procedure

```
TO TRIANGLE :SIDE
REPEAT 3[FD :SIDE RT 120]
END
```

can be invoked by typing TRIANGLE followed by a space and a value for N to produce these figures:

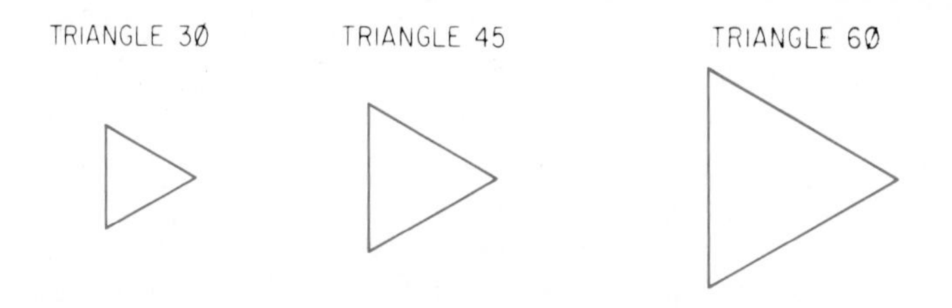

The use of variables allows for a more economical use of storage space in the computer's memory, since only one procedure needs to be stored. It can be adapted to handling the various requirements of an equilateral triangle by entering the length of the side needed when it is called. A similar program with turns can be used as needed.

The procedure

```
TO NGON :N :SIDE
REPEAT :N[FD :SIDE RT 360/:N]
END
```

involves two variables and can be used to produce a variety of regular n-gons with inputs stipulating the number of sides, :N, and the desired length of those sides, :SIDE. The command NGON 4 20 produces the square shown, NGON 5 30 gives the pentagon, and NGON 7 25 gives the heptagon.

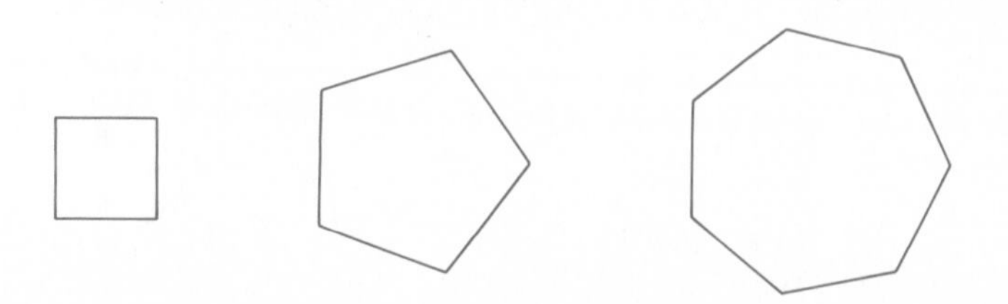

We are now in a position to use the Turtle Graphic's commands to create a variety of geometric designs.

Example 2 Write a procedure to draw the following four-square design.

Solution One possible procedure might be:

```
TO FOURSQUARE :N
REPEAT 4[REPEAT 4[FD :N RT 9Ø]LT 9Ø]
END
```

Another approach is to define a procedure using a procedure to draw the small squares first

```
TO SQUARE :N
REPEAT 4[FD :N RT 9Ø]
END
```

and then create a new procedure involving SQUARE :N. For example,

```
TO FOURSQR :N
REPEAT 4[SQUARE :N LT 9Ø]
END ■
```

The role of the variable N in the FOURSQR :N procedure defined in the solution of Example 2 is shown by these outputs for FOURSQR 2Ø, FOURSQR 3Ø, and FOURSQR 4Ø.

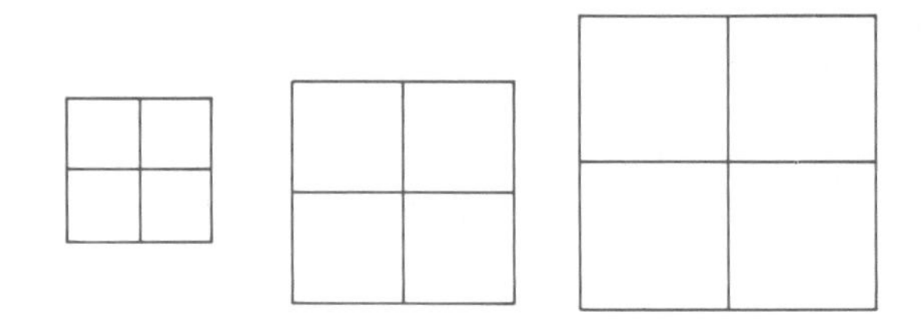

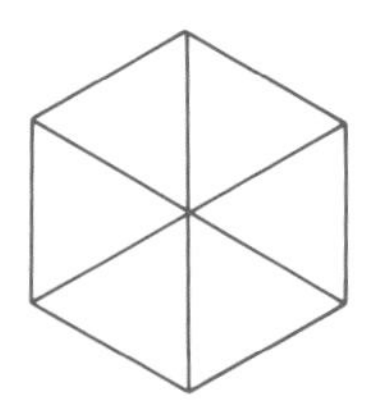

The next procedure can be used to draw a figure that we call a snowflake.

```
TO SNOWFLAKE :SIDE
REPEAT 6[TRIANGLE :SIDE RT 6Ø]
END
```

2 * :SIDE indicates a magnitude of twice the value of the variable :SIDE.

The SNOWFLAKE :SIDE procedure can then be used, as in the following procedure, as a building block for a more general snow flake by moving the snowflake about the screen and repeating it.

```
TO SUPERFLAKE :SIDE
REPEAT 6[FD 2 * :SIDE SNOWFLAKE :SIDE LT 60
         FD 2 * :SIDE SNOWFLAKE :SIDE RT 120]
END
```

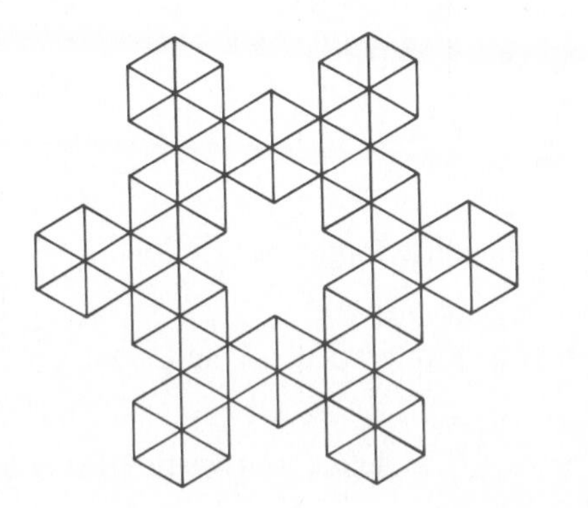

EXERCISES *Use pencil and paper to sketch the output of each procedure. Observe the final position of the turtle even if it is hidden. In Exercise 4 use the procedures that have been defined in Exercises 1 through 3.*

1.
```
TO LETTER
FD 100 RT 90 FD 30 BK 60 PU HT LT 90
END
```
2.
```
TO WRITE
PU FD 60 PD BK 60 RT 90 FD 30 LT 90 FD 60
END
```
3.
```
TO FINISH
FD 60 RT 90 FD 30 RT 90 PU FD 30 RT 90 FD 10 PD FD 20
END
```
4.
```
TO COMPOSE
LT 90 PU FD 40 RT 90 PD LETTER PU HOME PD WRITE
PU HOME RT 90 FD 60 LT 90 PD FINISH
END
```

Select appropriate distances and angles and write a Logo program for drawing a figure of the given shape.

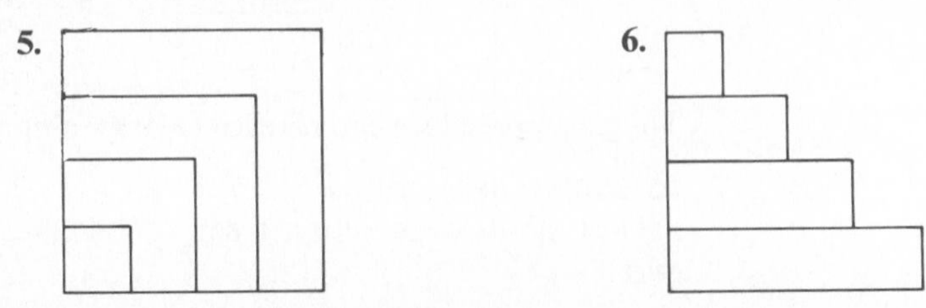

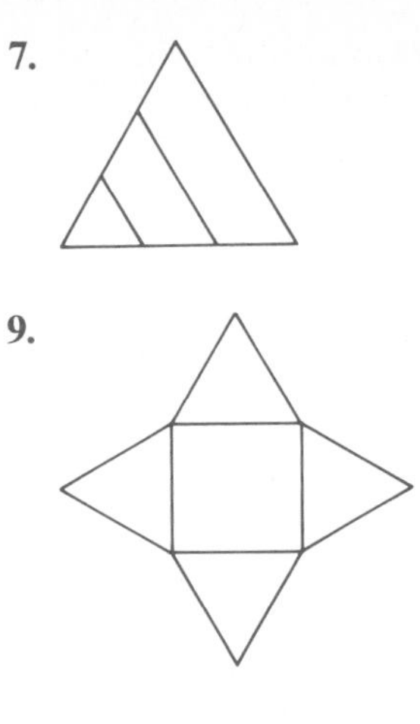

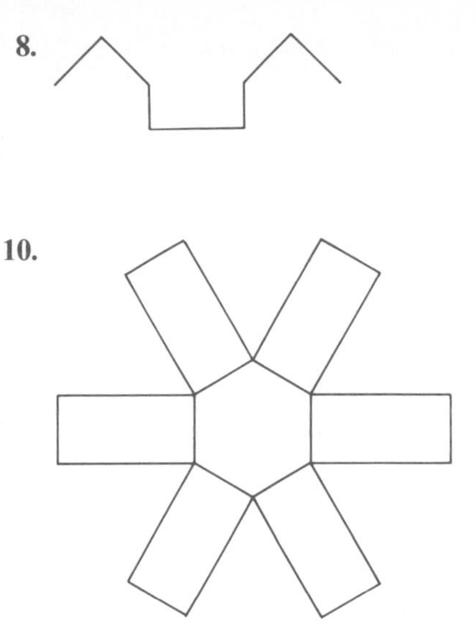

EXPLORATIONS

1. The following procedure provides an opportunity to study the repetition of a simple design to create more involved designs.

```
TO TURN :R :N :SIDE
REPEAT :R[NGON :N :SIDE RT 360/:R]
END
```

Consider the output associated with

```
TURN 8 3 60,    TURN 4 4 60,    TURN 24 3 60,
```

and some other possibilities.

2. The following procedure shows the use of a procedure within itself:

```
TO SQUAREALOT :SIDE
REPEAT 4[FD :SIDE RT 90]
RT 18
SQUAREALOT :SIDE
END
```

Consider SQUAREALOT and examine the drawing it defines. Note that it does not stop but continues to draw and draw until you interfere with the program, making it stop. This is because the program continues to refer to itself in executing the commands. The process by which a program refers to itself in executing the program is called *recursion* and is a very important concept in computing.

Chapter Review

Write each statement in BASIC.

1. $7(5 - 11)$
2. $5^2 + 13$
3. $35 \div 2$
4. $72 \div (3^3 - 3)$

List, in order, the operations that you would perform to evaluate each expression.

5. $29 - 2 \times 5$
6. $11 + 3^2$
7. $190 \div (3^3 - 17)$
8. $4^2 \div 2 \times 5^2 + 3$

Evaluate each of these BASIC commands.

9. `(5 * 3) ∧ 2`
10. `15 + (20/4) ∧ 3`
11. `INT(1Ø.5Ø23 – Ø.Ø5)`
12. `SQR(INT(36 + 64.Ø5))`

State in ordinary notation the printed output associated with each BASIC program.

13.
```
1Ø  FOR X = 1 TO 4
2Ø  PRINT X, 12 – X
3Ø  NEXT X
4Ø  END
```
14.
```
1Ø  READ A, B
2Ø  PRINT A, B, A * B
3Ø  GOTO 1Ø
4Ø  DATA 1, 2, 3, 4, 5, 6
```
15.
```
1Ø  FOR N = 1 TO 5
2Ø  PRINT N, N ∧ 2, N ∧ 3
3Ø  NEXT N
4Ø  END
```
16.
```
1Ø  REM PYTHAGORAS' DELIGHT
2Ø  READ A, B, C
3Ø  PRINT A, B, C
4Ø  IF A * A + B * B = C * C THEN GOTO 7Ø
5Ø  PRINT "ARE NOT SIDES OF A RIGHT TRIANGLE"
6Ø  GOTO 1Ø
7Ø  PRINT "ARE SIDES OF A RIGHT TRIANGLE"
8Ø  GOTO 1Ø
9Ø  DATA 7, 24, 25, 9, 17, 2Ø
```

Sketch the output associated with each Logo instruction.

17. `RT 9Ø FD 1Ø RT 9Ø FD 1Ø LT 9Ø FD 1Ø`
18. `REPEAT 4[FD 1Ø RT 9Ø FD 1Ø LT 9Ø FD 1Ø RT 9Ø]`
19. `NGON 3 3Ø`
20. `NGON 3 3Ø NGON 4 4Ø NGON 5 5Ø NGON 6 6Ø`

21. Write a BASIC program to find the sum of the first 100 natural numbers.
22. Write a BASIC program to find the sum of the first 100 odd natural numbers.

Select appropriate distances and angles and write a Logo program for drawing a figure of the given shape.

23.

24.

25.

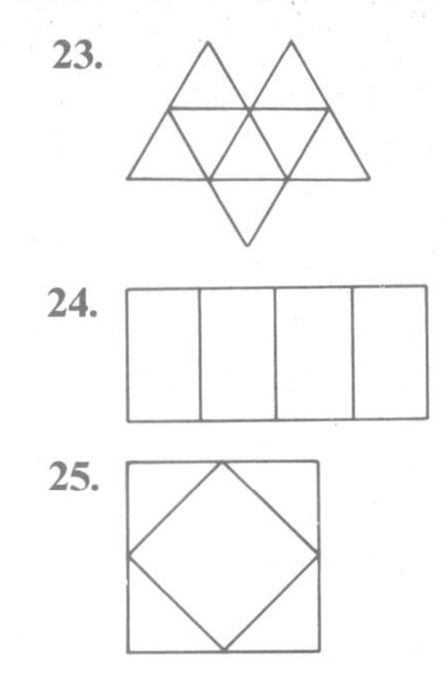

Answers to Odd-Numbered Exercises and Chapter Reviews

Chapter 1 An Introduction to Problem Solving

1-1 What is Mathematics?: page 6

1. (a) 25, 36, 49 (b) 14, 17, 20
3. (a) 12
   (b) Only halfway; then you start walking out.
   (c) *One* of them is not a nickel, but the other one is.
   (d) There is no dirt in a hole.
   (e) Brother-sister.
5. There are 11 trips needed. First one cannibal and one missionary go over; the missionary returns. Then two cannibals go over and one returns. Then two missionaries go over; one missionary and one cannibal return. Two missionaries go over next and one cannibal returns. Then two cannibals go over and one of them returns. Finally, the last two cannibals go over.
7. After 15 days the cat still has 3 feet to go. The cat does this the next day and is at the top after 16 days.
9.

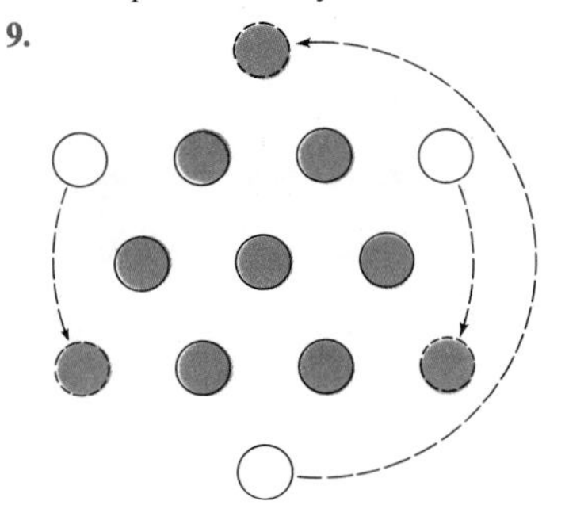

11. If the penny is in the left hand and the dime in the right hand, the computation will give $3 + 60 = 63$, an odd number. If the coins are reversed, we have $30 + 6 = 36$, an even number.
13. There really are no missing dollars. The computation may be done in one of two ways: $(60 - 6) - 4 = 50$, or $50 + 4 + 6 = 60$. In the problem the arithmetic was done in a manner that is not legitimate, that is, $(60 - 6) + 4$.
15. Use the matchsticks to form a triangular pyramid.

*17. Both *A*'s and *B*'s will say that they are *B*'s. Therefore, when the second man said that the first man said he was a *B*, the second man was telling the truth. Thus the first two men told the truth and the third man lied.

19. Eight moves are needed. The coins are identified in the following diagram as well as the squares which may be used. The moves are as follows, where the first numeral indicates the position of the coin and the second one tells you where to move it: D_1: 4-3; P_2: 2-4; P_1: 1-2; D_1: 3-1; D_2: 5-3; P_2: 4-5; P_1: 2-4; D_2: 3-2.

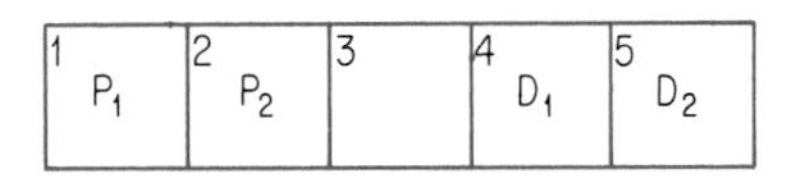

21. E, N. These are the first letters of the names (one, two, three, etc.) of the natural numbers.

23. Use H for the half-dollar, Q for the quarter, and N for the nickel. Assume that each coin is on a larger coin. The seven moves may be made in the order listed.

N: A to C	Q: A to B	N: C to B
H: A to C	N: B to A	Q: B to C
N: A to C		

For the four coins use P for the penny. Then the moves are

P: A to B	N: A to C	P: B to C
Q: A to B	P: C to A	N: C to B
P: A to B	H: A to C	P: B to C
N: B to A	P: C to A	Q: B to C
P: A to B	N: A to C	P: B to C

For a discussion of the number of moves required for 64 disks see page 171 of *Mathematics and the Imagination* by Edward Kasner and James Newman, Simon & Schuster, 1940. They estimate that it would take more than 58 billion centuries to complete the task.

25. The sum of the values on two opposite faces of any ordinary die is always 7. Therefore for three dice this sum is 21. Thus the difference between 21 and value on the top face of the top die is the sum of the values on the other specified faces.

27. **(a)** 3/4 **(b)** 7/8 **(c)** 15/16
In general: $(2^n - 1)/2^n$

29. Fred—red; Joe—green; Kathy—yellow; Linda—pink

1-2 Problem Solving with Arithmetic Patterns: page 17

1. 1×9, first; 2×9, second; 3×9, third; 4×9, fourth; 5×9, fifth; 6×9, sixth; 7×9, seventh; 8×9, eighth; 9×9, ninth.

3. $6^2 = 1 + 2 + 3 + 4 + 5 + 6 + 5 + 4 + 3 + 2 + 1$

$7^2 = 1 + 2 + 3 + 4 + 5 + 6 + 7 + 6 + 5 + 4 + 3 + 2 + 1$

$8^2 = 1 + 2 + 3 + 4 + 5 + 6 + 7 + 8 + 7 + 6 + 5 + 4 + 3 + 2 + 1$

$9^2 = 1 + 2 + 3 + 4 + 5 + 6 + 7 + 8 + 9 + 8 + 7 + 6 + 5 + 4 + 3 + 2 + 1$

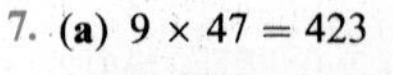
7. **(a)** $9 \times 47 = 423$

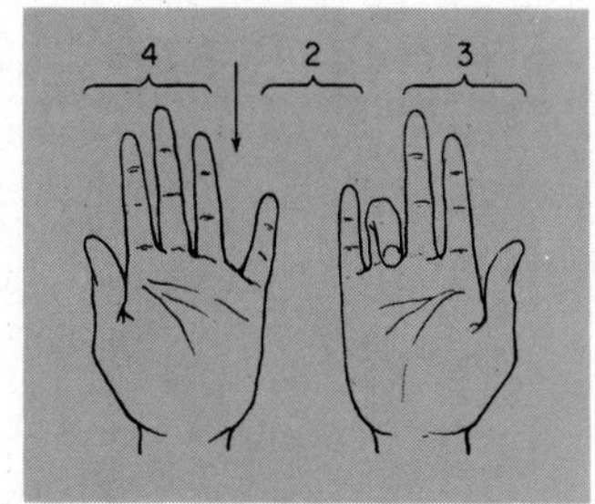

(b) $9 \times 39 = 351$

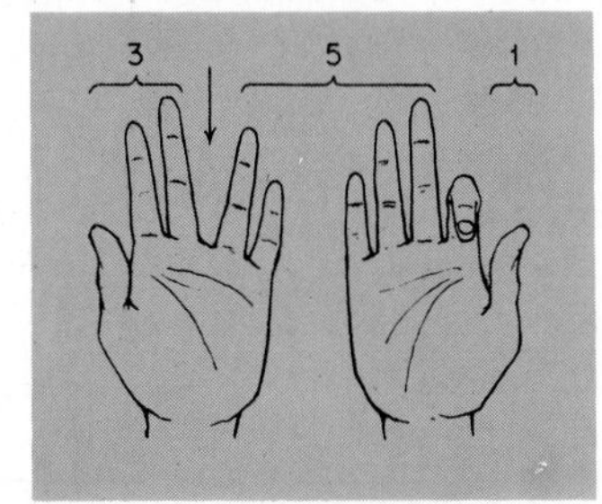

(c) $9 \times 18 = 162$

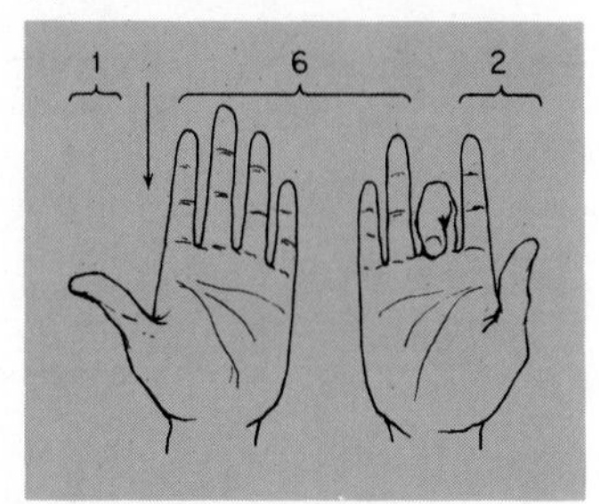

(d) $9 \times 27 = 243$

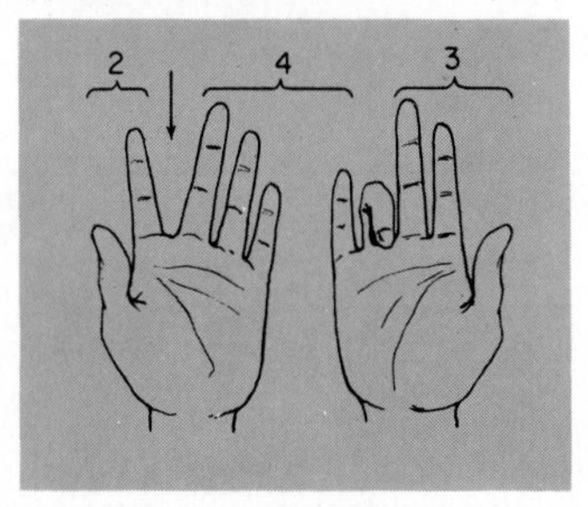

9. (a) 45×91, that is, 4095.
   (b) 150×301, that is, 45,150.
   (c) 10×40, that is, 400.
   (d) 75×300, that is, 22,500.
   (e) 125×502, that is, 62,750.
11. The sum appears to be nine times the number in the center of the array.
13. The sum of the two outer numbers appears to be equal to the sum of the two inner numbers.
15. (a) The conjecture is false.
    (b) The conjecture appears to be true.
17. There are only three such ways: $1 + 1 + 1 + 7$, $1 + 1 + 3 + 5$, and $1 + 3 + 3 + 3$.

*19. Impossible. The sum of three odd numbers must be an odd number.

1-3 Problem Solving with Logical Reasoning: page 27

1. If you pass the examination, then you pass the course.
3. If you do not pass the examination, then you do not pass the course.
5. (a) I like this book and I like mathematics.
   (b) I do not like mathematics.
   (c) It is not true that I do not like this book, that is, I like this book.
   (d) I do not like this book and I do not like mathematics.
7. 5(a), 5(c), 6(b), and 6(d).
9. Among others:
   (a) $(\sim p) \wedge (\sim q)$
   (b) $(\sim p) \wedge q$
   (c) $(\sim p) \wedge q$
   (d) $\sim[p \wedge (\sim q)]$
   (e) $p \vee (\sim q)$.
11. Among others:
    (a) $p \wedge (\sim q)$
    (b) $p \vee (\sim q)$
    (c) $(\sim p) \wedge q$
    (d) $\sim[(\sim p) \wedge q]$
    (e) $\sim[(\sim p) \wedge q]$.
13. It is not true that today is Monday. Today is not Monday.
15. It is not true that these two cars are not made by the same company. These two cars are made by the same company.
17. It is not true that I am young and I am happy. I am not young or I am not happy.
19. It is not true that Pedro has $50 and he has two tickets to the game. Pedro does not have $50 or he does not have two tickets to the game.
21. It is not true that the textbook is expensive or it is not used. The textbook is not expensive and it is used.
23. Every college is expensive.
25. (a) i (b) ii (c) i (d) ii
27. (a) True
    (b) True
    (c) False
    (d) True
29. (a) True
    (b) False
    (c) True
    (d) True
31. Negation: $x = 1$ and $x = 2$

    Converse: If $x \neq 2$, then $x = 1$.

    Inverse: If $x \neq 1$, then $x = 2$.

    Contrapositive: If $x = 2$, then $x \neq 1$.
33. Negation: We can afford a new car and we shall not buy it.

    Converse: If we buy a new car, then we can afford it.

    Inverse: If we cannot afford it, then we do not buy a new car.

    Contrapositive: If we do not buy a new car, then we cannot afford it.

1-4 Solving Verbal Problems with Logical Reasoning: page 36

Among others:

1. If a piece of fruit is an apple, then it is red.
3. If an animal is a dog, then it is a good watchdog.
5. If a geometric figure is a square, then it is a polygon.
7. If two people are ball players, then they are competitors.
9. If an object is an automobile, then it is expensive.
11. If you like this book, then you like mathematics.
13. If you like mathematics, then you like this book.
15. If you like mathematics, then you like this book.

17. If $12 + 4 = 15$, then $12 - 4 = 7$; true.

19. If $7 + 4 = 11$, then $7 \times 4 = 20$; false.

21. If $7 \times 5 = 75$, then $15 \times 5 \neq 75$; true.

23. I am happy and I pass the test. I am not happy and I do not pass the test.

25. You pass the test and you study hard. You do not pass the test and you study hard. You do not pass the test and you do not study hard.

27. I put on a sweater and I felt chilly. I do not put on a sweater.

29. For p: Elliot is a freshman.
q: Elliot takes mathematics.
The argument has the form
$[(p \rightarrow q) \wedge p] \rightarrow q$ and is valid.

31. For p: The Braves win the game.
q: The Braves win the pennant.
The argument has the form
$[(p \rightarrow q) \wedge (\sim q)] \rightarrow (\sim p)$ and is valid.

33. For p: You work hard.
q: You are a success.
The argument has the form
$[(p \rightarrow q) \wedge (\sim q)] \rightarrow (\sim p)$ and is valid.

35. For p: You are reading this book.
q: You like mathematics.
The argument has the form
$[(p \rightarrow q) \wedge (\sim p)] \rightarrow (\sim q)$ and is not valid.

37. This argument is of the form
$[(p \rightarrow q) \wedge (q \rightarrow r)] \rightarrow (r \rightarrow p)$
and is not valid.

*39. If I were to ask you "Is this the way to the river?" would you say "Yes"?

1-5 Problem Solving with Algebraic Patterns and Geometric Patterns: page 43

1. $(n - 11) + (n - 10) + (n - 9) + (n - 1) + (n) + (n + 1) + (n + 9) + (n + 10) + (n + 11) = 9n$, which is nine times the center number n.

3. The numbers in the shaded regions can be represented as shown. The sum of the quantities shown is $8n$. Note that $-11 + 11 = 0$, $-10 + 10 = 0$, $-9 + 9 = 0$, and $-2 + 2 = 0$.

	$n - 11$	$n - 10$	$n - 9$	
$n - 2$		n		$n + 2$
	$n + 9$	$n + 10$	$n + 11$	

5. The array has the form shown below. Each sum of numbers in opposite corners is $2a + 16$.

a	$a + 1$	$a + 2$
$a + 7$	$a + 8$	$a + 9$
$a + 14$	$a + 15$	$a + 16$

7. Use the array for Exercise 5. The products are $a(a + 16) = a^2 + 16a$ and $(a + 2)(a + 14) = a^2 + 16a + 28$. The constant difference is 28.

9. 0; n, $n + 7$, $3n + 21$, $3n$, n, 0.

11. There will be 4 holes produced by 4 folds, and 2^{n-2} holes with n folds.

13. $V + R = A + 2$. The sum of the number of vertices and the number of regions is 2 more than the number of arcs.

*15. Many answers are possible.

(a)

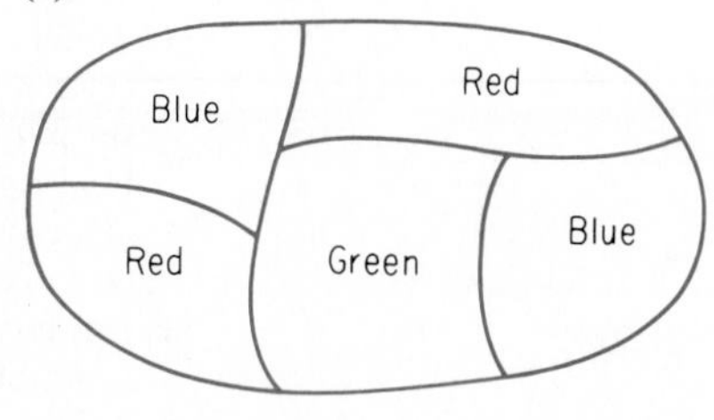

(b)

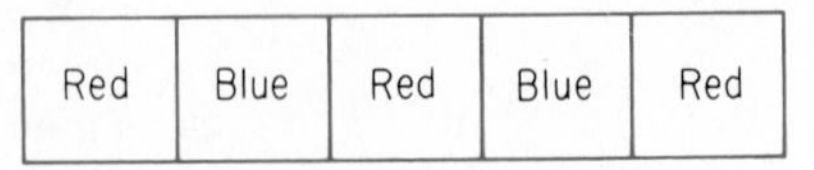

*17. 10; the faces would be all red, 1 red and 5 blue, 1 blue and 5 red, 2 adjacent blue and 4 red, 2 opposite blue and 4 red, 2 adjacent red and 4 blue, 2 opposite red and 4 blue, 3 red including two opposite faces and 3 blue, 3 red with no two opposite and 3 blue, or all blue.

*19. Represent the four cyclic numbers as follows and add:

$$\begin{array}{l} 1000a + 100b + 10c + d \\ 1000b + 100c + 10d + a \\ 1000c + 100d + 10a + b \\ \underline{1000d + 100a + 10b + c} \end{array}$$

The sum is $1000(a + b + c + d) + 100(a + b + c + d) + 10(a + b + c + d) + (a + b + c + d) = (1000 + 100 + 10 + 1)(a + b + c + d) = 1111(a + b + c + d)$. Dividing this sum by $a + b + c + d$ gives 1111.

*21. From the top of the deck n cards were removed by your friend and 20 more cards were removed to be spread out. Then your friend counted backward (toward the top of the deck) starting with the displayed card that had been furthest from the top of the deck. Since $n + 20 - n = 20$, the last card counted will be the predetermined card that had been placed 21st from the top of the deck.

Chapter Review: page 48

1. $1 + 4 + 9 + 16 + 25 + 36 + 49 + 64$, that is, 204

2.
1, 1, 72	1, 2, 36	1, 3, 24
1, 4, 18	1, 6, 12	1, 8, 9
2, 2, 18	2, 3, 12	2, 4, 9
2, 6, 6	3, 4, 6	3, 3, 8

3. There are 75 pairs of numbers. The sum of each pair is 151. Thus the total sum is 75×151, that is, 11,325.

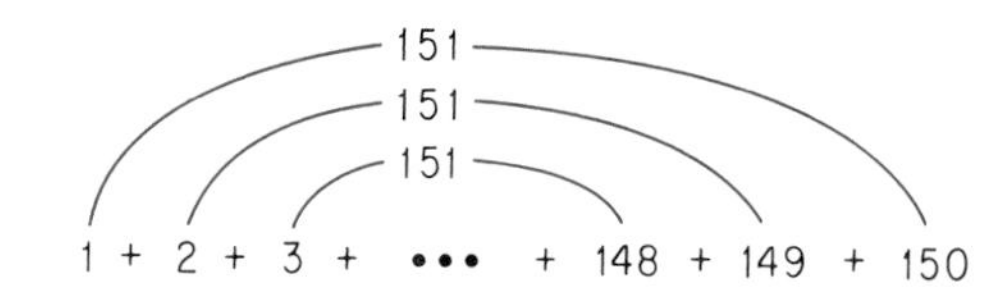

4. There are 50 pairs of numbers. The sum of each pair is 202. Thus the total sum is 50×202, that is, 10,100.

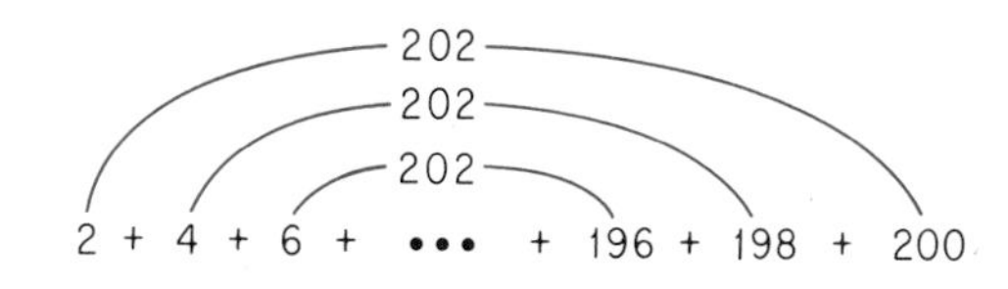

5. Among others:

   The numbers increase by 10 in each vertical column.

   The multiples of 10 are in the last column.

   There is a diagonal line that contains the multiples of 11. (See the third paragraph of Section 1-2.)

   There is a diagonal line that contains the multiples of 9. (See the fourth paragraph of Section 1-2.)

   For any square array of nine numbers the sums of the numbers in the opposite corners are equal.

6. Among others:

   The multiples of 9 are in the last column at the right.

   The sum of the digits of each number in the ninth row, or column, is 9.

   There is a diagonal line that contains the squares of the first nine natural numbers.

   Columns and rows with the same headings have the same elements.

   For any rectangular array of numbers the products of the numbers in opposite corners are equal.

7. Consider the array shown. The sums of the numbers in opposite corners are $n + (n + 40) = 2n + 40$ and $(n + 4) + (n + 36) = 2n + 40$ and thus are equal.

(n)	$n + 1$	$n + 2$	$n + 3$	($n + 4$)
$n + 12$	$n + 13$	$n + 14$	$n + 15$	$n + 16$
$n + 24$	$n + 25$	$n + 26$	$n + 27$	$n + 28$
($n + 36$)	$n + 37$	$n + 38$	$n + 39$	($n + 40$)

(Other forms of the array are possible.)

8. (a) Each sum is 1 less than the next power of 2, that is, $2^0 + 2^1 + 2^2 + \cdots + 2^n = 2^{n+1} - 1$.
   (b) $2^0 + 2^1 + 2^2 + 2^3 + 2^4 + 2^5 + 2^6 + 2^7 + 2^8 = 2^9 - 1 = 512 - 1 = 511$

9. Let $n = 8$. Then $[n(n - 3)]/2 = [8(5)]/2 = 20$.

10. Let $n = 8$. Then $[n(n - 1)]/2 = [8(7)]/2 = 28$. Alternatively, in terms of the numbers of diagonals and sides of an octagon, $[n(n - 3)]/2 + n = 20 + 8 = 28$. The answer can also be found by making a list of all possibilities.

(a) The indicated sums, 1, 3, 6, 10, have differences that increase by 1 each time. That is,

$$\underbrace{1 \quad 3}_{2} \quad \underbrace{\;}_{3} \quad 6 \quad \underbrace{\;}_{4} \quad 10$$

1 3 6 10
2 3 4

(b) To find the next three sums extend the array; that is, add 5, 6, and 7 to each preceding sum. Thus $10 + 5 = 15$, $15 + 6 = 21$, $21 + 7 = 28$, and the next three sums are 15, 21, and 28.

12. **(a)** $p \wedge q$ **(b)** $(\sim p) \wedge (\sim q)$
13. **(a)** $p \vee (\sim q)$ **(b)** $\sim[p \wedge (\sim q)]$
14. 13(a)
15. **(a)** I work hard and I shall not pass the course.
    (b) If I do not pass the course, then I have not worked hard.
16. *Negation*: The apples are red and they are not ripe.

    Converse: If the apples are ripe, then they are red.

    Inverse: If the apples are not red, then they are not ripe.

    Contrapositive: If the apples are not ripe, then they are not red.
17. **(a)** True **(b)** True **(c)** False **(d)** True
18. **(a)** If an animal is a horse, then it is a quadruped.
    (b) If you know Judy, then you like her.
19. **(a)** $p \rightarrow q$ **(b)** $p \rightarrow q$
20. **(a)** Valid **(b)** Not valid.

Chapter 2 Counting and Classifying—Sets and Reasoning

2-1 Counting and Classifying: page 56

1. 17
3. 27
5. 4
7. 15
9. **(a)** 8 **(b)** 12 **(c)** 6 **(d)** 1
11. **(a)** 0 **(b)** 8 **(c)** 44 ***(d)** 56 ***(e)** 12
13. **(a)** 1 **(b)** 5 **(c)** 14
15. 10
17. 14
19. 27
21. 40

*23. 12

2-2 Sets and Set Notation: page 65

1. **(a)** Yes **(b)** No **(c)** Yes
3. **(a)** Yes **(b)** Yes **(c)** Yes
5. Among others: the set of natural numbers less than 100.
7. Among others: the set of natural numbers that are multiples of 5.
9. Among others: the set of perfect squares, 1 through 36.
11. $\{x | (x \in W) \text{ and } (x < 10)\}$
13. Among others: $\{10x | (x \in N) \text{ and } (x < 20)\}$
15. **(a)** Equivalent **(b)** Equal
17. **(a)** Equivalent **(b)** Not equal
19. **(a)** 99 **(b)** 6
21. **(a)** $\varnothing$
    (b) $\{s, t, u\}$, $\{r, t, u\}$, $\{r, s, u\}$, $\{r, s, t\}$, $\{r, s\}$, $\{r, t\}$, $\{r, u\}$, $\{s, t\}$, $\{s, u\}$, $\{t, u\}$, $\{r\}$, $\{s\}$, $\{t\}$, $\{u\}$, $\varnothing$
23. **(a)** $\{1, 3, 4, 5, 7\}$ **(b)** $\{1, 3\}$
25. **(a)** $\{2, 4, 6, 7, 8\}$ **(b)** $\{4, 6, 8\}$
27. **(a)** $\{1, 2, 3, \ldots\}$ **(b)** $\varnothing$
29. **(a)** $\{3, 4, 5\}$ **(b)** $\{2, 4\}$
    (c) $\{2, 3, 4, 5\}$ **(d)** $\{4\}$
31. **(a)** $\{1, 2, 3, 4, 5, 6, 7\}$ **(b)** $\varnothing$
    (c) $\{1, 2, 3, 4, 5, 6, 7\}$ **(d)** $\varnothing$
33. $\{(0, 3), (0, 4), (1, 3), (1, 4), (2, 3), (2, 4)\}$
35. $\{(3, 3), (3, 4), (4, 3), (4, 4)\}$
37. $\varnothing$
39. **(a)** False **(b)** $n(B) \neq n(B')$
41. **(a)** True **(b)** $\varnothing$ is a subset of any set.
43. $A = \{1, 2, 3\}$, $B = \{2, 3\}$
45. $A = \{n, a, t, e, w, p\}$, $B = \{z\}$
47. **(a)** Always true
    (b) Not always true. For example, it is false if $A = \{1, 2\}$ and $B = \{1\}$. In general, the statement is false if $B \subseteq A$.
49. **(a)** Not always true. For example, if $A = \{1\}$ and $B = \{2\}$, the statement is false. In general, the statement is false if $A \cap B = \varnothing$.
    (b) Always true

*51. $A \subseteq B$

*53. Always true

*55. Always true

*57. $A = B$

*59. Always true

2-3 Venn Diagrams: page 73

1. **(a)** 3 **(b)** 7
   (c) 5 **(d)** 12

3. A' is shaded with horizontal lines; B is shaded with vertical lines. The union of these two sets is the subset of $\mathcal{U}$ that is shaded with horizontal lines, vertical lines, or lines in both directions.

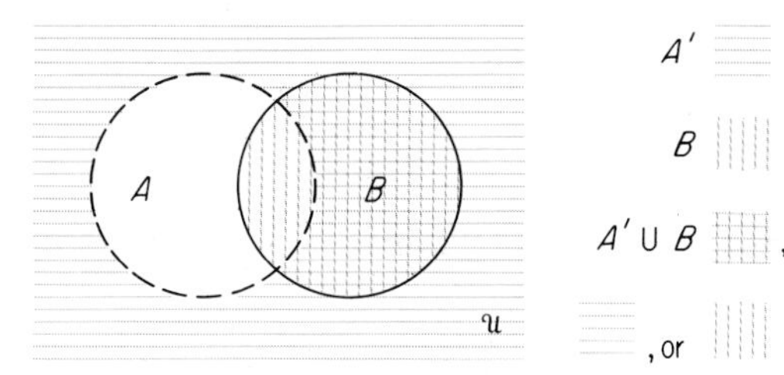

5. A is shaded with vertical lines; B' is shaded with horizontal lines. The intersection of these two sets is the subset of $\mathcal{U}$ that is shaded with lines in both directions.

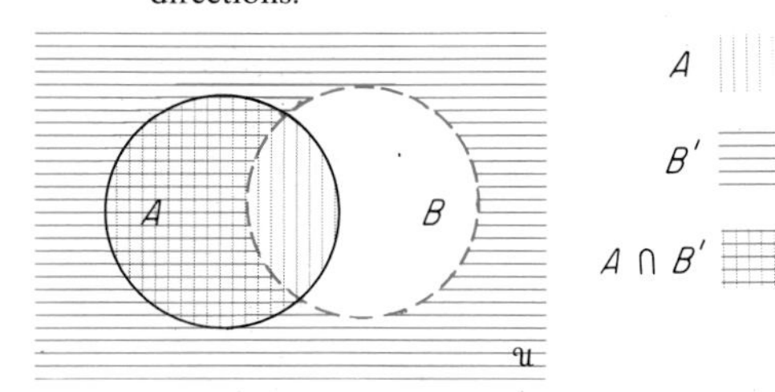

7. In the following pair of diagrams, the final result is the same, showing the equivalence of the statements given. The set $(A \cap B)$ is shaded with horizontal lines. Its complement, $(A \cap B)'$, is the remaining portion of $\mathcal{U}$ shaded with vertical lines.

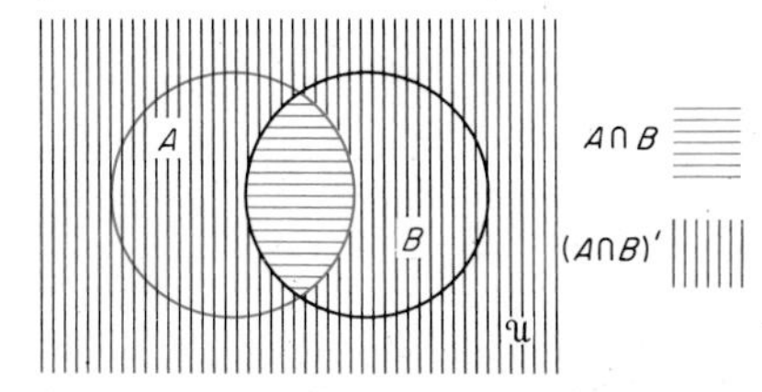

The set A' is shaded with vertical lines; the set B' is shaded with horizontal lines. Their union, $A' \cup B'$, is the portion of $\mathcal{U}$ shaded with vertical lines, horizontal lines, or lines in both directions.

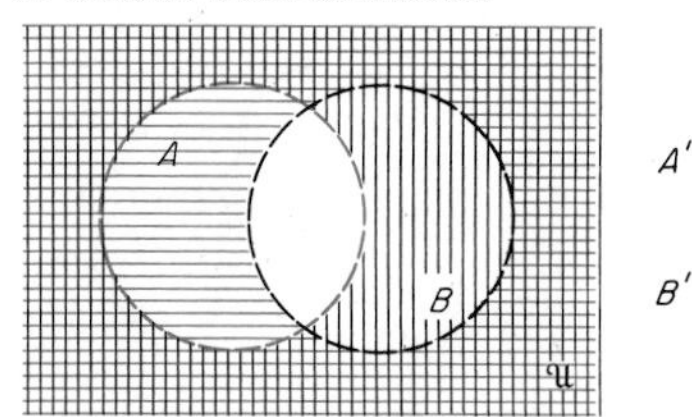

9. **(a)** 2 **(b)** 4 **(c)** 6
   (d) 15 **(e)** 28 **(f)** 29

11. **(a)** 33 **(b)** 10 **(c)** 10
    (d) 39 **(e)** 46 **(f)** 8

13. **(a)**

$A \cap B \cap C$

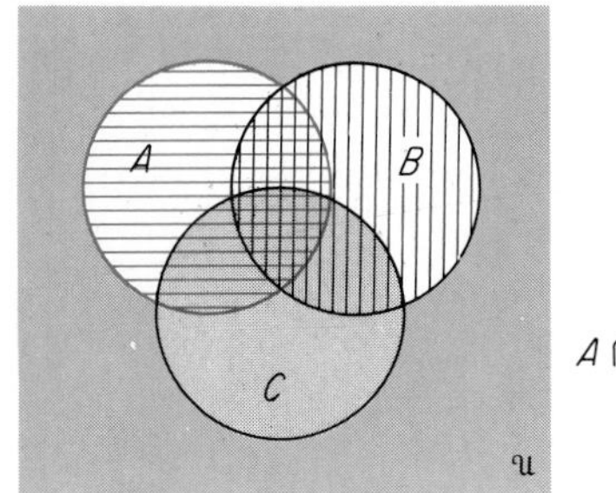

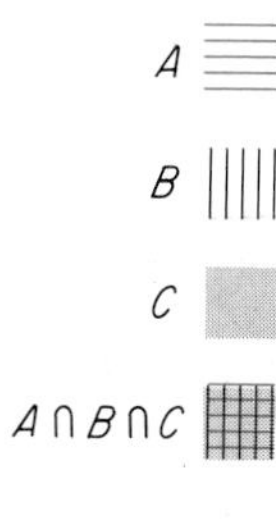

(b)

$A \cap B \cap C'$

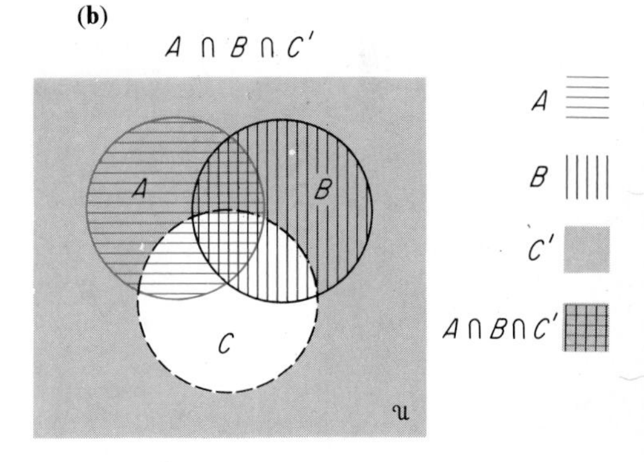

(c)

$A \cap B' \cap C$

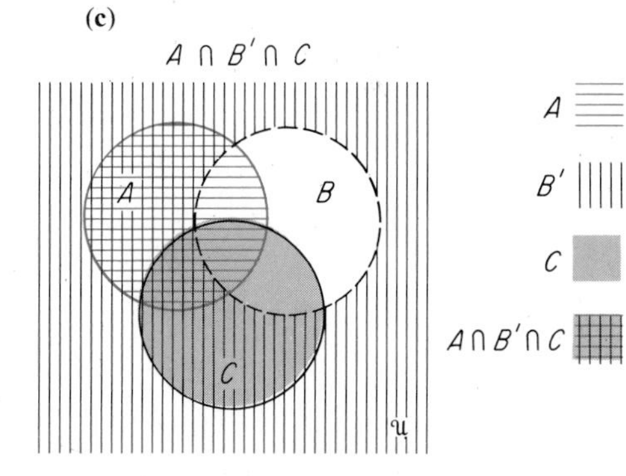

(d)

$A \cap B' \cap C'$

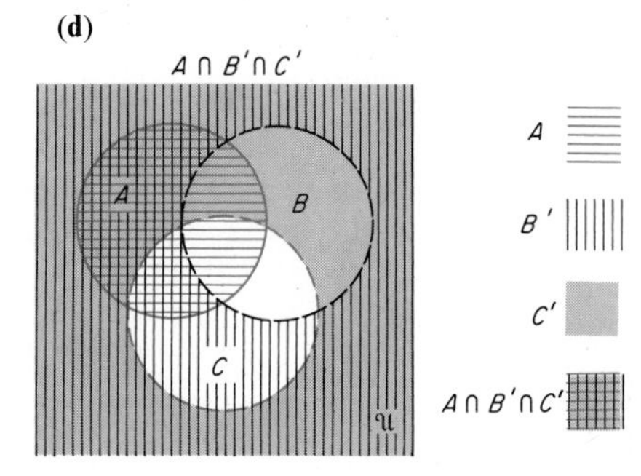

15. $(A \cup B) \cap C'$

17. $A \cap B' \cap C$

19. As shown in the following diagram, **(a)** 7 **(b)** 11 **(c)** 23.

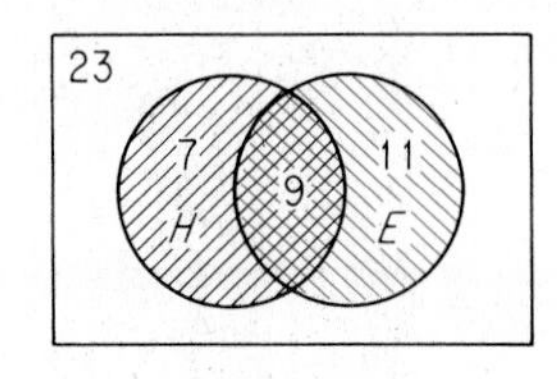

21. As shown in the following diagram, **(a)** 12 **(b)** 9 **(c)** 3.

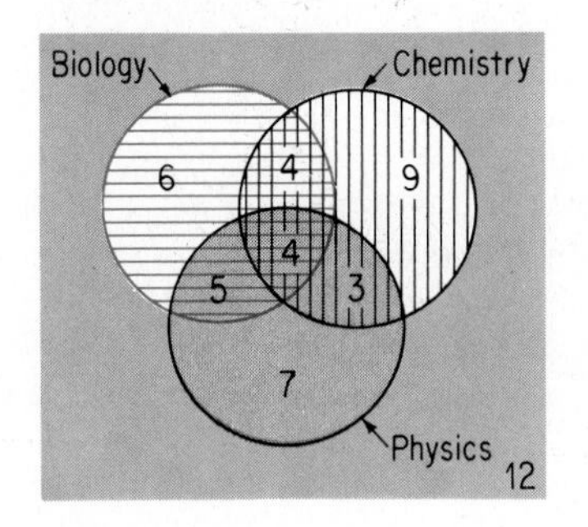

23. As shown in the following diagram, **(a)** 4 **(b)** 0.

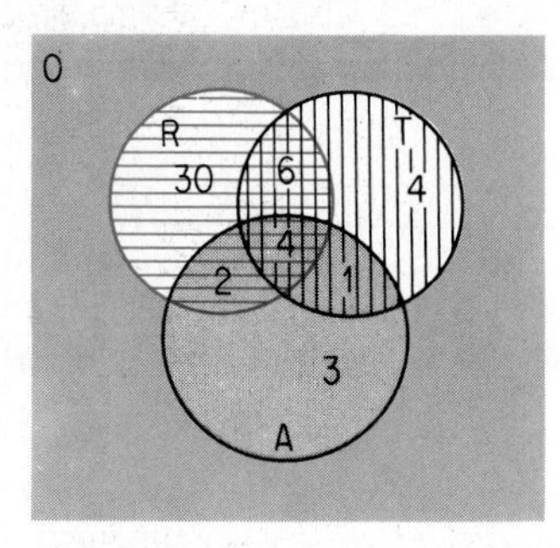

25. As shown in the following diagram, the data would require -2 students in $A' \cap B' \cap C$, which is impossible.

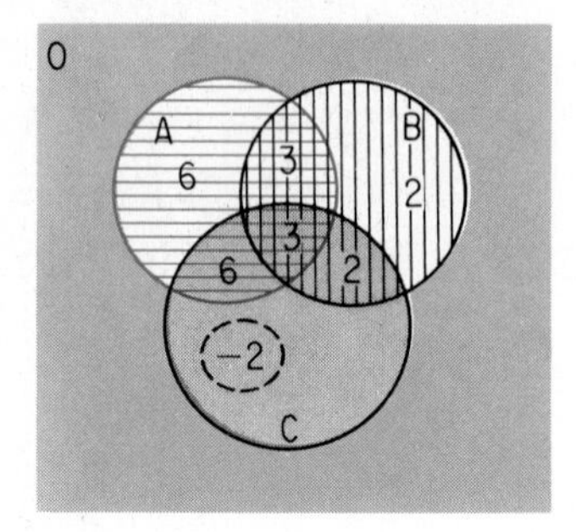

27. 150

2-4 Properties of Set Operations: page 81

1.

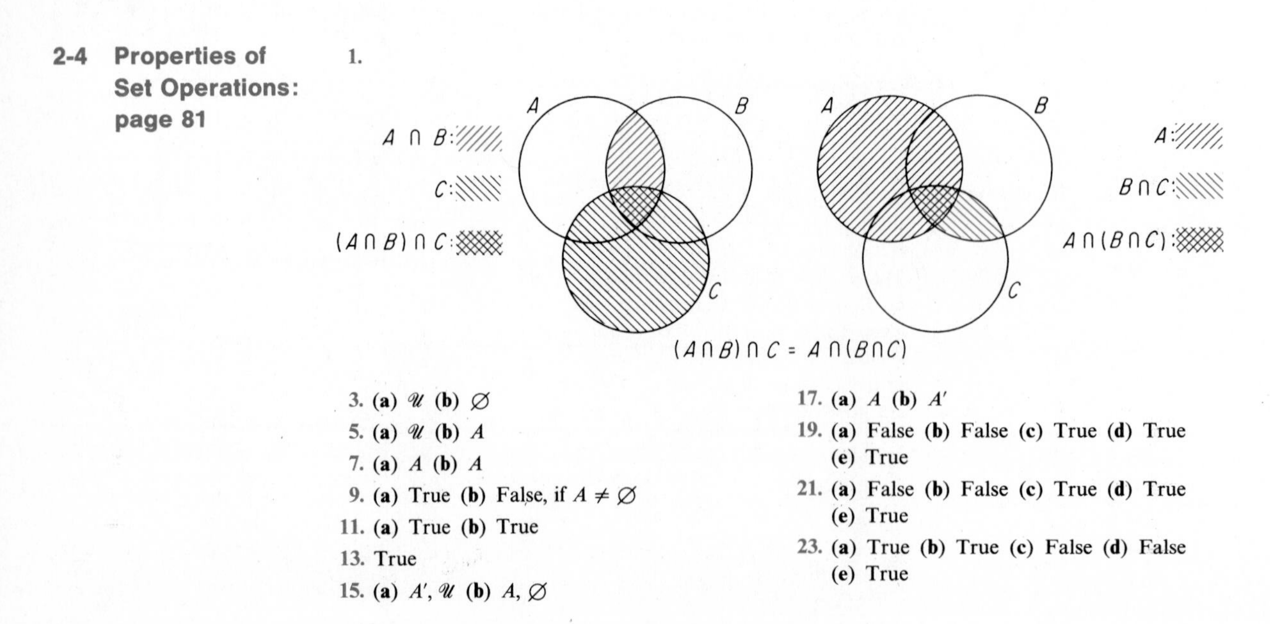

3. **(a)** $\mathscr{U}$ **(b)** $\varnothing$

5. **(a)** $\mathscr{U}$ **(b)** A

7. **(a)** A **(b)** A

9. **(a)** True **(b)** False, if $A \neq \varnothing$

11. **(a)** True **(b)** True

13. True

15. **(a)** $A', \mathscr{U}$ **(b)** $A, \varnothing$

17. **(a)** A **(b)** A'

19. **(a)** False **(b)** False **(c)** True **(d)** True **(e)** True

21. **(a)** False **(b)** False **(c)** True **(d)** True **(e)** True

23. **(a)** True **(b)** True **(c)** False **(d)** False **(e)** True

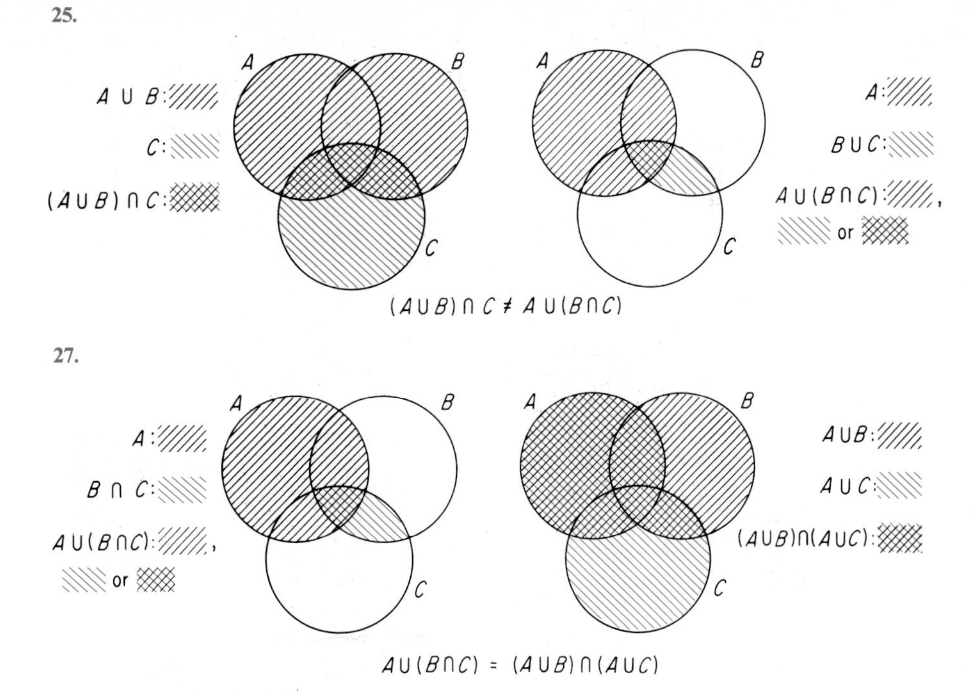

29. False, for example, if $A = \{1\}$, $B = \{2\}$, $C = \varnothing$, then $A \cup (B \times C) = A$ but $(A \cup B) \times (A \cup C) = \{(1, 1)\}$.

2-5 Use of Classifications in Reasoning: page 88

1. Karen is a gardener; Chad is an educator.
3. Dennis (first), Jim, John
5. Ha teaches statistics; Edge teaches analysis.
7. The fudge box contained macaroons; the macaroon box contained some chocolate chip cookies and some fudge; the chocolate chip box contained fudge.
9. Rose drives the pickup, Sue drives the panel truck, and Theresa drives the van.
11. Babette was first with an iris arrangement, Andrea was second with a rose arrangement, and Chandra was third with a daisy arrangement.

Chapter Review: page 90

1. 63, 127, 255, 511, 1023
2. **(a)** 6 **(b)** 12 **(c)** 8
3. 11 (*between*)
4. 16
5. Among others: the set of odd natural numbers less than 10.
6. **(a)**

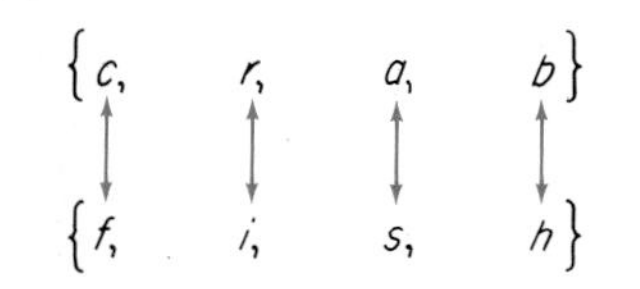

The sets are equivalent since they can be placed in one-to-one correspondence, that is, they have the same cardinal number.

(b) The sets are not equal since they have different elements.

7. $\varnothing$, $\{4\}$, $\{5\}$, $\{6\}$, $\{4, 5\}$, $\{4, 6\}$, $\{5, 6\}$
8. $A' = \{6, 8, 10\}$
9. $B' = \{1, 3, 5, 7, 9, 10\}$
10. $A - B = \{1, 3, 5, 7, 9\}$
11. $B - B = \varnothing$
12. $X \cap Y = \{4, 5\}$

13. $A' \cap B' = \{8, 9, 10\}$

14. $A \cup B = \{1, 2, 3, 4, 5, 7, 9\}$

15. $B \times A = \{(r, 1), (r, 2), (r, 3), (s, 1), (s, 2), (s, 3)\}$

16.

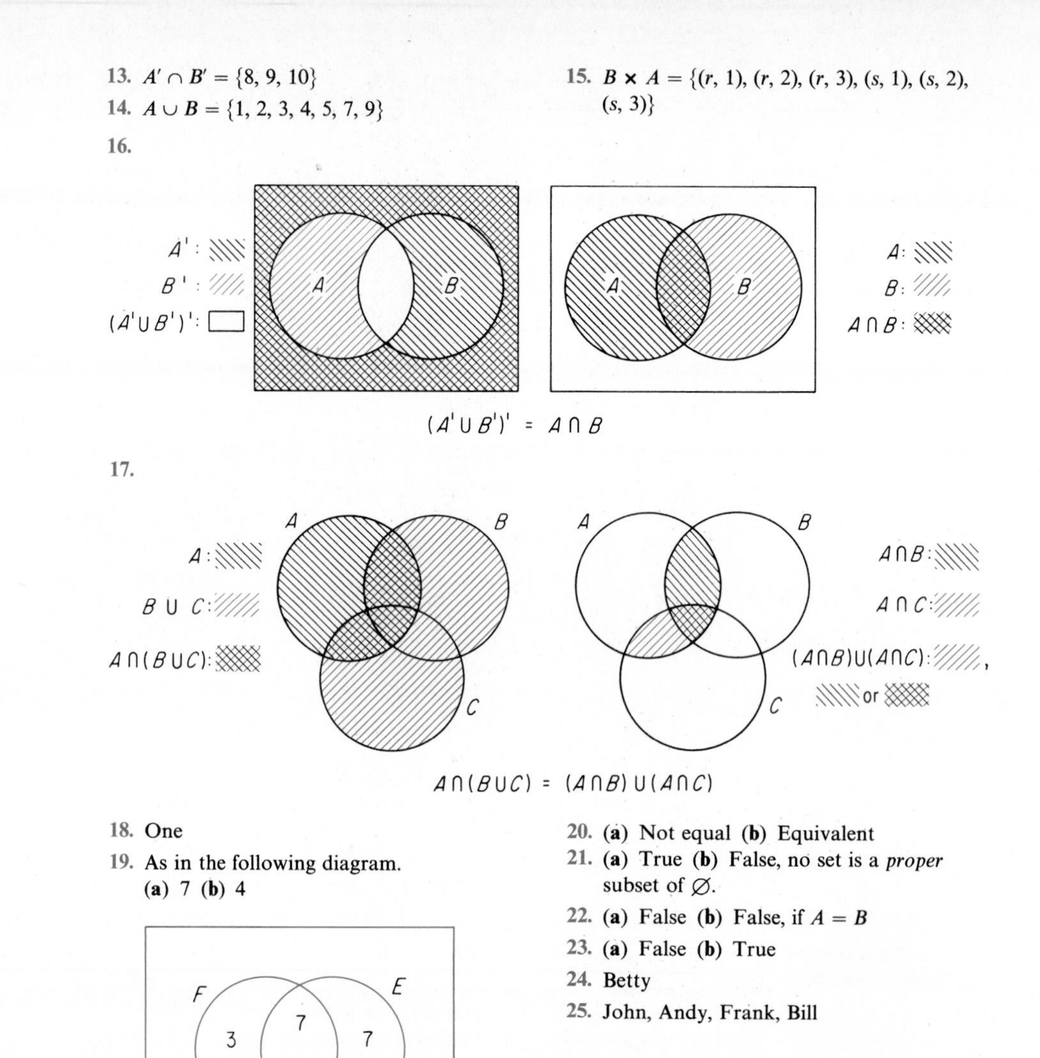

17.

18. One

19. As in the following diagram.
(**a**) 7 (**b**) 4

20. (**a**) Not equal (**b**) Equivalent

21. (**a**) True (**b**) False, no set is a *proper* subset of $\varnothing$.

22. (**a**) False (**b**) False, if $A = B$

23. (**a**) False (**b**) True

24. Betty

25. John, Andy, Frank, Bill

Chapter 3 Numeration and Whole Numbers

3-1 Base Ten Notation: page 98

1. ∩∩∩|||||

3. 𓆼𓆼𓆼𓍢𓍢𓍢𓍢∩|||||||

5. 𓂭𓆼𓆼𓍢𓍢𓍢|||||||

7. 22

9. 1102

11. 1212

13. $(2 \times 10^2) + (5 \times 10^1) + (7 \times 10^0)$

15. $(3 \times 10^3) + (5 \times 10^2) + (0 \times 10^1) + (4 \times 10^0)$

17. $(2 \times 10^5) + (3 \times 10^4) + (5 \times 10^3) + (1 \times 10^2) + (0 \times 10^1) + (0 \times 10^0)$

19. $(5 \times 10^5) + (0 \times 10^4) + (0 \times 10^3) + (2 \times 10^2) + (0 \times 10^1) + (0 \times 10^0)$

21. 8165

23. 609,502

25. 60,000,000

27. 700,000,000

29. 37,019

31. 300,023

33. Five thousand three hundred seventy

35. Two hundred five thousand thirty

37. Twenty-five million two hundred three thousand five hundred

39. 345

41. 2300

43.

45.

47.

49.

51.

$(1) \times 45 = (45)$
$(2) \times 45 = (90)$
$4 \times 45 = 180$
$8 \times 45 = 360$
$(16) \times 45 = (720)$
$19 = 1 + 2 + 16$
$19 \times 45 = (1 + 2 + 16) \times 45$
$= 45 + 90 + 720 = 855$

53.

$(1) \times 55 = (55)$
$2 \times 55 = 110$
$4 \times 55 = 220$
$8 \times 55 = 440$
$(16) \times 55 = (880)$
$17 = 1 + 16$
$17 \times 55 = (1 + 16) \times 55$
$= 55 + 880 = 935$

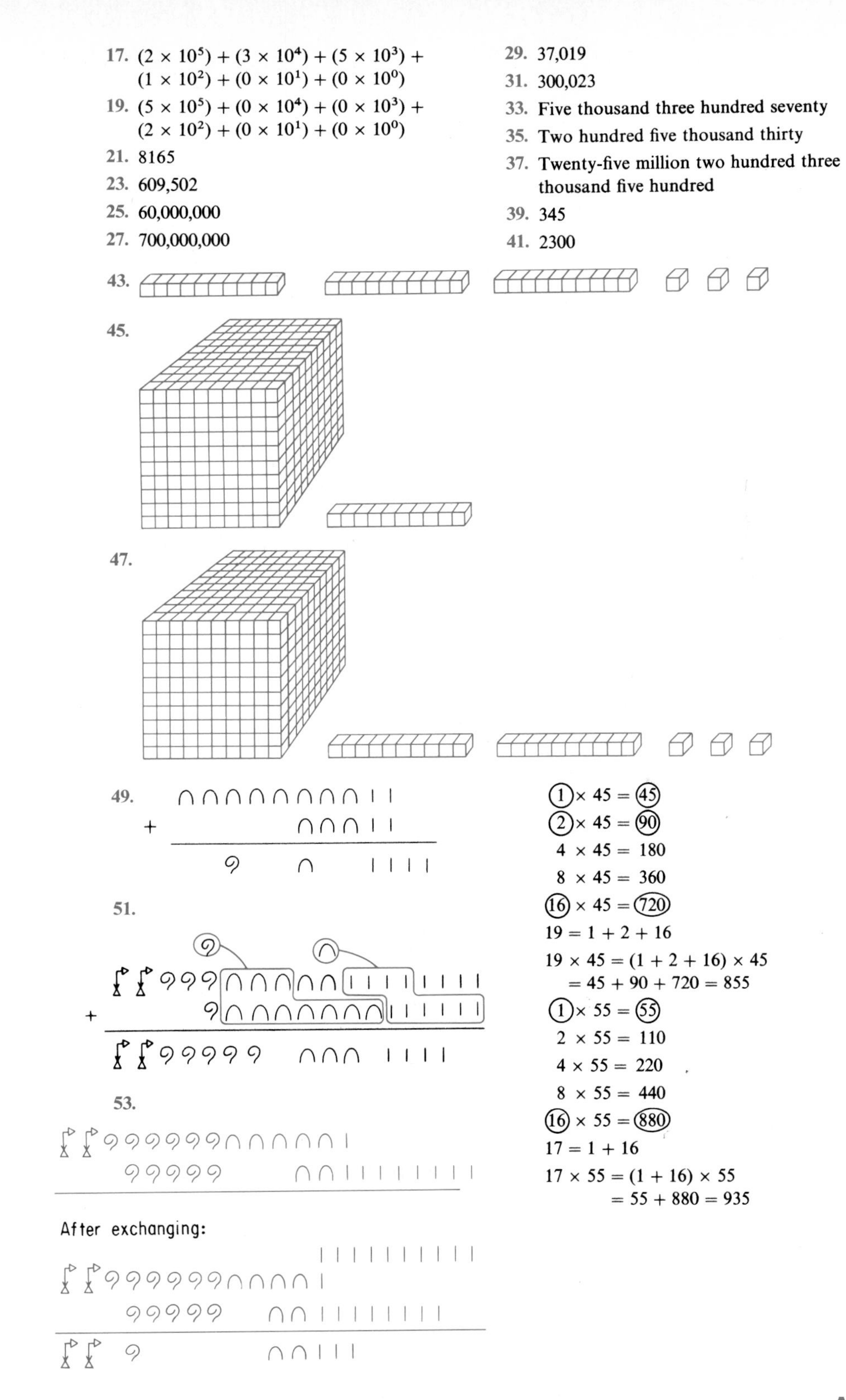

59. $(1) \times 29 = (29)$
$(2) \times 29 = (58)$
$4 \times 29 = 116$
$(8) \times 29 = (232)$
$16 \times 29 = 464$
$(32) \times 29 = (928)$
$43 = 1 + 2 + 8 + 32$
$$43 \times 29 = (1 + 2 + 8 + 32) \times 29 = 29 + 58 + 232 + 928 = 1247$$

1. Cardinal number
3. Ordinal number
5. Identification
7. 1
9. 11
11. 10
13. $\{(1,1), (1, 2), (1, 3), (1, 4), (2, 1), (2, 2), (2, 3), (2, 4)\}$.

15. 0 1 2 3 4 5 6 7

17. 0 1 2 3 4 5 6 7 8

19. 0 1 2 3 4 5 6 7

21. 0 1 2 3 4 5 6 7

23. The graph is the empty set; there are no points in the graph.
25. Identity, +
27. Closure, + and zero, ×
29. Identity, +
31. (**a**) Either $a = 1$ and $b = 2$ or $a = 2$ and $b = 1$.
(**b**) Either $a = 1$ and $b = 3$ or $a = 3$ and $b = 1$.
(**c**) Either $a = 1$ and $b = 4$, or $a = 2$ and $b = 2$, or $a = 4$ and $b = 1$.
33. Use any counterexample; for instance, $3 - 2 \neq 2 - 3$.
35. No, $8 - (3 - 2) = 8 - 1 = 7$, $(8 - 3) - 2 = 5 - 2 = 3$; no.
37. Use any counterexample; for instance, $2 + (3 \times 5) \neq (2 + 3) \times (2 + 5)$; $17 \neq 5 \times 7$.
39. $7 \times (80 - 1) = (7 \times 80) - (7 \times 1) = 560 - 7 = 553$
41. $8 \times (90 + 2) = (8 \times 90) + (8 \times 2) = 720 + 16 = 736$

43.
$$\begin{array}{rl} 45 & (4 \times 10) + (5 \times 1) \\ +\underline{38} & \underline{(3 \times 10) + (8 \times 1)} \\ & (7 \times 10) + (13 \times 1) = (8 \times 10) + (3 \times 1) = 83 \end{array}$$

45.
$$\begin{array}{rl} 375 & (3 \times 10^2) + (7 \times 10) + (5 \times 1) \\ +\underline{287} & \underline{(2 \times 10^2) + (8 \times 10) + (7 \times 1)} \\ & (5 \times 10^2) + (15 \times 10) + (12 \times 1) \\ & = (6 \times 10^2) + (6 \times 10) + (2 \times 1) = 662 \end{array}$$

47.
$$\begin{array}{rl} 1309 & (1 \times 10^3) + (3 \times 10^2) + (0 \times 10) + (9 \times 1) \\ +\underline{2578} & \underline{(2 \times 10^3) + (5 \times 10^2) + (7 \times 10) + (8 \times 1)} \\ & (3 \times 10^3) + (8 \times 10^2) + (7 \times 10) + (17 \times 1) \\ & = (3 \times 10^3) + (8 \times 10^2) + (8 \times 10) + (7 \times 1) = 3887 \end{array}$$

49.
$$\begin{array}{rl} 95 & (9 \times 10) + (5 \times 1) \\ -\underline{32} & \underline{(3 \times 10) + (2 \times 1)} \\ & (6 \times 10) + (3 \times 1) = 63 \end{array}$$

51.
$$\begin{array}{rll} 304 & (3 \times 10^2) + (0 \times 10) + (4 \times 1) = & (2 \times 10^2) + (9 \times 10) + (14 \times 1) \\ -\underline{128} & \underline{(1 \times 10^2) + (2 \times 10) + (8 \times 1)} = & \underline{(1 \times 10^2) + (2 \times 10) + (8 \times 1)} \\ & & (1 \times 10^2) + (7 \times 10) + (6 \times 1) = 176 \end{array}$$

53.
$$\begin{array}{rll} 5023 & (5 \times 10^3) + (0 \times 10^2) + (2 \times 10) + (3 \times 1) = & (4 \times 10^3) + (10 \times 10^2) + (1 \times 10) + (13 \times 1) \\ -\underline{2709} & \underline{(2 \times 10^3) + (7 \times 10^2) + (0 \times 10) + (9 \times 1)} = & \underline{(2 \times 10^3) + (7 \times 10^2) + (0 \times 10) + (9 \times 1)} \\ & & (2 \times 10^3) + (3 \times 10^2) + (1 \times 10) + (4 \times 1) = 2314 \end{array}$$

3-3 Early Procedures for Multiplication: page 116

1. 5 × 785, that is, 3925
3. 7 × 387, that is, 2709
5. 9 × 279, that is, 2511

7.

9.

11.

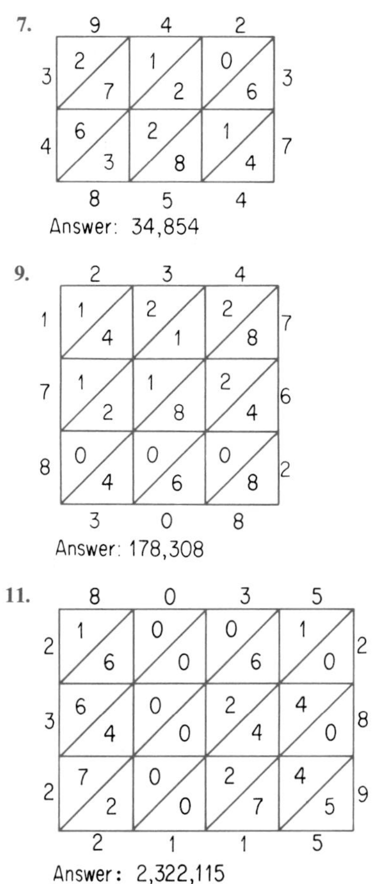

13.

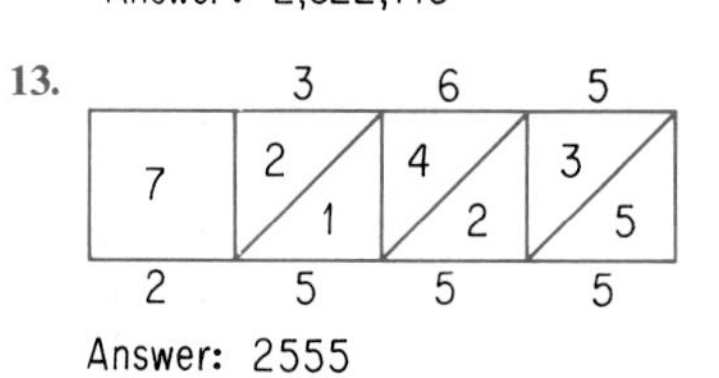

15.

17.

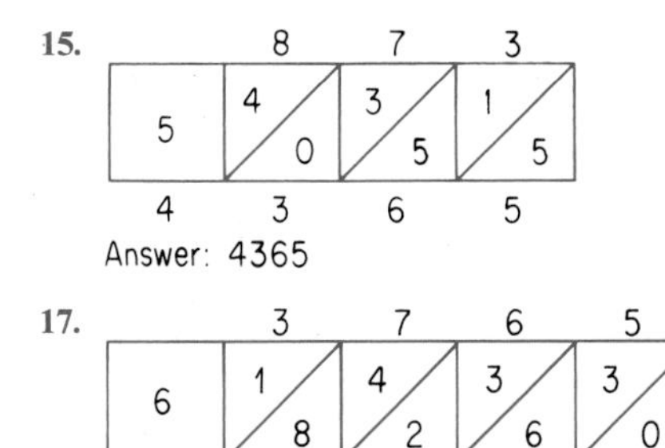

19.

21.

23.

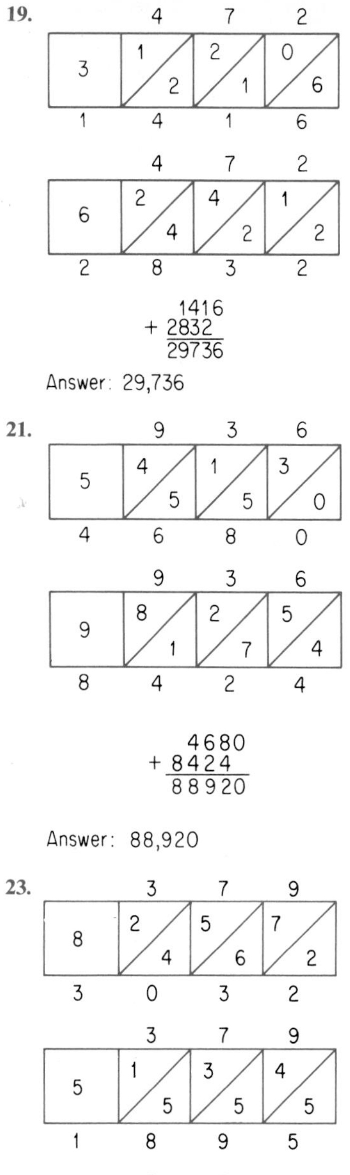

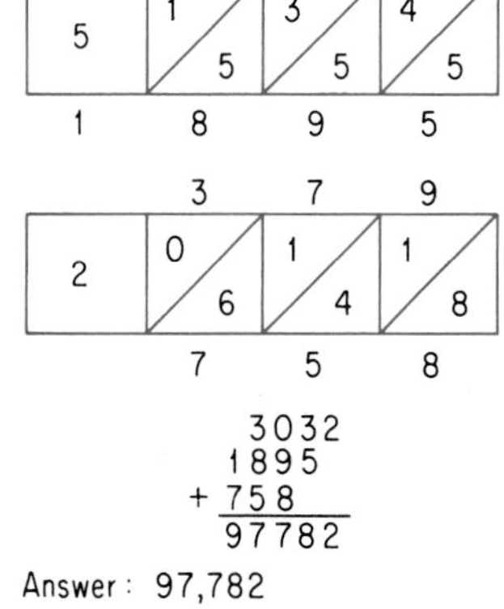

3-4 Multiplication and Division of Whole Numbers: page 124

1. 20 + 20 + 20 + 20 + 20 = 100

3. 5 + 5 + 5 + 5 + 5 + 5 + 5 + 5 = 40

5. 4 + 4 + 4 + 4 + 4 + 4 + 4 = 28

7. 40 − 5 = 35; 35 − 5 = 30; 30 − 5 = 25; 25 − 5 = 20; 20 − 5 = 15; 15 − 5 = 10; 10 − 5 = 5; 5 − 5 = 0; therefore, 40 ÷ 5 = 8.

9. 40 − 4 = 36; 36 − 4 = 32; 32 − 4 = 28; 28 − 4 = 24; 24 − 4 = 20; 20 − 4 = 16; 16 − 4 = 12; 12 − 4 = 8; 8 − 4 = 4; 4 − 4 = 0; therefore, 40 ÷ 4 = 10.

11. 105 − 15 = 90; 90 − 15 = 75; 75 − 15 = 60; 60 − 15 = 45; 45 − 15 = 30; 30 − 15 = 15; 15 − 15 = 0; therefore 105 ÷ 15 = 7.

13. $$\begin{aligned} 8 \times 15 &= 8 \times (10 + 5) \\ &= (8 \times 10) + (8 \times 5) \\ &= 80 + 40 \\ &= 120 \end{aligned}$$

15. $$\begin{aligned} 6 \times 45 &= 6 \times (40 + 5) \\ &= (6 \times 40) + (6 \times 5) \\ &= 240 + 30 = 270 \end{aligned}$$

17. $$\begin{aligned} 5 \times 435 &= 5 \times (400 + 30 + 5) \\ &= (5 \times 400) + (5 \times 30) + (5 \times 5) \\ &= 2000 + 150 + 25 \\ &= 2175 \end{aligned}$$

19. $$\begin{aligned} 12 \times 15 &= (10 + 2) \times (10 + 5) \\ &= (10 \times 10) + (10 \times 5) \\ &\quad + (2 \times 10) + (2 \times 5) \\ &= 100 + 50 + 20 + 10 \\ &= 180 \end{aligned}$$

21. $$\begin{aligned} 42 \times 57 &= (40 + 2) \times (50 + 7) \\ &= (40 \times 50) + (40 \times 7) \\ &\quad + (2 \times 50) + (2 \times 7) \\ &= 2000 + 280 + 100 + 14 \\ &= 2394 \end{aligned}$$

23. $$\begin{array}{r} 1 \\ 20 \\ 300 \\ 43\overline{)13803} \\ \underline{12900} \\ 903 \\ \underline{860} \\ 43 \\ \underline{43} \end{array} \quad \left.\right\} 321 \text{ (quotient)}$$

25. $$\begin{array}{r} 2 \\ 30 \\ 400 \\ 54\overline{)23328} \\ \underline{21600} \\ 1728 \\ \underline{1620} \\ 108 \\ \underline{108} \end{array} \quad \left.\right\} 432 \text{ (quotient)}$$

27. $$\begin{array}{r} 7 \\ 50 \\ 200 \\ 37\overline{)9509} \\ \underline{7400} \\ 2109 \\ \underline{1850} \\ 259 \\ \underline{259} \end{array} \quad \left.\right\} 257 \text{ (quotient)}$$

29. $$\begin{array}{r} 1 \\ 40 \\ 400 \\ 18\overline{)7943} \\ \underline{7200} \\ 743 \\ \underline{720} \\ 23 \\ \underline{18} \\ 5 \end{array} \quad \left.\right\} 441 \text{ (quotient)}, \quad 5 \text{ (remainder)}$$

31. The division is correct; (48 × 23) + 3 = 1107.

33. The division is incorrect; 3163 ÷ 37 = 85, remainder 18.

3-5 Other Systems of Numeration: page 130

1. 22_5

3. 30_4

5. (* * * * *)
(* * * * *)
(* * * * *)
* * * *

7. (* * * * * * *)
(* * * * * * *)
* * * * *

9. * * * *   * * * *   * *
* * * *
* * * *
* * * *

11. 19
13. 35
15. 23
17. 108
19. 99
21. 236
23. 517
25. 194
27. 3433_5
29. 4341_5
31. 2322_5
33. 4412_5
35. 1323_6
37. 10002_4
39. 223
41. 601
43. 694

3-6 Computation in Other Bases: page 136

1.

×	0	1	2	3	4
0	0	0	0	0	0
1	0	1	2	3	4
2	0	2	4	11_5	13_5
3	0	3	11_5	14_5	22_5
4	0	4	13_5	22_5	31_5

3. 113_5
5. 1431_5
7. 14121_5
9. 12_5
11. 123_5
13. 1142_5
15. 434_5
17. 2131_5
19. 10332_5
21. 3113_5
23. 24333_5
25. 1133131_5
27. 42_5
29. 341_5
31. 120_5
33. 243_5

35.

+	0	1	2	3
0	0	1	2	3
1	1	2	3	10_4
2	2	3	10_4	11_4
3	3	10_4	11_4	12_4

×	0	1	2	3
0	0	0	0	0
1	0	1	2	3
2	0	2	10_4	12_4
3	0	3	12_4	21_4

37. 1232_8
39. 214_8
41. 113_4
43. 14344_6
*45. 46_8
*47. 121_4 R 2

3-7 Binary Notation: page 140

1. $100\ 110_2$
3. $11\ 101_2$
5. $1\ 011\ 101_2$
7. $10\ 011\ 100_2$
9. $11\ 001\ 000_2$
11. $110\ 110\ 101_2$
13. 14
15. 27
17. 59
19. 43
21. 106
23. 185
25. $11\ 010_2$
27. $101\ 000_2$

29. 100_2
31. 1110_2
33. 101_2
35. $1\ 110\ 011_2$
37. (a) $1\ 000\ 010_2$
(b) $1\ 000\ 100_2$
(c) $1\ 000\ 111_2$
39. $234 = 352_8 = 11\ 101\ 010_2$; when placed in groups of three, starting with the units digit, the binary representation can be translated into the octal system, and conversely.
41. 3531_8
43. $22\ 533_8$
45. $1\ 531\ 046_8$
47. $101\ 000\ 010\ 011_2$
49. $100\ 110\ 010\ 100_2$
51. $11\ 110\ 101\ 100\ 011_2$

Chapter Review: page 142

1. $(5 \times 10^3) + (2 \times 10^2) + (8 \times 10^1) + (0 \times 10^0)$
2. 40,200
3. 4,030,020
4. (1) × 43 = (43)
2 × 43 = 86
(4) × 43 = (172)
8 × 43 = 344
(16) × 43 = (688)
$21 = 1 + 4 + 16$
$21 \times 43 = (1 + 4 + 16) \times 43$
$= 43 + 172 + 688 = 903$
5. (a) Cardinal number
(b) Ordinal number
6. (a) 8 (b) 17
7. (a)

-1 0 1 2 3 4 5 6

(b)

-1 0 1 2 3 4 5 6

8. (a) Addition and multiplication
(b) Addition and multiplication
9.

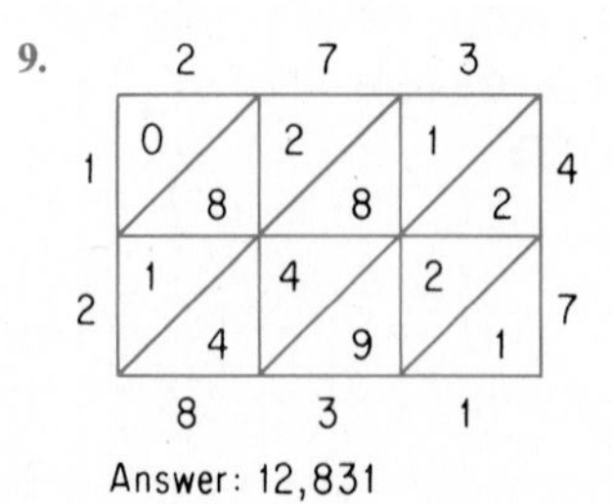

10. 8×483
11. $55 - 17 = 38$, $38 - 17 = 21$, $21 - 17 = 4$; $55 \div 17$ is 3 with remainder 4.
12. $7 \times (200 + 8) = (7 \times 200) + (7 \times 8) = 1400 + 56 = 1456$
13. 97
14. 346
15. 2331_5
16. $223\ 034_5$
17. 412_5
18. 21_5
19. 2411_5
20. 41_5
21. 175
22. $1\ 001\ 101_2$
23. $11\ 110\ 001_2$
24. 1010_2
25. $110\ 111_2$

Chapter 4 Integers

4-1 The Set of Integers: page 149

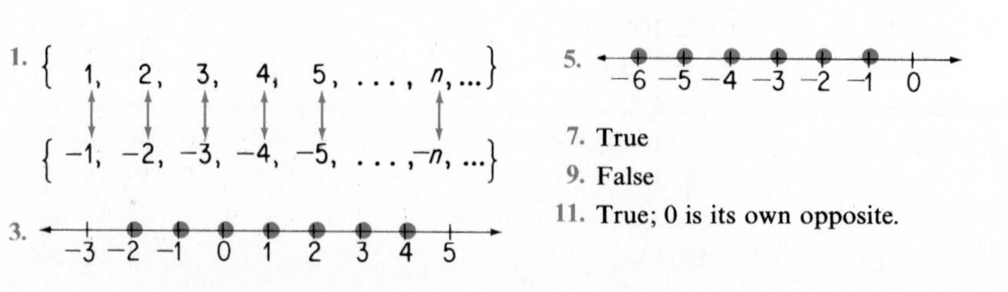

7. True
9. False
11. True; 0 is its own opposite.

13. False; the opposite of 0 is not a negative integer.

15. False; for example, $4 \div 3$ is not an integer.

17. False; the set is not closed since the sum of two odd integers is an even integer; also there is no identity element.

19. No, zero is an integer but is neither positive nor negative and thus is not in the union of the set of positive integers and the set of negative integers.

21. Among others:

$$\{1, 2, 3, 4, 5, 6, 7, \ldots, 2n, 2n+1, \ldots\}$$
$$\updownarrow \quad \updownarrow \quad \updownarrow \quad \updownarrow \quad \updownarrow \quad \updownarrow \quad \updownarrow \quad \updownarrow \quad \updownarrow$$
$$\{0, 1, -1, 2, -2, 3, -3, \ldots, n, -n, \ldots\}$$

23. Let $2k$ and $2m$ represent two even integers. Then $2k \times 2m = 2 \times 2 \times k \times m = 2(2km)$ where $2km$ is an integer. Thus $2(2km)$ is an even integer.

*25. Any integer is either even or odd. If an integer is even, then its square is even (as in Example 4). If an integer is odd, then its square is odd (as in Exercise 24). If the square of an integer is odd, the integer cannot be even and thus must be odd. If the square of an integer is even, the integer cannot be odd and thus must be even.

4-2 Addition and Subtraction of Integers: page 153

1.

$(+5) + (-7) = -2$

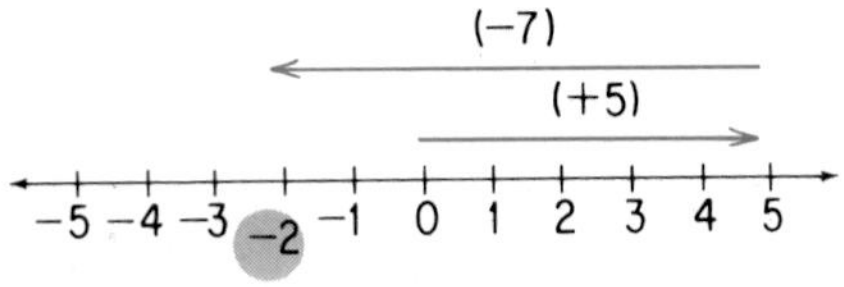

(−7)

(+5)

−5 −4 −3 −2 −1 0 1 2 3 4 5

3.

$(+3) + (+4) = +7$

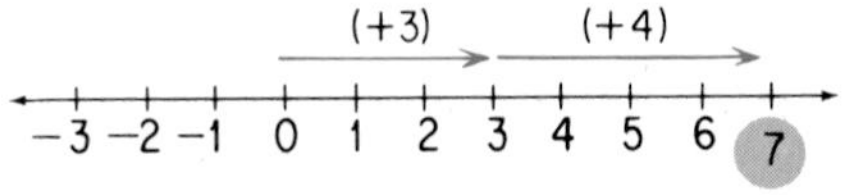

(+3) (+4)

−3 −2 −1 0 1 2 3 4 5 6 7

5. -4

7. $+4$

9. -27

11. -15

13. -9

15. 0

17. -3

19. $+12$

21. $+3$

23. $+7$

25. -27

27. -10

29. $+16$

31. -1

33. -31

35. -11

37. -11

39. $+7$

4-3 Multiplication and Division of Integers: page 159

1. -45

3. -45

5. -96

7. $+625$

9. -170

11. -105

13. 0

15. $+24$

17. -8

19. -8

21. $+2$

23. $+12$

25. -1

27. -5

29. $+8$

31. -2

33. -20

35. $+40$

37. -4

39. Division of integers is not associative; for example, $24 \div (6 \div 2) = 24 \div 3 = 8$ but $(24 \div 6) \div 2 = 4 \div 2 = 2$.

4-4 Clock Arithmetic: page 163

1.

×	1	2	3	4	5	6	7	8	9	10	11	12
1	1	2	3	4	5	6	7	8	9	10	11	12
2	2	4	6	8	10	12	2	4	6	8	10	12
3	3	6	9	12	3	6	9	12	3	6	9	12
4	4	8	12	4	8	12	4	8	12	4	8	12
5	5	10	3	8	1	6	11	4	9	2	7	12
6	6	12	6	12	6	12	6	12	6	12	6	12
7	7	2	9	4	11	6	1	8	3	10	5	12
8	8	4	12	8	4	12	8	4	12	8	4	12
9	9	6	3	12	9	6	3	12	9	6	3	12
10	10	8	6	4	2	12	10	8	6	4	2	12
11	11	10	9	8	7	6	5	4	3	2	1	12
12	12	12	12	12	12	12	12	12	12	12	12	12

3. 5
5. 7
7. 12
9. 7
11. 4
13. 12
15. 4
17. 4
19. 12
21. 9
23. 7
25. 1, 5, 9
27. 11

*29. An impossible equation; that is, there is no value of t for which this equation is true.

*31. An identity; that is, this equation is true for all possible replacements of t.

4-5 Modular Arithmetic: page 166

1. 5 (mod 12)
3. 4 (mod 12)
5. 0 (mod 12)
7. 2 (mod 5)
9. 3 (mod 5)
11. 3 (mod 5)
13. 4 (mod 5)
15. 3 (mod 5)
17. 2 (mod 5)
19. 4 (mod 5)
21. 4 (mod 5)
23. 3 (mod 5)
25. 3 (mod 5)
27. 3 (mod 5)
29. 2 (mod 5)
31. 2 (mod 7)
33. 5 (mod 7)
35. 3 (mod 9)
37. 3 (mod 6)
39. 4 (mod 8)
41. An impossible equation; that is, there is no value of x for which this equation is true.

43.

×	0	1	2	3	4	5	6	7	8	9	10	11
0	0	0	0	0	0	0	0	0	0	0	0	0
1	0	1	2	3	4	5	6	7	8	9	10	11
2	0	2	4	6	8	10	0	2	4	6	8	10
3	0	3	6	9	0	3	6	9	0	3	6	9
4	0	4	8	0	4	8	0	4	8	0	4	8
5	0	5	10	3	8	1	6	11	4	9	2	7
6	0	6	0	6	0	6	0	6	0	6	0	6
7	0	7	2	9	4	11	6	1	8	3	10	5
8	0	8	4	0	8	4	0	8	4	0	8	4
9	0	9	6	3	0	9	6	3	0	9	6	3
10	0	10	8	6	4	2	0	10	8	6	4	2
11	0	11	10	9	8	7	6	5	4	3	2	1

45. $2 \times 6 \equiv 0 \pmod{12}$; $3 \times 4 \equiv 0 \pmod{12}$; $3 \times 8 \equiv 0 \pmod{12}$; $4 \times 6 \equiv 0 \pmod{12}$; $4 \times 9 \equiv 0 \pmod{12}$; $6 \times 6 \equiv 0 \pmod{12}$; $6 \times 8 \equiv 0 \pmod{12}$; $6 \times 10 \equiv 0 \pmod{12}$; $8 \times 9 \equiv 0 \pmod{12}$. Thus the zero divisors in arithmetic modulo 12 are: 2 and 6, 3 and 4, 3 and 8, 4 and 6, 4 and 9, 6 and 6, 6 and 8, 6 and 10, and 8 and 9.

Chapter Review: page 168

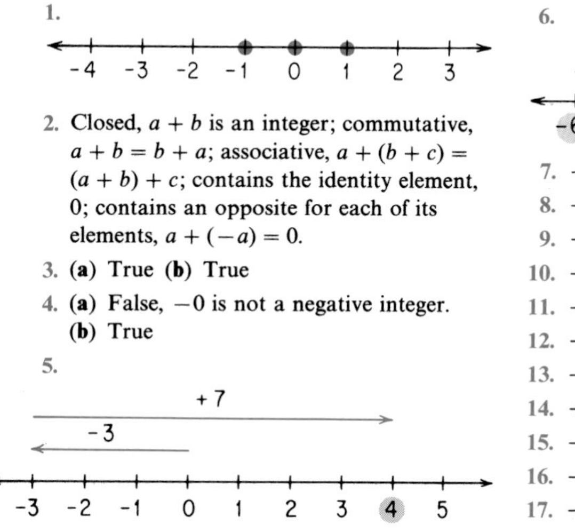

1.

2. Closed, $a + b$ is an integer; commutative, $a + b = b + a$; associative, $a + (b + c) = (a + b) + c$; contains the identity element, 0; contains an opposite for each of its elements, $a + (-a) = 0$.

3. **(a)** True **(b)** True

4. **(a)** False, -0 is not a negative integer. **(b)** True

5.

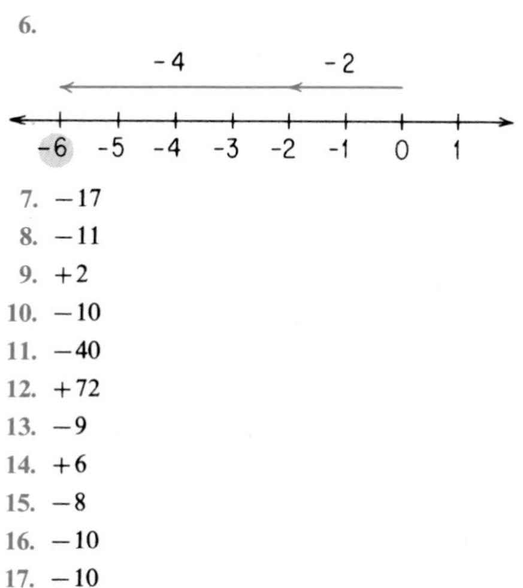

6.

7. -17

8. -11

9. $+2$

10. -10

11. -40

12. $+72$

13. -9

14. $+6$

15. -8

16. -10

17. -10

18. +4
19. Use any counterexample; for instance, $5 \div 2$ is not an integer.
20. (**a**) 10 (mod 12)
(**b**) 11 (mod 12)
21. 5 (mod 7)
22. 2 (mod 5)
23. (**a**) 3 (mod 12)
(**b**) 8 (mod 12)
(**c**) 6 (mod 12)
24. (**a**) 3 (mod 5)
(**b**) 2 (mod 5)
(**c**) 4 (mod 5)
25. Thursday

Chapter 5 Elements of Number Theory

5-1 Factors, Multiples, and Divisibility Rules: page 176

1. 3, 6, 9, 12, 15
3. 7, 14, 21, 28, 35
5. 11, 22, 33, 44, 55
7. 25, 50, 75, 100, 125
9. 1, 2, 4, 5, 10, 20
11. 1, 2, 3, 4, 6, 9, 12, 18, 36
13. 1, 2, 4, 8, 16, 32, 64
15. 1, 2, 4, 5, 8, 10, 16, 20, 40, 80
17. 1, 2, 4, 23, 46, 92
19. 1, 3, 37, 111
21. A natural number is divisible by 5 if its units digit is 0 or 5.
23. A natural number is divisible by 8 if the number formed by the last three digits, in order, of the given number is divisible by 8.
25. (**a**) Yes (**b**) Yes (**c**) Yes (**d**) Yes (**e**) Yes (**f**) Yes (**g**) No
27. (**a**) Yes (**b**) Yes (**c**) Yes (**d**) No (**e**) Yes (**f**) Yes (**g**) Yes
29. (**a**) Yes (**b**) No (**c**) Yes (**d**) Yes (**e**) No (**f**) No (**g**) No
31. (**a**) Yes (**b**) Yes (**c**) Yes (**d**) Yes (**e**) Yes (**f**) Yes (**g**) Yes
33. (**a**) Yes (**b**) No (**c**) Yes (**d**) Yes (**e**) No (**f**) No (**g**) No
35. (**a**) Yes (**b**) Yes (**c**) Yes (**d**) Yes (**e**) Yes (**f**) Yes (**g**) No

Among others:

37. If 3×5 divides 60, then 3 divides 60.
If 17×2 divides 680, then 17 divides 680.
If 11×7 divides 77, then 11 divides 77.
39. If 5 divides 40 and 5 divides 15, then 5 divides 55 and 25.

If 7 divides 70 and 7 divides 21, then 7 divides 91 and 49.

If 11 divides 99 and 11 divides 55, then 11 divides 154 and 44.
41. If 7 divides 3×21, then 7 divides 21; true.
If 7 divides 14×5, then 7 divides 5; false.
If 12 divides 4×6, then 12 divides 6; false.
43. If 6 does not divide 3 and 6 does not divide 7, then 6 does not divide 21; true.

If 6 does not divide 3 and 6 does not divide 4, then 6 does not divide 12; false.

If 6 does not divide 14 and 6 does not divide 15, then 6 does not divide 210; false.

5-2 Prime Numbers: page 180

1. Composite
3. Prime
5. Prime
7. Composite
9. Prime
11. Prime
13. 1×16, 2×8, 4×4
15. 1×21, 3×7
17. 1×31
19. 1×54, 2×27, 3×18, 6×9
21. 1×100, 2×50, 4×25, 5×20, 10×10
23. 1×125, 5×25
25. 2, 3, 5, 7, 11, 13, 17, 19
27. 41, 43, 47
29. 83, 89, 97
31. 4, 6, 8, 9
33. 32, 33, 34, 35, 36, 38, 39, 40, 42, 44, 45, 46, 48, 49
35. 81, 82, 84, 85, 86, 87, 88
37. No, for example, 9 is not a prime number.
No, 2 is a prime number.
39. There are other possible answers in many cases. $4 = 2 + 2$; $6 = 3 + 3$; $8 = 3 + 5$; $10 = 3 + 7$; $12 = 5 + 7$; $14 = 7 + 7$; $16 = 3 + 13$; $18 = 5 + 13$; $20 = 7 + 13$; $22 = 5 + 17$; $24 = 7 + 17$; $26 = 3 + 23$; $28 = 5 + 23$; $30 = 7 + 23$; $32 = 3 + 29$; $34 = 5 + 29$; $36 = 7 + 29$; $38 = 7 + 31$; $40 = 3 + 37$.

41. Any three consecutive odd numbers includes 3 or a multiple of 3. Each multiple of 3 that is greater than 3 is composite, and 1 is by definition not a prime. Therefore 3, 5, 7 is a set of three consecutive odd numbers that are all prime numbers, and, since any other such set contains a composite number that is a multiple of 3, this is the only prime triplet.

43. **(a)** 13 **(b)** 19 **(c)** 31

5-3 Fundamental Theorem of Arithmetic: page 185

1. 825
3. 864
5. 2400
7. 75,803
9. 3×5^2
11. 7×11
13. 79 is a prime number.
15. 257 is a prime number.
17. $2 \times 3 \times 103$
19. $7 \times 11 \times 13$
21. $2^3 \times 3 \times 5^3$
23. $5 \times 11 \times 89$
25. Using the factors one at a time: 2, 5, 7; two at a time: 2×5, 2×7, 5×7; three at a time: $2 \times 5 \times 7$; factors: {1, 2, 5, 7, 10, 14, 35, 70}
27. Using the factors one at a time: 2, 5; two at a time: 2×2, 2×5; three at a time: $2 \times 2 \times 5$; factors: {1, 2, 4, 5, 10, 20}
29. Using the factors one at a time: 2, 3, 5; two at a time: 2×3, 2×5, 3×3, 3×5; three at a time: $2 \times 3 \times 3$, $2 \times 3 \times 5$, $3 \times 3 \times 5$; four at a time: $2 \times 3 \times 3 \times 5$; factors: {1, 2, 3, 5, 6, 9, 10, 15, 18, 30, 45, 90}
31. 90
33. 140

5-4 Greatest Common Divisor: page 188

1. {1, 2}
3. {1, 2, 4, 8}
5. {1}
7. {1, 2, 5, 10}
9. {1, 5, 25}
11. {1, 2, 101, 202}
13. $42 = 2 \times 3 \times 7$; $60 = 2^2 \times 3 \times 5$; GCD(42, 60) $= 2 \times 3$, that is, 6.
15. $123 = 3 \times 41$; $287 = 7 \times 41$; GCD(123, 287) = 41
17. $123 = 3 \times 41$; $615 = 3 \times 5 \times 41$; GCD(123, 615) $= 3 \times 41$, that is, 123.
19. $68 = 2^2 \times 17$; $112 = 2^4 \times 7$; GCD(68, 112) $= 2^2$, that is, 4.
21. $600 = 2^3 \times 3 \times 5^2$; $800 = 2^5 \times 5^2$; GCD(600, 800) $= 2^3 \times 5^2$, that is, 200.
23. $2450 = 2 \times 5^2 \times 7^2$; $3500 = 2^2 \times 5^3 \times 7$; GCD(2450, 3500) $= 2 \times 5^2 \times 7$, that is, 350.
25. $12 = 2^2 \times 3$; $18 = 2 \times 3^2$; $21 = 3 \times 7$; GCD(12, 18, 21) = 3
27. $15 = 3 \times 5$; $25 = 5^2$; $40 = 2^3 \times 5$; GCD(15, 25, 40) = 5
29. $10 = 2 \times 5$; $20 = 2^2 \times 5$; $35 = 5 \times 7$; GCD(10, 20, 35) = 5
31. 2/5
33. 2/5
35. 5/6
37. 17/28
39. 2/5
41. 7/10
43. False, for example, 9 and 10 are relatively prime but neither 9 nor 10 is a prime number.
45. False, for example, 9 and 35 are relatively prime and both numbers are odd.
47. True

5-5 Least Common Multiple: page 193

1. 12, 24, 36
3. 15, 30, 45
5. 9, 18, 27
7. 30, 60, 90
9. 140, 280, 420
11. 96, 192, 288
13. $14 = 2 \times 7$; $40 = 2^3 \times 5$; LCM(14, 40) $= 2^3 \times 5 \times 7$, that is, 280.
15. $123 = 3 \times 41$; $287 = 7 \times 41$; LCM(123, 287) $= 3 \times 7 \times 41$, that is, 861.
17. $123 = 3 \times 41$; $615 = 3 \times 5 \times 41$; LCM(123, 615) $= 3 \times 5 \times 41$, that is, 615.
19. $68 = 2^2 \times 17$; $112 = 2^4 \times 7$; LCM(68, 112) $= 2^4 \times 7 \times 17$, that is, 1904.
21. $600 = 2^3 \times 3 \times 5^2$; $800 = 2^5 \times 5^2$; LCM(600, 800) $= 2^5 \times 3 \times 5^2$, that is, 2400.
23. $2450 = 2 \times 5^2 \times 7^2$; $3500 = 2^2 \times 5^3 \times 7$; LCM(2450, 3500) $= 2^2 \times 5^3 \times 7^2$, that is, 24,500.

25. $12 = 2^2 \times 3$; $18 = 2 \times 3^2$; $21 = 3 \times 7$; LCM(12, 18, 21) $= 2^2 \times 3^2 \times 7$, that is, 252.

27. $15 = 3 \times 5$; $25 = 5^2$; $40 = 2^3 \times 5$; LCM(15, 25, 40) $= 2^3 \times 3 \times 5^2$, that is, 600.

29. $10 = 2 \times 5$; $20 = 2^2 \times 5$; $35 = 5 \times 7$; LCM(10, 20, 35) $= 2^2 \times 5 \times 7$, that is, 140.

31. 29/24

33. 29/30

35. 9/20

37. 199/280

39. 38/615

41. −1/2400

43. True

45. True

In Exercises 47 and 49 any one of the stated conditions is a sufficient condition; all of the conditions are necessarily true.

47. n is a divisor of m, GCD$(m, n) = n$, LCM$(m, n) = m$

49. m is a divisor of n, GCD$(m, n) = m$, LCM$(m, n) = n$

5-6 Euclidean Algorithm and Applications: page 198

1. True

3. True

5. GCD(66, 96) = 6, LCM(66, 96) = 1056

7. GCD(154, 462) = 154, LCM(154, 462) = 462

9. GCD(1236, 1545) = 309, LCM(1236, 1545) = 6180

11. 5 = 3(15) − 40

13. 4 = 3(108) − 5(64)

15. 17 = 25(3961) − 104(952)

17. Every 12 days

19. 30 days

Chapter Review: page 199

1. **(a)** 1, 2, 3, 6, 7, 14, 21, 42 **(b)** 1, 71

2. **(a)** Yes **(b)** No **(c)** Yes **(d)** Yes

3. 1×98, 2×49, 7×14

4. 71, 73, 79

5. 23

6. 9000

7. 3850

8. $2^2 \times 3 \times 5$

9. $2^2 \times 3 \times 5^3$

10. {1, 3}

11. {1, 2, 4, 8}

12. 20

13. 6

14. 2/9

15. 26/39

16. 45, 90, 135

17. 4500

18. 120

19. 43/48

20. 16/205

21. 5/36

22. GCD(48, 120) = 24; LCM(48, 120) = 240

23. 24 = 120 − 3(48)

24. $m = n$

25. Impossible

Chapter 6 Rational Numbers

6-1 Concept of a Rational Number: page 210

1. **(a)** 2 **(b)** 1 **(c)** 2 **(d)** 0

3. 57/8

5. (−14)/3

7. $3\frac{1}{4}$

9. $-2\frac{1}{7}$

Among others for Exercises 11 through 21:

11. 2 + 1/3

13. 1/2 + 1/8

15. 1/3 + 1/9

17. 6 + 1/4

19. 5

21. 7 + 1/5 + 1/10

23. 3/10

25. (−3)/4

27. 8

29. 10

31. 35
33. 25
35. 8
37. 30
39. 7/4
41. 3/5
43. (−6)/7
45. (−4)/3
47. 8/12 and 9/12
49. 6/16 and 5/16
51. 10/24 and 9/24

6-2 Equivalence and Order Relations: page 217

1. **(a)** No **(b)** No **(c)** No **(d)** No
3. **(a)** Yes **(b)** No **(c)** Yes **(d)** No
5. **(a)** Yes **(b)** Yes **(c)** No **(d)** No
7. **(a)** No **(b)** Yes **(c)** No **(d)** No
9. $<$
11. $>$
13. $=$
15. $>$
17. $=$
19. $<$
21. $<$
23. $>$
25. $>$
27. $<$
29. $<$
31. **(a)** True, one-third less than Susan.
    (b) True, one-half more than Mary.
33. True
35. True
*37. False, $1^3 = 1^2$ but $3 \neq 2$.

6-3 Addition and Subtraction of Rational Numbers: page 223

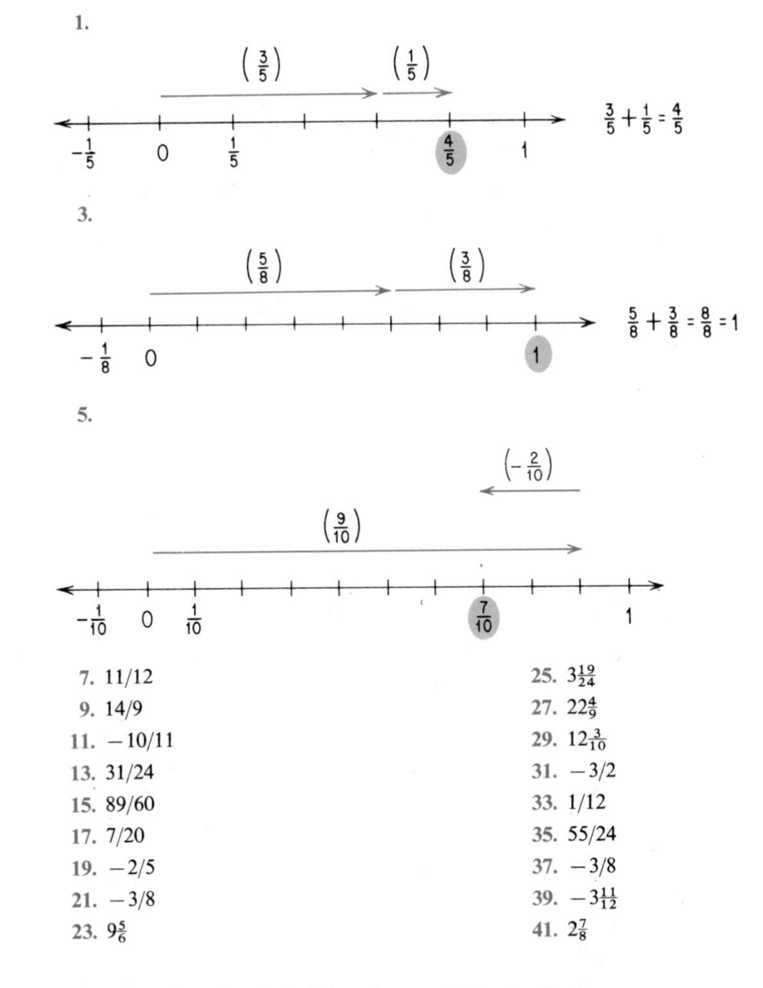

7. 11/12
9. 14/9
11. −10/11
13. 31/24
15. 89/60
17. 7/20
19. −2/5
21. −3/8
23. $9\frac{5}{6}$
25. $3\frac{19}{24}$
27. $22\frac{4}{9}$
29. $12\frac{3}{10}$
31. −3/2
33. 1/12
35. 55/24
37. −3/8
39. $-3\frac{11}{12}$
41. $2\frac{7}{8}$

*43. $(adf + bcf + bde)/bdf$

*45. $(adf + bcf + bde)/bdf$

47. (a) 2 (b) 2 *(c) 2

*49. Among others, depending in part on the model of calculator:

Enter: [a] [×] [d] [=] [Store] [1]

[b] [×] [d] [=] [Store] [2]

[b] [×] [c] [+] [Recall] [1] [=] [Clear] [1] [Store] [1]

The numerator $ad + bc$ is stored in position 1.
The denominator bd is stored in position 2.

6-4 Multiplication and Division of Rational Numbers: page 231

1. True
3. True
5. $\frac{1}{2} \times \frac{1}{3} = \frac{1}{6}$
7. $\frac{2}{5} \times \frac{3}{4} = \frac{6}{20} = \frac{3}{10}$
9. 11/28
11. 9/2
13. 3/10
15. 1
17. 40/21
19. −6/25
21. −3/10
23. $-1\frac{1}{3}$
25. $-2\frac{1}{2}$
27. 32
29. $1\frac{1}{2}$
31. $1\frac{11}{25}$
33. 9
35. 2/3
37. $1\frac{4}{7}$
39. 2/5
41. 5/12
43. $1\frac{1}{6}$
45. $1\frac{5}{9}$
47. 13/14
49. 4/5
51. $1\frac{1}{5}$
53. 24/35
55. (a) 2/3 (b) Yes (c) 36
57. Among others:
$\frac{2}{3} \div \left(\frac{1}{4} \div \frac{1}{2}\right) \neq \left(\frac{2}{3} \div \frac{1}{4}\right) \div \frac{1}{2}$; $\frac{4}{3} \neq \frac{16}{3}$.

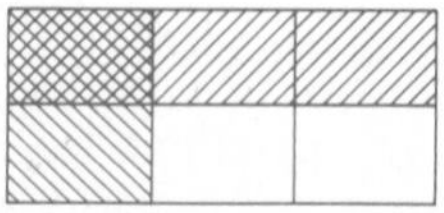

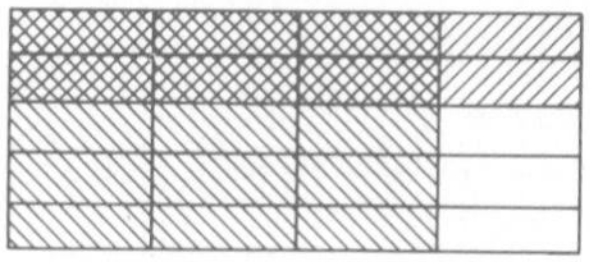

*59. Among others:
$\frac{2}{3} \div \left(\frac{1}{4} \times \frac{1}{2}\right) \neq \left(\frac{2}{3} \div \frac{1}{4}\right) \times \frac{1}{2}$; $\frac{16}{3} \neq \frac{4}{3}$.
In general:
$\frac{a}{b} \div \left(\frac{c}{d} \times \frac{e}{f}\right) = \frac{adf}{bce}$ and
$\left(\frac{a}{b} \div \frac{c}{d}\right) \times \frac{e}{f} = \frac{ade}{bcf}$.

	Natural numbers	Whole numbers	Integers	Positive rationals	Rational numbers
61.	√	√	√	√	√
63.	×	×	√	×	√
65.	×	×	√	×	√
67.	√	√	√	√	√
69.	×	×	×	√	√
71.	×	×	×	√	√

6-5 Decimal Fractions: page 241

1. $(4 \times 10^1) + (9 \times 10^0) + (6 \times 10^{-1})$
3. $(5 \times 10^0) + (0 \times 10^{-1}) + (1 \times 10^{-2}) + (0 \times 10^{-3}) + (3 \times 10^{-4})$
5. $(3 \times 10^3) + (0 \times 10^2) + (0 \times 10^1) + (0 \times 10^0) + (0 \times 10^{-1}) + (3 \times 10^{-2})$
7. 90.053
9. 6901.9
11. 0.30
13. 1.311
15. 0.236
17. 0.004
19. 0.00015
21. 170
23. **(a)** 9.9015 **(b)** 10.0191
25. **(a)** 43.7 **(b)** 0.39
27. 16.18
29. 7.83564
31. 41.5875
33. 10.93
35. (a)
37. (c)
39. (a)
41. (a)
43. **(a)** 0.65 **(b)** 0.085
45. **(a)** $0.\overline{714285}$ **(b)** $1.\overline{857142}$
47. 8/11
49. 47/111
51. 59/111
53. 55/9
55. 311/99
57. 107/333
59. **(a)** 0.4545454545 **(b)** 0.3572572572 **(c)** 0.8000000000 **(d)** 0.4289898989
61. \$2.08, \$2.37, \$2.59, \$2.65, \$2.89
63. $1.77\overline{7}$, 1.780, $1.7\overline{87}$, $1.78\overline{8}$, $1.88\overline{8}$

*65. Among others: 0.2343

*67. Among others: 0.2348

6-6 Ratio and Proportion: page 247

1. 2
3. 5/3
5. 5
7. 12
9. 7.5
11. 50
13. **(a)** 2/3 **(b)** 1 to 2 **(c)** 40
15. **(a)** m to f **(b)** $p = f/(m + f)$ **(c)** $m/(m + f)$ **(d)** $1 - p$
17. **(a)** 25 **(b)** 120 miles
19. **(a)** 9 years **(b)** 7 years
21. **(a)** 66 feet **(b)** 99 feet **(c)** 6 atmospheres **(d)** 59 pounds per square inch

*23. **(a)** 18 minutes **(b)** 24 minutes **(c)** $(k - 1)(6)$ minutes

6-7 Percent and Applications: page 254

1. 0.57
3. 0.03
5. 0.0095
7. 2.50
9. 1.00
11. 1/2
13. 19/20
15. 99/100
17. 3/2
19. 2/25
21. **(a)** 35% **(b)** 101% **(c)** 0.2%
23. **(a)** 145% **(b)** 0.6% **(c)** 100%
25. **(a)** 450% **(b)** 80% **(c)** 7.6%
27. 0.3; 30%
29. 11/20; 0.55
31. 0.85; 85%
33. 1/20; 0.05
35. 1.1; 110%
37. 0.715; 71.5%
39. 90
41. 60
43. 48
45. $40/100 = n/60$; 24
47. $35/100 = n/80$; 28
49. $20/160 = n/100$; $12\frac{1}{2}\%$
51. $120/160 = n/100$; 75%
53. $25/n = 20/100$; 125
55. $80/n = 125/100$; 64
57. (b)
59. (b)
61. (a)
63. (b)
65. \$180
67. \$144
69. \$54

71. $33\frac{1}{3}\%$ of \$120 is \$40; the correct selling price should be \$120 – \$40, that is, \$80. The merchant considered the discount, \$30, as $33\frac{1}{3}\%$ of the selling price, \$90.

73. $14\frac{2}{7}\%$

*75. $23\frac{1}{2}\%$

*77. In both cases, no action. That is, a 20% cut followed by a 25% increase returns one to the original salary; also a 25% increase followed by a 20% cut returns one to the original salary.

Chapter Review: page 258

1. **(a)** True **(b)** False
2. **(a)** True **(b)** True
3. **(a)** Yes **(b)** No **(c)** Yes **(d)** No
4. **(a)** Yes **(b)** No **(c)** Yes **(d)** No
5.

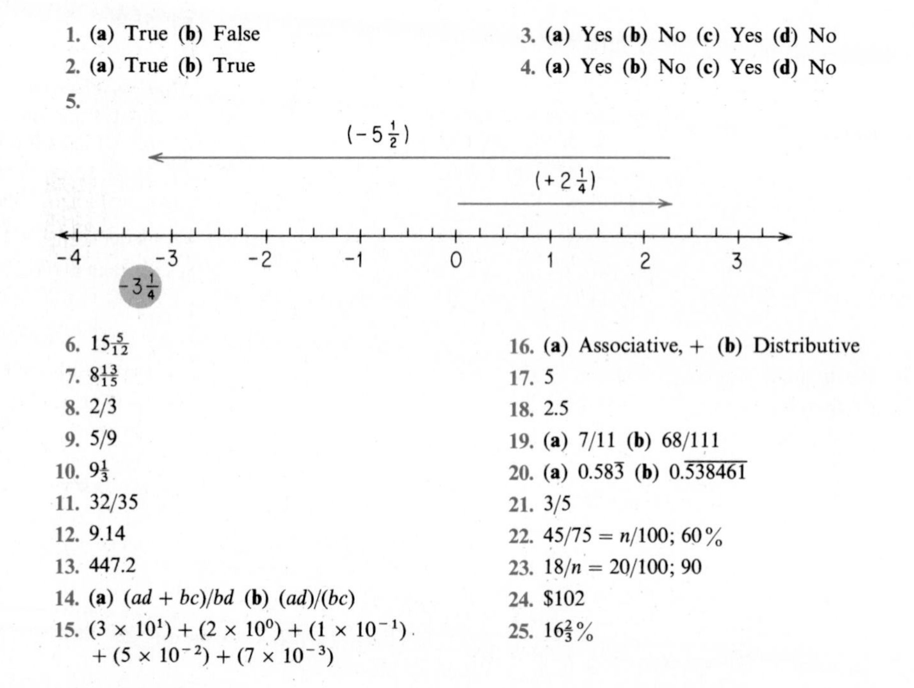

6. $15\frac{5}{12}$
7. $8\frac{13}{15}$
8. 2/3
9. 5/9
10. $9\frac{1}{3}$
11. 32/35
12. 9.14
13. 447.2
14. **(a)** $(ad + bc)/bd$ **(b)** $(ad)/(bc)$
15. $(3 \times 10^1) + (2 \times 10^0) + (1 \times 10^{-1}) + (5 \times 10^{-2}) + (7 \times 10^{-3})$
16. **(a)** Associative, + **(b)** Distributive
17. 5
18. 2.5
19. **(a)** 7/11 **(b)** 68/111
20. **(a)** $0.58\overline{3}$ **(b)** $0.\overline{538461}$
21. 3/5
22. $45/75 = n/100$; 60%
23. $18/n = 20/100$; 90
24. \$102
25. $16\frac{2}{3}\%$

Chapter 7 Real Numbers

7-1 Concept of Real Number: page 271

1. True
3. True
5. True
7. True
9. True
11. **(a)** Irrational **(b)** Irrational
13. **(a)** Rational **(b)** Irrational
15. **(a)** Rational **(b)** Rational
17. **(a)** Rational **(b)** Rational
19. **(a)** Rational **(b)** Irrational

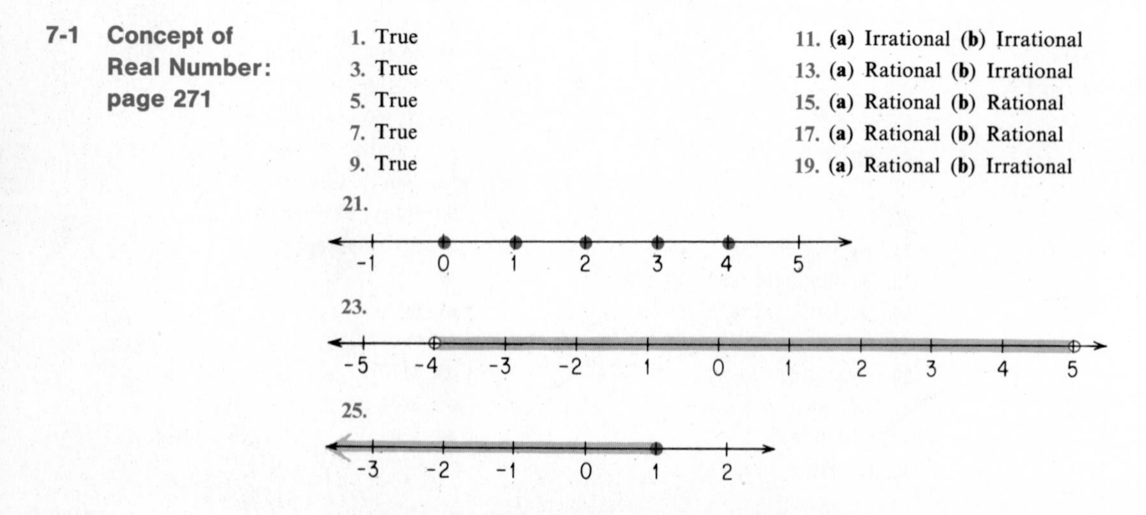

27.

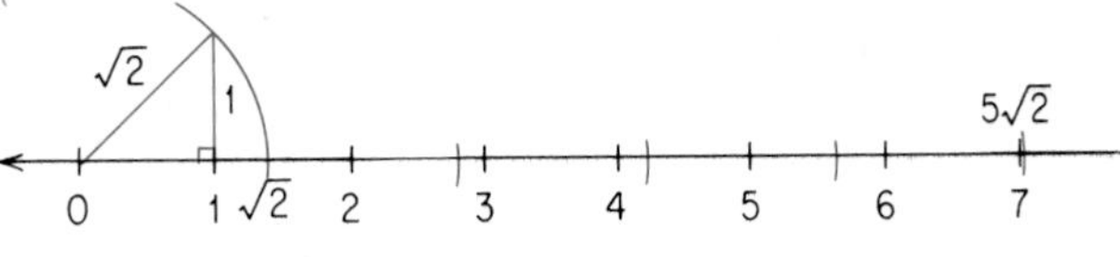

29.

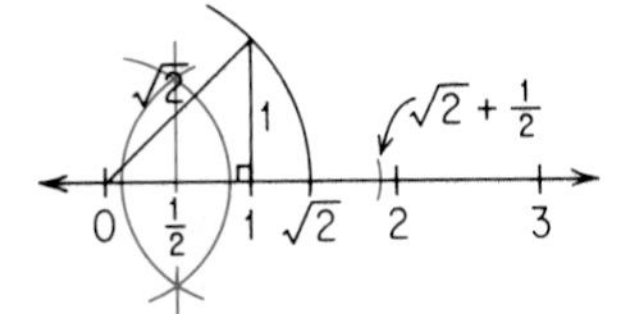

31.

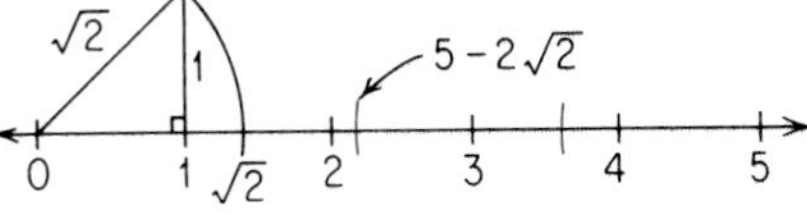

33. **(a)** Terminating; 0.375
  (b) Nonterminating, repeating; $0.41\overline{6}$
  (c) Nonterminating, nonrepeating

35. **(a)** 3 **(b)** 3

37. **(a)** 9 **(b)** 1

39. 0.45, $0.\overline{45}$, 0.45455, 0.454554555..., $0.4\overline{5}$

41. 0.06, $0.0\overline{6}$, 0.067, 0.67677677767..., $0.06\overline{7}$

43. **(a)** Not irrational
  (b) Not irrational
  (c) Irrational and between 0.234 and 0.235
  (d) Not between 0.234 and 0.235

45. Among others: 0.484484448..., 0.486486648666...

47. Among others: $\sqrt{2} + (-\sqrt{2})$, $2\sqrt{3} + (-2\sqrt{3})$, $3\sqrt{5} + (-3\sqrt{5})$

49. Among others: $(3 + \sqrt{2}) - (2 + \sqrt{2})$, $(\sqrt{3} - 5) - (\sqrt{3} - 6)$, $(8 + 4\sqrt{5}) - (7 + 4\sqrt{5})$

*51. Among others: $\sqrt{3} \times \sqrt{5}$, $\sqrt{8} \times \sqrt{12}$, $\sqrt{7} \times \sqrt{14}$

*53. Among others: $\sqrt{14} \div \sqrt{7}$, $\sqrt{54} \div \sqrt{18}$, $\sqrt{63} \div \sqrt{27}$

*55. If $3 + \sqrt{2}$ were a rational number a/b, then $\sqrt{2}$ would be a rational number $a/b - 3$ (impossible). Therefore $3 + \sqrt{2}$ cannot be a rational number and must be an irrational number.

*57. If $\sqrt{2} - 1$ were a rational number a/b, then $\sqrt{2}$ would be a rational number $a/b + 1$ (impossible). Therefore $\sqrt{2} - 1$ cannot be a rational number and must be an irrational number.

*59. If $5/2 + \sqrt{2}$ were a rational number a/b, then $\sqrt{2}$ would be a rational number $a/b - 5/2$ (impossible). Therefore $5/2 + \sqrt{2}$ cannot be a rational number and must be an irrational number.

*61. If $p + q\sqrt{11}$ were a rational number a/b, then $\sqrt{11}$ would be a rational number $(a - bp)/bq$ (impossible). Therefore $p + q\sqrt{11}$ cannot be a rational number and must be an irrational number.

7-2 Decimal Representations: page 277

1. **(a)** 3.1 **(b)** 3.142 **(c)** 3.141 592 7

3. **(a)** 3.1416 **(b)** 3.141 593
  (c) 3.141 592 653 589 8

5. The assertion was correct.

7. 4

9. 30

11. 200

13. 0.2

15. 0.4

17. 0.1

19. $3 < \sqrt{11} < 4$
  $3.3 < \sqrt{11} < 3.4$
  $3.31 < \sqrt{11} < 3.32$
  $\sqrt{11} \approx 3.3$

21. $6 < \sqrt{45} < 7$
  $6.7 < \sqrt{45} < 6.8$
  $6.70 < \sqrt{45} < 6.71$
  $\sqrt{45} \approx 6.7$

23. $9 < \sqrt{93} < 10$
  $9.6 < \sqrt{93} < 9.7$
  $9.64 < \sqrt{93} < 9.65$
  $\sqrt{93} \approx 9.6$

25. 2, 1.75, 1.73214..., 1.7320508...

27. 3, $2.\overline{6}$, $2.6458\overline{3}$, 2.6457513...

29. 4, 4.125, 4.123106..., 4.1231056...

31. 9, $9.0\overline{5}$, 9.055385..., 9.055385...

33. 10, 10.35, 10.344082..., 10.344080...

35. 20, 20.125, 20.124611..., 20.124611...

37. 0.4, 0.3875, 0.3872983..., 0.3872983...
39. 0.12, 0.1225, 0.12247448..., 0.12247448...
41. 0.015, 0.0155, 0.0154919..., 0.0154919...
43. **(a)** 1.140 **(b)** 36.05 **(c)** 114.0 **(d)** 0.1140
45. 209 feet. Note that no tables are needed; one application of the divide and average method suffices.
47. **(a)** 12 kilometers **(b)** 17 kilometers

7-3 Integers as Exponents: page 283

1. **(a)** 32 **(b)** 4 **(c)** 32
3. **(a)** 10,000,000 **(b)** 1000 **(c)** 100,000
5. **(a)** $20x^9$ **(b)** $6x^{12}$ **(c)** $6a^7b^7$
7. **(a)** $4a^2b^3$ **(b)** $7n^4$ **(c)** $4b^3$
9. **(a)** $16x^{12}$ **(b)** $25x^6$ **(c)** $x^4/4$
11. **(a)** 1 **(b)** $4x^6$ **(c)** x^6/y^3
13. **(a)** 9×10^4 **(b)** 7.5×10^6 **(c)** 4.5×10^5
15. **(a)** 1.235×10 **(b)** 4.516×10^4 **(c)** 4×10^{-3}
17. **(a)** 9.82×10^{-5} **(b)** 2.5×10^{-3} **(c)** 2.65×10^{-2}
19. 3.15×10^{-8}
21. 1×10^{-8}
23. 240,000
25. 24,500,000,000,000
27. 5,880,000,000,000
29. **(a)** 6×10^9 **(b)** 3×10^2
31. **(a)** 1.2×10^{-11} **(b)** 1.5×10^{-2}
33. **(a)** 3×10^{-6} **(b)** 7.5
35. 9.071848×10^2
37. 1.087×10^{-3}

7-4 Simple and Compound Interest: page 290

1. $630
3. $1400
5. $4130
7. $2630
9. $1893.75
11. $1007.93
13. $282.68
15. $6606.60
17. 12
19. **(a)** 8.16% **(b)** 8.24%
21. **(a)** $7.20 **(b)** $11.00
23. $19,326.12

*25. $8269.02

*27. **(a)** $5512.68 **(b)** $4563.92 **(c)** $948.76

7-5 Powers and Roots: page 295

1. $11^{1/2}$
3. $5^{2/3}$
5. $2 + 7^{1/3}$
7. $\sqrt{19}$
9. $\sqrt[3]{7^2}$
11. $6 - \sqrt[3]{3^2}$
13. Among others: $(1/2)^3 = 1/8$, $5^2 = 25$, $(7/8)^0 = 1$
15. Among others: $(\sqrt{17})^0 = 1$, $(\sqrt{17})^2 = 17$, $(\sqrt[3]{5})^6 = 25$
17. Among others: $\sqrt{2/3}$, $\sqrt[3]{4}$, $\sqrt[3]{5/8}$, that is, $(1/2)\sqrt[3]{5}$
19. $2\sqrt{2}$
21. $10\sqrt{2}$
23. $30\sqrt{5}$
25. $2\sqrt[3]{2}$
27. $2\sqrt[3]{5^2}$
29. $(2/3)\sqrt{2}$
31. $(1/2)\sqrt{10}$
33. $(1/4)\sqrt{14}$
35. $(1/2)\sqrt[3]{7}$
37. -2
39. $(-2/5)\sqrt[3]{75}$
41. 15
43. 8
45. $\sqrt{2}$
47. $(1/4)\sqrt{2}$
49. $\sqrt{3}$
51. $\sqrt[3]{3}$
53. $\sqrt[3]{9}$
55. 3.27
57. 7.07
59. 4.83
61. 3.35
63. 8.29

*65. 2.61

*67. **(a)** 0.746 **(b)** 0.645

7-6 Complex Numbers: page 301

1. Among others:

$$(2 + 3i) + (5 - 3i) = 7,$$
$$2i + (-2i) = 0,$$
$$(1 + i) + (2 - i) = 3$$

3. Among others:

$$i + (-i) = 0,$$
$$(1 + i) + (-1 - i) = 0,$$
$$(2 - 3i) + (-2 + 3i) = 0$$

5. Among others: $i^2 = -1$, $(3i)^2 = -9$, $(-2i)^2 = -4$

	Numbers	Set					
		Natural numbers	Rational numbers	Irrational numbers	Real numbers	Imaginary numbers	Complex numbers
7.	1	√	√	×	√	×	√
9.	$\sqrt{121}$	√	√	×	√	×	√
11.	$-3\sqrt{36}$	×	√	×	√	×	√
13.	$\sqrt[3]{-8}$	×	√	×	√	×	√
15.	$1 - \sqrt{8}$	×	×	√	√	×	√

17. $2\sqrt{5} + 0i$

19. $6\sqrt{3} - (\sqrt{3})i$

21. $3 + 5i$

23. $1 + 0i$

25. $3 - i$

27. $7 + 0i$

	Sentence	Replacement set					
		Natural numbers	Whole numbers	Integers	Rational numbers	Real numbers	Complex numbers
29.	$n + 5 = 0$	None	None	-5	-5	-5	-5
31.	$3n - 5 = 0$	None	None	None	5/3	5/3	5/3
33.	$n^2 - 5 = 0$	None	None	None	None	$\sqrt{5}, -\sqrt{5}$	$\sqrt{5}, -\sqrt{5}$

35. Among others:

$$(1 + i) + (i) = 1 + 2i,$$
$$(1 + i) + (1 - i) = 2 + 0i,$$
$$(2 - 3i) + (3 + 2i) = 5 - i$$

37. Among others:

$$[(i)(1 + i)](1 - i) = 2i = i[(1 + i)(1 - i)],$$
$$[(2i)(3i)](-4i) = 24i = (2i)[(3i)(-4i)],$$
$$[(3 - i)(3 + i)](1 + i) = 10 + 10i = (3 - i)[(3 + i)(1 + i)]$$

39. Among others:

$$i + 2i = 2i + i,$$
$$(1 - i) + (2 + 3i) = (2 + 3i) + (1 - i),$$
$$(2 + i) + (2 - i) = (2 - i) + (2 + i)$$

41. Among others:

$$i[i + (2 - i)] = 2i = i(i) + i(2 - i),$$
$$(2 + 3i)[(1 - i) + (-1 + i)] = 0 = (2 + 3i)(1 - i) + (2 + 3i)(-1 + i)$$
$$-i[(3 + i) + (-i)] = -3i = (-i)(3 + i) + (-i)(-i)$$

43. Among others:

$$(i)(-i) = 1,$$
$$(5i)[(-1/5)i] = 1,$$
$$(1 + i)[(1/2) - (1/2)i] = 1$$

Chapter Review: page 303

1. **(a)** True **(b)** False

2. **(a)** True **(b)** True

3.

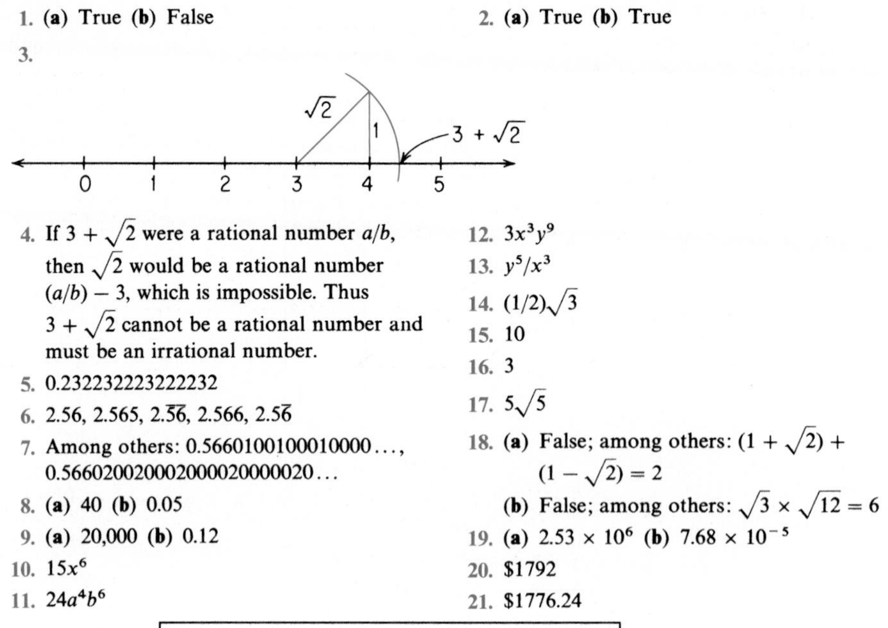

4. If $3 + \sqrt{2}$ were a rational number a/b, then $\sqrt{2}$ would be a rational number $(a/b) - 3$, which is impossible. Thus $3 + \sqrt{2}$ cannot be a rational number and must be an irrational number.
5. 0.232232223222232
6. 2.56, 2.565, $2.\overline{56}$, 2.566, $2.5\overline{6}$
7. Among others: 0.5660100100010000..., 0.566020020002000020000020...
8. **(a)** 40 **(b)** 0.05
9. **(a)** 20,000 **(b)** 0.12
10. $15x^6$
11. $24a^4b^6$
12. $3x^3y^9$
13. y^5/x^3
14. $(1/2)\sqrt{3}$
15. 10
16. 3
17. $5\sqrt{5}$
18. **(a)** False; among others: $(1 + \sqrt{2}) + (1 - \sqrt{2}) = 2$
    (b) False; among others: $\sqrt{3} \times \sqrt{12} = 6$
19. **(a)** 2.53×10^6 **(b)** 7.68×10^{-5}
20. \$1792
21. \$1776.24

		Replacement set			
	Sentence	Integers	Rational numbers	Irrational numbers	Real numbers
22.	$n + 3 = 3$	0	0	None	0
23.	$5n = 8$	None	8/5	None	8/5
24.	$n^2 = 3$	None	None	$\sqrt{3}, -\sqrt{3}$	$\sqrt{3}, -\sqrt{3}$
25.	$n^2 + 9 = 0$	None	None	None	None

Chapter 8 An Introduction to Geometry

8-1 Geometry and Arithmetic: page 313

1. Among others: an orange, a tennis ball, the sun.
3. Among others: cans (such as soup cans), some lamp shades, new pencils.
5. Among others: rooms, boxes, large books.
7. Among others: toy pyramids, special dice, space in corner of a room.
9. Among others: dice, sugar cubes, some boxes.
11. Among others: key ring, O-ring gaskets, hoops.
13. Cube
15. Cube
17. Triangular pyramid
19. True
21. False
23. 10 cm by 10 cm
25. 10 cm, 20 cm, and 20 cm
27. **(a)** Rectangular regions 3 m long and 1 m wide
    (b) Circular regions of diameter 1 m

*29. Yes; 30 cm

31. Among others: a circular region; an ice cream cone shape, as in the figure.

8-2 Points and Lines on a Plane: page 321

1. Among others:

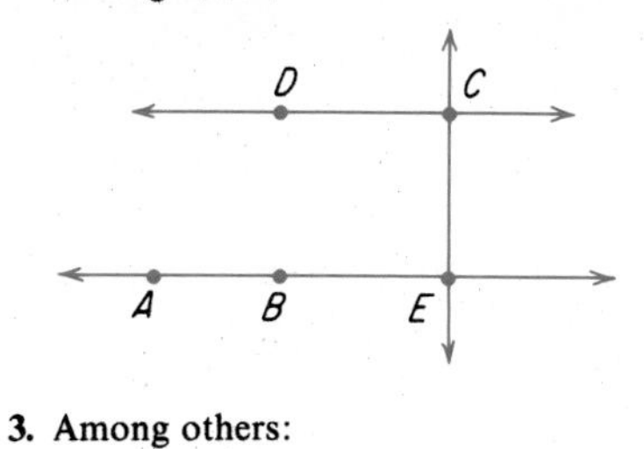

3. Among others:

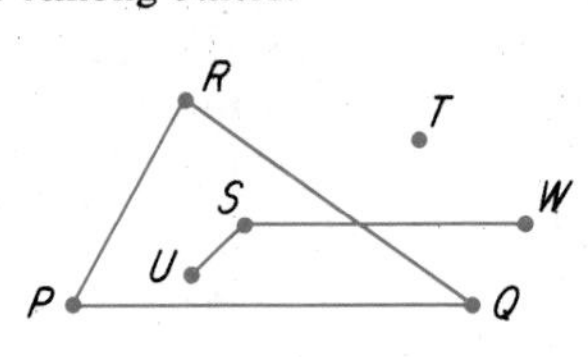

5. **(a)** False **(b)** True
7. **(a)** True **(b)** True
9. **(a)** True **(b)** True

11 through *27. (Guidelines are provided in the exercises.)

8-3 Constructions: page 329

5. Construct a right angle and bisect it.
7. Bisect an angle of 45°.
9. Among others: Construct a right angle, bisect it, and bisect one of its halves.

11.

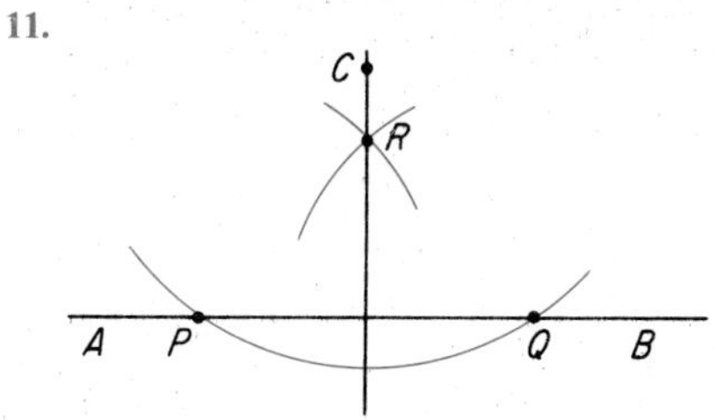

13. Copy the given line segments on the sides of a right angle.
15. Construct a right angle *B*. Copy the given line segment *TW* on *AB* as one side and copy *XY* as the hypotenuse *AC* of the right triangle *ABC*.
17. The triangles may be superimposed.

8-4 Polygons: page 337

1. Among others:
(a)
(b)

3. Among others:
(a)
(b)

5. Among others:
(a) **(b)**

7. Among others:

9. Among others:

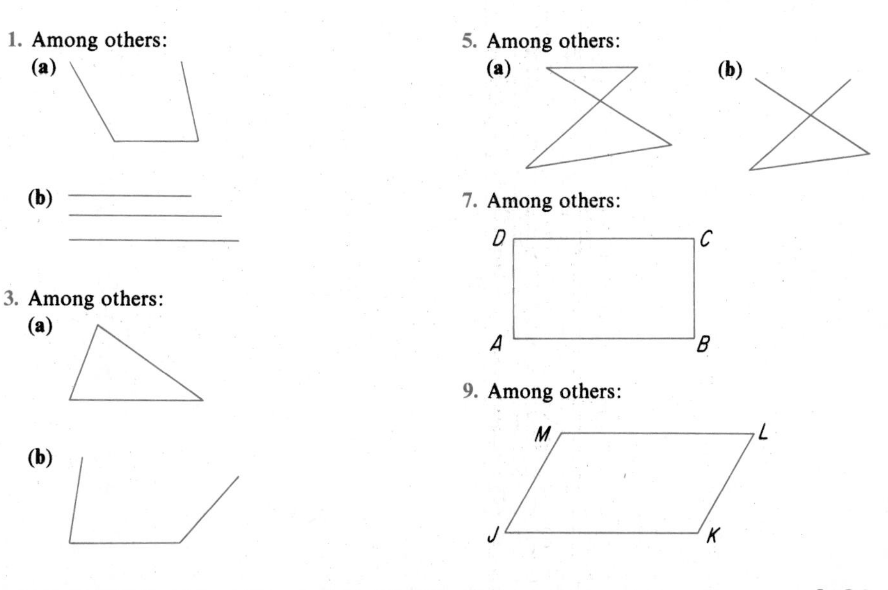

11. Among others:

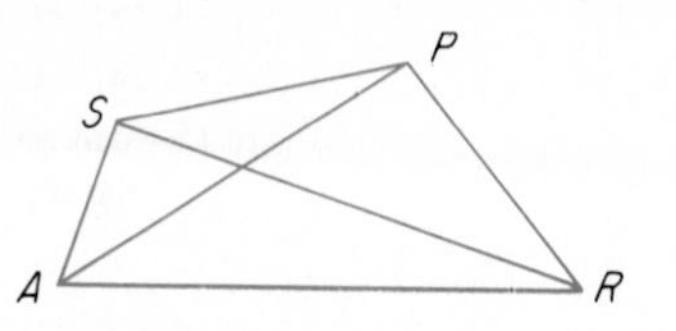

13. Among others: Fold a line m, lines p and q perpendicular to m, and a line t perpendicular to p.

15. Among others: Fold a line m, a line t perpendicular to m, a line p perpendicular to t, and a line q that intersects m but is not perpendicular to m and does not

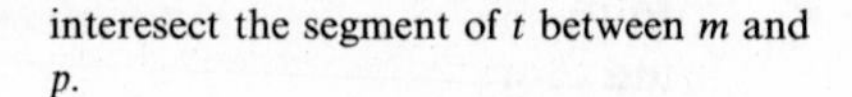
interesect the segment of t between m and p.

17. (c) $BCFE$ is a trapezoid.

19. Make folds for the altitudes AA', BB', and CC'. Make folds a, b, and c perpendicular to AA' at A, BB' at B, and CC' at C, respectively. Then a and c intersect at D; b and c intersect at F.

23. A line is separated into $n + 1$ disjoint parts by n distinct points.

25. Four, if n is odd; three, if n is even.

*27. The conjecture fails when $n = 6$ and only 31 regions are determined.

8-5 Polyhedra: page 343

1. **(a)** M, N, O, P
   (b) $\overline{MN}, \overline{MO}, \overline{MP}, \overline{ON}, \overline{NP}, \overline{PO}$
   (c) The triangular regions MNO, PMN, PNO, PMO.

3. **(a)** A, B, C, D, E, F, G, H
   (b) $\overline{AB}, \overline{BC}, \overline{CD}, \overline{AD}, \overline{AE}, \overline{BF}, \overline{CG}, \overline{DH}, \overline{EF}, \overline{FG}, \overline{GH}, \overline{HE}$
   (c) The square regions $ABCD$, $ABFE$, $BCGF$, $CDHG$, $DAEH$, $EFGH$.

5. (b) and (c)

7. (a)

9. Among others:

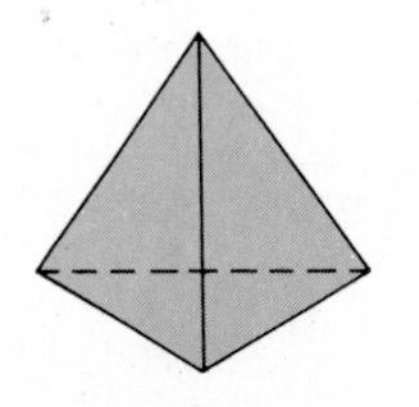

11. Among others:

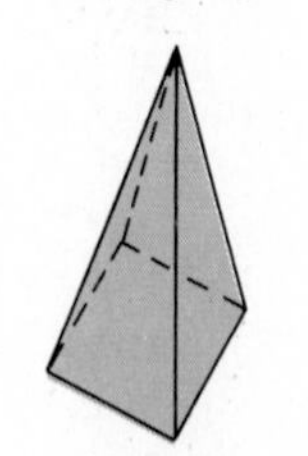

13. Among others:

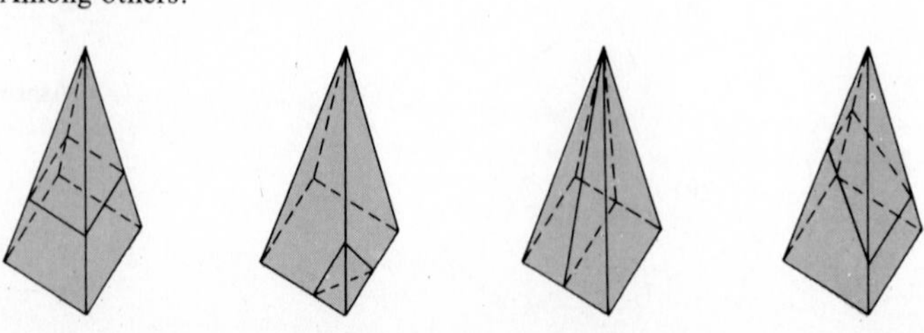

*15. Among others:

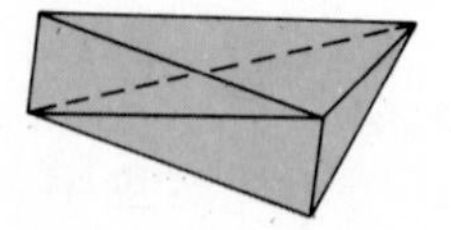

17. True

19. False, there are infinitely many such lines.

21. False, there are infinitely many such planes.

23. False, there are infinitely many such lines.

25. False, there are infinitely many such planes.

27. True

29. **(a)** True **(b)** True **(c)** False

31. **(a)** True **(b)** True **(c)** True

33. **(a)**

Figure	*F*	*V*	*E*
Triangular pyramid	4	4	6
Square pyramid	5	5	8
Pentagonal pyramid	6	6	10
Cube	6	8	12

(b) $F + V - 2 = E$

8-6 Networks and Graphs: page 352

1. (**a**) 4 (**b**) 0
   (**c**) Traversable starting at any vertex.

3. (**a**) 2 (**b**) 2
   (**c**) Traversable starting at K or M.

5. (**a**) 4 (**b**) 0
   (**c**) Traversable starting at any vertex.

7. (**a**) 1 (**b**) 4
   (**c**) Not traversable.

9. (**a**) 4 (**b**) 2
   (**c**) Traversable starting at A or B.

11. (**a**) 0 (**b**) 4
    (**c**) Not traversable.

13. (**a**) Yes; among others: $ABFEAFCEDCB$.
    (**b**) No

15. (**a**) Yes; among others: $CBEGIBDEFGHIAB$.
    (**b**) No

17. None or two.

19. (**a**) Yes; among others: $ABCDEF$.
    (**b**) Yes; among others: $ABCDEFA$.

21. (**a**) Yes; among others: $CBDEFGHIA$.
    (**b**) No

23. The inspector can use a map for the highways involved as a network, determine the number of odd vertices, and know that each section can be traversed exactly once in a single trip if there are at most two odd vertices.

25.

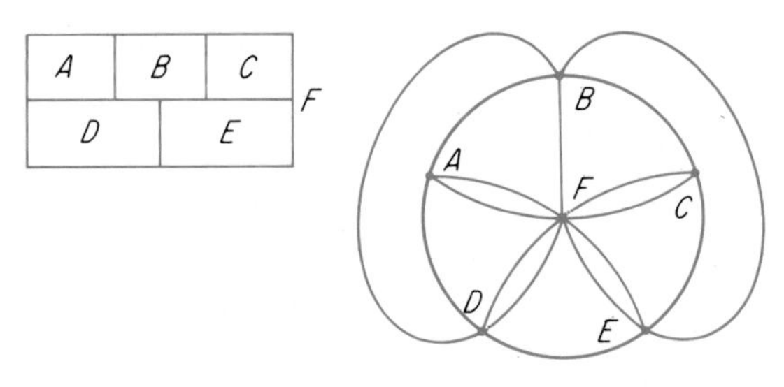

Think of the six regions as labeled and note the line segments that are needed as represented by arcs in the network. Since the network has four odd vertices (B, D, E, and F), the network is not traversable and the desired broken line cannot be drawn.

27.

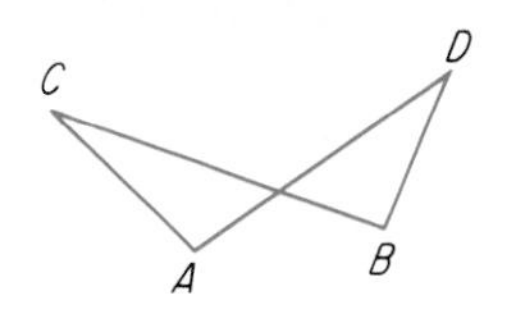

29.

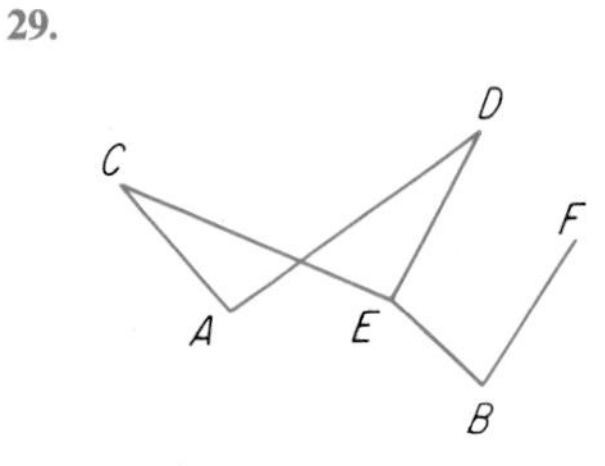

31. Among others:
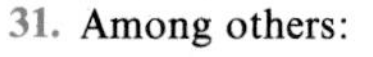

(**a**)

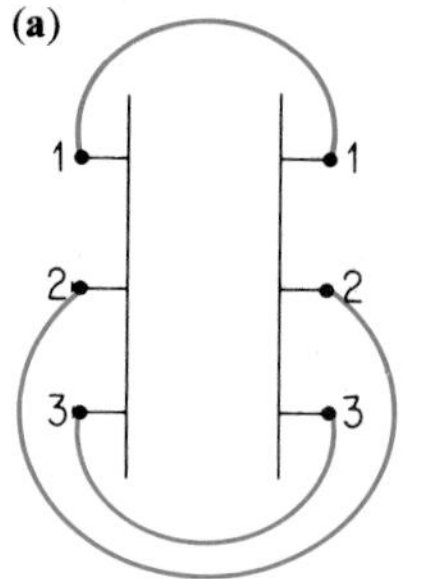

(**b**)

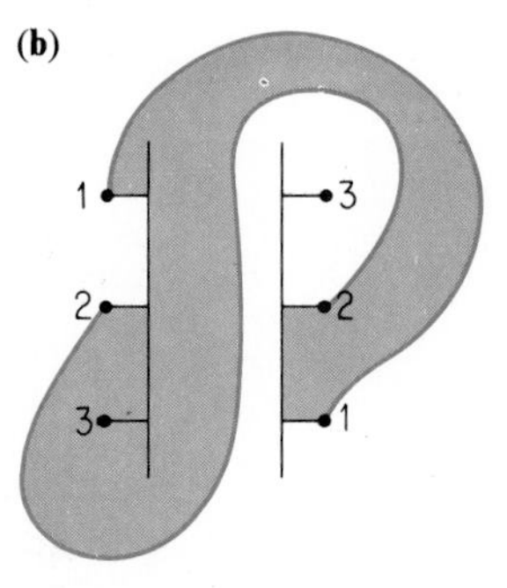

Impossible, by the Jordan curve theorem.

Chapter Review: page 357

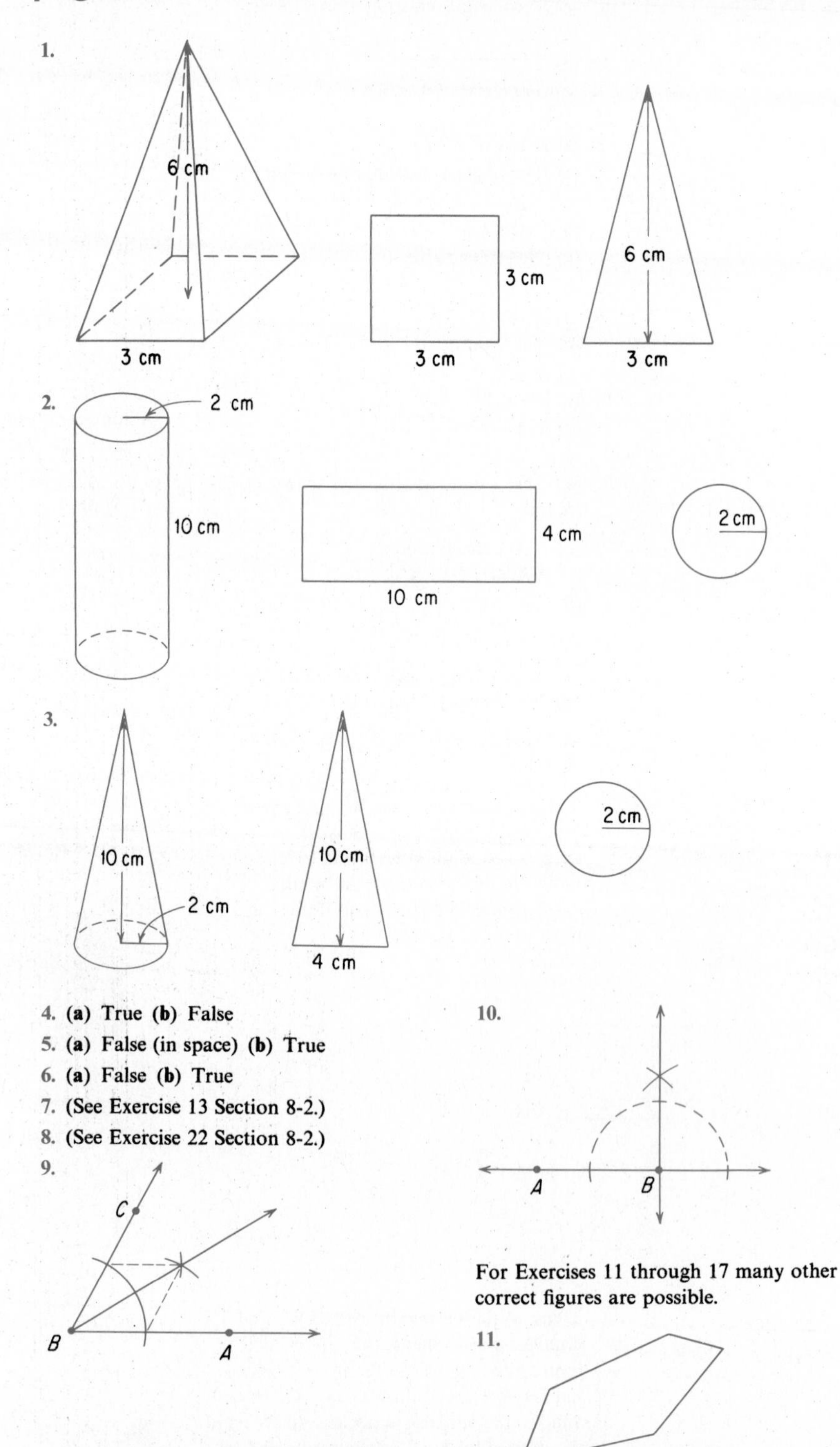

4. (a) True (b) False
5. (a) False (in space) (b) True
6. (a) False (b) True
7. (See Exercise 13 Section 8-2.)
8. (See Exercise 22 Section 8-2.)

For Exercises 11 through 17 many other correct figures are possible.

12.

13.

14.

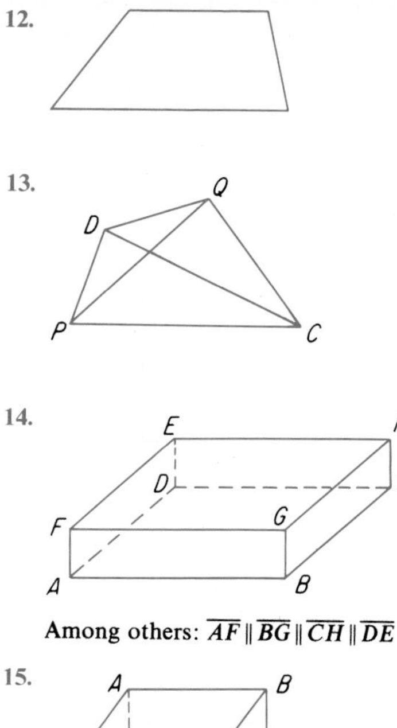

Among others: $\overline{AF} \parallel \overline{BG} \parallel \overline{CH} \parallel \overline{DE}$

15.

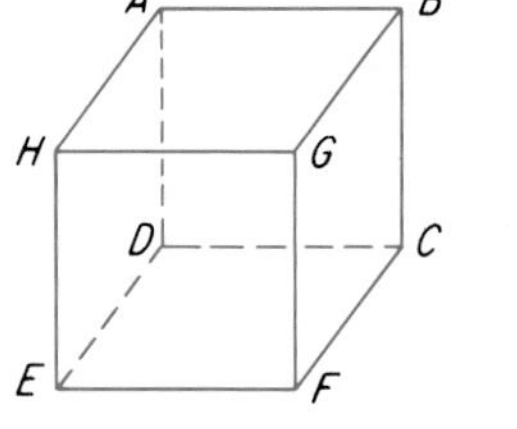

Among others: $\overleftrightarrow{EH} \perp$ plane EFC

16. Use the figure for Exercise 15. Among others: plane $EFC \perp$ plane CFG.

17.

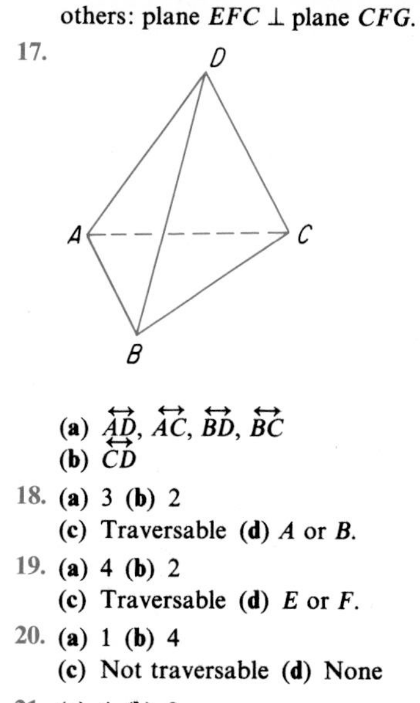

(a) $\overleftrightarrow{AD}$, $\overleftrightarrow{AC}$, $\overleftrightarrow{BD}$, $\overleftrightarrow{BC}$
(b) $\overleftrightarrow{CD}$

18. **(a)** 3 **(b)** 2
(c) Traversable **(d)** A or B.

19. **(a)** 4 **(b)** 2
(c) Traversable **(d)** E or F.

20. **(a)** 1 **(b)** 4
(c) Not traversable **(d)** None

21. **(a)** 4 **(b)** 2
(c) Traversable **(d)** A or B.

22. In the figure for Exercise 20, add an arc joining any two odd vertices; for example, an arc joining A and B.

23. **(a)** Yes **(b)** No **(c)** Yes **(d)** Yes

24. **(a)** Yes **(b)** Yes **(c)** Yes **(d)** Yes

25. **(a)** No **(b)** No **(c)** Yes **(d)** Yes

Chapter 9 Measurement and Geometry

9-1 The Metric System: page 368

1. False
3. False
5. True
7. False
9. True
11. Unlikely
13. Likely
15. Unlikely
17. Likely
19. Likely
21. (b)
23. (c)
25. (b)
27. 48.2 cm
29. 7.500 m
31. 982 cm
33. 3500 mm
35. 7000 g
37. 2.500 liters
39. **(a)** 35 mm **(b)** 3.5 cm
41. **(a)** 80 mm **(b)** 8.0 cm

*43. **(a)** 1 000 000 **(b)** 1 000 000
(c) 1000 **(d)** 5

9-2 Concept of Measurement: page 377

1. **(a)** $2\frac{1}{2}$ inches **(b)** 6 centimeters
3. **(a)** 1 kilometer **(b)** 1/10 millimeter
5. **(a)** 1/2 liter **(b)** 1/10 liter
7. **(a)** Exact **(b)** Approximate
9. **(a)** Approximate **(b)** Exact
11. **(a)** Three-digit **(b)** Five-digit
(c) Two-digit
13. **(a)** Two-digit **(b)** One-digit
(c) One-digit
15. 76.4, nearest tenth, three-digit accuracy

17. 1.00, nearest hundredth, three-digit accuracy

19. 4.0, nearest tenth, two-digit accuracy

21. 5.00, nearest hundredth, three-digit accuracy

23. 3.7×10^3, nearest hundred, two-digit accuracy

25. 1.3×10^6, nearest hundred thousand, two-digit accuracy

27. Among others:

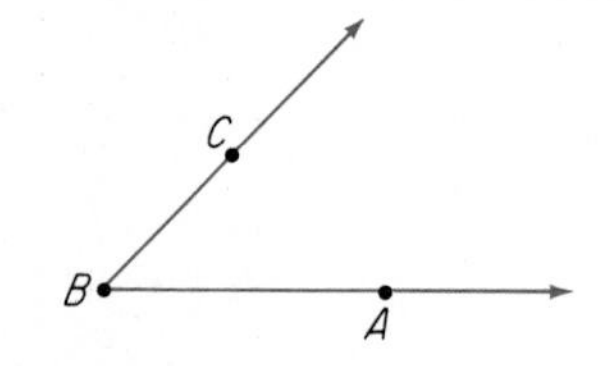

29. Among others:

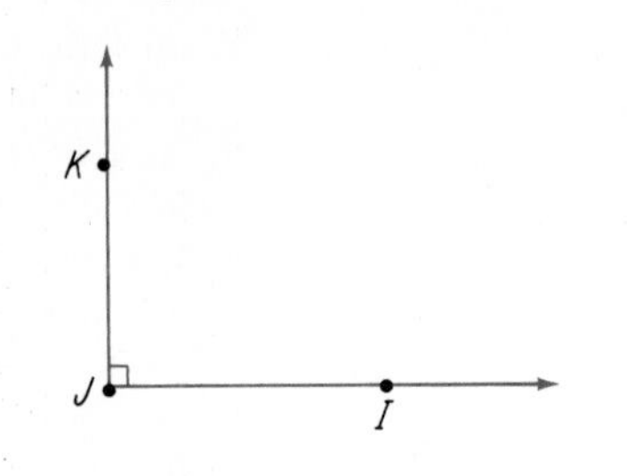

31. Among others:

33. $\pi/4$ radian

35. π radians

37. 2π radians

39. $5\pi/4$ radians

41. $-7\pi/4$ radians

43. 4π radians

45. 90°

47. 270°

49. 315°

51. −360°

53. −270°

55. −540°

9-3 Measures of Plane Figures: page 384

1. **(a)** 5 **(b)** 0 **(c)** 6

3. **(a)** 8 **(b)** 0 **(c)** 8

5. **(a)** 5 **(b)** 5 **(c)** 9

7. **(a)** 36 cm **(b)** 80 cm^2

9. **(a)** 48 cm **(b)** 45 cm^2

11. **(a)** 42 cm **(b)** 68 cm^2

13. **(a)** $26 + \sqrt{85} + 3\sqrt{5}$ cm **(b)** 78 cm^2

15. 28 cm^2

17. $3 \times 10^2\ \text{mm}^2$

19. $1.2 \times 10^2\ \text{km}^2$

21. **(a)** $60 + 20\pi$ units
(b) $600 + 100\pi$ square units

23. 184 cm^2

*25. $240 + 96\pi\ \text{m}^2$

*27. $3A$, $5A$, and $7A$

9-4 Geoboards: page 395

1.

3. $1, \sqrt{2}, 2, \sqrt{5}, \sqrt{8}, 3, \sqrt{10}, \sqrt{13}, 4, \sqrt{17}, \sqrt{18}, \sqrt{20}, 5, \sqrt{26}, \sqrt{29}, \sqrt{32}, \sqrt{34}, \sqrt{41}, \sqrt{50}$

5. **(a)** $\sqrt{2}$
(b) $\sqrt{2}, 2\sqrt{2}, 3\sqrt{2}, 4\sqrt{2}, 5\sqrt{2}$
(c) $x = y$

7.

	P	Q	R	S
(1)	(1, 1)	(4, 2)	(3, 5)	(0, 4)
(2)	(1, 0)	(4, 1)	(3, 4)	(0, 3)
(3)	(2, 1)	(5, 2)	(4, 5)	(1, 4)
(4)	(2, 0)	(5, 1)	(4, 4)	(1, 3)
(5)	(1, 1)	(4, 0)	(5, 3)	(2, 4)
(6)	(0, 1)	(3, 0)	(4, 3)	(1, 4)
(7)	(1, 2)	(4, 1)	(5, 4)	(2, 5)
(8)	(0, 2)	(3, 1)	(4, 4)	(1, 5)

9. 1, 2, 3, 4, 5, 6, 8, 9, 10, 12, 15, 16, 20, 25

11. **(a)** Five **(b)** 1, 2, 3, 4, 5

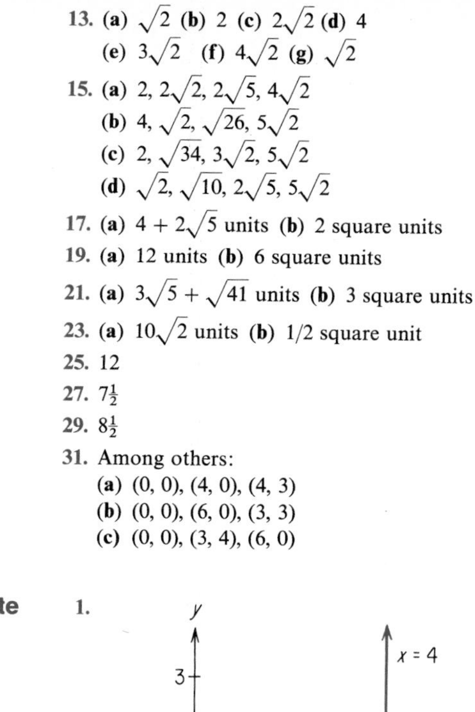

13. **(a)** $\sqrt{2}$ **(b)** 2 **(c)** $2\sqrt{2}$ **(d)** 4
    (e) $3\sqrt{2}$ **(f)** $4\sqrt{2}$ **(g)** $\sqrt{2}$
15. **(a)** $2, 2\sqrt{2}, 2\sqrt{5}, 4\sqrt{2}$
    (b) $4, \sqrt{2}, \sqrt{26}, 5\sqrt{2}$
    (c) $2, \sqrt{34}, 3\sqrt{2}, 5\sqrt{2}$
    (d) $\sqrt{2}, \sqrt{10}, 2\sqrt{5}, 5\sqrt{2}$
17. **(a)** $4 + 2\sqrt{5}$ units **(b)** 2 square units
19. **(a)** 12 units **(b)** 6 square units
21. **(a)** $3\sqrt{5} + \sqrt{41}$ units **(b)** 3 square units
23. **(a)** $10\sqrt{2}$ units **(b)** 1/2 square unit
25. 12
27. $7\frac{1}{2}$
29. $8\frac{1}{2}$
31. Among others:
    (a) (0, 0), (4, 0), (4, 3)
    (b) (0, 0), (6, 0), (3, 3)
    (c) (0, 0), (3, 4), (6, 0)
33. Among others:
    (a) (0, 0), (2, 0), (5, 4), (3, 4)
    (b) (0, 0), (6, 0), (9, 4), (3, 4)
    (c) (0, 0), (4, 0), (6, 3), (2, 3)
35. **(a)** C: (6, 5), perimeter 22, altitude 5, area 30.
    (b) C: (9, 4), perimeter 22, altitude 4, area 24.
    (c) C: (10, 3), perimeter 22, altitude 3, area 18.
    Conjecture: Among parallelograms with a given base and perimeter, the rectangle has the greatest area.
37. **(a)** Among others: (0, 3), (1, 3), (2, 3), (3, 3), (4, 3).
    (b) The line $y = 3$.
    (c) $2 + 2\sqrt{10}$
    (d) There is no largest possible perimeter.

9-5 The Coordinate Plane: page 404

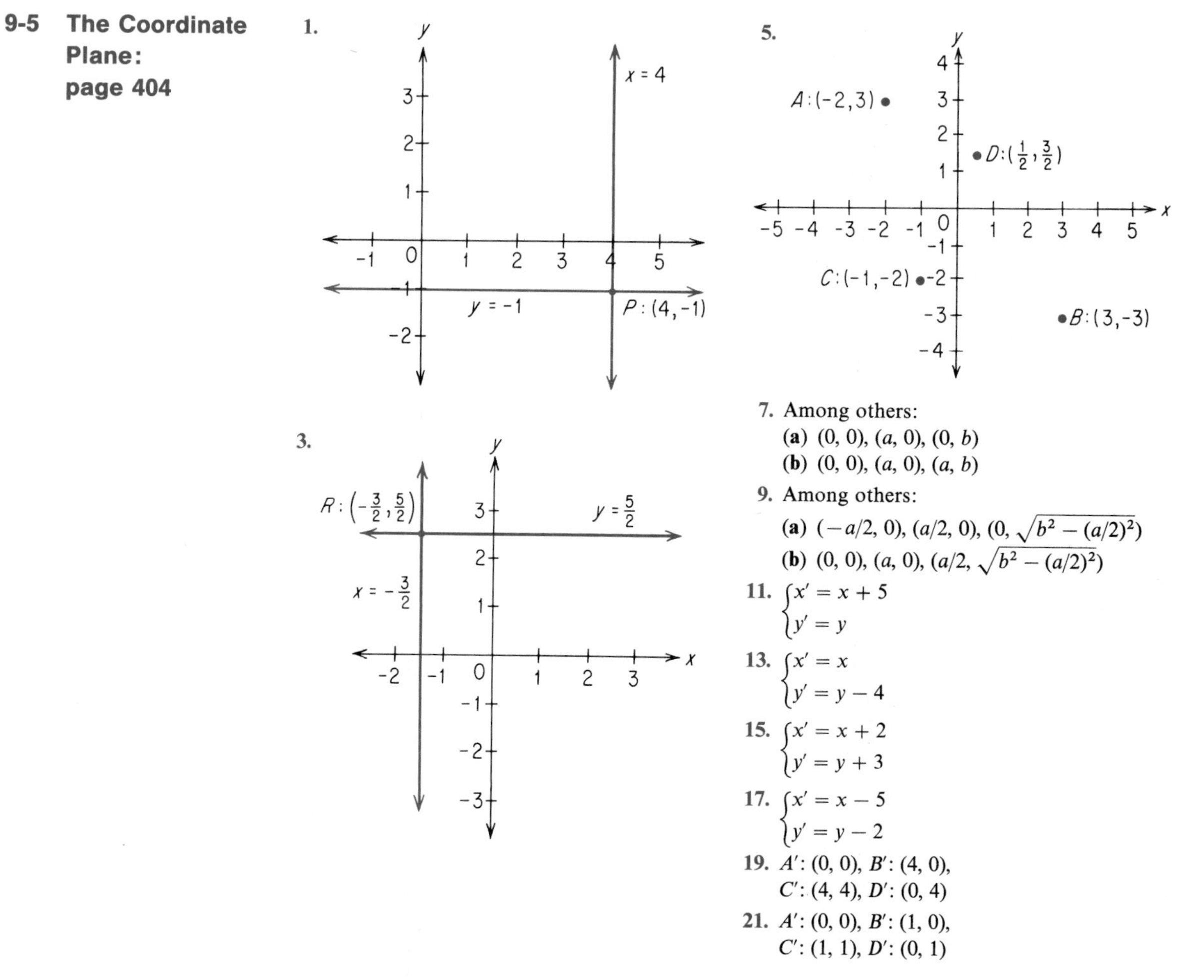

7. Among others:
   (a) (0, 0), $(a, 0)$, $(0, b)$
   (b) (0, 0), $(a, 0)$, (a, b)
9. Among others:
   (a) $(-a/2, 0), (a/2, 0), (0, \sqrt{b^2 - (a/2)^2})$
   (b) $(0, 0), (a, 0), (a/2, \sqrt{b^2 - (a/2)^2})$
11. $\begin{cases} x' = x + 5 \\ y' = y \end{cases}$
13. $\begin{cases} x' = x \\ y' = y - 4 \end{cases}$
15. $\begin{cases} x' = x + 2 \\ y' = y + 3 \end{cases}$
17. $\begin{cases} x' = x - 5 \\ y' = y - 2 \end{cases}$
19. A': (0, 0), B': (4, 0), C': (4, 4), D': (0, 4)
21. A': (0, 0), B': (1, 0), C': (1, 1), D': (0, 1)

23. **(a)** 4 to 1

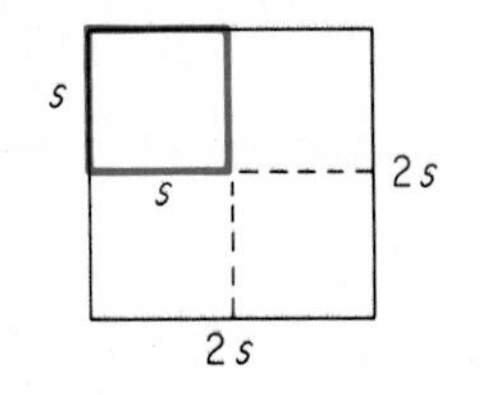

(b) The area of the larger square will be four times the area of the smaller square.

25. The ratio of the area of the larger figure to the area of the smaller figure will be k^2 to 1 since the area is being measured in unit squares and each unit square corresponds to a square of side k.

27. **(a)** **(b)** **(c)** **(d)**

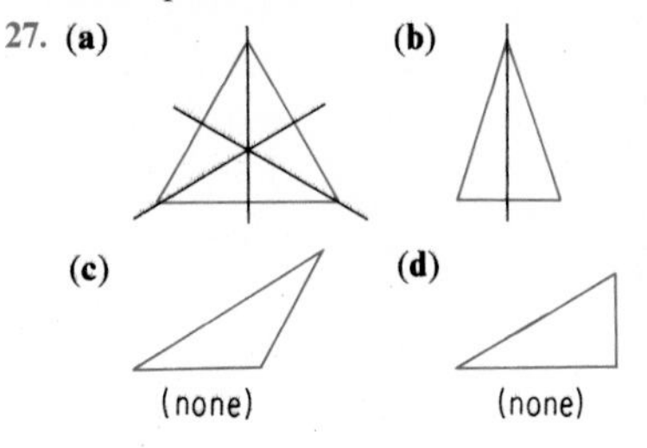

29. **(a)** **(b)** **(c)** **(d)**

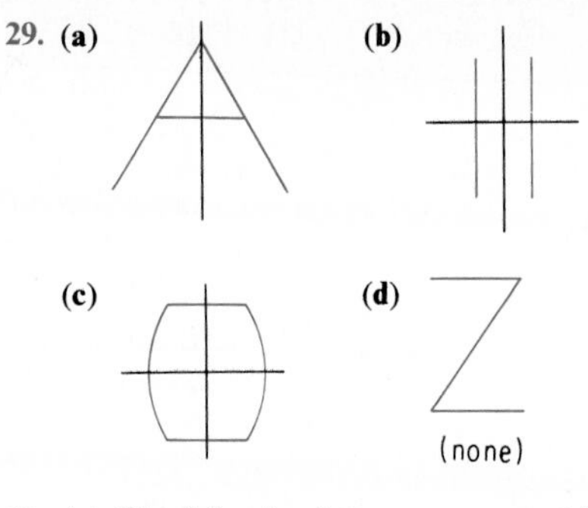

31. **(a)** The following letters are symmetric with respect to the y-axis when centered on the y-axis: A, H, I, M, O, T, U, V, W, X, Y.
(b) The following letters are symmetric with respect to the x-axis when centered on the x-axis: B, C, D, E, H, I, K, O, X.

9-6 Measures of Space Figures: page 410

1. **(a)** 125 cm^3 **(b)** 150 cm^2
3. **(a)** 343 cm^3 **(b)** 294 cm^2
5. **(a)** 3375 mm^3 **(b)** 1350 mm^2
7. **(a)** 84 cm^3 **(b)** 122 cm^2
9. **(a)** 80 m^3 **(b)** 132 m^2
11. 100 cm^3
13. **(a)** 128π cm^3 **(b)** 96π cm^2
15. **(a)** $4000\pi/3$ cm^3 **(b)** 400π cm^2
17. **(a)** 100π cm^3 **(b)** 90π cm^2
19. The volume is multiplied by
(a) 8 **(b)** 27 **(c)** 1/8 **(d)** k^3
The surface area is multiplied by
(a) 4 **(b)** 9 **(c)** 1/4 **(d)** k^2
21. The volume is multiplied by
(a) 8 **(b)** 27 **(c)** 1/8 **(d)** k^3
The surface area is multiplied by
(a) 4 **(b)** 9 **(c)** 1/4 **(d)** k^2
23. **(a)** 9 cm^2 **(b)** 144 cm^2
25. **(a)** \$1000 **(b)** \$64,000

*27. $B_1 = e^2, B_2 = e^2, M = e^2, h = e;$
$V = (e/6)(e^2 + 4e^2 + e^2) = e^3$

*29. $B_1 = 0, B_2 = 0, M = \pi r^2, h = 2r;$
$V = (2r/6)(0 + 4\pi r^2 + 0) = (4/3)\pi r^3$

*31. $B_1 = \pi r^2, B_2 = 0, M = (1/4)\pi r^2;$
$V = (h/6)[\pi r^2 + 4(1/4)\pi r^2 + 0]$
$= (1/3)\pi r^2 h$

Chapter Review: page 414

1. **(a)** 20 millimeters **(b)** 5 kilometers **(c)** 12,000 milligrams **(d)** 1.5 liters
2. **(a)** 1 millimeter **(b)** 1/2 centimeter
3. **(a)** Approximate **(b)** Exact
4. **(a)** $2\pi/3$ radians **(b)** 30°
5.

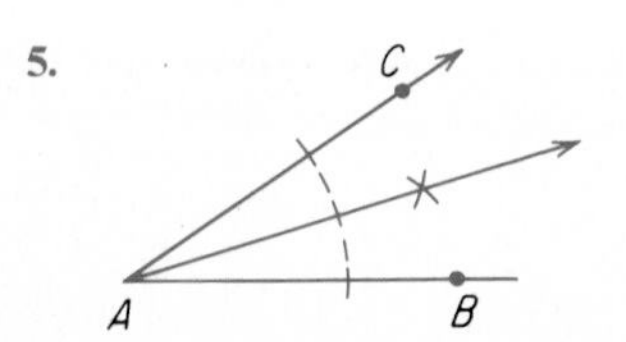

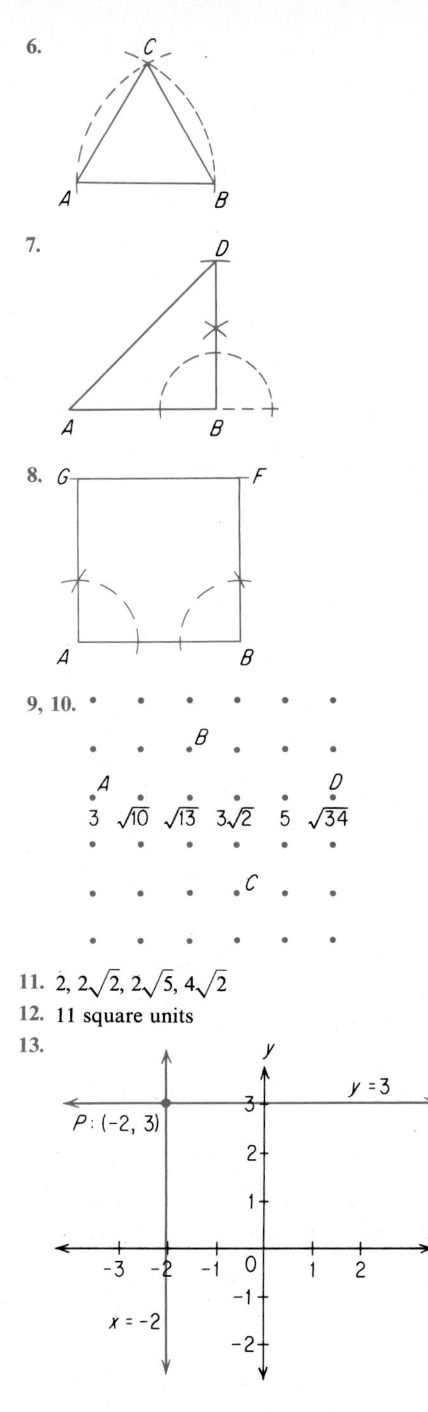

11. $2, 2\sqrt{2}, 2\sqrt{5}, 4\sqrt{2}$

12. 11 square units

14. C: (4, 3)

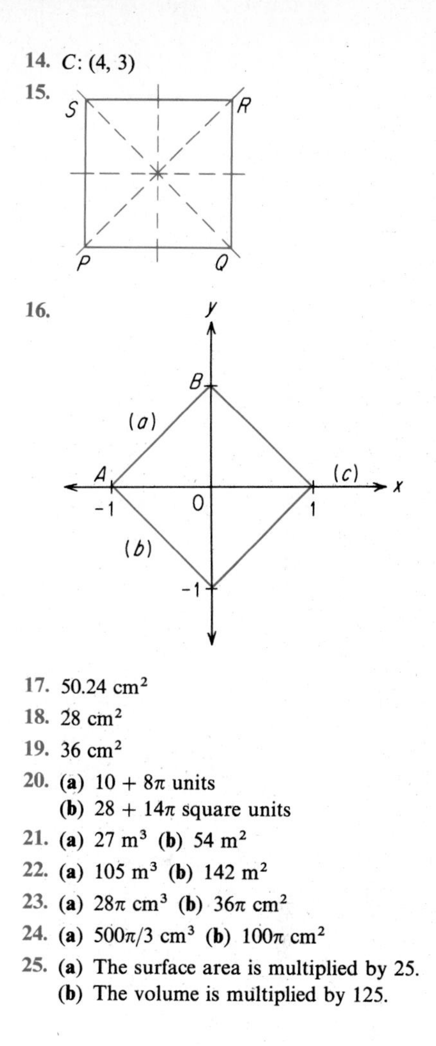

17. 50.24 cm^2

18. 28 cm^2

19. 36 cm^2

20. **(a)** $10 + 8\pi$ units
    (b) $28 + 14\pi$ square units

21. **(a)** 27 m^3 **(b)** 54 m^2

22. **(a)** 105 m^3 **(b)** 142 m^2

23. **(a)** 28π cm^3 **(b)** 36π cm^2

24. **(a)** $500\pi/3$ cm^3 **(b)** 100π cm^2

25. **(a)** The surface area is multiplied by 25.
    (b) The volume is multiplied by 125.

Chapter 10 An Introduction to Probability

10-1 Concept of Probability: page 423

1. **(a)** 24 **(b)** 120

3. **(a)** 64 **(b)** 216

5. **(a)** 8 **(b)** 15 **(c)** 18

7. **(a)** 24 **(b)** 30 **(c)** 72

9. 1/2

11. 2/3

13. 5/6
15. 0
17. 1/13
19. 1/4
21. **(a)** 12 **(b)** 16
23. **(a)** 30 **(b)** 36
25. **(a)** 56 **(b)** 64
27. **(a)** 48 **(b)** 100
29. **(a)** 180 **(b)** 294
31. **(a)** 448 **(b)** 648
33. 15
35. 1/5
37. **(a)** 12 **(b)** 3
39. 1/12
41. 7/8
*43. **(a)** 180 **(b)** 36
*45. **(a)** 72 **(b)** 360
*47. **(a)** 250
(b) $3 \times 8 \times 4 + 2 \times 8 \times 5$, that is, 176
*49. $26 \times 25 \times 9 \times 9 \times 8 \times 7$, that is, 2,948,400
*51. **(a)** 205,320 **(b)** 410,640
*53. **(a)** 6/13 **(b)** 5/26

10-2 Sample Spaces: page 430

1. **(a)** $P(L) = 1/2$, $P(R) = 1/2$
(b) $P(A) = 1/4$, $P(B) = 1/2$, $P(C) = 1/4$
(c) $P(D) = P(E) = P(F) = P(G) = P(H) = P(I) = 1/6$
3. 1/10
5. 2/5
7. 3/5
9. 1/2
11.

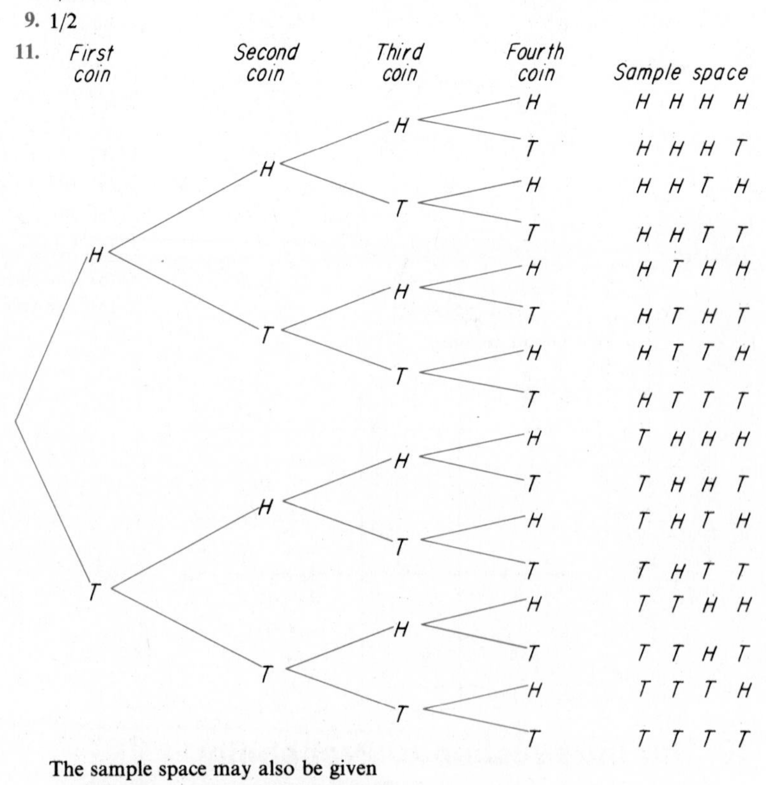

The sample space may also be given as: {*HHHH*, *HHHT*, *HHTH*, *HHTT*, *HTHH*, *HTHT*, *HTTH*, *HTTT*, *THHH*, *THHT*, *THTH*, *THTT*, *TTHH*, *TTHT*, *TTTH*, *TTTT*}.

13. **(a)** 5/16 **(b)** 5/16

15.

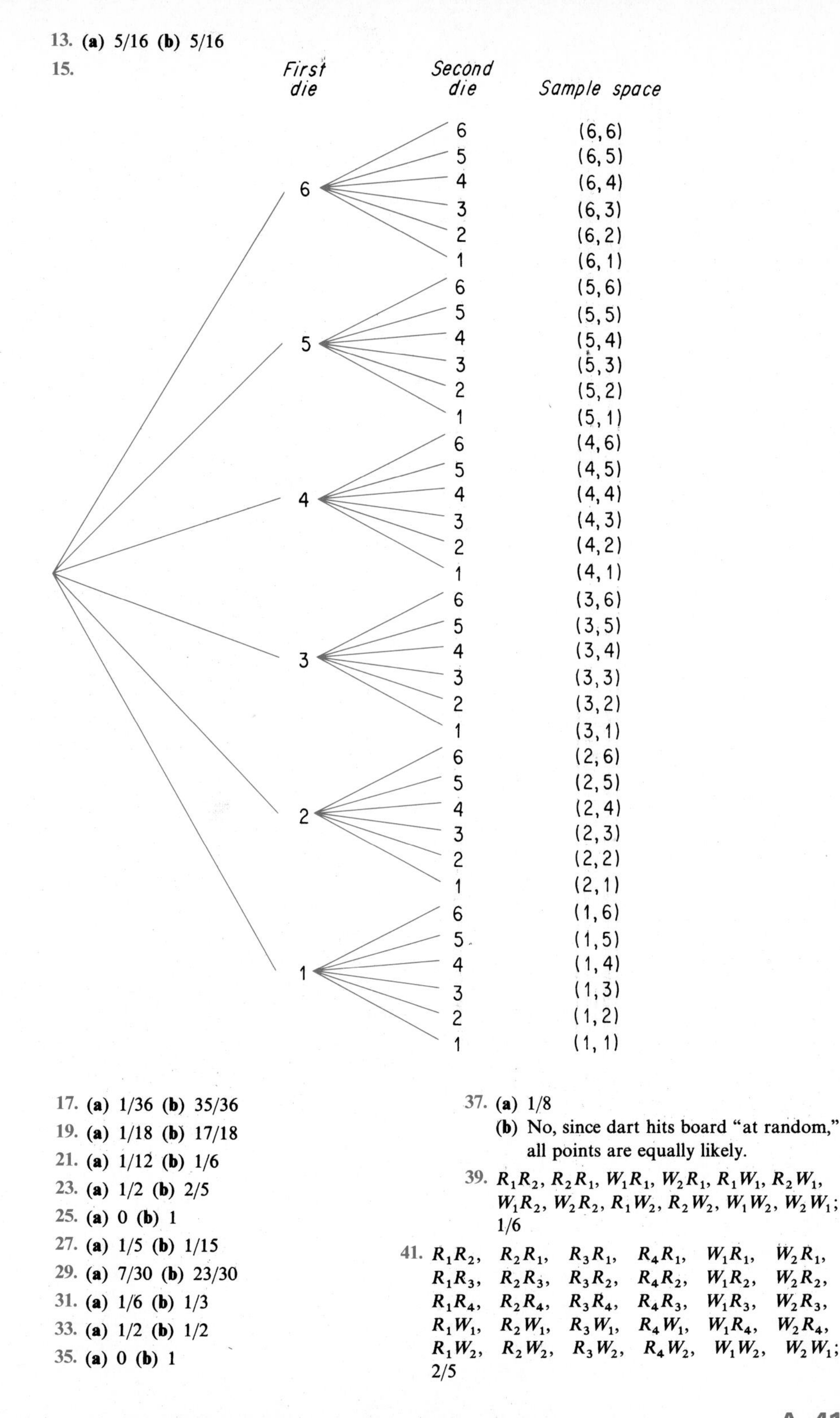

17. **(a)** 1/36 **(b)** 35/36

19. **(a)** 1/18 **(b)** 17/18

21. **(a)** 1/12 **(b)** 1/6

23. **(a)** 1/2 **(b)** 2/5

25. **(a)** 0 **(b)** 1

27. **(a)** 1/5 **(b)** 1/15

29. **(a)** 7/30 **(b)** 23/30

31. **(a)** 1/6 **(b)** 1/3

33. **(a)** 1/2 **(b)** 1/2

35. **(a)** 0 **(b)** 1

37. **(a)** 1/8
(b) No, since dart hits board "at random," all points are equally likely.

39. R_1R_2, R_2R_1, W_1R_1, W_2R_1, R_1W_1, R_2W_1, W_1R_2, W_2R_2, R_1W_2, R_2W_2, W_1W_2, W_2W_1; 1/6

41. R_1R_2, R_2R_1, R_3R_1, R_4R_1, W_1R_1, W_2R_1, R_1R_3, R_2R_3, R_3R_2, R_4R_2, W_1R_2, W_2R_2, R_1R_4, R_2R_4, R_3R_4, R_4R_3, W_1R_3, W_2R_3, R_1W_1, R_2W_1, R_3W_1, R_4W_1, W_1R_4, W_2R_4, R_1W_2, R_2W_2, R_3W_2, R_4W_2, W_1W_2, W_2W_1; 2/5

43.

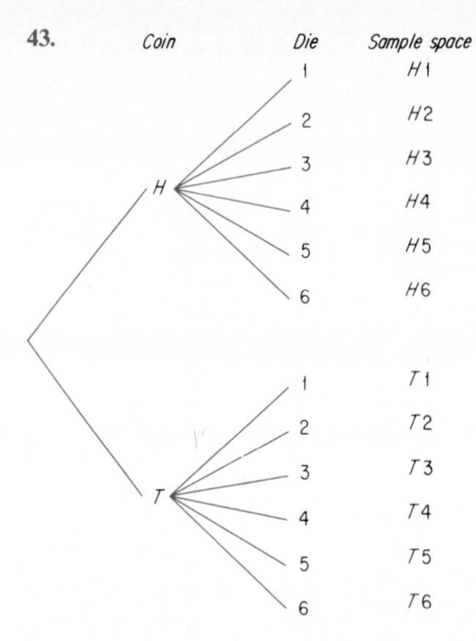

45. **(a)** 1/12 **(b)** 5/12

10-3 Computation of Probabilities: page 437

1. 2/3
3. 2/3
5. 1/2
7. 1
9. 1/2
11. 1/52
13. 19/52
15. 0
17. **(a)** 13/204 **(b)** 1/663
19. **(a)** 1/16 **(b)** 1/2704
21. 1/64
23. **(a)** 1/12 **(b)** 1/4 **(c)** 7/12 **(d)** 3/4
25. **(a)** 7/12 **(b)** 7/22
27. 33/66,640
29. **(a)** 27/2197 **(b)** 72/2197 **(c)** 343/2197 **(d)** 307/2197
31. **(a)** 2/143 **(b)** 6/143 **(c)** 35/286 **(d)** 25/286
33. **(a)** 1/8 **(b)** 1/8 **(c)** 1/8 **(d)** 3/8

*35. **(a)** 1/32 **(b)** 5/32 **(c)** 10/32

10-4 Odds, Mathematical Expectation, and Simulation: page 444

1. 1 to 3
3. 1 to 3
5. 3 to 1
7. 11 to 2
9. 2 to 7
11. $8
13. $(1/6)(10 - 2) + (5/6)(0 - 2) = -1/3$, that is, a loss of $0.33.
15. $2.00

*17. The probability that both of the bills drawn will be tens is 1/10. The probability that both will be fives is 3/10. The probability that one will be a five and one a ten is 3/5. The mathematical expectation is then found to be $14.

*19. The probability of selecting 2¢ is 2/30; the probabilities of 6¢, 11¢, 26¢, and 51¢ are 4/30; and the probabilities of 15¢, 30¢, 55¢, 35¢, 60¢, and 75¢ are each 2/30. Thus the mathematical expectation is $30\frac{2}{3}$¢.

*21. **(a)** p to $(1 - p)$ **(b)** $(1 - p)$ to p

*23. $(1/n)(4999) + [1/(n - 1)](999) + \{1 - [1/n + 1/(n - 1)]\}(-1)$, that is, $(4999/n) + [999/(n - 1)] + \{(2n - 1)/[n(n - 1)]\}(-1)$ dollars.

25. HH—16, HT—30, TH—24, TT—30

10-5 Permutations: page 449

1. **(a)** Bat, ball
   (b) (bat, ball), (ball, bat) (*Note*: Parentheses are often used for ordered sets, that is, permutations.)
3. **(a)** A, B, C, D
   (b) $AB, AC, AD, BA, BC, BD, CA, CB, CD, DA, DB, DC$
   (c) $ABC, ACB, ABD, ADB, ACD, ADC, BAC, BCA, BAD, BDA, BCD, BDC, CAB, CBA, CAD, CDA, CBD, CDB, DAB, DBA, DAC, DCA, DBC, DCB$
   (d) (*Note*: Simply add the remaining element at the end of each permutation in part **(c)**.) $ABCD$,

ACBD, ABDC, ADBC, ACDB, ADCB, BACD, BCAD, BADC, BDAC, BCDA, BDCA, CABD, CBAD, CADB, CDAB, CBDA, CDBA, DABC, DBAC, DACB, DCAB, DBCA, DCBA.

5. 720
7. 7920
9. 210
11. 10!, that is, 3,628,800
13. 1320
15. 604,800
17. $n!$
19. $n!/2$
21. $n!/n!$, that is, 1
23. 21
25. 5
27. 6
29. **(a)** 120 **(b)** 48 **(c)** 2/5 **(d)** 3/5
31. 210
33. 6: PAPA, APAP, PAAP, PPAA, APPA, AAPP
35. 34,650
37. 12,600

10-6 Combinations: page 454

1. 35
3. 120
5. 190
7. 28
9. 56
11. 165
13. ${}_3P_2$: $(r, t)(t, r)$ $(r, s)(s, r)$ $(s, t)(t, s)$
    ${}_3C_2$: $\{r, t\}$ $\{r, s\}$ $\{s, t\}$; 3
15. ${}_4C_3$: $\{a, b, c\}$ $\{a, b, d\}$ $\{a, c, d\}$ $\{b, c, d\}$
    ${}_4C_1$: $\{d\}$ $\{c\}$ $\{b\}$ $\{a\}$
17. ${}_nC_0 = 1$; the only possible combination of n things 0 at a time is obtained when none are selected.
19. ${}_5C_0 + {}_5C_1 + {}_5C_2 + {}_5C_3 + {}_5C_4 + {}_5C_5 = 1 + 5 + 10 + 10 + 5 + 1 = 32 = 2^5$
21. 64
23. 1848
25. 84
27. 80
29. **(a)** ${}_{16}C_4$, that is, 1820
    (b) ${}_{10}C_3 \times {}_6C_1$, that is, 720
31. $({}_4C_4 \times {}_{48}C_2)/{}_{52}C_6$, that is, 3/54,145
33. ${}_{12}C_4$, that is, 495
35. ${}_{12}C_2$, that is, 66
37. **(a)** ${}_{52}C_4$, that is, 270,725
    (b) ${}_{52}C_7$, that is, 133,784,560
39. $({}_{13}C_2)({}_{13}C_3) = 22,308$
41. **(a)** ${}_{25}C_2$, that is, 300
    (b) ${}_{26}C_2$, that is, 325

*43. $(1/2)_8C_4$, that is, 35

Chapter Review: page 458

1. **(a)** 57,120 **(b)** 10,100
2. **(a)** 380 **(b)** 190
3.

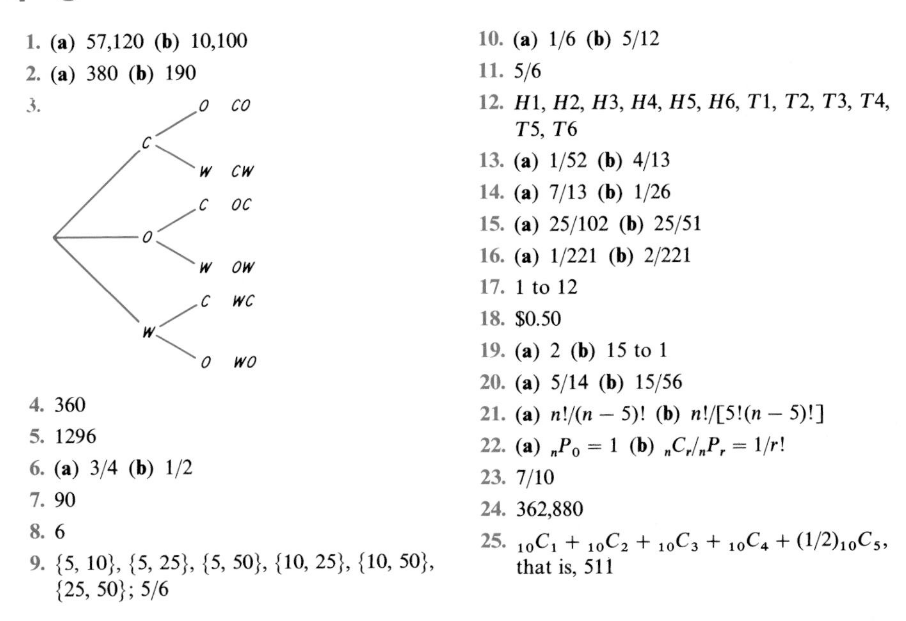

4. 360
5. 1296
6. **(a)** 3/4 **(b)** 1/2
7. 90
8. 6
9. {5, 10}, {5, 25}, {5, 50}, {10, 25}, {10, 50}, {25, 50}; 5/6
10. **(a)** 1/6 **(b)** 5/12
11. 5/6
12. *H*1, *H*2, *H*3, *H*4, *H*5, *H*6, *T*1, *T*2, *T*3, *T*4, *T*5, *T*6
13. **(a)** 1/52 **(b)** 4/13
14. **(a)** 7/13 **(b)** 1/26
15. **(a)** 25/102 **(b)** 25/51
16. **(a)** 1/221 **(b)** 2/221
17. 1 to 12
18. $0.50
19. **(a)** 2 **(b)** 15 to 1
20. **(a)** 5/14 **(b)** 15/56
21. **(a)** $n!/(n-5)!$ **(b)** $n!/[5!(n-5)!]$
22. **(a)** ${}_nP_0 = 1$ **(b)** ${}_nC_r/{}_nP_r = 1/r!$
23. 7/10
24. 362,880
25. ${}_{10}C_1 + {}_{10}C_2 + {}_{10}C_3 + {}_{10}C_4 + (1/2)_{10}C_5$, that is, 511

Chapter 11 An Introduction to Statistics

11-1 Uses and Abuses of Statistics: page 466

1. The fact that most accidents occur near home probably means that most of the miles driven are near the home of the driver. It does not necessarily mean that long trips are safer than short trips.
3. The conclusion that mathematics is very popular is not justified by the previous statement even though the conclusion may be true. The fact that more students are studying mathematics reflects the fact that colleges are requiring more mathematics.
5. The 100% sale probably indicates that the book was the required text in a course that all students had to take, but the given evidence does not justify the stated conclusion.
7. The survey may not have covered a representative sample of the voting population. Details of the sampling procedure would be needed to supply confidence in the conclusion.
9. One wonders how such a count could possibly have been made. Details of the sampling procedure would be needed to supply confidence in the conclusion.
11. Many people who have asthma go to Arizona because of the climate. Probably a larger proportion of the people in Arizona have asthma than in any other state. Thus the conclusion is not justified.
13. People do not necessarily vote for the candidate of the party in which they are registered.
15. It should be stated what *average* means in this statement. If average denotes the median, then there would always be 50% at or below average.
17. How short? How tall? What is meant by aggressive? How is aggressiveness measured?
19. Effective for what conditions? How does brand A compare to other brands? What is meant by effectiveness? How is effectiveness determined?
21. Fewer than whom? What age groups are under consideration? How much swimming is considered?
23. What is meant by success? What is implied by "student"?
25. Are those who marry during their last year of college included? What is the source of this information?
27. The small changes in temperature appear very large due to the fact that the temperature scale does not start at zero.
29. No; a scale is needed before any such conclusions can be made.
31. No, the amount of increase of NEWS was greater.
33. The absence of a scale to indicate the number of papers in circulation for each newspaper; also, an indication that the same types (paid, free, and so on) of circulation are being compared.

11-2 Descriptive Statistics: page 472

7.

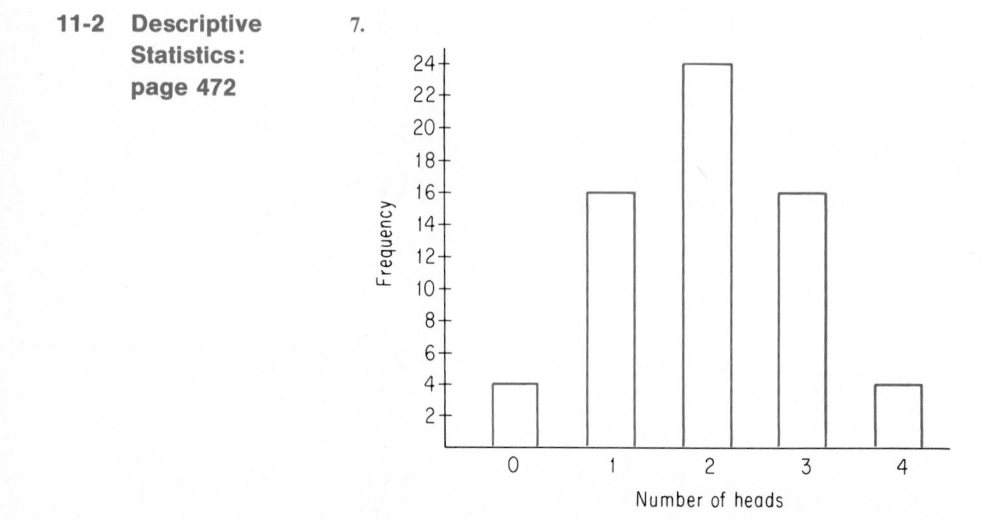

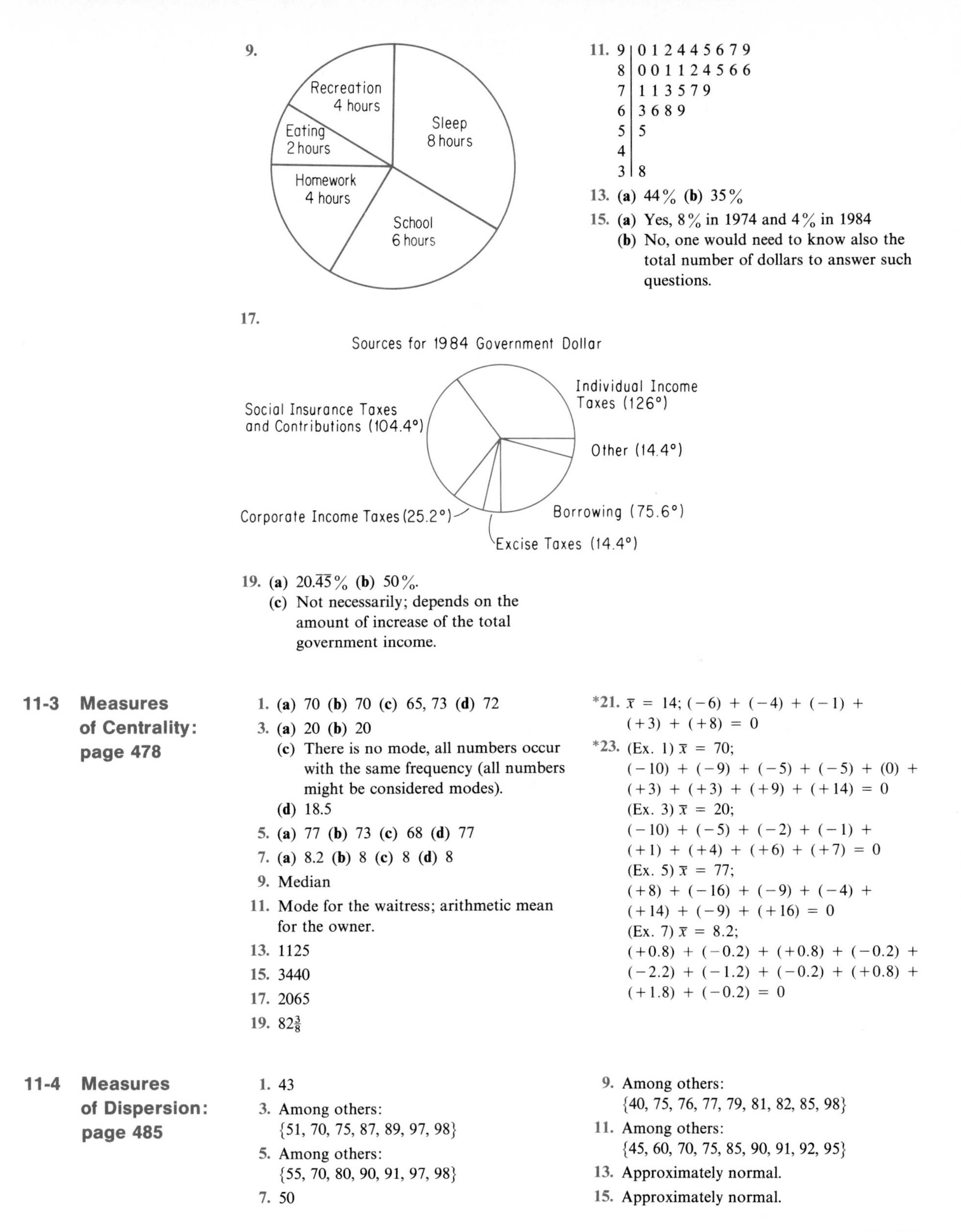

9.

11.

9	0 1 2 4 4 5 6 7 9
8	0 0 1 1 2 4 5 6 6
7	1 1 3 5 7 9
6	3 6 8 9
5	5
4	
3	8

13. **(a)** 44% **(b)** 35%

15. **(a)** Yes, 8% in 1974 and 4% in 1984
(b) No, one would need to know also the total number of dollars to answer such questions.

17.

19. **(a)** $20.\overline{45}\%$ **(b)** 50%.
(c) Not necessarily; depends on the amount of increase of the total government income.

11-3 Measures of Centrality: page 478

1. **(a)** 70 **(b)** 70 **(c)** 65, 73 **(d)** 72
3. **(a)** 20 **(b)** 20
(c) There is no mode, all numbers occur with the same frequency (all numbers might be considered modes).
(d) 18.5
5. **(a)** 77 **(b)** 73 **(c)** 68 **(d)** 77
7. **(a)** 8.2 **(b)** 8 **(c)** 8 **(d)** 8
9. Median
11. Mode for the waitress; arithmetic mean for the owner.
13. 1125
15. 3440
17. 2065
19. $82\frac{3}{8}$

*21. $\bar{x} = 14$; $(-6) + (-4) + (-1) + (+3) + (+8) = 0$

*23. (Ex. 1) $\bar{x} = 70$;
$(-10) + (-9) + (-5) + (-5) + (0) + (+3) + (+3) + (+9) + (+14) = 0$
(Ex. 3) $\bar{x} = 20$;
$(-10) + (-5) + (-2) + (-1) + (+1) + (+4) + (+6) + (+7) = 0$
(Ex. 5) $\bar{x} = 77$;
$(+8) + (-16) + (-9) + (-4) + (+14) + (-9) + (+16) = 0$
(Ex. 7) $\bar{x} = 8.2$;
$(+0.8) + (-0.2) + (+0.8) + (-0.2) + (-2.2) + (-1.2) + (-0.2) + (+0.8) + (+1.8) + (-0.2) = 0$

11-4 Measures of Dispersion: page 485

1. 43
3. Among others: {51, 70, 75, 87, 89, 97, 98}
5. Among others: {55, 70, 80, 90, 91, 97, 98}
7. 50
9. Among others: {40, 75, 76, 77, 79, 81, 82, 85, 98}
11. Among others: {45, 60, 70, 75, 85, 90, 91, 92, 95}
13. Approximately normal.
15. Approximately normal.

17. **(a)** 2.5% **(b)** 16% **(c)** 95%

19. ≈3.9

21. ≈4.6

*23. ≈9.96

11-5 Measures of Position: page 490

1. 495
3. 275
5. 137.5
7. 99
9. 97
11. 94
13. Charles
15. Charles
17. 665
19. **(a)** 50 **(b)** 95 **(c)** 40 **(d)** 20 **(e)** 5 **(f)** 99

11-6 Distributions and Applications: page 495

1. 120
3. 150
5. 100 to 140
7. 45 to 55
9. 35 to 65
11. 30
13. 15
15. No; it would appear that her SAT scores should be much closer to 700, that is, about two standard deviations above the mean.
17. Among other things emphasize that *no single class should be expected to have a normal distribution.* Also clarify the meaning of the term *average.*
19. 16 to 24

Chapter Review: page 498

1. Among others: Failure to label scales, the use of scales that do not start at 0 or some other clearly designated reference point, the misuse of areas or volumes to represent linear measures.

2.

Grade	A	B	C	D	F
Frequency	8	23	4	1	1

3.

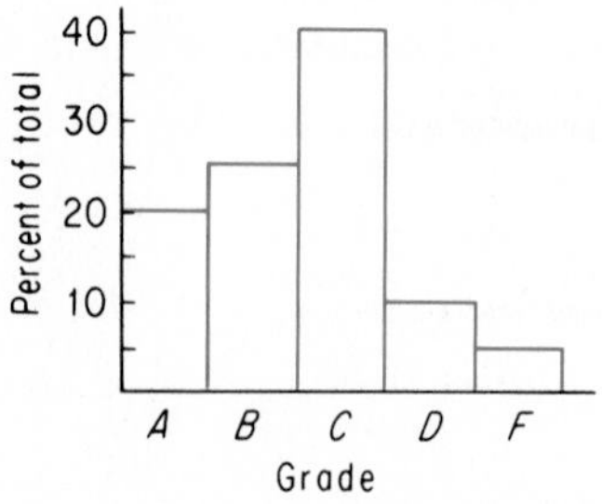

4.

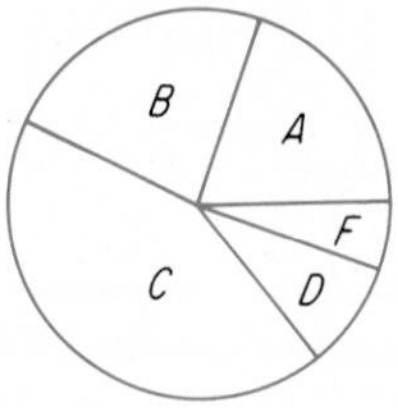

5.

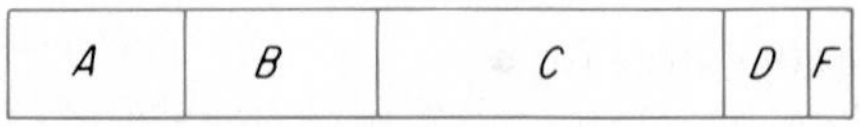

6. 22%
7. 80°
8. 35 mm
9. 69
10. 72
11. 74
12. **(a)** 34 **(b)** 81
13. **(a)** 55 **(b)** 72.5
14. 2.5
15. 84
16. 16
17. 81.5
18. Almost 0
19. 8th
20. 95
21. 32
22. 4
23. Between 20 and 44
24. 84
25. 84

Chapter 12 An Introduction to Algebra

12-1 Algebra and Arithmetic: page 507

1. Open sentence
3. Open sentence
5. Identity
7. False statement
9. Identity
11. False statement
13. $\{5\}$
15. $\{0, 1, 2, 3\}$
17. $\{0, 1, 2, 3, 4, 5, 6\}$
19. $\{3, 2, 1, 0, -1, -2, -3, \ldots\}$
21. $\{2, 1, 0, -1, -2, \ldots\}$
23. $\{-2, -1, 0, 1, 2, 3, \ldots\}$
25. x is any real number greater than 2.
27. x is any real number less than 6.
29. x is any real number other than 5.
31. $2n + 3$
33. $5(n - 2)$
35. 33
37. 2
39. 4
41. 4
43. 162

12-2 Algebra and Geometry: page 513

1. A point.
3. A ray.
5. A line segment.
7. A line.
9. A half-line.
11. A ray.

13.

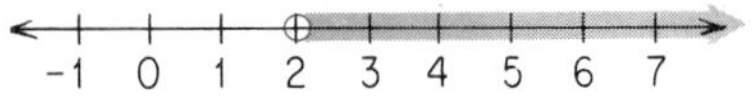

-1 0 1 2 3 4 5 6 7

15.

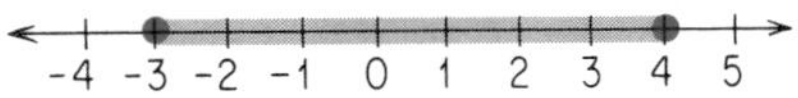

-4 -3 -2 -1 0 1 2 3 4 5

17.

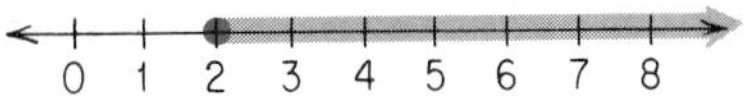

0 1 2 3 4 5 6 7 8

19.

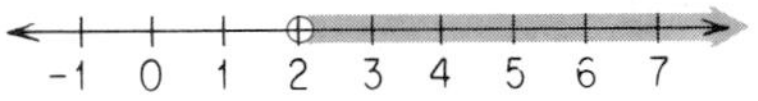

-1 0 1 2 3 4 5 6 7

21.

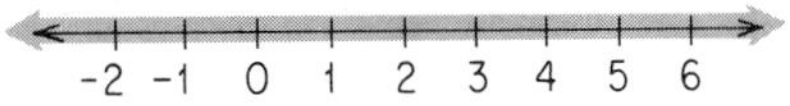

-2 -1 0 1 2 3 4 5 6

23.

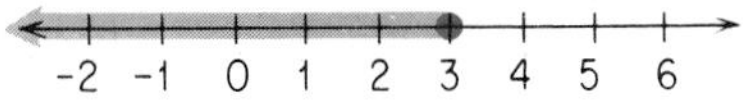

-2 -1 0 1 2 3 4 5 6

*25.

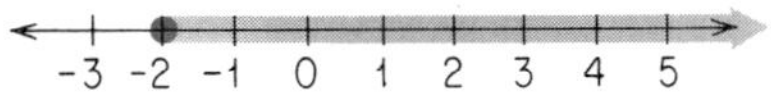

-3 -2 -1 0 1 2 3 4 5

*27.

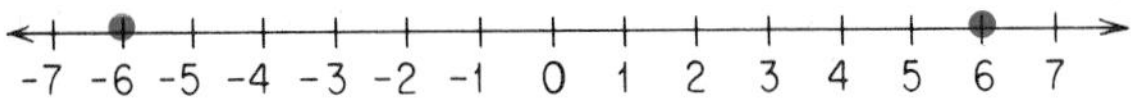

-7 -6 -5 -4 -3 -2 -1 0 1 2 3 4 5 6 7

*29.

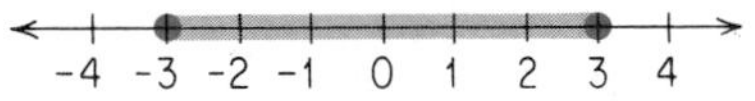

-4 -3 -2 -1 0 1 2 3 4

31. $\{1, 2, 3, 4, 5\}$
33. $\{2, 3, 4\}$
35. $\{-1, -2, -3, \ldots\} \cup \{1, 2, 3, \ldots\}$

37.

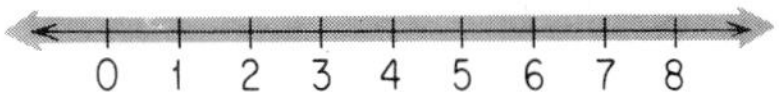

0 1 2 3 4 5 6 7 8

39.

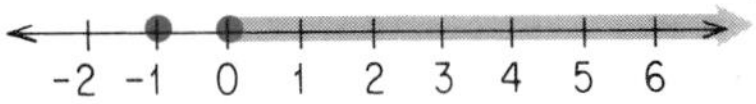

-2 -1 0 1 2 3 4 5 6

41. $\varnothing$
43. $\varnothing$

45.

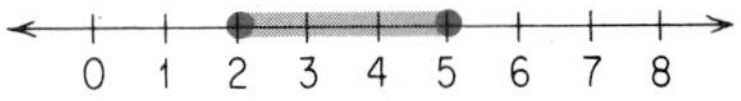

0 1 2 3 4 5 6 7 8

47.

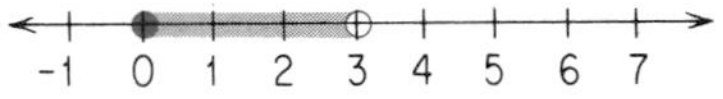

-1 0 1 2 3 4 5 6 7

49.

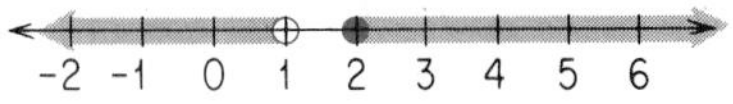

-2 -1 0 1 2 3 4 5 6

51. $\varnothing$
53. $\varnothing$

55.

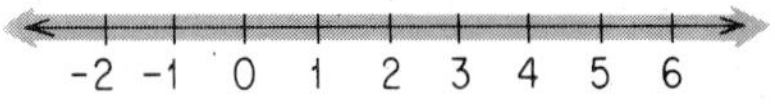

-2 -1 0 1 2 3 4 5 6

57.

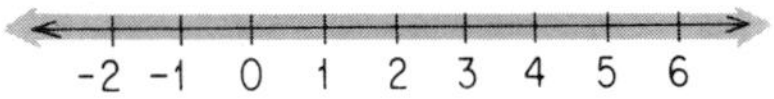

-2 -1 0 1 2 3 4 5 6

*59.

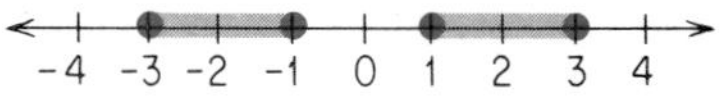

-4 -3 -2 -1 0 1 2 3 4

12-3 Equations and Inequalities: page 516

1. (a) Given
   (b) Subtract 3
   (c) Divide by 2
3. (a) Given
   (b) Add 4
   (c) Divide by 2.
5. (a) Given
   (b) Subtract 5
   (c) Divide by -3.
7. $x = 4$
9. $x = 5$
11. $x = -3$
13. $x = -4$
15. $x = 2$
17. $x < 4$
19. $x > 3$
21. $x > -4$
23. $x < 3$
25. $x = 22$
27. $x = 12$
29. $x > -8$

*31. Always true.

*33. Not always true; among others: $2 - 7 \nless 3 - 10$.

*35. Not always true; among others: $2(-4) \nless 3(-4)$.

12-4 Algebra and Problem Solving: page 521

1. 32 °F
3. 77 °F
5. 140 °F
7. 0 °C
9. 30 °C
11. 50 °C
13. 12
15. 7
17. Length, 20 cm; width, 10 cm
19. Length, 10 cm; perimeter, 30 cm
21. 12
23. 8 people, $39
25. 525 miles

12-5 Sentences Involving Absolute Value: page 525

1. $\{5, -5\}$
3. $\{-2, -1, 0, 1, 2\}$
5. $\varnothing$
7. $\{-3, 5\}$
9. $\{-4, 0\}$
11. $\{-4, -2\}$
13. 9
15. 8
17. 36
19. Line segment
21. Union of two rays
23. Line segment.
25.

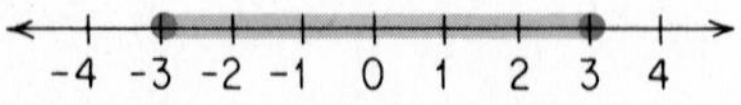

-4 -3 -2 -1 0 1 2 3 4

27.

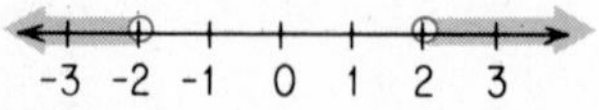

-3 -2 -1 0 1 2 3

29.

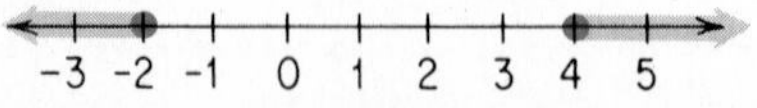

-3 -2 -1 0 1 2 3 4 5

31. $\varnothing$
33.

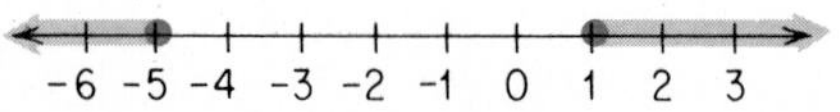

-6 -5 -4 -3 -2 -1 0 1 2 3

35.

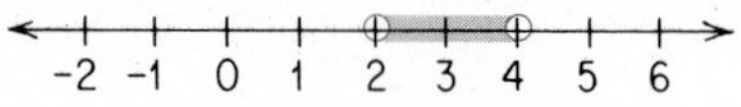

-2 -1 0 1 2 3 4 5 6

*37.

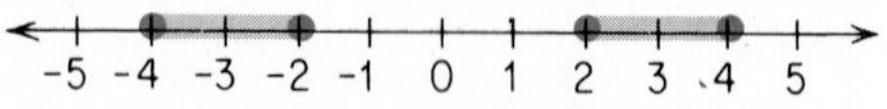

-5 -4 -3 -2 -1 0 1 2 3 4 5

*39.

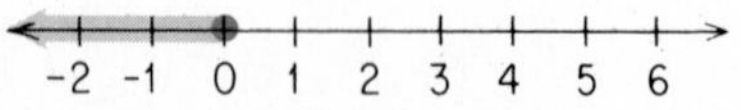

-2 -1 0 1 2 3 4 5 6

*41.

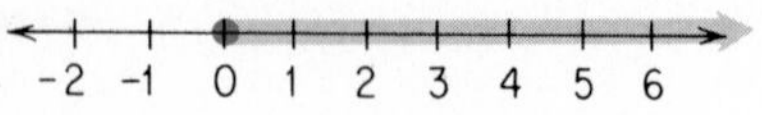

-2 -1 0 1 2 3 4 5 6

*43. $\{5/2\}$

*45. $\{9, 18\}$

12-6 Quadratic Equations: page 528

1. $\{-8, 8\}$
3. $\{-25, 25\}$
5. $\{-5, 5\}$
7. $\{-6, 6\}$
9. $\{-3, 3\}$
11. $\{-3, 3\}$

13. **(a)** 1/2 second **(b)** 5 seconds

15. **(a)** 28,400 feet **(b)** 23,600 feet

17. $\{-3, 4\}$

19. $\{-5/2, 1\}$

21. $\{-2, 3/5\}$

23. $x^2 + 0x - d^2 = 0$, $x = \dfrac{0 \pm \sqrt{0 + 4d^2}}{2} = \dfrac{\pm 2d}{2} = d$ or $-d$

*25. **(a)** Multiplication, $=$
(b) Zero, $\times$
(c) Addition, $=$
(d) Associative, $+$
(e) Addition
(f) Zero, $+$
(g) Addition, $=$
(h) Commutative, $+$
(i) Equivalent fractions, $-c/a = -4ac/(4a^2)$
(j) Addition of fractions
(k) Addition, $=$
(l) Addition of fractions
(m) Definition of subtraction

Chapter Review: page 531

1. **(a)** $\{0, 1, 2, 3\}$ **(b)** $\{0, 1, 2, 3\}$

2. **(a)** x is any real number less than 5.
(b) x is any real number greater than or equal to -2.

3. 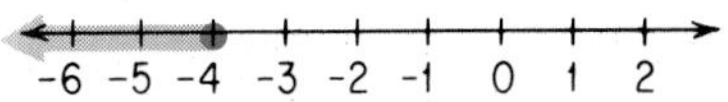
-6 -5 -4 -3 -2 -1 0 1 2

4.
-1 0 1 2 3 4 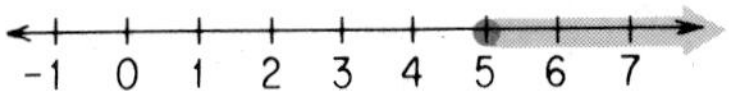 5 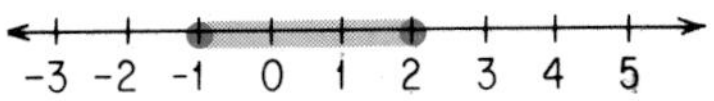 6 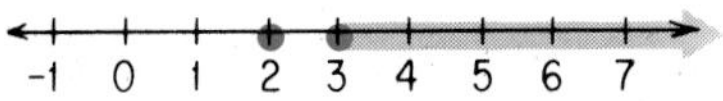 7

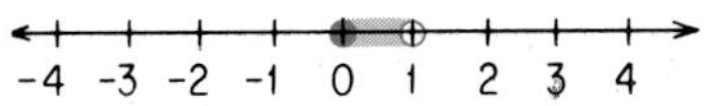

5.
-3 -2 -1 0 1 2 3 4 5

6.
-1 0 1 2 3 4 5 6 7

7.
-4 -3 -2 -1 0 1 2 3 4

8.
-3 -2 -1 0 1 2 3 4 5

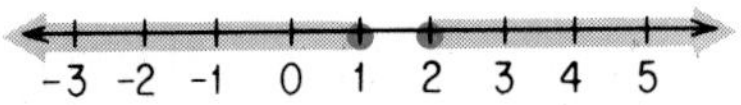

9.
-4 -3 -2 -1 0 1 2 3 4

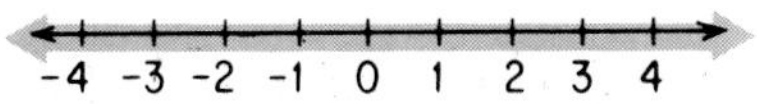

10.
-4 -3 -2 -1 0 1 2 3 4

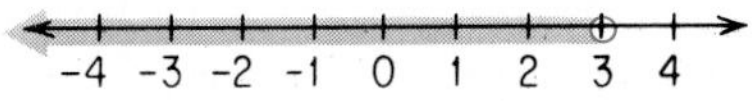

11.
-4 -3 -2 -1 0 1 2 3 4

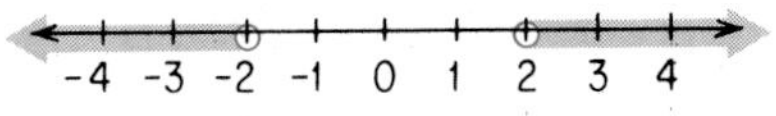

12.
-2 -1 0 1 2 3 4 5 6

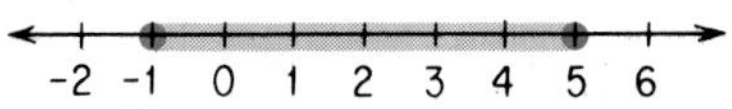

13.
-5 -4 -3 -2 -1 0 1 2 3

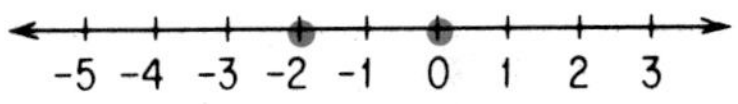

14.
-6 -5 -4 -3 -2 -1 0 1 2

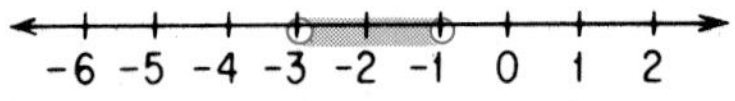

15. $\{4\}$

16. $\{-1\}$

17. $x > 20$

18. $x > 3$

19. $\{-1\}$

20. $\{-2, 8\}$

21. $\{-2, 2\}$

22. $\{-1, 5/2\}$

23. $\{9, 16\}$

24. $\{15\}$

25. **(a)** 59 °F **(b)** 95 °F

Chapter 13 Algebra and Geometry

13-1 Sentences in Two Variables: page 540

1. {(1, 3), (2, 2), (3, 1)}
3. {(1, 1), (1, 2), (2, 1)}
5. ∅
7. {(1, 5), (2, 4), (3, 3), (4, 2), (5, 1)}
9. {(1, 1), (1, 2), (1, 3), (2, 1), (2, 2), (3, 1)}
11. {(1, 1), (1, 2), (2, 1), (3, 1)}
13. {(−3, 3), (−2, 2), (−1, 1), (0, 0), (1, 1), (2, 2), (3, 3)}
15. {(−3, −3), (−2, −2), (−1, −1), (0, 0), (1, −1), (2, −2), (3, −3)}
17. {(−1, 2), (0, 1), (1, 2)}
19. 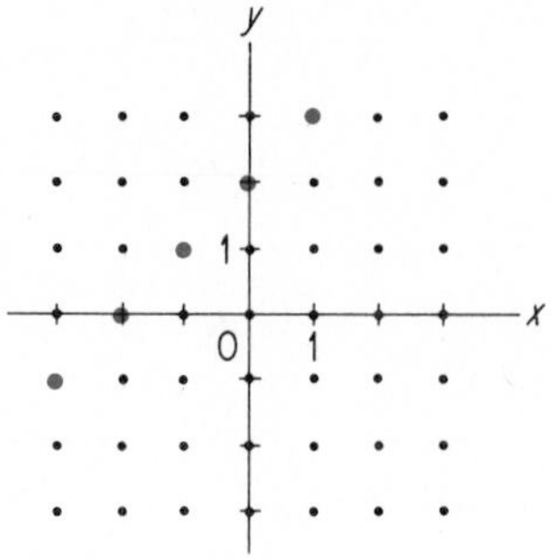
21. 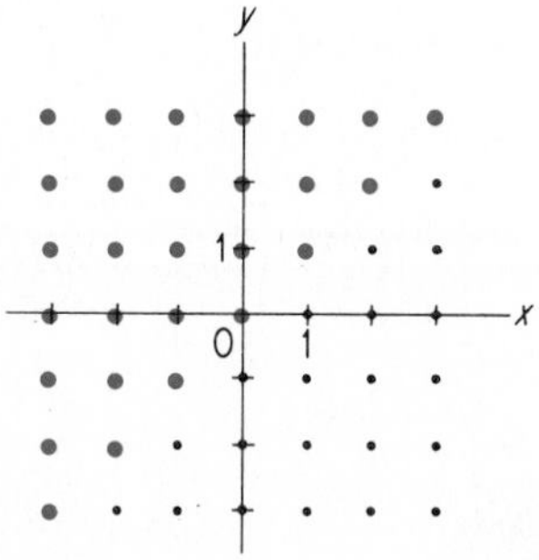
23. 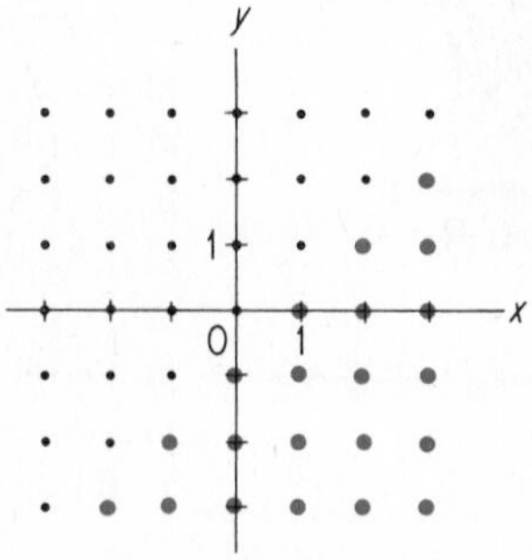
25. 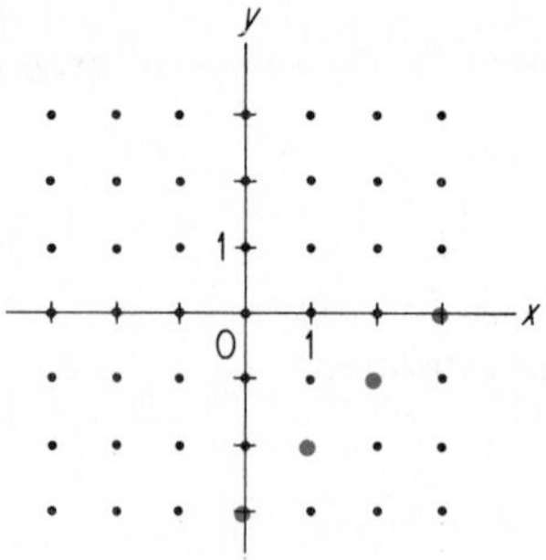
27. 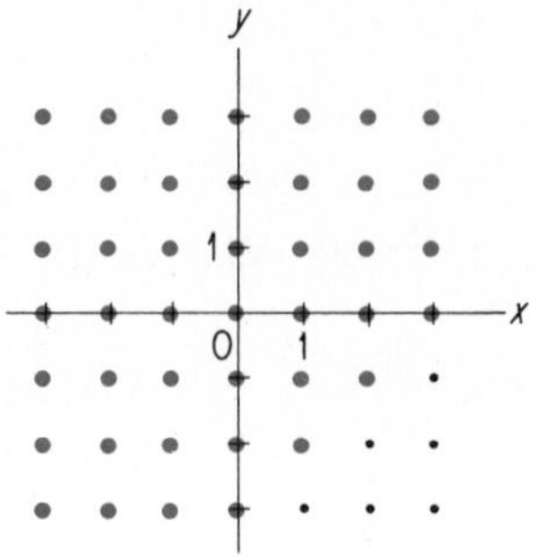
29. 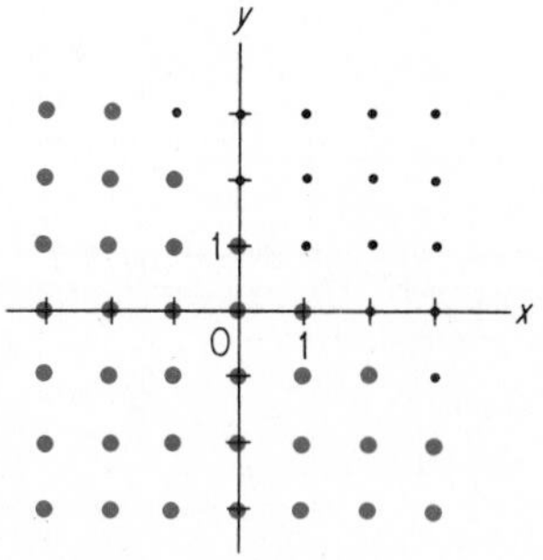
31. Function
33. Not a function
35. Function
37. Domain: {1, 2, 3, 4}; range: {1, 2, 3, 4}; a function.
39. Domain: {1, 2, 3, 4}; range: {1, 2, 3, 4}; not a function.
41. Domain: {1}; range: {1}; a function.
43. Domain: 𝒰; range: 𝒰; a function.
45. Domain: 𝒰; range: 𝒰; a function.
47. Domain: 𝒰; range: 𝒰; a function.
49. Domain: 𝒰; range: real numbers ≥ 0; a function.
51. Domain: 𝒰; range: real numbers ≥ 2; a function.

53. Domain: $\mathscr{U}$; range: $\mathscr{U}$; not a function.

55. Domain: $\mathscr{U}$; range: $\mathscr{U}$; not a function.

57. Domain: $\mathscr{U}$; range: real numbers ≤ 0; a function.

59. **(a)** -2 **(b)** 0 **(c)** 7

61. **(a)** 3 **(b)** -13 **(c)** -32

63. **(a)** -2 **(b)** 9 **(c)** -46

13-2 Linear Equations and Inequalities: page 550

1. **(a)** 3 **(b)** -3 **(c)** 1
3. **(a)** 4 **(b)** -6 **(c)** $3/2$
5. **(a)** -4 **(b)** 8 **(c)** 2
7. $y = 4x + 5$
9. $y = -x - 3$
11. $y = -2$
13. $y = x + 2$
15. $y = 1$
17. $y = -2x + 10$
19. 0, undefined, 0, undefined; a parallelogram
21. 0, $3/2$, 0, $3/2$; a parallelogram
23. $-2/3$, $3/2$, $-2/3$, $3/2$; a parallelogram
25. $-5/2$, $2/5$, $-5/2$, $2/5$; a parallelogram

*27. k/h, $(d - b)/(c - a)$, k/h, $(d - b)/(c - a)$; a parallelogram

*29. $(f - b)/(e - a)$, $(h - d)/(g - c)$, $(f - b)/(e - a)$, $(h - d)/(g - c)$; a parallelogram

31. $y = -2x + 9$
33. $y \geq -2x + 8$
35. $y \geq 2x - 8$
37. (f)
39. (g)
41. (c)
43.

45.

47.

49.

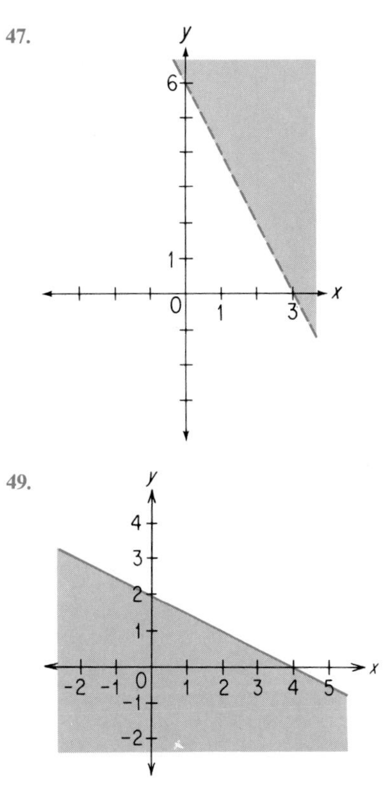

51.

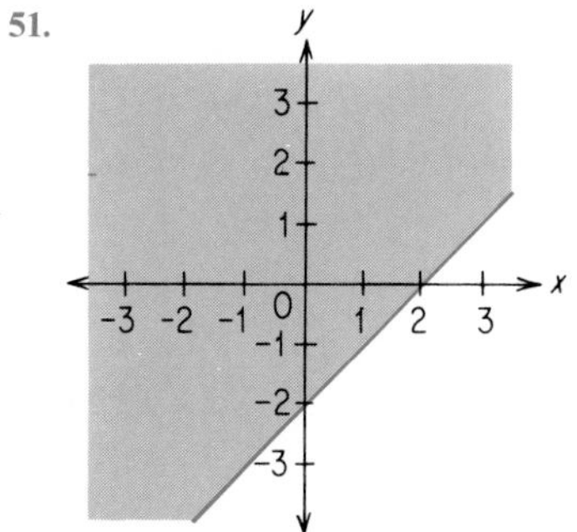

53.

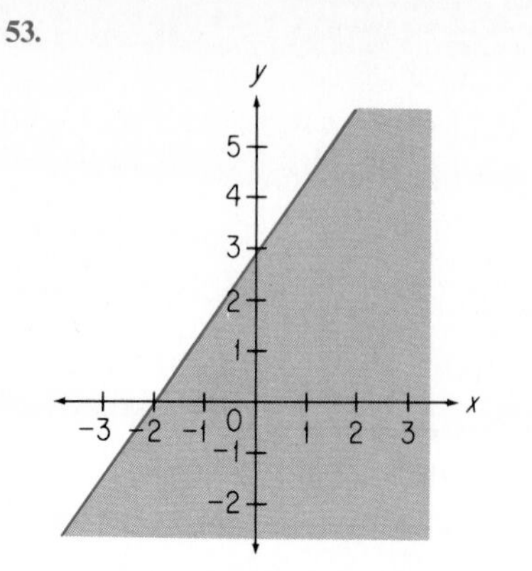

13-3 Linear Systems: page 564

1.

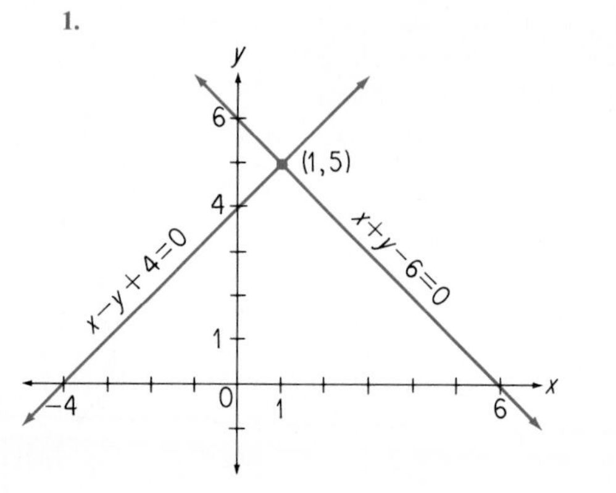

3.

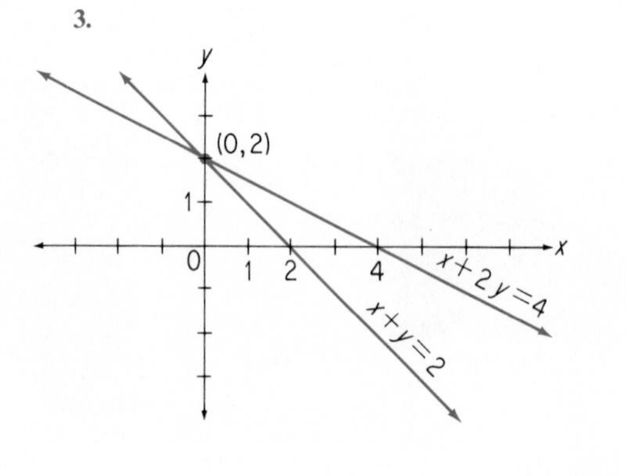

5.

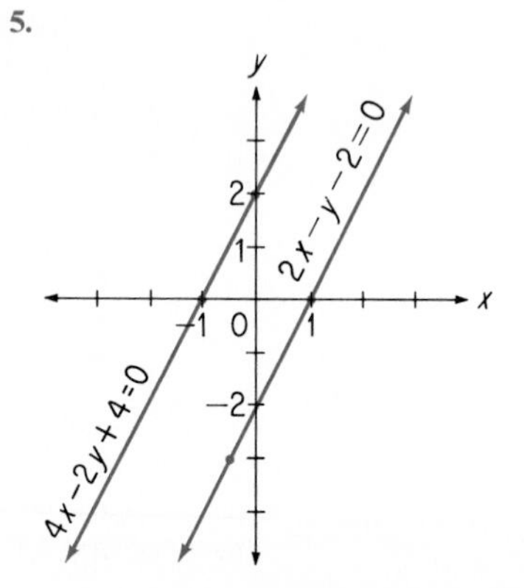

The lines are parallel; the solution set is the empty set.

7.

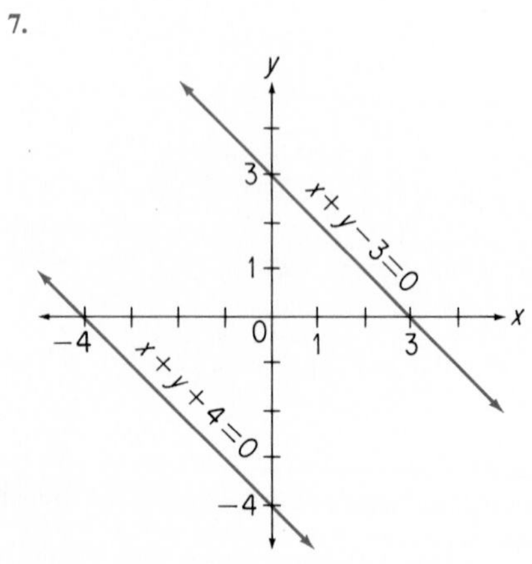

The graph consists of union of the two lines.

9.

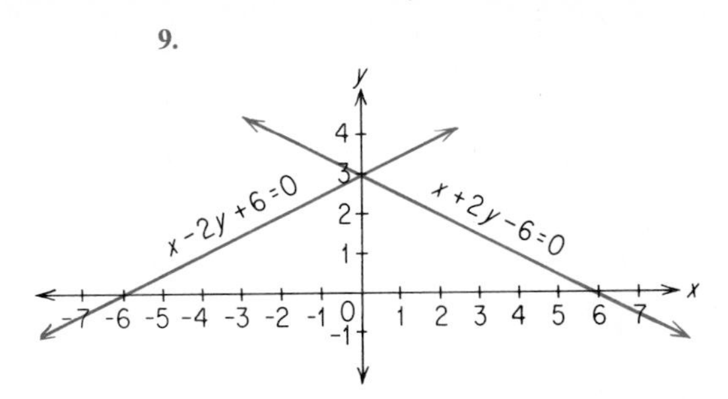

The graph consists of the union of the two lines.

11.

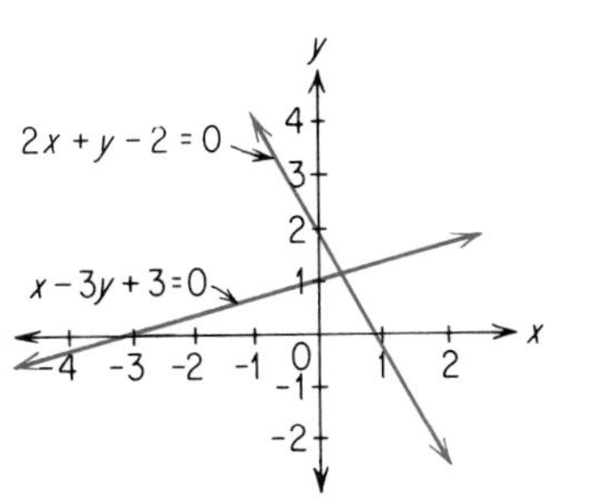

The graph consists of the union of the two lines.

13. $\begin{cases} x + y = 19 \\ x - y = 7 \end{cases}$

$x = 13,\ y = 6$

15. $\begin{cases} x + y = 25 \\ 2x = 3y \end{cases}$

$x = 15,\ y = 10$

17. $\begin{cases} x + 2y = 12 \\ x - y = 3 \end{cases}$

$x = 6,\ y = 3$

19.

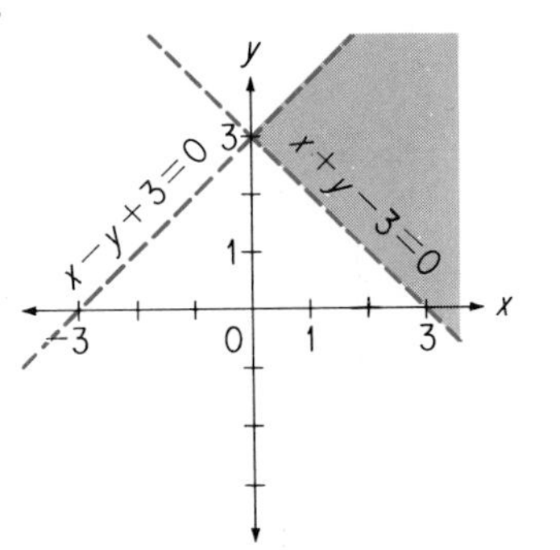

21.

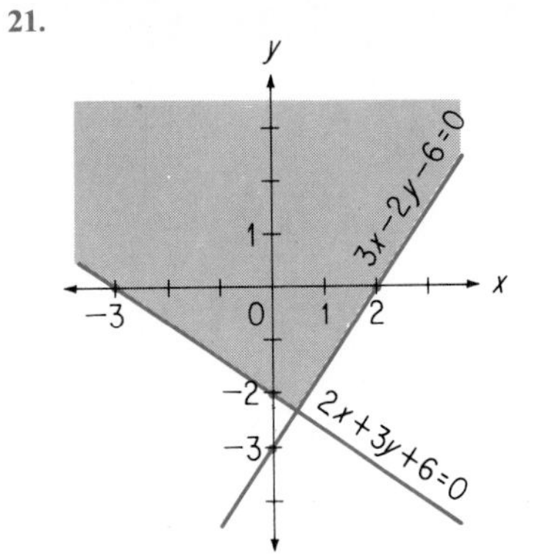

23.

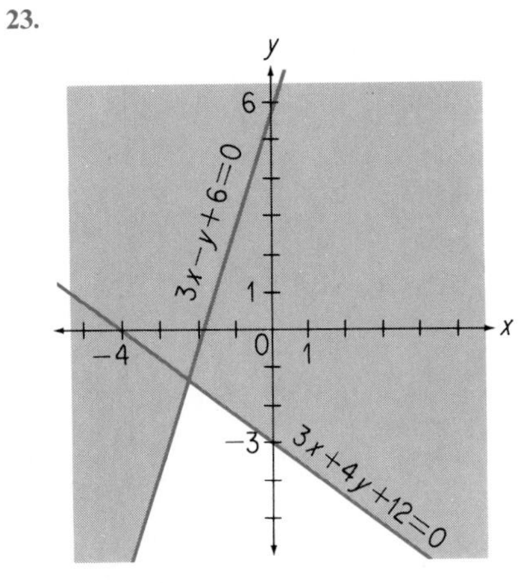

25.

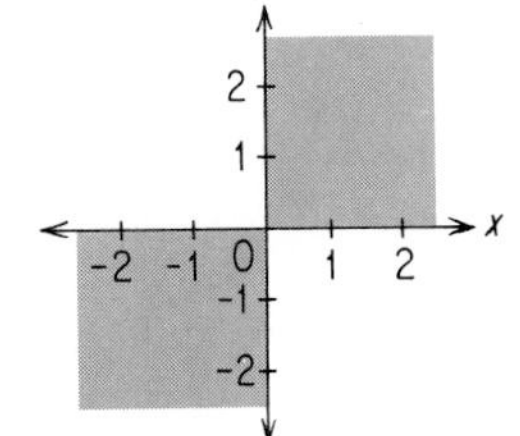

27.

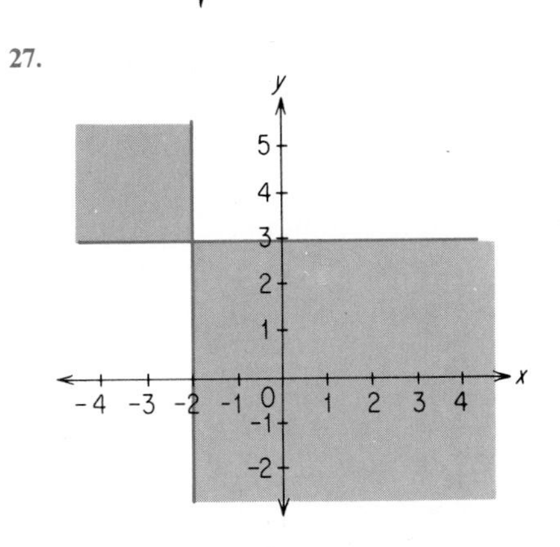

29.

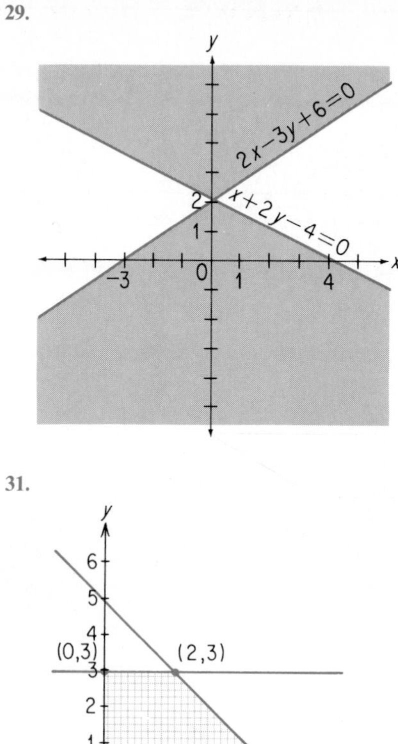

31.

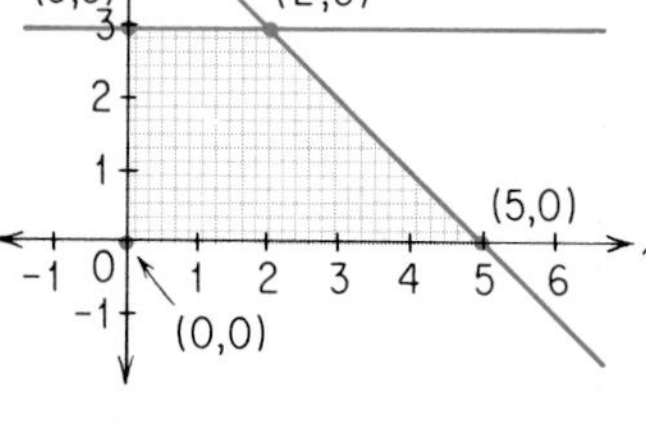

33.

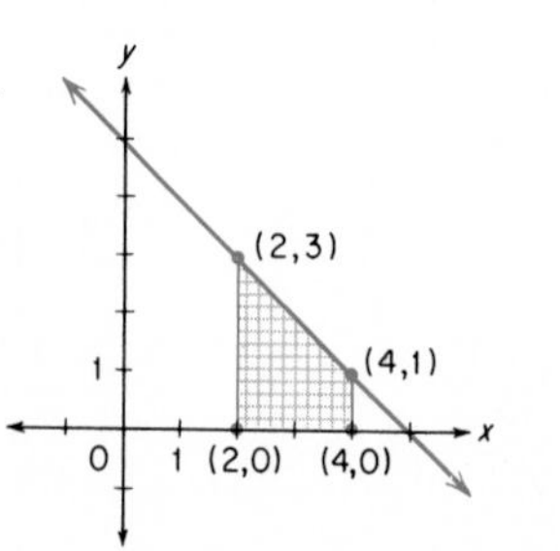

35.

37. **(a)** Maximum value of 8 at (2, 3).
(b) Minimum value of 2 at (2, 0).

39. **(a)** Maximum value of 13 at (4, 1).
(b) Minimum value of 6 at (2, 0).

41. **(a)** Maximum value of 11 at (4, 1).
(b) Minimum value of 3 at (0, 1).

43. The conditions are $0 \le x$, $0 \le y$, $x + y \le 180$, and $5x + 3y \le 750$. The maximum of $3x + 2y$ occurs at (105, 75); thus under the assumptions of this exercise, 105 minutes of regular teaching and 75 minutes of televised instruction would be best for the students.

45. Replace the condition $0 \le x$ by $30 \le x$ and retain all of the other conditions of Exercise 43. Then the maximum of $x + 2y$ occurs at (30, 150); thus under the assumptions of this exercise, 30 minutes of regular teaching and 150 minutes of televised instruction would be best for the students.

*47. $\begin{cases} x \ge 0 \\ y \ge 0 \\ y \le -x + 1 \end{cases}$

*49. $\begin{cases} x \ge 1 \\ x \le 5 \\ y \ge -2 \\ y \le 1 \end{cases}$

*51. $\begin{cases} x + y \le 2 \\ x - y \le 2 \\ -x + y \le 2 \\ -x - y \le 2 \end{cases}$

13-4 Figures on a Coordinate Plane: page 574

1. 4; (2, 3)
3. 4; (2, 3)
5. 5; (−3, 9/2)
7. (3, −2)
9. (−5, 3)
11. (2, 5)
13. (3, 7)
15. (−1, −8)
17. (a, b)
19. **(a)** $(a, 6b)$
(b) $(a, 4b)$
21. 5
23. 25
25. $3\sqrt{2}$
27. A circle with center (0, 0) and radius 4.
29. The interior points of the circle with center (0, 0) and radius 6.

31. The circular region with center (1, 2) and radius 2.

33. $(x - 2)^2 + (y - 5)^2 = 9$

35. $(x + 3)^2 + (y + 1)^2 \leq 16$

37. $(x + 4)^2 + (y - 3)^2 > 16$

39. $(x + 1)^2 + (y - 2)^2 \leq 25$

41. $9 \leq (x - 1)^2 + (y + 2)^2 \leq 16$

43. **(a)** $|x - 2| = 3$ **(b)** $|y + 1| = 4$

45. The line segment has endpoints $(b/2, c/2)$ and $((a + b)/2, c/2)$, is on the line $y = c/2$, which is parallel to AB on the x-axis, and has length $a/2$, which is half the length of AB.

*47. Any parallelogram $ABCD$ has a base b and height h and may be represented on a coordinate plane with vertices A: (0, 0), B: (b, 0), C: (b + c, h), D: (c, h). The diagonals AC and BD have the same midpoint $((b + c)/2, h/2)$ and thus bisect each other.

13-5 Figures in a Coordinate Space: page 582

1.

3.

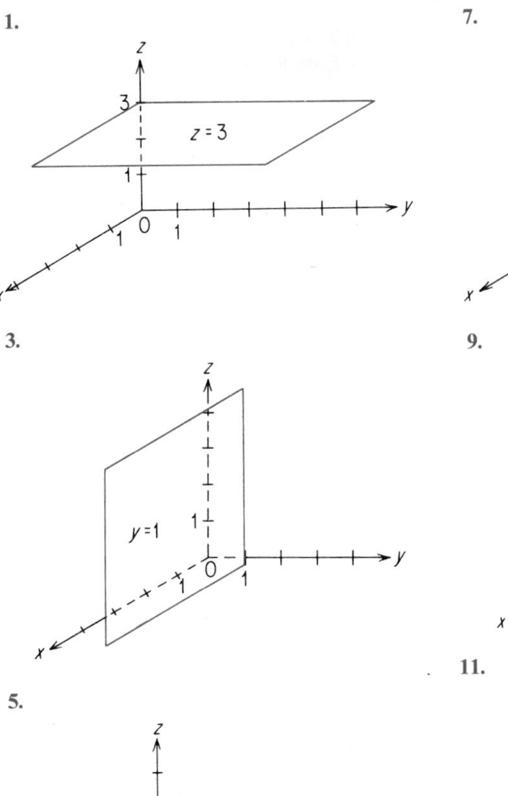

5.

z = −1

7.

9.

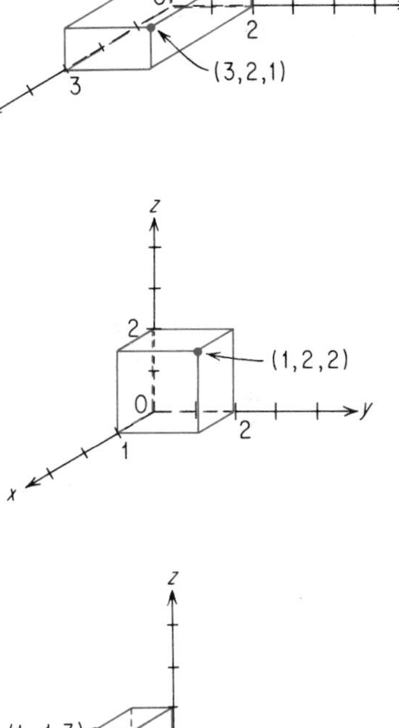

11.

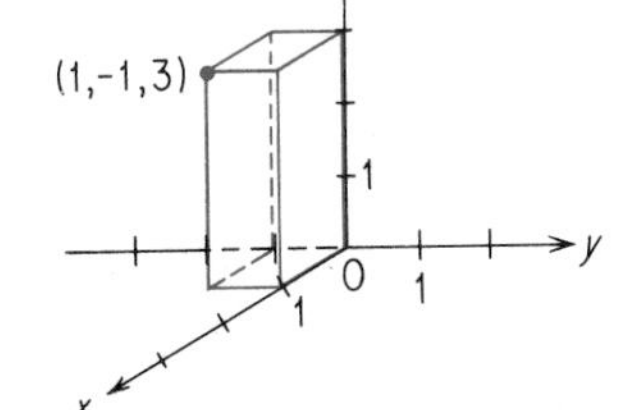

13. The yz-plane

15. The xy-plane

17. A plane parallel to the xz-plane and at a directed distance of b units from it.

19. The z-axis

21. The y-axis

23. A line parallel to the z-axis and through the point $(2, 3, 0)$.

25. A: $(4, 0, 0)$, B: $(4, 2, 0)$, C: $(4, 2, 3)$, D: $(4, 0, 3)$, E: $(0, 0, 0)$, F: $(0, 2, 0)$, G: $(0, 2, 3)$, H: $(0, 0, 3)$

27. A: $(4, -2, 0)$, B: $(4, 0, 0)$, C: $(4, 0, 3)$, D: $(4, -2, 3)$, E: $(0, -2, 0)$, F: $(0, 0, 0)$, G: $(0, 0, 3)$, H: $(0, -2, 3)$

29. A: $(7, 0, 0)$, B: $(7, 5, 0)$, C: $(7, 5, 2)$, D: $(7, 0, 2)$, E: $(0, 0, 0)$, F: $(0, 5, 0)$, G: $(0, 5, 2)$, H: $(0, 0, 2)$

31. A: $(7, -5, 0)$, B: $(7, 0, 0)$, C: $(7, 0, 2)$, D: $(7, -5, 2)$, E: $(0, -5, 0)$, F: $(0, 0, 0)$, G: $(0, 0, 2)$, H: $(0, -5, 2)$

33. $\sqrt{(1-1)^2 + (4-0)^2 + (5-2)^2}$; that is, 5.

35. $\sqrt{(3-2)^2 + (1+1)^2 + (2-0)^2}$; that is, 3.

37. $\sqrt{(1+1)^2 + (-5-5)^2 + (7-2)^2}$; that is, $\sqrt{129}$

39. $(x-1)^2 + (y+2)^2 + (z-5)^2 = 4$

41. $(x+1)^2 + (y+2)^2 + (z-3)^2 = 16$

43. $x^2 + (y+2)^2 + (z-4)^2 = 49$

45. A sphere with center at the origin and radius $\sqrt{5}$.

47. A spherical region with center at the origin and radius 3.

49. A sphere with center at $(1, 2, 3)$ and radius 2.

51. The interior of a sphere with center at $(-3, 1, -1)$ and radius 1.

53. A spherical region with center at $(5, -7, -2)$ and radius 4.

55. The exterior of a sphere with center at $(-1, 3, -4)$ and radius $\sqrt{10}$.

57. $z = 3$

59. $z \geq 2$

61. $x^2 + y^2 + z^2 = 9$

63. $x^2 + y^2 + z^2 \geq 25$

65. $(x-2)^2 + (y-3)^2 + (z-4)^2 \leq 4$

67. $4 \leq (x-2)^2 + (y+1)^2 + (z-3)^2 \leq 9$

69. $25 \leq (x-2)^2 + (y+3)^2 + (z-5)^2 \leq 49$

*71. $|y| \leq 3$

Chapter Review: page 585

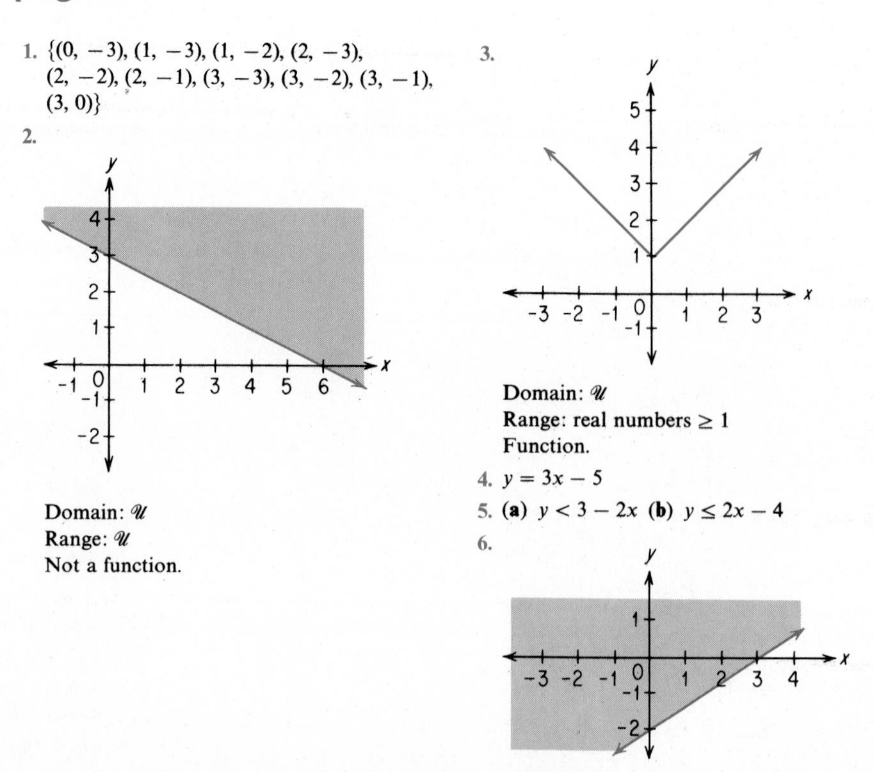

1. $\{(0, -3), (1, -3), (1, -2), (2, -3), (2, -2), (2, -1), (3, -3), (3, -2), (3, -1), (3, 0)\}$

2.

Domain: $\mathscr{U}$
Range: $\mathscr{U}$
Not a function.

3.

Domain: $\mathscr{U}$
Range: real numbers ≥ 1
Function.

4. $y = 3x - 5$

5. **(a)** $y < 3 - 2x$ **(b)** $y \leq 2x - 4$

6.

7.

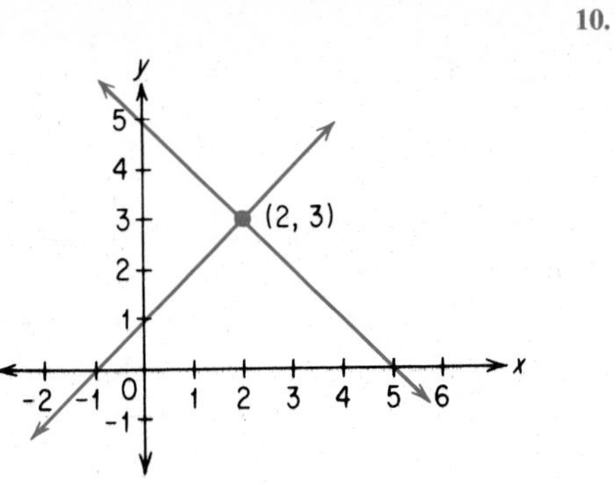

8.

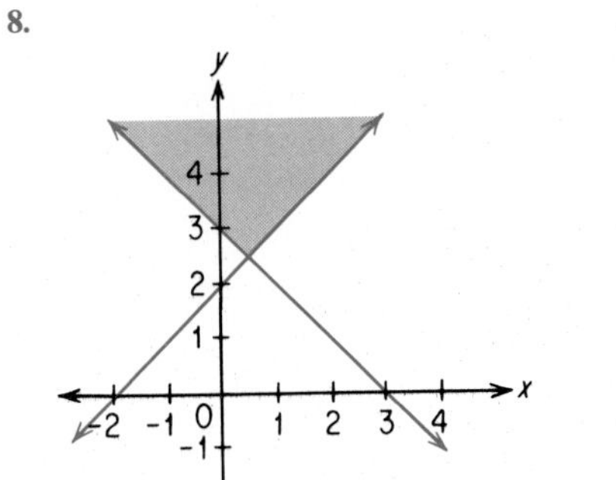

9.

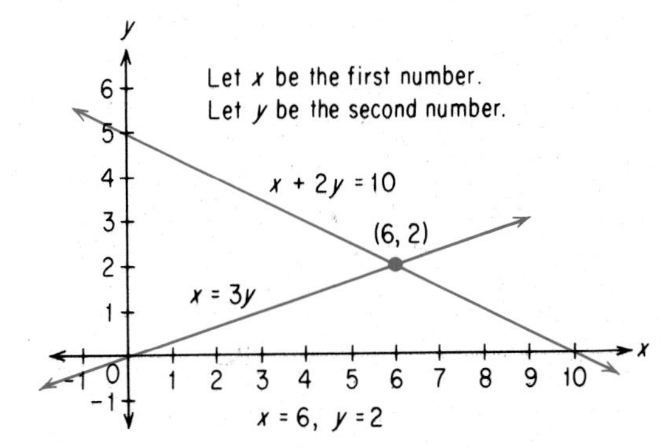

10.

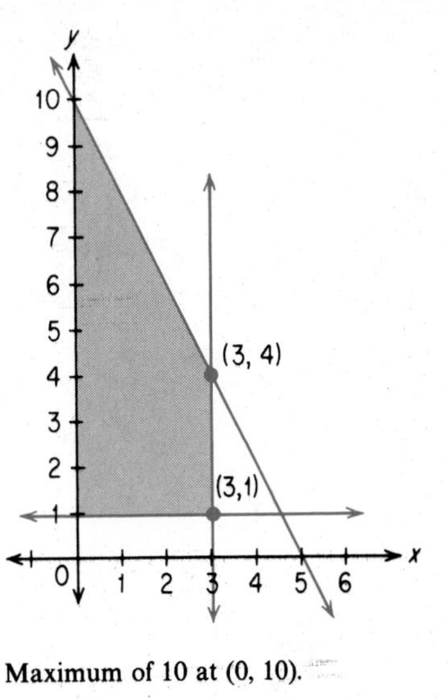

Maximum of 10 at (0, 10).

11. **(a)** $6\sqrt{5}$ **(b)** (5, 1)
12. (3, 2)
13. $(x - 5)^2 + (y + 2)^2 = 9$
14. **(a)** The interior points (between the circles) of a circular ring with center (0, 0), inner radius 2, and outer radius 4.
    (b) The circular region with center (1, −2) and radius 3.
15. **(a)** $|x| = 3$ **(b)** $|x - 2| = 2$
16. $1 \leq (x - 2)^2 + (y - 3)^2 \leq 4$
17.

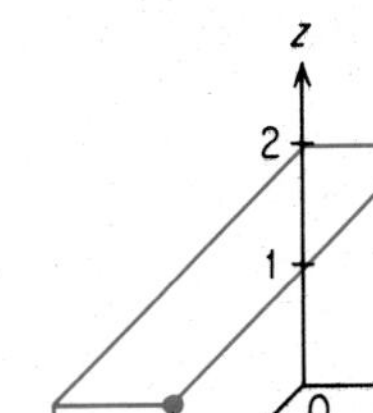

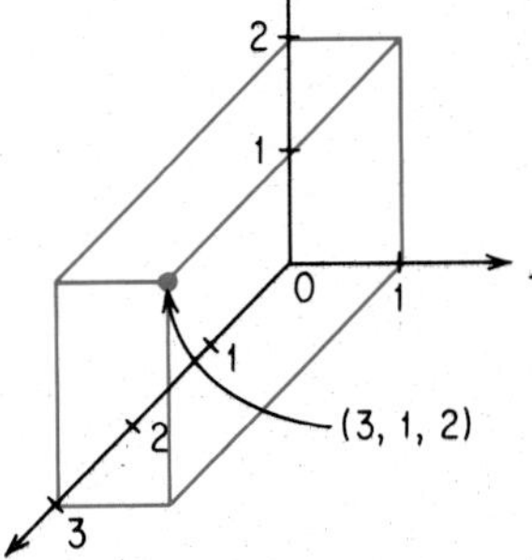

18. The plane parallel to the xz-plane and at a directed distance of 2 units from it.
19. The line parallel to the z-axis and through the point (2, 3, 0).
20. The spherical region with center (0, 0, 0) and radius 1.
21. The sphere with center (1, −2, 0) and radius 3.

22. $\begin{cases} y = 0 \\ z = 0 \end{cases}$

23. $z < 0$

24. $(x-2)^2 + (y+1)^2 + (z-5)^2 = 9$

25. $1 \le (x-1)^2 + y^2 + (z+2)^2 \le 9$

Chapter 14 Computers in Elementary Mathematics

14-1 Introduction: page 594

1. 9
3. 1
5. 4
7. 3
9. −1
11. 27
13. 2
15. 10/3
17. 200
19. `5*6/3`; 10
21. `6*SQR (25+24)`; 42
23. `8ØØØ/(2^5)`; 250
25. $7 \times 2 = 14$, $6 + 14 = 20$; 20
27. $6 \times 7 = 42$; $42 \div 3 = 14$; 14
29. $12^2 = 144$, $144 - 100 = 44$; 44
31. $3^2 = 9$, $7 \times 9 = 63$, $5 + 63 = 68$; 68
33. $\sqrt{169} = 13$, $11 + 13 = 24$; 24
35. $12^2 = 144$, $5^2 = 25$, $144 + 25 = 169$, $\sqrt{169} = 13$, $17 + 13 = 30$; 30

14-2 BASIC Programs: page 599

1.
```
2   1
4   2
6   3
8   4
```

3.
```
1    1
2    8
3   27
4   64
```

5.
```
1   2    3
3   4    7
5   6   11
```

7.
```
 1     2     3
 6

 4     5     6
15

 7     8     9
24

1Ø    11    12
33
```

9.
```
1Ø   1ØØ
11   121
12   144
13   169
14   196
15   225
```

11.
```
1   2    2
2   3    6
3   4   12
4   5   2Ø
5   6   3Ø
6   7   42
```

Among others:

13.
```
1Ø  FOR X = 1 TO 2Ø
2Ø  LET Y = X ^ 5
3Ø  PRINT X, Y
4Ø  NEXT X
5Ø  END
```

15.
```
1Ø  FOR X = 1 TO 1ØØØ
2Ø  LET Y = X ^ 2
3Ø  LET Z = X ^ 3
4Ø  PRINT X, Y, Z
5Ø  NEXT X
6Ø  END
```

17.
```
1Ø  FOR X = 1 TO 1Ø
2Ø  LET A = X ^ 2
3Ø  LET B = X ^ 3
4Ø  LET C = SQR (X)
5Ø  PRINT X, A, B
6Ø  PRINT C
61  PRINT
7Ø  NEXT X
8Ø  END
```

*19.
```
1Ø  FOR X = 1 TO 2Ø
2Ø  LET Y = X*0.12
3Ø  LET Z = Y + 1
4Ø  LET W = 1ØØØ*Z
5Ø  PRINT X, W
6Ø  NEXT X
7Ø  END
```

*21.
```
1Ø  READ E, G1, G2, G3
2Ø  DATA...
3Ø  LET A = E/2 + (G1 + G2 + G3)/6
4Ø  PRINT E
5Ø  PRINT G1, G2, G3
6Ø  PRINT A
7Ø  PRINT
8Ø  GOTO 1Ø
9Ø  END
```

14-3 Improving BASIC Programs: page 606

1. 35
3. 321.2
5. 3.29
7. 3.285
9.
```
ANNE
ANNE
ANNE
ANNE
ANNE
```
11.
```
0
2
4
8
12
```
13.
```
1
2
5
10
```
15.
```
N     Y
1     6
2    66
3   222
4   522
5  1014
```
Among others:

17.
```
10  FOR N = 1 TO 50
20  LET Y = 3 * N
30  IF Y / 2 = INT(Y / 2) THEN PRINT Y
40  NEXT N
50  END
```

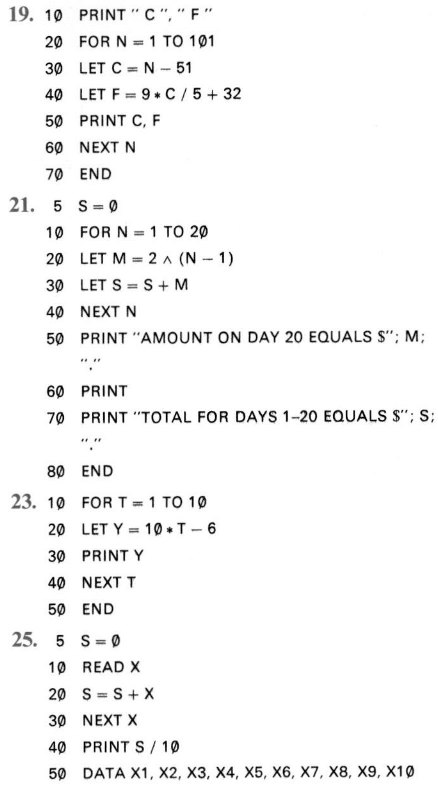

19.
```
10  PRINT " C ", " F "
20  FOR N = 1 TO 101
30  LET C = N - 51
40  LET F = 9 * C / 5 + 32
50  PRINT C, F
60  NEXT N
70  END
```
21.
```
 5  S = 0
10  FOR N = 1 TO 20
20  LET M = 2 ^ (N - 1)
30  LET S = S + M
40  NEXT N
50  PRINT "AMOUNT ON DAY 20 EQUALS $"; M;
    "."
60  PRINT
70  PRINT "TOTAL FOR DAYS 1-20 EQUALS $"; S;
    "."
80  END
```
23.
```
10  FOR T = 1 TO 10
20  LET Y = 10 * T - 6
30  PRINT Y
40  NEXT T
50  END
```
25.
```
 5  S = 0
10  READ X
20  S = S + X
30  NEXT X
40  PRINT S / 10
50  DATA X1, X2, X3, X4, X5, X6, X7, X8, X9, X10
60  END
```

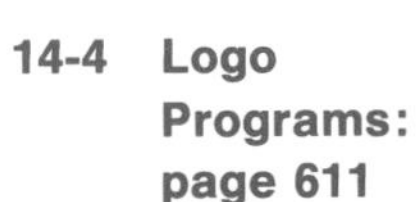

14-4 Logo Programs: page 611

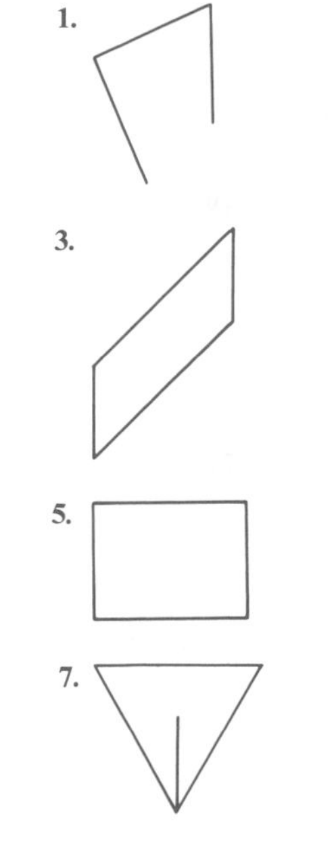

1.

3.

5.

7.

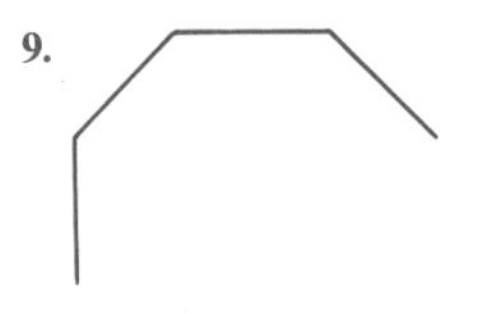

9.

Among others:

11. REPEAT 2[FD 10 RT 90 FD 20 RT 90]
13. RT 45 REPEAT 2[FD 30 RT 45 FD 50 RT 135]
15. REPEAT 4[RT 90 FD 15]
    RT 30 FD 15 RT 120 FD 15
17. FD 30 RT 135 FD 15 LT 90 FD 15 RT 135 FD 30
19. REPEAT 4[FD 20 RT 90]
    PU RT 90 FD 10 LT 90 FD 10 RT 90 PD
    REPEAT 4[FD 20 RT 90]

14-5 Logo Procedures: page 616

1.

3.

Among others:

5.
```
TO SQUARE :SIDE
REPEAT 4[FD :SIDE RT 9Ø]
END
SQUARE 1Ø SQUARE 2Ø SQUARE 3Ø SQUARE 4Ø
```

7.
```
TO TRIANGLE :SIDE
REPEAT 3[FD :SIDE RT 12Ø]
END
RT 3Ø TRIANGLE 1Ø TRIANGLE 2Ø TRIANGLE 3Ø
```

9.
```
TO TRIANGLE1Ø
REPEAT 3[FD 1Ø RT 12Ø]
END
REPEAT 4[LT 6Ø TRIANGLE1Ø RT 6Ø FD 1Ø RT 9Ø]
```

Chapter Review: page 618

1. 7 * (5 − 11)
2. 5 ∧ 2 + 13
3. 35/2
4. 72/(3 ∧ 3 − 3)
5. Multiply 5 by 2. Subtract 10 from 29.
6. Square 3. Add 11 and 9.
7. Raise 3 to the third power. Subtract 17 from 27. Divide 190 by 10.
8. Square 5. Square 4. Divide 16 by 2. Multiply 8 by 25. Add 200 and 3.
9. 225
10. 140
11. 10
12. 10
13.
```
1   11
2   10
3    9
4    8
```
14.
```
1   2    2
3   4   12
5   6   3Ø
```
15.
```
1    1     1
2    4     8
3    9    27
4   16    64
5   25   125
```
16.
```
7   24   25
ARE SIDES OF A RIGHT TRIANGLE
9   17   2Ø
ARE NOT SIDES OF A RIGHT TRIANGLE
```
17. 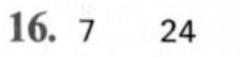

18. 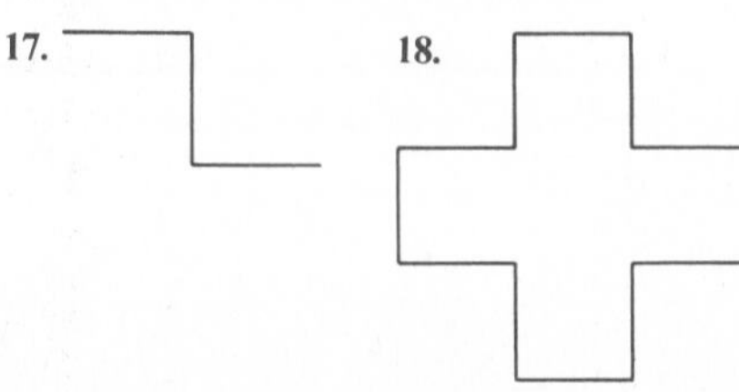

19. 

20. 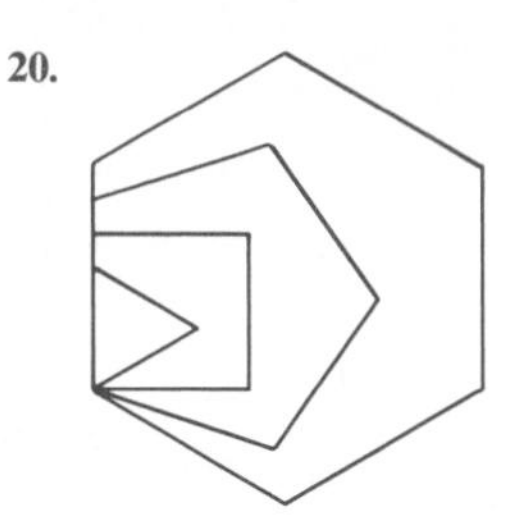

Among others:

21.
```
1Ø  S = Ø
2Ø  FOR N = 1 TO 1ØØ
3Ø  LET S = S + N
4Ø  NEXT N
5Ø  PRINT S
6Ø  END
```
22.
```
1Ø  S = Ø
2Ø  FOR N = 1 TO 1ØØ
3Ø  LET S = S + 2 * N − 1
4Ø  NEXT N
5Ø  PRINT S
6Ø  END
```
23.
```
TO TRIANGLE4Ø
REPEAT 3[FD 4Ø RT 12Ø]
END
RT 3Ø TRIANGLE4Ø RT 6Ø FD 2Ø LT 6Ø
TRIANGLE4Ø LT 6Ø FD 2Ø RT 12Ø TRIANGLE4Ø
```
24.
```
REPEAT 4[FD 2Ø RT 9Ø FD 1Ø RT 9Ø FD 2Ø RT 9Ø
FD 1Ø BK 1Ø RT 9Ø]
```
25.
```
REPEAT 4[FD 4Ø RT 9Ø]
FD 2Ø RT 45
REPEAT 4[FD 28.28 RT 9Ø]
END
```

Index